ADVANCES IN
MOLECULAR AND CELL BIOLOGY
VOLUME 32

Molecular and Cellular Insights to Ion Channel Biology

ADVANCES IN
MOLECULAR AND CELL BIOLOGY
VOLUME 32

Molecular and Cellular Insights to Ion Channel Biology

Series Editor:

E. Edward Bittar
University of Wisconsin - Madison
Madison, Wisconsin,
USA

Volume Editor:

Robert A. Maue
Hanover,
USA

2004

ELSEVIER

Amsterdam - Boston - Heidelberg - London - New York - Oxford
Paris - San Diego - San Francisco - Singapore - Sydney - Tokyo

ELSEVIER B.V.
Sara Burgerhartstraat 25
P.O. Box 211, 1000 AE
Amsterdam, The Netherlands

ELSEVIER Inc.
525 B Street, Suite 1900
San Diego, CA 92101-4495
USA

ELSEVIER Ltd
The Boulevard, Langford Lane
Kidlington, Oxford OX5 1GB
UK

ELSEVIER Ltd
84 Theobalds Road
London WC1X 8RR
UK

First edition 2004

Library of Congress Cataloging in Publication Data
A catalog record is available from the Library of Congress.

British Library Cataloguing in Publication Data
A catalogue record is available from the British Library.

ISBN: 0-444-50645-4
ISSN: 1569-2558 (series)

⊚ The paper used in this publication meets the requirements of ANSI/NISO Z39.48-1992 (Permanence of Paper).
Printed in The Netherlands.

TABLE OF CONTENTS

PREFACE

The roughly parallel emergence of primary sequence information acquired with molecular biological approaches (ca. 1982) and the functional resolution of currents through individual ion channels using patch clamp techniques (ca. 1976) began an exciting and explosive increase in our understanding of ion channel proteins. It opened our eyes to the remarkable diversity that exists, not only in the types of ion channels found in biological membranes, but in the mechanisms by which they are regulated. This trend toward molecular understanding has continued, with new structural information coming to light and new channel sequences uncovered in the complete genomic information now available for a variety of organisms, including the nematode, fruit fly, and human. However, these "reductionist" approaches have been joined by increasing efforts to understand the importance of specific ion channels to the behavior of the cells, and ultimately the organism, in which they are expressed. As a result, it is increasingly common to see geneticists, molecular biologists, and biochemists investigating ion channel function in collaboration with researchers interested in organ and tissue physiology from basic science and clinical perspectives.

One consequence of the increasingly broad-based investigations of ion channels is that those new to these remarkable proteins find themselves faced with the challenge of becoming familiar, if not well-versed, with widely diverse experimental approaches and a vast array of information. I've found this true at all levels, from the students in the "Integrated Approaches to Ion Channel Biology" summer course at Cold Spring Harbor that I had the privilege of teaching for several years, to the more experienced researchers who have become nascent "students of ion channels". Therefore, among the purposes of this volume is to serve as a resource for those new to ion channels by providing informative, up-to-date accounts of a wide variety of ion channel types. There has also been an effort to present, when possible, instances where information garnered from cellular and molecular approaches has revealed something about the "biology" of channels, and furthered our understanding of their physiological role(s) and/or involvement in diseases of cells, tissues, and organs. Hopefully, this will stimulate additional efforts to understand the cellular role(s) of specific ion channel proteins. Finally, to illustrate the rich diversity of approaches and viewpoints that have shaped our understanding of ion channels, contributions describing the results of genetic, molecular biological, biochemical, electrophysiological, immunocytochemical, and immunological approaches have been included in order to emphasize the value of all of these levels of inquiry. I hope that in all of these ways (though to different degrees depending upon the individual) this volume will be useful to a broad audience of undergraduate students, graduate students in physiology, biophysics, pharmacology, cell biology, and neuro-science, as well as medical students, clinicians, and basic scientists.

The goals of this book are addressed, in part, by its organization and scope. The book consists of individual chapters, with an equal number devoted to excitable and inexcitable cells and tissues. In general, each chapter focuses on a specific type of ion channel and includes background information. However, two chapters purposefully do not follow this pattern, and instead focus on the spectrum of channels found in membranes of particular interest (sperm membranes; membranes of intracellular organelles) These chapters were included to illustrate the importance of ion channels in all types of membranes and the efforts to understand the role(s) of these membrane proteins within a physiological context. By design, the contributions are diverse, and range from molecular biological approaches and biochemical analyses, to biophysical measures of ion channel function in specific membranes, cell types, and tissues. They include new perspectives (i.e. the immunology of the glutamate receptor; the aquaporins as ion channels) and original styles of presentation (i.e. the "natural history" of calcium-activated potassium channels). Examples were chosen to illustrate not only the success in uncovering the role and importance of specific channel types in disease, but to also highlight the importance of specific channels under normal physiological conditions. Something this book does not attempt to do is provide comprehensive or exhaustive coverage of ion channels, and I apologise on behalf of my authors to those who felt their work or their favorite type of ion channel was not well represented. For information about ion channels that is not presented here, there are a number of excellent books that may prove useful, including well-known "classics" (Hille (1992) *Ionic Channels of Excitable Membranes*, Sakmann and Neher (1995) *Single-channel Recording*), more recent, well-regarded works (Aidley and Stanfield (1996) *Ion Channels - Molecules in Action*, Ashcroft (2000) *Ion Channels and Disease*), and more narrowly focused volumes on channel dysfunction and disease (Lehmann-Horn and Jurkat-Rott (2000) *Channelopathies*). Finally, as with all previous works this book is not immune to the continual discovery of new ion channel subunits and new aspects of ion channel function. For example, during the assembly of this book new channel subunits were cloned (i.e. auxiliary subunits of Cl^- channels), new structural information was obtained for several channel types (Cl^- channels; K^+ channels; acetylcholine-gated channels; aquaporin channels), and new information was obtained about the roles of VR-1 channel family members in sensing temperature. Overall, it has been a stimulating endeavor to assemble a book on this subject, and I have learned a great deal about ion channels in the process – I hope that my readers do too.

As with any undertaking like this, many thanks are in order to those who helped shape this project and make it a reality. For my perspective on ion channels, I would like to thank my thesis advisor Vince Dionne and my postdoctoral advisors Irwin Levitan and Gail Mandel for their patience with me as a student of ion channels, and for imparting not only their superb biophysical, biochemical, and molecular biological approaches to ion channels, but also instilling in me by example an integrative, inclusive, and encompassing view of these amazing proteins. Their discussions and advice with regard to this project were also appreciated. I would also like to thank Rock Levinson, John Caldwell, Paul Gardner, and Leslie Henderson for engendering a fearless and fun approach to studying ion channels, particularly during the Cold Spring Harbor courses. The excitement of the courses and our late night discussions had a definite influence on this project. Of course, I would like to thank the more than forty contributing authors from more than a half dozen countries around the world for their perspective, their patience, and the quality of their efforts, which made my task much easier and enjoyable. Similarly, thanks go to the several dozen reviewers that commented on drafts of the manuscripts. I was told by a colleague that in taking on such a project I would find out who my friends really were, and it would test the best of my friendships. I am happy to report that I still have all of my original friends and have made some new ones in the process. I would like to thank JAI Press for their original interest in ion channels and E. Edward Bittar for asking me to put this book together. Special thanks go to Netty Vreugdenhil at Elsevier Press, who took over this project and exhibited an infinite supply of patience. Without her encouragement and support this project would not have come to fruition. Thanks also go to Terry Hall in the

Physiology Department at Dartmouth for her assistance with correspondence and the preparation of manuscripts. Finally, I would especially like to thank my family – my spouse and colleague, Leslie Henderson, for her help with nearly all aspects of this project, and my sons Tyler and Casey, for their patience and understanding – I appreciate the time they gave to me to do this.

Robert A. Maue, Ph.D.
Departments of Physiology and of Biochemistry
Dartmouth Medical School
Hanover, New Hampshire

ATP-sensitive potassium channels and insulin secretion diseases

C.G. Nichols,[a,*] S.-L. Shyng,[b] B. Marshall[a] and J.C. Koster[a]

[a]Department of Cell Biology and Physiology, Washington University School of Medicine,
660 South Euclid Avenue, St. Louis, MO 63110, USA
[b]Center for Research on Occupational and Environmental Toxicology,
Oregon Health Sciences University, Portland, Oregon 97201, USA
*Correspondence address: Tel.: +1-314-362-6630; fax: +1-314-362-7463
E-mail: cnichols@cellbio.wustl.edu

1. Introduction: a paradigm for control of insulin secretion

Regulated secretion of insulin from pancreatic β-cells and the actions of this peptide hormone on its target tissues maintains glucose homeostasis. The concerted action of nutrients (glucose and amino acids) together with the effect of hormones and of neuro-transmitters acting on G-protein coupled receptors leads to exocytotic release of insulin from β-cells. Exocytosis proceeds through distinct steps that result in the fusion of secretory granules with the plasma membrane and release of insulin into the extracellular space (Fig. 1). Only a small proportion of insulin is released even under maximal stimulatory conditions, and circulating levels therefore depend mainly on the regulation of secretion, rather than on the rate of synthesis. As we discuss, inappropriate control of insulin secretion results in human diseases: insufficient insulin results in various forms of diabetes, in particular in pre-type I (insulin- dependent diabetes mellitus) and in many forms of the more common type II non-insulin-dependent diabetes (NIDDM); constitutive insulin release causes persistent hyperinsulinemic hypoglycemia of infancy (PHHI).

In response to elevated glucose levels, the currently best understood metabolic coupling event is a change in the intracellular [ATP]/[ADP] ratio, which induces closure of ATP-sensitive K^+ (K_{ATP}) channels (Fig. 1), leading to membrane depolarization and opening of voltage-dependent calcium channels. The subsequent increase in cytosolic free Ca^{2+} then triggers exocytosis. Although the role of the K_{ATP} channel in the coupling of metabolism to insulin release [1] is the essential topic of this article, it is important to realize that important regulatory steps also occur distally, and that insulin release may be induced by glucose through a mechanism distinct from closure of K_{ATP} channels or

Advances in Molecular and Cell Biology, Vol. 32, pages 1–14
ISSN: 1569-2558 / DOI: 10.1016/S1569-2558(03)32001-6

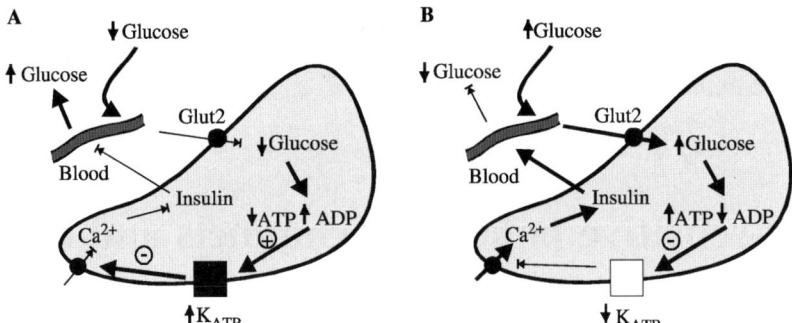

Fig. 1. Role of K_{ATP} channel in coupling glucose levels to insulin secretion. Decreased blood glucose leads to decrease in β-cell [glucose] and hence decreased [ATP]:[ADP]. This opens K_{ATP} channels causing hyperpolarization of the cell and consequent closure of calcium (Ca) channels, block of Ca entry, and suppression of insulin secretion. This decreases glucose uptake in peripheral tissues, and raises blood glucose.

alterations in Ca^{2+} [2], which may imply metabolic coupling factors other than intracellular nucleotides.

2. Molecular regulation of K_{ATP} channel activity

2.1. The 'Classical view' of nucleotide regulation

ATP-sensitive potassium channels are unique amongst K channels in being rapidly, and reversibly, inhibited by the non-hydrolytic binding of cytoplasmic adenosine nucleotides, as well as being activated in complex ways by nucleotide tri- and diphosphates [3,4]. Inhibition by nucleotides requires at least one phosphate, and the potency increases significantly from AMP ($K_{1/2}$, [nucleotide] causing half-maximal inhibition ∼ 10 mM) to ADP ($K_{1/2}$ ∼ 250 μM), to ATP ($K_{1/2}$ ∼ 10–25 μM) in native K_{ATP} channels. The adenosine moiety seems to be critical since GTP and pyrimidine triphosphates are essentially without inhibitory effects [3,4]. The early efforts to describe nucleotide activation of the channel were rather less successful at defining the underlying mechanisms, and this is because multiple overlapping mechanisms are likely to be responsible (see below). In addition to inhibition, ATP could activate channels in the presence of Mg^{2+} which indicates that hydrolysis is necessary [5,6]. However, MgADP and MgGDP could also activate channels by unknown mechanisms that led to an apparent antagonism of the inhibitory effect of ATP [7–11].

2.2. Molecular basis of K_{ATP} channel activity: SUR1 and Kir6.2 subunits generate β-cell K_{ATP} channels

K_{ATP} channels are blocked by sulfonylurea drugs [12,13] and it has long been recognized that the membrane binding affinity of different sulfonylureas closely correlates with their ability to inhibit pancreatic K_{ATP} channels [14]. The receptor protein was therefore long considered a candidate K_{ATP} channel constituent, and a dedicated effort to purify the 140 kDa sulfonylurea receptor from a β-cell tumor cell line, utilizing a radioiodinated derivative of glyburide [15], eventually resulted in the cloning of the SUR1 cDNA (Fig. 2). SUR1 encodes a 1582 amino acid protein, expressed in the pancreas, brain and heart [16]. At the same time as SUR1 was being cloned, a family of inwardly rectifying K (Kir) channels, with pore properties similar to those of native K_{ATP} channels were also being cloned [17] and coexpression of SUR1 with a novel Kir subunit (Kir6.2, see Fig. 2) finally resulted in the reconstitution of K_{ATP} channel activity [18].

Subsequent studies indicated that K_{ATP} channels are normally formed by a tetramer of Kir6.2 subunits, each Kir subunit being associated with one SUR1 subunit [19–21], although functional channels can be generated by C-terminally truncated Kir6.2 subunits in the absence of SUR1 [22]. It has recently been demonstrated that endoplasmic reticulum retention signals are present within the SUR1 subunit and within the Kir6.2 subunit C-terminus, leading to the current hypothesis that association of one subunit with the other is normally required for efficient trafficking of each to the plasma membrane [23].

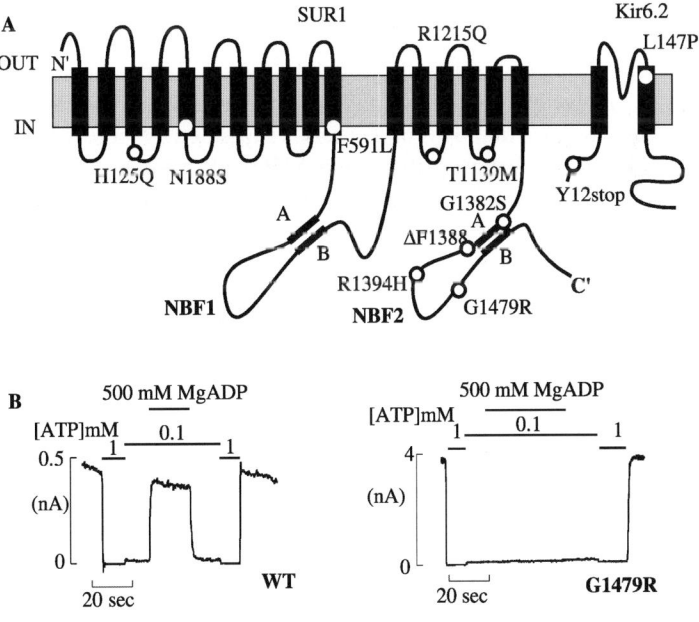

Fig. 2. K_{ATP} channel mutations underlie PHHI (A) Probable membrane topology of SUR1 and Kir6.2, indicating mutations associated with PHHI. (B) Currents from wild type and SUR1[G1479R] mutant K_{ATP} channels in response to applied inhibitory ATP, and ATP plus ADP (all in the presence of 1 mM free Mg^{2+}). The response to MgADP is specifically abolished in the disease mutant (unpublished).

2.3. Separable roles of the sulfonylurea receptor and Kir6 subunits: SUR controls nucleotide activation and drug sensitivity

Two SUR genes have been cloned. In addition to SUR1, two major splice variants of SUR2, differing in the sequence of the C-terminal 45 amino acids, are expressed predominantly in striated muscle (SUR2A) and in smooth muscle (SUR2B) [24,25]. The SUR isoform is the major determinant of channel pharmacology. Channels formed with SUR1 are activated by diazoxide and inhibited by glibenclamide, whereas channels formed with SUR2 isoforms are more readily activated by the cardioselective potassium channel opening drugs such as pinacidil and nicorndil, and inhibited by 5-hydroxyde-canoate [26]. SURs are members of the ATP binding cassette (ABC) family of membrane proteins. All ABC proteins contain two classical nucleotide binding folds [27] (NBFs) in the cytoplasmic loops, leading to the natural presumption that inhibitory ATP binding should occur at the NBFs of SUR1. However, while the NBFs of other ABC proteins bind and hydrolyze MgATP, the inhibitory effect of ATP on K_{ATP} channels is Mg^{2+}-independent and does not require hydrolysis [4,28]. The first insight to the role of the NBFs in regulating K_{ATP} channel activity came from examination of a mutation in NBF2 (G1479R, see Fig. 2) that causes PHHI [29] (see below). This mutation selectively abolished MgADP- and diazoxide-mediated stimulation of channel activity, with no effect on ATP inhibition. Similar results were obtained with multiple other introduced mutations in NBF2 [30], and Gribble et al. [31] showed that mutation of the conserved lysine residues in the Walker A motifs of either NBF1 (mutation K719A), or NBF2 (mutation K1384M), which are predicted to reduce ATP hydrolytic activity, can block the stimulatory effect of MgADP.

These electrophysiological results suggested the hypothesis that nucleotide hydrolysis at both NBFs is involved in channel stimulation by MgADP and by potassium channel opening drugs such as diazoxide. This hypothesis is supported by more recent biochemical studies. Ueda et al. [32] first showed that while ATP binds NBF1 in a Mg^{2+}-independent manner, ATP or ADP binding at NBF2 requires Mg^{2+}. The same group further demonstrated that MgADP stabilizes ATP binding at NBF1 either by direct binding to NBF2 or by hydrolysis of bound MgATP [33]. Mutations in the nucleotide binding motifs of NBF2 (K1385M and D1506N) abolish this cooperative binding of nucleotides to the two NBFs. Bienengraeber et al. [34] showed that both NBFs of SUR2A are capable of hydrolyzing ATP, with NBF2 having higher ATPase activity than NBF1. Interestingly, mutations in NBF2 that reduce the ATPase activity produced channels with increased ATP sensitivity, whereas drugs such as diazoxide, which reduce ATP sensitivity of the channel, promote ATPase activity. These results are consistent with our model that SUR acts as a "hypersensitivity switch" to increase ATP sensitivity of the channel, with the hypersensitizing switch being turned off by nucleotide hydrolysis at the NBFs [30].

2.4. Kir6.2 controls nucleotide inhibition

While the SUR1 subunit seemed the obvious candidate for providing nucleotide sensitivity, the Kir6.2 subunits generate the pore of the channel [35]. However, to our initial surprise, mutations in Kir6.2 also affected ATP-sensitivity [35]. Tucker et al. [22]

isolated many mutations that caused more dramatic reductions of ATP-sensitivity, and subsequent analyses have shown that truncations of the N-terminus [36–38], and point mutations throughout the N- and C-terminal regions of Kir6.2 can significantly decrease ATP sensitivity [39–43]. Most of these mutations seem to affect ATP-sensitivity allosterically by altering the intrinsic stability of the open state [40,41], but certain mutations suggest residues in the cytoplasmic domains (including residues 50, 182, 185 and 334–337) that are potentially associated with an ATP binding site [39,42–44].

ATP-sensitivity of K_{ATP} channels is modulated by several cytoplasmic factors, and by the lipid composition of the membrane. In particular, acidosis and negatively charged phospholipids enhance ATP sensitivity [45–47], by interaction with cytoplasmic residue in the channel. So far, there have been few studies looking at the potential for dynamic modulation by these agents, but in addition to setting the baseline of channel ATP sensitivity, the potential exists for "physiological" modulation of ATP sensitivity by them. There is also evidence for modification of the Kir6.2 subunit by protein kinases. While there remains some controversy as to the exact residues that are phosphorylated, two groups have reported the existence of protein kinase A (PKA)-mediated phosphorylation of Kir6.2 [48,49]. Both groups reported that the effect of PKA was to increase channel open probability [48,49] and decrease ATP sensitivity [49]. Neurotransmitters that couple to adenylate cyclase, and therefore PKA, are likely to affect Kir6.2 phosphorylation and hence channel activity. Similarly, protein kinase C modulates Kir6.2 [50], and here too, we may envision a route for dynamic modulation.

3. K_{ATP} channels and pancreatic diseases

Inappropriate control of insulin secretion results in human diseases. Insufficient insulin results in various forms of diabetes, in particular pre-type I (insulin-dependent diabetes mellitus) and many forms of the more common type II NIDDM. The etiology of diabetes is complex and involves multiple genes and causal factors. However, in many cases, an effective treatment for type II NIDDM is the administration of sulfonylurea drugs that inhibit K_{ATP} channels and thereby enhance insulin secretion. Overproduction of insulin results in PHHI, a disease of newborns that in the majority of cases results from defects in K_{ATP} subunit genes. These defects render the channels inactive in the pancreas, and as might be predicted, PHHI can sometimes be treated by the drug diazoxide, which enhances channel opening (see above). The importance of K_{ATP} channels in the regulation of insulin secretion has also been demonstrated in the laboratory, as mutations that enhance channel activity have been shown to cause experimental diabetes. These findings, and their implications, will be the focus of the remainder of this chapter.

3.1. The features of the PHHI disease

PHHI, also known as familial hyperinsulinism or pancreatic nesidioblastosis is a rare genetic disease characterized by inappropriately high levels of insulin in parallel with low blood glucose levels in newborns and infants [51,52]. The frequency of occurrence varies but tends to be highest in populations that are small and in which consanguineous marriages are common. Infants with PHHI may have convulsions, sleepiness, lethargy, lack of interest in feeding, breathing difficulty, excessive sweating, abnormal temperature, and blue appearance, all features associated with hypoglycemia from any cause. The most serious complication is brain damage and the limited treatment options are geared to avoiding this outcome and maintaining a normal blood sugar level. This can sometimes be achieved by administration of drugs (octreotide and/or diazoxide) which may activate K_{ATP} channels and hence inhibit the secretion of insulin, but in cases where there is no response, surgical removal of the pancreas becomes necessary, leading to insulin dependency. More background on the disease can be found at the PHHI web page (http://www.sur1.com/).

The molecular basis for the disease also seems to be variable, and currently there are four different genes that have been identified as responsible for producing PHHI. Two (glucokinase and glutamate dehydrogenase) genes cause a mild dominant form of the disease. Mutations in two genes (encoding glucokinase and glutamate dehydrogenase) cause enhanced enzyme activity and, by enhancing glycolytic flux, enhance the production of ATP and increase the cellular [ATP]/[ADP] ratio, leading to reduced K_{ATP} channel activity [53]. However, the majority of PHHI seems to result from mutations in the SUR1 and Kir6.2 genes (Fig. 2) mutations have various functional consequences for K_{ATP} channel activity, leading to both mild and severe recessive forms of the disease.

3.2. K_{ATP} genetic defects underlying PHHI

PHHI was first mapped to human chromosome 11p15.1 by linkage analysis [54–56]. Following the cloning of SUR1 [16] and Kir6.2 (BIR) [18], both genes were mapped to this region [18,57,58], with the Kir6.2 gene immediately 3-prime of the SUR1 gene. Mutation screening rapidly demonstrated 2 separate SUR gene splice site mutations that segregated with the disease phenotype in affected individuals from 9 different families, and which result in aberrant processing of the RNA sequence and disruption of the SUR1 protein [57].

Subsequent large scale screening has demonstrated a strong association of PHHI with mutations in the SUR1 gene [59–62]. For instance, screening of 45 familial hyperinsulinism probands of various ethnic origins revealed 17 novel and 3 previously described mutations in SUR1 [59]. Mutations have been found throughout SUR1 (Fig. 2). In recombinant studies, these mutations invariably cause either abnormal processing or truncation, or altered metabolite sensitivity of the expressed K_{ATP} channels.

3.3. K_{ATP} channel defects underlying PHHI

In patients with PHHI, treatments involving the use of agents that inhibit insulin secretion, such as diazoxide and somatostatin, and/or sub-total pancreatectomy are required in order to avoid permanent hypoglycemia-induced brain damage. Kane et al. [63] have reported that β-cells isolated from infants with sporadic PHHI undergoing therapeutic pancreatectomy lack any functional K_{ATP} channel activity. To date, more than 50 mutations in both channel subunits have been identified in patients with the recessive form of PHHI. The first mutation demonstrating altered function of K_{ATP} channels was a PHHI associated missense mutation (G1479R) in the second nucleotide binding fold of SUR1 [64]. Recombinant channels encoded by this mutation in hamster SUR1 behave essentially normally with respect to single channel conductance and ATP-sensitivity in inside-out membrane patches. However, these channels do not respond to stimulation by MgADP, a defect that would lead to inability of the channel to respond to glucose deprivation and, consequently, to hyperinsulinemia and hypoglycemia [64].

We subsequently extended the functional analysis [65] to include other mutations that were identified in extensive analyses of SUR1 genomic DNA from PHHI patients. Rb^+ efflux assays were used to determine the conductance of recombinant K_{ATP} channels when activated by metabolic inhibitors. Channel activity followed the order, WT $\sim$ N188S > G1382S > F591L > G1479R > R1215Q > ΔF1388 $\sim$ zero. In each case, channel activity was lower in the mutant than in wild type channels. Lack of, or a reduction of, K_{ATP} channel activity in intact cells would be expected to cause persistent hyperinsulinemia, since the hyperpolarization of the pancreatic β-cells and consequent cessation of insulin secretion that normally results from K_{ATP} channel activation would be absent or reduced.

Lack of, or decreased, activity of mutant channels in the Rb^+ efflux assay may result from decreased expression of channel protein, normal expression of channel proteins that are without function, channels with altered regulation by intracellular nucleotides, or combinations of all three. With the exception of ΔF1388, all of these point mutants generated active channels in the absence of ATP. All mutant channels had essentially wild-type ATP sensitivity, but most had reduced stimulatory effect of MgADP [65]. The fact that a large number of SUR1 mutations identified in PHHI patients to date result in channels with attenuated responses to MgADP, suggests that this is a common mechanism underlying the disease. Diazoxide acts on the SUR1 subunit and stimulates K_{ATP} channel activity in pancreatic β-cells and has been used successfully to treat some forms of PHHI. As with MgADP stimulation, most PHHI mutants also showed a reduced sensitivity to diazoxide [65], potentially explaining the variable responsiveness of patients to these drugs. Although it is not yet clear exactly how MgADP and potassium channel opening drugs act on SUR1 to affect channel activity, recent studies on another NBF2 mutation (R1420C) suggest that reduced cooperative nucleotide binding between the two NBFs may underlie the effects of mutations [62,66,67]. Bienengraeber et al. [34] have recently demonstrated that the NBF2 of SUR2A harbors an ATPase activity that determines the activating ability of nucleotides. It is possible that some mutations may alter the ATPase activity at the NBFs, thereby affecting the response of the channel to nucleotides and diazoxide.

A number of SUR1 mutations result in no K_{ATP} channel activity when coexpressed with Kir6.2 [65]. Sharma et al. [68] have identified an anterograde trafficking signal at the C-terminus of SUR1 that is required for surface expression of the channel complex, and proposed that some of the SUR1 C-terminus truncation mutations may cause defective trafficking of the channel. Cartier et al. [69] recently demonstrated that the ΔF1388 mutation in SUR1 causes defective trafficking and lack of surface expression of functional K_{ATP} channels. The mutant protein appears to be retained in the ER, similar to ΔF508, a common mutation of CFTR that causes cystic fibrosis [70]. These studies establish that defective trafficking of K_{ATP} channels is a molecular basis of PHHI. Future studies are likely to reveal more mutations with trafficking defects.

In contrast to SUR1, only two PHHI mutations have been reported in Kir6.2 ([58,71], Fig. 2): Leu 147 to Pro and a nonsense mutation that truncates the protein after 12 amino acids. The truncation mutation, as expected, resulted in no functional channels. The L147P mutation also resulted in no functional channels, although the exact mechanism is not yet clear.

3.4. Correlation between in vitro and clinical findings

For some patients, the in vitro properties of the mutant channels seem to adequately explain the clinical picture. For example, two siblings, homozygous for the ΔF1388 mutation, had very severe disease and were unresponsive to diazoxide treatment, as expected since this mutant channel is completely inactive in vitro [60]. Another patient, compound heterozygous for the H125Q mutation and the ΔF1388 mutation had clinically mild disease, consistent with functional activity of the H125Q mutation [65]. However, the patient did not respond with decreased insulin secretion after diazoxide treatment, nor with increased insulin secretion after glucose or tolbutamide treatment. One patient, heterozygous for a previously described splice-site mutation (3992−9 g to a) and the R1394H mutation had severe clinical disease which was unresponsive to diazoxide, as expected from the observation that the 3992−9 g to a mutation is usually associated with severe non drug-responsive disease [65]. In contrast, a patient who was compound heterozygous for the 3992−9 g to a mutation and R1215Q mutation did have some clinical response to diazoxide, consistent with the observation that the R1215Q mutation has reduced sensitivity to stimulation by diazoxide [65].

Although for some mutations the in vitro findings correlate well with the clinical observations, this is not always the case. It is possible that some mutations alter transcription, translation, trafficking, or stability of the protein that have not been examined in detail. Alternatively, it is possible that some mutations are in fact benign polymorphisms that are in linkage disequilibrium with an unidentified disease-causing mutation.

3.5. Animal models of PHHI

Knockout of SUR1 in mice [72] results in spontaneous action potentials in pancreatic β-cells, like those seen in patients with PHHI [63], but the mice are normoglycemic unless stressed. SUR1(-/-) islets lack first phase insulin secretion and exhibit an attenuated glucose-stimulated second phase secretion, which leads to mild glucose intolerance. Also consistent with the human PHHI disease, glucose-dependent Ca signalling is abolished in β-cells in which the Kir6.2 gene is disrupted [73]. Miki et al. [73] generated a transgenic mouse overexpressing a dominant-negative Kir6.2[G132S] in β-cells, abolishing K_{ATP} channel activity. However this had only minimal effects on glucose homeostasis. In addition, in our hands, mice expressing a similarly dominant-negative transgene (Kir6.2[AAA], in which the structure of the channel pore is disrupted, abolishing ion permeation), also develop apparently normally and are fertile [74]. Thus, it seems that while some of the features of PHHI are reiterated in mice with reduced or abolished β-cell K_{ATP} channel, the details of insulin secretion or action are not identical, and the functional consequences for the animal are not as severe as in the human disease.

4. K_{ATP} channels and diabetes

There are two main types of diabetes mellitus: Type-1 and Type-2. Type-1 (or juvenile) diabetes is typically considered to be the consequence of autoimmune destruction of pancreatic β-cells, and consequent complete lack of circulating insulin. Type-2 non-insulin dependent (or adult onset) diabetes is a complex disease with multiple potential mechanisms. However, in many cases, the disease is a consequence of altered regulation of insulin signalling, rather than insulin production, since effective treatment can be obtained, at least for a period, by causing β-cell K_{ATP} channel blockade, and enhanced secretion, by treatment with sulfonylurea drugs. This suggests that defects in the link between blood glucose and insulin secretion, via the K_{ATP} channel, could be causal.

In addition to reduced K_{ATP} channel activity being a causal mechanism of PHHI, some forms of PHHI have been shown to result from an *enhanced* activity of glucokinase [75], which catalyses the conversion of glucose to glucose-6-phosphate, the first reaction of glycolysis. Overactive glucokinase will therefore increase the glycolytic flux, providing a stronger inhibitory signal ([ATP]/[ADP]) to the K_{ATP} channel, and hence increasing insulin secretion for a given glucose level. In direct contrast, one form of maturity onset diabetes of the young (MODY) has been shown to be frequently associated with *reduced* glucokinase activity [76–81]. Reduced glucokinase activity will reduce the glycolytic flux, hence lowering the [ATP]/[ADP] ratio, increasing K_{ATP} channel activity and reducing insulin secretion [53,82]. Thus, it would appear that mutations which render K_{ATP} channels overactive, either by increased channel density, reduced ATP-sensitivity, or increased MgADP sensitivity, would lead to reduced insulin secretion in the face of a glucose challenge. However, genetic studies are yet to reveal any marked association between K_{ATP} channel mutations and diabetes. This is surprising, since as discussed above, numerous mutations with the appropriate phenotype can be generated experimentally in K_{ATP} channel subunits, and as discussed below, we have demonstrated the potential experimentally in vivo utilizing a transgenic strategy.

4.1. Overactive K_{ATP} channels as a mechanistic basis of diabetes in transgenic mice

We attempted to generate transgenic mice expressing mutant Kir6.2 subunits that generate K_{ATP} channels with ATP-sensitivity reduced either 10-fold, or 400-fold, when transfected into COSm6 cells. For the more severely shifted mutation, we failed to identify the transgene in 60 potential founder mice. However, five founder mice were generated that expressed the less severely ATP-insensitive transgene [74] (Kir6.2[ΔN30, K185Q]-GFP, see Fig. 3) All of these founder mice were euglycemic, developed apparently normally, and were fertile, yet all transgenic F1 progeny from four of the five lines were severely hyperglycemic with hypoinsulinemia and exhibited significantly elevated blood

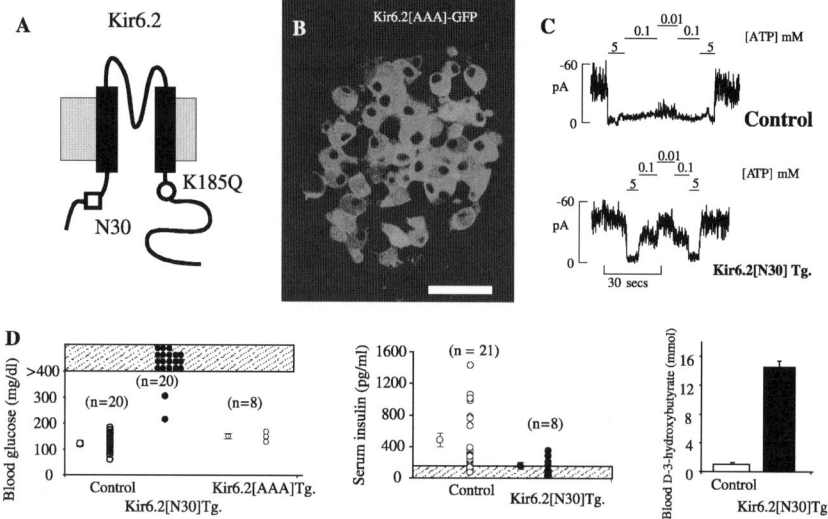

Fig. 3. Diabetic phenotype of Kir6.2[ΔN30] transgenic mice (A) Position of mutations in Kir6.2 transgene. (B) Confocal image of a slice through an islet from a transgenic animal expressing a GFP-tagged Kir6.2 subunit under control of the insulin promoter. Strong fluorescence is in most cells, and the fluorescence is excluded from the nucleus and localized most strongly at the cell periphery. (C) Currents in a control β-cell membrane patch, and in a patch from a transgenic β-cell expressing Kir6.2[ΔN30]-GFP. On average, the transgenic channels show ~5-fold reduction of ATP sensitivity [74]. (D) (left) Blood glucose levels in neonatal (day 2–4) control and Kir6.2[ΔN30] transgenic mice. Control mice show normal blood glucose levels (~120 mg/dl), whereas transgenic mice typically have blood glucose levels above the level of detection (>400 mg/dl). (middle) Serum insulin levels in neonatal (day 2–4) mice are normal (~500 pg/ml) in control mice, but at or below the level of detection (<200 pg/ml) in transgenic mice. (right) Blood ketone levels (D-3-hydroxybutyrate) are very low in control but highly elevated in transgenic neonatal mice. (Adapted from ref. [74] and unpublished.)

D-3-hydroxybutyrate levels. Neonatal lethality was uniformly observed in progeny from all four diabetic transgenic lines [74]. Most animals died by day five; of the few transgenic mice that survived to weaning, whole body weight was significantly reduced when compared to control mice and could only be increased by administering increasing doses of insulin.

The transgenic construct used in these studies included a green fluorescent protein (GFP) tag at the C-terminal end of the Kir6.2 coding sequence, and green fluorescence was present in β-cells in islets from transgenic mice but not in islets from control mice (Fig. 3). When Kir6.2 subunits are overexpressed in the transgenic β-cells, SUR1 is likely to be rate-limiting for heteromeric channel formation, such that the transgenic subunits will replace endogenous subunits, and an effective "functional knock-in" of the transgenic construct will be obtained. In accordance with this prediction, ATP sensitivity was reduced about 5-fold in transgenic β-cells, yet the total density of K_{ATP} channels was not obviously higher [74]. Importantly, islet morphology was normal in the trasnsgenic pancreas in the first few days after birth, and only after prolonged hyperglycemia (>3 days) was a characteristic, diabetes-induced, redistribution of α- and β-cells observed [74].

Taken together, these data indicate a dramatic effect of reduced K_{ATP} sensitivity of K_{ATP} channels on control of insulin secretion, and suggest a crucial requirement for appropriate control of K_{ATP} channel activity for normal regulation of insulin secretion. Importantly, the severely diabetic phenotype exhibited by the transgenic mice results from a comparatively small reduction of intrinsic ATP-sensitivity ($\sim$5-fold) and suggests that channel ATP-sensitivity is a critical determinant of the coupling between glucose and insulin secretion. As discussed above, site-directed mutational analyses show that mutations throughout the Kir6.2 structure can cause similar, and even far more significant, reductions of ATP sensitivity. Studies of diabetic patients are now gradually beginning to indicate linkage between diabetes and mutations of Kir6.2 or SUR1 [83–86], and it remains possible that untested diabetic populations suffer from mutations in these genes.

5. Conclusions and perspectives

Since the cloning of the genes encoding the constituent subunits of the K_{ATP} channel in 1995, there has been a rapid advance in understanding the link between K_{ATP} channel activity and insulin secretion. In particular, the mechanistic basis of PHHI has essentially been elucidated, and potential new treatment strategies are being developed in consequence. It is our own bias that mutations of K_{ATP}, or of genes regulating the expression of K_{ATP}, are also likely to be causally involved in Type II diabetes. Further linkage analysis, combined with expression analysis, and genetic manipulations in experimental animals, will be required to provide a full picture of the role of this channel in diseases of insulin secretion.

Acknowledgements

We are grateful to the numerous collaborators who have contributed to our own publications in this area. Our experimental work has been primarily supported by NIH grant DK55282 (to CGN), by NIH grant DK57699 and a Career Development grant from the ADA (to SLS), and by a Pilot and Feasibility grant from the Washington University NIH DRTC (to JCK).

References

[1] Ashcroft, F.M., Proks, P., Smith, P.A., Ammala, C., Bokvist, K., Rorsman, P., 1994. J. Cell. Biochem. 55, 54–65.
[2] Komatsu, M., Sharp, G.W., Aizawa, T., Hashizume, K., 1997. Jpn. J. Physiol. 47, S22–S24.
[3] Ashcroft, S.J., Ashcroft, F.M., 1990. Cell. Signal. 2, 197–214.
[4] Nichols, C.G., Lederer, W.J., 1991. Am. J. Physiol. 261, H1675–H1686.
[5] Findlay, I., 1988. Pflugers Arch. 412, 37–41.
[6] Findlay, I., Dunne, M.J., 1986. Pflugers Arch. Eur. J. Physiol. 407, 238–240.
[7] Dunne, M.J., Petersen, O.H., 1986. Pflugers Arch. Eur. J. Physiol. 407, 564–565.
[8] Dunne, M.J., Petersen, O.H., 1986. FEBS Lett. 208, 59–62.
[9] Findlay, I., 1988. Laboratoire de Physiologie Comparee (UA CNRS 1121), U.d.P.X., Orsay, France. J. Membr. Biol. 101(1), 83–92.

[10] Lederer, W.J., Nichols, C.G., 1989. J. Physiol. 419, 193–211.

[11] Kakei, M., Kelly, R., Ashcroft, S.J., Ashcroft, F.M. FEBS Lett. 208, 63–66.

[12] Dunne, M-J., Illot, M-C., Peterson, O-H., 1987. J. Membr. Biol. 99(3), 215–224.

[13] Trube, G., Rorsman, P., Ohno, S.T., 1986. Pflugers Arch. Eur. J. Physiol. 407, 493–499.

[14] Schmid-Antomarchi, H., De, W.J., Fosset, M., Lazdunski, M., 1987. J. Biol. Chem. 262, 15840–15844.

[15] Aguilar-Bryan, L., Nichols, C.G., Rajan, A.S., Parker, C., Bryan, J., 1992. J. Biol. Chem. 267, 14934–14940.

[16] Aguilar-Bryan, L., Nichols, C.G., Wechsler, S.W., Clement, J.P. IV, Boyd, A.E.r., Gonzalez, G., Herrera-Sosa, H., Nguy, K., Bryan, J., Nelson, D.A., 1995. Science 268, 423–426.

[17] Nichols, C.G., Lopatin, A.N., 1997. Annu. Rev. Physiol. 59, 171–191.

[18] Inagaki, N., Gonoi, T., Clement, J.P.t., Namba, N., Inazawa, J., Gonzalez, G., Aguilar-Bryan, L., Seino, S., Bryan, J., 1995. Science 270, 1166–1170.

[19] Shyng, S., Nichols, C.G., 1997. J. Gen. Physiol. 110, 655–664.

[20] Clement, J.P.t., Kunjilwar, K., Gonzalez, G., Schwanstecher, M., Panten, U., Aguilar-Bryan, L., Bryan, J., 1997. Neuron 18, 827–838.

[21] Inagaki, N., Gonoi, T., Seino, S., 1997. FEBS Lett. 409, 232–236.

[22] Tucker, S.J., Gribble, F.M., Zhao, C., Trapp, S., Ashcroft, F.M., 1997. Nature 387, 179–183.

[23] Zerangue, N., Schwappach, B., Jan, Y.N., Jan, L.Y., 1999. Neuron 22, 537–548.

[24] Chutkow, W.A., Samuel, V., Hansen, P.A., Pu, J., Valdivia, C.R., Makielski, J.C., Burant, C.F., 2001. Proc. Natl Acad. Sci. USA 98, 11760–11764.

[25] Babenko, A.P., Aguilar-Bryan, L., Bryan, J., 1998. Annu. Rev. Physiol. 60, 667–687.

[26] Hu, H., Sato, T., Seharaseyon, J., Liu, Y., Johns, D.C., O'Rourke, B., Marban, E., 1999. Mol. Pharmacol. 55, 1000–1005.

[27] Higgins, C.F., 1995. Cell 82, 693–696.

[28] Ashcroft, F.M., 1988. Annu. Rev. Neurosci. 11, 97–118.

[29] Nichols, C.G., Shyng, S.L., Nestorowicz, A., Glaser, B., Clement, J.P., Gonzalez, G., Aguilarbryan, L., Permutt, M.A., Bryan, J., 1996. Science 272, 1785–1787.

[30] Shyng, S., Ferrigni, T., Nichols, C.G., 1997. J. Gen. Physiol. 110, 643–654.

[31] Gribble, F.M., Tucker, S.J., Ashcroft, F.M., 1997. EMBO J. 16, 1145–1152.

[32] Ueda, K., Inagaki, N., Seino, S., 1997. J. Biol. Chem. 272, 22983–22986.

[33] Ueda, K., Komine, J., Matsuo, M., Seino, S., Amachi, T., 1999. Proc. Natl Acad. Sci. USA 96, 1268–1272.

[34] Bienengraeber, M., Alekseev, A.E., Abraham, M.R., Carrasco, A.J., Moreau, C., Vivaudou, M., Dzeja, P.P., Terzic, A., 2000. FASEB J. 14, 1943–1952.

[35] Shyng, S., Ferrigni, T., Nichols, C.G., 1997. J. Gen. Physiol. 110, 141–153.

[36] Koster, J.C., Sha, Q., Shyng, S., Nichols, C.G., 1999. J. Physiol. 515, 19–30.

[37] Reimann, F., Tucker, S.J., Proks, P., Ashcroft, F.M., 1999. J. Physiol. (Lond.) 518, 325–336.

[38] Babenko, A.P., Gonzalez, G., Bryan, J., 1999. Biochem. Biophys. Res. Commun. 255, 231–238.

[39] Tucker, S.J., Gribble, F.M., Proks, P., Trapp, S., Ryder, T.J., Haug, T., Reimann, F., Ashcroft, F.M., 1998. EMBO J. 17, 3290–3296.

[40] Enkvetchakul, D., Loussouarn, G., Makhina, E., Shyng, S.L., Nichols, C.G., 2000. Biophys. J. 78, 2334–2348.

[41] Loussouarn, G., Makhina, E.N., Rose, T., Nichols, C.G., 2000. J. Biol. Chem. 275, 1137–1144.

[42] Drain, P., Li, L., Wang, J., 1998. Proc. Natl Acad. Sci. USA 95, 13953–13958.

[43] Li, L., Wang, J., Drain, P., 2000. Biophys. J. 79, 841–852.

[44] Tanabe, K., Tucker, S.J., Matsuo, M., Proks, P., Ashcroft, F.M., Seino, S., Amachi, T., Ueda, K., 1999. J. Biol. Chem. 274, 3931–3933.

[45] Shyng, S.L., Nichols, C.G., 1998. Science 282, 1138–1141.

[46] Baukrowitz, T., Schulte, U., Oliver, D., Herlitze, S., Krauter, T., Tucker, S.J., Ruppersberg, J.P., Fakler, B., 1998. Science 282, 1141–1144.

[47] Xu, H., Wu, J., Cui, N., Abdulkadir, L., Wang, R., Mao, J., Giwa, L.R., Chanchevalap, S., Jiang, C., 2001. J. Biol. Chem. 276, 38690–38696.

[48] Beguin, P., Nagashima, K., Nishimura, M., Gonoi, T., Seino, S., 1999. EMBO J. 18, 4722–4732.

[49] Lin, Y.F., Jan, Y.N., Jan, L.Y., 2000. EMBO J. 19, 942–955.

[50] Light, P.E., Bladen, C., Winkfein, R.J., Walsh, M.P., French, R.J., 2000. Proc. Natl Acad. Sci. USA 97, 9058–9063.

[51] Aynsley-Green, A., Polak, J.M., Bloom, S.R., Gough, M.H., Keeling, J., Ashcroft, S.J., Turner, R.C., Baum, J.D., 1981. Arch. Dis. Child 56, 496–508.

[52] Aynsley-Green, A., 1982. Clin. Endocrinol. Metab. 11, 159–194.

[53] Sakura, H., Ashcroft, S.J., Terauchi, Y., Kadowaki, T., Ashcroft, F.M., 1998. Diabetologia 41, 654–659.

[54] Glaser, B., Chiu, K.C., Anker, R., Nestorowicz, A., Landau, H., Ben-Bassat, H., Shlomai, Z., Kaiser, N., Thornton, P.S., Stanley, C.A., 1994. Nat. Genet. 7, 185–188.

[55] Thomas, P.M., Cote, G.J., Hallman, D.M., Mathew, P.M., 1995. Am. J. Hum. Genet. 56, 416–421.

[56] Fantes, J.A., Oghene, K., Boyle, S., Danes, S., Fletcher, J.M., Bruford, E.A., Williamson, K., Seawright, A., Schedl, A., Hanson, I., Zehetrer, G., Bhogal, R., Lehrach, M., Gregory, S., Williams, J., Little, P.F.R., Sellar, G.C., Moovers, J., Mannens, H., Weissenbach, J., Junien, C., van Meyninger, V., Bickmore, W.A., 1995. Genomics 25, 447–461.

[57] Thomas, P.M., Cote, G.J., Wohllk, N., Haddad, B., Mathew, P.M., Rabl, W., Aguilar-Bryan, L., Gagel, R.F., Bryan, J., 1995. Science 268, 426–429.

[58] Thomas, P., Ye, Y., Lightner, E., 1996. Hum. Mol. Genet. 5, 1809–1812.

[59] Nestorowicz, A., Glaser, B., Wilson, B.A., Shyng, S.L., Nichols, C.G., Stanley, C.A., Thornton, P.S., Permutt, M.A., 1998. Hum. Mol. Genet. 7, 1119–1128.

[60] Nestorowicz, A., Wilson, B.A., Schoor, K.P., Inoue, H., Glaser, B., Landau, H., Stanley, C.A., Thornton, P.S., Clement, J.P.t., Bryan, J., Aguilar-Bryan, L., Permutt, M.A., 1996. Hum. Mol. Genet. 5, 1813–1822.

[61] Otonkoski, T., Ammala, C., Huopio, H., Cote, G.J., Chapman, J., Cosgrove, K., Ashfield, R., Huang, E.,Komulainen, J., Ashcroft, F.M., Dunne, M.J., Kere, J., Thomas, P.M., 1999. Diabetes 48, 408–415.

[62] Tanizawa, Y., Matsuda, K., Matsuo, M., Ohta, Y., Ochi, N., Adachi, M., Koga, M., Mizuno, S., Kajita, M., Tanaka, Y., Tachibana, K., Inoue, H., Furukawa, S., Amachi, T., Ueda, K., Oka, Y., 2000. Diabetes 49, 114–120.

[63] Kane, C., Shepherd, R.M., Squires, P.E., Johnson, P.R., James, R.F., Milla, P.J., Aynsley-Green, A., Lindley, K.J., Dunne, M.J., 1996. Nat. Med. 2, 1344–1347.

[64] Nichols, C.G., Shyng, S.L., Nestorowicz, A., Glaser, B., Clement, J.P.t., Gonzalez, G., Aguilar-Bryan, L., Permutt, M.A., Bryan, J., 1996. Science 272, 1785–1787.

[65] Shyng, S.L., Ferrigni, T., Shepard, J.B., Nestorowicz, A., Glaser, B., Permutt, M.A., Nichols, C.G., 1998. Diabetes 47, 1145–1151.

[66] Matsuo, M., Tanabe, K., Kioka, N., Amachi, T., Ueda, K., 2000. J. Biol. Chem. 275, 28757–28763.

[67] Matsuo, M., Trapp, S., Tanizawa, Y., Kioka, N., Amachi, T., Oka, Y., Ashcroft, F.M., Ueda, K., 2000. J. Biol. Chem. 275, 41184–41191.

[68] Sharma, N., Crane, A., Clement, J.P.t., Gonzalez, G., Babenko, A.P., Bryan, J., Aguilar-Bryan, L., 1999. J. Biol. Chem. 274, 20628–20632.

[69] Cartier, E., Conti, L.R., Vandenberg, C.A., Shyng, S.-L., 2003. Proc. Natl. Acad Sci. USA 98, 2882–2887.

[70] White, M.B., Amos, J., Hsu, J.M., Gerrard, B., Finn, P., Dean, M., 1990. Nature 344, 665–667.

[71] Nestorowicz, A., Inagaki, N., Gonoi, T., Schoor, K.P., Wilson, B.A., Glaser, B., Landau, H., Stanley, C.A., Thornton, P.S., Seino, S., Permutt, M.A., 1997. Diabetes 46, 1743–1748.

[72] Seghers, V., Nakazaki, M., DeMayo, F., Aguilar-Bryan, L., Bryan, J., 2000. J. Biol. Chem. 275, 9270–9277.

[73] Miki, T., Tashiro, F., Iwanaga, T., Nagashima, K., Yoshitomi, H., Aihara, H., Nitta, Y., Gonoi, T., Inagaki, N., Miyazaki, J., Seino, S., 1997. Proc. Natl Acad. Sci. USA 94, 11969–11973.

[74] Koster, J.C., Marshall, B.A., Ensor, N., Corbett, J.A., Nichols, C.G., 2000. Cell 100, 645–654.

[75] Glaser, B., Kesavan, P., Heyman, M., Davis, E., Cuesta, A., Buchs, A., Stanley, C.A., Thornton, P.S., Permutt, M.A., Matschinsky, F.M., Herold, K.C., 1998. N. Engl. J. Med. 338, 226–230.

[76] Vionnet, N., Stoffel, M., Takeda, J., Yasuda, K., Bell, G.I., Zouali, H., Lesage, S., Velho, G., Iris, F., Passa, P., Froguel, P., Cohen, D., 1992. Nature 356, 721–722.

[77] Froguel, P., Vaxillaire, M., Sun, F., Velho, G., Zouali, H., Butel, M.O., Lesage, S., Vionnet, N., Clement, K., Fougerousse, F., Tanizawa, Y., Weissenbach, J., Beckmann, S., Lathrop, G.M., Passa, P., Permutt, M.A., Cohen, D., 1992. Nature 356, 162–164.

[78] Velho, G., Froguel, P., Clement, K., Pueyo, M.E., Rakotoambinina, B., Zouali, H., Passa, P., Cohen, D., Robert, J.J., 1992. Lancet 340, 444–448.

[79] Sakura, H., Eto, K., Kadowaki, H., Simokawa, K., Ueno, H., Koda, N., Fukushima, Y., Akanuma, Y., Yazaki, Y., Kadowaki, T., 1992. J. Clin. Endocrinol. Metab. 75, 1571–1573.

[80] Guazzini, B., Gaffi, D., Mainieri, D., Multari, G., Cordera, R., Bertolini, S., Pozza, G., Meschi, F., Barbetti, F., 1998. Hum. Mutat. 12, 136.

[81] Ellard, S., Beards, F., Allen, L.I., Shepherd, M., Ballantyne, E., Harvey, R., Hattersley, A.T., 2000. Diabetologia 43, 250–253.

[82] Page, R., Hattersley, A., Turner, R., 1992. Lancet 340, 1162–1163.

[83] Inoue, H., Ferrer, J., Warren-Perry, M., Zhang, Y., Millns, H., Turner, R.C., Elbein, S.C., Hampe, C.L., Suarez, B.K., Inagaki, N., Seino, S., Permutt, M.A., 1997. Diabetes 46, 502–507.

[84] Hani, E.H., Boutin, P., Durand, E., Inoue, H., Permutt, M.A., Velho, G., Froguel, P., 1998. Diabetologia 41, 1511–1515.

[85] Ohta, Y., Tanizawa, Y., Inoue, H., Hosaka, T., Ueda, K., Matsutani, A., Repunte, V.P., Yamada, M., Kurachi, Y., Bryan, J., Aguilar-Bryan, L., Permutt, M.A., Oka, Y., 1998. Diabetes 47, 476–481.

[86] Gloyn, A.L., Hashim, Y., Ashcroft, S.J., Ashfield, R., Wiltshire, S., Turner, R.C., 2001. Diabet. Med. 18, 206–212.

The biology of voltage-gated sodium channels

John H. Caldwell[a] and S. Rock Levinson[b,*]

[a]Department of Cellular and Structural Biology, University of Colorado Health Sciences Center,
Denver, CO 80262, USA
[b]Department of Physiology and Biophysics, University of Colorado Health Sciences Center,
Denver, CO 80262, USA
[*]Correspondence address: Campus Box C 240, Department of Physiology and Biophysics,
University of Colorado Health Sciences Center, Denver, CO 80262, USA

1. Introduction

Voltage-gated sodium channels are the primary mediators of the propagating action potential of nerve and muscle, the fundamental phenomenon by which various elements of the nervous system communicate. In addition, sodium channels are the major targets of a variety of natural toxins as well as several clinically important drug classes, such as local anesthetics, antiarrhythmics, and antiseizure drugs. Finally, genetic defects in the structure of these moieties give rise to a number of human diseases, e.g. long Q–T syndrome and the periodic paralyses of muscle. For these reasons, sodium channels have been well studied in the past and will continue to be a focus of major investigative efforts for some time to come.

As with other ion channels, within a given organism sodium channels are encoded by a number of genes, while there are indications that RNA processing events that vary the protein sequences encoded by individual genes may generate further diversity. Consideration of the biological significance of such sodium channel isoforms is the major theme of this chapter. However, before doing this we first review the basic functional properties of sodium channels and their role in various excitation phenomena in nerve and muscle; this is followed by a brief account of the current state of knowledge of the molecular mechanism of these sodium channel functions. We next describe the molecular diversity of the sodium channel family and the various transcriptional, post-transcriptional, and post-translational mechanisms that are responsible. Finally we discuss the role of such diversity in the differential function, differential gene expression, and differential localization of sodium channel isoforms.

Advances in Molecular and Cell Biology, Vol. 32, pages 15–50
© 2004 Elsevier B.V. All rights of reproduction in any form reserved.
ISSN: 1569-2558 / DOI: 10.1016/S1569-2558(03)32002-8

2. Basic physiology of sodium channels

2.1. Role in generation of propagating action potentials

The basic functional properties of sodium channels have been known for nearly 50 years and are well described in a number of texts (e.g. Ref. [1]). For the naïve reader only a brief description is given here.

Simply defined, the action potential is a brief, propagating reversal in the transmembrane voltage. As originally described by the pioneering work of Hodgkin and Huxley (see Ref. [1]), this waveform is generated by the concerted flow of sodium and potassium ions across the cell membrane (Fig. 1). Decades of subsequent investigation have established that discrete molecular pores known as voltage-gated sodium and potassium ion channels mediate these flows. In particular, these channels are closed in the resting state, but are opened by any stimulus (e.g. by a postsynaptic potential at a synapse or a "generator" potential in a sensory receptor) that causes the resting membrane potential to decrease towards zero (i.e. to "depolarize"), exceeding a threshold level. Initially, this causes sodium channels to open or "*activate*", allowing sodium ions to be driven into the cell by their electrochemical gradient, resulting in the rising phase of the action potential which goes toward V_{Na} (approximately as given by the Nernst relation for sodium). In reality, V_{Na} is never quite reached because of two subsequent limiting processes: (1) the *inactivation* of sodium channels, thus shutting off the further influx of sodium ions, and (2) the *delayed opening* of potassium channels, which allows the outward flow of potassium ions from the cell driven by their own electrochemical gradient, thus returning ("repolarizing") the membrane potential to its original resting level. After a brief delay (the "refractory period"), during which sodium channels recover from inactivation and potassium channels close, another action potential may be generated.

The basic description above allows one to explain certain fundamental properties of the action potential, namely the "all-or-none" nature of its amplitude and its propagation. Both phenomena are related to the fact that in excitable tissues the influx of sodium ions through the first few opening sodium channels causes an increased local depolarization that accelerates the opening of some of the surrounding channels; influx of sodium through

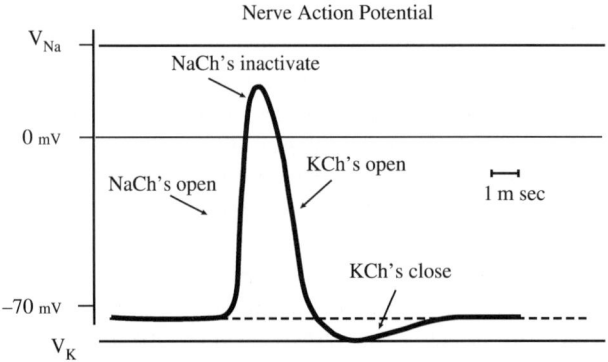

Fig. 1. Ionic flows involved in generation of the action potential.

these channels in turn further depolarizes the membrane, thus increasing the opening rate of more sodium channels, and so on. This "regenerative" effect quickly becomes the dominant stimulus for sodium channel activation, hence overwhelming the influence of the original stimulus. As a result, all subsequent processes in the action potential (e.g. sodium channel inactivation, potassium channel opening) are automatically entrained by the regenerative phenomenon, and thus the action potential amplitude and shape are independent of the initial depolarizing stimulus. This explosively regenerative upswing in the membrane potential also propagates in domino-like fashion, as channels adjacent to this activation zone are recruited into a conducting state. In the simplest case of impulse propagation along nerve axons, it appears that the density and gating properties of ion channels are constant along the axon length; thus all supra-threshold stimuli at one end of the axon evoke propagating action potentials of the same size and velocity, while sub-threshold stimuli fail to evoke any propagating response.

2.2. Conduction modes in the nervous system: continuous versus saltatory

The basic description above applies to impulse conduction in which channels are more or less uniformly and diffusely distributed over the cell membrane. Such a scheme applies mainly to action potential propagation in unmyelinated axons. However, in vertebrates a much faster and more energetically efficient mode of conduction has evolved based on the insulation of long stretches of inexcitable axon by myelinating glia. In this mode action potentials are generated only at small, regularly spaced gaps in the myelin called nodes of Ranvier. The influx of sodium at such nodes, by virtue of the insulating and capacitance-reducing properties of myelin, is capable of rapidly activating the sodium channels at the next node of Ranvier, roughly 1 mm or some 100 axon diameters away. Thus in myelinated fibers impulses jump very rapidly from one node to the next in this *saltatory* mode of propagation. The role of myelin makes this mode of conduction susceptible to a number of pathological demyelinating conditions that result in compromised action potential propagation (e.g. multiple sclerosis, Guillain–Barré syndrome [2]).

2.3. Action potentials of skeletal muscle and cardiac tissue

Propagating action potentials in skeletal muscle and cardiac tissues are more complex in terms of the ion channels involved in their generation. While the basic action potential in skeletal muscle is mediated by voltage-gated sodium and potassium channels much as in nerve axons, the muscle surface action potential is further shaped by the flow of chloride currents through non-gated chloride channels. Further, when this action potential invades the t-tubular system, it becomes modified by the opening of calcium channels, which are involved in the process of excitation–contraction coupling. In the heart, the nature of the action potential depends on the cell-type in which it arises. Thus the contractile cells of the atria and ventricles and the specialized His Purkinje conduction cells have fast-rising, plateau-phased action potentials mediated by sodium, calcium, and a number of potassium channels, while pacemaker cells do not express significant sodium currents at all but have slow rising, rounded action potentials whose rising phase is mediated by calcium channels.

2.4. Role of sodium channels in the initiation of action potentials and in integrating multiple synaptic signals at dendrites

In contrast to the all-or-none properties of the propagating action potential, sodium channels also are involved in initiating action potentials and in responding to more graded signals such as stimuli transduced by sensory receptors or synaptic potentials in neuron dendrites.

One essential role of sodium channels is to reinitiate the action potential on the postsynaptic side as a result of synaptic transmission. Thus in the motor endplate sodium channels are located in the folds of the postsynaptic membrane [3]. Both the high density of channels [4] and the geometry of the folds themselves [5] allow endplate potentials to initiate reliably a propagating action potential in the postsynaptic membrane. In neurons, sodium channels are present at high density in the initial segment of the axon ("axon hillock"). These channels initiate an action potential only when synaptic inputs at dendritic synapses summate sufficiently to depolarize the membrane potential at the hillock. Clearly, in both of these cases the voltage-sensitivity of the sodium channels will determine the strength of synaptic signal required to fire an action potential in the postsynaptic cell. In addition, a very low and nonuniform density of sodium channels has been inferred to exist in the dendrites themselves. These channels are thought to be part of the mechanism by which dendrites perform signal processing of multiple synaptic inputs to produce an appropriate output at the axon hillock [6–9]. Finally, sodium channels are found in certain sensory receptors, e.g. those for pain [10–12], taste, and sound, where they are part of the transduction mechanisms that amplify weak sensory stimuli.

In summary, recent work shows that sodium channels play a number of important yet different roles in neuronal communication and sensory transduction. The multiplicity of such functions further suggests that a number of functionally distinct sodium channel types will exist to subserve these different roles.

3. The structural basis for ion channel function

3.1. Basic functional properties of sodium channels

The microscopic functional properties of individual sodium ion channels that underlie the above macroscopic phenomena may be categorized as follows. First, sodium channels consist of a *transmembrane aqueous pore* that allows for the passage of ions across the hydrophobic barrier of the plasma membrane. Second, within this pore reside the mechanisms of *ion selectivity* that allow the channel to translocate sodium over the various other ions in physiological solutions (e.g. potassium, chloride, calcium). Finally, there are *voltage-sensitive activation and inactivation gates* that allow channels to assume the various closed, open, and inactivated states in response to changes in membrane voltage.

3.2. Functional characterization of voltage-dependent gating

Membrane potential may be controlled with a suitable voltage-clamp apparatus, allowing gating to be analyzed in terms of its *steady state* or *kinetic properties*. For both steady state activation and inactivation a sigmoidal relationship is obtained between membrane potential and the resultant number of open channels of a population (or the fractional open time of a single channel). It has been traditional to interpret these curves as reflecting gating transitions between two energy levels (i.e. those between the closed/resting and open or inactivated states). The Boltzmann fits to such data yield both a voltage midpoint "$V_{1/2}$" (i.e. the voltage at which opening is 50% maximal) and a "z" parameter related to the steepness of the sigmoidal curve. While in theory these fit parameters reflect the thermodynamics of the underlying gating process (e.g. the effective charge of a gating sensor), in practice they are better used in a relative sense to measure *changes* in gating behavior or differences between sodium channel isoforms.

The *rate* at which channels activate as a function of voltage is thought to reflect the activation energy barrier to the transition between open and closed states. A primary observation in kinetic measurements lies in the delayed onset of currents in response to activating voltages: this has been interpreted to reflect the need for transitions of multiple gating mechanisms in a single channel (see Ref. [1]). Fits of this data to the original formalism developed by Hodgkin and Huxley suggested models in which from 2–4 gating particles must move in a quasi-independent manner for channel opening. Studies using single channel and gating current analysis also support the notion of multiple gating sensors [13].

We next consider the molecular structure of voltage-gated ion channels and the domains that mediate pore formation, ion selectivity, and voltage gating. Progress in this area has largely been made by a combination of biochemical, genetic, and nucleic acid manipulation techniques [14]. In these endeavors the main goal has been the description of the subunit composition of these channels and the amino acid sequence of their subunits. However, an immediate (and perhaps surprising) finding from these studies was that ion channels are very diverse at the molecular level. We will discuss the potential significance of this diversity later.

3.3. Subunit composition and amino acid sequence

Biochemically, many voltage-gated ion channels are hetero-oligomeric assemblies of subunits. However, it is now known that for sodium channels the basic functional structure (i.e. that which forms voltage-gated, ion selective pores) consists of one subunit type, usually designated as "α". This is a very large polypeptide of approximately 2000 amino acid residues in length and the amino acid sequence of these proteins has been determined from their cloned cDNAs. With this information, predictive models of the secondary and higher order structure have been made using a combination of hydropathy and secondary structure prediction analysis. Thus, predicted transmembrane domains are assigned to sequences of appropriate length (i.e. about 20 residues) that contain a large proportion of hydrophobic amino acids with a high probability of forming α-helices

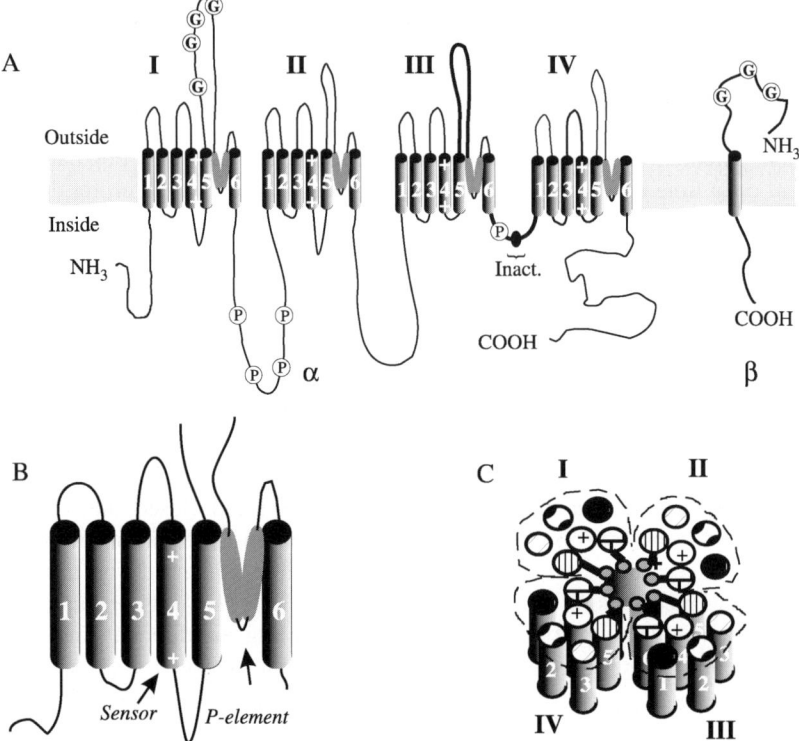

Fig. 2. A. Predicted secondary structure of typical sodium channel α and β subunits. Shown are approximate locations and numbers of potential phosphorylation sites ("P" circles), glycosylation sites ("G" circles) and the inactivation domain ("inact."). B. Enlarged view of a transmembrane homology domain, showing segments postulated to comprise the voltage sensor and pore lining/selectivity filter ("p-element"). C. Staves of a barrel tertiary structure involved in pore formation by α subunit.

(a thermodynamically stable protein structure in a hydrophobic environment). In addition, the sequences themselves have been analyzed for homology with one another and for the presence of internal repeats.

The results of such predictive models for the first sodium channel clones produced are shown in Fig. 2 (see Ref. [14]). First, based on hydropathy and secondary structure predictions sodium channels consist of four transmembrane domains (I, II, III, and IV), each of which consists of six hydrophobic α-helices (S1–S6). Second, this architecture suggests that the four domains might contain segments (i.e. the α-helices) of highly (but not completely) conserved sequence homology to one another. These "internal repeats" have been confirmed by homology analysis.

In addition to these primary α subunits, many channels are associated with one or several "accessory" β subunits [15]. While not directly part of the actual pore, selectivity, or gating structures, there is increasing evidence that such subunits play important roles in regulating channel numbers, localization, and gating, as is discussed below.

3.4. Functional models of ion channel mechanisms based on amino acid sequence analysis

Given the primary structure (amino acid sequence) and secondary structure predicted as above, what conjectures may be formed regarding the basic ion channel functions of pore formation, selectivity, and gating?

3.4.1. Pore formation

The 4-fold pseudosymmetry of sodium channel domains has been taken to imply that transmembrane pores are formed by a "staves of a barrel" arrangement in which each internal repeat domain forms one quarter of the pore structure (Fig. 2, see Ref. [14]). This architecture is highly similar to that of many other ion channels as described in other chapters, e.g. potassium channels that are formed by a 4-fold association of individual subunits shown. In fact, it has been proposed that a potassium channel gene was prototypic, and that evolutionarily this element underwent a series of gene duplications to form the large, four-domain sodium and calcium channels. Further, the staves of a barrel architecture has been confirmed in a number of studies using functionally altered mutant domains [16].

3.4.2. Ion selectivity

Previous biophysical studies had suggested that the ion selectivity apparatus lay within the lining of the transmembrane pore itself, and consisted of a narrowing of the pore to form a "selectivity filter" (see Ref. [1]). An early hypothesis for ion selectivity was that fully hydrated ions would be too large to pass through the filter by themselves, but that upon binding to the filter they would shed their associated water and become small enough to permeate past the barrier (Fig. 3, [1]). Selectivity among ions would thus occur through

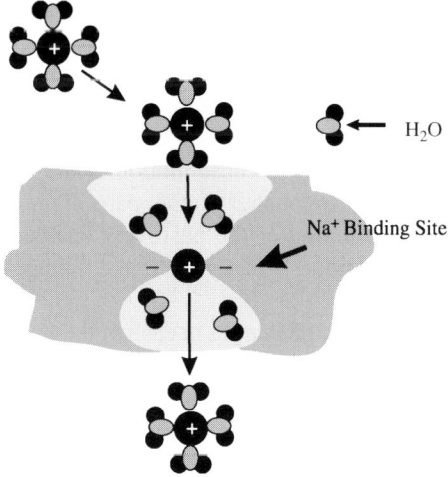

Fig. 3. "Selectivity filter" model of ion selectivity in the pore of sodium channels.

differences in their relative affinity for binding to chemical groups on the filter, such as negative charges and dipoles.

To identify domains that might contain such a selectivity filter, more sophisticated hydropathy analysis has been used to identify "amphipathic" segments, i.e. helices in which hydrophilic and hydrophobic amino acids are arranged on opposite sides of the long axis [17]. This chemical separation of side chains would allow the helix to associate with the major hydrophobic transmembrane helices while the hydrophilic side chains would form a polar environment highly conducive to the movement of ions. Such segments were thus identified in calcium, potassium, and sodium channels, residing in the sequences between helices S5 and S6, and have been called variously as "SS1/SS2", "H5" and more recently the "P-element" or "P-loop" (Fig. 2). Thus channels are postulated to contain four such loops, each one of which forms a quarter-sector of the cylindrical pore lining.

Evidence for a crucial role in permeation and selectivity has been obtained through mutagenesis in the P-loop regions, in which permeation may be reduced or eliminated and/or selectivity altered. For example, the P-loop is rather homologous between sodium and calcium channels, the main difference being substitutions of neutral or positively charged amino acids in sodium channels for conserved negatively charged glutamates in two of the P-loops in calcium channels. When negatively charged, these amino acids have been postulated to form a second calcium binding site that was suggested by earlier electrophysiological experiments. Thus replacing the appropriate uncharged amino acids in sodium channels with glutamates results in a channel that becomes calcium selective [18].

3.4.3. Voltage-sensitive gating

Perhaps the most strikingly conserved motif among voltage-sensitive ion channels is that of helix S4, consisting of a repeated triad of two very hydrophobic amino acids (usually leucine, isoleucine, or valine) followed by a positively charged amino acid (arginine or lysine) (Fig. 4, [19]). This structure is highly suggestive of a voltage sensor lying within the transmembrane electrical field [20]. The basic idea is that sufficient changes in transmembrane potential will cause motion of the channel in the membrane, and this motion is then mechanically coupled to other parts of the channel structure to open the pore. Thus mutagenic substitution of the positively charged amino acids for either neutral or negatively charged residues usually alters the voltage-dependence of channel opening [16,21].

For sodium channels, the number of charges in each of the four S4 regions in individual domains varies between 4 and 8, suggesting that they contribute differentially to the overall voltage-sensitive gating behavior of sodium channels [13]. In fact, recent advanced biophysical studies suggest that S4 segments in domains I and II seem to determine the kinetics and voltage sensitivity of activation, while those in domains III and IV appear to be more involved in the inactivation process (e.g. perhaps in creating the "receptor" pocket for the inactivation "ball" discussed below). In any case, the exact mechanisms by which

The S4 Motif (HH+)

		+	+	+	+	+	+	+	+	+
Electric eel NaCh IV-S4		L F R	V I R	L A R	I A R	V L R	L I R	A A K	G I R
Rat brain NaCh IV-S4		L F R	V I R	L A R	I G R	I L R	L I K	G A K	G I R
Rat skeletal muscle NaCh IV-S4		L F K	V I R	L A R	I G R	V L R	L I R	G A K	G I R
Drosophila NaCh S4		L L R	V V R	V F R	I G R	I L R	L I K	A A K	G I R
Rabbit skel. muscle CaCh IV-S4		S S A	F F R	L F R	V M R	L I R	L L S	R A E	G V R
Mouse brain KCh S4		I L R	V I R	L V R	V F R	I F K	L S R	H S K	
Drosophila Shaker S4		I L R	V I R	L V R	V F R	I F K	L S R	H S K	

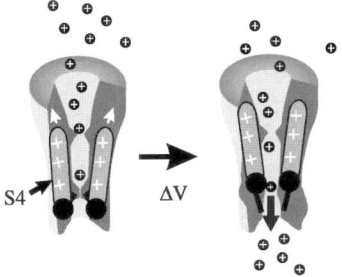

Fig. 4. The S4 voltage sensor of voltage-gated channels. A. Conservation of the basic hydrophobic–hydrophobic-charge motif among voltage gated channels from widely different species. B. Cartoon of postulated S4 sensor movement during voltage-induced gating.

S4 and other channel elements create voltage-dependent activation are still unknown and under intense investigation [13].

The mechanism of channel inactivation seems more certain. Evidence from previous electrophysiological studies suggested that inactivation for ion channels generally occurred through a "ball and chain" mechanism (Fig. 5, see Ref. [1]). For sodium channels, this segment lies in a highly conserved cytoplasmic sequence between domains III and IV, in which the inactivation ball lies within a chain that is tethered at both ends [20]. The role of this region in inactivation has been confirmed in a series of mutagenesis experiments, which have identified critical residues that form the ball and chain structure [22,23]. Mutagenesis has also identified candidate sites within the channel to which the ball binds to occlude the pore ([24–26]).

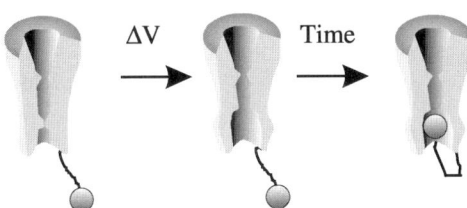

Fig. 5. The "ball and chain" model of voltage-dependent inactivation. The single-tethered ball shown is found on the N-terminal in inactivating potassium channels; for sodium channels, this structure lies within the sequence between domains III and IV, hence is a ball tethered by two chains (see Fig. 1).

4. Molecular diversity

As described earlier, sodium channels consist of an α subunit that forms the pore and accessory β subunits. Ten genes that code for sodium channel α subunits and three for β subunits have been identified in mammals. The nomenclature that evolved for the different α isoforms is complex, and Goldin et al. [27] have proposed a simpler nomenclature similar to that used for calcium and potassium channels. That convention will be used here. There are two families of α subunits, Na_v1 and Na_v2. All but one of the α subunits fall into the Na_v1 family. Na_v2 is a separate, "atypical" sodium channel family. None of the atypical sodium channels has yet been successfully expressed in heterologous systems, raising questions as to whether it is a functional sodium channel in vivo. Each of the three β subunits has a single transmembrane domain with a short cytoplasmic tail and an extracellular domain that has Ig-like motifs [15]. Although there has been no systematic study describing all the molecular isoforms generated by each sodium channel gene, it is likely that each gene gives rise to a geometric increase in protein isoforms due to alternative splicing and editing of mRNA, to post-translational modifications, and to different combinations of subunits (Fig. 6). As described below, we are now aware of the enormous molecular diversity of sodium channels without knowing what unique role is played by each isoform.

Alternative splicing of sodium channel mRNA has been reported for many isoforms of rodent sodium channels [28–34] although the only study focused specifically upon the extent of splicing has been in Drosophila [35]. In one region of the Drosophila *para* sodium channel there are 48 theoretically possible splice isoforms; 19 have been observed and it is likely that a more extensive search would reveal more of these possible isoforms [36]. Moreover, only a fraction of the coding region of *para* was examined for splicing, implying that many more splice isoforms exist. In addition, the type and extent of splicing varies developmentally in *para* and in rat $Na_v1.2$ [32,35]. Splicing has also been found for one of the β subunits [37]. Several basic questions remain about the extent and consequences of alternative splicing in sodium channels. The possibility that the splicing is spatially regulated, e.g. differing between tissues and between cell types within a tissue, has not been tested. In addition, splice isoforms could alter not only kinetic behavior but also associations with other proteins that regulate targeting or modulation, and these

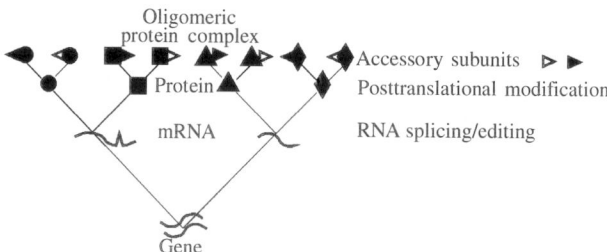

Fig. 6. Geometric expansion of molecular isoforms of sodium channels due to alternative splicing and/or editing of RNA, posttranslational modifications of the protein (glycosylation, fatty acylation, and phosphorylation), and co-assembly with various beta subunits.

functional consequences of splicing have yet to be determined. Some reported splice variants would result in truncated proteins or in proteins that are likely to be nonfunctional [28,33], and these also have not been tested for function. Another potential concern stems from the existence of multiple splice isoforms for one gene. Since published cDNAs are often pieced together from overlapping PCR amplification reactions or from partial sequences from cDNA library screening, it is conceivable that some published full length cDNAs never exist in vivo because some of these fragments may include splicing isoforms that are mutually exclusive.

A second mechanism for diversification of mRNA is pre-mRNA editing. This occurs at four sites in the *para* sodium channel of Drosophila and results in amino acid changes [38,39]. The editing mechanism is probably similar to that found for other channels because they are all a change of adenosine to inosine, which is interpreted as guanosine by tRNA, and because the predicted secondary structure of the RNA is similar to that required for other RNA editases. Editing has not been reported for mammalian sodium channels.

Post-translational modifications of sodium channels include both permanent changes (glycosylation and fatty acylation) as well as more transient changes (phosphorylation). All sodium channels have consensus sites for glycosylation in extracellular domains. Most of the channels when separated on an SDS gel appear as diffuse bands of approximately 250 kD, which is larger than the molecular weight predicted from the amino acid sequence, presumably due to the post-translational addition of sugar. Functional consequences of glycosylation have been studied and are described below. The eel electroplax sodium channel (the ortholog of the rodent skeletal muscle channel) is heavily acylated with about 100 fatty acid molecules covalently linked to the channel. Thus, the eel channel provides its own immediate lipid environment. Acylation has also been described in mammalian brain channels, but has not been quantified [40]. Phosphorylation of sodium channels is mediated by both protein kinases A and C [41]. Multiple PKA sites exist on the I–II cytoplasmic linker and a PKC site is found on the III–IV linker. Phosphorylation of some PKA sites is dependent on phosphorylation of a PKC site [41]. The functional effects of phosphorylation are dependent on the kinase, and include changes in amplitude and kinetics (see below).

Most α subunit isoforms are thought to be associated with one or more β subunits. $\beta 2$ is covalently linked to the α subunit by a disulfide bond, and the stoichiometry suggests 1:1 binding. Since $\beta 2$ is not expressed in skeletal muscle (see below), $Na_v 1.4$ only associates with $\beta 1$ and/or $\beta 3$. The binding of $\beta 1$ is also thought to be 1:1 [15]. However, it is not clear whether each α subunit always binds one, two, or three different β subunits. It is also possible that binding of $\beta 1$ and 3 is reversible, which would affect channel kinetic behavior on a moment-to-moment time scale. Based on the somewhat complementary distributions of $\beta 1$ and $\beta 3$, it has been suggested that the α subunit binds either $\beta 1$ or $\beta 3$ [42]; however, this has not been tested. Further, it has been suggested from co-immunoprecipitation experiments and functional studies that not all α subunits are associated with β's ([43, 44], and below). Finally, it has been reported that during brain development Nav1.2 α subunits are initially not complexed to $\beta 2$, but only later become associated with accessory subunits [43]. These speculations about β subunit composition illustrate the current level of uncertainty about which β subunits associate with which α subunits.

5. Physiological significance of sodium channel diversity

A major current research thrust concerns the reasons for ion channel diversity. Most investigators agree that there are three basic possibilities: (1) Functional variations may exist among isoforms that provide for the specific excitation requirements of different cell types. For example, fast kinetic channels are suited for the transmission of high frequencies of action potentials in myelinated fibers, while slow kinetic gating of certain isoforms may be required for the long duration action potentials in smooth muscle or cardiac tissue. (2) Amino acid sequence variations may encode signals that allow channel isoforms to be selectively localized to different membrane areas or specializations (e.g. nodes of Ranvier, neuron dendrites, muscle fiber t-tubules) through interactions with transport, targeting, or clustering molecules. (3) The noncoding region of channel isoform genes may contain unique regulatory elements that would allow cell, tissue or developmentally specific expression of a given isoform. For example, the regulation of cardiac function probably requires sodium channel expression to respond to very different processes and factors than the development of excitability in the nervous system. Note that these three different explanations are not exclusive of one another. At present, there is good evidence that all three explanations of ion channel diversity apply in at least one example. Below we discuss these examples.

5.1. Functional variations among sodium currents in cells and tissues in vivo

In addition to their central role in mediating the propagating action potential, sodium channels have been recently recognized to play an increasing number of roles in sensory transduction and in signal processing events (see above). Hence one might expect that functional differences among sodium channel isoforms would be found that reflect the varied requirements of these roles. Before considering the evidence that functional differences among isoforms exist, we first discuss the variations in the functional properties of endogenous sodium currents among various tissues.

In fact, in classical studies voltage-dependent sodium currents have shown somewhat modest functional differences among a range of tissues and species (see discussion in Ref. [1]). Kinetically, some tissues show fairly fast activating and inactivating currents, while others are slower; however the time constants for such processes are usually within a factor of two of one another. This contrasts with the much larger variations seen for potassium current kinetics (for example see other chapters). The voltage-dependences of activation and inactivation were also rather similar among a variety of tissues, while differences in the ion selectivity of sodium channels have been found to be relatively minor.

However, striking differences were seen in the sensitivities of various endogenous currents to toxins and certain drugs. Thus the guanidinium toxins tetrodotoxin and saxitoxin strongly inhibited sodium channels in mature mammalian skeletal muscle, peripheral nerve fibers, and most CNS neurons, while sodium currents in heart and neonatal muscle were found to be about 3 orders of magnitude less sensitive to these compounds. Furthermore, sensory neurons from the dorsal root ganglia of several species have been found to express both tetrodotoxin-sensitive and resistant currents in the same cells, and these resistant channels in the DRG are much more tetrodotoxin-insensitive than those of the heart [45,46]. Finally, heart sodium currents are about 100 times more sensitive to block by the drug lidocaine than those in other tissues; this differential sensitivity is the basis for the selectivity of lidocaine as an antiarrhythmic agent. These pharmacological differences were the first clues for the existence of sodium channel isoforms. However, it is not clear whether these differences are relevant to normal physiological differences among isoforms.

More recently, with the advent of advanced recording and preparation techniques, several examples of novel sodium currents have been described. For example, certain cerebellar neurons display fast sodium currents that appear to inactivate normally on membrane depolarization. However, when the membrane is repolarized, these Purkinje cell currents transiently reappear instead of remaining inactivated during the refractory period [165]. Such "resurgent currents" are thought to reflect the unusual ability of the underlying channels to reopen from an inactivated state during their return to the resting closed state. The functional role of such currents is not presently understood. Another recent example is the discovery of extremely fast-gating sodium channels in the presynaptic terminals of synapses that convey sensory information in the auditory system. This system must encode intensity, frequency, and spatial information about sounds in the environment. Highly sophisticated methods allow one to study these currents in presynaptic terminals of large, cap-like synapses. It has been found here that the action potentials are of very short duration; this is due to the ability of sodium channels in or near the calyx to inactivate about an order of magnitude faster than normal channels do ([47], R. Leao and H. von Gersdorff, personal communication). It is thought that this rapid gating allows fast conduction and subsequent comparison of the time of arrival of sounds in one ear versus the other, thus allowing sounds to be located in space.

5.2. Differences in gating among sodium channel isoforms

Are the unusual properties of sodium currents in the two above examples due to the unique functional characteristics of individual isoforms? Alternatively, are they caused by cellular modifications to the "stereotypic" manner in which all sodium channels gate, or could they be due to a combination of these two mechanisms? To address the above questions one must be able to characterize the functional properties of individual sodium channel isoforms in isolation. However, such studies are problematic for two basic reasons. First, there are very few examples of tissues or cells in which a homogenous population of a single isoform is known to occur naturally (perhaps the only examples are those of mature skeletal and cardiac muscles, which express relatively pure populations of $Na_v1.4$ and 1.5 isoforms, respectively). Thus, at present one must resort to comparing functional properties for isoforms expressed exogenously. However, this approach is limited in turn by observations that sodium channels function differently when expressed in exogenous systems, such as *Xenopus* oocytes, when compared to their function in parent tissues in situ. The most well-known examples of this are the dramatically slowed kinetics of inactivation of sodium currents resulting from mRNA injections into oocytes [48]. Another example is the often-striking difference in the voltage sensitivity of gating in expression systems when compared with parent tissues [49,50]. Such differences may reflect variations in the modulation of channel function by post-translational processes, the differential association with "accessory" subunit proteins (described below), or variations in the cellular environment (e.g. membrane lipid composition).

Thus, while there are numerous reports in the literature on sodium channel isoform gating characteristics in the literature, there are often large discrepancies in gating properties between reports, and few have directly compared gating between isoforms in the same system. As an exception, the major skeletal muscle isoform ($Na_v1.4$) and the cardiac muscle isoform ($Na_v1.5$) have indeed been comparatively studied by a number of investigators (e.g. Ref. [51,52]). All have found that $Na_v1.5$ steady state gating behaviors occur at much more negative potentials (i.e. ~ 20 mV) than $Na_v1.4$. Significant differences in the voltage and time-dependence of gating of the neuronal isoforms $Na_v1.1$, 1.2, and 1.6 were also found in another thorough comparative study using oocyte expression [53]. Significantly, when these isoforms were co-expressed with the $\beta1$ subunit (see below), these differences were largely eliminated. Finally, although not directly comparative, the kinetics of gating of $Na_v1.7$ and $Na_v1.8$ isoforms expressed in commonly used mammalian cell lines have been noted to be characteristically slower than other isoforms [46,54].

5.3. The effect of β subunits on sodium channel function

The *Xenopus* oocyte expression system has been widely used to study sodium channel mechanisms. When mRNAs encoding sodium channel α subunits are injected into oocytes, they induce sodium currents that display unusually slow inactivation kinetics [48,53]. Subsequent work has shown that fast inactivation can be restored in oocytes if $\beta1$ subunits are co-expressed with α [15,55–57]. In addition, the steady state voltage-dependence of activation and inactivation are also displaced toward more negative membrane potentials with $\beta1$ co-expression. Such observations suggested that $\beta1$ subunits were required for normal sodium channel gating (see Ref. [15]).

In addition, there have been reports that different isoforms may be differentially affected by co-expression with the $\beta1$ subunit. For example, the heart isoform ($Na_v1.5$) shows relatively normal inactivation kinetics when expressed alone in oocytes, and its gating properties are affected relatively little upon $\beta1$ co-expression [58]. Further, in the comparative study of $Na_v1.1$, 1.2, and 1.6 cited above [53] and in a separate comparative study of $Na_v1.2$ and 1.3 [56], $\beta1$ had differential effects on the gating properties among all four isoforms. Most recently, the normally slow inactivation kinetics of the $Na_v1.8$ isoform in oocytes were found to be unaffected by $\beta1$ co-expression [46]. These differential affects of $\beta1$ on the gating properties of sodium channel isoforms suggest a possible manner of sodium current modulation through selective $\beta1$ expression and association. Further support for such an idea comes from co-immunoprecipitation studies that show that $\beta1$ physically associates with some isoforms ($Na_v1.2$, 1.4, and 1.5) but not others ($Na_v1.7$) [44].

However, recent developments have challenged this notion of $\beta1$ modulation. In particular, it has been demonstrated that one can restore normal fast inactivation to α-alone expressing oocytes by stretching the oocyte plasma membrane [59,60].

The voltage-sensitivities of gating were also identical to those of $\alpha-\beta1$ induced currents. In these studies, evidence was found that disruption of the oocyte cytoskeleton caused inactivation kinetics to become normal [60]. In addition, in mammalian cells α subunits expressed without $\beta1$ display fast inactivation kinetics, which co-expression with $\beta1$ only modestly changes [54,61–63]. Overall, it has been proposed that in the oocyte expression system a unique interaction of α subunits with the cytoskeletal elements results in abnormal gating. In this view $\beta1$ does not directly confer normal properties on α, but rather *reverses the abnormal influence* of the oocyte cytoskeleton. On the other hand, in mammalian-based expression systems such interactions might not occur, and thus $\beta1$ contributes relatively modest effects to channel function, possibly via a passive electrostatic mechanism (see below).

Other β subunits (e.g. $\beta2$, $\beta3$) have been found to have similar but lesser effects on sodium channel function in the oocyte expression system. However, whether β accessory subunits play a major role in shaping sodium channel function in vivo is not clear, and instead investigations now focus more on roles for these moieties in sodium channel surface expression, clustering, or targeting (see below).

5.4. Sodium channel isoforms may be differentially modulated by intracellular second messenger cascades

A general property of ion channels is the ability of their functional properties to be acutely modified by intracellular signaling processes ([64], other chapters). For sodium channels, numerous studies have shown that sodium channels are biochemical substrates for the protein kinases PKA and PKC (reviewed in Ref. [20]). Further, in expression systems the resultant phosphorylation of channels has been shown to have several functional consequences. In particular, PKA-induced phosphorylation of the $Na_v1.2$ isoform reduces the magnitudes of sodium currents without affecting gating properties [65]. Further, it has been shown that the channel-associated sites involved in this phenomenon lie on the I–II intracellular interdomain loop ([66], see Fig. 2) On the other hand, PKC-associated phosphorylation of $Na_v1.2$ both reduces currents and slows the kinetics of inactivation. The sites of PKC phosphorylation for the two effects are distinct, being on the I–II loop and the inactivation domain in the III–IV loop respectively [67]. Sodium currents can also be modulated in cell lines via tyrosine kinase pathways, which cause a hyperpolarizing shift in the voltage-dependence of inactivation [68].

More recent studies have suggested that these modulatory processes may act in vivo. Thus activation of dopamine receptors on neurons can reduce sodium currents via a cAMP-dependent mechanism [69,70]. More recently, sodium channels in neurons were found to be directly associated with the receptor protein tyrosine phosphatase (RPTP) β complex [71]. Binding to this complex was through α and $\beta1$ subunits, and when RPTPβ was co-expressed with $Na_v1.2$, significant slowing of inactivation along with positive-going shifts in steady state inactivation was seen. Thus it is proposed that sodium channel gating may be dynamically modulated by the opposing actions of tyrosine kinases and closely bound, ligand-activated tyrosine phosphatases.

Modulatory effects on several other sodium channel isoforms have been investigated. Thus PKA-dependent modulation of the cardiac $Na_v1.5$ isoform has been reported, but in contrast to the $Na_v1.2$, such modification increases rather than decrease sodium conductance [72]. Similar increases that were G-protein dependent have also been seen in isolated hippocampal neurons and cardiac myocytes [73,74]. On the other hand, the skeletal muscle isoform $Na_v1.4$, despite the fact that it is capable of phosphorylation by PKA, is not functionally affected by such modifications [75]. Lastly, PKA-induced functional changes for the $Na_v1.8$ isoform have been recently reported [76]. Here also the functionally relevant phosphorylation sites were located on the I–II loop, but again in contrast with $Na_v1.2$, the modification was associated with changes to both activation and inactivation gating.

Overall, it appears that sodium channel isoforms may indeed be differentially modified by a host of mechanisms. Thus, there is great potential for the dynamic regulation of sodium channel function by these processes. However, it remains to be determined whether physiologically meaningful regulation of sodium channels occurs in vivo.

5.5. Differences in gating among isoforms may be affected by differential glycosylation

Sodium channels are significantly glycosylated, a post-translational modification that occurs in the endoplasmic reticulum and in the Golgi apparatus [77,78]. A large fraction of this carbohydrate is sialic acid, a negatively charged sugar ([79–81]), and in particular, purified brain and skeletal muscle channels may have over 100 of these residues per molecule. In general, it has been thought that these modifications might be involved in protein folding or targeting to the plasma membrane during the biosynthetic process. However, it may be that these domains also play a more active role in channel mechanisms. It has been considered possible that such charges might influence the electrical field around gating elements of the channel. In fact, the existence of substantial negative charge near the external surface of sodium channels has long been inferred from the studies of Frankenhauser and Hodgkin [82] and others (see Ref. [1]), who observed that when external calcium concentration was elevated, substantial positive-going shifts in channel voltage-dependence occurred. To explain this phenomenon, it was proposed that fixed negative charges effectively depolarized the transmembrane voltage near channel voltage sensors, thus biasing them toward the open state. In this hypothesis calcium acted by neutralizing these fixed charges, either by direct interaction or a bulk solution "screening" effect, hence hyperpolarizing the juxta-channel membrane potential, i.e. making the transchannel potential more negative. Thus in elevated external calcium, sodium channels would require a larger depolarization in applied membrane potential to affect the same degree of channel activation seen at lower calcium levels, resulting in positively shifted gating behaviors.

In support of this idea, it has been found that enzymatic removal of these sugars from purified sodium channels from the electric organ of the electric eel caused significant positive-going changes in the voltage-sensitivity of their activation when they were reconstituted into lipid bilayers [83]. More recent studies extended these results using

Na$_v$1.4 and a mammalian cell expression system [84]. In these experiments, in addition to enzymatic desialylation, the sialic acid content of the channel was reduced in a number of different ways, such as mutagenic elimination of the sites of Na$_v$1.4 glycosylation. Regardless of the manner in which sialic acids were reduced, all voltage-dependent gating parameters shifted toward positive voltages and by the same magnitude. Furthermore, the ability of calcium to cause such shifts was dramatically reduced, thus identifying these sugars as the sites at which calcium interacts to cause the gating shifts described classically. Thus such sugars might represent a means for the cell to control the gating behavior of its channels appropriate to its excitation needs.

More recent studies have focused on the role of glycosylation among other channel isoforms. Recent studies reported that neuraminidase exposure of *Xenopus* oocytes expressing Na$_v$1.2 and Na$_v$1.5 channels shifted steady state activation of both isoforms in a positive direction [85]. Similar results were obtained for Na$_v$1.4 and 1.5 in a mammalian cell line using either neuraminidase or glycosylation processing inhibitors [86]. In both studies activation gating shift sensitivity to external divalents was dramatically reduced with decreased glycosylation or sialylation. Thus the authors also concluded that the effects of reduced sialic acid reflected an alteration to surface charge.

The above results suggest that sialic acid can contribute via a surface charge effect to gating behavior in sodium channels. The next question is whether variations in sialic acid content can account for gating differences between isoforms. This question arises because the basic gating sensors, i.e. the S4 segments, are highly conserved in sequence among isoforms; hence variations in voltage sensitivity of gating must originate from other structures. In fact, channel isoforms are known to vary considerably both in glycosylation and in sialic acid content [79,87–89], presumably because the number of potential asparagine-linked glycosylation sites on individual isoforms varies considerably. In the above-cited study comparing Na$_v$1.4 and 1.5, few substantial differences between the isoforms were found. However, in a more recent study comparing the poorly glycosylated isoform Na$_v$1.7 with the highly sialylated Na$_v$1.4, large differences in sialylation-related gating behavior were observed, with Na$_v$1.7 being relatively unchanged by sialic acid reduction manipulations [89,90]. Furthermore, chimeric replacement of the sites on Na$_v$1.7 with those of Na$_v$1.4 cause the mutant 1.7 isoform to assume the voltage-dependent gating properties and sensitivity to sialic acid removal and calcium elevation of the Na$_v$1.4 donor isoform. Thus it was suggested that gating variations among sodium channel isoforms are due in large part to variations in sialic acid-dependent surface potential effects.

Lastly, as β1 subunits have also been shown to be heavily sialylated, the contribution of β1 sialic acid to voltage-dependent gating parameters has been studied using the same approaches as for the α subunit [89,91]. Perhaps surprisingly, evidence was found that β1 sialic acid contributed significantly to the surface potential affecting gating sensors for both Nav1.4 and Nav1.7 isoforms. Overall, as sialic acid related contributions to gating are similar in magnitude to those of acute modulatory processes, it would appear that variable glycosylation represents another potent mechanism for the regulation of sodium channel function.

5.6. Even subtle changes to channel function may have major physiological consequences

Finally, a brief comment on the issue of what constitutes a physiologically meaningful functional difference seems appropriate. While the answer to this question is not known with certainty, nonetheless it is highly possible that very slight gating changes or differences may have dramatic physiological consequences when viewed in context of the whole organism. For example, changes to the voltage-dependence of channel activation or inactivation of just a few millivolts could significantly affect the macroscopic properties of a neural circuit or network, perhaps resulting in phenomena such as epileptiform discharges (i.e. epilepsy) or diminished synaptic function such as long-term potentiation (hence cognitive defects). Such functional changes would be difficult or impossible to detect on the cellular level using current methods; thus the "lack" of functional differences seen among isoforms in some studies may not be a definitive answer to the question of whether functional differences among isoforms are required for specialized physiological roles.

6. Differential expression

Studies of expression utilize methods to identify either mRNA or protein. Differing levels of sensitivity and spatial resolution are provided by PCR, Northern blots, and in situ hybridization to detect mRNAs. Each method has advantages and disadvantages, and all of them assume that detection of mRNA implies that protein is being made. Thus, discrepancies between observations made using different techniques can arise for many reasons. The exquisite sensitivity of PCR requires a certain amount of caution, e.g. the levels of mRNA may not be functionally significant. Both RT-PCR and Northern blot analysis utilize tissue homogenization that may obscure localized expression in a small region. In situ hybridization would detect locally discrete regions of expression between brain regions or between cell types within a brain region. Differences in control of translation from mRNA to protein have not been studied within the nervous system, and thus, it is not clear that an abundance of mRNA means an abundance of protein. Proteins can be directly detected with antibodies by immunocytochemistry. In most cases, subcellular localization in the nervous system can only be determined with immunocyto-chemistry (described below). Results with antibodies also need to be interpreted with caution since they can miss a low level of expression (false negative) and may cross react with other proteins (false positive). In a restricted number of examples, especially in muscle and large invertebrate neurons, subcellular channel distribution can be determined electrophysiologically by loose patch voltage clamp recording.

6.1. Tissue and Regional Differences

The pattern of expression of each sodium channel isoform in different tissues provides a useful means for grouping them (Fig. 7). Four sodium channels are expressed in both the CNS and PNS ($Na_v1.1-1.3$, and $Na_v1.6$). Three sodium channels ($Na_v1.7-1.9$) are expressed almost exclusively in sensory neurons in the PNS. Two sodium channels are

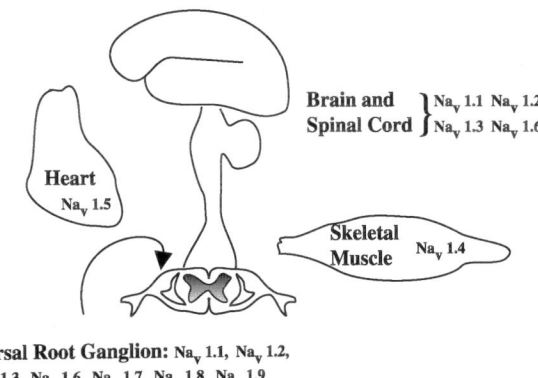

Brain and } Na$_V$ 1.1 Na$_V$ 1.2
Spinal Cord } Na$_V$ 1.3 Na$_V$ 1.6

Heart
Na$_V$ 1.5

Skeletal Muscle Na$_V$ 1.4

Dorsal Root Ganglion: Na$_V$ 1.1, Na$_V$ 1.2, Na$_V$ 1.3, Na$_V$ 1.6, Na$_V$ 1.7, Na$_V$ 1.8, Na$_V$ 1.9

Fig. 7. Tissue distribution of various sodium channel isoforms.

primarily muscle channels: Na$_V$1.4 in adult skeletal muscle and Na1.5 in cardiac muscle. The tissue expression of the sodium channels within the second family (Na$_V$2) is widespread and very different from the first family, being present in heart, uterus, glia, and dorsal root ganglion neurons (in the latter case detected by in situ hybridization but absent by immunocytochemistry (Levinson, personal communication)). Although it is common to refer to an isoform, for example, as a "brain" channel or a "cardiac" channel, multiple exceptions to the patterns described above have been reported. In most cases the presence of these channels is inferred from detection of mRNA; confirmation by immunocyto-chemistry and electrophysiology is desirable. The cardiac channel (Na$_V$1.5) is expressed in developing and denervated skeletal muscle and has the same subcellular distribution as the adult skeletal muscle channel [92]. The cardiac channel also is expressed in neurons in the limbic cortex [93]. Transcripts for two neuronal channels (Na$_V$1.1 and Na$_V$1.3) are expressed in skeletal and cardiac muscles [30, 94] A PNS channel (Na$_V$1.8) is expressed in the CNS in brains of mice with experimental allergic encephalomyelitis and humans with multiple sclerosis [95]. Functional consequences of these exceptions to the primary expression pattern (for example, a "muscle" channel expressed in neurons) are unknown.

The distribution of the three β subunit isoforms has been reported. β1 and β3 are expressed in brain, heart and skeletal muscle, and it has been suggested that they tend to have a complementary distribution, implying that sodium channels associate with either β1 or β3. β2 is expressed in both brain and heart, but not skeletal muscle. Human β3 is also expressed in lung, kidney and pancreas, raising the possibility that β3 is associated with proteins other than voltage-gated sodium channels [96,97].

Expression of sodium channels in different tissues and cells is likely to be controlled at the level of DNA transcription into RNA. One simple prediction is that a factor present only in neurons would turn on transcription of sodium channel genes. Contrary to this expectation, transcriptional regulation of the sodium channel Na$_V$1.2 has been shown to be due to a protein that is present in non neuronal cells and binds to a silencer element in the 5$'$ flanking region of the gene. This transcription factor has been designated REST or NRSF [98–100]. All tissues and cells make the REST/NRSF protein except neurons, and the $Na_v1.2$ gene is thus constitutively turned off in all cells except in neurons where there is a release from inhibition. Therefore, we can anticipate that combinations of positive and negative regulatory elements are required for control of sodium channel transcription.

6.2. Development

In the rat CNS the earliest sodium channel to appear is $Na_v1.3$. This is followed by $Na_v1.1$ (which increases postnatally and subsequently declines) and $Na_v1.2$ (which increases postnatally and remains highly expressed by Northern blot analysis) [101]. $Na_v1.6$ is expressed primarily postnatally [102] although there is a report of embryonic expression [103]. In the adult rat $Na_v1.3$ mRNA is very low, leading to the concept that this is primarily an embryonic channel. However, $Na_v1.3$ is abundantly expressed in the adult human [104]. Developmental expression of the CNS channels in different brain regions has been studied with Northern blot [101], in situ hybridization [103], PCR [102] and immunocytochemistry [105]. There are distinct patterns of expression for each isoform in each region.

In rat dorsal root ganglion neurons, all the neuronal channels are expressed (at the mRNA level) at birth and, with the exception of $Na_v1.3$, increase in expression postnatally [88,103,106,107]. Both $Na_v1.4$ and $Na_v1.5$ are expressed in embryonic and neonatal skeletal muscles. Expression of the adult cardiac channel $Na_v1.5$ in skeletal muscle is downregulated during the first few weeks after birth [108]. In cardiac muscle there is evidence that $Na_v1.1$ is expressed in neonatal sinoatrial myocytes and later disappears, with no sodium channels expressed in these cells in the adult [94].

6.3. Expression in glia

Glial cells are generally considered to be electrically inexcitable, and it is therefore surprising that voltage-gated sodium channels are expressed in astrocytes and Schwann cells. What is equally surprising is that glial expression is not restricted to one or two sodium channel isoforms. Although the particular isoform expressed differs for different glial cells and for glial cells in different brain regions, the general conclusion is that most of the isoforms are expressed in glia [102,105,109–112], including the cardiac muscle channel $Na_v1.5$ [113]. The role of sodium channels in glia is not understood, but it has been suggested that glia might synthesize sodium channels and transfer them to neurons at the node of Ranvier [114], that they might allow a sodium influx that would keep intracellular sodium in a range that would maintain Na/K pump activity when K is elevated outside the neuron [115,116], and finally that glial sodium channels might trigger action potentials and play a bigger role during development than in the adult [117].

7. Regulation of sodium channel localization and clustering

Ion channels are often nonhomogeneously expressed and distributed among excitable tissues (e.g. sodium channels at nodes of Ranvier, calcium channels in muscle fiber t-tubules, no sodium channels in SA node cells). As a field of much current research interest, it is now clear that highly selective molecular mechanisms are responsible for regulating the synthesis and localization of individual ion channel types through control of transcriptional, post-translational, protein targeting, and clustering processes. At the protein level, sodium channel isoforms are themselves highly selectively expressed, both in specific cell types as well as within specific micro-domains within these cells.

At the outset, it is important to state that the expression and distribution of isoforms is best determined by the presence of the actual isoform protein rather than by the presence of its specific mRNA. This is said for two reasons. First, the presence of message does not necessarily correspond to the level of isoform protein nor even that it is expressed at all. Second, since mRNA for sodium channels is thought to be restricted to the cell soma, the location of message does not tell one anything about the ultimate targeting or density of channels at the target site. On the other hand, the level of isoform-specific mRNA is indeed important in assessing the activity of the actual sodium channel gene, e.g. in response to trophic influences. With these basic considerations in mind, we now focus on what is known about the isoform-specific localization of sodium channels.

The basic approach to study isoform localization is straightforward, and consists of using isoform-specific antibodies generated by the immunization of host animals with synthetic peptides encoding short peptide segments unique to various isoforms. Using these reagents it has been possible to observe the specific expression of almost all of the existing isoforms.

However, despite the availability of isoform-specific labeling reagents, few detailed studies of the differential distribution of isoforms have been published. On the level of whole tissue, it has long been known that adult skeletal muscle expresses a single isoform, $Na_v1.4$ (see below), while cardiac myocytes appear to express only the $Na_v1.5$ isoform [118]. In the CNS, it has been reported that $Na_v1.2$ is enriched in the brain over $Na_v1.1$, but that these abundances are reversed in spinal cord [119,120]. In these studies the relative expression of these two isoforms appeared to change during development, with $Na_v1.2$ appearing earlier. Within the brain, a more recent study found evidence for the differential localization of these two subtypes among various regions; for example, $Na_v1.1$ is more abundant in brainstem and cortex and $Na_v1.2$ more abundant in hippocampus and thalamus [43]. Developmentally, this same study reported that $Na_v.1.1$ protein expression peaked and started to decline at one month postpartum, while $Na_v1.2$ levels continued to rise throughout development. Also at the macroscopic level, $Na_v1.7$ has been observed to be restricted in its expression to peripheral nerve tracts and sensory ganglia [88].

At the cellular level, there is evidence for differential localization of some isoforms in specialized processes and membrane domains (Fig. 8). In two studies, $Na_v1.1$ was found to be restricted to the cell bodies of neurons in several areas of the brain and spinal cord, with $Na_v1.2$ usually present in unmyelinated fiber layers in the brain [43,120]. Recently, several studies have identified $Na_v1.1$, 1.2, and 1.6 in the dendrites of CNS neurons ([43, 121, 105], Fig. 9B,C). There is less information regarding the localization of isoforms to subcellular membrane domains. Perhaps the most striking examples known are those of high density clusters of $Na_v1.4$ in the postsynaptic folds of the neuromuscular junction (see below), and of $Na_v1.6$ at the nodes of Ranvier in the CNS and PNS ([121], Fig. 9A).

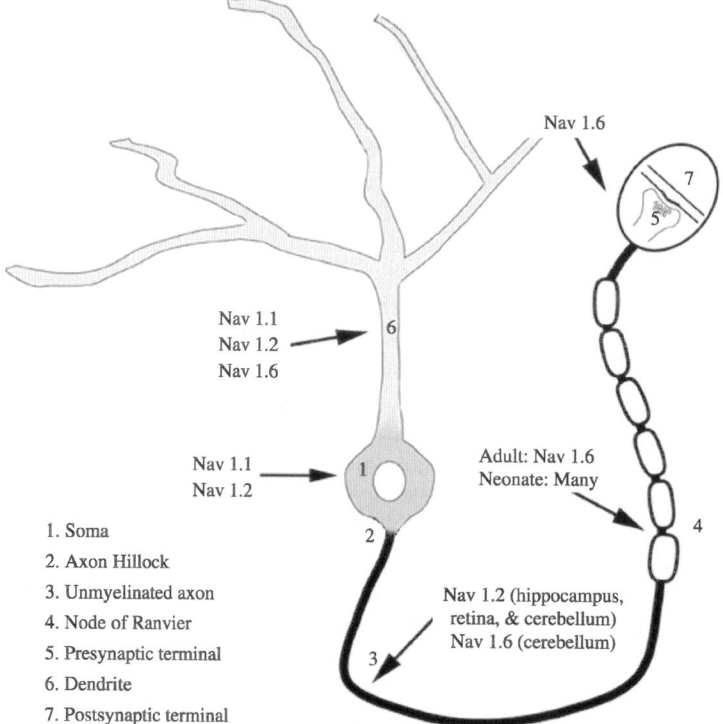

Nav 1.6

Nav 1.1
Nav 1.2   6
Nav 1.6

Nav 1.1   1   Adult: Nav 1.6
Nav 1.2   Neonate: Many

1. Soma   2
2. Axon Hillock
3. Unmyelinated axon   Nav 1.2 (hippocampus,
4. Node of Ranvier   retina, & cerebellum)
5. Presynaptic terminal   Nav 1.6 (cerebellum)
6. Dendrite   3
7. Postsynaptic terminal

4

7
5

Fig. 8. Cartoon of neuron showing typical subcellular localization of isoforms. This pattern may very from one type of neuron to another.

Based largely on the above studies and the in situ hybridization results discussed in the previous section, it appears very likely that many neurons will synthesize more than one isoform and that these are to some degree differentially targeted to specific subcellular domains. Recently, a striking example of such specific subcellular targeting was described for adult retinal ganglion cells, whose axons consist of a long initial

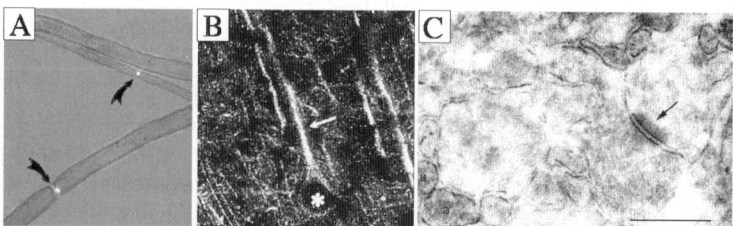

Fig. 9. Immunocytochemical images of the subcellular localization of the Na$_v$1.6 isoform. A. Nodes of Ranvier of sciatic nerve. B. Dendrites of cortical pyramidal cells. (Asterisk indicates the soma. Arrow indicates apical dendrite.) C. Dendritic synapses (electronmicrograph. Arrow indicates a synapse; note staining is both pre- and postsynaptic). Fig. A from Ref. [105] and B and C from Ref. [121].

unmyelinated segment within the retina, followed by myelinated axons that form the optic nerve [122]. In adult axons, $Na_v1.2$ was almost exclusively localized to the unmyelinated zone, while $Na_v1.6$ was restricted to the nodal membrane. Thus it would appear that such neurons have the capability of selectively targeting these two isoforms to different membrane regions in the same cell.

Fig. 8 schematically summarizes the results discussed above. It should be stressed that this diagram is very general and preliminary and thus is very likely to be modified by future findings. While it may be tempting to conclude from early studies that each isoform is targeted to a unique membrane domain, recent results suggest that such a view is overly simplified. It seems more probable that isoform distributions will depend on the individual cell in which they are expressed. Further, one cannot yet rule out that different isoforms may even co-exist in the same membrane domains in some cases.

In any case, by what mechanisms might these specialized distributions arise? At present, it appears that at least three separately regulated processes may play a role. Thus, there is the selective control of the *synthesis* of each isoform, e.g. by regulation of gene transcription and/or translation of the isoform mRNA. The regulation of $Na_v1.2$ transcription by negative enhancer elements in its gene promoter region has been described above. Second, cells have the ability to *target* individual isoforms via selective transport processes to specific membrane domains (e.g. proximal dendrites ($Na_v1.1$) versus nodes of Ranvier ($Na_v1.6$)). At present little is known of the mechanism of these processes except that they probably involve interaction of the selective targeting machinery with unique structures on each isoform. Third, once the isoforms arrive at target sites, they are specifically *clustered* into a local membrane domain, e.g. at the axon hillock, node of Ranvier, or muscle postsynaptic membrane. There has been considerable progress in understanding this process in recent years, as described next.

7.1. Sodium channel distribution in skeletal and cardiac muscle

Sodium channels in skeletal muscles are not uniformly distributed over the surface membrane. Sodium channels in adult muscles are maintained at high density at the neuromuscular junction and are kept at low density near the tendon (Fig. 10). Even in extrajunctional membrane, channel density varies several fold over short distances [123], and these variations are not associated with underlying nuclei or with morphological specializations [124].

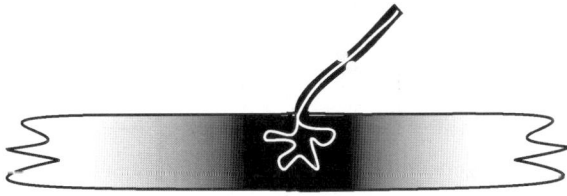

Fig. 10. Schematic showing distribution of $Na_v1.4$ in a skeletal muscle fiber.

The density of sodium channels in the postsynaptic membrane was estimated to be approximately 2000 sodium channels/μm^2 [125,126] using the loose-patch voltage clamp technique; this is approximately the density measured at the node of Ranvier. The density at the endplate is 10 to 20-fold greater than in the extrajunctional membrane where it is a few hundred sodium channels/μm^2. The density of sodium channels falls off with distance from the endplate, reaching extrajunctional values $100-200$ μm from the endplate [92,108]. In the mammalian muscle a further decrease in density occurs near the tendon. Similar results have been reported for human, mouse, rat, and snake muscles. Differences in sodium channel density and distribution occur between various muscle types. For example, the sodium channel density at endplates and perijunctional membrane of a fast-twitch muscle is greater than that of a slow-twitch muscle [126,127].

Ultrastructural studies showed that postsynaptic membrane of muscle is divided into separate domains with AChRs located on the crests of the postsynaptic folds and sodium channels found in the troughs of the folds [3,128]. For the skeletal muscle $Na_v1.4$ isoform, it appears that direct interactions with the cytoskeletal components syntrophin and ankyrin occur. Ankyrin and syntrophin are molecules that bind both the cytoskeleton and sodium channels [129,130], and isoforms of ankyrin and syntrophin are concentrated at the neuromuscular junction. Further, interactions of $Na_v1.4$ with syntrophin occur in part through a so-called PDZ domain on syntrophin with the C-terminus of $Na_v1.4$ [130]. Syntrophin is then thought to interact with other cytoskeletal elements to help cluster and anchor the channels at the postsynaptic membrane, in particular deep within the folds of the synapse. In these folds, the interaction of channels with ankyrin has also been suggested [3]. However, neither ankyrin nor syntrophin has been shown to be required for the formation and maintenance of the high density of sodium channels in muscle. On the other hand, ankyrin G is required for concentration of sodium channels in some neurons. For example, sodium channels are not concentrated at the axon hillock of cerebellar Purkinje cells in mice missing some isoforms of ankyrin [131].

An increased density of sodium channels has been observed at the site of nerve contact in nerve–muscle co-cultures [132,133]. These in vitro results suggest that sodium channels begin to cluster shortly after nerve contact. On the other hand, clustering of sodium channels in vivo is associated with synapse maturation rather than synapse formation. Neuromuscular junctions in the rodent begin to form about a week before birth. Acetylcholine receptors aggregate almost immediately after nerve contact. Electrophysiological recordings [108] and immunocytochemical measurements [4] showed that sodium channels begin to be concentrated in the synaptic and perijunctional regions during the first two weeks after birth. This time course of sodium channel concentration correlates with the formation of postsynaptic junctional folds. Thus, the aggregation of sodium channels and AChRs is likely to depend on different extracellular and cytoplasmic interactions.

7.2. Clustering of sodium channels at nodes of Ranvier

While the molecular players in stable clusters of channels are being identified, it is less certain how such assemblies are formed dynamically and how they are regulated. A much-studied phenomenon has been the formation of the high-density clusters of channels at nodes of Ranvier in the peripheral and central nervous systems. Two competing theories of nodal cluster formation have been proposed. In the first, the neuron itself contains the machinery that specifies where the nodes are to be located during development, and channels are inserted at this point and glia are directed to myelinate up to this nodal gap. Alternatively, myelinating glia (Schwann cells or oligodendrocytes) themselves define the site of the node by their pattern of attachment and subsequent extension along the developing axon. Although nodal formation itself and sodium channel clustering at this domain may be independent processes, this question has been addressed by studying the development of nodal sodium channel clusters. Although there is some experimental support for both theories, the weight of the evidence appears to strongly favor a determinant role of myelinating glia in nodal cluster formation. Thus, glia appear to form high density "heminodal" clusters at their extending ends almost as soon as they attach and start to myelinate axons ([134,135], see Ref. [136]). Further, such clusters are inferred to move just ahead of the extending myelinating glia. Finally, mature nodal clusters are stabilized when adjacent moving clusters from extending glial processes converge. Such a phenomenon has been demonstrated in both the PNS and CNS [137] and has been supported by data from myelination-delayed mutant mice [138] as well as by the de novo formation of nodal clusters during recovery from various manipulations that cause acute demyelination [139–141]. On the other hand, there is strong evidence that in cultured retinal ganglion cells a factor secreted by oligodendrocytes can induce node-like clusters in neuronal processes without glial contact nor even glial presence in the culture [142]. Nonetheless, it appears that sodium channel clusters at nodes are formed primarily by glial-directed processes without neuronal specification.

However, there is recent evidence that neurons do play an important role in determining the isoform that populates nodes of Ranvier. Thus in developing rodent optic nerve $Na_v1.2$ is first found at nodes, to be later replaced by $Na_v1.6$ as the animal develops [122]. In the mature nerve $Na_v1.2$ is then specifically localized to the initial unmyelinated zone of the optic nerve. It was suggested that the process of myelination itself instructs the cell to differentially target these isoforms to their respective domains.

Finally, the complexes that form sodium channels into nodal clusters are beginning to be elucidated. In addition to the intracellular protein ankyrin [131,143–145], it appears that the β1 and β2 subunits may interact with the extracellular recognition proteins tenascin-R [146] and contactin (Isom, personal communication) respectively. These latter molecules are proposed to interact with glial recognition/ nodal formation molecules such as Nr-Cam and neurofascin [147]. Hence it appears that formation and/or stabilization of the high-density sodium channel cluster at nodes of Ranvier will involve a complex of cytoplasmic axonal proteins and extracellularly projecting membrane proteins from both axon and glia.

8. Sodium channelopathies

Sodium channel mutations that cause diseases in humans, horses and mice have been identified over the past 10 years. The mutations not only reveal the primary cause of the disease, permitting physicians to devise rational therapies, but also provide insight into the basic properties and structural features of the channel. Each of the missense mutations can be studied in heterologous expression systems to determine the specific defect in the channel, with the expectation that the defect will explain the phenotype. In addition, these diseases, as well as experimentally produced null mice, have answered one concern that is often raised about multigene families: are the protein isoforms functionally interchangeable or redundant? In other words, is it correct to assume that each isoform serves a unique role?

Deletions of sodium channel isoforms indicate that if there is compensation for loss of one isoform, it is incomplete. If the isoforms serve unique roles, one would predict that mutations that cause a loss of a CNS channel or a muscle channel will be lethal, while loss of PNS isoforms will not be lethal. Mice that are null for CNS neuronal sodium channel $Na_v1.2$ die at birth [148] and mice null for another CNS channel, $Na_v1.6$, die at 3–4 weeks postnatal [149]. Thus, the remaining isoforms in these mice do not substitute for the missing sodium channels. Mice lacking $Na_v1.8$, which is restricted to the PNS and especially to small neurons that are primarily responsible for pain and temperature sensation, are behaviorally normal to the casual observer. Threshold for action potential generation and speed of conduction of the small neurons indicated some compensation by other sodium channel isoforms. However, compensation for the loss of $Na_v1.8$ (by the increased expression of other channel isoforms) was either incomplete or abnormal since the mice had a deficit in mechanoreception and thermoreception as well as a diminished response to inflammatory pain [106]. Overall, it appears likely from these three examples that sodium channel isoforms serve unique functions.

Within the past ten years many single amino acid mutations in the skeletal muscle sodium channel have been identified that produce skeletal muscle diseases, especially the periodic paralyses (hyperkalemic periodic paralysis, paramyotonia congenita, and some forms of hypokalemic periodic paralysis [150]; reviewed recently by Lehmann-Horn et al. [151]). Mutations of the cardiac sodium channel are responsible for some cases of long Q–T syndrome, a cause of sudden cardiac arrest. The fact that skeletal and cardiac muscle express different sodium channel isoforms and that these isoforms are not found in neurons means that mutations in a sodium channel in either skeletal or cardiac muscles do not produce defects in other tissues. For example, people with periodic paralysis of skeletal muscles do not have cardiac disorders or cognitive deficits. Most of the muscle sodium channel mutations produce a gain-of-function, and this is typically an alteration in sodium channel inactivation. Thus, channels stay open longer than normal and produce additional depolarization and electrical excitation. If the depolarization is sufficiently large and prolonged, sodium channels enter a slow inactivation state; the cells then become inexcitable, producing muscle paralysis.

Point mutations in the neuronal isoforms produce neurological deficits. A point mutation in the mouse $Na_v1.6$ causes the *jolting* phenotype, a cerebellar ataxia (unco-ordinated locomotion) with a progressive loss of cerebellar Purkinje cells [149,152]. The first human neuronal sodium channel mutations have been recently identified. Mutations of a β subunit [153] and of $Na_v1.1$ [154] each produce a rare form of epilepsy, termed generalized epilepsy with febrile seizures-plus (GEFS +). This form of epilepsy is linked to fevers during childhood, and usually disappears after 6 years of age. However, in some families the seizures can continue into adulthood, and this distinction adds the "plus" to the GEFS category. By studying large families with this type of epilepsy, it was shown that the type of seizure varies from one family member to another (e.g. atonic, absence, or generalized tonic clonic epilepsy). Variable penetrance is the term used to describe the fact that the same mutation can produce different symptoms, including the lack of symptoms. Variable penetrance within one family is not often reported and could be due to differences in genes that modify the function of the sodium channel.

Finally, although not caused by structural defects in sodium channels, there are pathological conditions in which the regulation of sodium channel numbers and distributions may play an important part. In particular, there is great interest in the role of sodium channels in the establishment and maintenance of chronic pain states [11,12,155]. In general, there appear to be two general types of chronic pain, i.e. those cause by tissue inflammations (e.g. burns or cuts) and those resulting from direct damage to peripheral nerves, such as crushes (neuropathic pain) [156]. In the inflammatory response, an acute cascade of cytokines and other inflammatory agents is followed by the maintenance of a hyperalgesic state in which the injury site and its immediate uninjured surrounds are very sensitive to stimuli. This produces a guarding response that is conducive to wound healing. In experimental models of inflammation in rodents, it has been shown that after 24 h peripheral injury produces a dramatic upregulation in the synthesis of sodium channels in the sensory neurons that innervate the inflamed site [157,158]. Further, this upregulation tends to be specific to $Na_v1.7$ but not $Na_v1.8$ [159]. However, it is not known what role these increases in sodium channel abundance play in the mechanism of inflammatory hyperalgesia. On the other hand, in pain of neuropathic origin, there is evidence to suggest that sodium channels accumulate at the site of injury and make the nerve highly sensitive to mechanical stimulation at this point [160–163]. Such stimulation (by touch for example) might then generate action potentials in sensory pain fibers, resulting in the sensation of pain. In any case, knowing whether specific sodium channel isoforms play a role in the generation or maintenance of chronic pain conditions could be very helpful in the development of new pain drugs that could be specifically targeted to these isoforms [155,164].

9. Future directions

As of this writing, the biological significance of sodium channel diversity remains highly unclear. We know that sodium channel isoforms do display functional differences, that they are differentially expressed, and that they tend to be differentially localized within given cell. What conclusions can we make about features that unify or separate voltage-gated sodium channels?

Each member of the family has preserved the voltage-dependence of activation and the subsequent rapid inactivation of the open state. This conservation of function is reflected in the high conservation of the amino acid sequence of both membrane domains and the fast inactivation cytoplasmic domain (III–IV linker) that blocks the pore. Naturally occurring mutations in these domains in humans and domesticated animals cause severe phenotypes that would be lethal in the wild. Thus, when sodium channel genes were duplicated hundreds of millions of years ago, only mutations that created minor functional changes were permitted. On the other hand, once we look past the broad similarities of the isoforms, it is remarkable how diverse the sodium channel isoforms are. Each isoform is unique in its developmental expression, tissue distribution, subcellular location, response to modulation, and functional properties, such as the absolute voltage dependence or rate of inactivation.

It is not yet clear how the functional properties of each sodium channel isoform are important for each neuronal cell or subcellular domain that contains that isoform. All but two of the isoforms ($Na_v1.7$ and $Na_v1.8$) have similar rapid kinetics of activation and inactivation. Differences in the voltage-dependence of activation and inactivation are likely to be important distinctions between the isoforms, and it is important to keep in mind that differences of only a few millivolts in voltage dependence can have drastic effects on threshold for action potential firing and for the number of channels that are available to open.

Additionally, ion channels need to be targeted to a specific subcellular site, tethered to that site, perhaps modulated during their lifetime, and finally removed for degradation. All of these functions require associations with other proteins, which are expected to be identified in the near future. Targeting of the channels could involve specific vesicles and directed insertion of these vesicles or could be mediated by random insertion, diffusion, and trapping at specific sites. Since the lifetime of a sodium channel in vivo is not known for any isoform in any tissue, it is difficult to rule out random insertion and diffusion. Proteins that link sodium channels to the cytoskeleton, such as ankyrin and syntrophin, are essential for maintaining a high concentration. The specific sites of interaction and additional proteins involved in these interactions need to be determined. There is evidence that sodium channels are modulated by kinases and by nitric oxide. More modulators will be identified and their physiological roles in synaptic transmission, action potential initiation and action potential conduction will be characterized. A complete understanding of sodium channel behavior will require the integration of all this information.

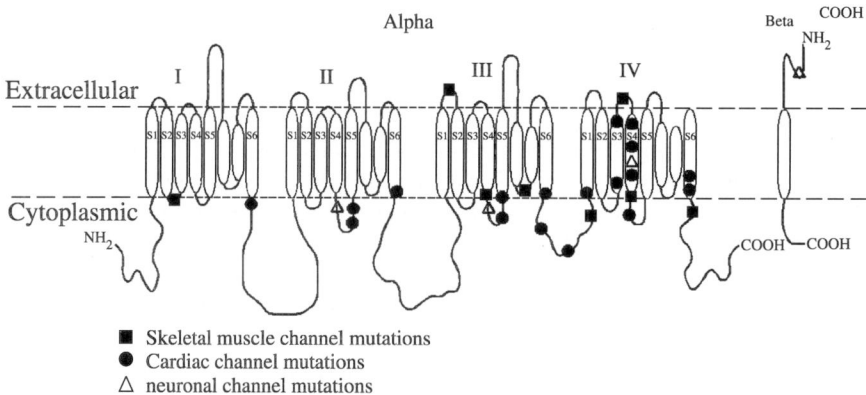

Fig. 11. Naturally occurring sodium channel mutations causing disease in skeletal muscle, cardiac muscle, and the CNS.

There is no doubt that pathological mutations in each α and β sodium channel isoform exist in humans and are responsible for neurological diseases. Therefore, many more diseases will be identified that are caused by defective sodium channels. Most of the identified mutations in muscle and neuronal sodium channels lie in three general regions: the transmembrane domains (especially domains III and IV), the regions close to the membrane, or the cytoplasmic domain that is important for fast inactivation (the III–IV linker). It is highly likely that deleterious but nonlethal mutations occur in the large intracellular or extracellular domains and that these mutations interfere with channel targeting and/or modulation (see Fig. 11).

Finally, now that detailed information is accruing regarding the functional diversity among sodium channel isoforms and its differential regulation, it becomes essential that this diversity be studied in the context of meaningful biological roles. Ultimately, we need to be able to explain how the unique properties of each isoform shape the response of a neuron or muscle cell to its synaptic input and local environment. Thus future studies will (or should, at least) focus on the part sodium channels play in dynamically regulated phenomena, such as synaptic integration on dendrites (e.g. LTP) and sensory adaptation (e.g. the development of hyperalgesia after injury).

References

[1] Hille, B., 2001. Ion channels of excitable membranes, 3rd ed. Sinauer Associates, Sunderland, MA.

[2] Raine, C., Waxman, S. (Eds.), 1978. Physiology and Pathobiology of Axons. Raven Press, New York, pp. 283–311.

[3] Flucher, B.E., Daniels, M.P., 1989. Distribution of Na^+ channels and ankyrin in neuromuscular junctions is complementary to that of acetylcholine receptors and the 43 kd protein. Neuron 3(2), 163–175.

[4] Wood, S.J., Shewry, K., Young, C., Slater, C.R., 1998. An early stage in sodium channel clustering at developing rat neuromuscular junctions. Neuroreport 9(9), 1991–1995.

[5] Martin, A.R., 1994. Amplification of neuromuscular transmission by postjunctional folds. Proc. R. Soc. Lond. [Biol.] 258, 321–326.

[6] Stuart, G., Sakmann, B., 1995. Amplification of EPSPs by axosomatic sodium channels in neocortical pyramidal neurons. Neuron 15(5), 1065–1076.

[7] Stuart, G., Schiller, J., Sakmann, B., 1997. Action potential initiation and propagation in rat neocortical pyramidal neurons. J. Physiol. (Lond.) 505(Pt 3), 617–632.

[8] Stuart, G., Spruston, N., Sakmann, B., Hausser, M., 1997. Action potential initiation and backpropagation in neurons of the mammalian CNS. Trends Neurosci. 20(3), 125–131 (Review, 101 refs).

[9] Stuart, G.J., Sakmann, B., 1994. Active propagation of somatic action potentials into neocortical pyramidal cell dendrites. Nature 367(6458), 69–72.

[10] Brock, J.A., McLachlan, E.M., Belmonte, C., 1998. Tetrodotoxin-resistant impulses in single nociceptor nerve terminals in guinea-pig cornea. J. Physiol. (Lond.) 512(Pt 1), 211–217.

[11] Eglen, R.M., Hunter, J.C., Dray, A., 1999. Ions in the fire: recent ion-channel research and approaches to pain therapy. Trends Pharmacol. Sci. 20(8), 337–342 (Review, 45 refs).

[12] Waxman, S.G., Dib-Hajj, S., Cummins, T.R., Black, J.A., 1999. Sodium channels and pain. Proc. Natl Acad. Sci. USA 96(14), 7635–7639 (Review, 53 refs).

[13] Bezanilla, F., 2000. The voltage sensor in voltage-dependent ion channels. Physiol. Rev. 80, 555–592.

[14] Levinson, S.R., Sather, W.A. In Sperelakis, N., (Ed.), 2001. Structure and mechanism of voltage-gated ion channels. Cell Physiology Sourcebook, 27. Academic Press, San Diego, pp. 455–478.

[15] Isom, L.L., De Jongh, K.S., Catterall, W.A., 1994. Auxiliary subunits of voltage-gated ion channels. Neuron 12(6), 1183–1194 (Review, 125 refs).

[16] Stuhmer, W., Conti, F., Suzuki, H., Wang, X.D., Noda, M., Yahagi, N., Kubo, H., Numa, S., 1989. Structural parts involved in activation and inactivation of the sodium channel. Nature 339(6226), 597–603.

[17] Guy, H.R., Durell, S.R., 1995. Structural models of Na^+, Ca^{2+}, and K^+ channels. Soc. Gen. Physiol. Ser. 50, 1–16 (Review, 46 refs).

[18] Heinemann, S.H., Terlau, H., Stuhmer, W., Imoto, K., Numa, S., 1992. Calcium channel characteristics conferred on the sodium channel by single mutations. Nature 356, 441–443.

[19] Noda, M., Numa, S., 1993. Structure and function of sodium channels. Ann. NY Acad. Sci. 707, 20–37.

[20] Catterall, W.A., 2000. From ionic currents to molecular mechanisms: the structure and function of voltage-gated sodium channels. Neuron 26(1), 13–25 (Review, 120 refs).

[21] Papazian, D.M., Timpe, L.C., Jan, Y.-N., Jan, L.Y., 1991. Alteration of voltage-dependence of Shaker potassium channel by mutations in the S4 sequence. Nature 349, 305–310.

[22] Eaholtz, G., Scheuer, T., Catterall, W.A., 1994. Restoration of inactivation and block of open sodium channels by an inactivation gate peptide. Neuron 12, 1041–1048.

[23] West, J.W., Patton, D.E., Scheuer, T., Wang, Y., Goldin, A.L., Catterall, W.A., 1992. A cluster of hydrophobic amino acid residues required for fast Na(+)-channel inactivation. Proc. Natl Acad. Sci. USA 89(22), 10910–10914.

[24] McPhee, J.C., Ragsdale, D.S., Scheuer, T., Catterall, W.A., 1995. A critical role for transmembrane segment IVS6 of the sodium channel α subunit in fast inactivation. J. Biol. Chem. 270, 12025–12034.

[25] McPhee, J.C., Ragsdale, D.S., Scheuer, T., Catterall, W.A., 1998. A critical role for the S4–S5 intracellular loop in domain IV of the sodium channel α-subunit in fast inactivation. J. Biol. Chem. 273(2), 1121–1129.

[26] Smith, M.R., Goldin, A.L., 1997. Interaction between the sodium channel inactivation linker and domain III S4–S5. Biophys. J. 73, 1885–1895.

[27] Goldin, A.L., Barchi, R.L., Caldwell, J.H., Hofmann, F., Howe, J.R., Hunter, J.C., Kallen, R.G., Mandel, G., Meisler, M.H., Netter, Y.B., et al., 1901. Nomenclature of voltage-gated sodium channels. Neuron 28(2), 365–368.

[28] Oh, Y., Waxman, S.G., 1998. Novel splice variants of the voltage-sensitive sodium channel alpha subunit. Neuroreport 9(7), 1267–1272.

[29] O'Dowd, D.K., Gee, J.R., Smith, M.A., 1995. Sodium current density correlates with expression of specific alternatively spliced sodium channel mRNAs in single neurons. J. Neurosci. 2(5 Pt 2), 4005–4012.

[30] Schaller, K.L., Krzemien, D.M., McKenna, N.M., Caldwell, J.H., 1992. Alternatively spliced sodium channel transcripts in brain and muscle. J. Neurosci. 12(4), 1370–1381.

[31] Gustafson, T.A., Clevinger, E.C., O'Neill, T.J., Yarowsky, P.J., Krueger, B.K., 1993. Mutually exclusive exon splicing of type III brain sodium channel alpha subunit RNA generates developmentally regulated isoforms in rat brain. J. Biol. Chem. 268(25), 18648–18653.

[32] Sarao, R., Gupta, S.K., Auld, V.J., Dunn, R.J., 1991. Developmentally regulated alternative RNA splicing of rat brain sodium channel mRNAs. Nucleic Acids Res. 19(20), 5673–5679.

[33] Plummer, N.W., McBurney, M.W., Meisler, M.H., 1997. Alternative splicing of the sodium channel SCN8A predicts a truncated two-domain protein in fetal brain and non-neuronal cells. J. Biol. Chem. 272(38), 24008–24015.

[34] Dietrich, P.S., McGivern, J.G., Delgado, S.G., Koch, B.D., Eglen, R.M., Hunter, J.C., Sangameswaran, L., 1998. Functional analysis of a voltage-gated sodium channel and its splice variant from rat dorsal root ganglia. J. Neurochem. 70(6), 2262–2272.

[35] Thackeray, J.R., Ganetzky, B., 1994. Developmentally regulated alternative splicing generates a complex array of Drosophila para sodium channel isoforms. J. Neurosci. 14(5 Pt 1), 2569–2578.

[36] Thackeray, J.R., Ganetzky, B., 1995. Conserved alternative splicing patterns and splicing signals in the Drosophila sodium channel gene para. Genetics 141(1), 203–214.

[37] Kazen-Gillespie, K.A., Ragsdale, D.S., D'Andrea, M.R., Mattei, L.N., Rogers, K.E., Isom, L.L., 2000. Cloning, localization, and functional expression of sodium channel beta1A subunits. J. Biol. Chem. 275(2), 1079–1088.

[38] Hanrahan, C.J., Palladino, M.J., Ganetzky, B., Reenan, R.A., 2000. RNA editing of the Drosophila para Na(+) channel transcript. Evolutionary conservation and developmental regulation. Genetics 155(3), 1149–1160.

[39] Reenan, R.A., Hanrahan, C.J., Barry, G., 2000. The mle(napts) RNA helicase mutation in drosophila results in a splicing catastrophe of the para Na⁺ channel transcript in a region of RNA editing. Neuron 25(1), 139–149.

[40] Schmidt, J.W., Catterall, W.A., 1987. Palmitylation, sulfation, and glycosylation of the alpha subunit of the sodium channel. Role of post-translational modifications in channel assembly. J. Biol. Chem. 262(28), 13713–13723.

[41] Catterall, W.A., 1997. Modulation of sodium and calcium channels by protein phosphorylation and G proteins. Adv. Second Messenger Phosphoprot. Res. 31, 159–181 (Review, 136 refs).

[42] Morgan, K., Stevens, E.B., Shah, B., Cox, P.J., Dixon, A.K., Lee, K., Pinnock, R.D., Hughes, J., Richardson, P.J., Mizuguchi, K., et al., 2000. beta 3: an additional auxiliary subunit of the voltage-sensitive sodium channel that modulates channel gating with distinct kinetics. Proc. Natl Acad. Sci. USA 97(5), 2308–2313.

[43] Gong, B., Rhodes, K.J., Bekele-Arcuri, Z., Trimmer, J.S., 1999. Type I and type II Na(+) channel alpha-subunit polypeptides exhibit distinct spatial and temporal patterning, and association with auxiliary subunits in rat brain. J. Comp. Neurol. 412(2), 342–352.

[44] Shah, R.M., 1999. Biochemical and electrophysiological analysis of the interaction between sodium channel α and β subunits. Dartmouth College, University Microfilms, Ann Arbor.

[45] Akopian, A.N., Sivilotti, L., Wood, J.N., 1996. A tetrodotoxin-resistant voltage-gated sodium channel expressed by sensory neurons. Nature 379(6562), 257–262.

[46] Sangameswaran, L., Delgado, S.G., Fish, L.M., Koch, B.D., Jakeman, L.B., Stewart, G.R., Sze, P., Hunter, J.C., Eglen, R.M., Herman, R.C., 1996. Structure and function of a novel voltage-gated, tetrodotoxin-resistant sodium channel specific to sensory neurons. J. Biol. Chem. 271(11), 5953–5956.

[47] Taschenberger, H., von Gersdorff, H., 2000. Fine-tuning an auditory synapse for speed and fidelity: developmental changes in presynaptic waveform, EPSC kinetics, and synaptic plasticity. J.Neurosci. 20(24), 9162–9173.

[48] Krafte, D.S., Goldin, A.L., Auld, V.J., Dunn, R.J., Davidson, N., Lester, H.A., 1990. Inactivation of cloned Na channels expressed in Xenopus oocytes. J. Gen. Physiol. 96(4), 689–706.

[49] Patel, V.V., 1994. The contribution of nonprotein domains to the function of the voltage-dependent Na⁺ channel. University of Colorado Health Sciences Center, University Microfilms, Ann Arbor.

[50] Patel, V.V., Levinson, S.R., Caldwell, J.H., 1994. Surface charge variations account for differences in the gating of rSkM1 sodium channels in different cells. Biophys. J. 66, A102 (Abstract).

[51] Pappone, P.A., 1980. Voltage-clamp experiments in normal and denervated mammalian skeletal muscle fibres. J. Physiol. (Lond.) 306, 377–410.

[52] Chahine, M., Deschene, I., Chen, L.Q., Kallen, R.G., 1996. Electrophysiological characteristics of cloned skeletal and cardiac muscle sodium channels. Am. J. Physiol. 271(2 Pt 2), H498–H506.

[53] Smith, M.R., Smith, R.D., Plummer, N.W., Meisler, M.H., Goldin, A.L., 1998. Functional analysis of the mouse Scn8a sodium channel. J. Neurosci. 18(16), 6093–6102.

[54] Klugbauer, N., Lacinova, L., Flockerzi, V., Hofmann, F., 1995. Structure and functional expression of a new member of the tetrodotoxin-sensitive voltage-activated sodium channel family from human neuroendocrine cells. EMBO J. 14, 1084–1090.

[55] Isom, L.L., De Jongh, K.S., Patton, D.E., Reber, B.F., Offord, J., Charbonneau, H., Walsh, K., Goldin, A.L., Catterall, W.A., 1992. Primary structure and functional expression of the beta 1 subunit of the rat brain sodium channel. Science 256(5058), 839–842.

[56] Patton, D.E., Isom, L.L., Catterall, W.A., Goldin, A.L., 1994. The adult rat brain β₁ subunit modifies activation and inactivation gating of multiple sodium channel α subunits. J. Biol. Chem. 269, 17649–17655.

[57] Cannon, S.C., McClatchey, A.I., Gusella, J.F., 1993. Modification of the Na⁺ current conducted by the rat skeletal muscle alpha subunit by coexpression with a human brain beta subunit. Pflugers Arch. Eur. J. Physiol. 423(1-2), 155–157.

[58] Makielski, J.C., Limberis, J.T., Chang, S.Y., Fan, Z., Kyle, J.W., 1996. Coexpression of beta 1 with cardiac sodium channel alpha subunits in oocytes decreases lidocaine block. Mol. Pharmacol. 49(1), 30–39.

[59] Tabarean, I.V., Juranka, P., Morris, C.E., 1999. Membrane stretch affects gating modes of a skeletal muscle sodium channel. Biophys. J. 77(2), 758–774.

[60] Shcherbatko, A., Ono, F., Mandel, G., Brehm, P., 1999. Voltage-dependent sodium channel function is regulated through membrane mechanics. Biophys. J. 77(4), 1945–1959.

[61] Scheuer, T., Auld, V.J., Boyd, S., Offord, J., Dunn, R., Catterall, W.A., 1990. Functional properties of rat brain sodium channels expressed in a somatic cell line. Science 247(4944), 854–858.

[62] West, J.W., Scheuer, T., Maechler, L., Catterall, W.A., 1992. Efficient expression of rat brain type IIA Na$^+$ channel alpha subunits in a somatic cell line. Neuron 8(1), 59–70.

[63] Bennett, E., Urcan, M.S., Tinkle, S.S., Koszowski, A.G., Levinson, S.R., 1997. Contribution of sialic acid to the voltage dependence of sodium channel gating – A possible electrostatic mechanism. J. Gen. Physiol. 109(3), 327–343.

[64] Levitan, I.B., 1999. Modulation of ion channels by protein phosphorylation. How the brain works. Adv. Second Messenger Phosphoprot. Res. 33, 3–22 (Review, 89 refs).

[65] Li, M., West, J.W., Lai, Y., Scheuer, T., Catterall, W.A., 1992. Functional modulation of brain sodium channels by cAMP-dependent phosphorylation. Neuron 8(6), 1151–1159.

[66] Smith, R.D., Goldin, A.L., 1996. Phosphorylation of brain sodium channels in the I–II linker modulates channel function in *Xenopus* oocytes. J. Neurosci. 16(6), 1965–1974.

[67] Cantrell, A.R., Ma, J.Y., Scheuer, T., Catterall, W.A., 1996. Muscarinic modulation of sodium current by activation of protein kinase C in rat hippocampal neurons. Neuron 16(5), 1019–1026.

[68] Hilborn, M.D., Vaillancourt, R.R., Rane, S.G., 1998. Growth factor receptor tyrosine kinases acutely regulate neuronal sodium channels through the Src signaling pathway. J. Neurosci. 18(2), 590–600.

[69] Surmeier, D.J., Kitai, S.T., 1993. D1 and D2 dopamine receptor modulation of sodium and potassium currents in rat neostriatal neurons. Prog. Brain Res. 99, 309–324 (Review, 56 refs).

[70] Cantrell, A.R., Smith, R.D., Goldin, A.L., Scheuer, T., Catterall, W.A., 1997. Dopaminergic modulation of sodium current in hippocampal neurons via cAMP-dependent phosphorylation of specific sites in the sodium channel alpha subunit. J. Neurosci. 17(19), 7330–7338.

[71] Ratcliffe, C.F., Qu, Y., McCormick, K.A., Tibbs, V.C., Dixon, J.E., Scheuer, T., Catterall, W.A., 2000. C.F., Qu, Y., McCormick, K.A., Tibbs, V.C., Dixon, J.E., A sodium channel signaling complex: modulation by associated receptor protein tyrosine phosphatase beta. Nat. Neurosci. 3(5), 437–444 (see comments).

[72] Frohnwieser, B., Chen, L.Q., Schreibmayer, W., Kallen, R.G., 1997. Modulation of the human cardiac sodium channel α-subunit by cAMP-dependent protein kinase and the responsible sequence domain. J. Physiol. (Lond.) 498(2), 309–318.

[73] Ma, J.Y., Li, M., Catterall, W.A., Scheuer, T., 1994. Modulation of brain Na$^+$ channels by a G-protein-coupled pathway. Proc. Natl Acad. Sci. USA 91(25), 12351–12355.

[74] Lu, T., Lee, H.C., Kabat, J.A., Shibata, E.F., 1999. Modulation of rat cardiac sodium channel by the stimulatory G protein alpha subunit. J. Physiol. (Lond.) 518(Pt 2), 371–384.

[75] Smith, R.D., Goldin, A.L., 1992. Protein kinase A phosphorylation enhances sodium channel currents in *Xenopus* oocytes. Am. J. Physiol. 263(3 Pt 1), C660–C666.

[76] Fitzgerald, E.M., Okuse, K., Wood, J.N., Dolphin, A.C., Moss, S.J., 1999. cAMP-dependent phosphorylation of the tetrodotoxin-resistant voltage-dependent sodium channel SNS. J. Physiol. (Lond.) 516(Pt 2), 433–446.

[77] Thornhill, W.B., Levinson, S.R., 1987. Biosynthesis of electroplax sodium channels in Electrophorus electrocytes and *Xenopus* oocytes. Biochemistry 26(14), 4381–4388.

[78] Thornhill, W.B., Levinson, S.R., 1992. Biosynthesis of ion channels in cell-free and metabolically labeled cell systems. Methods Enzymol. 207, 659–670.

[79] Miller, J.A., Agnew, W.S., Levinson, S.R., 1983. Principal glycopeptide of the tetrodotoxin/saxitoxin binding protein from Electrophorus electricus: isolation and partial chemical and physical characterization. Biochemistry 22(2), 462–470.

[80] Messner, D.J., Catterall, W.A., 1985. The sodium channel from rat brain. Separation and characterization of subunits. J. Biol. Chem. 260(19), 10597–10604.

[81] Roberts, R.H., Barchi, R.L., 1987. The voltage-sensitive sodium channel from rabbit skeletal muscle. Chemical characterization of subunits. J. Biol. Chem. 262(5), 2298–2303.

[82] Frankenhaeuser, B., Hodgkin, A.L., 1957. The action of calcium on the electrical properties of the squid axon. J. Physiol. (Lond.) 137, 218–244.

[83] Recio-Pinto, E., Thornhill, W.B., Duch, D.S., Levinson, S.R., Urban, B.W., 1990. Neuraminidase treatment modifies the function of electroplax sodium channels in planar lipid bilayers. Neuron 5(5), 675–684.

[84] Bennett, E., Urcan, M.S., Tinkle, S.S., Koszowski, A.G., Levinson, S.R., 1997. Contribution of sialic acid to the voltage dependence of sodium channel gating. A possible electrostatic mechanism. J. Gen. Physiol. 109(3), 327–343.

[85] Hauser, T.H., Fozzard, H.A., Satin, J., 1995. Surface potential differences between heart and brain sodium channels expressed in *Xenopus* oocytes. Biophys. J. 68, A156 (Abstract).

[86] Zhang, Y., Hartmann, H.A., Satin, J., 1999. Glycosylation influences voltage-dependent gating of cardiac and skeletal muscle sodium channels. J. Membr. Biol. 171(3), 195–207.

[87] Cohen, S.A., Levitt, L.K., 1993. Partial characterization of the rH1 sodium channel protein from rat heart using subtype-specific antibodies. Circ. Res. 73, 735–742.

[88] Toledo-Aral, J.J., Moss, B.L., He, Z.J., Koszowski, A.G., Whisenand, T., Levinson, S.R., Wolf, J.J., Silos-Santiago, I., Halegoua, S., Mandel, G., 1997. Identification of PN1, a predominant voltage-dependent sodium channel expressed principally in peripheral neurons. Proc. Natl Acad. Sci. USA 94(4), 1527–1532.

[89] Whisenand, T., 2000. The contribution of alpha and beta subunit sialylation to the function of the voltage-dependent sodium channel. University of Colorado Health Sciences Center, University Microfilms.

[90] Whisenand, T., Levinson, S.R., 2001. The effect of I-S5/S6 sialylation on sodium channel gating. Biophys. J. 80, 232a–233a (Abstract).

[91] Whisenand, T., Maue, R.A., Shah, R.M., 2001. The effect of β1 sialylation on sodium channel gating. Biophys. J. 80, 232a (Abstract).

[92] Caldwell, J.H., Milton, R.L., 1988. Sodium channel distribution in normal and denervated rodent and snake skeletal muscle. J. Physiol. (Lond.) 401, 145–161.

[93] Hartmann, H.A., Colom, L.V., Sutherland, M.L., Noebels, J.L., 1999. Selective localization of cardiac SCN5A sodium channels in limbic regions of rat brain. Nat. Neurosci. 2(7), 593–595 (letter).

[94] Baruscotti, M., Westenbroek, R., Catterall, W.A., DiFrancesco, D., Robinson, R.B., 1997. The newborn rabbit sino-atrial node expresses a neuronal type I-like Na$^+$ channel. J. Physiol. (Lond.) 498(Pt 3), 641–648.

[95] Black, J.A., Dib-Hajj, S., Baker, D., Newcombe, J., Cuzner, M.L., Waxman, S.G., 2000. Sensory neuron-specific sodium channel SNS is abnormally expressed in the brains of mice with experimental allergic encephalomyelitis and humans with multiple sclerosis. Proc. Natl Acad. Sci. USA 97(21), 11598–11602.

[96] Stevens, E.B., Cox, P.J., Shah, B.S., Dixon, A.K., Richardson, P.J., Pinnock, R.D., Lee, K., 2001. Tissue distribution and functional expression of the human voltage-gated sodium channel β3 subunit. Pflügers Arch. Eur. J. Physiol. 441, 481–488.

[97] Malhotra, J.D., Chen, C., Rivolta, I., Abriel, H., Malhotra, R., Mattei, L.N., Brosius, F.C., Kass, R.S., Isom, L.L., 2001. Characterization of sodium channel α- and β-subunits in rat and mouse cardiac myocytes. Circulation 103, 1303–1310.

[98] Maue, R.A., Kraner, S.D., Goodman, R.H., Mandel, G., 1990. Neuron-specific expression of the rat brain type II sodium channel gene is directed by upstream regulatory elements. Neuron 4(2), 223–231.

[99] Chong, J.A., Tapia-Ramirez, J., Kim, S., Toledo-Aral, J.J., Zheng, Y., Boutros, M.C., Altshuller, Y.M., Frohman, M.A., Kraner, S.D., Mandel, G., 1995. REST: a mammalian silencer protein that restricts sodium channel gene expression to neurons. Cell 80(6), 949–957.

[100] Schoenherr, C.J., Anderson, D.J., 1995. The neuron-restrictive silencer factor (NRSF): a coordinate repressor of multiple neuron-specific genes. Science 267(5202), 1360–1363.

[101] Beckh, S., Noda, M., Lubbert, H., Numa, S., 1989. Differential regulation of three sodium channel messenger RNAs in the rat central nervous system during development. EMBO J. 8(12), 3611–3616.

[102] Schaller, K.L., Caldwell, J.H., 2000. Developmental and regional expression of sodium channel isoform NaCh6 in the rat central nervous system. J. Comp. Neurol. 420(1), 84–97.

[103] Felts, P.A., Yokoyama, S., Dib-Hajj, S., Black, J.A., Waxman, S.G., 1997. Sodium channel alpha-subunit mRNAs I, II, III, NaG, Na6 and hNE (PN1): different expression patterns in developing rat nervous system. Brain Res. Mol. Brain Res. 45(1), 71–82.

[104] Whitaker, W.R., Clare, J.J., Powell, A.J., Chen, Y.H., Faull, R.L., Emson, P.C., 2000. Distribution of voltage-gated sodium channel alpha-subunit and beta-subunit mRNAs in human hippocampal formation, cortex, and cerebellum. J. Comp. Neurol. 422(1), 123–139.

[105] Krzemien, D.M., Schaller, K.L., Levinson, S.R., Caldwell, J.H., 2000. Immunolocalization of sodium channel isoform NaCh6 in the nervous system. J. Comp. Neurol. 420(1), 70–83.

[106] Akopian, A.N., Souslova, V., England, S., Okuse, K., Ogata, N., Ure, J., Smith, A., Kerr, B.J., McMahon, S.B., Boyce, S., et al., 1999. The tetrodotoxin-resistant sodium channel SNS has a specialized function in pain pathways. Nat. Neurosci. 2(6), 541–548.

[107] Dib-Hajj, S.D., Tyrrell, L., Black, J.A., Waxman, S.G., 1998. NaN, a novel voltage-gated Na channel, is expressed preferentially in peripheral sensory neurons and down-regulated after axotomy. Proc. Natl Acad. Sci. USA 95(15), 8963–8968.

[108] Lupa, M.T., Krzemien, D.M., Schaller, K.L., Caldwell, J.H., 1993. Aggregation of sodium channels during development and maturation of the neuromuscular junction. J. Neurosci. 13(3), 1326–1336.

[109] Oh, Y., Black, J.A., Waxman, S.G., 1994. Rat brain Na$^+$ channel mRNAs in non-excitable Schwann cells. FEBS Lett. 350(2-3), 342–346.

[110] Reese, K.A., Caldwell, J.H., 1999. Immunocytochemical localization of NaCh6 in cultured spinal cord astrocytes. Glia 26(1), 92–96.

[111] Black, J.A., Westenbroek, R., Minturn, J.E., Ransom, B.R., Catterall, W.A., Waxman, S.G., 1995. Isoform-specific expression of sodium channels in astrocytes in vitro: immunocytochemical observations. Glia 14(2), 133–144.

[112] Gautron, S., Dos, S.G., Pinto-Henrique, D., Koulakoff, A., Gros, F., Berwald-Netter, Y., 1992. The glial voltage-gated sodium channel: cell- and tissue-specific mRNA expression. Proc. Natl Acad. Sci. USA 89(15), 7272–7276.

[113] Black, J.A., Dib-Hajj, S., Cohen, S., Hinson, A.W., Waxman, S.G., 1998. Glial cells have heart: rH1 Na$^+$ channel mRNA and protein in spinal cord astrocytes. Glia 23(3), 200–208.

[114] Belcher, S.M., Zerillo, C.A., Levenson, R., Ritchie, J.M., Howe, J.R., 1995. Cloning of a sodium channel alpha subunit from rabbit Schwann cells. Proc. Natl Acad. Sci. USA 92(24), 11034–11038.

[115] Sontheimer, H., Black, J.A., Waxman, S.G., 1996. Voltage-gated Na$^+$ channels in glia: properties and possible functions. Trends Neurosci. 19(8), 325–331 (Review, 67 refs).

[116] Sontheimer, H., Fernandez-Marques, E., Ullrich, N., Pappas, C.A., Waxman, S.G., 1994. Astrocyte Na$^+$ channels are required for maintenance of Na$^+$/K$^+$-ATPase activity. J. Neurosci. 14, 2464–2475.

[117] Gritti, A., Rosati, B., Lecchi, M., Vescovi, A.L., Wanke, E., 2000. Excitable properties in astrocytes derived from human embryonic CNS stem cells. Eur. J. Neurosci. 12(10), 3549–3559.

[118] Cohen, S.A., 1996. Immunocytochemical localization of rH1 sodium channel in adult rat heart atria and ventricle – presence in terminal intercalated disks. Circulation 94(12), 3083–3086.

[119] Gordon, D., Merrick, D., Auld, V., Dunn, R., Goldin, A.L., Davidson, N., Catterall, W.A., 1987. Tissue-specific expression of the RI and RII sodium channel subtypes. Proc. Natl Acad. Sci. USA 84(23), 8682–8686.

[120] Westenbroek, R.E., Merrick, D.K., Catterall, W.A., 1989. Differential subcellular localization of the RI and RII Na$^+$ channel subtypes in central neurons. Neuron 3(6), 695–704.

[121] Caldwell, J.H., Schaller, K.L., Lasher, R.S., Peles, E., Levinson, S.R., 2000. Sodium channel Na(v)1.6 is localized at nodes of Ranvier, dendrites, and synapses. Proc. Natl Acad. Sci. USA 97(10), 5616–5620.

[122] Boiko, T., Rasband, M.N., Levinson, S.R., Caldwell, J.H., Mandel, G., Trimmer, J.S., Matthews, G., 2001. Compact myelin dictates the differential targeting of two sodium channel isoforms in the same axon. Neuron 30, 91–104.

[123] Almers, W., Stanfield, P.R., Stuhmer, W., 1983. Slow changes in currents through sodium channels in frog muscle membrane. J. Physiol. (Lond.) 339, 253–271.

[124] Roberts, W.M., 1987. Sodium channels near end-plates and nuclei of snake skeletal muscle. J. Physiol. (Lond.) 388, 213–232.

[125] Caldwell, J.H., Campbell, D.T., Beam, K.G., 1986. Na channel distribution in vertebrate skeletal muscle. J. Gen. Physiol. 87(6), 907–932.

[126] Milton, R.L., Lupa, M.T., Caldwell, J.H., 1992. Fast and slow twitch skeletal muscle fibres differ in their distribution of Na channels near the endplate. Neurosci. Lett. 135(1), 41–44.

[127] Ruff, R.L., 1992. Na current density at and away from end plates on rat fast- and slow-twitch skeletal muscle fibers. Am. J. Physiol. 262(1 Pt 1), C229–C234.

[128] Boudier, J.L., Le Treut, T., Jover, E., 1992. Autoradiographic localization of voltage-dependent sodium channels on the mouse neuromuscular junction using 125I-alpha scorpion toxin. II. Sodium distribution on postsynaptic membranes. J. Neurosci. 12(2), 454–466.

[129] Srinivasan, Y., Elmer, L., Davis, J., Bennett, V., Angelides, K., 1988. Ankyrin and spectrin associate with voltage-dependent sodium channels in brain. Nature 333(6169), 177–180.

[130] Gee, S.H., Madhavan, R., Levinson, S.R., Caldwell, J.H., Sealock, R., Froehner, S.C., 1998. Interaction of muscle and brain sodium channels with multiple members of the syntrophin family of dystrophin-associated proteins. J. Neurosci. 18(1), 128–137.

[131] Zhou, D.X., Lambert, S., Malen, P.L., Carpenter, S., Boland, L.M., Bennett, V., 1998. Ankyrin$_G$ is required for clustering of voltage-gated Na channels at axon initial segments and for normal action potential firing. J. Cell Biol. 143(5), 1295–1304.

[132] Angelides, K.J., 1986. Fluorescently labelled Na$^+$ channels are localized and immobilized to synapses of innervated muscle fibres. Nature 321(6065), 63–66.

[133] Fry, M., Moody-Corbett, F., 1999. Localization of sodium and potassium currents at sites of nerve-muscle contact in embryonic Xenopus muscle cells in culture. Pflugers Arch. Eur. J. Physiol. 437(6), 895–902.

[134] Dugandzija-Novakovic, S., Koszowski, A.G., Levinson, S.R., Shrager, P., 1995. Clustering of Na$^+$ channels and node of Ranvier formation in remyelinating axons. J. Neurosci. 15(1 Pt 2), 492–503.

[135] Vabnick, I., Novakovic, S.D., Levinson, S.R., Schachner, M., Shrager, P., 1996. The clustering of axonal sodium channels during development of the peripheral nervous system. J. Neurosci. 16(16), 4914–4922.

[136] Vabnick, I., Shrager, P., 1998. Ion channel redistribution and function during development of the myelinated axon. J. Neurobiol. 37(1), 80–96 (Review, 90 refs).

[137] Rasband, M.N., Peles, E., Trimmer, J.S., Levinson, S.R., Lux, S.E., Shrager, P., 1999. Dependence of nodal sodium channel clustering on paranodal axoglial contact in the developing CNS. J. Neurosci. 19(17), 7516–7528.

[138] Koszowski, A.G., Owens, G.C., Levinson, S.R., 1998. The effect of the mouse mutation *claw paw* on myelination and nodal frequency in sciatic nerves. J. Neurosci. 18(15), 5859–5868.

[139] Dugandzija-Novakovic, S., Koszowski, A.G., Levinson, S.R., Shrager, P., 1995. Clustering of Na$^+$ channels and node of Ranvier formation in remyelinating axons. J. Neurosci. 15(1 Pt 2), 492–503.

[140] Novakovic, S.D., Deerinck, T.J., Levinson, S.R., Shrager, P., Ellisman, M.H., 1996. Clusters of axonal Na$^+$ channels adjacent to remyelinating Schwann cells. J. Neurocytol. 25(6), 403–412.

[141] Novakovic, S.D., Levinson, R., Schachner, M., Shrager, P., 1998. Disruption and reorganization of sodium channels in experimental allergic neuritis. Muscle Nerve 21(8), 1019–1032.

[142] Kaplan, M.R., Meyer-Franke, A., Lambert, S., Bennett, V., Duncan, I.D., Levinson, S.R., Barres, B.A., 1997. Induction of sodium channel clustering by oligodendrocytes. Nature 386(6626), 724–728.

[143] Davis, J.Q., Lambert, S., Bennett, V., 1996. Molecular composition of the node of Ranvier: identification of ankyrin-binding cell adhesion molecules neurofascin (mucin + /third FNIII domain-) and NrCAM at nodal axon segments. J. Cell Biol. 135(5), 1355–1367.

[144] Lambert, S., Davis, J.Q., Bennett, V., 1997. Morphogenesis of the node of Ranvier: co-clusters of ankyrin and ankyrin-binding integral proteins define early developmental intermediates. J. Neurosci. 17(18), 7025–7036.

[145] Malhotra, J.D., Kazen-Gillespie, K., Hortsch, M., Isom, L.L., 2000. Sodium channel beta subunits mediate homophilic cell adhesion and recruit ankyrin to points of cell–cell contact. J. Biol. Chem. 275(15), 11383–11388.

[146] Xiao, Z.C., Ragsdale, D.S., Malhotra, J.D., Mattei, L.N., Braun, P.E., Schachner, M., Isom, L.L., 1999. Tenascin-R is a functional modulator of sodium channel beta subunits. J. Biol. Chem. 274(37), 26511–26517.

[147] Ching, W., Zanazzi, G., Levinson, S.R., Salzer, J.L., 1999. Clustering of neuronal sodium channels requires contact with myelinating Schwann cells. J. Neurocytol. 28(4-5), 295–301.

[148] Planells-Cases, R., Caprini, M., Zhang, J., Rockenstein, E.M., Rivera, R.R., Murre, C., Masliah, E., Montal, M., 2000. Neuronal death and perinatal lethality in voltage-gated sodium channel alpha(II)-deficient mice. Biophys. J. 78(6), 2878–2891.

[149] Duchen, L.W., Stefani, E., 1971. Electrophysiological studies of neuromuscular transmission in hereditary 'motor end-plate disease' of the mouse. J. Physiol. (Lond.) 212(2), 535–548.

[150] Jurkat-Rott, K., Mitrovic, N., Hang, C., Kouzmekine, A., Iaizzo, P., Herzog, J., Lerche, H., Nicole, S., Vale-Santos, J., Chauveau, D., et al., 2000. Voltage-sensor sodium channel mutations cause hypokalemic periodic paralysis type 2 by enhanced inactivation and reduced current. Proc. Natl Acad. Sci. USA 97(17), 9549–9554.

[151] Lehmann-Horn, F., Jurkat-Rott, K., 1999. Voltage-gated ion channels and hereditary disease. Physiol. Rev. 79(4), 1317–1372 (Review, 598 refs).

[152] Burgess, D.L., Kohrman, D.C., Galt, J., Plummer, N.W., Jones, J.M., Spear, B., Meisler, M.H., 1995. Mutation of a new sodium channel gene, Scn8a, in the mouse mutant 'motor endplate disease'. Nat. Genet. 10(4), 461–465.

[153] Wallace, R.H., Wang, D.W., Singh, R., Scheffer, I.E., George, A.L.J., Phillips, H.A., Saar, K., Reis, A., Johnson, E.W., Sutherland, G.R., et al., 1998. Febrile seizures and generalized epilepsy associated with a mutation in the Na$^+$-channel beta1 subunit gene SCN1B. Nat. Genet. 19(4), 366–370.

[154] Escayg, A., MacDonald, B.T., Meisler, M.H., Baulac, S., Huberfeld, G., An-Gourfinkel, I., Brice, A., LeGuern, E., Moulard, B., Chaigne, D., et al., 2000. Mutations of SCN1A, encoding a neuronal sodium channel, in two families with GEFS + 2. Nat. Genet. 24(4), 343–345.

[155] Waxman, S.G., Wood, J.N., 1999. Sodium channels: from mechanisms to medicines? Brain Res. Bull. 50(5-6), 309–310.

[156] Millan, M.J., 1999. The induction of pain: an integrative review. Prog. Neurobiol. 57(1), 1–164 (Review, 1895 refs).

[157] Gould, H.J. III, England, J.D., Liu, Z.P., Levinson, S.R., 1998. Rapid sodium channel augmentation in response to inflammation induced by complete Freund's adjuvant. Brain Res. 802(1-2), 69–74.

[158] Gould, H.J., Gould, T.N., Paul, D., England, J.D., Liu, Z.P., Reeb, S.C., Levinson, S.R., 1999. Development of inflammatory hypersensitivity and augmentation of sodium channels in rat dorsal root ganglia. Brain Res. 824(2), 296–299.

[159] Gould, H.J., Gould, T.N., England, J.D., Paul, D., Liu, Z.P., Levinson, S.R., 2000. A possible role for nerve growth factor in the augmentation of sodium channels in models of chronic pain. Brain Res. 854(1-2), 19–29.

[160] England, J.D., Gamboni, F., Ferguson, M.A., Levinson, S.R., 1994. Sodium channels accumulate at the tips of injured axons. Muscle Nerve 17, 593–598.

[161] England, J.D., Happel, L.T., Kline, D.G., Gamboni, F., Thouron, C.L., Liu, Z.P., Levinson, S.R., 1996. Sodium channel accumulation in humans with painful neuromas. Neurology 47(1), 272–276.

[162] Cummins, T.R., Waxman, S.G., 1997. Downregulation of tetrodotoxin-resistant sodium currents and upregulation of a rapidly repriming tetrodotoxin-sensitive sodium current in small spinal sensory neurons after nerve injury. J. Neurosci. 17(10), 3503–3514.

[163] Dib-Hajj, S.D., Fjell, J., Cummins, T.R., Zheng, Z., Fried, K., LaMotte, R., Black, J.A., Waxman, S.G., 1999. Plasticity of sodium channel expression in DRG neurons in the chronic constriction injury model of neuropathic pain. Pain 83(3), 591–600.

[164] Cummins, T.R., Dib-Hajj, S.D., Black, J.A., Waxman, S.G., 2000. Sodium channels and the molecular pathophysiology of pain. Prog. Brain Res. 129, 3–19 (Review, 76 refs).

[165] Raman, I.M., Bean, B.P., 1997. Resurgent sodium current and action potential formation in dissociated cerebellar Purkinje neurons. J. Neurosci. 17(12), 4517–4526.

Towards a natural history of calcium-activated potassium channels

David P. McCobb

Department of Neurobiology and Behavior, Cornell University, Ithaca, New York, USA

This review is a "molecular safari" [1], where the focus is a handful of genes that have been central to the survival and evolution of diverse life-forms in diverse ecological contexts. We take a somewhat scenic route, hoping for glimpses that might spur novel insights. The K_{Ca} channels, both by refining and preserving useful structural and functional features, and by devising a startling array of mechanisms for adapting themselves for varied use in different tissues and species, have made themselves fascinating elements in the grand history of life on the planet. Trying to summarize the highlights of just one class of channels impresses on one how many things have been evolving in parallel. This fact gives an unending, fractal image-like quality to the learning process. Pursuing specific channels into a few of their different frames of existence underscores the interconnectedness of different cellular phenomena (from ciliary beating to sensory signal integration) as well as the enormous resourcefulness of Mother Nature. To convey the scope of K_{Ca} channel natural history, which is vast even if still mostly uncharted, one has to almost arbitrarily select which of many recounts to zoom-in on. While this review emphasizes big-conductance K^+ ("BK") channels, the goal is to understand K_{Ca} channels as a class by drawing comparisons with small ("SK") and intermediate ("IK")-conductance K_{Ca} channels, in favor of including many specifics regarding BK channels.

1. Ca^{2+}-sensitivity in K channels: ancestral overlap, independent inventions

In the pre-cloning era, patch clamp studies established a dichotomy between BK and SK channels that was near absolute, but for scattered reports of IK channels [2]. All were K^+ selective and activated by intracellular calcium. The BK channels, of gargantuan conductance (~ 280 pS), were less sensitive to calcium, but co-activated by depolarizing voltages, and were blocked by TEA^+ and the scorpion-derived peptides charybdotoxin and iberiotoxin. The 4–20 pS SK channels were voltage-insensitive and mostly blocked by the bee-derived apamin. Attempts at more detailed functional taxonomies were rudely interrupted by molecular studies [3], forcing what would be an astonishingly fruitful interplay between molecular and electrophysiological camps.

Advances in Molecular and Cell Biology, Vol. 32, pages 51–71
ISSN: 1569-2558/DOI: 10.1016/S1569-2558(03)32003-X

It is now clear that BK, SK, IK, and perhaps even calcium-activated cation-selective channels are all descended from ancient, probably calcium-independent, precursors with six-transmembrane domains (6-TMs) [4–13]. 6-TM channels are themselves derived from a 2-TM archetype that remains at the heart of all channels in the superfamily, which stretches from inward rectifiers to Na^+ and Ca^{2+} channels. TMs 5 and 6 in the 6-TM channels correspond to the two alpha helices that flank the "pore" domain in the "structurally solved" 2-TM bacterial K channel [14]. In all of them, four copies are arranged around the central canal to form the channel. Structural details responsible for the huge conductance differences between BK and SK channels remain unclear. Calcium-sensitivity clearly evolved multiple times. The calcium-sensitivity of animal SK channels, and their close cousins the IK channels, is conferred by the intimate association of calmodulin with the basic 6-TM subunit [15,16]. Tandem duplication of an ancestral 2-TM gene has given rise to a widespread family with 4-TMs and two pore domains (two copies needed per channel). In higher plants, one representative of the 4-TM family has an EF hand calcium-binding domain and a very respectable calcium-dependent activation [17]. Calcium-sensitivity of BK channels has yet another origin.

2. BK channels: dual gating

The first BK channel must have arisen nearly half a billion years ago [18], in a slimy tide pool perhaps, when a genetic recombination event shuffled DNA so that a voltage-gated potassium channel gene ended up upstream of what may have been a calcium-dependent serine protease (or related sequence) [19]. Eventually a large multi-exon domain was successfully spliced, in-frame, onto the channel mRNA. The result was a roughly 1200 amino acid protein represented by Slo (for the *slowpoke* fly it was discovered in Fig. 1). Slo shows a very high level of conservation between *Drosophila* and man, with about 60% amino acid identity over 900 residues, excluding a non-conserved "linker" region [20]. The functional novelty of this K channel's dual sensitivity to internal calcium and voltage rendered it one of the most useful. The appended tail was thus a ticket to immortality, and not in a restricted lineage; this ticket was valid on phylogenetic lines that would radiate out in virtually all directions from its ancestral hub. The Slo protein would become indispensable in such diverse roles as regulating blood flow, producing tears, regulating graded contractility of crayfish skeletal muscle, regulating visceral peristalsis, and even making computations "upstairs" [21–26].

Unlike the SK and IK channels, the 6-TM "core" of Slo (minus the calcium-binding tail) retains the essential voltage-sensing mechanics shared with products of more than 30 (in mammals) calcium-insensitive 6-TM K channels. Slo has a unique 7th TM at the N-terminus ("TM-0") [27]. TM-4 (or S4) is thought to be the heart of the essential dipole that translocates and transduces between voltage and gating [28–30]. TM-4 in mammalian Slo has three basic residues, fewer than most voltage-gated K channels, perhaps explaining the relatively modest steepness of its voltage-sensitivity. Surprisingly, Slo-homologues Slo2 in *C. elegans* and Slack in mammals (discussed further below) retain none of the three regularly spaced basic residues in TM-4, and yet retain voltage-sensitivity [10,31]. They have three additional basic residues at irregular intervals in TM-4, but their

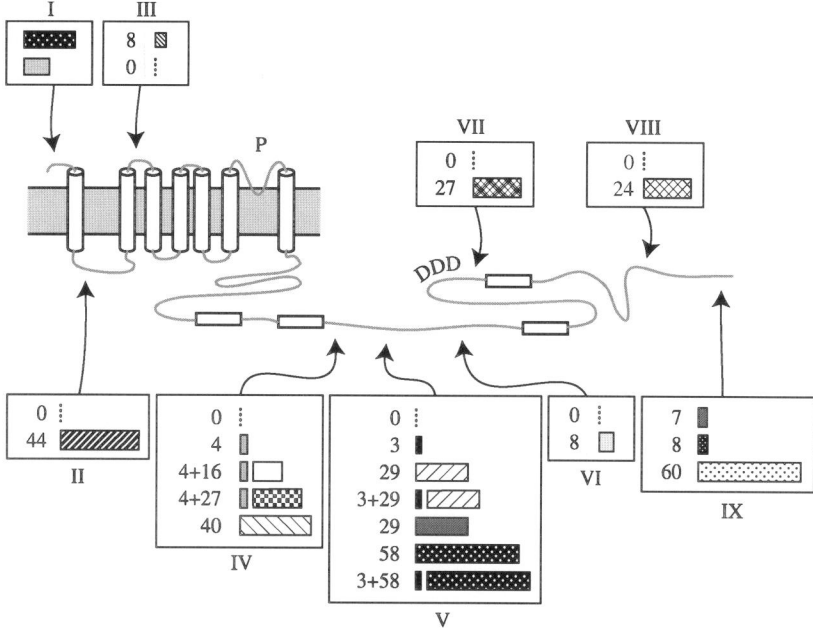

Fig. 1. Alternative splicing in the vertebrate Slo gene, encoding the channel-forming subunit of BK potassium channels. Compiling all exonic variation reported from mammalian genomes (at nine sites, labelled I–IX, exon sizes given in number of amino acid residues) reveals a potentially enormous structural heterogeneity. If all possible combinations of configurations can occur, i.e. the exons assort independently, then 6720 variants are possible. Even if many exon combinations do not occur, or are not functionally distinct, structural heterogeneity will be enormous, since heterotetrameric BK channels can form from different combinations of splice variant subunits. The cylinders above represent the 7 transmembrane alpha helices (TM-0 through TM-6), P stands for the pore domain, DDD represents the aspartate-rich region at the heart of the calcium-binding domain, and the 4 white boxes represent hydrophobic stretches in the C-terminal tail of the subunit protein. The 58-amino acid exon at site V (black with white stippling) is the STREX exon referred to in the text.

functional importance is unclear. Also surprisingly, SK and IK channels, which lack any appreciable voltage sensitivity, do have the three regularly spaced basic residues in TM-4 that Slo1 has [6,32]. Other residues scattered through the other TMs, and even the pore domain [33], clearly can influence TM-4's function, including the steepness and the absolute range over which the sensor responds.

How remote calcium-binding is functionally coupled to channel gating is not fully understood. For BK channels, depolarization and internal calcium act synergistically on the measurable quantity, current flowing through the channels. This raises a semantic issue; neither "calcium-sensitivity" nor "voltage-sensitivity" of BK channels can be measured without reference to the other. The synergistic interaction has been variously explained as either calcium modulating the voltage-dependent gating or the inverse, voltage modulating the calcium-dependent gating. However, rigorous analyses in the absence of calcium and in saturating calcium now argue for separate and independent influences of voltage and calcium on the gating mechanism of the channel [34–37].

In the conserved region of Slo's tail, the aspartate-rich "calcium bowl" is clearly at the heart of Ca^{2+} binding, though not solely responsible for it [38,39]. Schreiber et al. [40] have exploited the fact that an interesting relative of Slo, Slo3, lacks calcium-sensitivity, though it shares general tail structure [41]. (Also exploited was the fact that functionally normal BK channels arise from the co-expression in oocytes of split Slo transcripts, i.e. co-injection of two engineered transcripts, one encoding TM-0 through TM-6 plus hydrophobic segments S7 and S8, and the second encoding S9 and S10 [42].) Surprisingly, co-expression of the corresponding Slo1 tail with the Slo3 core produced channels that were voltage-sensitive with little or no calcium-sensitivity, even though the Slo1 core by itself is apparently non-functional. Swapping a small stretch including the calcium bowl of Slo1 into the Slo3 tail restored substantial calcium-sensitivity. A second stretch downstream that excluded the calcium bowl also restored significant calcium-sensitivity, though less. Together the two restored most of it. Even more surprisingly, when a 91-residue stretch just upstream of Slo's calcium bowl was swapped into the Slo3 tail, the voltage-dependence was shifted substantially in the positive direction, meaning stronger depolarizations were required to open the channel. This effect was observed whether or not the calcium-sensitivity domains were also transplanted. This region of Slo1, as compared with its counterpart in Slo3, somehow favors the closed state of the channel over the open. Presumably this negative effect on opening increases the necessity for calcium binding to achieve an open probability appropriate in a physiological range of voltage. If this region is also involved in communicating the binding of calcium to an upstream gate, the corresponding region of Slo3 seems to be just as effective in that capacity. Further structure–function experiments and structural details from crystallography should further illuminate the mechanics of the interaction between ligand binding and gating.

In another fascinating twist, Slo2, a *C. elegans* paralog of Slo1, has replaced many of the negatively charged residues in Slo's calcium bowl with positively charged residues [31]. Referred to as the "chloride bowl", charge-substitution studies on Slo2 confirm a role in chloride sensitivity. The Slo2 channels in native form retain significant calcium-sensitivity, in addition to the chloride-sensitivity, and surprisingly, the calcium-sensitivity is also negatively affected by the same charge substitutions that disrupt chloride-sensitivity. Evidently this one domain is interacting with multiple ion types.

3. Functional niches for K_{Ca} channels: similarities, differences, and multiplicities

The 6-TM precursor of the 3 known SK channel genes, representing paralogs approximately 60% identical to each other, also gave rise to a cousin with about 40% identity to the others, the IK gene (BK channels are less than 20% identical to SK and IK genes) [12,13,32,43]. IK is expressed in inexcitable cells, particularly those of hematopoietic origin. The first-studied K_{Ca} channel, characterized in red blood cells by Gardos [44], belongs to this family. IK channels are also prominent in T-lymphocytes, where they are particularly important in cell proliferation [45]. IK channels appear to predominate in immature, proliferating vascular smooth muscle cells, while BK channels dominate in more mature, contractile cells [12].

SK channels are present in most excitable cells and many inexcitable cells. While an essential role may be homeostatic, hyperpolarizing the membrane to limit calcium influx through depolarization-activated channels, SK channels have also taken on more dynamic roles [6,8,9]. SK channels are responsible for spike frequency adaptation in many neurons [2,46–48]. Functional suppression of SK channels is seen in various neurons by neuromodulators acting through metabotropic receptors, including receptors to glutamate, norepinephrine, acetylcholine, serotonin, dopamine, histamine and several peptides. The negative modulation of SK channel is important in promoting wide-scale CNS arousal, and apparently plays more locally restricted roles in attention, emotion, memory and other cognitive functions, as well. For example, adenosine's role in promoting sleep may derive in part from its ability to increase SK channel activation, and reduce spike frequency adaptation [6,49]. Apamin injections into the hippocampus have been reported to enhance memory and promote immediate early gene expression [50,51], and endogenous β-adrenergic modulation may similarly enhance memory by reducing spike frequency adapation [52].

By analogy with spike frequency adaptation, SK channels can participate in terminating regular bursts of action potentials [48,53,54]. However, BK-type channels are perhaps just as likely to play this role in some cells [55–57], and slowly activating Ca-independent K channels can also participate [56,58]. In sharp contrast to the burst-terminating role, SK channels suppress burst firing in hippocampal CA1 pyramidal neurons and a variety of other cells [58–60]. Thus based on the experimental augmentation of bursting behavior by apamin, SK currents have been attributed with increasing the interspike interval, which dampens the spike frequency and obviates burst termination. Reduced accumulation of Ca that would activate a burst-terminating K_{Ca} channel (BK channels or apamin-insensitive SK channels?) and/or improved recovery of Na and/or Ca channels from inactivation between spikes could be involved. Clearly, definitive resolution of the molecular players underlying specific K_{Ca} currents (and their modulation) are needed to better appreciate the contributions of the various types to excitability in specific contexts, and to design better therapeutic approaches to electrical malfunction.

The ancestor of Slo has given rise to apparently just two additional paralogs in mammals, despite the extremely varied roles played by BK channels. Slo3 is known only in spermatocytes, and exhibits pH- and voltage-sensitive gating but no sensitivity to calcium. Regulation of sperm motility and the fertilization process through pH is important in most species, but *C. elegans* is an exception, perhaps explaining its lack of a Slo3 paralog. Slo2 was discovered in *C. elegans*, but shares 41% homology with mammalian Slack [10,31,61]. The nematode channel has chloride-, calcium-, and voltage-sensitive gating, the utility of which is unknown. Slo2 is even more widely expressed in the worm than Slo1. The rat Slack gene is widely expressed in the brain, and occurs in kidney as well. When expressed in *Xenopus* oocytes, Slack forms voltage-gated channels with an apparent negative regulation by calcium and a single channel conductance of around 50 pS, making them unlikely to be confused with classic BK channels. Whether Slack is chloride-sensitive is unknown. Intriguingly, co-expression of Slo and Slack results in novel, presumed-heteromultimeric channels [10]. These channels exhibit a positive regulation by calcium and a very flickery gating, which makes the single-channel conductance difficult to measure. This is still unlike Slo channels alone or any native BK currents characterized to date. The extent to which Slack homomultimers or Slack/Slo heteromultimers underlie K_{Ca} currents in vivo, or underlie channels that can be confused with channels comprised solely of Slo1 subunits, remains to be explored. Despite this, it seems likely from their similarity with heterologously expressed Slo products, that many of what are functionally identified as BK channels in diverse tissues arise from alpha subunits encoded by Slo itself. Considerable diversity is achieved through (1) alternative splicing of Slo, and (2) differential association of Slo with accessory subunits unrelated to Slo.

BK channels strongly resembling Slo homomultimers are widespread, though most of the details of structure, association, localization, and regulation that tailor the channels for diverse function remain to be discovered. BK channels in exocrine epithelia are concentrated at the lumenal cell surface, where a primary function is to provide a conduit for moving potassium, which assists with water movement osmotically [23,62]. At the vertebrate neuromuscular junction, Slo channels are tightly interspersed with pre-synaptic calcium channels, where their modulation shapes action potential duration and neurotransmitter release [63–65]. Differential coupling of K_{Ca} channels to distinct calcium channel types is an important variable, both between BK and SK channels, and between cell types [66–68]. BK channels are very important as negative regulators of smooth muscle contractility, as discussed further below, and may play anti-excitatory roles in neurons as well. Also elaborated on below, BK channels shape voltage oscillations in vertebrate auditory hair cells, and can play a positive role in repetitive firing in adrenal chromaffin cells by recovering inactivated Na channels following spikes. BK channels seem to be essential in burst firing of cerebrocortical neurons [69]. Further illustrating just how counterintuitive K_{Ca} channel functions can be, rapid BK channel activation has been attributed with enabling pituitary somatotropes to fire very long duration plateau potentials. Thus by clipping action potential peak-amplitude, the BK channels limit spike amplitude. This prevents activation of delayed rectifier K channels which would otherwise drive a rapid action potential repolarization, and enables the long plateau [70].

4. Tailoring Slo properties through alternative splicing

An astronomical 1500 distinct exon configurations of *Drosophila* Slo have been postulated, and many have been documented to occur [71]. As illustrated in Fig. 1, vertebrate Slo genes have roughly an order of magnitude power potential configurations. However, the genomic structure is not yet fully charted, and surprises remain to be explained, including the " − 26" splice variant in turtles that lacks sequence thought to belong to constitutive exons [72]. For both *Drosophila* and mammals, most differential splicing is in the tail region. What the variation accomplishes is largely unknown, but several configurations confer differences in calcium- and voltage-dependent gating [73–77].

One approach to understanding the function of splice variants in vivo has exploited the power of *Drosophila* as an experimental system. This approach involves introducing pre-spliced Slo transgenes into a Slo-null background [78]. A muscle-specific variant transgene adequately rescues flight muscle action potential repolarization and the ability to fly when the insect is dropped. However, it fails to rescue the hyperexcitability of foot motoneurons that leaves the fly, when it is stimulated in a standing position, stuck in-place by excessive gripping of the substrate. It remains to be seen whether introduction of a motoneuron-native variant will rescue this "sticky-feet" behavior, and perhaps leave an aberration in flight muscle function. This would further suggest specific functional roles for individual splice variants.

There are several other fascinating glimpses of Slo splice variant function in vivo. Vertebrate cochlear hair cells specialize in being most sensitive to a narrow segment of the sound frequency spectrum [79]. "Tonotopy" refers to the orderly basal to apical cochlear distribution of high to low frequency specialists. The specialization depends on frequency-restricted accentuation, which requires simultaneous involvement of multiple mechanisms. In reptiles, birds, and fishes, differences in intrinsic electrical resonance properties along the cochlear axis provide one important mechanism. In these cells, steady current injections through an electrode trigger voltage oscillations with a characteristic frequency. Sound of this frequency is maximally effective at driving voltage oscillations in the cell. The oscillations are driven by feedback between calcium channels and BK channels, and the kinetics of this active process are determined chiefly by the voltage-dependence and kinetics of gating of the more rate-limiting BK channels. Interestingly, several Slo splice variants are expressed in the cochlea, as described below.

At least seven exons appear to exhibit differential distributions along the cochlea that are consistent with their influence on BK function [80–82; Rosenblatt, 1997, 5968]. Insertion of an optional 61 amino acids at one splice site facilitates opening at negative voltages, increasing the activation kinetics at a specific voltage, and potentially lowering the characteristic oscillation frequency for the cell [83]. Inserts of various lengths at nearby sites have similar effects. Intriguingly, Jones et al. [80] noted a rough correlation between the collective length of inserts and an increasing calcium-sensitivity. Effects on the binding affinity for calcium could be involved, but other explanations are plausible. Exploiting the apparent inhibitory, or positive-shifting, effect of the 91-residue region upstream of the calcium bowl described above [40], one could speculate that the optional inserts facilitate gating by interfering with the region's inhibitory effect. Whatever the mechanism, BK channels in hair cells clearly represent a very intriguing case of precise "channel-tweaking" through alternative splicing, with important behavioral significance. Beta subunits are also thought to contribute to the cochlear heterogeneity, as discussed further below. We can look forward to more complete characterization of the players, better resolution of their distributions, and further mechanistic analysis of structure–function relations behind the frequency tuning. Further over the horizon loom the mysteries of how hair cells know their position, and control how many and which combinations of variants and subunits to express accordingly.

The 61-amino acid Slo insert identified in hair cells is generated from the 174 base pair "STREX" exon known from chromaffin cells [75,84], plus an additional 9 base pair mini-exon inserted in front of STREX at the same site [83]. STREX is an acronym for stress-axis regulated exon. On the theory that pituitary hormones, particularly ACTH (adrenocorticotropic hormone), might regulate chromaffin BK channel expression, Xie and McCobb made comparisons in Slo expression between normal and pituitary-ablated (hypophysectomized) rats. Though total transcript levels were not different, the relative abundance of the two predominant splice forms was. With primers bracketing this splice site, PCR products representing the configuration with no insert ("ZERO"), and the STREX configuration were generated from chromaffin tissue. In normal rats, transcripts with STREX account for approximately 40% of the total, with ZERO transcripts accounting for the remainder. The STREX representation was well below 20% in hypophysectomized animals [84].

A variety of evidence suggests bidirectional regulation of STREX splicing involving multiple steroid hormones. The ability of ACTH injections to prevent the STREX decline suggests that CORT, the synthesis of which is ACTH-driven, might mediate the pituitary effect. STREX drops rapidly over the 2 weeks after surgery, a time course paralleling the drop in mRNA for the enzyme phenylethanolamine-*N*-methyltransferase (PNMT) that occurs with hypophysectomy. PNMT transcription is promoted by the glucocorticoid receptor–CORT complex. Surprisingly, however, experiments involving direct application of steroids to bovine chromaffin cells in culture indicated a negative regulation of STREX inclusion by cortisol and dexamethasone. In contrast, a lesser-known adrenal corticosteroid, dehydroepiandrosterone (DHEA), had the opposite effect, increasing representation of STREX transcripts relative to ZERO [85]. A number of studies have shown DHEA synthesis to be stimulated by stress or ACTH, raising the possibility that DHEA is a key player in the effect of pituitary ablation on STREX. Though the biological role of adrenal DHEA is largely unknown, it is considered a weak androgen. Application of testosterone to bovine cells in vitro, like DHEA, raised STREX levels [85]. Behavioral experiments to assess the effects of stress on STREX levels have been done, exploiting a paradigm involving chronic exposure of subordinate male tree shrews to a dominant conspecific. This robust chronic stress resulted in a significant decline in adrenal STREX levels [86]. While CORT levels were chronically elevated in the stressed animals, androgen levels were depressed. Moreover, unstressed female tree shrews had lower STREX levels than stressed or unstressed males. Further evidence for a role for sex steroids in STREX splicing comes from the rat pituitary gland. Pre-pubertal castration of males significantly reduced pituitary STREX expression, and this can be prevented with testosterone implants [87]. Further studies are needed to better understand links between steroids and Slo splicing, which appear to differ meaningfully between cell types, sexes, and species.

Mechanistic studies by Xie and Black indicate that STREX splicing of a Slo construct introduced into a cell line is negatively regulated by calcium [88]. This is thought to be mediated by interaction of calcium/calmodulin-dependent kinase IV (CaMKIV) with a response element in the intronic sequence upstream of the STREX exon. Additional exonic sequences also play a role. A role for calcium in chromaffin cells requires further study, but preliminary evidence from bovine cells in vitro suggests positive regulation of STREX by calcium, rather than negative (Lai and McCobb, unpublished data). Though one wonders whether calcium could be a factor in STREX splicing in cochlear or uterine smooth muscle cells as well, signaling pathways in splicing regulation are likely to differ.

Of what physiological consequence is hormonal regulation of Slo splicing in chromaffin cells? We have hypothesized that making the BK channels more accessible for activation will make the cells more, rather than less, excitable, and thereby increase catecholamine responses to autonomic inputs [89,90]. In *Xenopus* oocytes STREX shifts the half-activation voltage approximately 25 mV in the hyperpolarizing direction, as compared with the otherwise equivalent ZERO form, and simultaneously increases and decreases rates of activation and deactivation respectively [75,84]. Together these features are expected to speed repolarization and augment the brief afterhypolarization. This will facilitate repetitive firing by minimizing the accumulative entry of Na^+ and Ca^{2+} channels into, and maximizing recovery from, the inactivated state. Such a role for a K channel exploits the nearly universal tendency for Na^+ channels to inactivate rapidly, which itself protects against prolonged or overly frequent depolarization. Na^+ channel deinactivation is faster at more negative potentials, thus rapid activation of K channels will facilitate deinactivation. BK enhancement of repetitive firing is supported by the ability of pharmacological block of BK channels to reduce it [90–92]. Interestingly, the STREX exon has recently been identified as a direct target for phosphorylation, serving to switch the effects of PKA phosphorylation of the channel from activation-promoting to activation-inhibiting [93]. This adds an unexplored dimension to the role of splicing regulation.

As one test of the above scenario, chromaffin cells from hypophysectomized rats were compared with their counterparts from normal rats [90]. Gating of BK currents from hypophysectomized rats differed from that in normal rats as predicted from the reduction in STREX transcripts. Moreover, cells from hypophysectomized animals were indeed less excitable than normal. With perforated patch recording, hypophysectomized rat cells could generate fewer than half the number of action potentials during sustained depolarizing stimuli. Their action potentials also had smaller afterhyperpolarizations, and a more rapid decline in action potential amplitude and increase in breadth over successive action potentials. We cannot rule out that changes in expression of Slo Beta subunits or a number of other channel types, including Na^+ or Ca^{2+} channels, contribute to the changes in excitability observed. The rapid inactivation characteristic of rat chromaffin BK channels was not altered by hypophysectomy, implying a lack of hormonal regulation of inactivation-conferring Beta subunits [90].

Hormonal control of STREX splicing may underlie bovine–rat species differences, differences between rat strains, and changes in STREX abundance during development. STREX transcripts comprise about 15% of the total in bovine chromaffin cells, compared with 40–55% in rat (depending on the strain). This may be related to the fact that there are approximately 10-fold lower levels of serum cortisol (compared with its analog in rats, corticosterone) and 5-fold lower levels of ACTH [90]. As in hypophysectomized rat cells, bovine chromaffin cell BK channels require more positive voltages to activate, activate less rapidly, and deactivate more rapidly. These results argue that evolution has exploited hormonal regulation of Slo splicing to adapt autonomic function for different life styles [89]. Bovine cells differ from both normal and hypophysectomized rat cells in having markedly slower and often incomplete BK inactivation. In rats a postnatal decline in STREX of approximately 50% coincides temporally with the stress-hyporesponsive period of development, during which CORT levels drop dramatically before returning to pre-natal levels [94,95].

To what extent Slo splicing is responsive to natural stress-axis activation, and whether there are significant health consequences associated with it remain to be addressed.

Individual and species differences in stress-readiness may be attributable to genetic differences in stress hormone, receptor expression, or differences in the regulatory mechanism associated with the Slo gene. Several other cell types in which STREX occurs are intimately involved with stress responses, including the pituitary, the hippocampus, and pancreatic β-islet cells. This raises the possibility of more general stress-related regulation of excitability through BK splicing [96–98].

In uterine smooth muscle, the relative abundance of several Slo splice variants, including STREX and ZERO, change as pregnancy progresses [99]. Specifically, the levels of STREX-containing transcripts declines relative to ZERO transcripts, reaching a minimum at parturition. A decline in accessible BK channels, at least in part reflecting a positive shift in the voltage-dependence of gating, is observed over pregnancy, despite a puzzling increase in Slo mRNA abundance in the tissue. Splicing-related changes in channel gating could therefore help explain the apparent paradox. A more complete understanding of this awaits better temporal and species-specific resolution of the complement of factors regulating excitability and contractility. However, it is tantalizing to consider potential roles for BK currents. Leading up to parturition, changes in BK number or properties could conceivably reduce contractility to facilitate uterine expansion, increase tension to support a growing brood, or more complexly alter the dynamics or sensitivity of contractility to hormonal or neural control.

5. Companions of Slo: the Beta subunits

Though our BK binoculars have thus far been trained on Slo, associated creatures are emerging from the shadows. Thanks to genome databases, a family of "Beta subunits" comprised of at least four distinct members in mammalian genomes has been rapidly cataloged, following the breakthrough cloning of the first member, now known as Beta1 [100]. Expressed in greatest abundance in smooth muscles of various sources, Beta1 shifts the voltage/calcium sensitivity as much as 50 mV in the negative direction, making the channels much more accessible for activation [83,101,102]. It also slows channel activation, and even more dramatically slows deactivation. The second Beta cloned was from quail, and though it has been variously referred to, its relationship to mammalian Betas is still unclear [103]. Beta1 is its closest known mammalian homolog, but its fairly scant ~45% identity make it more divergent than would be expected for a true ortholog. Almost certainly there are other avian Beta genes, and perhaps a truer mammalian ortholog as well. Intriguingly, the avian Beta is preferentially expressed in the apical one third of the cochlea, and is believed to be a key factor in conferring the slowest oscillation frequencies on those cells [104]. The early but relatively gradual activation, combined with delayed but gradual deactivation would seem to provide an ideal mechanism for slowing the innate oscillatory drive of the lower frequency cells. Its effects may be even more than additive with those of Slo splicing configurations that have similar effects; though mechanistically unexplained, the 61 amino acid insert mentioned above seems to exaggerate the Beta-induced slowing in a more than simply additive fashion [104]. At a physiological level, this would extend the frequency range at the more slowly oscillating, lower frequency end. Mammalian Beta1 has effects on Slo that are similar, but not identical to those of avian Beta1. Perhaps the apparent absence of a closer mammalian ortholog of the avian Beta reflects the reduced importance of electrical tuning in mammalian audition.

Much slower voltage oscillations than those in the cochlea drive smooth muscle contractions [25]. Nevertheless, a rhythm-slowing influence of the Slo–Beta1 combination could be analogous to that postulated in hair cells. However, smooth muscle varies dramatically between, for example, circular and longitudinal colonic muscle, bronchi, and between capillary beds in different tissues. Electrical balance and dynamic modulation are achieved by fine-regulation of different combinations of several ion channel types in different tissues [21]. The dual sensitivity of BK channels to voltage and calcium must provide for a uniquely dynamic influence that is likely to be more important in some contexts than others. In some vascular smooth muscle, a "reactive" tonus generated reflexively in response to supernormal tension (from rapidly elevated blood pressure perhaps) may be a more important role of BK channels than maintaining a "passive" tonus. On the other hand, the line between passive and reactive is fine and clearly context dependent. Calcium released from smooth muscle SR during sporadic "sparks" can trigger BK-driven hyperpolarizing transients that may continually promote a relatively relaxed state [105]. While the cell would be exploiting the calcium sensitivity of the BK channel primarily, voltage would presumably also play a role, for example, in limiting the duration of the hyperpolarizing transient.

The protein referred to as Slo Beta2 has effects on activation and deactivation gating that are similar to those of Beta1, but it also has the striking ability to confer fairly rapid inactivation on the channel [106,107]. The inactivation may be mediated by a tethered "ball" provided by the N-terminus of the Beta2 protein that swings into a position occluding the inner channel mouth. Like the N-terminal balls of alpha and Beta subunits of some Shaker-related channel variants, this ball can be functionally removed by trypsin digestion, or by N-terminal truncation of the Beta expression construct. It can also be transplanted to an otherwise non-inactivating Beta. Note that the still unexplained inability of TEA$^+$ to competitively interfere with inactivation in Slo contrasts with its ability to influence shaker N-terminal inactivation, and serves to remind us that there are differences between these two types of K channels [106].

Slo Beta4, widely expressed in neurons, shifts the voltage-dependent gating in the positive direction at low calcium concentrations, but in the negative direction in higher concentrations. This will effectively increase the steepness of the relationship between calcium concentration and open probability [108,109]. Another family member, Beta3, is alternatively spliced, and two of its four known variants confer inactivation on Slo [108, 110,111]. This inactivation, though considerably faster than that of Beta2, is decidedly incomplete, despite long depolarizing voltage clamp pulses. An interesting suggestion has been put forward; perhaps what this achieves is to raise the apparent rate of BK current activation, while avoiding a subsequent BK current of overwhelming amplitude. Thus the rapid inactivation allows rapid activation to be achieved by simply "exaggerating" the number of channels expressed.

Neither is the "raison d'être" of Beta2 inactivation certain. Co-expression of Beta2 with Slo in *Xenopus* oocytes produces BK inactivation more closely resembling that in rat adrenal chromaffin cells than does co-expression of Beta3 and Slo. The slower but more complete inactivation of Beta2 might simply serve to remove BK channels from availability. The calcium-sensitivity of this inactivation, and the fact that it accrues too slowly to have much impact during infrequent spikes, would suggest that it might figure most prominently during rapid firing and/or muscarinic acetylcholine receptor-mediated elevations of cytoplasmic calcium [92,112].

The biological role of BK inactivation must vary depending on the role of BK channel activation in specific cellular contexts. In hippocampal neurons BK inactivation is credited with frequency-dependent spike broadening [98]. The net effect of this on synaptic function and behavior is unclear. If, as is undoubtedly true in many cases, the BK channels' primary role is to dampen excitability or slow the dynamics of oscillatory voltage changes, then BK inactivation should enhance excitability or speed dynamics. If the role of the BK channels is to facilitate repetitive firing, as postulated for chromaffin cells above, then channel inactivation should serve to limit repetitive firing. Perhaps this is an important barrier to over-stimulation of chromaffin cells. The subset of rat chromaffin cells with a prominent inactivating component are also apparently the ones with greater repetitive firing ability, as well as the BK activation and deactivation properties suited for promoting rapid firing, as discussed above [113]. Though it may seem ironic that BK inactivation should play a negative feedback role, this does not rule out the possibility that BK activation plays a negative feedback role by dampening excitability, for example, in the subset of chromaffin cells with the non-inactivating BK channels. Who is to say that Mother Nature cannot have it both ways at once? The complexity of BK channel gating could enable them to define both the lower and upper limits of a cell's responsiveness. This should be testable in future experiments involving a fast and well-programmed dynamic clamp.

6. Dynamic modulation of BK channel gating behavior

Rapid modulation of the gating properties of K_{Ca} channels is extraordinarily diverse and critical in many contexts [114,115]. Direct phosphorylation or dephosphorylation of Slo channels can increase or decrease open probabilities of K_{Ca} channels [116–120; 93,121]. Gating can also be directly effected by changes in the redox state of cysteine residues [122]. Nitric oxide may directly affect redox state, but can also interact with Slo channels via G-proteins, which can themselves interact with Slo channels either directly or indirectly via phosphorylation pathways [123–125]. In vascular smooth muscle, phospholipid metabolites regulate BK channels [126,127]. Estradiol activates smooth muscle BK channels by binding to Beta1 subunits of Slo directly [128]. Acute application of the steroid DHEA promotes BK activation in airway smooth muscle [129,130]. One can thus speculate that stress-enhanced adrenal DHEA secretion enhances air intake during stress responses. Similarly-acute cortisol exposure promotes BK channel activation and facilitates repetitive firing in bovine adrenal chromaffin cells, which should augment catecholamine secretion in response to stress [131]. Participation of Beta subunits in these modulatory responses is being tested. Glucocorticoids have been shown to modulate BK channels in pituitary-type cells through two less-acute mechanisms, involving gene transcription [132] and modulation of the phosphorylation state [93,119]. Ethanol appears to affect BK channels rather directly [133].

Definitive identification of the full complement of molecular players involved in modulation, and analysis of their importance in specific contexts, still has a long way to go. Newly discovered proteins like Slob, 14-3-3, dSLIP1, PDZ proteins, and the Beta subunits discussed above act as modulators, modulatory targets, and/or key links in "scaffolding" that couples BK channels to functional partners like calcium channels and G-proteins [109,134–136]. At the *Drosophila* neuromuscular junction, Slob may directly promote Slo opening, or participate in the inhibition of Slo through an interaction with 14-3-3 protein and calcium-calmodulin kinase II that results in the phosphorylation of serine residues on Slob [135,136].

7. Regulation of K_{Ca} channel expression

The genomes of *C. elegans* and *H. sapiens* each include on the order of 100 K channel genes, each with some potential functional overlap with the K_{Ca} subset [5,7,137]. The selective expression of these subtypes, including not only which but also how many of each are expressed, will determine the dynamic repertoire of cellular excitability, potentially uniquely in each cell. Generating a net K_{Ca} current that just suits demand depends just as importantly on the number of copies of a channel type as on their calcium- or voltage-sensitive properties. In fact, somehow these variables must be mutually coordinated. Cloning and characterizing the channels and their accessory proteins is just the foundation for understanding the mechanisms governing the qualitative and quantitative deployment of channels in cells.

Transcriptional regulation of the *Drosophila* Slo gene involves at least five transcription start sites [138–140]. The mammalian counterpart has at least three, and probably more [141]. Some meaningful control elements within them can be identified by similarity to known consensus sequences, although experiments to decipher the combinatorial workings of control elements and factors suggest that, like most genes analyzed to date, transcriptional control of these genes is not simple. Comparisons between fly species have proven useful for identifying novel putative control elements. The relative ease of promoter analysis in *Drosophila*, and the likelihood of shared elements, factors, and principles across taxa, clearly justify detailed study in *Drosophila*, and similarly, in *C. elegans* and probably zebrafish, as well as mouse. Spread over roughly 8 kb, *Drosophila* promoters have been shown to direct expression in tissue specific fashion. In this regard, the viability of the Slo null mutant has been advantageous. Transgenic flies with reporter genes driven by various promoter constructs reveal two distinct neuronal promoters, one muscle and trachea promoter, and two midgut promoters. Roughly 10-fold quantitative differences in transcription from the two neuronal promoters occur in different cells, and probably derive in part from differences in TATA box sequence. Differential use of distinct elements in the muscle promoter leads to selective expression in larval muscle, adult muscle, and tracheal cells. A deletion mutant lacking only the neuronal promoters will complement the flight problems of the null mutant, but not the sticky-feet phenotype mentioned above.

Behavioral effects of transgenic expression driven by different promoters, and expression of different splice variants and mutants will be of interest for years to come.

The regulation of BK functional expression also appears to occur at post-transcriptional levels. The chick iris releases a soluble growth factor, identified as the avian TGFβ1 homolog that somehow unmasks BK expression in ciliary ganglion neurons both via a transient, protein synthesis-independent pathway and via a longer-lasting protein synthesis-dependent pathway [142,143]. Both involve the MAP kinase Erk, and neither alters transcription of Slo itself. Clearly Mother Nature misses no opportunity to fine-tune BK functional expression, and employs a variety of means to do so.

8. Summary

K_{Ca} channels have provided strategic advantages for many cells. Those with and without additional voltage dependence are co-deployed in many if not most cases. All probably participate in negative-feedback protection against calcium overload in some contexts; however, they have many other roles, some even contradictory to this role. Slo gene-encoded BK channels dampen excitation, secretion, and contraction, facilitate rapid spiking and secretion, enhance auditory frequency discrimination, modulate pre-synaptic output, shape rhythmicity, and transport potassium in bulk. Barely studied paralogs of Slo (Slo2, Slack, and Slo3) have mysterious chloride-, calcium-, or pH-sensitive aspects to their gating. SK channels from several genes mediate spike frequency adaptation, help pace rhythmic bursting, prevent rhythmic bursting, fine-tune threshold, and influence wakefulness and memory formation. Related IK channels provide a driving force for calcium, and help regulate immune function and hematopoiesis. Some of the classically described K_{Ca} channels, e.g. those that help regulate ciliary beating in *Paramecium*, and those that produce the quiet phase in elegant molluscan bursting rhythms, remain uncharacterized at the molecular level [144–147].

Slo channel proteins do not do it alone; dedicated Beta subunits and more promiscuous signaling and scaffolding proteins assist in many ways, including some yet to be discovered. Identifying the factors and interpreting the "code" regulating proper expression of the players is a challenge that extends even beyond understanding their interactive mechanics. Understanding the process by which all this fantastic complexity evolved obviously extends further still.

Analogies are imperfect, but thinking of genes as if they were species, each with a long and unique history of exposure to mutational and selective pressures that drive adaptation and diversification, has an appeal similar to that of gaining an in-depth knowledge of the evolutionary ecology of a species. Many episodes remain to be written to cover the natural history of the K_{Ca} channels.

References

[1] Wei, A.D., 1994. Personal Communication.
[2] Hille, B., 2001. Ionic Channels of Excitable Membranes, 3rd ed. Sinauer Associates, Inc, Sunderland, Massachusetts, p. 814.

[3] Atkinson, N.S., Robertson, G.A., Ganetzky, B., 1991. A component of calcium-activated potassium channels encoded by the *Drosophila* Slo locus. Science 253(5019), 551–555.

[4] Meera, P., Wallner, M., Toro, L., 2001. Molecular biology of high-conductance, Ca^{2+}-activated potassium channels. In: Rusch, A.a. (Ed.), Potassium Channels in Cardiovascular Biology. Kluwer academic/Plenum, Dordrecht/New York, pp. 49–70.

[5] Coetzee, W.A., Amarillo, Y., Chiu, J., Chow, A., Lau, D., McCormack, T., Moreno, H., Nadal, M.S., Ozaita, A., Pountney, D., Saganich, M., Vega-Saenz de Miera, E., Rudy, B., 1999. Molecular diversity of K+ channels. Ann. NY Acad. Sci. 868, 233–285.

[6] Bond, C.T., Maylie, J., Adelman, J.P., 1999. Small-conductance calcium-activated potassium channels. Ann. NY Acad. Sci. 868, 370–378.

[7] Wei, A., Jegla, T., Salkoff, L., 1996. Eight potassium channel families revealed by the *C. elegans* genome project. Neuropharmacology 35(7), 805–829.

[8] Vergara, C., Latorre, R., Marrion, N.V., Adelman, J.P., 1998. Calcium-activated potassium channels. Curr. Opin. Neurobiol. 8(3), 321–329.

[9] Sah, P., Davies, P., 2000. Calcium-activated potassium currents in mammalian neurons [in process citation]. Clin. Exp. Pharmacol. Physiol. 27(9), 657–663.

[10] Joiner, W.J., Tang, M.D., Wang, L.Y., Dworetzky, S.I., Boissard, C.G., Gan, L., Gribkoff, V.K., Kaczmarek, L.K., 1998. Formation of intermediate-conductance calcium-activated potassium channels by interaction of Slack and Slo subunits. Nat. Neurosci. 1(6), 462–469.

[11] Jorgensen, T.D., Jensen, B.S., Strobaek, D., Christophersen, P., Olesen, S.P., Ahring, P.K., 1999. Functional characterization of a cloned human intermediate-conductance Ca(2+)-activated K+ channel. Ann. NY Acad. Sci. 868, 423–426.

[12] Neylon, C.B., Lang, R.J., Fu, Y., Bobik, A., Reinhart, P.H., 1999. Molecular cloning and characterization of the intermediate-conductance Ca(2+)-activated K(+) channel in vascular smooth muscle: relationship between K(Ca) channel diversity and smooth muscle cell function. Circ. Res. 85(9), e33–e43.

[13] Logsdon, N.J., Kang, J., Togo, J.A., Christian, E.P., Aiyar, J., 1997. A novel gene, *hKCa4*, encodes the calcium-activated potassium channel in human T lymphocytes. J. Biol. Chem. 272(52), 32723–32726.

[14] Doyle, D.A., Morais Cabral, J., Pfuetzner, R.A., Kuo, A., Gulbis, J.M., Cohen, S.L., Chait, B.T., MacKinnon, R., 1998. The structure of the potassium channel: molecular basis of K+ conduction and selectivity. Science 280(5360), 69–77 (see comments).

[15] Xia, X.M., Fakler, B., Rivard, A., Wayman, G., Johnson-Pais, T., Keen, J.E., Ishii, T., Hirschberg, B., Bond, C.T., Lutsenko, S., Maylie, J., Adelman, J.P., 1998. Mechanism of calcium gating in small conductance calcium-activated potassium channels. Nature 395(6701), 503–507.

[16] Fanger, C.M., Ghanshani, S., Logsdon, N.J., Rauer, H., Kalman, K., Zhou, J., Beckingham, K., Chandy, K.G., Cahalan, M.D., Aiyar, J., 1999. Calmodulin mediates calcium-dependent activation of the intermediate conductance KCa channel, IKCa1. J. Biol. Chem. 274(9), 5746–5754.

[17] Czempinski, K., Zimmermann, S., Ehrhardt, T., Muller-Rober, B., 1997. New structure and function in plant K+ channels: KCO1, an outward rectifier with a steep Ca^{2+} dependency [published erratum appears in EMBO J. 16(22), 1997, 6896]. Embo J. 16(10), 2565–2575.

[18] Freeman, S., Herron, J.C., 1998. Evolutionary Analysis. Prentice Hall, Upper Saddle River, NJ.

[19] Moss, G.W., Marshall, J., Morabito, M., Howe, J.R., Moczydlowski, E., 1996. An evolutionarily conserved binding site for serine proteinase inhibitors in large conductance calcium-activated potassium channels. Biochemistry 35(50), 16024–16035.

[20] Butler, A., Tsunoda, S., McCobb, D.P., Wei, A., Salkoff, L., 1993. Mslo a complex mouse gene encoding maxi calcium-activated potassium channels. Science 261(5118), 221–224.

[21] Brayden, J.E., 1996. Potassium channels in vascular smooth muscle. Clin. Exp. Pharmacol. Physiol. 23(12), 1069–1076.

[22] Rosenfeld, C.R., White, R.E., Roy, T., Cox, B.E., 2000. Calcium-activated potassium channels and nitric oxide coregulate estrogen-induced vasodilation. Am. J. Physiol. Heart Circ. Physiol. 279(1), H319–H328.

[23] Brink, P.R., Roemer, E.J., Walcott, B., 1990. Maxi-K channels in plasma cells. Pflugers Arch. 417(3), 349–351.

[24] Araque, A., Buno, W., 1999. Fast BK-type channel mediates the Ca(2+)-activated K(+) current in crayfish muscle. J. Neurophysiol. 82(4), 1655–1661.

[25] Carl, A., Bayguinov, O., Shuttleworth, C.W., Ward, S.M., Sanders, K.M., 1995. Role of Ca(2+)-activated K+ channels in electrical activity of longitudinal and circular muscle layers of canine colon. Am. J. Physiol. 268(3 Pt 1), C619–C627.

[26] Golding, N.L., Jung, H.Y., Mickus, T., Spruston, N., 1999. Dendritic calcium spike initiation and repolarization are controlled by distinct potassium channel subtypes in CA1 pyramidal neurons. J. Neurosci. 19(20), 8789–8798.

[27] Wallner, M., Meera, P., Toro, L., 1996. Determinant for beta-subunit regulation in high-conductance voltage-activated and Ca(2+)-sensitive K+ channels: an additional transmembrane region at the N terminus. Proc. Natl Acad. Sci. USA 93(25), 14922–14927.

[28] Baker, O.S., Larsson, H.P., Mannuzzu, L.M., Isacoff, E.Y., 1998. Three transmembrane conformations and sequence-dependent displacement of the S4 domain in shaker K+ channel gating. Neuron 20(6), 1283–1294.

[29] Larsson, H.P., Baker, O.S., Dhillon, D.S., Isacoff, E.Y., 1996. Transmembrane movement of the shaker K+ channel S4. Neuron 16(2), 387–397.

[30] Diaz, L., Meera, P., Amigo, J., Stefani, E., Alvarez, O., Toro, L., Lattore, R., 1998. Role of the S4 segment in a voltage-dependent calcium-sensitive potassium (hSlo) channel. J. Biol. Chem. 273(49), 32430–32436.

[31] Yuan, A., Dourado, M., Butler, A., Walton, N., Wei, A., Salkoff, L., 2000. SLO-2, a K+ channel with an unusual Cl− dependence. Nat. Neurosci. 3(8), 771–779.

[32] Ishii, T.M., Silvia, C., Hirschberg, B., Bond, C.T., Adelman, J.P., Maylie, J., 1997. A human intermediate conductance calcium-activated potassium channel. Proc. Natl Acad. Sci. USA 94(21), 11651–11656.

[33] Kim, M., Baro, D.J., Lanning, C.C., Doshi, M., Farnham, J., Moskowitz, H.S., Peck, J.H., Olivera, B.M., Harris-Warrick, R.M., 1997. Alternative splicing in the pore-forming region of shaker potassium channels. J. Neurosci. 17(21), 8213–8224.

[34] Talukder, G., Aldrich, R.W., 2000. Complex voltage-dependent behavior of single unliganded calcium-sensitive potassium channels. Biophys. J. 78(2), 761–772.

[35] Horrigan, F.T., Aldrich, R.W., 1999. Allosteric voltage gating of potassium channels II. Mslo channel gating charge movement in the absence of Ca(2+). J. Gen. Physiol. 114(2), 305–336.

[36] Horrigan, F.T., Cui, J., Aldrich, W., 1999. Allosteric voltage gating of potassium channels I. Mslo ionic currents in the absence of Ca(2+). J. Gen. Physiol. 114(2), 277–304.

[37] Rothberg, B.S., Magleby, K.L., 2000. Voltage and Ca2+ activation of single large-conductance Ca2+-activated K+ channels described by a two-tiered allosteric gating mechanism. J. Gen. Physiol. 116(1), 75–99.

[38] Schreiber, M., Salkoff, L., 1997. A novel calcium-sensing domain in the BK Channel. Biophys. J. 73, 1355–1363.

[39] Reinhart, P., Personal Communication, 2000.

[40] Schreiber, M., Yuan, A., Salkoff, L., 1999. Transplantable sites confer calcium sensitivity to BK channels. Nat. Neurosci. 2(5), 416–421.

[41] Schreiber, M., Wei, A., Yuan, A., Gaut, J., Saito, M., Salkoff, L., 1998. Slo3, a novel pH-sensitive K+ channel from mammalian spermatocytes. J. Biol. Chem. 273(6), 3509–3516.

[42] Wei, A., Solaro, C., Lingle, C., Salkoff, L., 1994. Calcium sensitivity of BK-type KCa channels determined by a separable domain. Neuron 13(3), 671–681.

[43] Joiner, W.J., Wang, L.Y., Tang, M.D., Kaczmarek, L.K., 1997. hSK4, a member of a novel subfamily of calcium-activated potassium channels. Proc. Natl Acad. Sci. USA 94(20), 11013–11018.

[44] Gardos, G., 1958. The function of calcium in the potassium permeability of human erythrocytes. Biochem. Biophys. Acta 30, 653–654.

[45] Ghanshani, S., Wulff, H., Miller, M.J., Rohm, H., Neben, A., Gutman, G.A., Cahalan, M.D., Chandy, K.G., 2000. Up-regulation of the IKCa1 potassium channel during T-cell activation. Molecular mechanism and functional consequences. J. Biol. Chem. 275(47), 37137–37149.

[46] Yarom, Y., Sugimori, M., Llinas, R., 1985. Ionic currents and firing patterns of mammalian vagal motoneurons in vitro. Neuroscience 16(4), 719–737.

[47] Barrett, E.F., Barret, J.N., 1976. Separation of two voltage-sensitive potassium currents, and demonstration of a tetrodotoxin-resistant calcium current in frog motoneurones. J. Physiol. 255(3), 737–774.

[48] Grillner, S., Wallen, P., Hill, R., Cangiano, L., El Manira, A., 2001. Ion channels of importance for the locomotor pattern generation in the lamprey brainstem–spinal cord. J. Physiol. 533(Pt 1), 23–30.

[49] Gorelova, N., Reiner, P.B., 1996. Role of the afterhyperpolarization in control of discharge properties of septal cholinergic neurons in vitro. J. Neurophysiol. 75(2), 695–706.

[50] Messier, C., Mourre, C., Bontempi, B., Sif, J., Lazdunski, M., Destrade, C., 1991. Effect of apamin, a toxin that inhibits Ca(2+)-dependent K+ channels, on learning and memory processes. Brain Res. 551(1–2), 322–326.

[51] Heurteaux, C., Messier, C., Destrade, C., Lazdunski, M., 1993. Memory processing and apamin induce immediate early gene expression in mouse brain. Brain. Res. Mol. Brain. Res. 18(1–2), 17–22.

[52] Dunwiddie, T.V., Taylor, M., Heginbotham, L.R., Proctor, W.R., 1992. Long-term increases in excitability in the CA1 region of rat hippocampus induced by beta-adrenergic stimulation: possible mediation by cAMP. J. Neurosci. 12(2), 506–517.

[53] D'Angelo, E., Nieus, T., Maffei, A., Armano, S., Rossi, P., Taglietti, V., Fontana, A., Naldi, G., 2001. Theta-frequency bursting and resonance in cerebellar granule cells: experimental evidence and modeling of a slow k+-dependent mechanism. J. Neurosci. 21(3), 759–770.

[54] del Negro, C.A., Hsiao, C.F., Chandler, S.H., 1999. Outward currents influencing bursting dynamics in guinea pig trigeminal motoneurons. J. Neurophysiol. 81(4), 1478–1485.

[55] Crest, M., Gola, M., 1993. Large conductance Ca(2+)-activated K+ channels are involved in both spike shaping and firing regulation in Helix neurones. J. Physiol. 465, 265–287.

[56] Lara, J., Acevedo, J.J., Onetti, C.G., 1999. Large-conductance Ca2+-activated potassium channels in secretory neurons. J. Neurophysiol. 82(3), 1317–1325.

[57] Jin, W., Sugaya, A., Tsuda, T., Ohguchi, H., Sugaya, E., 2000. Relationship between large conductance calcium-activated potassium channel and bursting activity. Brain Res. 860(1–2), 21–28.

[58] Azouz, R., Jensen, M.S., Yaari, Y., 1996. Ionic basis of spike after-depolarization and burst generation in adult rat hippocampal CA1 pyramidal cells. J. Physiol. 492(Pt 1), 211–223.

[59] Bennett, B.D., Callaway, J.C., Wilson, C.J., 2000. Intrinsic membrane properties underlying spontaneous tonic firing in neostriatal cholinergic interneurons. J. Neurosci. 20(22), 8493–8503.

[60] Gu, X., Blatz, A.L., German, D.C., 1992. Subtypes of substantia nigra dopaminergic neurons revealed by apamin: autoradiographic and electrophysiological studies. Brain Res. Bull. 28(3), 435–440.

[61] Lim, H.H., Park, B.J., Choi, H.S., Park, C.S., Eom, S.H., Ahnn, J., 1999. Identification and characterization of a putative *C. elegans* potassium channel gene (Ce-slo-2) distantly related to Ca(2+)-activated K(+) channels. Gene 240(1), 35–43.

[62] Peterson, O.II., 1986. Calcium activated Potassium Channels and Fluid Secretion by Exocrine Glands. American Physiological Society, Bethesda, MD.

[63] Robitaille, R., Charlton, M.P., 1992. Presynaptic calcium signals and transmitter release are modulated by calcium-activated potassium channels. J. Neurosci. 12(1), 297–305.

[64] Robitaille, R., Garcia, M.L., Kaczorowski, G.J., Charlton, M.P., 1993. Functional colocalization of calcium and calcium-gated potassium channels in control of transmitter release. Neuron 11(4), 645–655.

[65] Robitaille, R., Bourque, M.J., Vandaele, S., 1996. Localization of L-type Ca2+ channels at perisynaptic glial cells of the frog neuromuscular junction. J. Neurosci. 16(1), 148–158.

[66] Marrion, N.V., Tavalin, S.J., 1998. Selective activation of Ca2+-activated K+ channels by co-localized Ca2+ channels in hippocampal neurons. Nature 395(6705), 900–905.

[67] Wisgirda, M.E., Dryer, S.E., 1994. Functional dependence of Ca2+-activated K+ current on L- and N-type Ca2+- channels: differences between chicken sympathetic and parasympathetic neurons suggest different regulatory mechanisms. Proc. Natl Acad. Sci. USA., 91.

[68] Prakriya, M., Lingle, C.J., 2000. Activation of BK channels in rat chromaffin cells requires summation of Ca(2+) influx from multiple Ca(2+) channels. J. Neurophysiol. 84(3), 1123–1135.

[69] Jin, W., Sugaya, A., Tsuda, T., Ohguchi, H., Sugaya, E., 2000. Relationship between large conductance calcium-activated potassium channel and bursting activity. Brain Res. 860(1–2), 21–28.

[70] Van Goor, F., Li, Y.X., Stojilkovic, S.S., 2001. Paradoxical role of large-conductance calcium-activated K+ (BK) channels in controlling action potential-driven Ca2+ entry in anterior pituitary cells. J. Neurosci. 21(16), 5902–5915.

[71] Atkinson, N.S., Brenner, R., Bohm, R.A., Yu, J.Y., Wilbur, J.L., 1998. Behavioral and electrophysiological analysis of Ca-activated K-channel transgenes in *Drosophila*. Ann. NY Acad. Sci. 860, 296–305.

D.P. McCobb

[72] Jones, E.M., Gray-Keller, M., Art, J.J., Fettiplace, R., 1999. The functional role of alternative splicing of Ca(2+)-activated K+ channels in auditory hair cells. Ann. NY Acad. Sci. 868, 379–385.

[73] Ha, T.S., Jeong, S.Y., Cho, S.W., Jeon, H., Roh, G.S., Choi, W.S., Park, C.S., 2000. Functional characteristics of two BKCa channel variants differentially expressed in rat brain tissues. Eur. J. Biochem. 267(3), 910–918.

[74] Tseng-Crank, J., Foster, C., Krause, J., Mertz, R., Godinot, N., DiChiara, T., Reinhart, P., 1994. Cloning, expression, and distribution of functionally distinct Ca2+-activated K+ channel isoforms from human brain. Neuron 13, 1315–1330.

[75] Saito, M., Nelson, C., Salkoff, L., Lingle, C.J., 1997. A cysteine-rich domain defined by a novel exon in a Slo variant in rat adrenal chromaffin cells and PC12 cells. J. Biol. Chem. 272(18), 11710–11717.

[76] Lagrutta, A., Shen, K.Z., North, R.A., Adelman, J.P., 1994. Functional differences among alternatively spliced variants of slowpoke, a *Drosophila* calcium-activated potassium channel. J. Biol. Chem. 269(32), 20347–20351.

[77] Korovkina, V.P., Fergus, D.J., Holdiman, A.J., England, S.K., 2001. Characterization of a novel 132-bp exon of the human maxi-K channel. Am. J. Physiol. Cell Physiol. 281(1), C361–C367.

[78] Brenner, R., Yu, J.Y., Srinivasan, K., Brewer, L., Larimer, J.L., Wilbur, J.L., Atkinson, N.S., 2000. Complementation of physiological and behavioral defects by a slowpoke Ca(2+)-activated K(+) channel transgene. J. Neurochem. 75(3), 1310–1319.

[79] Fettiplace, R., Fuchs, P.A., 1999. Mechanisms of hair cell tuning. Annu. Rev. Physiol. 61, 809–834.

[80] Jones, E.M., Gray-Keller, M., Fettiplace, R., 1999. The role of Ca2+-activated K+ channel spliced variants in the tonotopic organization of the turtle cochlea. J. Physiol. (Lond.) 518(Pt 3), 653–665 (see comments).

[81] Navaratnam, D.S., Bell, T.J., Tu, T.D., Cohen, E.L., Oberholtzer, J.C., 1997. Differential distribution of Ca2+-activated K+ channel splice variants among hair cells along the tonotopic axis of the chick cochlea. Neuron 19(5), 1077–1085.

[82] Jiang, G.J., Zidanic, M., Michaels, R.L., Michael, T.H., Griguer, C., Fuchs, P.A., 1997, CSlo encodes calcium-activated potassium channels in the chick's cochlea. Proc. R. Soc. Lond. B Biol. Sci. 264(1382), 731–737.

[83] Ramanathan, K., Michael, T.H., Jiang, G.J., Hiel, H., Fuchs, P.A., 1999. A molecular mechanism for electrical tuning of cochlear hair cells. Science 283(5399), 215–217.

[84] Xie, J., McCobb, D.P., 1998. Control of alternative splicing of potassium channels by stress hormones. Science.

[85] Lai, G.J., McCobb, D.P., 2002. Opposing actions of adrenal androgens and glucocorticoids on alternative splicing of Slo potassium channels in bovine chromaffin cells. Proc. Natl. Acad. Sci. U.S.A. 99(11), 7722–7727.

[86] McCobb, D.P., Hara, Y., Lai, G.J., Mahmoud, S.F., Flugge, G., 2003. Subordination stress alters alternative splicing of the Slo gene in tree shrew adrenals. Horm. Behav., 43(1), 180–186.

[87] Mahmoud, S.F., Yu, Y., Lee, J., and McCobb, D.P., 2002. Regulation of potassium channel alternative splicing in rat pituitary by testosterone. In Preparation.

[88] Xie, J., Black, D.L., 2001. A CaMK IV responsive RNA element mediates depolarization-induced alternative splicing of ion channels. Nature 410(6831), 936–939.

[89] Lovell, P.V., James, D.G., McCobb, D.P., 2000. Bovine versus rat adrenal chromaffin cells: big differences in BK potassium channel properties. J. Neurophysiol. 83(6), 3277–3286.

[90] Lovell, P.V., McCobb, D.P., 2001. Pituitary control of BK potassium channel function and intrinsic firing properties of adrenal chromaffin cells. J. Neurosci. 21(10), 3429–3442.

[91] Solaro, C.R., Prakriya, M., Ding, J.P., Lingle, C.J., 1995. Inactivating and non-inactivating Ca2+- and voltage-dependent K+ current in rat adrenal chromaffin cells. J. Neurosci. 15, 6110–6123.

[92] Lingle, C.J., Solaro, C.R., Prakriya, M., Ding, J.P., 1996. Calcium-activated potassium channels in adrenal chromaffin cells. Ion Channels 4, 261–301.

[93] Tian, L., Duncan, R.R., Hammond, M.S., Coghill, L.S., Wen, H., Rusinova, R., Clark, A.G., Levitan, I.B., Shipston, M.J., 2001. Alternative splicing switches potassium channel sensitivity to protein phosphorylation. J. Biol. Chem. 276(11), 7717–7720.

[94] Lai, G., McCobb, D.P., 2000. Alternative splicing of the Slo potassium channel in adrenal chromaffin cells changes during the stress hyporesponsive period of development. Soc. Neurosci. Abstr., 26.

[95] Sapolsky, R.M., Meaney, M.J., 1986. Maturation of the adrenocortical stress response: neuroendocrine control mechanisms and the stress hyporesponsive period. Brain Res. Rev. 11, 65–76.

[96] Shipston, M.J., Duncan, R.R., Clark, A.G., Antoni, F.A., Tian, L., 1999. Molecular components of large conductance calcium-activated potassium (BK) channels in mouse pituitary corticotropes. Mol. Endocrinol. 13(10), 1728–1737.

[97] Ferrer, J., Wasson, J., Salkoff, L., Permutt, M.A., 1996. Cloning of human pancreatic islet large conductance Ca2+ -activated K+ channel (hSlo) cDNAs: evidence for high levels of expression in pancreatic islets and identification of a flanking genetic marker. Diabetologia 39, 891–898.

[98] Shao, L.R., Halvorsrud, R., Borg-Graham, L., Storm, J.F., 1999. The role of BK-type Ca2+-dependent K+ channels in spike broadening during repetitive firing in rat hippocampal pyramidal cells. J. Physiol. (Lond.) 521(Pt 1), 135–146.

[99] Benkusky, N.A., Fergus, D.J., Zucchero, T.M., England, S.K., 2000. Regulation of the Ca2+-sensitive domains of the maxi-K channel in the mouse myometrium during gestation [in process citation]. J. Biol. Chem. 275(36), 27712–27719.

[100] Knaus, H.G., Folander, K., Garcia-Calvo, M., Garcia, M.L., Kaczorowski, G.J., Smith, M., Swanson, R., 1994. Primary sequence and immunological characterization of beta-subunit of high conductance Ca(2+)-activated K+ channel from smooth muscle. J. Biol. Chem. 269(25), 17274–17278.

[101] Dworetzky, S.I., Boissard, C.G., Lum Ragan, J.T., McKay, M.C., Post Munson, D.J., Trojnacki, J.T., Chang, C.P., Gribkoff, V.K., 1996. Phenotypic alteration of a human BK (hSlo) channel by hSlo-beta subunit coexpression: changes in blocker sensitivity, activation-relaxation and inactivation kinetics, and protein kinase A modulation. J. Neurosci. 16(15), 4543–4550.

[102] McCobb, D.P., Fowler, N.L., Featherstone, T., Lingle, C.J., Saito, M., Krause, J.E., Salkoff, L., 1995. A human calcium-activated potassium channel gene expressed in vascular smooth muscle. Am. J. Physiol. 269(3 Pt 2), H767–H777.

[103] Oberst, C., Weiskirchen, R., Hartl, M., Bister, K., 1997. Suppression in transformed avian fibroblasts of a gene (CO6) encoding a membrane protein related to mammalian potassium channel regulatory subunits. Oncogene 14(9), 1109–1116.

[104] Ramanathan, K., Michael, T.H., Fuchs, P.A., 2000. Beta subunits modulate alternatively spliced, large conductance, calcium-activated potassium channels of avian hair cells. J. Neurosci. 20(5), 1675–1684.

[105] Nelson, M.T., Cheng, H., Rubart, M., Santana, L.F., Bonev, A.D., Knot, H.J., Lederer, W.J., 1995. Relaxation of arterial smooth muscle by calcium sparks. Science 270(5236), 633–637 (see comments).

[106] Xia, X.-M., Ding, J.P., Lingle, C.J., 1999. Molecular basis for the inactivation of Ca2+- and voltage-dependent BK channels in adrenal chromaffin cells and rat insulinoma tumor cells. J. Neurosci. 19(13), 5255–5264.

[107] Wallner, M., Meera, P., Toro, L., 1999. Molecular basis of fast inactivation in voltage and Ca2+-activated K+ channels: a transmembrane beta-subunit homolog [in process citation]. Proc. Natl Acad. Sci. USA 96(7), 4137–4142.

[108] Brenner, R., Jegla, T.J., Wickenden, A., Liu, Y., Aldrich, R.W., 2000. Cloning and functional characterization of novel large conductance calcium-activated potassium channel beta subunits, hKCNMB3 and hKCNMB4. J. Biol. Chem. 275(9), 6453–6461.

[109] Weiger, T.M., Holmqvist, M.H., Levitan, I.B., Clark, F.T., Sprague, S., Huang, W.J., Ge, P., Wang, C., Lawson, D., Jurman, M.E., Glucksmann, M.A., Silos-Santiago, I., DiStefano, P.S., Curtis, R., 2000. A novel nervous system beta subunit that downregulates human large conductance calcium-dependent potassium channels. J. Neurosci. 20(10), 3563–3570.

[110] Xia, X.M., Ding, J.P., Zeng, X.H., Duan, K.L., Lingle, C.J., 2000. Rectification and rapid activation at low Ca2+ of Ca2+-activated, voltage-dependent BK currents: consequences of rapid inactivation by a novel beta subunit. J. Neurosci. 20(13), 4890–4903.

[111] Uebele, V.N., Lagrutta, A., Wade, T., Figueroa, D.J., Liu, Y., McKenna, E., Austin, C.P., Bennett, P.B., Swanson, R., 2000. Cloning and functional expression of two families of beta-subunits of the large conductance calcium-activated K+ channel. J. Biol. Chem. 275(30), 23211–23218.

[112] Neely, A., Lingle, C.J., 1992. Effects of muscarine on single rat adrenal chromaffin cells. J. Physiol. 453(0), 133–166.

[113] Solaro, C.R., Prakriya, M., Ding, J.P., Lingle, C.J., 1995. Inactivating and noninactivating Ca(2+)- and voltage-dependent K+ current in rat adrenal chromaffin cells. J. Neurosci. 15(9), 6110–6123.

[114] Levitan, I.B., 1999. Modulation of ion channels by protein phosphorylation. How the brain works. Adv. Second Messenger Phosphoprotein Res. 33, 3–22.

[115] Toro, L., Stefani, E., 1991. Calcium-activated K+ channels: metabolic regulation. J. Bioenerg. Biomembr. 23(4), 561–576.

[116] Chung, S., Reinhart, P.H., Martin, B.L., Brautigan, D., Leitan, I.B., 1991. Protein Kinase activity closely associated with a reconstituted calcium-activated potassium channel. Science 253(5019), 560–562.

[117] Esguerra, M., Wang, J., Foster, C.D., Adelman, J.P., North, R.A., Levitan, I.B., 1994. Cloned Ca-2+-dependent K+ channel modulated by a functionally associated protein kinase. Nature (Lond.) 369(6481), 563–565.

[118] Reinhart, P.H., Levitan, I.B., 1995. Kinase and phosphatase activities intimately associated with a reconstituted calcium-dependent potassium channel. J. Neurosci. 15(6), 4572–4579.

[119] Tian, L., Knaus, H.G., Shipston, M.J., 1998. Glucocorticoid regulation of calcium-activated potassium channels mediated by serine/threonine protein phosphatase. J. Biol. Chem. 273(22), 13531–13536.

[120] Nara, M., Dhulipala, P.D., Wang, Y.X., Kotlikoff, M.I., 1998. Reconstitution of beta-adrenergic modulation of large conductance, calcium-activated potassium (maxi-K) channels in Xenopus oocytes. Identification of the camp-dependent protein kinase phosphorylation site. J. Biol. Chem. 273(24), 14920–14924.

[121] Nara, M., Dhulipala, P.D., Ji, G.J., Kamasani, U.R., Wang, Y.X., Matalon, S., Kotlikoff, M.I., 2000. Guanylyl cyclase stimulatory coupling to K(Ca) channels. Am. J. Physiol. Cell Physiol. 279(6), C1938–C1945.

[122] DiChiara, T.J., Reinhart, P.H., 1997. Redox modulation of hslo Ca2+-activated K+ channels. J. Neurosci. 17(13), 4942–4955.

[123] Bolotina, V., Najibi, S., Palacino, J., Pagano, P., Cohen, R., 1994. Nitric oxide directly activates calcium-dependent potassium channels in vascular smooth muscle. Nature 368, 850–853.

[124] Scornik, F.S., Codina, J., Birnbaumer, L., Toro, L., 1993. Modulation of coronary smooth muscle KCa channels by Gs alpha independent of phosphorylation by protein kinase A. Am. J. Physiol. 265(4 Pt 2), H1460–H1465.

[125] Roy-Contancin, L., Garcia, M.L., Galvez, A., Kaczorowski, G.J., Katz, G.M., Williams, D., Reuben, J.P., 1990. Ca2(+)-activated K+ channels in bovine aortic smooth muscle and GH3 cells: properties and regulation by guanine nucleotides. Prog. Clin. Biol. Res. 334, 145–170.

[126] Ordway, R.W., Walsh, J.V. Jr, Singer, J.J., 1989. Arachidonic acid and other fatty acids directly activate potassium channels in smooth muscle cells. Science 244(4909), 1176–1179.

[127] Zou, A.P., Fleming, J.T., Falck, J.R., Jacobs, E.R., Gebremedhin, D., Harder, D.R., Roman, R.J., 1996. Stereospecific effects of epoxyeicosatrienoic acids on renal vascular tone and K(+)-channel activity. Am. J. Physiol. 270(5 Pt 2), F822–F832.

[128] Valverde, M.A., Rojas, P., Amigo, J., Cosmelli, D., Orio, P., Bahamonde, M.I., Mann, G.E., Vergara, C., Latorre, R., 1999. Acute activation of Maxi-K channels (hSlo) by estradiol binding to the beta subunit. Science 285(5435), 1929–1931 (see comments).

[129] Farrukh, I.S., Peng, W., Orlinska, U., Hoidal, J.R., 1998. Effect of dehydroepiandrosterone on hypoxic pulmonary vasoconstriction: a Ca(2+)-activated K(+)-channel opener. Am. J. Physiol. 274(2 Pt 1), L186–L195.

[130] Peng, W., Hoidal, J.R., Farrukh, S., 1999. Role of a novel KCa opener in regulating K+ channels of hypoxic human pulmonary vascular cells. Am. J. Respir. Cell Mol. Biol. 20(4), 737–745.

[131] Lovell, P.V., King, J.T., McCobb, D.P., 2002. Acute modulation of BK channel function and cellular excitability by glucocorticoids in bovine and rat chromaffin cells. In Preparation.

[132] Tian, L., Hammond, M.S., Florance, H., Antoni, F.A., Shipston, M.J., 2001. Alternative splicing determines sensitivity of murine calcium-activated potassium channels to glucocorticoids. J. Physiol. 537(Pt 1), 57–68.

[133] Dopico, A.M., Chu, B., Lemos, J.R., Treistman, S.N., 1999. Alcohol modulation of calcium-activated potassium channels. Neurochem. Int. 35(2), 103–106.

[134] Xia, X.-M., Hirschberg, B., Smolik, S., Forte, M., Adelman John, P., 1998. dSLo interacting protein 1, a novel protein that interacts with large-conductance calcium-activated potassium channels. J. Neurosci. 18(7), 2360–2369.

[135] Zhou, Y., Schopperle, W.M., Murrey, H., Jaramillo, A, Dagan, D., Griffith, L.C., Levitan, I.B., 1999. A dynamically regulated 14-3-3, Slob, and Slowpoke potassium channel complex in *Drosophila* presynaptic nerve terminals. Neuron 22(4), 809–818.

[136] Schopperle, W.M., Holmqvist, M.H., Zhou, Y., Wang, J., Wang, Z., Griffith, L.C., Keselman, I., Kusinitz, F., Dagan, D., Levitan, I.B., 1998. Slob, a novel protein that interacts with the Slowpoke calcium-dependent potassium channel. Neuron 20(3), 565–573.

[137] Jan, L.Y., Jan, Y.N., 1997. Voltage-gated and inwardly rectifying potassium channels. J. Physiol. (Lond.) 505(Pt 2), 267–282.

[138] Chang, W.M., Bohm, R.A., Strauss, J.C., Kwan, T., Thomas, T., Cowmeadow, R.B., Atkinson, N.S., 2000. Muscle-specific transcriptional regulation of the slowpoke Ca(2+)-activated K(+) channel gene. J. Biol. Chem. 275(6), 3991–3998.

[139] Bohm, R.A., Wang, B., Brenner, R., Atkinson, N.S., 2000. Transcriptional control of Ca(2+)-activated K(+) channel expression: identification of a second, evolutionarily conserved, neuronal promoter. J. Exp. Biol. 203(Pt 4), 693–704.

[140] Atkinson, N.S., Brenner, R., Chang, W., Wilbur, J., Larimer, J.L., Yu, J., 2000. Molecular separation of two behavioral phenotypes by a mutation affecting the promoters of a Ca-activated K channel. J. Neurosci. 20(8), 2988–2993.

[141] Dhulipala, P.D., Kotlikoff, M.I., 1999. Cloning and characterization of the promoters of the maxiK channel alpha and beta subunits. Biochim. Biophys. Acta 1444(2), 254–262.

[142] Lhuillier, L., Dryer, S.E., 2000. Developmental regulation of neuronal KCa channels by TGFbeta 1: transcriptional and posttranscriptional effects mediated by Erk MAP kinase. J. Neurosci. 20(15), 5616–5622.

[143] Dourado, M.M., Brumwell, C., Wisgirda, M.E., Jacob, M.H., Dryer, S.E., 1994. Target tissues and innervation regulate the characteristics of K+ currents in chick ciliary ganglion neurons developing in situ. J. Neurosci. 14(5 Pt 2), 3156–3165.

[144] Satow, Y., 1978. Internal calcium concentration and potassium permeability in Paramecium. J. Neurobiol. 9(1), 81–91.

[145] Preston, R.R., Kink, J.A., Hinrichsen, R.D., Saimi, Y., Kung, C., 1991. Calmodulin mutants and Ca2(+)-dependent channels in Paramecium. Annu. Rev. Physiol. 53, 309–319.

[146] Meech, R.W., Standen, N.B., 1974. Calcium-mediated potassium activation in Helix neurones. J. Physiol. (Lond.) 237(2), 43P–44P.

[147] Crest, M., Jacquet, G., Gola, M., Zerrouk, H., Benslimane, A., Rochat, H., Mansuelle, P., Martin-Eauclaire, M.F., 1992. Kaliotoxin, a novel peptidyl inhibitor of neuronal BK-type Ca(2+)-activated K+ channels characterized from *Androctonus mauretanicus mauretanicus* venom. J. Biol. Chem. 267(3), 1640–1647.

Cystic fibrosis transmembrane conductance regulator

J.W. Hanrahan*

Department of Physiology, McGill University, 3655 Prom. Sir-William-Osler, Montréal, Québec, Canada, H3G 1Y6
*Correspondendce address: Tel.: +1-514-398-8320; fax: +1-514-398-7452
E-mail: hanrahan@med.mcgill.ca

1. Introduction

The cystic fibrosis transmembrane conductance regulator (CFTR) is a 1480 amino acid membrane glycoprotein, which is mutated in the autosomal recessive disease cystic fibrosis (CF) [1–3]. Although details of pathogenesis remain controversial, CF is characterized by abnormal ion transport across exocrine epithelia [4]. Defects in CFTR processing or function lead to viscous secretions in the pancreatic ducts, airways, and intestine. Chronic bacterial infections and ensuing airway inflammation lead to respiratory failure by age 32 in about 2/3 of CF patients.

CFTR has two membrane domains (TMD1 and TMD2), each comprising six trans-membrane segments that probably form alpha helices (see Fig. 1). A nucleotide binding domain follows each TMD in the linear sequence (NBD1 and NBD2, respectively), placing CFTR in a large superfamily of ATP binding cassette (ABC) transport proteins present in all organisms from bacteria to humans. Members of the ABC superfamily transport a diverse set of substrates including peptides (e.g. STE6 [5], TAP1/TAP2 [6]), amino acids and sugars (e.g. HisP [7] and MalK [8], respectively), lipophilic drugs and xenobiotics (e.g. MDR [9] and MRP1 [10], respectively), and phospholipids (e.g. MDR2 [11]). CFTR is the only known ion channel in the superfamily, and the only one that has a central regulatory (R) domain. This chapter provides an overview of CFTR structure and function, emphasizing recent advances in our understanding of anion permeation, channel regulation, and interactions between various domains of CFTR and between CFTR and other proteins.

2. The CFTR chloride channel

When the *cftr* gene was functionally expressed in mammalian and insect cells, it resulted in a cAMP-stimulated conductance [12] that was mediated by ohmic

Advances in Molecular and Cell Biology, Vol. 32, pages 73–94
ISSN: 1569-2558 / DOI: 10.1016/S1569-2558(03)32004-1

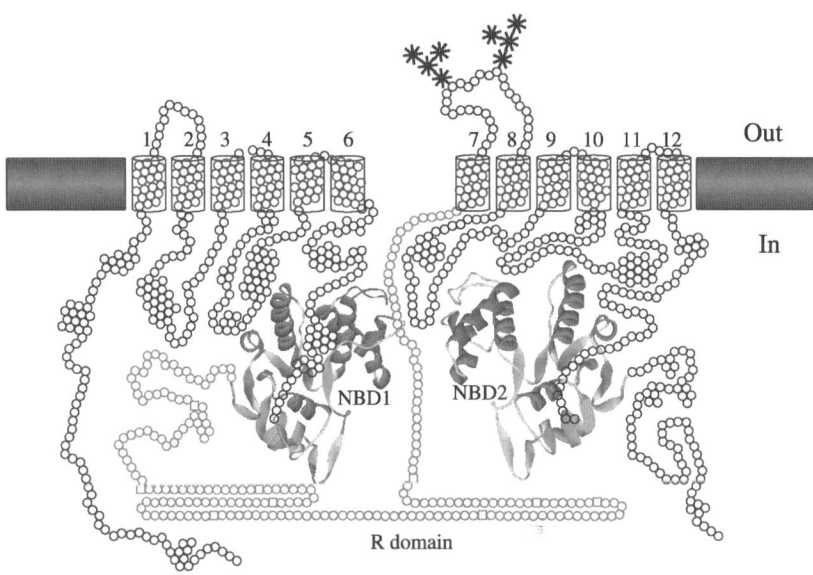

Fig. 1. Cartoon illustrating the domains of CFTR. Transmembrane domains and the lipid bilayer are shown in blue. The regulatory domain is shown in red, with squares indicating locations of dibasic consensus sites for PKA phosphorylation. The fourth extracellular loop has two N-linked glycosylation sites, both of which are utilized. The nucleotide binding folds (NBD1, NBD2) are drawn from the crystal structure of a bacterial ABC protein, the maltose transporter MalK (see text). The NBDs are presumed to be tightly associated but are shown separated in this drawing for clarity. Alpha helices elsewhere in the cytoplasmic domains of CFTR were predicted using the program nnpredict (www.cmphar.ucsf.edu/cgi-bin/nnpredict.pl). (*For a colored version of this figure, see plate section, page 460.*)

Cl^- channels having relatively low (7–10 pS) conductance [13,14]. These and other single channel properties (slow gating, "flickery" openings at negative membrane potentials, weak sensitivity to external disulfonic stilbene inhibition) were reminiscent of endogenous Cl^- channels characterized previously in epithelia [15–17]. It was therefore proposed that CFTR was by itself a non-rectifying, low-conductance Cl^- channel [13,14, 18–20]. Mutations in predicted transmembrane segments altered the selectivity of macroscopic CFTR currents, providing further evidence that CFTR forms a pore [21]. The channel activity of CFTR was formally demonstrated when it was purified to homogeneity and reconstituted into planar bilayers [22]. These early studies have been reviewed in detail elsewhere [23–28].

3. Searching for the CFTR channel pore

Amino acids that may line the CFTR pore have been inferred by comparing the functional properties of wild-type and mutant CFTR channels (i.e. their anion permeability ratios, conductances, and sensitivity to blockers). Based on such evidence, a strong case can be made for involvement of the sixth transmembrane segment (TM6), since mutations there affect anion selectivity [21,29], conductance [30,31], multi-ion pore behaviour [31], and blocker sensitivity [32,33]. Many mutagenesis/patch clamp results are difficult to interpret, however, and need confirmation using other (e.g. biochemical) methods. For example, deleting the N-terminus and TMs 1, 2, 3, and 4 has little effect on channel function when the construct is expressed in *Xenopus* oocytes [34], yet several disease-associated mutations that alter channel properties (e.g. R117H) occur in this region. Mutations at K95 and G91 in TM1 affect anion selectivity [21] and rectification [35], respectively, and cysteines inserted into TM1 are accessible to extracellular hydrophilic sulfhydryl reagents, consistent with access to an aqueous pore [36]. Interestingly, CFTR molecules lacking their N-termini due to alternate translation start sites constitute about 10% of the CFTR expressed in mammalian cells [37]. These variants do not reach the surface of mammalian cells, although they might be functional if expressed in *Xenopus* oocytes where the quality control seems less stringent. Paradoxical results have also been reported for the C-terminal half of CFTR. Mutations in TM12 affect sensitivity to the open channel blocker diphenylamine-2-carboxylic acid (DPC), suggesting it lines the pore [32], yet normal-looking channel activity can be recorded when only the N-terminal half of CFTR is expressed (i.e. lacking TMs 7–12 and NBD2) [38]. In the latter case, the N-terminal halves might dimerize to form a functional unit, but this has not yet been shown. Regardless, if TM12 does line the pore, its contribution must be very different from that of TM6, since alanine substitutions at T1134, M1137, N1138, S1141 and T1142 (amino acids in TM12 that align with important residues in TM6) have little or no effect on permeation [39]. Finally, the flickery, voltage-dependent block of the CFTR by the pH buffer 3-(*N*-morpholino)propanesulfonic acid (MOPS) from the intracellular, but not extracellular, side of the channel suggests the presence of a site within the pore that is accessible from a wide cytoplasmic mouth of the pore but not a narrower extracellular entrance [40].

Mutations at G314 in TM5 and S1118 in TM11 affect both ion permeation and channel gating, suggesting these segments also contribute to the pore [41–43]. A model in which amino acids immediately after TM6 fold back into the membrane rather like the P loop in cation channels has been proposed based on the accessibility of cysteines engineered in this region to externally applied hydrophilic sulfhydryl reagents [44]. Notwithstanding the caveats of substituted-cysteine accessibility for deducing topology [35,45,46], mutations at R358 in the proposed reentrant loop dramatically reduce anion:cation selectivity, consistent with the model [47].

4. Permeation properties of CFTR

Selectivity of the CFTR pore was first studied in channels endogenously expressed in pancreatic duct [48] and thyroid epithelial cells [16]. In cell attached membrane patches, complete replacement of external Na^+ with K^+ or tetramethyl ammonium, or partial replacement with *N*-methyl-D-glucamine had little effect on the reversal potential (E_{rev}), whereas replacement of most extracellular Cl^- with gluconate or sulfate shifted E_{rev} by $+26$ mV, indicating high Cl^- selectivity [15,16,49]. Lower limits for P_{Cl}/P_{cation} of >8–14 were reported using channels expressed in human colonic T84 and intestinal CaCo2 cell lines [17,50].

Among the halides, permeability ratios (P_X/P_{Cl}) generally follow the (inverse) lyotropic series, with large, weakly hydrated ions having the highest values. The exception is I^-, which gives inconsistent results even within laboratories. P_I/P_{Cl} was near unity when determined at the single channel level [48], but was 0.4 in whole cell recordings, a value similar to macroscopic estimates with sweat ducts [51] and T84 cells [52,53]. By contrast, I^- currents were not detected in a study of CFTR in cultured thyroid cells [16], and iodide block of Cl^- currents was also reported [54]. Low P_I/P_{Cl} was obtained in whole cell [21] and transepithelial [53] studies, and has been widely used as a signature of CFTR-mediated Cl conductance. We observed high P_I/P_{Cl} immediately after channels were exposed to I^- solutions under biionic conditions (~ 1.2; (13)), but this ratio quickly switched during the course of an experiment to lower values (~ 0.4). This behaviour caused an apparent hysteresis in the single channel current–voltage relationship [55]. Attempts to study the iodide effect were frustrated because it was not evident in macroscopic recordings, where P_I/P_{Cl} was always about 0.8 (e.g. [56]). To explain the high initial P_I/P_{Cl} in single channel experiments, we have speculated that I^- alters the permeation pathway, producing two distinct permeability states with high initial and low steady-state I^- permeabilities, respectively [55]. Chimeras between human CFTR and the *Xenopus* homolog, which has relatively high P_I/P_{Cl}, suggest that amino acids influencing P_I/P_{Cl} may be situated in the first TMD [57]. However, it remains unclear whether the same determinants are responsible for variable ratios in human CFTR channels and the high P_I/P_{Cl} in *Xenopus* channels.

When determined from the extracellular side using mixed solutions, the permeability ratios for polyatomic anions followed the sequence $NO_3(1.73) > Cl(1.0) > HCO_3(0.25) \gg$ gluconate(0.03) [48]. A similar sequence $NO_3^- > Cl^- > HCO_3^- >$ formate $>$ acetate was obtained under biionic conditions, although permeability to large kosmotropic anions was higher from the cytoplasmic than extracellular side [29]. Indeed, external pyruvate, propanoate, methane sulfonate, ethane sulfonate and gluconate were not measurably permeant (i.e. $P_X/P_{Cl} < 0.06$) when macroscopic current was measured using excised patches exposed to PKA and ATP, yet currents carried by these anions were detectable from the cytoplasmic side [58]. The relationship between macroscopic permeability ratio and ion diameter suggested a minimum functional diameter of 5.3 Å from the outside [29], and ~ 12 Å from the inside [29,58]. Macroscopic currents corresponding to glutathione (GSH) efflux were induced by cAMP using excised, inside–out patches exposed to high concentrations of GSH in the bath [59]. The effects of anion channel blockers and CFTR mutations on GSH-mediated currents suggested that CFTR can mediate GSH permeation. However, CFTR-dependent GSH effluxes from intact cells are small [60] and may depend on cytoplasmic Cl^- concentration rather than CFTR per se [61], thus it remains to be established whether GSH permeates through CFTR at significant rates under physiological conditions.

The role of CFTR in HCO_3^- secretion has recently aroused much attention since this is the major ion transported in pancreas and may be important in the airway and colonic secretions [62]. While the data are not completely concordant, HCO_3^- efflux accounts for significant anion secretion across the airway submucosal gland cell line Calu3 under resting [63] or cAMP stimulated conditions [64]. Early calculations suggested that a fraction of HCO_3^- secretion must be electrogenic to achieve the high concentrations found in pancreatic juice (e.g. Ref. [65]). Subsequent patch clamp experiments revealed that bicarbonate does permeate through the CFTR channel pore [24,29,48,66], although the physiological significance of this route for HCO_3^- has been challenged [67]. External HCO_3^- inhibits CFTR-mediated chloride currents with an apparent K_i of 7 mM [68]). This inhibition may reduce back-leakage of HCO_3^- through CFTR channels when the luminal concentration approaches 150 mM in the pancreatic duct. Other mechanisms by which CFTR may regulate HCO_3^- transport are discussed below.

Several lines of evidence suggest that the CFTR pore can hold more than one ion simultaneously. First, the biionic permeability ratio P_I/P_{Cl} is concentration-dependent [54], a common feature of channels having multi-ion pores [69,70]. Second, at least two distinct electrical distances have been calculated for voltage-dependent blockers, and block by intracellular gluconate can be relieved by raising external Cl^- concentration. Third, when wild-type CFTR channels were bathed in symmetrical mixtures of Cl^- and SCN^-, single channel conductance decreased from 7 to 2 pS as the mole fraction of SCN^- (which is more permeant than Cl^- when studied under biionic conditions) in the mixture was increased from 0 to 7%, and then increased again as the SCN^- mole fraction was increased further to 97% [31]. The anomalous mole fraction effect (AMFE) and voltage-dependent block of Cl^- currents by cytoplasmic SCN^- were abolished when R347 was mutated to aspartate, and became pH dependent in the R347H mutant, suggesting positive charge at this site is necessary for the AMFE and binding by internal SCN^-. The AMFE was not observed when measuring macroscopic currents in *Xenopus* oocytes bathed with SCN^-/Cl^- mixtures on the extracellular side, however SCN and other lyotropes can inhibit gating, which might complicate interpretation of macroscopic CFTR currents.

Permeability and conductance ratios differ for some anions (e.g. SCN^-, I^-, and Br^-). Dawson and colleagues have emphasized the importance of anion binding in CFTR (e.g. Ref. [71]). They showed that conductance ratios are more strongly affected than permeability ratios by mutations in TMs 1, 5, and 6 and therefore may be a more sensitive indicator of pore structure [35]. Permeability ratios, which depend on barrier heights and relative ease with which anions enter the pore from the bulk solution, are dominated by anion–water interactions (i.e. hydration energy) whereas conductance ratios should provide a measure of relative binding of anions in the pore. A four barrier, three site (4B3S) Eyring rate theory model has been developed for CFTR that reproduces current–voltage relationships quantitatively under a wide range of conditions, i.e. single channel conductances, reversal potentials, block by intracellular gluconate, and AMFEs in mixtures of SCN^- and Cl^- [72]. Loss of AMFE (as seen in the R347D mutant) could be simulated using the 4B3S model by reducing the depth of the well that corresponds to intracellular SCN^- block. Whether the positive charge at R347 contributes directly to SCN^- binding or indirectly affects pore architecture is not known [73]. The low conductance of R347H and R347P mutant channels may help explain the mild phenotype associated with these mutations in CF patients.

5. Phosphorylation regulates CFTR channel gating

Phosphorylation of the R domain stimulates CFTR channel activity [18,74] and may enhance its interactions with other proteins. Regulation by PKA and PKC are well established although other serine–threonine and tyrosine kinases may also have a role.

5.1. cAMP- and cGMP-dependent protein kinases

Elevating cAMP stimulates Cl^- secretion across many epithelia through its effects on apical Cl^- conductance [75,76]. In COS cells, forskolin increases in vivo phosphorylation of CFTR by 1.8-fold [74] to an apparent stoichiometry of ~5 mol PO_4/mol protein [77], although this is bound to be an underestimate if all intracellular ATP is not radiolabeled during incubation of cells with $[^{32}P]PO_4$. Sites that become conspicuously phosphorylated under in vivo conditions are $RRNS_{660}I$, $KRKNS_{700}I$, $RKVS_{795}L$ and $RRLS_{813}Q$. Phosphorylation of additional sites (i.e. $RKFS_{712}I$, $RIS_{753}V$, and $RRQS_{768}V$) has been detected in vitro, presumably because kinase activity is higher (or phosphatase activity lower) than in intact cells [78]. Sites that do not appear to be phosphorylated in vivo cannot be dismissed, as they can support channel activity (~50%) and are highly conserved between species despite sequence divergence in the R domain generally. At least one PKA site (at S_{768}) may be phosphorylated in resting cells and downregulate basal channel activity [79,80]. This "parking brake" function is consistent with results obtained with the mutant $\Delta769-783$, the smallest deletion identified so far that allows constitutive channel activity [81]. Mutating all nine dibasic (R–R/K–X–S/T) and five monobasic (R/K–X–S/T) consensus sequences for PKA on the R domain along with the dibasic site $RKTS_{422}N$ near the beginning of NBD1 abolishes channel responses to PKA when the mutant is incorporated into planar bilayers [82]. Whether all functionally important PKA sites are situated on the R domain is still debated, however, since "split" channels lacking the R domain (defined as amino acids 634–836) respond to stimulation by PKA in patch clamp experiments, albeit weakly [83]. The Type II isoform of cyclic GMP-dependent protein kinase (cGKII) activates CFTR channels in intestinal epithelium and is implicated in the secretory diarrhea induced by heat-stable enterotoxins [84]. The activation of channels by cGKII in excised patches is somewhat slower than by PKA, although very similar phosphopeptide maps suggest the same sites are phosphorylated by both kinases [85].

Phosphorylation may enable opening of the channel through enhanced binding of the nucleotide, by facilitating the formation of a transition state complex with bound nucleotide, or by allowing transduction of the associated conformational change from the NBD to the channel gate [86–90]. Exogenous phospho-R domain weakly stimulates mutant channels that lack most of the R domain when added to excised patches [91]. However, the significance of this stimulation remains uncertain because only a fraction of the wild-type response can be restored. There is also evidence that phospho-R domain is not essential for full stimulation, since split channels lacking the R domain have constitutive activity which is comparable to that of maximally stimulated split channels bearing an R domain [83]. Phosphorylation does alter the secondary structure of recombinant R domain protein (aa 595–831) [92], probably in a region now considered to be the distal end of NBD1 [93] where disease-associated mutations have been shown to cause misprocessing and to reduce channel opening rate [94].

5.2. Protein kinase C

While protein kinase C (PKC) is about 10% as effective as PKA in activating CFTR channels, the main effect of PKC is to enhance CFTR responses to PKA [18].

PKC increases both the rate and magnitude of activation by PKA in excised patches [18] and maintains CFTR in a PKA-responsive state [95]. Without PKC, channels become refractory to PKA ~ 10 min after excision, presumably because critical PKC sites become dephosphorylated by the membrane-associated phosphatase (Fig. 2). Loss of PKA responsiveness can be reversed by adding PKC + ATP to the bath, and prevented if PKC is continuously present [95]. PKC belongs to a family of 11 related, lipid-dependent serine/threonine kinases having homologous catalytic domains. The isotypes are not specific in their ability to stimulate CFTR in vitro [96], nevertheless PKCξ has been identified as the one regulating CFTR in the airway submucosal gland cell line Calu3 [97]. Mutant channels lacking all nine PKC consensus sequences on the R domain respond to PKA in the same manner as wild-type channels treated with the PKC inhibitor chelerythrine, thus PKC acts primarily by phosphorylating CFTR itself rather than some accessory protein [98]. The relative importance of individual PKC sites in potentiating

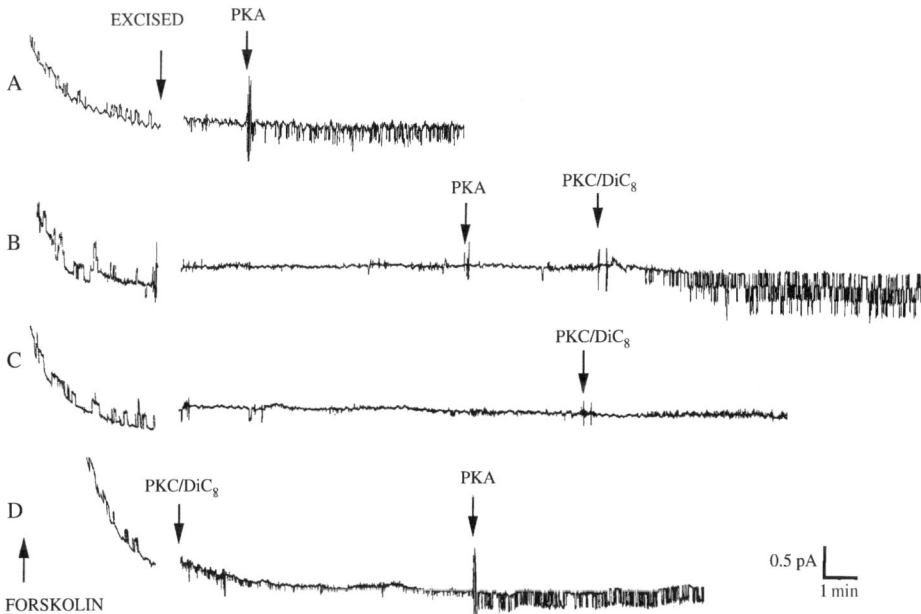

Fig. 2. PKC dependence of CFTR channel activation by PKA. Channels recorded in four patches, beginning in the cell-attached configuration with forskolin (5 μM) and MgATP (1 mM) in the bath and Vpipette = − 30 mV. The interruption in each trace indicates when it was excised and the holding potential switched to +30 mV. Downward deflections indicate inward currents carried by Cl⁻ ions flowing into the pipette. PKA (200 units/mL), PKC (3.78 nM) + DiC$_8$ (5 μM) were added to the cytoplasmic side at the arrow. A, stimulation of CFTR activity elicited by the addition of PKA within 2 min of rundown. B, inability of PKA to stimulate CFTR channels when added 10 min after excision and subsequent stimulation after addtion of PKC and DiC$_8$. C, trace showing that PKC and DiC$_8$ had no effect on channel activity in the absence of PKA when added after the same time interval as in B. Note that at least two channels were active in this patch in the cell-attached configuration. D, presence of PKC and DiC$_8$ during the 10 min period after excision prevents the loss of responsiveness to PKA (compare with trace B). Adapted from Ref. [95].

PKA responses is not known, nor is the mechanism of regulation by converging kinases. However, the latter may have broad significance as PKC modulation of PKA responses has also been observed in voltage-gated Na^+ [99] and K_{ATP} channels [100]. In the latter, PKC modulation could be attributed to phosphorylation of a single site [101].

What is the physiological significance of such hierarchial regulation of CFTR by multiple kinases? Basal PKC activity in the cell is apparently sufficient for most of the CFTR response to PKA, however the regulation may only appear constitutive because the cultured cells are maintained in vitro under conditions that stimulate PKC (i.e. in the presence of serum and other media supplements). Since regulation by PKC may be more dynamic in vivo, studies need to be carried out using epinephrine and other physiological agonists that may activate both PKA and PKC. Multiple agonists that could act synergistically from opposite sides of the epithelium should also be considered; e.g. apical ATP or adenosine may stimulate PKC locally, allowing channels to respond to global elevation of cAMP induced by basolateral agonists.

5.3. Src family tyrosine kinases

Ion channels regulated by src family tyrosine kinases include NMDA receptors, Ca^{2+}-activated K^+ channels, and voltage-gated Ca^{2+} channels [102–105]. Tyrosine phosphorylation of CFTR has received little attention; probably because no phosphotyrosine was detected on CFTR in early studies of resting or forskolin-stimulated COS cells [74], and because there are no strong consensus sequences on CFTR for tyrosine phosphorylation. Nevertheless $p60^{c\text{-}src}$, a ubiquitous membrane-associated tyrosine kinase, does induce CFTR channel flickering and increases responsiveness to PKA when added to excised patches [106]. Src also activates CFTR channels in vitro [107] and directly phosphorylates CFTR at unidentified sites. Src activation is independent of PKA since a mutant lacking mono- and dibasic consensus PKA sites on the R domain responds to $p60^{c\text{-}src}$ as well as does wild-type CFTR. The physiological role of tyrosine kinase regulation of CFTR is not known but may involve regulation by purinergic receptors, as suggested above for PKC. In this regard, it is interesting that $p60^{c\text{-}src}$ and the Src family tyrosine kinase Fyn are both activated by purinergic receptors and phosphorylate the anion exchanger AE1 in cardiac cells [108]. A Src family tyrosine kinase ($p62^{c\text{-}yes}$) is expressed at the apical membrane of airway epithelium, where it interacts with yes-associated-protein of 65 kD (YAP65) and ezrin–radixin–moesin binding phosphoprotein of 50 kDa (EBP-50) [109]. The latter binds to CFTR and could potentially tether regulatory proteins, although that remains to be demonstrated [110–112]. Interestingly, the effect of $p62^{c\text{-}yes}$ on CFTR channel activity has not yet been examined.

5.4. Membrane-associated protein phosphatase activity

CFTR channels "run down" soon after they are excised from cAMP-stimulated cells [18,49]. This spontaneous decline in open probability (P_o) has been attributed to phosphatase activity in the excised patch because channels could be reactivated by exposure to MgATP plus PKA [18]. Rundown was not inhibited significantly by high concentrations of okadaic acid, and did not require Ca or calmodulin, indicating that it was not mediated by PP1, PP2A or PP2B. Rundown was sensitive to the alkaline phosphatase inhibitors levamisole and bromotetramisole [113], but at concentrations (>200 μM) that we now know inhibit all four protein phosphatase types [114]. Reducing Mg^{2+} concentration slows rundown, consistent with the dependence of PP2C on millimolar concentrations of Mg^{2+} [115]. In T_{84} monolayers, deactivation of cAMP-stimulated currents was insensitive to okadaic acid and calyculin A, suggesting most CFTR deactivation in epithelia is mediated by PP2C [115,116]. As discussed below, PP2C is the only phosphatase that is physically associated with CFTR [117]. It is not yet known whether this interaction is direct or mediated by other proteins.

PP2C is a monomeric enzyme of ~ 44 kD. There are many mammalian isoforms (e.g., α, β, δ and γ), with the α isoform having at least two splice variants and the β isoform at least five. The variant associated with CFTR has not been identified, although γ can be excluded because it is not recognized by the antibody used to probe Western blots of proteins coimmunoprecipitated with CFTR. The N-terminal 310 amino acids of PP2C form a highly conserved catalytic domain while the C-terminal 80 amino acids influence substrate specificity and cellular location of the enzyme. PP2C activity requires two Mg^{2+} ($EC_{50} \sim 1.5$ mM; [118]) or Mn^{2+} ions, probably to coordinate the target phosphoryl moiety and provide a nucleophile acid for the reaction.

The phosphatase that regulates CFTR has been suggested as a therapeutic target in CF because phosphatase inhibitors activate mutant CFTR channels [113,114]. However, specific inhibitors of PP2C are not presently available, and the phenylimidazothiazoles that have been used to activate CFTR in patch clamp experiments were not sufficiently potent to stimulate G551D mutant channels in a mouse model of CF [119].

6. Nucleotides open the CFTR pore

ATP hydrolysis by the NBDs was proposed to energize CFTR channel gating because, in early studies, non-hydrolyzable ATP analogs did not activate macroscopic currents once the contaminating ATP had been removed [120,121]. Moreover a strict dependence on Mg^{2+}, which is required for ATPase activity, was observed [120,122], although not in all laboratories [123,124]. More recently, CFTR activity was recorded in nominally Mg-free solutions [89,90] and with high concentrations of AMP–PNP [90]. AMP–PNP is widely used as a non-hydrolyzable analog of ATP, although it may be hydrolyzed by some enzymes such as F_1-ATPase [125]. The temperature dependence of CFTR gating in planar bilayers suggests that open and closed states of CFTR have similar free energies and therefore ATP hydrolysis energy is not required for closing the channel [126]. It has been proposed that the energy comes instead from the interaction of MgATP with CFTR as it enters a transition state for hydrolysis, perhaps through an "induced fit" mechanism [90]. Although some uncertainties remain, (e.g. why the activation energy for closing is comparable to that of aqueous diffusion [126], while the closing rate appears strongly temperature-dependent in other studies [90]), non-hydrolytic binding is clearly sufficient for channel opening. Hydrolysis may allow efficient reversibility under physiological conditions [126]. This scheme is analogous with those proposed for other ABC proteins such as the bacterial $MalEFGK_2$ maltose transporter, where strain induced by ATP binding has been proposed to energize a conformational transition that opens an aqueous channel for maltose after the docking of a periplasmic maltose binding protein [8]. Opening of the channel is inhibited by low ionic strength ($EC_{50} = 80$ mM) when MgATP is kept constant [127]. The mechanism of this electrolyte effect is not known, but it is independent of the particular ionic species present and mediated by changes in the opening rate.

Different functions have been ascribed to NBD1 and NBD2 based on electrophysiological studies of mutant channels. It was proposed that hydrolysis at one of them (presumably NBD1) opens the channel and subsequent hydrolysis at the other NBD allows the channel to close [128–130]. According to this scheme, ATP binds at NBD2 only if CFTR is strongly phosphorylated and once bound there, keeps the channel open until it, or its hydrolysis products, are released. Patch clamp and bilayer data support this model insofar as the channel enters a prolonged open state in mixtures of ATP and AMP–PNP or pyrophosphate [129,131]. Moreover high concentrations of the transition state phosphate analog orthovanadate, which stabilizes (i.e. "traps") ADP in the catalytic site by binding tightly in place of Pi, can also lock the channel open [128]. Finally, the mutation K1250A, which should interfere with hydrolysis at NBD2, results in occasional long-lived open bursts [130–132]. However, the opposing functions of NBD1 and NBD2 are not completely dissociable, since mutations in NBD2 also affect channel opening rate and the converse is true for mutations at NBD1 [130]. A simple scheme in which phosphorylation increases open probability (P_o) by a "locking open" mechanism seems unlikely, since phosphorylation acts in large part by shortening the interburst duration (i.e. by increasing the opening rate) [87,133,134]. Indeed, locking in open bursts still occurs after all strong PKA consensus sequences have been removed so that in vitro phosphorylation is reduced by more than 90% [133]. Photoaffinity labeling studies with 8-azido-[α-^{32}P]ATP indicate that nucleotides interact predominantly at NBD1 and can become occluded there in a Mg-dependent manner [135]. In addition to locking the channel in an open burst state (Fig. 3A), AMP–PNP also delays channel opening (Fig. 3B; [136,137]) and suppresses photolabeling of NBD1 by azido-ATP, indicating competition there with ATP [138]. Phosphorylation reduces the apparent EC_{50} for ATP [91,133,139], however biochemical measurements of ATP binding are needed to assess if this is due to an increase in nucleotide affinity. Co-operativity between the NBDs is suggested by the Hill coefficient of ATP dependence (1.8) reported for WT CFTR [139] and a low-phosphorylation mutant [133], and also by the ability of AMP–PNP binding at NBD1 to enhance vanadate trapping (and presumably hydrolysis) of nucleotide at NBD2 [138]. Interactions between NBDs have been demonstrated for other ABC transporters (e.g. Ref. [140]). Indeed, the asymmetric structure of the malK dimer has identified a "lid" region between the Walker A and B motifs of each monomer that may insert into the apposed NBD and hydrogen bond with the β phosphate of ATP that is bound there [8].

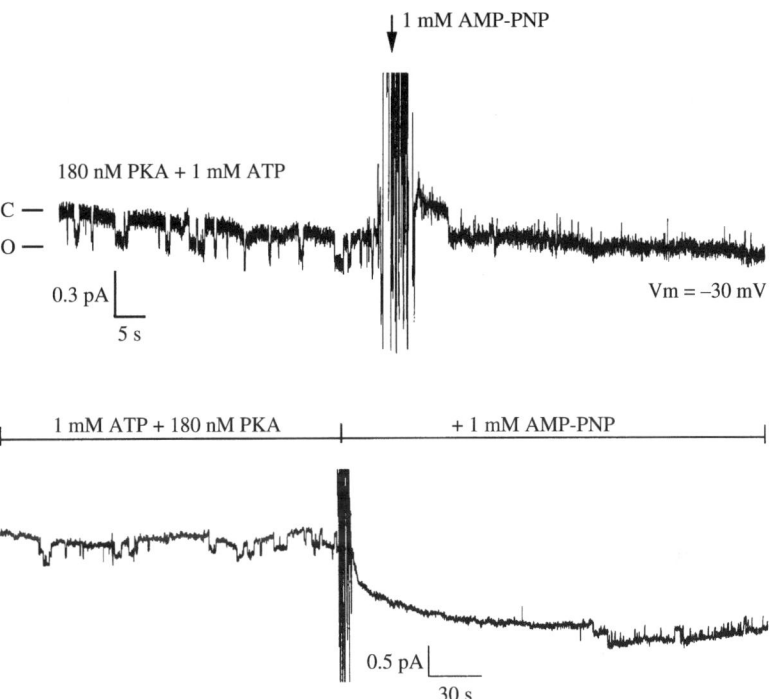

Fig. 3. Two different effects of the ATP analog AMP–PNP on CFTR channel gating by ATP. A, Excised patch recording of a single CFTR channel before and after addition of AMP–PNP to the cytoplasmic side in the continued presence of ATP and PKA. Channel openings are downwards and there is an artifactual positive baseline shift after AMP–PNP addition. Note the prolonged open bursts in the presence of AMP–PNP. B, Effect of AMP–PNP on a patch containing two mutant CFTR channels which have reduced open probability due to removal of all 10 dibasic PKA sites. Note the < 2 min delay after adding AMP–PNP before channels reopen into the "locked open state". Both effects may be mediated by AMP–PNP binding at the first nucleotide binding domain (NBD1). Adapted from Ref. [136].

7. Protein interactions involving CFTR

CFTR may be associated in a regulatory complex with other proteins at the plasma membrane, where it has been reported to influence other transport proteins such as the epithelial sodium channel (ENaC). Much effort is presently being directed towards understanding these interactions since they could contribute to the pathobiology of CF.

7.1. Syntaxin 1A

Perhaps the best characterized interactions are those involving the N-terminal tail (N-tail) of CFTR and syntaxin 1A [141,142]. Syntaxin 1A is part of the vesicle fusion apparatus at neuronal synapses and is also expressed in airway epithelium [143]. It binds to the N-tail of CFTR and inhibits its channel activity, apparently by disrupting interaction of the N-tail with the R domain of CFTR. The N-tail/R domain interaction depends on a cluster of acidic amino acids between aa 46 and 63 of the N-tail that are predicted to lie on one side of an alpha helix. Interaction between the N-tail (or a recombinant N-tail-GST fusion protein) and the R domain increases open probability by prolonging the open state. The N-tail interacts preferentially with the N-terminal region of the R domain between aa 595 and 740, with a more distal region of the R domain apparently stabilizing the interaction. The N-tail also influences the phosphorylation-mediated control of CFTR, as mutating the acidic residues to alanines slows channel activation by forskolin and IBMX and accelerates deactivation after forskolin and IBMX washout [144]. Another membrane protein, MUNC18, may compete with CFTR for interaction with syntaxin 1A and thereby relieve the inhibition of CFTR by syntaxin 1A (see reviews [145,146]).

7.2. EBP50 (and p62$^{c\text{-}yes}$)

As discussed above in connection with Src family kinases, the C-terminal amino acids of CFTR (DTRL) form a motif that binds PDZ (Post synaptic density 95/ Discs large/ ZO-1) domains in scaffolding/adaptor proteins such as EBP50. The structure of the PDZ1 domain from EBP50 bound to the C-terminal five amino acids of CFTR has been solved at 1.7 Å resolution, and indicates there are two hydrogen bonds and two salt bridges between the penultimate arginine and PDZ1 residues [147]. Mutating threonine or leucine in the D*T*RL motif of CFTR leads to its partial redistribution from the apical to lateral plasma membrane in polarized epithelial cells, and abolishes binding to PDZ1 of EBP-50 in vitro [110,111]. PDZ2 of EBP50 binds YAP65, which in turn targets the Src family tyrosine kinase p62$^{c\text{-}yes}$ to the apical membrane, as discussed above [109]. However E3KARP, another PDZ-domain protein related to EBP50, is also situated at the apical membrane of epithelia and can be immunoprecipitated with CFTR [148] and a PDZ domain protein called CAL (CFTR associated ligand) has been proposed to interact with CFTR during its maturation in the trans-Golgi compartment [149]. A protein with four PDZ domains called CAP70 was identified using an affinity column with the C-tail of CFTR as bait [150]. Adding recombinant CAP70 to excised patches had a biphasic effect on P_o, which increased at low CAP70 concentrations and decreased at high concentrations. It was proposed that dimerization of CFTR monomers is enhanced by low concentrations of CAP70, but this bivalent interaction declines when CAP70 is in excess because each CFTR molecule interacts with a single CAP70 molecule at the PDZ domain having highest affinity (i.e. PDZ3). Similar results were obtained with fragments of EBP50 containing the two PDZ domains [151].

7.3. Ezrin, AKAPs and other putative adaptor proteins

PKA is often localized near target substrates by A-kinase anchoring proteins (AKAPs; [152]). Recently, Huang et al. [153] found that CFTR channels in patches excised from Calu-3 cells could sometimes be activated by cpt-cAMP and ATP, indicating that the PKA holoenzyme must already be present. PKA catalytic and type II (RII) regulatory subunits both co-immunoprecipitate with CFTR [148], and endogenous kinase activity in the immunoprecipitates is abolished by Ht31, a peptide derived from the N-terminus of RII that specifically disrupts interaction with AKAPs. Ezrin has been proposed as an AKAP for CFTR because it binds to RII in overlay experiments, is expressed at the apical membrane of Calu-3 and T84 cells, and can be co-immunoprecipitated with CFTR. It also interacts with E3KARP, and thus could provide a link between E3KARP-CFTR and PKA [148]. The cytoskeleton may also be involved since PKA activation depends on the actin cross-linking protein ABP-280x [154]. PKC may also be anchored (e.g. by "regulators of C kinase" or RACKS), although these proteins have not yet been shown to co-immunoprecipitate with CFTR [155]. The importance of targeting in CFTR regulation is best underlined by comparing activation by the soluble and membrane-bound froms of cGMP-dependent protein kinase (Types I vs. II, respectively; [156]). Type I does not activate CFTR, but a chimera containing the Type I enzyme with Type II N-terminal anchoring domain associates with the membrane and activates CFTR channels.

In addition to PDZ domain proteins, CFTR associates with $\alpha 1$ and $\alpha 2$ catalytic subunits of AMP-activated protein kinase (AMPK) [157]. AMPKs are stimulated by increases in the AMP:ATP ratio and thus may serve as metabolic sensors. $\alpha 1$-AMPK binds to the C-tail of CFTR between residues 1420 and 1457. This interaction is disrupted by mutations at either of two protein trafficking motifs located in this region. AMPK phosphorylates CFTR in vitro and inhibits CFTR currents by 35–50% when coexpressed in *Xenopus* oocytes. Whether this is due to direct phosphorylation of CFTR or altered metabolism remains to be established, however AMPK and CFTR are co-localized at the apical membrane of nasal epithelium.

8. CFTR effects on other transporters

Activating CFTR with forskolin causes inhibition of amiloride-sensitive short-circuit current across MDCK cell monolayers made to co-express CFTR and ENaC channels [158]. Qualitatively similar results were obtained in whole cell patch recordings, although the effects were small compared to the variation in basal current. Loss of downregulation by CFTR in cystic fibrosis airway epithelium was proposed as an explanation for sodium hyperabsorption reported previously [159]. Inhibition of ENaC by CFTR has also been reported in *Xenopus* oocytes injected with ENaC and CFTR cRNA [160–164], although the opposite relationship (i.e. a dependence on CFTR) was observed in sweat duct [165]. CFTR regulation of ENaC in oocytes was recently challenged and attributed to inadequate voltage clamping of the membrane potential due to the low resistance of the oocyte membrane during cAMP stimulation relative to the series resistance [166]. The series resistance is mainly due to chamber geometry and the bathing solution resistance (5–7 KΩ; R. Grygorczyk, personal communication). However, it is unlikely that such artifacts could explain CFTR effects on ENaC in other preparations. Indeed, CFTR has been reported to influence virtually all transport proteins examined, including an outwardly rectifying Cl$^-$ channel [167], rat outer medullary K channel ROMK2 [168,169], aquaporin 3 water [170], Ca-activated Cl$^-$ channels [171,172], ATP release channels (e.g. Ref. [173,174]), intracellular calcium signaling [175], and plasma membrane anion exchangers [176] (see Ref. [177]).

Regulation of anion exchange by CFTR was inferred by measuring forskolin-enhanced intracellular alkalinization when CFTR-expressing cells were exposed to bicarbonate-buffered gluconate solutions. The anion exchanger isoform involved and the precise mechanism are not known, however the regulation is apparently independent of Cl^- conductance and is correlated with the severity of pancreatic dysfunction when different disease-associated mutants are compared in vitro [67,178,179].

9. Conclusions

Although much progress has been made since cloning of the CFTR gene, major questions remain concerning the location of the CFTR channel pore, the action of nucleotides during ATP-dependent gating, and control by phosphorylation. As the list of membrane transporters influenced by CFTR grows, it becomes increasingly likely that their dependence will be secondary to some global function or protein association involving CFTR rather than individual, direct associations. Identifying physiologically relevant interacting proteins should provide clues concerning the regulatory mechanisms.

Acknowledgements

This article reviews the literature as to work in the author's laboratory was supported by the Canadian Institutes of Health Research (CIHR), National Institutes of Health (NIDDK; DK54075-03), and Canadian Cystic Fibrosis Foundation (CCFF). The author is a CIHR senior scientist.

References

[1] Rommens, J.M., Iannuzzi, M.C., Kerem, B.-S., Drumm, M.L., Melmer, G., Dean, M., Rozmahel, R., Cole, J.L., Kennedy, D., Hidaka, N., et al., 1989. Identication of the cystic fibrosis gene: chromosome walking and jumping. Science 245, 1059–1065.

[2] Kerem, B.-S., Rommens, J.M., Buchanan, J.A., Markiewicz, D., Cox, T.K., Chakravarti, A., Buchwald, M., Tsui, L.-C., 1989. Identification of the cystic fibrosis gene: genetic analysis. Science 245, 1073–1080.

[3] Riordan, J.R., Rommens, J.M., Kerem, B.-S., Alon, N., Rozmahel, R., Grzelczak, Z., Zielenski, J., Lok, S., Plavsic, N., Chou, J.-L., et al., 1989. Identification of the cystic fibrosis gene: cloning and characterization of complementary DNA. Science 245, 1066–1073.

[4] Quinton, P.M., 1990. Cystic fibrosis: a disease in electrolyte transport. FASEB J. 4, 2709–2717.

[5] Berkower, C., Taglicht, D., Michaelis, S., 1996. Functional and physical interactions between partial molecules of STE6, a yeast ATP-binding cassette protein. J. Biol. Chem. 271, 22983–22989.

[6] Russ, G., Esquivel, F., Yewdell, J.W., Cresswell, P., Spies, T., Bennink, J.R., 1995. Assembly, intracellular localization, and nucleotide binding properties of the human peptide transporters TAP1 and TAP2 expressed by recombinant vaccinia virus. J. Biol. Chem. 270, 21312–21318.

[7] Hung, L.-W., Wang, I.X., Nikaido, K., Liu, P.Q., Ames, G.F.L., Kim, S.-H., 1998. Crystal structure of the ATP-binding subunit of an ABC transporter, the histidine permease of Salmonella typhimurium. Nature 396, 703–707.

[8] Diederichs, K., Diez, J., Greller, G., Muller, C., Breed, J., Schnell, C., Vonrhein, C., Boos, W., Welte, W., 2000. Crystal structure of MalK, the ATPase subunit of the trehalose/maltose ABC transporter of the archeon Thermococcus litoralis. EMBO J. 19, 5951–5961.

[9] Ambudkar, S.V., Dey, S., Hrycyna, C.A., Ramachandra, M., Pastan, I., Gottesman, M.M., 1999. Biochemical, cellular, and pharmacological aspects of the multidrug transporter. Annu. Rev. Pharmacol. Toxicol. 39, 361–398.

[10] Cornwell, M.M., Tsuruo, T., Gottesman, M.M., Pastan, I., 1987. ATP-binding properties of P glycoprotein from multidrug- resistant KB cells. FASEB J. 1, 51–54.

[11] Reutz, S., Gros, P., 1994. Phosphatidylcholine translocase: a physiological role for mdr2. Cell 77, 1071–1081.

[12] Anderson, M.P., Rich, D.P., Gregory, R.J., Smith, A.E., Welsh, M.J., 1991. Generation of cAMP-activated chloride currents by expression of CFTR. Science 251, 679–682.

[13] Kartner, N., Hanrahan, J.W., Jensen, T.J., Naismith, A.L., Sun, S., Ackerley, C.A., Reyes, E.F., Tsui, L.-C., Rommens, J.M., Bear, C.E., et al., 1991. Expression of the cystic fibrosis gene in non-epithelial invertebrate cells produces a regulated anion conductance. Cell 64, 681–691.

[14] Berger, H.A., Anderson, M.P., Gregory, R.J., Thompson, S., Howard, P.W., Maurer, R.A., Mulligan, R., Smith, A.E., Welsh, M.J., 1991. Identification and regulation of the cystic fibrosis transmembrane conductance regulator-generated chloride channel. J. Clin. Invest. 88, 1422–1431.

[15] Gray, M.A., Harris, A., Coleman, L., Greenwell, J.R., Argent, B.E., 1989. Two types of chloride channel on duct cells cultured from human fetal pancreas. Am. J. Physiol. 257, C240–C251.

[16] Champigny, G., Verrier, B., Gérard, C., Mauchamp, J., Lazdunski, M., 1990. Small conductance chloride channels in the apical membrane of thyroid cells. FEBS Lett. 259, 263–268.

[17] Tabcharani, J.A., Low, W., Elie, D., Hanrahan, J.W., 1990. Low-conductance chloride channel activated by cAMP in the epithelial cell line T_{84}. FEBS Lett. 270, 157–164.

[18] Tabcharani, J.A., Chang, X.-B., Riordan, J.R., Hanrahan, J.W., 1991. Phosphorylation-regulated Cl^- channel in CHO cells stably expressing the cystic fibrosis gene. Nature 352, 628–631.

[19] Bear, C.E., Duguay, F., Naismith, A.L., Kartner, N., Hanrahan, J.W., Riordan, J.R., 1991. Cl^- channel activity in *Xenopus* oocytes expressing the cystic fibrosis gene. J. Biol. Chem. 266, 19142–19145.

[20] Rommens, J.M., Dho, S., Bear, C.E., Kartner, N., Kennedy, D., Riordan, J.R., Tsui, L.-C., Foskett, K., 1991. cAMP-inducible chloride conductance in mouse fibroblast lines stably expressing human cystic fibrosis transmembrane conductance regulator. Proc. Natl Acad. Sci. USA 88, 7500–7504.

[21] Anderson, M.P., Gregory, R.J., Thompson, S., Souza, D.W., Paul, S., Mulligan, R.C., Smith, A.E., Welsh, M.J., 1991. Demonstration that CFTR is a chloride channel by alteration of its anion selectivity. Science 253, 202–205.

[22] Bear, C.E., Li, C., Kartner, N., Bridges, R.J., Jensen, T.J., Ramjeesingh, M., Riordan, J.R., 1992. Purification and functional reconstitution of the cystic fibrosis transmembrane conductance regulator (CFTR). Cell 68, 809–818.

[23] Riordan, J.R., Chang, X.-B., 1992. CFTR, a channel with the structure of a transporter. Biochim. Biophys. Acta Bio-Energetics 1101, 221–222.

[24] Hanrahan, J.W., Tabcharani, J.A., Grygorczyk, R., Dodge, J.A., Brock, D.J.H., Widdicombe, J.H., (Eds.), 1993. Patch clamp studies of apical membrane chloride channels. Cystic Fibrosis-Current Topics, John Wiley and Sons, Ltd., London, pp. 93–137.

[25] Gadsby, D.C., Nagel, G., Hwang, T.-C., 1995. The CFTR chloride channel of mammalian heart. Annu. Rev. Physiol. 57, 387–416.

[26] Hanrahan, J.W., Tabcharani, J.A., Becq, F., et al., 1995. Function and dysfunction of the CFTR chloride channel. In: Dawson, D.C., Frizzell, R.A. (Eds.), Ion Channels and Genetic Diseases. Rockefeller Univ. Press, New York, pp. 125–137.

[27] Quinton, P.M., 1999. Physiological basis of cystic fibrosis: a historical perspective. Physiol. Rev. 79, S3–S22.

[28] Sheppard, D.N., Welsh, M.J., 1999. Structure and function of the CFTR chloride channel. Physiol. Rev. 79, S23–S45.

[29] Linsdell, P., Tabcharani, J.A., Rommens, J.M., Hou, Y.-X., Chang, X.-B., Tsui, L.-C., Riordan, J.R., Hanrahan, J.W., 1997. Permeability of wild-type and mutant cystic fibrosis transmembrane conductance regulator chloride channels to polyatomic anions. J. Gen. Physiol. 110, 355–364.

[30] Sheppard, D.N., Rich, D.P., Ostedgaard, L.S., Gregory, R.J., Smith, A.E., Welsh, M.J., 1993. Mutations in CFTR associated with mild-disease-form Cl^- channels with altered pore properties. Nature 362, 160–164.

[31] Tabcharani, J.A., Rommens, J.M., Hou, Y.-X., Chang, X.-B., Tsui, L.-C., Riordan, J.R., Hanrahan, J.W., 1993. Multi-ion pore behaviour in the CFTR chloride channel. Nature 366, 79–82.

[32] McDonough, S., Davidson, N., Lester, H.A., McCarty, N.A., 1994. Novel pore-lining residues in CFTR that govern permeation and open-channel block. Neuron 13, 623–634.

[33] Linsdell, P., Hanrahan, J.W., 1996. Disulphonic stilbene block of cystic fibrosis transmembrane conductance regulator Cl⁻ channels expressed in a mammalian cell line and its regulation by a critical pore residue. J. Physiol. 496, 687–693.

[34] Carroll, T.P., Morales, M.M., Fulmer, S.B., Allen, S.S., Flotte, T.R., Cutting, G.R., Guggino, W.B., 1995. Alternate translation initiation codons can create functional forms of cystic fibrosis transmembrane conductance regulator. J. Biol. Chem. 270, 11941–11946.

[35] Mansoura, M.K., Smith, S.S., Choi, A.D., Richards, N.W., Strong, T.V., Drumm, M.L., Collins, F.S., Dawson, D.C., 1998. Cystic fibrosis transmembrane conductance regulator (CFTR) anion binding as a probe of the pore. Biophys. J. 74, 1320–1332.

[36] Akabas, M.H., Kaufmann, C., Cook, T.A., Archdeacon, P., 1994. Amino acid residues lining the chloride channel of the cystic fibrosis transmembrane conductance regulator. J. Biol. Chem. 269, 14865–14868.

[37] Pind, S., Mohamed, A., Chang, X.-B., et al., 1996. Multiple initiation sites are used during translation of the mRNA encoding CFTR. Ped. Pulmonol.(Suppl. 12), 180.

[38] Sheppard, D.N., Ostedgaard, L.S., Rich, D.P., Welsh, M.J., 1994. The amino-terminal portion of CFTR forms a regulated Cl⁻ channel. Cell 76, 1091–1098.

[39] Gupta, J., Evagelidis, A., Hanrahan, J.W., Linsdell, P., 2003. Asymmetric structure of the cystic fibrosis transmembrane conductance regulator chloride channel pore suggested by mutagenesis of the twelfth transmembrane region. Biochemistry 40, 6620–6627.

[40] Ishihara, H., Welsh, M.J., 1997. Block by MOPS reveals a conformational change in the CFTR pore produced by ATP hydrolysis. Am. J. Physiol. (Cell Physiol.) 273, C1278–C1289.

[41] Mansoura, M.K., Strong, T.V., Collins, F.S., et al., 1994. A disease-related mutation in the fifth putative transmembrane segment alters the conduction and gating properties of CFTR. J. Gen. Physiol. 104, 36a (Abstract).

[42] Zhang, Z.-R., McDonough, S.I., McCarty, N.A., 2000. Interaction between permeation and gating in a putative pore domain mutant in the cystic fibrosis transmembrane conductance regulator. Biophys. J. 79, 298–313.

[43] McCarty, N.A., 2000. Permeation through the CFTR chloride channel. J. Exp. Biol. 203, 1947–1962.

[44] Cheung, M., Akabas, M.H., 1997. Locating the anion-selectivity filter of the cystic fibrosis transmembrane conductance regulator (CFTR) chloride channel. J. Gen. Physiol. 109, 289–299.

[45] Holmgren, M., Liu, Y., Xu, Y., Yellen, G., 1996. On the use of thiol-modifying agents to determine channel topology. Neuropharmacology 35(7), 797–804.

[46] Yang, N., George, J.r.A.L., Horn, R., 1996. Molecular basis of charge movement in voltage-gated sodium channels. Neuron 16, 113–122.

[47] Guinamard, R., Akabas, M.H., 1999. Arg352 is a major determinant of charge selectivity in the the cystic fibrosis transmembrane conductance regulator chloride channel. Biochemistry, 385528–385537.

[48] Gray, M.A., Pollard, C.E., Harris, A., Coleman, L., Greenwell, J.R., Argent, B.E., 1990. Anion selectivity and block of the small conductance chloride channel on pancreatic duct cells. Am. J. Physiol. (Cell Physiol.) 259, C752–C761.

[49] Gray, M.A., Greenwell, J.R., Argent, B.E., 1988. Secretin-regulated chloride channel on the apical plasma membrane of pancreatic duct cells. J. Membr. Biol. 105, 131–142.

[50] Bear, C.E., Reyes, E.F., 1992. cAMP-activated chloride conductance in the colonic cell line, Caco-2. Am. J. Physiol. (Cell Physiol.) 262, C251–C256.

[51] Yang, I.C.H., Cheng, T.-H., Wang, F., Price, E.M., Hwang, T.-C., 1997. Modulation of CFTR chloride channels by calyculin A and genistein. Am. J. Physiol. (Cell Physiol.) 272, C142–C155.

[52] Cliff, W.H., Frizzell, R.A., 1990. Separate Cl⁻ conductances activated by cAMP and Ca²⁺ in Cl⁻ secreting epithelial cells. Proc. Natl Acad. Sci. USA 87, 4956–4960.

[53] Bell, C.L., Quinton, P.M., 1992. T84 cells: anion selectivity demonstrates expression of Cl⁻ conductance affected in cystic fibrosis. Am. J. Physiol. (Cell Physiol.) 262, C555–C562.

[54] Tabcharani, J.A., Chang, X.-B., Riordan, J.R., Hanrahan, J.W., 1992. The cystic fibrosis transmembrane conductance regulator chloride channel: Iodide block and permeation. Biophys. J. 62, 1–4.

[55] Tabcharani, J.A., Linsdell, P., Hanrahan, J.W., 1997. Halide permeation in wild-type and mutant cystic fibrosis transmembrane conductance regulator chloride channels. J. Gen. Physiol. 110, 341–354.

[56] Linsdell, P., Evagelidis, A., Hanrahan, J.W., 2000. Molecular determinants of anion selectivity in the cystic fibrosis transmembrane conductance regulator chloride channel pore. Biophys. J. 78, 2973–2982.

[57] Price, M.P., Ishihara, H., Sheppard, D.N., Welsh, M.J., 1996. Function of *Xenopus* cystic fibrosis transmembrane conductance regulator (CFTR) Cl⁻ channels and use of human-*Xenopus chimeras* to investigate the pore properties of CFTR. J. Biol. Chem. 271, 25184–25191.

[58] Linsdell, P., Hanrahan, J.W., 1998. Adenosine triphosphate-dependent asymmetry of anion permeation in the cystic fibrosis transmembrane conductance regulator chloride channel. J. Gen. Physiol. 111, 601–614.

[59] Linsdell, P., Hanrahan, J.W., 1998. Glutathione permeability of CFTR. Am. J. Physiol. (Cell Physiol.) 275, C323–C326.

[60] Gao, L., Kim, K.J., Yankaskas, J.R., Forman, H.J., 1999. Abnormal glutathione transport in cystic fibrosis airway epithelia. Am. J. Physiol. (Lung Cell Mol. Physiol.) 277, L113–L118.

[61] Gao, L., Broughman, J.R., Iwamaoto, T., et al., 2000. Correction of glutathione secretion in cystic fibrosis airway epithelia by a chloride channel forming peptide and chlorzoxazone. Ped. Pulmonol.(Suppl. 20), 248 (Abstract).

[62] Quinton, P.M., 2001. The neglected ion: HCO₃⁻. Nat. Med. 7, 292–293.

[63] Lee, M.C., Penland, C.M., Widdicombe, J.H., Wine, J.J., 1998. Evidence that Calu-3 human airway cells secrete bicarbonate. Am. J. Physiol. (Lung Cell Mol. Physiol.), 274L450–274L453.

[64] Devor, D.C., Singh, A.K., Lambert, L.C., DeLuca, A., Frizzell, R.A., Bridges, R.J., 1999. Bicarbonate and chloride secretion in Calu-3 human airway epithelial cells. J. Gen. Physiol., 113743–113760.

[65] Hanrahan, J.W., Tabcharani, J.A., 1989. Possible role of outwardly rectifying anion channels in epithelial transport. Ann. NY Acad. Sci. 574, 30–43.

[66] Poulsen, J.H., Fischer, H., Illek, B., Machen, T.E., 1994. Bicarbonate conductance and pH regulatory capability of cystic fibrosis transmembrane conductance regulator. Proc. Natl Acad. Sci. USA 91, 5340–5344.

[67] Lee, M.G., Choi, J.Y., Luo, X., Strickland, E., Thomas, P.J., Muallem, S., 1999. Cystic fibrosis transmembrane conductance regulator regulates luminal Cl⁻/HCO₃⁻ exchange in mouse submandibular and pancreastic ducts. J. Biol. Chem. 274, 14670–14677.

[68] O'Reilly, C.M., Winpenny, J.P., Rgent, B.E., Ray, M.A., 2000. Cystic fibrosis transmembrane conductance regulator currents in guinea pig pancreatic duct cells: inhibition by bicarbonate ions. Gastroenterology 118, 1187–1196.

[69] Eisenman, G., Horn, R., 1983. Ionic selectivity revisited: the role of kinetic and equilibrium processes in ion permeation through channels. J. Membr. Biol. 76, 197–225.

[70] Hille, B., 1992. Ionic channels of excitable membranes, 2nd ed. Sinauer Associates Incorporated, Sunderland, Massachusetts.

[71] Dawson, D.C., Smith, S.S., Mansoura, M.K., 1999. CFTR: mechanism of anion conduction. Physiol. Rev. 79, S47–S75.

[72] Linsdell, P., Tabcharani, J.A., Hanrahan, J.W., 1997. A multi-ion mechanism for ion permeation and block in the cystic fibrosis transmembrane conductance regulator chloride channel. J. Gen. Physiol. 110, 365–377.

[73] Cotten, J.F., Welsh, M.J., 1999. Cystic fibrosis-associated mutations at arginine 347 alter the pore architecture of CFTR. J. Biol. Chem. 274, 5429–5435.

[74] Cheng, S.H., Rich, D.P., Marshall, J., Gregory, R.J., Welsh, M.J., Smith, A.E., 1991. Phosphorylation of the R domain by cAMP-dependent protein kinase regulates the CFTR chloride channel. Cell 66, 1027–1036.

[75] Klyce, S.D., Wong, R.K.S., 1977. Site and mode of adrenaline action on chloride transport across the rabbit corneal epithelium. J. Physiol. 266, 777–799.

[76] Frizzell, R.A., Field, M., Schultz, S.G., 1979. Sodium-coupled chloride transport by epithelial tissues. Am. J. Physiol. (Renal, Fluid Electrolyte Physiol.) 236, F1–F8.

[77] Picciotto, M.R., Cohn, J.A., Bertuzzi, G., Greengard, P., Nairn, A.C., 1992. Phosphorylation of the cystic fibrosis transmembrane conductance regulator. J. Biol. Chem. 267, 12742–12752.

[78] Gadsby, D.C., Nairn, A.C., 1999. Control of CFTR channel gating by phosphorylation and nucleotide hydrolysis. Physiol. Rev. 79, S77–S107.

[79] Wilkinson, D.J., Strong, T.V., Mansoura, M.E., Wood, D.L., Smith, S.S., Collins, F.S., Dawson, D.C., 1997. CFTR activation: additive effects of stimulatory and inhibitory phosphorylation sites in the R domain. Am. J. Physiol. (Lung Cell Mol. Physiol.) 273, L127–L133.

[80] Chan, K.W., Smith, S.S., Csanady, L., Nairn, A.C., Dawson, D.C., Gadsby, D.C., 1998. PKA- and Ca^{2+} dependent regulation of WT and S768A mutant CFTR Cl^- channels in *Xenopus* oocytes. Biophys. J. 74, A396.

[81] Baldursson, O., Ostedgaard, L.S., Rokhlina, T., Cotten, J.F., Welsh, M.J., 2001. Cystic fibrosis transmembrane conductance regulator Cl^- channels with R domain deletions and translocations show phosphorylation-dependent and -independent activity. J. Biol. Chem. 276, 1904–1910.

[82] Seibert, F.S., Chang, X.-B., Aleksandrov, A.A., Clarke, D.M., Hanrahan, J.W., Riordan, J.R., 2000. Influence of phosphorylation by protein kinase A on CFTR at the cell surface and endoplasmic reticulum. Biochim. Biophys. Acta 1461, 275–283.

[83] Chan, K.M., Csanády, L., Seto-Young, D., Nairn, A.C., Gadsby, D.C., 2000. Severed molecules functionally define the boundaries of the cystic fibrosis transmembrane conductance regulator's NH_2-terminal nucleotide binding domain. J. Gen. Physiol. 116(2), 163–180.

[84] Pfeifer, A., Aszódi, A., Seidler, U., Ruth, P., Hofmann, F., Fässler, R., 1996. Intestinal secretory defects and dwarfism in mice lacking cGMP-dependent protein kinase II. Science 274, 2082–2086.

[85] French, P.J., Bijman, J., Edixhoven, M., Vaandrager, A.B., Scholte, B.J., Lohmann, S.M., Nairn, A.C., Dejonge, H.R., 1995. Isotype-specific activation of cystic fibrosis transmembrane conductance regulator-chloride channels by cGMP-dependent protein kinase II. J. Biol. Chem. 270, 26626–26631.

[86] Ma, J., Tasch, J.E., Tao, T., Zhao, J., Xie, J., Drumm, M.L., Davis, P.B., 1996. Phosphorylation-dependent block of cystic fibrosis transmembrane conductance regulator chloride channel by exogenous R domain protein. J. Biol. Chem. 271, 7351–7356.

[87] Csanády, L., Chan, K.W., Seto-Young, D., Kopsco, D.C., Nairn, A.C., Gadsby, D.C., 2000. Severed channels probe regulation of gating of cystic fibrosis transmembrane conductance regulator by its cytoplasmic domains. J. Gen. Physiol. 116(3), 477–500.

[88] Ma, J., Zhao, J., Drumm, M.L., Xie, J., Davis, P.B., 1997. Function of the R domain in the cystic fibrosis transmembrane conductance regulator chloride channel. J. Biol. Chem. 272, 28133–28141.

[89] Ikuma, M., Welsh, M.J., 2000. Regulation of CFTR Cl^- channel gating by ATP binding and hydrolysis. Proc. Natl Acad. Sci. USA 97, 8675–8680.

[90] Aleksandrov, A.A., Chang, X.-B., Aleksandrov, L., Riordan, J.R., 2000. The non-hydrolytic pathway of cystic fibrosis transmembrane conductance regulator ion channel gating. J. Physiol. 528, 259–265.

[91] Winter, M.C., Welsh, M.J., 1997. Stimulation of CFTR activity by its phosphorylated R domain. Nature 389, 294–296.

[92] Dulhanty, A.M., Riordan, J.R., 1994. Phosphorylation by cAMP-dependent protein kinase causes a conformational change in the R domain of the cystic fibrosis transmembrane conductance regulator. Biochemistry 33, 4072–4079.

[93] Ostedgaard, L.S., Baldursson, O., Vermeer, D.W., Welsh, M.J., Robertson, A.D., 2000. A functional R domain from cystic fibrosis transmembrane conductance regulator is predominantly unstructured in solution. Proc. Natl Acad. Sci. USA 97, 5657–5662.

[94] Pasyk, E.A., Morin, X.K., Zeman, P., Garami, E., Galley, K., Huan, L.J., Wang, Y., Bear, C.E., 1998. A conserved region of the R domain of cystic fibrosis transmembrane conductance regulator is important in processing and function. J. Biol. Chem. 273, 31759–31764.

[95] Jia, Y., Mathews, C.J., Hanrahan, J.W., 1997. Phosphorylation by protein kinase C is required for acute activation of cystic fibrosis transmembrane conductance regulator by protein kinase A. J. Biol. Chem. 272, 4978–4984.

[96] Berger, H.A., Travis, S.M., Welsh, M.J., 1993. Regulation of the cystic fibrosis transmembrane conductance regulator Cl^- channel by specific protein kinases and phosphatases. J. Biol. Chem. 268, 2037–2047.

[97] Liedtke, C.M., Cole, T.S., 1998. Antisense oligonucleotide to PKC-ε alters cAMP-dependent stimulation of CFTR in Calu-3 cells. Am. J. Physiol. (Cell Physiol.) 275, C1357–C1364.

[98] Hinkson, D.A.R., Riordan, J.R., Chang, X.B., et al., 2000. Regulation of CFTR by protein kinase C. Pediatr. Pulmonol.(Suppl. 20), 176 (Abstract).

[99] Li, M., West, J.W., Numann, R., Murphy, B.J., Scheuer, T., Catterall, W.A., 1993. Convergent regulation of sodium channels by protein kinase C and cAMP-dependent protein kinase. Science 261, 1439–1442.

[100] Light, P.E., Sabir, A.A., Allen, B.G., Walsh, M.P., French, R.J., 1996. Protein kinase C-induced changes in the stoichiometry of ATP binding activate cardiac ATP-sensitive K^+ channels. Circ. Res. 79, 399–406.

[101] Light, P.E., Bladen, C., Winkfein, R.J., Walsh, M.P., French, R.J., 2000. Molecular basis of protein kinase C-induced activation of ATP-sensitive potassium channels. Proc. Natl Acad. Sci. USA 97, 9058–9063.

[102] Wang, Y.T., Salter, M.W., 1994. Regulation of NMDA receptors by tyrosine kinases and phosphatases. Nature 369, 233–235.

[103] Yu, X.-M., Askalan, R., Keil, G.J. II, Salter, M.W., 1997. NMDA channel regulation by channel-associated protein tyrosine kinase Src. Science 275, 674–678.

[104] Köhr, G., Seeburg, P.H., 1996. Subtype-specific regulation of recombinant NMDA receptor-channels by protein tyrosine kinases of the src family. J. Physiol. 492, 445–452.

[105] Wijetunge, S., Hughes, A.D., 1996. Activation of endogenous c-Src or a related tyrosine kinase by intracellular (PY)EEI peptide increases voltage-operated calcium channel currents in rabbit ear artery cells. FEBS Lett. 399(1–2), 63–66.

[106] Fischer, H., Machen, T.E., 1996. The tyrosine kinase $p60^{c-src}$ regulates the fast gate of the cystic fibrosis transmembrane conductance regulator chloride channel. Biophys. J. 71, 3073–3082.

[107] Jia, F., Seibert, F., Chang, X.-B., et al., 1997. Activation of CFTR chloride channels by tyrosine phosphorylation. Pediatr. Pulmonol.(Suppl. 14), 214 (Abstract).

[108] Pucéat, M., Roche, S., Vassort, G., 1998. Src family tyrosine kinase regulates intracellular pH in cardiomyocytes. J. Cell Biol. 141, 1637–1646.

[109] Mohler, P.J., Kreda, S.M., Boucher, R.C., Sudol, M., Stutts, M.J., Milgram, S.L., 1999. Yes-associated protein 65 localizes $p62^{c-Yes}$ to the apical compartment of airway epithelia by association with EBP50. J. Cell Biol. 147, 879–890.

[110] Wang, S., Raab, R.W., Schatz, P.J., Guggino, W.B., Li, M., 1998. Peptide binding consensus of the NHE-RF-PDZ1 domain matches the C-terminal sequence of cystic fibrosis transmembrane conductance regulator (CFTR). FEBS Lett. 427, 103–108.

[111] Short, D.B., Trotter, K.W., Reczek, D., Kreda, S.M., Bretscher, A., Boucher, R.C., Stutts, M.J., Milgram, S.L., 1998. An apical PDZ protein anchors the cystic fibrosis transmembrane conductance regulator to the cytoskeleton. J. Biol. Chem. 273, 19797–19801.

[112] Hall, R.A., Ostedgaard, L.S., Premont, R.T., Blitzer, J.T., Rahman, N., Welsh, M.J., Lefkowitz, R.J., 1998. A C-terminal motif found in the β_2-adrenergic receptor, P2Y1 receptor and cystic fibrosis transmembrane conductance regulator determines binding to the Na^+/H^+ exchanger regulatory factor family of PDZ proteins. Proc. Natl Acad. Sci. USA 95, 8496–8501.

[113] Becq, F., Jensen, T.J., Chang, X.-B., Savoia, A., Rommens, J.M., Tsui, L.-C., Buchwald, M., Riordan, J.R., Hanrahan, J.W., 1994. Phosphatase inhibitors activate normal and defective CFTR chloride channels. Proc. Natl Acad. Sci. USA 91, 9160–9164.

[114] Luo, J., Zhu, T., Evagelidis, A., Pato, M., Hanrahan, J.W., 2000. Role of protein phosphatases in the activation of CFTR (ABCC7) by genistein and bromotetramisole. Am. J. Physiol. (Cell Physiol.) 279, C108–C119.

[115] Luo, J., Pato, M.D., Riordan, J.R., Hanrahan, J.W., 1998. Differential regulation of single CFTR channels by PP2C, PP2A, and other phosphatases. Am. J. Physiol. (Cell Physiol.) 274, C1397–C1410.

[116] Travis, S.M., Berger, H.A., Welsh, M.J., 1997. Protein phosphatase 2C dephosphorylates and inactivates cystic fibrosis transmembrane conductance regulator. Proc. Natl Acad. Sci. USA 94, 11055–11060.

[117] Zhu, T., Dahan, D., Evaglelidis, A., Zheng, S.-X., Luo, J., Hanrahan, J.W., 1999. Association of cystic fibrosis transmembrane conductance regulator and protein phosphatase 2C. J. Biol. Chem. 274, 29102–29107.

[118] Cohen, P., Schelling, D.L., Stark, M.J., 1989. Remarkable similarities between yeast and mammalian protein phosphatases. FEBS Lett. 250, 601–606.

[119] Smith, S.N., Delaney, S.J., Dorin, J.R., Farley, J.M., Geddes, D.M., Porteous, D.J., Wainwright, B.J., Alton, E.W., 1998. Effect of IBMX and alkaline phosphatase inhibitors on Cl^- secretion in G551D cystic fibrosis mutant mice. Am. J. Physiol. (Cell Physiol.) 274, C492–C499.

[120] Anderson, M.P., Berger, H.A., Rich, D.P., Gregory, R.J., Smith, A.E., Welsh, M.J., 1991. Nucleoside triphosphates are required to open the CFTR chloride channel. Cell 67, 775–784.

[121] Carson, M.R., Welsh, M.J., 1993. 5'-adenylylimidodiphosphate (AMP–PNP) does not activate CFTR chloride channels in cell-free patches of membrane. Am. J. Physiol. 265, L27–L32.

[122] Dousmanis, A.G., Nairn, A.C., Gadsby, D.C., 1996. [Mg^{2+}] governs CFTR Cl$^-$ channel opening and closing rates, confirming hydrolysis of two ATP molecules per gating cycle. Biophys. J. 70, A127 (Abstract).

[123] Gunderson, K.L., Kopito, R.R., 1995. Conformational states of CFTR associated with channel gating: the role of ATP binding and hydrolysis. Cell 82, 231–239.

[124] Schultz, B.D., Bridges, R.J., Frizzell, R.A., 1996. Lack of conventional ATPase properties in CFTR chloride channel gating. J. Membr. Biol. 151, 63–75.

[125] Tomaszek, T.A. Jr, Schuster, S.M., 2001. Hydrolysis of adenyl-5-yl imidodiphosphate by beef heart mitochondrial ATPase. J. Biol. Chem. 261, 2264–2269.

[126] Aleksandrov, A.A., Riordan, J.R., 1998. Regulation of CFTR ion channel gating by MgATP. FEBS Lett. 431, 97–101.

[127] Wu, J.V., Joo, N.S., Krause, M.E., Wine, J.J., 2001. Cystic fibrosis transmembrane conductance regulator gating requires cytosloic electrolytes. J. Biol. Chem. 276, 6473–6478.

[128] Baukrowitz, T., Hwang, T.-C., Nairn, A.C., Gadsby, D.C., 1994. Coupling of CFTR Cl$^-$ channel gating to an ATP hydrolysis cycle. Neuron 12, 473–482.

[129] Hwang, T.-C., Nagel, G., Nairn, A.C., Gadsby, D.C., 1994. Regulation of the gating of cystic fibrosis transmembrane conductance regulator Cl channels by phosphorylation and ATP hydrolysis. Proc. Natl Acad. Sci. USA 91, 4698–4702.

[130] Carson, M.R., Travis, S.M., Welsh, M.J., 1995. The two nucleotide-binding domains of CFTR have distinct functions in controlling channel activity. J. Biol. Chem. 270, 1711–1717.

[131] Gunderson, K.L., Kopito, R.R., 1994. Effects of pyrophosphate and nucleotide analogs suggest a role for ATP hydrolysis in cystic fibrosis transmembrane regulator channel gating. J. Biol. Chem. 269, 19349–19353.

[132] Carson, M.R., Welsh, M.J., 1995. Structural and functional similarities between the nucleotide-binding domains of CFTR and GTP-binding proteins. Biophys. J. 69, 2443–2448.

[133] Mathews, C.J., Tabcharani, J.A., Chang, X.-B., Jensen, T.J., Riordan, J.R., Hanrahan, J.W., 1998. Dibasic protein kinase A sites regulate bursting rate and nucleotide sensitivity of the cystic fibrosis transmembrane conductance regulator chloride channel. J. Physiol. 508, 365–377.

[134] Wang, F., Zeltwanger, S., Hu, S.H., Hwang, T.C., 2000. Deletion of phenylalanine 508 causes attenuated phosphorylation-dependent activation of CFTR chloride channels. J. Physiol. 524(3), 637–648.

[135] Szabó, K., Szakás, G., Hegedüs, T., Sarkadi, B., 1999. Nucleotide occlusion in the human cystic fibrosis transmembrane conductance regulator. J. Biol. Chem. 274, 12209–12212.

[136] Mathews, C.J., Tabcharani, J.A., Hanrahan, J.W., 1998. The CFTR chloride channel: nucleotide interactions and temperature-dependent gating. J. Membr. Biol. 163, 55–66.

[137] Weinreich, F., Riordan, J.R., Nagel, G., 1999. Dual effects of ADP and adenylylimidodiphosphate on CFTR channel kinetics show binding to two different nucleotide binding sites. J. Gen. Physiol. 114, 55–70.

[138] Aleksandrov, L., Mengos, A., Chang, X.-B., Aleksandrov, A.A., Riordan, J.R., 2001. Differential interactions of nucleotides at the two nucleotide binding domains of CFTR. J. Biol. Chem. 276(16), 12918–12923.

[139] Li, C., Ramjeesingh, M., Wang, W., Garami, E., Hewryk, M., Lee, D., Rommens, J.M., Galley, K., Bear, C.E., 1996. ATPase activity of the cystic fibrosis transmembrane conductance regulator. J. Biol. Chem. 271, 28463–28468.

[140] Qu, Q., Sharom, F.J., 2001. FRET analysis indicates that the two ATPase active sites of the P-glycoprotein multidrug transporter are closely associated. Biochemistry 40(5), 1413–1422.

[141] Naren, A.P., Nelson, D.J., Xie, W.W., Jovov, B., Pevsner, J., Bennett, M.K., Benos, D.J., Quick, M.W., Kirk, K.L., 1997. Regulation of CFTR chloride channels by syntaxin and Munc18 isoforms. Nature 390, 302–305.

[142] Naren, A.P., Quick, M.W., Collawn, J.F., Nelson, D.J., Kirk, K.L., 1998. Syntaxin 1A inhibits CFTR chloride channels by means of domain-specific protein-protein interactions. Proc. Natl Acad. Sci. USA 95, 10972–10977.

[143] Naren, A.P., Di, A., Cormet-Boyaka, E., Boyaka, P.N., McGhee, J.R., Zhou, W., Akagawa, K., Fujiwara, T., Thome, U., Engelhardt, J.F., et al., 2000. Syntaxin 1A is expressed in airway epithelial cells, where it modulates CFTR Cl$^-$ currents. J. Clin. Invest. 105, 377–386.

[144] Naren, A.P., Cormet-Boyaka, E., Fu, J., Villain, M., Blalock, J.E., Quick, M.W., Kirk, K.L., 1999. CFTR chloride channel regulation by an interdomain interaction. Science 286, 544–548.

[145] Naren, A.P., Cormet-Boyaka, E., Fu, J., Villain, M., Blalock, J.E., Quick, M.W., Kirk, K.L., 1999. CFTR chloride channel regulation by an interdomain interaction. Science 286, 544–548.

[146] Naren, A.P., Kirk, K.L., 2000. CFTR chloride channels: Binding partners and regulatory networks. News Physiol. Sci. 15, 57–61.

[147] Karthikeyan, S., Leung, T., Ladias, J.A.A., 2001. Structural basis of the NHERFPDZ1 interaction with the carboxyl-terminal region of the cystic fibrosis transmembrane conductance regulator. J. Biol. Chem. 276(23), 19683–19686.

[148] Sun, F., Hug, M.J., Bradbury, N.A., Frizzell, R.A., 2000. Protein kinase A associates with cystic fibrosis transmembrane conductance regulator via an interaction with ezrin. J. Biol. Chem. 275, 14360–14366.

[149] Cheng, J., Moyer, B.D., Milewski, M., et al., 1999. CFTR associates with the PDZ domain protein CAL at the trans-Golgi network. Pediatr. Pulmonol.(Suppl. 19), 168–169 (Abstract).

[150] Wang, S., Yue, H., Derin, R.B., Guggino, W.B., Li, M., 2000. Accessory protein facilitated CFTR–CFTR interaction, a molecular mechanism to potentiate the chloride channel activity. Cell 103, 169–179.

[151] Raghuram, V., Mak, D.-O.D., Foskett, J.K., 2001. Regulation of cystic fibrosis transmembrane conductance regulator single-channel gating by bivalent PDZ-domain-mediated interaction. Proc. Natl Acad. Sci. USA 98, 1300–1305.

[152] Gray, P.C., Scott, J.D., Catterall, W.A., 2000. Regulation of ion channels by cAMP-dependent protein kinase and A-kinase anchoring proteins. Curr. Opin. Neurobiol. 8, 330–334.

[153] Huang, P., Trotter, K., Boucher, R.C., Milgram, S.L., Stutts, M.J., 2000. PKA holoenzyme is functionally coupled to CFTR by AKAPs. Am. J. Physiol. (Cell Physiol.) 278, C417–C422.

[154] Prat, A.G., Cunningham, C.C., Jackson, G.R.J.r., Borkan, S.C., Wang, Y., Ausiello, D.A., Cantiello, H.F., 1999. Actin filament organization is required for proper cAMP-dependent activation of CFTR. Am. J. Physiol. (Cell Physiol.) 277, C1160–C1169.

[155] Yarwood, S.J., Steele, M.R., Scotland, G., Houslay, M.D., Bolger, G.B., 1999. The RACK1 signaling scaffold protein selectively interacts with the cAMP-specific phosphodiesterase PDE4D5 isoform. J. Biol. Chem. 274, 14909–14917.

[156] Vaandrager, A.B., Smolenski, A., Tilly, B.C., Houtsmuller, A.B., Ehlert, E.M., Bot, A.G.M., Edixhoven, M., Boomaars, W.E.M., Lohmann, S.M., De Jonge, H.R., 1998. Membrane targeting of cGMP-dependent protein kinase is required for cystic fibrosis transmembrane conductance regulator Cl⁻ channel activation. Proc. Natl Acad. Sci. USA 95, 1466–1471.

[157] Hallows, K.R., Raghuram, V., Kemp, B.E., Witters, L.A., Foskett, J.K., 2000. Inhibition of cystic fibrosis transmembrane conductance regulator by novel interaction with the metabolic sensor AMP-activated protein kinase. J. Clin. Invest. 105, 1711–1721.

[158] Stutts, M.J., Canessa, C.M., Olsen, J.C., Hamrick, M., Cohn, J.A., Rossier, B.C., Boucher, R.C., 1995. CFTR as a cAMP-dependent regulator of sodium channels. Science 269, 847–850.

[159] Knowles, M., Gatzy, J., Boucher, R., 1983. Relative ion permeability of normal and cystic fibrosis nasal epithelium. J. Clin. Invest. 71, 1410–1417.

[160] Mall, M., Hipper, A., Greger, R., Kunzelmann, K., 1998. Wild type but not deltaF508 CFTR inhibits Na⁺ conductance when coexpressed in *Xenopus* oocytes. FEBS Lett. 381, 47–52.

[161] Briel, M., Greger, R., Kunzelmann, K., 1998. Cl⁻ transport by cystic fibrosis transmembrane conductance regulator (CFTR) contributes to the inhibition of epithelial Na⁺ channels (ENaCs) in Xenopus oocytes coexpressing CFTR and ENaC. J. Physiol. 508, 825–836.

[162] Chabot, H., Vives, M.F., Dagenais, A., Grygorczyk, C., Berthiaume, Y., Grygorczyk, R., 1999. Downregulation of epithelial sodium channel (ENaC) by CFTR co-expressed in *Xenopus* oocytes is independent of Cl⁻ conductance. J. Membr. Biol. 169, 175–188.

[163] Ji, H.L., Chalfant, M.L., Jovov, B., Lockhart, J.P., Parker, S.B., Fuller, C.M., Stanton, B.A., Benos, D.J., 2000. The cytosolic termini of the β- and γ-ENaC subunits are involved in the functional interactions between CFTR and ENaC. J. Biol. Chem. 275, 27947–27956.

[164] Jiang, Q., Li, J., Dubroff, R., Ahn, Y.J., Foskett, J.K., Engelhardt, J., Kleyman, T.R., 2000. Epithelial sodium channels regulate cystic fibrosis transmembrane conductance regulator chloride channels in *Xenopus* oocytes. J. Biol. Chem. 275, 13266–13274.

[165] Reddy, M.M., Light, M.J., Quinton, P.M., 1999. Activation of the epithelial Na⁺ channel (ENaC) requires CFTR Cl⁻ channel function. Nature 402, 301–304.

[166] Nagel, G., Szellas, T., Riordan, J.R., Friedrich, T., Hartung, K., 2001. Non-specific activation of the epithelial sodium channel by the CFTR chloride channel. EMBO Rep. 21, 249–254.

[167] Egan, M., Flotte, T., Afione, S., Solow, R., Zeitlin, P.L., Carter, B.J., Guggino, W.B., 1992. Defective regulation of outwardly rectifying Cl⁻ channels by protein kinase A corrected by insertion of CFTR. Nature 358, 581–584.

[168] McNicholas, C.M., Guggino, W.B., Schwiebert, E.M., Hebert, S.C., Giebisch, G., Egan, M.E., 1996. Sensitivity of a renal K⁺ channel (ROMK2) to the inhibitory sulfonylurea compound glibenclamide is enhanced by coexpression with the ATP-binding cassette transporter cystic fibrosis transmembrane regulator. Proc. Natl Acad. Sci. USA 93(15), 8083–8088.

[169] McNicholas, C.M., Nason, M.W. Jr, Guggino, W.B., Schwiebert, E.M., Hebert, S.C., Giebisch, G., Egan, M.E., 1997. A functional CFTR-NBF1 is required for ROMK2-CFTR interaction. Am. J. Physiol. Renal Physiol. 273, F843–F848.

[170] Schreiber, R., Nitschke, R., Greger, R., Kunzelmann, K., 1999. The cystic fibrosis transmembrane conductance regulator activates aquaporin 3 in airway epithelial cells. J. Biol. Chem. 274, 11811–11816.

[171] Vennekens, R., Trouet, D., Vankeerberghen, A., Voets, T., Cuppens, H., Eggermont, J., Cassiman, J.J., Droogmans, G., Nilius, B., 1999. Inhibition of volume-regulated anion channels by expression of the cystic fibrosis transmembrane conductance regulator. J. Physiol. (Lond.) 515(1), 75–85.

[172] Wei, L., Vankeerberghen, A., Cuppens, H., Eggermont, J., Cassiman, J.J., Droogmans, G., Nilius, B., 1999. Interaction between calcium-activated chloride channels and the cystic fibrosis transmembrane conductance regulator. Pflugers Arch. 438, 635–641.

[173] Pasyk, E.A., Foskett, J.K., 1997. Cystic fibrosis transmembrane conductance regulator-associated ATP and adenosine 3′-phosphate 5′-phosphosulfate channels in endoplasmic reticulum and plasma membranes. J. Biol. Chem. 272, 7746–7751.

[174] Sugita, M., Yue, Y., Foskett, J.K., 1998. CFTR Cl⁻ channel and CFTR-associated ATP channel: distinct pores regulated by common gates. EMBO J. 17, 898–908.

[175] Reinlib, L., Jefferson, D.J., Marini, F.C., Donowitz, M., 1992. Abnormal secretagogue-induced intracellular free Ca²⁺ regulation in cystic fibrosis nasal epithelial cells. Proc. Natl Acad. Sci. USA 89, 2955–2959.

[176] Grubman, S.A., Perrone, R.D., Lee, D.W., et al., 1997. Regulation of chloride/bicarbonate exchanger activity by wild-type and mutant CFTR. Pediatr. Pulmonol.(Suppl.14), 277 (Abstract).

[177] Schwiebert, E.M., Benos, D.J., Egan, M.E., Stutts, M.J., Guggino, W.B., 1999. CFTR is a conductance regulator as well as a chloride channel. Physiol. Rev. 79, S145–S166.

[178] Lee, M.G., Wigley, W.C., Zeng, W., Noel, L.E., Marino, C.R., Thomas, P.J., Muallem, S., 1999. Regulation of the Cl⁻/HCO₃⁻ exchange by cystic fibrosis transmembrane conductance regulator expressed in NIH 3T3 and HEK 293 cells. J. Biol. Chem. 274, 3414–3421.

[179] Choi, J.Y., Muallem, D., Kiselyov, K., Lee, M.G., Thomas, P.J., Muallem, S., 2001. Aberrant CFTR-dependent HCO₃⁻ transport in mutations associated with cystic fibrosis. Nature 410, 94–97.

Molecular insights into acetylcholine receptor structure and function revealed by mutations causing congenital myasthenic syndromes

Steven M. Sine,[a],[*] Andrew G. Engel,[b] Hai-Long Wang[a] and Kinji Ohno[b]

[a]Receptor Biology Laboratory, Department of Physiology and Biophysics, Mayo Foundation,
Rochester, MN 55905, USA
[b]Muscle Research Laboratory, Department of Neurology, Mayo Foundation,
Rochester, MN 55905, USA
[*]Corresponding author

Congenital myasthenic syndromes (CMS) are neither new nor uncommon disorders. In 1937, Rothbart [1] described four brothers under the age of 2 years suffering from a myasthenic disorder, and by 1972 Sarah Bundey [2] collected 97 familial cases of myasthenia with onset before the age of 2. In the 1970s, myasthenia gravis was traced to an autoimmune response to the muscle nicotinic acetylcholine receptor (AChR), and a decade later the cause of the Lambert–Eaton myasthenic syndrome was shown to be autoimmunity to the voltage-gated calcium channel of the motor nerve terminal. These findings implied that myasthenic disorders occurring in a familial or congenital setting have a different etiology [3]. Subsequent ultrastructural, cytochemical, and in vitro microelectrode studies of congenital myasthenic patients revealed a number of distinct syndromes: a presynaptic disorder associated with a paucity of synaptic vesicles and decreased quantal release by nerve impulse [4]; a presynaptic disease attributed to a defect in the resynthesis or vesicular packaging of acetylcholine (ACh) [5,6]; synaptic acetylcholinesterase (AChE) deficiency [7]; postsynaptic AChR deficiency [8,9]; the slow-channel syndrome, attributed to prolonged activation episodes of the AChR [10]; and a fast-channel syndrome attributed to a defect in AChR activation by ACh [11].

The above discoveries pointed to defects in endplate (EP)-specific proteins, but analysis at the molecular level was hindered by lack of knowledge of their primary sequences in humans. Three further developments paved the way for molecular analysis. First, by the early 1990s, the cDNA sequences of the human AChR subunits [12–15], the catalytic subunit of AChE [16,17], and of choline acetyltransferase (ChAT) [18] were determined; by 1998, the cDNA sequence and genomic structure of the collagenic tail subunit of AChE was reported [19], and the cholinergic gene locus for the vesicular ACh transporter and ChAT was identified [20]; and by 2000, the entire genomic structure of human ChAT

Advances in Molecular and Cell Biology, Vol. 32, pages 95–119
© 2004 Elsevier B.V. All rights of reproduction in any form reserved.
ISSN: 1569-2558/DOI: 10.1016/S1569-2558(03)32005-3

was described [21]. Second, in the early 1990s, Milone and coworkers [22] succeeded in patch-clamping EPs in human intercostal muscles to allow resolution of single channel currents through EP AChRs. Third, the advent of mammalian expression systems allowed detailed analysis of engineered mutant AChRs, AChEs, and ChATs yielding further insights into the mechanisms that compromise the safety margin of neuromuscular transmission. As a result of these advances, the cause of the CMS attributed to defective resynthesis or vesicular packaging of ACh was traced to missense mutations of ChAT that reduce the enzyme's catalytic efficiency [21]; EP AChE deficiency was traced to mutations in the collagenic tail subunit of AChE that prevent the anchorage of the catalytic subunit of AChE in the synaptic basal lamina [19]; the slow-channel syndrome is now known to be due to gain-of-function mutations in different AChR subunits that alter affinity for ACh, gating kinetics, or both [23–29]; and the fast-channel CMS attributed to defective activation of AChR by ACh stems from state-specific reduced affinity of AChR for ACh caused by a mutation in the AChR ε subunit [30].

1. Investigation of the CMS

The investigation of CMS proceeds from a generic clinical diagnosis to defining the morphologic and electrophysiologic phenotype. If these studies point to a candidate gene whose sequence is known, then mutation analysis becomes feasible. Further, if a mutation in a relevant protein is identified, then appropriate expression studies are designed [31] (Table 1).

A generic clinical diagnosis of a CMS is based on the history of myasthenic symptoms from birth or early childhood, similarly affected relatives, a decremental electromyographic (EMG) response of the compound muscle fiber action potential on low-frequency (2–3 Hz) stimulation, and negative test for anti-AChR antibodies. Some CMS cases, however, are sporadic or present in later life, and the decremental EMG response may not be present in all muscles or at all times.

Conventional microelectrode studies on muscle specimens excised from origin to insertion readily reveal whether the defect of neuromuscular transmission is presynaptic, synaptic, or postsynaptic, and define the factors that impair the safety margin of neuromuscular transmission. These studies determine the amplitude of the miniature EP potential (MEPP), the amplitude and time course of the miniature EP current (MEPC), the amplitude of the EP potential (EPP), the number of quanta released by nerve impulse (m), the number of readily releasable quanta (n), and the probability of quantal release (p). The amplitude of the MEPP depends on the number of ACh molecules per synaptic vesicle, the geometry of the synaptic space, the functional state of AChE in the synaptic space, the density of AChR on the junctional folds, the single channel potential amplitude and the input resistance of the muscle fiber. The amplitude of the MEPC is independent of the input resistance of the muscle fiber, but otherwise is determined by the same factors that determine the amplitude of the MEPP. With AChE normally active at the EP, the decay time constant of the MEPC approximates the duration of the activation episodes of the AChR. The amplitude of the EPP is a function of the amplitude of the MEPP and the number of transmitter quanta released by nerve impulse (m).

Table 1
Investigation of congenital myasthenic syndromes

Clinical data
  History, examination, response to AChE inhibitor
  EMG: conventional needle EMG, repetitive stimulation, SFEMG
  Serologic tests (AChR antibodies, tests for botulism)

Morphologic studies
  Routine histochemical studies
  Cytochemical and immunocytochemical localization of AChE, AChR, agrin, β_2-laminin, utrophin, and rapsyn
    at the EP
  Estimate of the size, shape, and configuration of AChE-reactive EPs or EP regions on teased muscle fibers
  Quantitative electron microscopy; electron cytochemistry

Endplate-specific ^{125}I-α-bungarotoxin binding sites

In vitro electrophysiology studies
  Conventional microelectrode studies: MEPP, MEPC, evoked quantal release (*m*, *n*, *p*)
  Noise analysis: channel kinetics
  Single-channel patch-clamp recordings: channel types and kinetics

Molecular genetic studies
  Mutation analysis (if candidate gene or protein identified)
  Linkage analysis (if no candidate gene or protein recognized)
  Expression studies (if mutation identified)

AChE = acetylcholinesterase; AChR = acetylcholine receptor; α-bgt = α-bungarotoxin; EP = endplate; EMG = electromyography; MAC = C5b–9 complement membrane attack complex; MEEP = miniature endplate potential; MEPC = miniature endplate current; m = number of ACh quanta released by nerve impulse; n = number of readily releasable ACh quanta; p = probability of quantal release; SFEMG = single fiber EMG.

The value of *m*, in turn, is a function of the probability of release (*p*) and the number of readily releasable quanta (*n*). When the EPP amplitude exceeds the activation threshold (V_T) of the perijunctional Na channels, the EPP initiates a muscle fiber action potential. *The safety margin of neuromuscular transmission is defined as* EPP-V_T. *In each CMS, the safety margin of transmission is compromised by one or more specific mechanisms.*

Ultrastructural analysis of the EP defines the size of the nerve terminal, and the numerical density and dimensions of synaptic vesicles; it can also reveal destructive changes involving the postsynaptic region resulting from cholinergic over activity. Other EP studies involve determining the number of AChRs per EP with ^{125}I-α-bungarotoxin, and single-channel patch-clamp recording from EP AChRs. These studies establish whether there is a deficiency or kinetic abnormality of AChR.

2. Congenital myasthenic syndromes caused by defects in the acetylcholine receptor

All postsynaptic CMS identified to date have been traced to a deficiency, a kinetic abnormality, or both, of AChR.

2.1. Deficiency of AChR with or without minor kinetic abnormality

Severe EP AChR deficiency results from different types of recessive mutations in AChR subunit genes. The mutations are either homozygous or, more frequently, heterozygous; most reside in the ε subunit, whereas fewer reside in α or β subunits. Different types of recessive mutations causing severe EP AChR deficiency have been identified:

(1) Mutations causing premature termination of the translational chain. These mutations are frameshifting [32–36], occur at a splice site [32,36,37], or produce a stop codon directly [33].
(2) Missense mutation in a signal peptide region (εG-8R) [30].
(3) Missense mutations in residues essential for assembly of the pentameric receptor. Mutations of this type were observed in the ε subunit at an N-glycosylation site (εS143L) [30]; in cysteine 128 (εC128S), which forms the essential C128–C142 disulfide loop in the extracellular domain [38]. In the β subunit, a 3-codon deletion (β426delEQE) in the long cytoplasmic loop impairs AChR assembly by disrupting a specific interaction between β and δ subunits [39].
(4) Missense mutations affecting both AChR expression and kinetics. For example, εR311W in the long cytoplasmic loop between M3 and M4 decreases, whereas εP245L in the M1 domain increases the open duration of channel events [33]. In the case of εR311W and εP245L, the kinetic consequences are modest and are likely overshadowed by the reduced expression of the mutant gene.

EP AChR deficiency could also result from defects in neurally secreted AChR-inducing activity (ARIA) or its postsynaptic receptors that govern AChR transcription by junctional nuclei, or from mutations in neurally secreted agrin and its postsynaptic receptors and effectors that regulate AChR aggregation at the junction; CMS-causing mutations in these proteins have not been detected to date.

2.2. Kinetic abnormality of AChR with or without AChR deficiency

Since 1995, two major kinetic syndromes stemming from mutations in AChR have emerged: the slow-channel syndrome [23–29]; and the fast-channel syndrome [30,38, 40]. The slow-channel syndrome is caused by gain-of-function dominant mutations that increase the synaptic response to ACh; the fast-channel syndrome is due to recessive, loss-of-function mutations that decrease the synaptic response to ACh. The two syndromes are physiological opposites, have different effects on endplate morphology, and respond to different modalities of therapy (Fig. 1; Table 2). It is also noteworthy that choline, present in serum at $10-20\ \mu M$, has been reported to activate the AChR channel [82], and this mechanism may contribute to the cationic overloading of the postsynaptic region and the EP myopathy in slow-channel CMS. The fast-channel syndrome responds to measures that increase the synaptic response to ACh, namely 3,4-diaminopyridine, which increases the number of quanta released by nerve impulse,

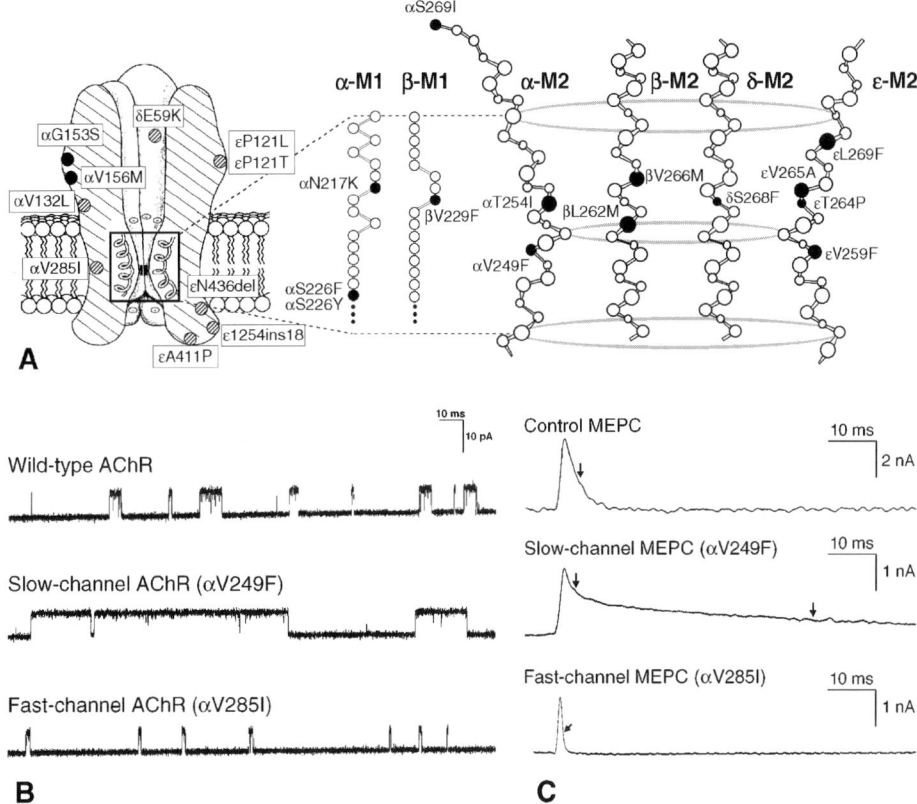

Fig. 1. (A) Schematic diagram of 14 slow-channel (solid circles) and 7 fast-channel (shaded circles) mutations. The drawing on the left shows a section through the acetylcholine receptor with two slow-channel mutations, αG153S and αV156M, in the extracellular domain near the ACh binding site of the α subunit, and 3 fast channel mutations, εP121L, εP121T and δE59K near the ACh binding site of the ε or δ subunits, αV132L in the disulfide loop just outside the membrane, αV285I in the M3 domain of the α subunit, and ε1254ins18 and εA411P in the long-cytoplasmic loop of the ε subunit. In the drawing on the right, dotted lines delimit transmembrane domains. Slow-channel mutations appear in the M2 domains of the α, β, δ and ε subunits, and in the M1 domain of the α and β subunits. The αS269I slow-channel mutation above the dotted line is in the extracellular M2–M3 linker. (B) Examples of single channel currents from wild-type, slow-channel (αV249F), and fast-channel (αV285I) AChRs expressed in HEK cells. (C) Miniature endplate currents (MEPC) recorded from endplates of a control subject, a patient harboring the αV249F slow-channel mutation, and a patient harboring the αV285I fast-channel mutation. Arrows indicate decay time constants. The slow-channel MEPC decays biexponentially due to expression of both wild-type and mutant AChRs at the EP, with one decay time constant that is normal and one that is markedly prolonged.

and cholinesterase inhibitors, which increase the number of AChRs activated by each quantum [30]. Conversely, the slow-channel syndrome is improved by decreasing the synaptic response: quinidine, a long-lived open-channel blocker of AChR, normalizes the duration of channel opening events at concentrations attainable in clinical practice [41,83] (see Fig. 2).

Table 2
Kinetic abnormalities of AChR

	Slow-channel syndromes	Fast-channel syndromes
Endplate currents	Slow decay	Fast decay
Channel opening events	Prolonged	Brief
Open states	Stabilized	Destabilized
Closed states	Destabilized	Normal
Mechanisms[a]	Increased affinity	Reduced ACh affinity for open state
	Increased β	Decreased β
	Decreased α	Increased α
		Mode-switching kinetics
Endplate morphology	Endplate myopathy[b]	No abnormality
Therapy	Long-lived open-channel blockade by quinidine[c]	AChE inhibitor drugs and 3,4-diaminopyridine[d]

β = channel opening rate; α = channel closing rate.
[a]Different combinations of mechanisms operate in the individual slow- and fast-channel syndromes.
[b]Cationic overloading of the postsynaptic region results in destruction of the junctional folds with loss of AChR, widening of the synaptic space, and degeneration of muscle fiber organelles near the endplate.
[c]Micromolar concentrations of quinidine, attainable in clinical practice, normalize the duration of slow-channel activation episodes and improve the disease [41,42].
[d]3,4-Diaminopyridine increases the number of quanta released by nerve impulse and AChE inhibitors increase the number of AChR activated by each quantum.

3. Mechanistic implications of AChR mutations

Mechanistic studies of AChR mutations that cause CMS have provided unique insights relating the structure of the AChR to the channel activation process. This insight emerges because nature, in causing the CMS, selects residues vitally important for AChR activation, the very process researchers are working to understand. Also, by modifying discrete reaction steps, the mutations delineate how each step contributes to endplate potential time course and amplitude, parameters that directly impact safety margin of neuromuscular transmission. Finally, results of the mechanistic studies lay a sound foundation for treatment of the CMS.

4. Overview of AChR structure

The muscle AChR is a pentamer of homologous subunits with compositions $\alpha_2\beta\gamma\delta$ in embryonic and denervated muscles and $\alpha_2\beta\epsilon\delta$ in adult muscles. The subunits are arranged as barrel staves around a central pore, with the counter clockwise order of subunits $\alpha\gamma\alpha\delta\beta$ or $\alpha\epsilon\alpha\delta\beta$ when viewed from the synaptic cleft. This order of subunits was deduced by site-directed labeling and mutagenesis studies [43–45], and was confirmed by the crystal structure of an acetylcholine binding protein (AChBP) homologous to the major extracellular domain of the AChR [81]. In two-dimensional crystals of *Torpedo* AChR, electron density resolved at 9 Å reveals symmetrical profiles in each of the five subunits relative to the central axis of the pentamer [47], consistent with evolution of the subunits

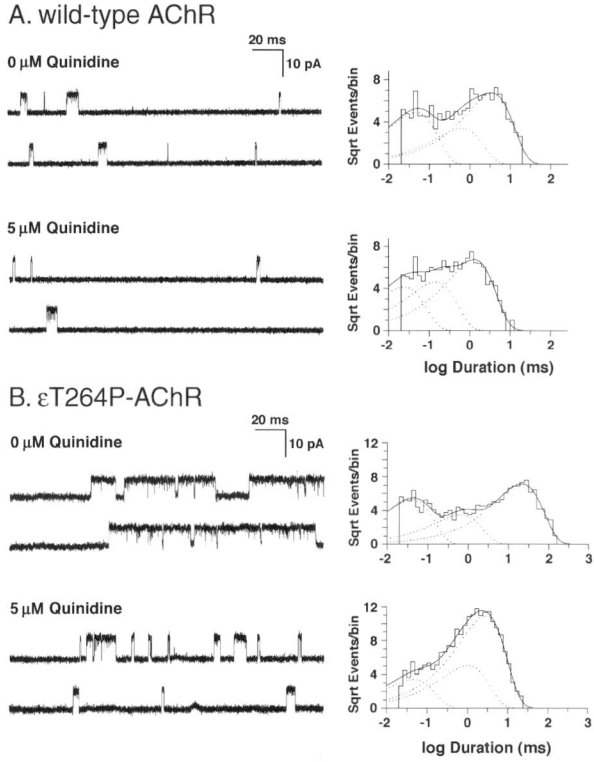

Fig. 2. The effect of 5 μM quinidine on channel events of wild-type- (A, left) and εT264P-AChR (B, left). The abnormally long bursts characteristic of the slow-channel mutant become much shorter in the presence of 5 μM quinidine. The burst duration histograms (A and B right) shows three components of bursts for both types of receptors. Note that quinidine shifts the duration of the longest burst component (τ_{bs}) to the left and that the shift is greater for mutant (B, right) than wild-type (A, right) AChR. ACh concentration was 50 nM for wild-type and 250 nM for mutant AChR. Membrane potential; −80 mV; temperature; 22 °C. Channel openings appear as upward deflections.

from a common ancestor, and equivalent positioning of the γ, ε and δ subunits with respect to the α subunit [45,48–50]. Approximately half of each subunit is extracellular, where the ACh binding sites are located. The remainder forms transmembrane domains (TMD) 1 through 4, plus a large cytoplasmic domain between TMD3 and TMD4.

Unwin and colleagues recently resolved electron density in two-dimensional crystals of *Torpedo* AChR to a resolution of 4.6 Å [51]. The extracellular domains of each subunit appear as crescents of predominantly β sheet, running normal to the membrane, and the five crescents align to form a turbine-like structure. The crystal structure of AChBP confirms that the extracellular domain is composed predominantly of β-sheet, and reveals the location of the ACh binding sites near the periphery of the oligomer where ACh has ready access [81]. Upon rapid application of ACh, contours of the extracellular domains change and electron densities within them tilt, suggesting rearrangement of secondary structural elements accompanying AChR activation [47].

At the level of the membrane, the electron density reveals five rods typical of α-helices that twist upon activation to allow flow of permeant ions. These rods correspond to TMD2, with possible contributions from TMD1, based on substituted cysteine accessibility, mutagenesis and functional measurements. The collective studies indicate that the extracellular half of TMD2 forms an α-helix, the middle three residues form an extended structure, and the remainder is β-strand that forms the channel gate and ion selectivity filter near the cytoplasmic limit of TMD2 [52–54]. Electron densities corresponding to TMD3 and TMD4 have not been assigned, but the long cytoplasmic domain between them likely contributes to a multi-subunit assembly that resembles a fenestrated basket-like structure extending into the cytoplasm [51].

The two ACh binding sites are formed at interfaces between subunits: α–γ and α–δ in embryonic and denervated muscles, and α–ε and α–δ in adult muscles. Assuming each subunit positions equivalently within the pentamer, this pairing of subunits provided evidence for the subunit order, αγαδβ (or αεαδβ), which was recently confirmed by the AChBP structure [81]. Residues from both α and non-α subunits contribute profoundly to binding of ACh and competitive antagonists, indicating that the ACh binding sites are formed at interfaces between subunits [53,55,56]. Probing with ligands that select between the two binding sites, mutagenesis studies reveal seven linearly distinct loops among α and non-α subunits that contribute to the site. Site-directed labeling studies physically localized all but one of the seven loops to the binding site, further supporting the multi-loop hypothesis [57–60]. Loops from opposing α and non-α subunits come into exceptionally close proximity, as revealed by analysis of pair-wise interactions between peptide toxins and binding site residues [61,62]. The subunit interface and multiloop hypotheses were confirmed in elegant detail by the AChBP structure [81].

CMS mutations have been found in all AChR subunits and in several domains of the subunits, including the major extracellular domain contributing to the ACh binding site, TMDs 1 through 3 and the long cytoplasmic domain between TMDs 3 and 4 (Fig. 1A) [63]. Despite this diversity of pathogenic targets, kinetic consequences of CMS mutations fall into two broad categories: increasing response to ACh in slow channel syndromes or decreasing response in fast channel syndromes.

5. Mechanistic description of AChR activation

Molecular schemes provide a mechanistic framework for understanding consequences of CMS mutations of the AChR. The simplest possible scheme incorporates ACh binding and channel gating into a single step,

$$A + R \underset{k-}{\overset{k+}{\rightleftharpoons}} AR^* \qquad \text{(Scheme 1)}$$

where A is the agonist, R is the resting state and AR* is the open channel state. Inherent in Scheme 1 and its expanded forms is the existence of stable ground states, R and AR*, which represent deep wells in the energy landscape. The concept of a small number of stable ground states is supported by the observation that single channel dwell times distribute into a small number of exponential components. Implicit in Scheme 1 is a transition state between the R and AR* ground states, which corresponds to a saddle point in the energy landscape whose height is determined by the rate constants $k+$ and $k-$. Scheme 1 is too simple, however, as it does not account for full versus partial agonism, nor for the brief closed intervals during a single agonist occupancy. These features are readily explained by separating the agonist binding from the channel gating steps,

$$A + R \underset{k-}{\overset{k+}{\rightleftharpoons}} AR \underset{\alpha}{\overset{\beta}{\rightleftharpoons}} AR^* \qquad \text{(Scheme 2)}$$

where AR is the agonist-occupied resting state. Existence of two agonist binding sites per AChR, together with positive cooperativity in the ACh dose–response relationship, requires a second binding step followed by a gating step:

$$A + R \underset{k_{-1}}{\overset{k_{+1}}{\rightleftharpoons}} AR + A \underset{k_{-2}}{\overset{k_{+2}}{\rightleftharpoons}} A_2R \underset{\alpha}{\overset{\beta}{\rightleftharpoons}} A_2R^* \qquad \text{(Scheme 3)}$$

By designating distinct agonist association and dissociation rate constants, Scheme 3 allows the possibility of non-equivalent binding sites, as expected from different α–ε and α–δ binding site interfaces. Scheme 3 accounts for all essential features of activation for AChRs from a variety of species [64,65], and represents the starting point for analyzing the kinetics of mutant AChRs [27,30,40,66]. Scheme 3 is a subset of the classical Monod–Wyman–Changeux (MWC) allosteric description of oligomeric protein function [67],

$$
\begin{array}{ccccc}
R^* + A & \overset{K_1^*}{\rightleftharpoons} & AR^* + A & \overset{K_2^*}{\rightleftharpoons} & A_2R^* \\
\theta_0 \updownarrow & & \theta_1 \updownarrow & & \theta_2 \updownarrow \\
R + A & \underset{K_1}{\rightleftharpoons} & AR + A & \underset{K_2}{\rightleftharpoons} & A_2R
\end{array}
\qquad \text{(Scheme 4)}
$$

which contains resting (R) and active (R*) states that interconvert in the absence of agonist, and tighter binding of agonist to the active over the resting state (i.e. the equilibrium dissociation constants K^* are much smaller than K). Given the tighter binding to the active state, the equilibrium constant θ between resting and active states increases progressively with increasing agonist occupancy such that $\theta_2 \gg \theta_1 \gg \theta_0$; for muscle AChR, θ_0 is approximately 10^{-6}, θ_1 is approximately 10^{-2} and θ_2 ranges from 25 to 100. Thus the essential task of agonist is to bind more tightly to the active than to the resting state to overcome the unfavorable equilibrium constant θ_0 [68].

6. CMS mutations in TMD2

The first CMS mutation discovered in humans was described in a collaborative effort between the Engel and Sine laboratories [23]. That first case arose from a threonine to proline mutation in TMD2 of the ε subunit. Since that first case, additional mutations in TMD2 causing CMS have been described. Mutations have been found in α, β and ε subunits, and span the TMD2s from just inside their extracellular border to just beyond the three extended hydrophobic residues in their middle (Fig. 1A). When mapped on the current consensus model of TMD2 based on mutagenesis and cysteine accessibility studies [69], all but one mutation (αV249F) localize to the putative α-helical portion of the domain.

6.1. Functional consequences of mutations in TMD2

The most conspicuous effect of TMD2 mutations is a dramatic increase in the duration of ACh-induced openings. Mean open durations increase from 0.5 ms for wild-type AChR to approximately 50 ms for mutations in TMD2 [23,25,26]. Also conspicuous are frequent channel openings in the absence of ACh [23,26], which occur only rarely in wild-type AChR. In addition to increasing open duration, the CMS mutation αV249F increases affinity of ACh for the resting state of the AChR [26]. A complete set of activation rate constants has not been determined for receptors with mutations in TMD2 due to difficulty in assigning the multiple closed time components to steps in a kinetic scheme [26]. Finally, mutations in TMD2 enhance desensitization of the AChR, as shown by increased ACh affinity following equilibrium exposure to agonist [25,26].

6.2. Mechanistic consequences of mutations in TMD2

Increased spontaneous openings due to mutations in TMD2 indicate an increase of θ_0, which describes the equilibrium between the resting and active states in the absence of ACh. A small value of θ_0 is vital for minimizing cation influx at rest and for maximizing the range of the response to ACh. This ACh-independent equilibrium is the starting point for drawing the MWC scheme for AChR activation (Scheme 4). The first ACh binding step thus begins with a mutant receptor already prone to opening, and the enhanced affinity of ACh for the active over the resting state further promotes opening of the mono-liganded receptor according to $\theta_1 = \theta_0 K_1/K_1^*$. Analogously, the increase of θ_1 propagates to the doubly liganded AChR, increasing the corresponding opening equilibrium constant θ_2 according to $\theta_2 = \theta_1 K_2/K_2^*$. Rates of channel opening are also increased, the most obvious being the ACh-independent opening rate β_0. The doubly-liganded opening rate β_2 also increases with mutations in TMD2 [26], although the extent of the increase is uncertain because the opening rate for wild type AChR approaches resolution limits of the patch clamp. The doubly liganded closing rate α_2 slows with mutations in TMD2 owing to stabilization of the open state relative to the transition state between the open and closed states. Finally, desensitization is enhanced to varying degrees for the different mutations in TMD2. A likely mechanistic explanation is increase of the equilibrium constant governing desensitization in the absence of ACh, analogous to that just described for the channel opening process; enhanced affinity of ACh for the desensitized over the resting state promotes desensitization of singly- and doubly-liganded receptors. Thus for mutations in TMD2, channel activation is enhanced upon short-term exposure to ACh, whereas desensitization is enhanced upon long-term exposure.

6.3. Interpretation and implications of mutations in TMD2

Because TMD2 is well established to form the ion channel, functional consequences of mutating this domain likely result from direct perturbation of the gating apparatus. Although the gating apparatus comprises TMDs from all five subunits, and gating comprises synchronous movement of all five, TMD2 mutations in individual subunits appear independent of the other subunits [70]. The overall results of mutating TMD2 indicate that its structure is essential for stabilizing the channel in the closed state relative to the active and desensitized states; closed state stability prevents unwanted cation influx at rest, while optimizing the fraction of activatable receptors. The structure of TMD2 is also vital for tuning the stability of the open relative to the transition state to provide millisecond EPC decay times. Because TMD2 couples tightly to the ACh binding site, mutations in TMD2 that affect ACh binding affinity likely exert their actions through allosteric propagation of the initial perturbation in the channel via a coupling structure.

7. Contributions of other TMDs to channel gating

The four TMDs from each subunit are thought to fold in a complex of α-helix and β-sheet structures to form the gating machinery of each subunit [71]. Thus each TMD has the potential to affect functional counterparts of the ion permeation pathway, as just described for TMD2. TMD1 is physically closest to residues forming the ACh binding site, and being closely associated with TMD2, is a logical candidate for coupling ACh binding to gating. Although TMD4 forms the perimeter of the TMD complex, it contributes profoundly to channel gating kinetics [72]. The importance of TMD3, on the other hand, remained unknown until discovery and mechanistic analysis of the CMS mutation αV285I in the extracellular one third of this TMD [40].

7.1. Functional consequences of αV285I

By simply adding a methyl group, the mutation αV285I impairs the efficiency of channel gating [40]. The mutation slows the rate of channel opening and speeds the rate of channel closing, consequences opposite to those of TMD2 mutations. Agonist binding steps are largely unaffected by αV285I, indicating lack of allosteric effects at the remote binding site. Analysis of genetically engineered mutations of αV285 showed that effects on channel gating depend on the size of the side chain, with αV285L impairing gating beyond that of αV285I, and αV285A enhancing gating beyond that of the wild type (Fig. 3). Because the leucine side chain is the same size as that of isoleucine, stereochemical considerations were required to explain the effects on channel gating. Gating depended on the size of the moiety attached to the β-carbon, which is ethyl for isoleucine and isopropyl for leucine. Moreover, the volume of the β-carbon substitution altered the free energy of the channel gating equilibrium in a linear fashion, spanning ranges of 4 kcal/mol and 110 Å^3, indicating both stereochemical and volume contributions to channel gating. Mutations in equivalent positions of ε and β subunits affected gating similarly to those in

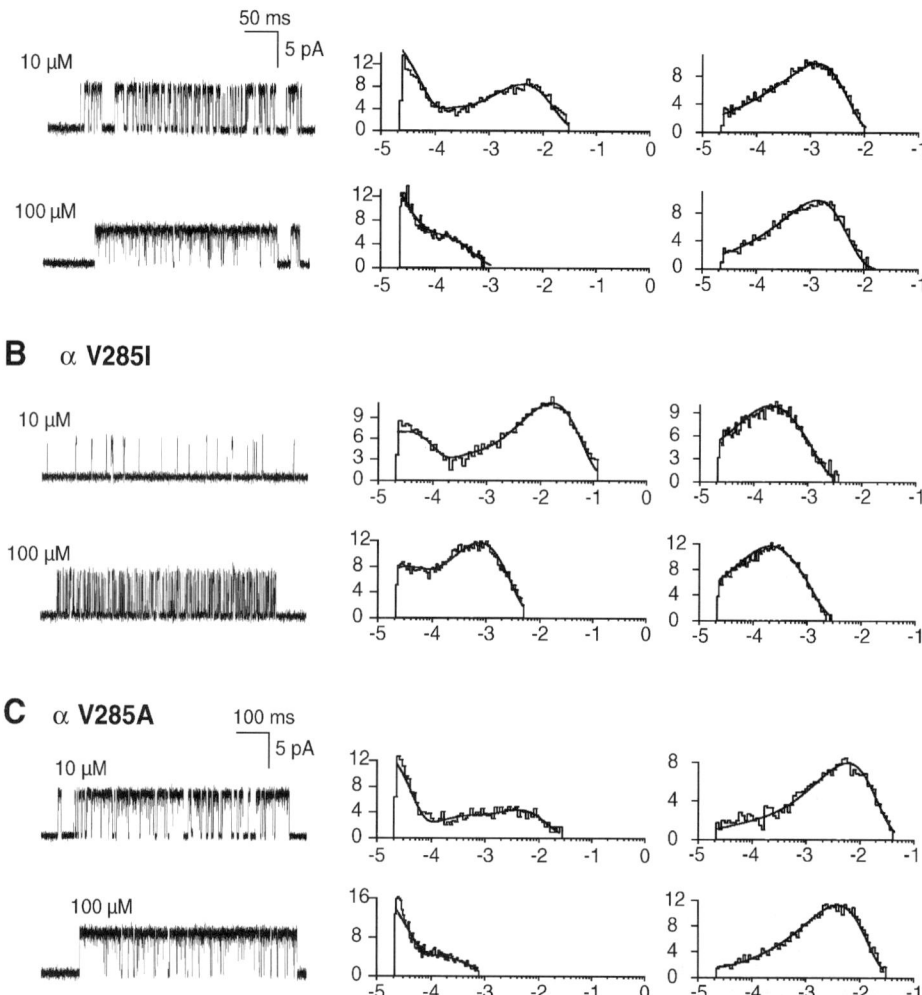

Fig. 3. Activation kinetics for receptors containing wild-type α (a), αV285I (b) and αV285A (c). Left column displays individual clusters of single channel currents recorded at the indicated ACh concentrations from HEK cells expressing adult human AChRs containing the indicated α subunits. Currents are displayed at a bandwidth of 10 kHz, with channel openings shown as upward deflections. Center and right columns display the corresponding closed and open duration histograms with fitted probability density functions superimposed. Fitted rate constants are given in Ref. [40]. Reproduced by permission from Nature Neuroscience.

the α subunit, while mutations in multiple subunits showed additive effects. However, mutations in the δ subunit were without effect, indicating differences in TMD3 contributions to gating between the δ and α, β and ε subunits. Finally, the αV285I mutation exhibited two open states even at saturating concentrations of ACh, unlike the wild type, which exhibited one open state.

7.2. Mechanistic consequences of the TMD3 mutation αV285I

Kinetic analysis of currents through AChRs containing αV285I provided a complete set of rate constants describing the activation process, as shown by fitting of a kinetic scheme to closed and open intervals (Fig. 3) [40]. The mutation impairs the channel gating step, most likely by destabilizing the doubly liganded open state. The possibility that ACh binds more tightly to the resting state cannot explain the impaired gating because ACh affinity for this state was only minimally affected and did not correlate with changes in gating produced by other side chains placed at αV285. The presence of two distinct open states in the mutant was described better by open states connected in series rather than open states branching from a common closed state, as follows.

$$A + R \underset{k_{-1}}{\overset{k_{+1}}{\rightleftharpoons}} AR + A \underset{k_{-2}}{\overset{k_{+2}}{\rightleftharpoons}} A_2R \underset{\alpha}{\overset{\beta}{\rightleftharpoons}} A_2R^* A_2R \underset{\alpha'}{\overset{\beta'}{\rightleftharpoons}} A_2R^{**} \qquad \text{(Scheme 5)}$$

Thus αV285 may unmask a metastable open state not normally detected in wild type AChRs; the free energy profile for channel opening in the mutant thus contains a distinct well, which in wild-type is much shallower and not detectable over the experimentally accessible bandwidth.

7.3. Interpretation and implications of mutations in TMD3

αV285 is located one third of the way along TMD3 from the extracellular side of the membrane. Hydrophobic photolabeling probes do not react with αV285, suggesting it faces the protein interior away from bulk lipid [73]. Mutating αV285 likely interferes with intermeshing of TMD3 in the α subunit with TMDs from the same or adjacent subunits, which in turn impairs gating, akin to a misshapen gear in a clock. The correlation between gating and side chain volume suggests TMD3 was fine-tuned by evolution to produce millisecond EPC decay times appropriate for neuromuscular transmission.

8. α and ε subunits contribute to the ACh binding site

The past decade saw a convergence of mutagenesis, chemical labeling and structural studies demonstrating that the ACh binding sites are formed at interfaces between α and non-α subunits [53–55,81,86]. Localization of the binding sites progressed from identifying key subunits, to identifying key regions in the subunits, to identifying key residues and finally to the atomic structure of the homologous AChBP. Structural modeling, based on scanning mutagenesis and the AChBP structure [86], provides precise positioning of residue side chains in the ligand binding sites of human AChR (Fig. 4). CMS mutations have been identified in key regions of both α and ε subunits, showing that both subunits are vital for correct function of the ACh binding site (Fig. 4).

8.1. Functional consequences of the CMS mutation αG153S

αG153S was the first CMS mutation discovered at the ACh binding site, and localized to one of three key regions of the α subunit that contribute to the site [24].

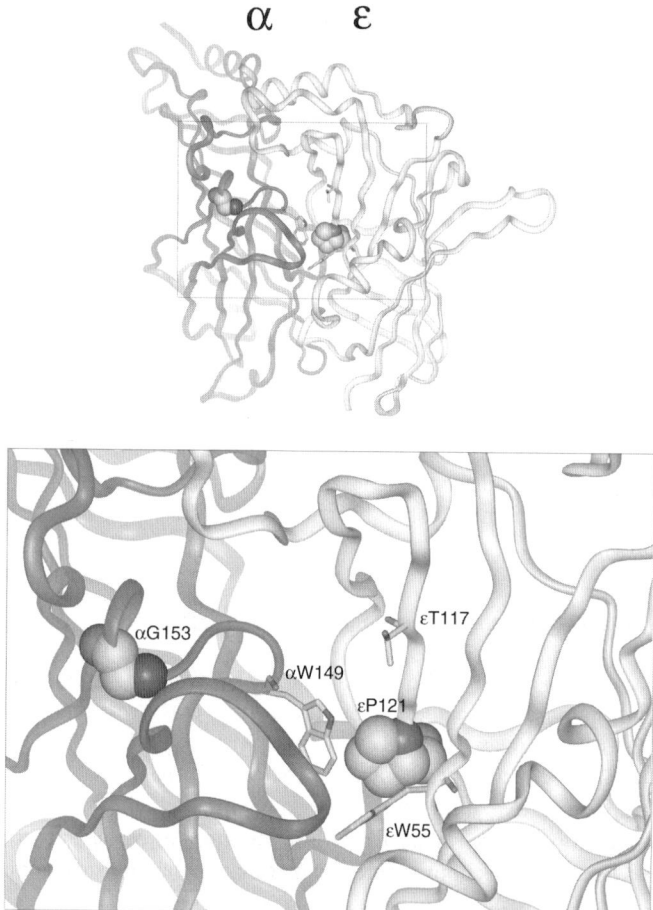

Fig. 4. Structural model of ligand binding regions of human α and ε subunits generated by scanning mutagenesis and homology modeling [86]. The α subunit portion of the ligand-binding site is shown in red ribbons, and the ε subunit portion is shown in yellow ribbons. Rendered in space-filling format are mutant residues in slow channel (αG153) and fast channel (εP121) congenital myasthenic syndromes. For orientation, key binding site residues (αW149, εW55 and εT117) are rendered in stick format. (*For a colored version of this figure, see plate section, page 461.*)

Flanking previously identified key aromatic residues αW149 and αY151 [74] (Fig. 4), αG153S greatly slowed the rate of ACh dissociation from the doubly liganded closed state of the AChR, increasing ACh affinity for this inactive state. In the mutant AChR, the doubly liganded closed state opened repeatedly during a single ACh occupancy because the rate of channel opening was some 46-fold greater than the competing rate of ACh dissociation, as opposed to 3.5-fold greater as in wild type AChR. αG153S also stabilized the open channel and desensitized states, suggesting increased ACh affinity for these functional states. Increased stability of the open state was suggested by slowing of the channel closing rate, while increased stability of the desensitized state was demonstrated by tighter binding of ACh to receptors desensitized by the local anesthetic proadifen. Thus αG153S enhances ACh affinity for resting, active and desensitized states of the AChR.

8.2. Mechanistic consequences of αG153S

By slowing the rate of ACh dissociation k_{-2} (Scheme 3), αG153S prolongs individual AChR activation episodes, known as bursts, according to $\tau_B = \tau_O(1 + \beta/k_{-2})$, where τ_B is mean burst duration and τ_O is mean open interval duration. Burst duration increases further due to the slower rate of channel closing α, which increases mean open duration according to $\tau_O = 1/\alpha$. Probability that a doubly liganded receptor will open increases according to $P = \beta/(\beta + k_{-2})$, predicting an increased peak response following instantaneous delivery of ACh. Desensitization is enhanced by αG153S, owing to tighter binding of ACh to this refractory state. Thus by increasing residence time of ACh at the binding site in all states of the AChR, αG153S produces a greater peak response, prolongs burst duration and enhances desensitization.

8.3. Functional consequences of the CMS mutation εP121L

A second mutation to affect ACh binding, εP121L, localizes to the opposite face of the ACh binding site (Fig. 4), within one of four key regions of the ε subunit that contribute to the site. Located at the C-terminal boundary of a series of residues that affect competitive antagonist binding [55,86], including εY111, εT117 and εD59, εP121L does not affect ACh affinity for the resting state of the AChR [30]. Instead, εP121L dramatically reduces probability of channel opening at saturating ACh concentrations. Accompanying this reduced opening probability, the rate of channel opening of the doubly liganded AChR slows by nearly 500-fold, while the rate of channel closing increases about 2-fold. Desensitization of single channel events, on the other hand, develops more slowly and to a lesser extent due to reduced ACh affinity for the α–ε site in the desensitized state. Reduced ACh affinity for the desensitized state results in nearly a steady stream of single channel events even at high concentrations of ACh. Thus εP121L impairs entry of the AChR into its two functional states, active and desensitized.

8.4. Mechanistic consequences of εP121L

Kinetic analysis of currents through AChRs containing εP121L provided a complete set of rate constants describing the activation process [30]. Thus the conclusion of normal ACh affinity for the resting state followed directly from the measured rate constants. Open state affinity was determined by detailed balancing, using the measured closed state affinity and gating steps for singly and doubly liganded receptors, in the following cycle derived from the MWC scheme for activation:

$$
\begin{array}{ccc}
 & K_2^* & \\
AR^* + A & \rightleftharpoons & A_2R^* \\
\theta_1 \big\Updownarrow & & \theta_2 \big\Updownarrow \\
R + A \rightleftharpoons AR + A & \rightleftharpoons & A_2R \\
\quad K_1 \qquad\qquad K_2 &
\end{array}
\qquad \text{(Scheme 6)}
$$

Open state affinity, given by $K_2^* = K_2 \theta_1 / \theta_2$, decreased from 35 nM for wild-type to 1.5 μM for εP121L [30]. Thus εP121 selectively stabilizes ACh bound to the open state of the AChR. Additionally, the dramatic slowing of the rate of opening of the doubly liganded AChR, β_2, could potentially result from stabilization of ACh bound to the resting state; however, resting state affinity was only slightly affected by εP121L. Instead, εP121 emerges as critical in forming the transition state in the path towards the open state. Impaired desensitization by εP121L is explained analogously to that just described for activation: loss of tight binding of ACh to the desensitized state allows the resting state to predominate even at high ACh concentrations. Thus the results from the εP121L mutation suggest that the proline is critical in contributing to a binding site structure that better complements ACh bound to open and desensitized states relative to the resting state. Results from these studies show that tighter binding of ACh to functional states is the fundamental driving force underlying agonist-induced activation and desensitization of the AChR.

8.5. Interpretation and implications of CMS mutations in α and ε subunits

Because αG153S and εP121L localize to well-established regions contributing to the ACh binding site (Fig. 4), their functional consequences result at least partly from direct perturbation of the binding site interface. However, for both the mutations the initial perturbation likely propagates beyond the altered side chain. Replacing the flexible glycine with serine at position αG153 may displace or alter mobility of the nearby aromatic residues αW149 and αY151, which are critical for ACh binding [55]. Similarly, for εP121L, loss of conformational restriction provided by the imino ring of proline may alter nearby residues upstream of the mutation that affect agonist and competitive antagonist binding (i.e. εT117), as well as alter the network of β-strands of which εP121 is part [81,86] (Fig. 4). Both αG153S and εP121L demonstrate that transition from resting to functional states is driven by tighter binding of ACh to the functional state, as embodied in the MWC scheme. Additionally, certain residues stabilize both ground and transition states (i.e. εP121L), whereas others stabilize primarily ground states (i.e. αG153S); residues stabilizing solely transition states remain to be discovered. Finally, the two mutations illustrate a divergence in functional contributions, as they define two classes of ACh-stabilizing residues: those that stabilize ACh in a state-independent manner (αG153S) and those that stabilize ACh in a state-dependent manner (εP121L).

9. Cytoplasmic domain between TMD3 and TMD4 affects channel gating

Since the initial cloning of AChR subunits, the long cytoplasmic domain between TMD3 and TMD4 has drawn considerable attention. Fourier analysis of residue hydrophobicity as a function of sequence strongly suggested an amphipathic α-helical secondary structure in the cytoplasmic domain [75]. More recent secondary structural analysis, based on multiple sequence alignments, divides the cytoplasmic domain into three predicted α-helices designated F, G and H, corresponding to ε subunit residues 320–333, 398–415, and 421–435, respectively [76]. The amphipathic composition initially suggested formation of the ion conductance pathway, but mutagenesis showed that the cytoplasmic domain could be deleted without loss of channel activity [77]. Clues about functional significance of the cytoplasmic domain emerged from studies investigating the structural basis of the fetal to adult kinetic switch, in which mean open duration decreases from about 10 to 1 ms, and AChRs containing the γ subunit are replaced by those containing the ε subunit. Multiple residues in the cytoplasmic domain of the ε and γ subunits were shown to account for about half of the kinetic switch, with the balance due to two residues in the neighboring TMD4 domain [78]. The overall results pointed to a specific role of the ε subunit in tuning the kinetics of the adult AChR, with functionally significant residues located in helices G and H of the cytoplasmic domain. Two CMS mutations further pointed to the cytoplasmic domain as important in governing the kinetics of AChR activation. The first case arose from tandem duplication of six residues, STRDQE, originating at codons 413–418 in helix G [38], and the second from the mutation εA411P just two residues upstream from the duplication [79].

9.1. Functional consequences of a six-residue duplication STRDQE in the ε subunit

The STRDQE duplication causes individual receptor channels to suddenly change kinetics, also known as mode-switching. Mode-switching is quite rare in wild-type AChR [79,84,85], but the fact that it occurs suggests that the STRDQE mutation amplifies a normal process. Mode-switches were readily observed during activation episodes elicited by high concentrations of ACh, which appeared as clusters of events in quick succession flanked by prolonged quiescent periods. Three distinct kinetic modes could be discerned within clusters, each with reduced probability of opening due to slower rates of channel opening and faster rates of channel closing. Also, kinetic analysis of the separated modes revealed two open states at saturating concentrations of ACh, rather than the single open state observed for wild type AChR. Thus studies of the STRDQE insertion provided the first hint that the cytoplasmic loop governs the uniformity of AChR gating kinetics.

9.2. Mechanistic consequences of the six residue insertion STRDQE in the ε subunit

The overall findings indicate at least three stable gating modes of the mutant AChR, which interconvert on a time scale that is long compared to individual gating events. This interconversion rate is similar to that of the desensitization process, which determines the duration of activation episodes elicited by high concentrations of ACh. Once a mutant receptor switched modes, however, the change was instantaneous, appearing in the span of one closed or open interval. The STRDQE duplication perturbs the energy landscape governing channel gating to either create additional wells or increase accessibility to pre-existing ones. Individual modes showed two distinct open states, with open states connected in series describing the data better than open states branching from a common closed state. Thus the additional open state unmasked by the mutation may correspond to a metastable open state in the wild type AChR, as suggested by the CMS mutation αV285I in TMD3. Thus the cytoplasmic domain emerges as a determinant of rates of opening and closing of the channel gate.

9.3. Functional consequences of the mutation εA411P

Unlike the STRDQE duplication beginning at codon 413 of the ε subunit, the nearby mutation εA411P does not increase frequency of mode switching within clusters, but instead causes individual clusters of activation episodes to span a wide spectrum of kinetics [79]. Current pulses through most individual mutant receptors appeared kinetically uniform, but each activation episode had a unique kinetic signature, spanning a wide range of open probability (Fig. 5). Hidden Markov modeling analysis of the kinetics of individual receptors revealed a Gaussian distribution for each rate constant in a kinetic description of receptor activation (i.e. Scheme 3). The distributions for agonist binding rate constants were unaltered by the mutation, but those for channel opening and closing steps showed remarkable broadening (Fig. 6). Proline mutations placed in positions flanking εA411 also produced a wide spectrum of kinetics similar to that produced by εA411P, whereas proline mutations placed in equivalent positions of β and δ subunits produced the usual narrow range of kinetics [79]. The possibility of folding heterogeniety of individual receptors was considered unlikely because in a few activation episodes several gating modes were readily detected. Thus the ε subunit again emerges as specific for governing the kinetics of

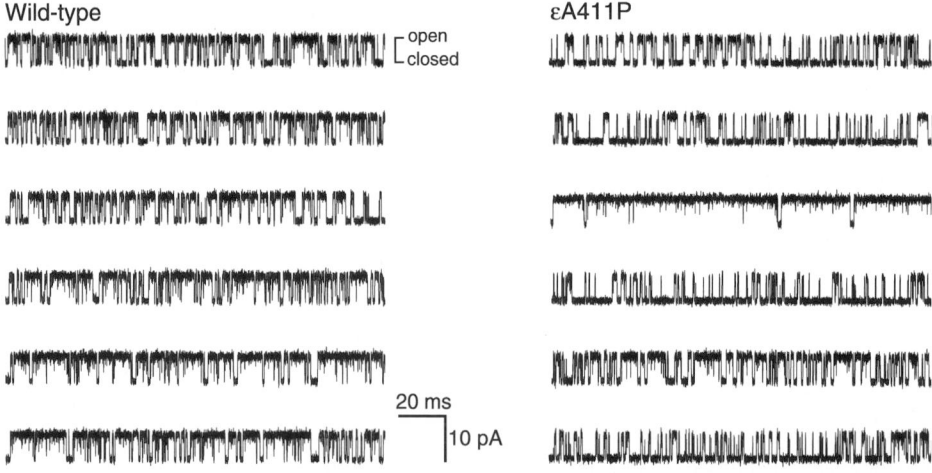

Fig. 5. Increased kinetic range for individual receptors caused by the εA411P mutation. Each trace is from a single cluster of activation episodes elicited by 30 μM ACh for wild-type or εA411P AChRs. The left-hand edge of each trace is the start of the cluster. Traces are displayed at 10 kHz bandwidth.

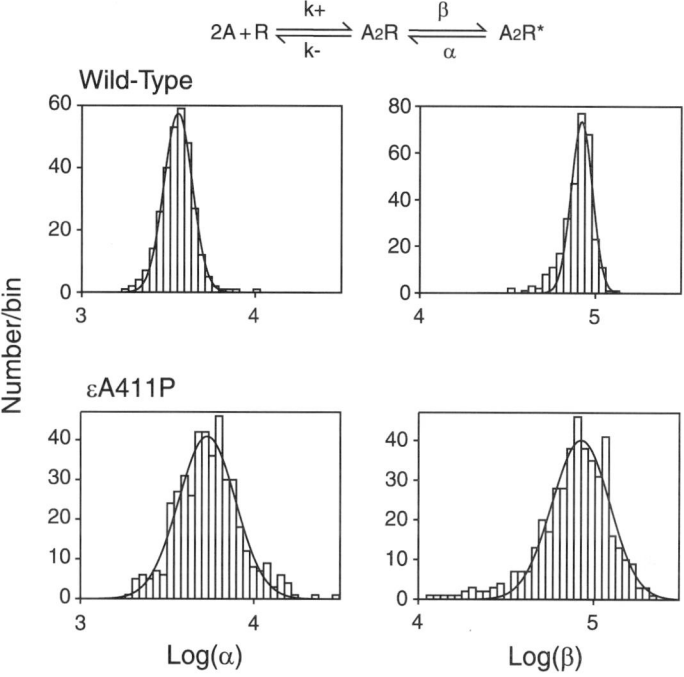

Fig. 6. Distribution of channel gating rate constants for individual activation episodes from wild-type of εA411P AChRs. For each rate constant histogram, one entry corresponds to the most likely rate constant for an individual episode of channel activity determined by hidden Markov modeling using the scheme at the top. The bell-shaped curves are fitted Gaussian distributions, with fitted parameters given in Ref. [79]. For each receptor type, data are from single patches. Note that in wild-type AChRs distributions of gating rate constants are much narrower than in εA411P AChRs.

channel gating where residues flanking εA411 maintain fidelity of the opening and closing rate constants.

9.4. Mechanistic consequences of the mutation εA411P

Analysis of the εA411P mutation provides a unifying explanation for the kinetic consequences of several structural changes in the cytoplasmic loop, including the fetal to adult kinetic switch, mode-switching by the STRDQE duplication and the wide spectrum of kinetics produced by εA411P. Each of these effects can be ascribed to a localized structural change in the cytoplasmic domain that affects global energetics governing channel gating. The observation that both wild-type and mutant AChRs shuttle among multiple stable states indicates that the energy landscape underlying gating of the AChR is corrugated [80]. Because the wild-type AChR gates in predominantly one mode, the corrugations superimpose upon a steep funnel-shaped foundation (Fig. 7). Receptors containing the εA411P mutation are subject to a similar corrugated energy landscape, but the corrugations superimpose on a much shallower

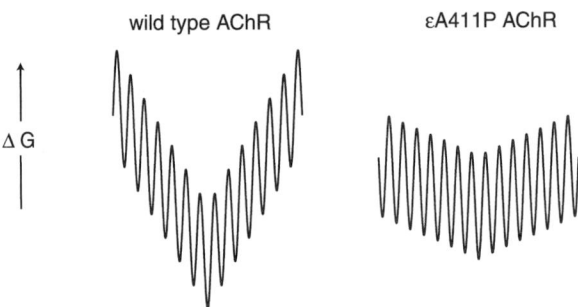

Fig. 7. Hypothetical free energy landscapes for channel gating. The profile for wild-type AChR, although corrugated, maintains gating rate constants in primarily one mode owing to the steep funnel-shaped foundation. By contrast, the corrugated profile for εA411P AChR allows multiple modes of gating rate constants owing to the shallow funnel-shaped foundation.

funnel-shaped foundation. Thus the local region flanking εA411 can be viewed as shaping the broad foundation upon which the corrugations superimpose. In the εA411P mutant, barriers separating energy wells are relatively high, similar to wild-type; so only rare-mode switches are detected during individual activation episodes. Thus in addition to affecting absolute rates of gating, the cytoplasmic loop of the ε subunit controls the fidelity of the gating rate constants.

9.5. Interpretation and implications of mutations in the long cytoplasmic domain of the ε subunit

Mutations in the cytoplasmic domain likely perturb a local α-helical structure, which in turn may alter interaction with cytoplasmic domains from the same or neighboring subunits. The predicted α-helical region containing εA411 likely contributes, along with analogous regions of the other subunits, to the fenestrated basket-like structure that protrudes into the cytoplasm [51]. Interplay between cytoplasmic domains from the five subunits thus appears to affect movement of distant structures, such as the ion channel gate, through changes in the global energy landscape. The link between cytoplasmic domains and the ion channel gate is not known, nor is the number of linking structures. However the ε subunit is known to play a specific role in affecting global energetics governing channel gating, and helix G in the cytoplasmic domain of the ε subunit tunes the fidelity of gating. Point mutations of helix G unmask the corrugated nature of the energy landscape governing gating [80]. A corrugated landscape is also present in wild-type AChRs, but it is overlaid upon a steep funnel-shaped foundation, which confines gating to predominantly one mode. The overall findings suggest that the primordial AChR was gated over a spectrum of kinetic modes, but that evolution introduced new structures and fine-tuned them to confine gating kinetics to primarily one mode.

10. Future prospects

Future studies of CMS mutations of the AChR will likely reveal new structures critical to AChR function. Still unknown is the identity of the molecular trigger that transduces occupancy of the ACh binding site to opening of the channel. Also unknown is the linkage structure that couples the binding site trigger to the channel gating apparatus. Residues governing movement of secondary structures during channel opening and desensitization also remain to be identified. Structures with unexpected functions may also emerge, as demonstrated by the discovery that the large cytoplasmic domain in the ε subunit governs fidelity of channel gating. Finally, current structure–function studies face inherent limitations because the atomic structure of the AChR is not known. The overall fold and residue placement in the major extracellular domain is provided by the homologous AChBP structure [81]. Precise placement of residue side chains in human AChR extracellular domains is provided by homology modeling based on mutagenesis and the AChBP structure [86]. The atomic structure of the entire AChR will allow unprecedented correlation between structure and function. Analysis of CMS mutations of the AChR coupled with the three dimensional structure will provide powerful synergies in understanding the molecular workings of this prototype of neurotransmitter-gated ion channels.

Acknowledgements

Supported by NIH grants to S.M. Sine (NS32744) and A.G. Engel (NS6277), and a research grant from the Muscular Dystrophy Association to A.G. Engel.

References

[1] Rothbart, H.B., 1937. Myasthenia gravis. Familial occurrence. JAMA 108, 715–717.
[2] Bundey, S., 1972. A genetic study of infantile and juvenile myasthenia gravis. J. Neurol. Neurosurg. Psychiatry 35, 41–51.
[3] Engel, A.G., 1994. Myasthenic syndromes. In: Engel, A.G., Franzini-Armstrong, C. (Eds.), Myology. 2nd ed, McGraw-Hill, New York, pp. 1798–1835.
[4] Walls, T.J., Engel, A.G., Nagel, A.S., Harper, C.M., Trastek, V.F., 1993. Congenital myasthenic syndrome associated with paucity of synaptic vesicles and reduced quantal release. Ann. NY Acad. Sci. 681, 461–468.
[5] Engel, A.G., Lambert, E.H., 1987. Congenital myasthenic syndromes. Electroencephalogr. Clin. Neurophysiol. Suppl. 39, 91–102.
[6] Mora, M., Lambert, E.H., Engel, A.G., 1987. Synaptic vesicle abnormality in familial infantile myasthenia. Neurology 37, 206–214.
[7] Engel, A.G., Lambert, E.H., Gomez, M.R., 1977. A new myasthenic syndrome with end-plate acetylcholinesterase deficiency, small nerve terminals, and reduced acetylcholine release. Ann. Neurol. 1, 315–330.
[8] Engel, A.G., Hutchinson, D.O., Nakano, S., Murphy, L., Griggs, R.C., Gu, Y., Hall, Z.W., Lindstrom, J., 1993. Myasthenic syndromes attributed to mutations affecting the epsilon subunit of the acetylcholine receptor. Ann. NY Acad. Sci. 681, 496–508.
[9] Vincent, A., Newsom-Davis, J., Wray, D., Shillito, P., Harrison, J., Betty, M., Beeson, D., Mills, K., Palace, J., Molenaar, P., Murray, N., 1993. Clinical and experimental observations in patients with congenital myasthenic syndromes. Ann. NY Acad. Sci. 681, 451–460.

[10] Engel, A.G., Lambert, E.H., Mulder, D.M., Torres, C.F., Sahashi, K., Bertorini, T.E., Whitaker, J.N., 1982. A newly recognized congenital myasthenic syndrome attributed to a prolonged open time of the acetylcholine-induced ion channel. Ann. Neurol. 11, 553–569.

[11] Uchitel, O., Engel, A.G., Walls, T.J., Nagel, A., Atassi, Z.M., Bril, V., 1993. Congenital myasthenic syndromes. II. A syndrome attributed to abnormal interaction of acetylcholine with its receptor. Muscle Nerve 16, 1293–1301.

[12] Noda, M., Furutani, Y., Takahashi, H., Toyosato, M., Tanabe, T., Shimizu, S., Kikyotani, S., Kayano, T., Hirose, T., Inayama, S., Numa, S., 1983. Cloning and sequence analysis of calf cDNA and human genomic DNA encoding alpha-subunit precursor of muscle acetylcholine receptor. Nature 305, 818–823.

[13] Beeson, D., Brydson, M., Newsom-Davis, J., 1989. Nucleotide sequence of human muscle acetylcholine receptor beta subunit. Nucleic Acids Res. 17, 4391.

[14] Luther, M.A., Schoepfer, R., Whiting, P., Casey, B., Blatt, Y., Montal, M.S., Montal, M., Lindstrom, J., 1989. A muscle acetylcholine receptor is expressed in the human cerebellar medulloblastoma cell line TE671. J. Neurosci. 9, 1082–1096.

[15] Beeson, D., Brydson, M., Betty, M., Jeremiah, S., Povey, S., Vincent, A., Newsom-Davis, J., 1993. Primary structure of the human muscle acetylcholine receptor: cDNA cloning of the gamma and epsilon subunits. Eur. J. Biochem. 215, 229–238.

[16] Soreq, H., Ben-Azis, R., Prody, C.A., Seidman, S., Gnatt, A., Neville, L., Lieman-Hurwitz, J., Lev Lehmann, E., Ginzberg, D., Lapidot-Lifson, Y., Zakut, H., 1990. Molecular cloning and construction of the coding region for human acetylcholinesterase reveals a G + C-rich attenuating structure. Proc. Natl Acad. Sci. USA 87, 9688–9692.

[17] Li, Y., Camp, S., Rachinsky, T.L., Getman, D., Taylor, P., 1991. Gene structure of mammalian acetylcholinesterase. Alternative exons dictate tissue specific expression. J. Biol. Chem. 266, 23083–23090.

[18] Oda, Y., Nakanishi, I., Deguchi, T., 1992. A complementary DNA for human choline acetyltransferase induces two forms of enzyme with different molecular weights in cultured cells. Brain Res. Mol. Brain Res. 16, 287–294.

[19] Ohno, K., Brengman, J.M., Tsujino, A., Engel, A.G., 1998. Human endplate acetylcholinesterase deficiency caused by mutations in the collagen-like tail subunit (ColQ) of the asymmetric enzyme. Proc. Natl Acad. Sci. USA 95, 9654–9659.

[20] Erickson, J.D., Varoqui, H., Eiden, L.E., Schafer, M.K., Modi, W., Diebler, M., Weihe, E., Rand, J., Bonner, T.I., Usdin, T.B., 1994. Functional identification of a vesicular acetylcholine transporter and its expression from a 'cholinergic' gene locus. J. Biol. Chem. 269, 21929–21932.

[21] Ohno, K., Tsujino, A., Brengman, J.M., Michel Harper, C.M., Bajzer, Z., Udd, B., Beyring, R., Robb, S., Kirkham F.J., Engel, A.G., 2001. Choline acetyltransferase mutations cause myasthenic syndrome in humans. Proc. Natl. Acad. Sci. USA 98, 2017–2022.

[22] Milone, M., Hutchinson, D.O., Engel, A.G., 1994. Patch-clamp analysis of the properties of acetylcholine receptor channels at the normal human endplate. Muscle Nerve 17, 1364–1369.

[23] Ohno, K., Hutchinson, D.O., Milone, M., Brengman, J.M., Bouzat, C., Sine, S.M., Engel, A.G., 1995. Congenital myasthenic syndrome caused by prolonged acetylcholine receptor channel openings due to a mutation in the M2 domain of the ε subunit. Proc. Natl Acad. Sci. USA 92, 758–762.

[24] Sine, S.M., Ohno, K., Bouzat, C., Auerbach, A., Milone, M., Pruitt, J.N., Engel, A.G., 1995. Mutation of the acetylcholine receptor α subunit causes a slow-channel myasthenic syndrome by enhancing agonist binding affinity. Neuron 15, 229–239.

[25] Engel, A.G., Ohno, K., Milone, M., Wang, H.-L., Nakano, S., Bouzat, C., Pruitt, J.N., Hutchinson, D.O., Brengman, J.M., Bren, N., Sieb, J.P., Sine, S.M., Engel, A.G., 1996. New mutations in acetylcholine receptor subunit genes reveal heterogeneity in the slow-channel congenital myasthenic syndrome. Hum. Mol. Genet. 5, 1217–1227.

[26] Milone, M., Wang, H.-L., Ohno, K., Fukudome, T., Pruitt, J.N., Bren, N., Sine, S.M., Engel, A.G., 1997. Slow-channel syndrome caused by enhanced activation, desensitization, and agonist binding affinity due to mutation in the M2 domain of the acetylcholine receptor alpha subunit. J. Neurosci. 17, 5651–5665.

[27] Wang, H.-L., Auerbach, A., Bren, N., Ohno, K., Engel, A.G., Sine, S.M., 1997. Mutation in the M1 domain of the acetylcholine receptor alpha subunit decreases the rate of agonist dissociation. J. Gen. Physiol. 109, 757–766.

[28] Croxen, R., Newland, C., Beeson, D., Oosterhuis, H., Chauplanaz, G., Vincent, A., Newsom-Davis, J., 1997. Mutations in different functional domains of the human muscle acetylcholine receptor α subunit in patients with the slow-channel congenital myasthenic syndrome. Hum. Mol. Genet. 6, 767–773.

[29] Gomez, C.M., Maselli, R., Gammack, J., Lasalde, J., Tamamizu, S., Cornblath, D.R., Lehar, M., McNamee, M., Kuncl, R., 1996. A beta-subunit mutation in the acetylcholine receptor gate causes severe slow-channel syndrome. Ann. Neurol. 39, 712–723.

[30] Ohno, K., Wang, H.-L., Milone, M., Bren, N., Brengman, J.M., Nakano, S., Quiram, P., Pruitt, J.N., Sine, S.M., Engel, A.G., 1996. Congenital myasthenic syndrome caused by decreased agonist binding affinity due to a mutation in the acetylcholine receptor ε subunit. Neuron 17, 157–170.

[31] Engel, A.G., Ohno, K., Sine, S.M., 1999. Congenital myasthenic syndromes. In: Engel, A.G. (Ed.), Myasthenia Gravis and Myasthenic Disorders. Oxford University Press, New York, pp. 251–297.

[32] Ohno, K., Anlar, B., Özdirim, E., Brengman, J.M., De Bleecker, J., Engel, A.G., 1998. Myasthenic syndromes in Turkish kinships due to mutations in the acetylcholine receptor. Ann. Neurol. 44, 234–241.

[33] Ohno, K., Quiram, P., Milone, M., Wang, H-L., Harper, C.M., Pruitt, J.N., Brengman, J.M., Pao, L., Fischbeck, K.H., Crawford, T.O., Sine, S.M., Engel, A.G.., 1997. Congenital myasthenic syndromes due to heteroallelic nonsense/missense mutations in the acetylcholine receptor ε subunit gene: identification and functional characterization of six new mutations. Hum. Mol. Genet. 6, 753–766.

[34] Ohno, K., Fukudome, T., Nakano, S., Milone, M., Feasby, T.E., Tyce, G.M., Engel, A.G., 1996. Mutational analysis in a congenital myasthenic syndrome reveals a novel acetylcholine receptor epsilon subunit mutation. Soc. Neurosci. Abstr. 22, 234.

[35] Engel, A.G., Ohno, K., Bouzat, C., Sine, S.M., Griggs, R.G., 1996. End-plate acetylcholine receptor deficiency due to nonsense mutations in the ε subunit. Ann. Neurol. 40, 810–817.

[36] Middleton, L., Ohno, K., Christodoulou, K., Brengman, J.M., Milone, M., Neocleous, V., Serdaroglu, P., Deymeer, F., Özdemir, C., Mubaidin, A., Horany, K., Al-Shehab, A., Mavromatis, I., Mylonas I., Tsingis, M., Zamba, E., Pantzaris, M., K. Kyriallis, K., Engel, A.G., 1999. Congenital myasthenic syndromes linked to chromosome 17p are caused by defects in acetylcholine receptor ε subunit gene. Neurology 53, 1076–1082.

[37] Ohno, K., Engel, A.G., Milone, M., Sieb, J.P., Ianacomne, S., 1995. A congenital myasthenic syndrome with severe acetylcholine receptor deficiency caused by heteroallelic frameshifting mutations in the epsilon subunit. Neurology 45(Suppl. 4), A283 (Abstract).

[38] Milone, M., Wang, H-L., Ohno, K., Prince, R.J., Shen, X-M., Brengman, J.M., Griggs, R.C., Engel, A.G., 1998. Mode switching kinetics produced by a naturally occurring mutation in the cytoplasmic loop of the human acetylcholine receptor ε subunit. Neuron 20, 575–588.

[39] Quiram, P., Ohno, K., Milone, M., Patterson, M.C., Pruitt, N.J., Brengman, J.M., Sine, S.M., Engel, A.G., 1999. Mutation causing congenital myasthenia reveals acetylcholine receptor β/δ subunit interaction essential for assembly. J. Clin. Invest. 104, 1403–1410.

[40] Wang, H.-L., Milone, M., Ohno, K., Shen, X.-M., Tsujino, A., Batocchi, A.P., Tonali, P., Brengman, J.M., Engel, A.G., Sine, S.M., 1998. Acetylcholine receptor M3 domain: stereochemical and volume contributions to channel gating. Nat. Neurosci. 2, 226–233.

[41] Fukudome, T., Ohno, K., Brengman, J.M., Engel, A.G., 1998. Quinidine normalizes the open duration of slow-channel mutants of the acetylcholine receptor. Neuroreport 9, 1907–1911.

[42] Harper, C.M., Engel, A.G., 1998. Quinidine sulfate therapy for the slow-channel congenital myasthenic syndrome. Ann. Neurol. 43, 480–484.

[43] Blount, P., Merlie, J.P., 1989. Molecular basis of the two non-equivalent ligand binding sites of the muscle nicotinic acetylcholine receptor. Neuron 3, 347–357.

[44] Pedersen, S., Cohen, J.B., 1990. *d*-Tubocurarine binding sites are located at alpha–gamma and alpha–delta subunit interfaces of the nicotinic acetylcholine receptor. Proc. Natl Acad. Sci. USA 87, 2785–2789.

[45] Sine, S.M., 1993. Molecular dissection of subunit interfaces in the acetylcholine receptor: identification of residues that determine curare selectivity. Proc. Natl Acad. Sci. USA 90, 9436–9440.

[46] Kubalek, E., Ralston, S., Lindstrom, J., Unwin, N., 1987. Location of subunits within the acetylcholine receptor by electron image analysis of tubular crystals from *Torpedo marmorata*. J. Cell Biol. 105, 9–18.

[47] Unwin, N., 1995. Acetylcholine receptor channel imaged in the open state. Nature 373, 37–43.

[48] Sine, S.M., Kreienkamp, H.J., Bren, N., Maeda, R., Taylor, P., 1995. Molecular dissection of subunit interfaces in the acetylcholine receptor: identification of determinants of α-Conotoxin M1 selectivity. Neuron 15, 205–211.

[49] Bren, N., Sine, S., 1997. Identification of residues in the adult nicotinic acetylcholine receptor that confer selectivity for curariform antagonists. J. Biol. Chem. 272, 30793–30798.

[50] Sine, S., 1997. Identification of equivalent residues in the γ, δ, and ε subunits of the nicotinic receptor that contribute to α-bungarotoxin binding. J. Biol. Chem. 272, 23521–23527.

[51] Miyazawa, A., Fujiyoshi, Y., Stowell, M., Unwin, N., 1999. Nicotinic acetylcholine receptor at 4.6 Å resolution: transverse tunnels in the channel wall. J. Mol. Biol. 288, 765–786.

[52] Akabas, M.H., Kaufmann, C., Archdeacon, P., Karlin, A., 1994. Identification of acetylchoine receptor channel-lining residues in the entire M2 segment of the alpha subunit. Neuron 13, 919–927.

[53] Karlin, A., Akabas, M.H., 1995. Toward a structural basis for the function of nicotinic acetylcholine receptors and their cousins. Neuron 15, 1231–1244.

[54] Corringer, P.J., Bertrand, S., Galzi, J.L., Devillers-Thiery, Changeux, J.P., Bertrand, D., 1999. Mutational analysis of the charge selectivity filter of the α7 nicotinic acetylcholine receptor. Neuron 22, 831–843.

[55] Prince, R., Sine, S.M., 2000. The ligand binding domains of the nicotinic acetylcholine receptor. In: Barrantes, F.J. (Ed.), The Nicotinic Acetylcholine Receptor: Current Views and Future Trends. Landes Bioscience, Austin, TX, pp. 31–59.

[56] Corringer, P.J., Le Novère, N., Changeux, J.P., 2000. Nicotinic receptors at the amino acid level. Annu. Rev. Pharmacol. Toxicol. 40, 431–458.

[57] Galzi, J.L., Revah, F., Black, D., Goeldner, M., Hirth, C., Changeux, J.P., 1990. Identification of a novel amino acid α-tyrosine 93 within the cholinergic ligands-binding sites of the acetylcholine receptor by photoaffinity labeling. J. Biol. Chem. 265, 10430–10437.

[58] Czajkowski, C., Karlin, A., 1995. Structure of the nicotinic receptor acetylcholine-binding site. J. Biol. Chem. 270, 3160–3164.

[59] Chiara, D.C., Middleton, R.E., Cohen, J.B., 1998. Identification of tryptophan 55 as the primary site of [³H]nicotine photoincorporation in the gamma-subunit of the *Torpedo* nicotinic acetylcholine receptor. FEBS Lett. 423, 223–226.

[60] Wang, D., Chirar, D.C., Xie, Y., Cohen, J.B., 2000. Probing the structure of the nicotinic acetylcholine receptor with 4-benzoylbenzoylcholine, a novel photoaffinity competitive antagonist. J. Biol. Chem. 275, 28666–28674.

[61] Fu, D.X., Sine, S.M., 1994. Competitive antagonists bridge the α–γ subunit interface of the acetylcholine receptor through quaternary ammonium-aromatic interactions. J. Biol. Chem. 269, 26152–26157.

[62] Bren, N., Sine, S.M., 2000. Hydrophobic pairwise interactions stabilize α-conotoxin MI in the muscle acetylcholine receptor binding site. J. Biol. Chem. 275, 12692–12700.

[63] Engel, A.G., Ohno, K., Sine, S.M., 1999. Congenital myasthenic syndromes: recent advances. Arch. Neurol. 56, 163–167.

[64] Sine, S.M., Claudio, T., Sigworth, F.J., 1990. Activation of Torpedo acetylcholine receptors expressed in mouse fibroblasts: single channel current kinetics reveal distinct agonist binding affinities. J. Gen. Physiol. 96, 395–437.

[65] Zhang, Y., Chen, J., Auerbach, A., 1995. Activation of recombinant mouse acetylcholine receptors by acetylcholine, carbamylcholine and tetramethylammonium. J. Physiol. 486, 189–206.

[66] Ohno, K., Quiram, P., Milone, M., Wang, H.L., Harper, M., Pruitt, J., Brengman, J., Pao, L., Fischbeck, K., Crawford, T., Sine, S., Engel, A., 1997. Congenital myasthenic syndromes due to heteroallelic nonsense/missense mutations in the acetylcholine receptor ε subunit gene: Identification and functional characterization of six new mutations. Hum. Mol. Genet. 6, 753–766.

[67] Monod, J., Wyman, J., Changeux, J.P., 1965. On the nature of allosteric transitions: a plausible model. J. Mol. Biol. 12, 88–118.

[68] Jackson, M.B., 1989. Perfection of a synaptic receptor: kinetics and energetics of the acetylcholine receptor. Proc. Natl Acad. Sci. USA 86, 2199–2203.

[69] Wilson, G.G., Karlin, A., 1998. The location of the channel gate in the acetylcholine receptor channel. Neuron 20, 1269–1281.

[70] Labarca, C., Nowak, M.W., Zhang, H., Tang, L., Deshpande, P., Lester, H.A., 1995. Channel gating governed symmetrically by conserved leucine residue in the M2 domain of nicotinic receptors. Nature 376, 514–516.

[71] Ortells, M., Lunt, G.A., 1996. Mixed helix-beta-sheet model of the transmembrane region of the nicotinic acetylcholine receptor. Protein Eng. 9, 51–59.

[72] Bouzat, C., Barrantes, F., Sine, S., 2000. Nicotinic receptor fourth transmembrane domain: hydrogen bonding by conserved threonine contributes to channel gating kinetics. J. Gen. Physiol. 115, 663–671.

[73] Blanton, M., Cohen, J., 1994. Identifying the lipid-protein interface of the Torpedo nicotinic acetylcholine receptor: secondary structure implications. Biochemistry 33, 2859–2872.

[74] Dennis, M., Giraudat, J., Kotzyba-Hibert, F., Hirth, C., Chang, J.Y., Lazure, C., Chretien, M., Changeux, J.P., 1988. Amino acids of the Torpedo marmorata acetylcholine receptor alpha subunit labeled by photoaffinity ligand for the acetylcholine binding site. Biochemistry 27, 2346–2357.

[75] Finer-Moore, J., Stroud, R.M., 1984. Amphipathic analysis and possible formation of the ionic channel in an acetylcholine receptor. Proc. Natl Acad. Sci. USA 81, 155–159.

[76] Le Novère, N., Corringer, P.J., Changeux, J.P., 1999. Improved secondary structure predictions for a nicotinic receptor subunit: incorporation of solvent accessibility and experimental data into a two-dimensional representation. Biophys. J. 76, 2329–2345.

[77] Mishina, M., Tobimatsu, T., Imoto, K., Tanaka, K., Fujita, Y., Fukuda, K., Kurasaki, M., Takahashi, H., Morimoto, Y., Hirose, T., Inayama, S., Takahashi, T., Kuno, M., Numa, S., 1895. Location of functional regions of acetylcholine receptor a-subunit by site-directed mutagenesis. Nature 313, 364–369.

[78] Bouzat, C., Bren, N., Sine, S.M., 1994. Structural basis of the different gating kinetics of fetal and adult acetylcholine receptors. Neuron 13, 1395–1402.

[79] Wang, H.L., Ohno, K., Milone, M., Brengman, J., Evoli, A., Batocchi, A.P., Middleton, L., Christodoulou, K., Engel, A.G., Sine, S.M., 2000. Fundamental gating mechanism of nicotinic receptor channel revealed by mutation causing a congenital myasthenic syndrome. J. Gen. Physiol. 116, 449–460.

[80] Frauenfelder, H., Lesson, D.T., 1998. The energy landscape in non-biological and biological molecules. Nat. Struct. Biol. 5, 757–759.

[81] Brejc, K., van Dijk, W., Klassen, R., Schuurmans, M., van der Oost, J., Smit, A., Sixma, T., 2001. Crystal structure of an ACh-binding protein reveals the ligand-binding domain of nicotinic receptors. Nature 411, 269–276.

[82] Zhou, M., Engel, A.G., Auerbach, A., 1999. Serum choline activates mutant acetylcholine receptors that cause slow channel congenital myasthenic syndrome. Proc. Natl Acad. Sci. USA 96, 10466–10471.

[83] Harper, C.M., Engel, A.G., 1998. Quinidine sulfate therapy for the slow-channel congenital myasthenic syndrome. Ann. Neurol. 43, 480–484.

[84] Auerbach, A., Lingle, C., 1986. Heterogeneous kinetic properties of acetylcholine receptor channels in Xenopus myocytes. J. Physiol. 378, 119–140.

[85] Naranjo, D., Brehm, P., 1993. Modal shifts in acetylcholine receptor channel gating confer subunit-dependent desensitization. Science 260, 1811–1814.

[86] Sine, S.M., Wang, H.-L., Bren, N., 2002. Lysine scanning mutagenesis delineates structural model of the nicotinic receptor ligand binding domain. J. Biol. Chem. 277, 29210–29223.

Store-operated calcium channels: properties, functions and the search for a molecular mechanism

Murali Prakriya* and Richard S. Lewis

Department of Molecular and Cellular Physiology, Stanford University School of Medicine, Stanford, CA 94305, USA

*Correspondence address: E-mail: prakriya@stanford.edu

1. Introduction

In most or all animal cells, stimulation of cell surface receptors that generate 1,4,5-inositol trisphosphate (IP_3) causes the release of Ca^{2+} from the endoplasmic reticulum (ER)-derived Ca^{2+} stores followed by Ca^{2+} influx across the plasma membrane. This ubiquitous Ca^{2+} entry process, first described by Putney and originally termed capacitative Ca^{2+} entry, was proposed as a feedback mechanism designed to replenish Ca^{2+} in the stores [1]. It was later renamed store-operated Ca^{2+} entry to reflect the consensus that it is regulated by the level of free Ca^{2+} in the lumen of the ER and that it provides a direct conduit for Ca^{2+} to enter the cell across the plasma membrane.

A large number of Ca^{2+} imaging studies indicate that store-operated channels (SOCs) are ubiquitous in nonexcitable cells, where they comprise the main Ca^{2+} entry pathway, and some studies suggest that they may exist in excitable cells as well [2–4]. Patch-clamp methods have offered the most detailed information about the biophysical properties of these channels. The first and best characterized SOC is the Ca^{2+} release-activated Ca^{2+} (CRAC) channel, expressed in mast cells, T lymphocytes, and related cell lines [5,6]. While much is known about the basic biophysical properties of these channels, a critical issue remains the identification of the molecular mechanism(s) underlying their activity. Despite considerable efforts over the past decade, the genes that encode SOCs and the mechanisms that link store depletion to SOC activation are unknown. A subpopulation of ER residing close to the plasma membrane is believed to signal to SOCs [7,8], but it has not been clearly identified. There is also disagreement about the relationship between the extent of store depletion and SOC activation, with proposals ranging from a graded response [9,10] to threshold behavior once a certain degree of depletion is reached [8,11]. The slow progress in resolving these issues probably reflects the inherent difficulties of identifying and studying this class of channels. Unlike most voltage- or ligand-activated channels, no potent or selective blockers of SOCs are known, the proximal stimulus

Advances in Molecular and Cell Biology, Vol. 32, pages 121–140
ISSN: 1569-2558 / DOI: 10.1016/S1569-2558(03)32006-5

for channel activation is a mystery, activation is slow (on the order of 1–10 s), and for the Ca^{2+}-selective SOCs, whole-cell currents are usually quite small (on the order of pA per cell) and the single-channel Ca^{2+} currents are probably below the resolution of patch-clamp measurements.

In this chapter, we discuss those aspects of SOCs that may be especially pertinent to the search for a molecular mechanism, along with current guesses as to what that mechanism might be. CRAC channels are highlighted here, as a prototypic example of an SOC, with the expectation that many of the issues related to the molecular mechanism will apply to other SOCs as well. Finally, we discuss the known biological functions of CRAC channels, and describe some of the intrinsic properties that are most important for carrying out these roles.

2. Identification of store-operated Ca^{2+} channels

The defining characteristic of SOCs is that they are activated as a direct result of a reduction of the lumenal Ca^{2+} concentration in the ER. Physiologically this stimulus most often involves agonist activation of surface receptors that are linked to phospholipase C, producing IP_3 that releases Ca^{2+} through IP_3 receptors in the ER membrane. It is important to recognize that although such agonists cause store depletion, the membrane currents they activate may not necessarily be store-operated. For instance, activation of metabotropic receptors is known to activate nonselective cation currents through a store-independent mechanism involving the activation of receptor-associated G proteins; such currents are therefore often classified as "receptor-activated" [12,13].

What criteria must be satisfied to identify a particular current as being store-operated? Several methods are commonly used to deplete stores and more directly activate SOCs independently of cell-surface receptors. For example, IP_3 applied intracellularly through the recording pipette, or lipophilic Ca^{2+} ionophores (ionomycin, A23187) added extracellularly directly release Ca^{2+} from the ER. Passive depletion of stores is possible with membrane-permeant inhibitors of SERCA pumps such as thapsigargin (TG) or internal dialysis with Ca^{2+} buffers such as EGTA or BAPTA, which interrupt the normal flow of Ca^{2+} from cytosol to ER and allow a passive Ca^{2+} leak to empty the stores. Another approach is to apply N,N,N',N'-tetrakis(2-pyridylmethyl)ethylenediamine (TPEN) a membrane-permeant Ca^{2+} chelator that reduces free Ca^{2+} levels in the ER without altering fluxes across the ER membrane [10]. But these methods are not without limitations. IP_3 is capable of directly activating ion channels, and the rise in $[Ca^{2+}]_i$ produced by some of these stimuli may activate or modulate channels independently of Ca^{2+} store content. For these reasons, it is important to demonstrate current activation by store depletion under conditions of constant $[Ca^{2+}]_i$, for example, in the presence of a high concentration of intracellular Ca^{2+} buffer. Even this method could be misleading, however, as reduction of resting $[Ca^{2+}]_i$ has been reported to activate SOCs independently of store content [14]. Thus, an even better approach to identify SOCs is to deplete stores under conditions where cytosolic $[Ca^{2+}]_i$ is highly buffered near its normal resting level. This test has only been reported in a limited number of cases [10,15]. The importance of ruling out alternate modes of activation is underscored by the fact that the channels associated with some *trp* genes were originally thought be store-operated, but were later shown to be receptor-, IP_3-, or Ca^{2+}-activated instead (see below). Thus, a collection of different approaches is required to unequivocally identify a current as being store-operated.

3. The defining characteristics of the CRAC channel, a prototypic SOC

SOCs have been identified using patch-clamp techniques in a variety of different cell types, including hematopoietic cells such as T lymphocytes, mast cells, and macrophages, as well as endothelial cells, hepatocytes, pancreatic acinar and β cells, and the MDCK cell line (see below). Imaging studies indicate that store-operated Ca^{2+} entry mechanisms exist in an even wider range of tissues [7,16] that may include excitable cells [2–4]. Nevertheless, the most detailed information about the properties of the channels underlying store-operated Ca^{2+} entry is available only in a few cell types such as T cells, mast cells, and related cell lines. The channel described in these cells is referred to as the CRAC channel, and the current through these channels as I_{CRAC}.

3.1. Activation by store depletion

I_{CRAC} is activated by a range of stimuli that share the ability to deplete Ca^{2+} stores. These include stimulation of endogenous antigen receptors in T cells [17–19] and Fc receptors on mast cells [20] as well as by heterologously expressed muscarinic receptors [8,18,21], all of which are linked to the production of IP_3. In addition, I_{CRAC} is effectively activated by ionomycin, TG, TPEN, or intracellular dialysis with Ca^{2+} chelators [10,15, 18,22–25]. Whereas these receptor-independent stimuli by themselves do not offer unequivocal proof of store dependence as discussed above, many of them are effective even in cells pre-loaded with exogenous Ca^{2+} buffers at normal $[Ca^{2+}]_i$ [10,15]. Thus, the evidence that I_{CRAC} is activated by depletion of Ca^{2+} stores is extensive.

3.2. Calcium selectivity

A hallmark of the CRAC channel is its lack of voltage dependence and an extremely high selectivity for Ca^{2+} over monovalent cations. Macroscopic I_{CRAC} displays an inwardly rectifying current–voltage relation, and the current does not reverse sign at potentials up to $\sim +50$ mV [17,22,24]. Using the fura-2 overload method [26], Hoth and Penner [23,27] estimated that CRAC channels conduct $Ca^{2+} > 1000$ times better than Na^+ under physiological conditions, placing them among the most Ca^{2+}-selective channels known. They are thought to achieve this high selectivity in a manner similar to voltage-gated Ca^{2+} channels, via ionic interactions within the pore. This idea is based on observations that CRAC channels, like L-type Ca^{2+} channels [28], conduct monovalent ions if extracellular divalent cations are reduced to submicromolar concentrations [29,30]. Their extremely high selectivity for Ca^{2+} makes the CRAC channel an efficient pathway for Ca^{2+} entry, causing a minimal amount of depolarization and hence preserving the electrical driving force on Ca^{2+}.

The selectivity of the CRAC channel among divalent ions is not easily compared with that of voltage-gated Ca^{2+} channels, because replacement of Ca^{2+} with Ba^{2+} or Sr^{2+} affects channel gating. Under steady-state conditions, the order of conductance is generally $Ca^{2+} > Ba^{2+} \geq Sr^{2+}$, with Ba^{2+} and Sr^{2+} conducting less than half as well as Ca^{2+} [18,22,24]. This contrasts with the behavior of many voltage-gated Ca^{2+} channels, for which the sequence is generally $Ba^{2+} > Sr^{2+} > Ca^{2+}$ [31]. However, two reports have shown that when exchanged rapidly with Ca^{2+}, Ba^{2+} initially conducts current at least as well as Ca^{2+}, and that the Ba^{2+} current subsequently declines over a period of seconds [27,32]. This current decline may be explained by a reversal of Ca^{2+}-dependent potentiation (CDP) (see below). Together, these data suggest that permeation through CRAC channels may actually resemble that of voltage-gated Ca^{2+} channels, implying that the pore regions of the two classes of channels could be related. More thorough studies of the divalent ion permeability of CRAC channels must be made to test this idea.

Interestingly, high Ca^{2+} selectivity is not a ubiquitous property of store-operated currents. SOCs in cell types such as pancreatic acinar cells, endothelial cells, and MDCK cells exhibit significantly lower Ca^{2+} selectivity than CRAC channels, and appear to have correspondingly higher unitary conductances [33–35]. Thus, in addition to Ca^{2+}, these channels conduct significant amounts of Na^+, which would enhance membrane depolarization and therefore reduce Ca^{2+} influx by diminishing the driving force for Ca^{2+} entry. In most cases, the true ability of these channels to produce changes in $[Ca^{2+}]_i$ has not been measured.

3.3. Unitary conductance

Another distinctive feature of the CRAC channel is its extremely small apparent unitary conductance for Ca^{2+}. Whole-cell I_{CRAC} produces no current noise visible by eye, and the activity of single CRAC channels conducting Ca^{2+} has not yet been measured. From noise analysis of whole-cell currents, Zweifach and Lewis [24] estimated that the unitary conductance of I_{CRAC} in Jurkat T cells is 9–24 fS in 2–110 mM Ca^{2+}. Similarly, Hoth and Penner [23] estimated a unitary conductance value $\ll 1$ pS in mast cells based on the lack of visible noise. By comparison, unitary Ca^{2+} conductances of voltage-gated Ca^{2+} channels range from 4 to 10 pS under comparable ionic conditions, i.e. about 1000-fold larger than CRAC channels [31]. These low estimates based on noise measurements were called into question when it was found that a significantly higher estimate of ~ 1.6 pS could be made by scaling the size of the Na^+ conductance through CRAC channels (~ 40 pS; measured after the removal of extracellular divalents) by the ratio of whole-cell Na^+ and Ca^{2+} currents (~ 25) [36]. However, more recent evidence [37–39] indicates that the 40-pS channels are in fact unrelated to CRAC channels; instead, they are store-independent cation channels regulated by intracellular Mg^{2+} and MgATP, and appear similar to the recently cloned LTRCP7 channels (see below). Under conditions that prevent the activation of LTRPC7 channels, the unitary conductance of CRAC channels to Na^+ is estimated by noise analysis to be extremely small, about 0.2 pS [37]. Thus, overall, these estimates suggest that the unitary conductance of CRAC channels to both Ca^{2+} and Na^+ is several

100-fold smaller than that of voltage-gated Ca^{2+} channels. It should be noted that noise measurements can underestimate the conductance if the single channels have a high open probability, or if the current fluctuations are brief enough to exceed the frequency limits of the recording apparatus. Thus, the question of the unitary Ca^{2+} conductance will be directly resolved only by experimental measurements of Ca^{2+} currents through single CRAC channels.

3.4. Pharmacology

A high-affinity, specific blocker for CRAC channels has yet to be identified. Nevertheless, a number of less specific inhibitors together provide a pharmacological profile that might be useful in identifying CRAC channels at a molecular level. The family of imidazole antimycotic compounds including econazole and SKF 96365, inhibit CRAC channels and other SOCs with IC_{50} values in the range of $0.6-15\ \mu M$ [40,41]. La^{3+} blocks I_{CRAC} with high affinity ($K_i = 20-60\ nM$) [42,43]. CRAC channels are insensitive to organic blockers of L-, N-, and P/Q-type voltage-gated Ca^{2+} channels, such as diltiazem, nifedipine, ω-conotoxin GVIIA or ω-agatoxin IVA [18]. In addition to these exogenous agents, I_{CRAC} is also blocked by endogenous lipid regulators of cell growth and proliferation such as sphingosine, ceramide, and related compounds with IC_{50}s in the range of $1-10\ \mu M$ [47]. In T cells, ceramide and sphingosine are generated by the crosslinking of Fas (CD95) on the cell surface; blockade of I_{CRAC} by these compounds may contribute to immunosuppression by tumor cells expressing high levels of Fas ligand, and may thereby enable them to evade immune surveillance in vivo [45].

Recently, 2-aminoethyldiphenyl borate (2-APB), a noncompetitive antagonist of the IP_3 receptor [46] was found to inhibit store-operated Ca^{2+} influx in human embryonic kidney (HEK 293) cells [47]. This was used as evidence for the idea that activation of SOCs occurs via conformational coupling with IP_3 receptors (see below). However, subsequent investigations in RBL cells and Jurkat T cells indicate that 2-APB is much more potent when applied extracellularly ($IC_{50} \sim 10\ \mu M$) than intracellularly, suggesting that it inhibits CRAC channels directly [48,49]. Surprisingly, at low concentrations ($\leq 5\ \mu M$) 2-APB enhances I_{CRAC} amplitude by up to 3-fold [49]. 2-APB may have only limited use as a selective I_{CRAC} blocker, as it also inhibits IP_3 receptors [46], *trp3* channels [47], and Ca^{2+} efflux from mitochondria [49]. However, its ability to potentiate I_{CRAC} may be useful in screening potential CRAC channel clones.

4. Regulation of CRAC channels by calcium

Ca^{2+} affects the function of CRAC channels in several ways, inducing fast inactivation, slow inactivation, and CDP. These modes of modulation may have important physiological effects, and serve as useful identifiers in testing putative CRAC-channel genes.

4.1. Fast inactivation

Like voltage-gated Ca^{2+} channels, CRAC channels inactivate in response to a local rise in $[Ca^{2+}]_i$. Fast inactivation is measured as a time-dependent decrease in current amplitude over tens of milliseconds during hyperpolarizing voltage steps [50–52]. Several observations indicate that fast inactivation is a local process, occurring as a result of Ca^{2+} flux through individual channels. First, inactivation is enhanced by increasing the driving force for Ca^{2+} entry, but is unaffected by changes in the extent of I_{CRAC} activation or global $[Ca^{2+}]_i$ [51]. Second, fast inactivation can be partially suppressed by intracellular BAPTA, a fast Ca^{2+} buffer, but not by EGTA, which binds Ca^{2+} ~ 100-fold more slowly [23,51]. A quantitative analysis of these results led to the conclusion that multiple Ca^{2+} ions bind to sites several nanometers from the pore to cause inactivation [51]. These sites may be located either directly on the channel itself or on a channel accessory protein. Berridge [7] has proposed that the site of Ca^{2+} inactivation is on IP_3 receptors that are physically coupled to the CRAC channel (the conformational coupling hypothesis, see below). This mechanism appears unlikely, however, given recent evidence that fast inactivation of I_{CRAC} occurs even in a DT40 B cell line [53] in which all three isoforms of the IP_3 receptor have been knocked out [49]. The fast kinetics of this form of inactivation imply that it will limit the magnitude of Ca^{2+} influx at negative potentials without affecting the slow Ca^{2+} dynamics typical of nonexcitable cells.

4.2. Slow inactivation

As global cell $[Ca^{2+}]_i$ rises due to influx through CRAC channels, SERCA-type Ca^{2+}-ATPases in the ER refill the stores, leading to a decline in I_{CRAC} [10,54]. This deactivation of the channels is not surprising given that store depletion triggers their activation. However, if refilling is prevented with TG, a slow inactivation of I_{CRAC} over tens of seconds is still observed, which is tightly correlated with an increase in $[Ca^{2+}]_i$ [54,55]. A high concentration of intracellular EGTA (12 mM), which prevents $[Ca^{2+}]_i$ from rising above ~ 200 nM greatly reduces slow inactivation. Thus, slow inactivation is governed by a rise in the global $[Ca^{2+}]_i$, in contrast to fast inactivation which is unaffected by global $[Ca^{2+}]_i$ or EGTA. The underlying mechanism of slow inactivation is unknown. Its sensitivity to a relatively slow Ca^{2+} buffer (EGTA) suggests that the site of slow inactivation is further from the channel than that of fast inactivation, perhaps > 100 nm based on the average capture distance of 12 mM EGTA [51]. Interestingly, recent evidence shows that active mitochondria function to limit or prevent slow inactivation in T cells and RBL cells, probably by sequestering Ca^{2+} near sites of influx [56–58]. These results raise the possibility that I_{CRAC} may be modulated by changes in the number, location, and metabolic state of mitochondria in intact cells.

4.3. Ca^{2+}-dependent potentiation

Extracellular Ca^{2+} appears to promote CRAC channel activation by a process known as CDP [32,59]. Readdition of external Ca^{2+} to store-depleted cells bathed in a Ca^{2+}-free solution causes I_{CRAC} to appear in two kinetically distinct stages. A small fraction of the current appears immediately after Ca^{2+} addition; presumably representing the current through which Ca^{2+} channels already open at the time of Ca^{2+} readdition. This instantaneous current is followed by a time-dependent, several-fold increase in I_{CRAC} amplitude plateauing eventually in 10–20 s. The extent of CDP increases with membrane hyperpolarization, suggesting that the site of CDP may reside within the pore of the channel. The site is not likely to be intracellular, as CDP is unaffected by intracellular BAPTA, and extracellular Ni^{2+} can substitute for Ca^{2+} in potentiating CRAC channels even though it is not thought to permeate the channels appreciably [59]. As noted above, Ba^{2+} does not support the potentiation process, so that CRAC currents decline over time after Ba^{2+} is substituted for Ca^{2+}. Although CDP has not been described explicitly in other cell types, time-dependent decline of Ba^{2+} currents and slow recovery of I_{CRAC} after extracellular Ca^{2+} is restored has been reported in mast and RBL cells [27]. Thus, CDP is likely to be a general characteristic of the CRAC channel. It imbues the CRAC channel with a slight voltage dependence, such that depolarization to 0 mV reduces the function of activatable channels by about half. In this way, CDP accentuates the ability of depolarization to reduce store-operated Ca^{2+} entry.

5. Biological roles of store-operated Ca^{2+} channels

In general, there are two ways in which store-operated Ca^{2+} channels contribute to cell physiology. First, by promoting store refilling, SOCs prevent the gradual loss of Ca^{2+} from stores that results from pumping of released Ca^{2+} across the plasma membrane. In this way, they enable the repetitive release of Ca^{2+} from stores that underlies sustained $[Ca^{2+}]_i$ oscillations in many cells [60]. Second, Ca^{2+} influx through SOCs can contribute directly to elevation of cytosolic $[Ca^{2+}]$, thus extending their signaling role significantly beyond the maintenance of Ca^{2+} store homeostasis. The roles of SOCs in cell physiology have been most often probed by examining the effect of removing extracellular Ca^{2+}, or of adding inhibitors like La^{3+}, Gd^{3+}, or organic blockers such as SKF 96365. These types of studies suggest that SOCs conduct the Ca^{2+} influx necessary for histamine and serotonin secretion by mast and RBL cells during allergic reactions [20,61,62]; that oscillations in SOC activity drive $[Ca^{2+}]_i$ oscillations in activated T lymphocytes [17,63] and parotid acinar cells [64]; and that Ca^{2+} entry through SOCs accelerates IP_3-triggered Ca^{2+} wave propagation in *Xenopus* oocytes [65]. SOCs appear to be closely coupled with multiple isoforms of Ca^{2+}-sensitive adenylate cyclase in glioma cells, contributing to crosstalk between Ca^{2+} and cAMP signaling pathways [66].

Many specific roles are to be expected for SOCs in light of their widespread tissue distribution. However, in many cases direct examination of the role of SOCs has been hampered by an unfortunate lack of specific, high-affinity pharmacological inhibitors. Ca^{2+} removal and heavy metals block Ca^{2+} flux through a variety of Ca^{2+} channels, and imidazole SOC blockers like SKF 96365 also inhibit cytochrome P450 [67] as well as other types of channels that may affect Ca^{2+} influx indirectly [40]. This lack of specificity, combined with the fact that only a very small Ca^{2+} current density (< 1 pA/pF) is usually required in a nonexcitable cell to account for Ca^{2+} signals, makes it difficult to rule out the action of additional Ca^{2+} conductances that have not been directly identified.

Genetic approaches provide an alternative method to assess the physiological roles of SOCs. Mutant Jurkat T cells have been isolated that display $<10\%$ of the normal density of I_{CRAC} [68,69]. Store refilling in these cells occurs at substantially reduced rates, demonstrating a role for SOCs in refilling Ca^{2+} stores [69]. Importantly, T cell activation as measured by the production of cytokines such as interleukin-2 is blocked in these mutant cells. This phenotype results from an inability of the mutant cells to generate a sustained $[Ca^{2+}]_i$ elevation [68,69] that is needed to drive nuclear translocation of NFAT, a critical transcription factor involved in cytokine gene expression [70]. Further genetic evidence indicates a critical role for CRAC channels in the activation of T cells by antigen in vivo. One type of severe immunodeficiency in humans has been causally linked to a lack of I_{CRAC} [19,71]. This central requirement for CRAC channels in T cell activation is consistent with genetic evidence that CRAC channels are the sole Ca^{2+} entry pathway activated by the antigen receptor [69], and with the inhibition of T cell activation by SKF 96365, La^{3+} or Ca^{2+} removal [43,72,73].

In excitable cells, specific functional roles for SOCs are generally less well known, which may be a consequence of the difficulty of identifying and studying SOCs in these cells. However, recent reports suggest that SOCs may play a role in the etiology of some forms of Alzheimer's disease related to mutations of presenilins 1 and 2 (PS1 and PS2). Knock-out of PS1 or expression of an inactive form of PS1 significantly elevates store-operated Ca^{2+} entry, suggesting that functional presenilin negatively regulates store-operated Ca^{2+} entry [74]. Most interestingly, mutant forms of PS1 and PS2 that result in Alzheimer's disease inhibit store-operated Ca^{2+} entry in multiple cell types, including neurons [74,75]. SKF 96365 mimicked the effects of the mutant proteins in elevating the production of toxic amyloid β peptide, thus supporting a causal connection between SOC inhibition and the pathogenesis of Alzheimer's disease [74]. It is unclear whether presenilins inhibit SOCs directly [74] or indirectly through an increase in ER Ca^{2+} store filling [75]. While the mechanistic details of these effects remain to be established, this example illustrates that by ensuring that Ca^{2+} stores are refilled and by serving as a conduit for Ca^{2+} entry, SOCs are almost certain to be involved in multiple aspects of normal Ca^{2+} homeostasis in neurons.

6. The activation mechanism of store-operated Ca^{2+} channels

The mechanism by which store depletion is linked to the activation of SOCs is currently the subject of extensive debate. A large number of possible coupling mechanisms have been proposed over the past decade, several of which have been abandoned in view of negative evidence (reviewed in Refs. [5,16]). Currently, the most popular hypotheses are activation by a diffusible factor, insertion of active channels through vesicle fusion, and activation of SOCs through physical coupling to the IP_3 receptor in the ER membrane.

6.1. Diffusible factor

According to this hypothesis, depletion of Ca^{2+} stores triggers the generation and/or release of a messenger from the stores which diffuses to the plasma membrane to trigger the activation of the store-operated current. Randriamampita and Tsien [76] first tested this idea by fractionating and partially purifying a 'Calcium Influx Factor' (CIF) from Jurkat T cells. Subsequent work showed this fraction to contain several activities, some of which appeared to operate through receptors on the cell surface [77,78]. However, interest in CIF has been revived by recent studies on more purified material that appears to activate or enhance the activity of SOCs when introduced into *Xenopus* oocytes and Jurkat T cells [79]. Cytosolic CIF activity in Jurkat cells is increased by store depletion, and is present in high concentrations in *pmr1* mutant yeast that are deficient for the vacuolar Ca^{2+}-ATPase and show an abnormally high uptake of extracellular Ca^{2+}. The picture is complicated somewhat by the fact that CIF also appears to activate a second Ca^{2+} entry pathway in oocytes and Jurkat cells, and it is difficult to be certain that it does not activate SOCs by facilitating store depletion. A direct effect of CIF on SOCs has been described in excised membrane patches from smooth muscle cells [80]. CIF prepared from *pmr1* yeast, or from activated Jurkat cells or platelets can activate a 3-pS weakly Ca^{2+}-permeable channel that is likely to be store-operated. Further purification of this factor will be an essential step in understanding its possible role in Ca^{2+} signaling.

6.2. Vesicle fusion

This model posits that SOCs are activated through depletion-induced insertion of new channels in the plasma membrane. The hypothesis was first proposed by Fasolato et al. [15] based on the observation that GTPγS, a nonselective inhibitor of small G proteins including those involved in vesicle trafficking, inhibits the induction but not the maintenance of I_{CRAC}. The fusion model might also explain the inhibition of store-operated Ca^{2+} entry by primaquine [81] and by reduced temperature [82], both of which are known to nonselectively inhibit vesicle trafficking. More recently, Yao et al. [25] reported evidence for a role of SNARE proteins in activation of I_{SOC} in oocytes. SNARE proteins are critical components of the vesicle fusion machinery and play important roles in the docking and fusion steps of exocytosis. Injection of oocytes with botulinum neurotoxin A (BoNT A), which is known to inhibit exocytosis in excitable cells by cleaving SNAP-25, significantly inhibited I_{SOC} activation. Moreover, C-terminal truncation mutants of SNAP-25 acted as dominant negative inhibitors of calcium entry, further supporting a role for SNAP-25 in SOC activation. Questions arising from this study include why BoNT E, which also inhibits exocytosis via cleavage of SNAP-25 (at a different site than BoNT A), and BoNT B, which cleaves synaptobrevin, did not affect I_{SOC}. Of particular interest is the issue of whether SNAREs are involved directly in SOC activation or are instead needed to establish or maintain the proper intracellular environment that permits SOC activation.

6.3. Conformational coupling

This hypothesis proposes that SOCs are gated by direct protein–protein interactions with Ca^{2+} release channels such as IP_3 receptors and ryanodine receptors in the ER [7]. Release of Ca^{2+} from stores is postulated to produce a conformational change in the Ca^{2+} release channels, which then triggers the activation of CRAC channels through direct contact. This idea has received indirect support from observations that the signal for SOC activation acts locally [83,84]. Moreover, promoting polymerization of cortical actin sometimes interferes with activation of SOCs, prompting the suggestion that an actin barrier between stores and channels disrupts communication between the two [85]. The strongest evidence for the conformational coupling model has come from studies describing direct interaction between the TRP proteins, suggested to be the molecular correlates of some store-operated Ca^{2+} channels, and IP_3 receptors or ryanodine receptors [86–89]. Coimmunoprecipitation and GST-pulldown experiments have provided biochemical evidence for binding of IP_3 receptors to TRP3, and regions of interaction have been identified [86,87]. Interestingly, overexpression of peptides derived from these regions can enhance or suppress TRP3 function [86–89]. However, the relationship of these results to activation of SOCs is not entirely clear. TRP3 generally does not behave as a SOC [47,90]. Importantly, IP_3 itself is required as a cofactor for activation of TRP3 by IP_3 receptors in microsomes, even when the microsomes are Ca^{2+}-depleted [86,87]. In contrast, CRAC channels in vivo do not require IP_3 and can even be activated by store depletion in the presence of high levels of heparin that would block binding of background levels of IP_3 to its receptor [22]. Most recently, the ability of 2-APB, an IP_3 receptor antagonist, to block the induction and maintenance of TRP3-mediated and native store-operated Ca^{2+} entry in HEK 293 cells has been cited in support of the conformational coupling model [47]. However, subsequent studies indicating that it probably blocks CRAC channels directly raises doubts about this interpretation [48,49].

Finally, two gene suppression studies suggest that IP_3 receptors are not obligatory for activation of store-operated currents. Antisense suppression of the type 1 IP_3 receptor in Jurkat T cells completely blocked Ca^{2+} release in response to stimulation of the T cell receptor, but did not prevent TG-stimulated Ca^{2+} entry [91]. Moreover, Sugawara et al. [53] knocked out all three known isoforms of the IP_3 receptor in the DT40 B cell line to completely disrupt the IP_3-induced Ca^{2+} release pathway, yet TG-induced Ca^{2+} entry in these cells was unaffected. I_{CRAC} measured electrophysiologically also appears to be normal in these mutant cells [49].

7. Molecular composition of SOCs

The molecular identity of SOCs remains a major unresolved issue that, together with the lack of selective pharmacological antagonists, has stymied the study of SOC function and regulation in vivo. Several problems have contributed to the difficulty in identifying the gene codings for SOCs. Expression cloning is difficult because most of the common expression systems express their own version of an endogenous SOC. Biochemical approaches to SOC protein isolation are hindered by the lack of specific high-affinity SOC ligands. Cloning by sequence homology to known Ca^{2+} channels is risky, given the unusual biophysical characteristics of the channel, although recent evidence for similarities in permeation may suggest more similarity than previously thought. Recent attention has been almost solely focused on the mammalian homologues of the *Drosophila* TRP channel, some of which appear to present good candidates for SOCs.

7.1. The origins of the TRP SOC hypothesis

Mutation of the *trp* gene in *Drosophila* abolishes the steady-state response of photoreceptors to light, resulting in a transient receptor potential [91]. Based on genetic evidence for the dependence of *Drosophila* vision on PLC and electrophysio-logical evidence that *trp* encodes a Ca^{2+}-selective channel, Hardie and Minke first proposed that *trp* might encode components of SOCs [93]. According to this view, light activates rhodopsin, which through a G protein stimulates PLC to produce IP_3, which in turn releases Ca^{2+} from stores to activate sustained inward current through TRP.

The best evidence that *trp* encodes a SOC comes from heterologous expression studies. TG stimulates a nonselective current in insect Sf9 cells transfected with *trp* [94] and in HEK 293 cells transfected with *trp* and its homolog, *trpl* [95]. However, extensive studies suggest that in its native environment, TRP does not behave as an SOC. First, there is no evidence that light-induced activation of PLC depletes the fly eye's internal Ca^{2+} store [91]. Second, TG, while producing a prominent rise in intracellular Ca^{2+} due to the depletion of internal stores, does not activate TRP or TRPL [96,97]. Finally, phototransduction is normal in flies with a mutation in the IP_3 receptor that effectively prevents Ca^{2+} release, as well as in flies in which IP_3 receptors are deleted [98,99]. Thus, the hypothesis that light activation of the TRP and TRPL channels is mediated by IP_3-sensitive Ca^{2+} store depletion appears untenable. A more likely alternative supported by recent observations is that PLC is needed to produce diacylglycerol (DAG), which is then metabolized to fatty acids that directly activate the channel. Both the native as well as heterologous TRP and TRPL channels can be activated rapidly in both whole-cell and in excised patches by DAG metabolites such as arachidonic acid and linolenic acid [100]. Further, inhibitory mutations in the diacylglycerol kinase (DGK), which inactivates DAG by converting it into an inactive metabolite, produce constitutive activity of the light-activated channels [101], suggesting that DGK normally helps turn off TRP and TRPL following a light flash. Collectively, these results imply that the light activates TRP/TRPL channels by a mechanism involving the DAG pathway rather than store depletion.

7.2. Classification and functions of mammalian TRPs

Work by many groups over the past 10 years has revealed that a large family of *trp*-related genes is expressed in mammalian cells, and there is avid interest in the possibility that some of these may contribute to SOCs [102–104]. The gene sequences share in common the typical organization of the cation channel family with six transmembrane segments and a hydrophobic pore region between the 5th and 6th transmembrane segments. They fall into three subfamilies, based on length and homology [105]. The long TRPs have a reading frame of ~1600 residues and carry out unknown functions. One member of this group is melanostatin, a protein expressed in melanocytes whose level decreases in metastatic cells. Two other notable members are the recently characterized ion channels LTRPC2 [106,107] and LTRPC7 [108]. LTRPC2 is a nonselective cation channel whose C-terminus functions as a specific ADPR pyrophosphatase [106]. Whole-cell and single-channel analysis of HEK-293 cells expressing LTRPC2 showed that LTRPC2 functions as a calcium-permeable cation channel gated by free adenosine 5'-diphosphoribose (ADPR) and nicotinamide adenine dinucleotide (NAD). It has a unitary conductance of 60 pS, is weakly Ca^{2+} permeable, and is expressed in a variety of cell types including thymocytes. It has been hypothesized that Ca^{2+} influx through LTRCP2 could play an important role in Ca^{2+} signaling in thymocytes. However, little is known about the natural stimuli that regulate endogenous ADPR production or the downstream effects of LTRPC2 activity on cell physiology.

LTRPC7 [108], also known as TRP-PLIK [109], is a widely expressed protein with a C-terminal domain that shares homology with the MHCK/EEF2 family of protein kinases. The exact mechanism of LTRPC7 channel activation is a matter of debate. Runnels and colleagues [109] have proposed that LTRPC7 functions both as an ion channel and a kinase, with the kinase domain phosphorylating the channel domain to promote channel activity. However, Nadler and colleagues [108] have contested this view, and have instead proposed that LTRCP7 channel activity is directly regulated by intracellular Mg^{2+} and MgATP, such that increasing the concentration of Mg^{2+} or MgATP results in a decrease in LTRCP7 activity. LTRCP7 channels have a unitary chord conductance of 40 pS in monovalent ions [108]. While channel activity is not strongly voltage-dependent, extracellular divalent ions block the channels at negative potentials, resulting in outwardly rectifying whole-cell currents. LTRCP7 channels are permeable to Ca^{2+}, although the extent of Ca^{2+} permeability has not been characterized [108,109]. Interestingly, both overexpression and selective knock-out of LTRCP7 in DT40 cells results in cell death [108], leading to speculation that Ca^{2+} signaling via LTRCP7 channels plays an important role in maintaining cellular homeostasis, perhaps in the regulation of ATP levels. Elucidation of the activation mechanism and direct measurements of the Ca^{2+} conductance of LTRPC7 channels will be important steps toward understanding their role in cell physiology.

The OTRP family, with a reading frame of ~900 residues, was named after the homology with *osm-9*, a *C. elegans* gene involved in responses to odorants, hyperosmotic conditions, and mechanical stimuli. Members of this family may be linked to various kinds of sensory transduction; for example, the mammalian members VR1 and VRL1 form channels that are activated by heat and other noxious stimuli [105]. The third subfamily consists of the short TRPs (~900 residues). Seven mammalian members, TRP1–7, have been cloned. Most of the short TRPs have been shown to be activated by receptors that are coupled via G proteins to PLC. However, it is currently unclear whether any of these truly form SOCs.

7.3. Do trp homologs encode store-operated channels?

Two general approaches, overexpression and suppression, have been applied to address the question of whether any of the known *trp* homologs encode the CRAC channel or a SOC in general. With the overexpression approach, the gene of interest is expressed heterologously in *Xenopus* oocytes or in a mammalian cell line (e.g. COS, CHO, or HEK 293). Typically, the cells are then tested for either divalent ion influx using fluorescence techniques, or for whole-cell currents in response to stimuli that deplete the stores. The electrophysiological studies indicate most TRPs produce nonselective cation currents [110–114], suggesting that they are unlikely by themselves to be the pore-forming subunits of CRAC channels. One exception is bovine TRP4, which was reported to form highly Ca^{2+}-selective channels when stably expressed in CHO cells [115]. Further, a mouse knock-out of TRP4 exhibited a dramatic deficit in store-operated Ca^{2+} currents in vascular endothelial cells [116]. These animals also displayed a significant reduction in vasorelaxation, suggesting a vital role for store-operated Ca^{2+} entry in setting vascular tone [116]. However, it is unknown whether store-operated currents were also affected in other tissues, notably in T cells and mast cells, where the properties of CRAC channels have been best characterized.

In addition to TRP4, two recently cloned TRP-related channels, CaT1 [117] and ECaC [118], are also highly Ca^{2+}-selective. These candidates share other properties with CRAC channels as well, including an ability to conduct a significant monovalent current in the absence of external Ca^{2+} [115,118] and fast inactivation [118]. Although initial reports concluded that ECaC and CaT1 are not activated by store depletion when over expressed in HEK cells and oocytes [117,118], a later study found that low levels of expression of recombinant CaT1 resulted in store-dependent Ca^{2+} selective currents in HEK cells [119]. This characteristic, along with the similarities in permeation and gating between CaT1 and CRAC channels led to the proposal that CaT1 comprises at least part of the CRAC channel pore [119]. However, more recent evidence based on a detailed comparison of the ion permeation and biophysical properties of CaT1 and CRAC channels has contested this view and concluded that CaT1 does not form the pore-forming subunit of CRAC channels [120]. Moreover, the reported single-channel conductance of CaT1 (55 ps) is >200 fold larger than the conductance of CRAC channels (~0.2 ps [40]). Thus, CAT1 by itself is unlikely to form the ion conduction pathway.

Very little is known about how the mammalian TRPs are activated. Most TRPs require stimulation of the PLC pathway for activation by external stimuli, and there is at least one report for each one of these members proposing that they are activated by store depletion [105]. Notably, TRP1, 3, 4, and 5 have been reported to generate currents that respond to store depletion when expressed in heterologous systems [112,121–123]. Other studies, however, report that these same channels are not store-operated [90,111, 114,124,125]. In addition, there is a growing body of literature for several TRP members (TRP1, 3, 6, and 7) indicating that they might not be activated by store depletion, but instead by intracellular messengers involving the DAG pathway much like TRP and TRPL [113,124,126].

The reasons for the discrepancies between these different studies are unknown, but it is generally suspected that the functional properties of the channels formed by the recombinant TRPs may depend critically on the presence of other subunits. Over expressing a particular TRP in a cell line might produce channels with functional properties different from SOCs because additional subunits that are needed to confer SOC-like properties may not be present naturally in sufficient quantities to interact with the vast excess of new TRP protein. Thus, the properties of a transfected TRP may depend on its level of expression as well as the cell type in which it is being expressed. For example, it has been suggested that the IP$_3$ receptor may be involved in coupling the activation of TRP3 to store depletion [86]. However, overexpression of recombinant TRP3 produces channels that are constitutively active and are not coupled to store depletion [87]. The presence of such store-independent cation channels has been attributed to an excess of TRP3 relative to endogenous IP$_3$ receptors, hence producing TRP3 channels uncoupled from IP$_3$ receptors [87]. In addition, despite experimental evidence that native TRP1 and TRP4 contribute to endogenous SOCs (see below), when expressed heterologously they are activated by PLC-linked stimuli but not by depletion [114,124]. If native SOCs are heteromultimers, a negative result in an over-expression study may not necessarily rule out a role in store-operated Ca^{2+} entry, as overexpression of a single subunit may lead to nonphysiological homomultimers or heteromultimers with incorrect stoichiometries.

An alternative approach to assess the role of TRP proteins in forming SOCs has been to use antisense constructs to knock down the expression of endogenous TRPs. For example, transfection of a human submandibular gland cell line with antisense *trp*1 significantly reduces the endogenous store-operated Ca^{2+} influx as well as TRP1 protein levels [127]. Similar results have been reported for TRP3 in human vascular endothelial cells [128] and for TRP4 in bovine adrenal cortical cells [129]. In the latter case, transfection with antisense *trp*4 significantly attenuated the endogenous CRAC-like current in parallel with the level of TRP4 protein [129]. If the roles of these TRP proteins in forming SOCs can be confirmed with other methods, these results imply that the usage of particular TRPs to make SOCs may be cell-specific.

Clearly the assignment of a particular gene or genes to any particular SOC in a given cell is a major challenge. The reliability of such an assignment will ultimately depend on a set of experimental tests that extend beyond showing that a particular combination of TRPs is expressed in a given cell with a well-defined SOC current, or that the current gets larger when one of these is overexpressed or smaller when one is suppressed. A truly decisive test should include biochemical evidence that specific TRP proteins are bound to one another in complexes, and that the selectivity or gating of the endogenous SOC current can be changed by introducing a modified TRP. Once a SOC has been unequivocally identified at the molecular level, it may be easier to tackle the nature of the activation signal, and finally approach a solution to the intriguing mystery of how SOCs are controlled.

Acknowledgements

We thank the members of the Lewis laboratory for helpful discussions. This work was supported by a post-doctoral fellowship from the Irvington Institute of Immunology to MP and an NIH grant to RL.

References

[1] Putney, J.W. Jr, 1990. Capacitative calcium entry revisited. Cell Calcium 11, 611–624.

[2] Fomina, A.F., Nowycky, M.C., 1999. A current activated on depletion of intracellular Ca^{2+} stores can regulate exocytosis in adrenal chromaffin cells. J. Neurosci. 19, 3711–3722.

[3] Bouron, A., 2000. Activation of a capacitative Ca^{2+} entry pathway by store depletion in cultured hippocampal neurones. FEBS Lett. 470, 269–272.

[4] Usachev, Y.M., Thayer, S.A., 1999. Ca^{2+} influx in resting rat sensory neurones that regulates and is regulated by ryanodine-sensitive Ca^{2+} stores. J. Physiol. 519(Pt 1), 115–130.

[5] Parekh, A.B., Penner, R., 1997. Store depletion and calcium influx. Physiol Rev. 77, 901–930.

[6] Lewis, R.S., 1999. Store-operated calcium channels. Adv. Second Messenger Phosphoprot. Res. 33, 279–307.

[7] Berridge, M.J., 1995. Capacitative calcium entry. Biochem. J. 312, 1–11.

[8] Parekh, A.B., Fleig, A., Penner, R., 1997. The store-operated calcium current I_{CRAC}: nonlinear activation by $InsP_3$ and dissociation from calcium release. Cell 89, 973–980.

[9] Jacob, R., 1990. Agonist-stimulated divalent cation entry into single cultured human umbilical vein endothelial cells. J. Physiol. 421, 55–77.

[10] Hofer, A.M., Fasolato, C., Pozzan, T., 1998. Capacitative Ca^{2+} entry is closely linked to the filling state of internal Ca^{2+} stores: a study using simultaneous measurements of I_{CRAC} and intraluminal $[Ca^{2+}]$. J. Cell Biol. 140, 325–334.

[11] Fierro, L., Parekh, A.B., 2000. Substantial depletion of the intracellular Ca^{2+} stores is required for macroscopic activation of the Ca^{2+} release-activated Ca^{2+} current in rat basophilic leukaemia cells. J. Physiol. 522, 247–257.

[12] Wickman, K., Clapham, D.E., 1995. Ion channel regulation by G proteins. Physiol. Rev. 75, 865–885.

[13] Barritt, G.J., 1999. Receptor-activated Ca^{2+} inflow in animal cells: a variety of pathways tailored to meet different intracellular Ca^{2+} signalling requirements. Biochem. J. 337, 153–169.

[14] Krause, E., Schmid, A., Gonzalez, A., Schulz, I., 1999. Low cytoplasmic $[Ca^{2+}]$ activates I_{CRAC} independently of global Ca^{2+} store depletion in RBL-1 cells. J. Biol. Chem. 274, 36957–36962.

[15] Fasolato, C., Hoth, M., Penner, R., 1993. A GTP-dependent step in the activation mechanism of capacitative calcium influx. J. Biol. Chem. 268, 20737–20740.

[16] Putney, J.W. Jr, Bird, G.S., 1993. The inositol phosphate-calcium signaling system in nonexcitable cells. Endocr. Rev. 14, 610–631.

[17] Lewis, R.S., Cahalan, M.D., 1989. Mitogen-induced oscillations of cytosolic Ca^{2+} and transmembrane Ca^{2+} current in human leukemic T cells. Cell Regul. 1, 99–112.

[18] Premack, B.A., McDonald, T.V., Gardner, P., 1994. Activation of Ca^{2+} current in Jurkat T cells following the depletion of Ca^{2+} stores by microsomal Ca^{2+}-ATPase inhibitors. J. Immunol. 152, 5226–5240.

[19] Partiseti, M., Le Deist, F., Hivroz, C., Fischer, A., Korn, H., Choquet, D., 1994. The calcium current activated by T cell receptor and store depletion in human lymphocytes is absent in a primary immunodeficiency. J. Biol. Chem. 269, 32327–32335.

[20] Zhang, L., McCloskey, M.A., 1995. Immunoglobulin E receptor-activated calcium conductance in rat mast cells. J. Physiol. 483, 59–66.

[21] McDonald, T.V., Premack, B.A., Gardner, P., 1993. Flash photolysis of caged inositol 1,4,5-trisphosphate activates plasma membrane calcium current in human T cells. J. Biol. Chem. 268, 3889–3896.

[22] Hoth, M., Penner, R., 1992. Depletion of intracellular calcium stores activates a calcium current in mast cells. Nature 355, 353–356.

[23] Hoth, M., Penner, R., 1993. Calcium release-activated calcium current in rat mast cells. J. Physiol. 465, 359–386.

[24] Zweifach, A., Lewis, R.S., 1993. Mitogen-regulated Ca^{2+} current of T lymphocytes is activated by depletion of intracellular Ca^{2+} stores. Proc. Natl. Acad. Sci. USA 90, 6295–6299.

[25] Yao, Y., Ferrer-Montiel, A.V., Montal, M., Tsien, R.Y., 1999. Activation of store-operated Ca^{2+} current in *Xenopus* oocytes requires SNAP-25 but not a diffusible messenger. Cell 98, 475–485.

[26] Neher, E., 1995. The use of fura-2 for estimating Ca buffers and Ca fluxes. Neuropharmacology 34, 1423–1442.

[27] Hoth, M., 1995. Calcium and barium permeation through calcium release-activated calcium (CRAC) channels. Pflügers Arch. 430, 315–322.

[28] Tsien, R.W., Hess, P., McCleskey, E.W., Rosenberg, R.L., 1987. Calcium channels: mechanisms of selectivity, permeation, and block. Annu. Rev. Biophys. Biophys. Chem. 16, 265–290.

[29] Lepple-Wienhues, A., Cahalan, M.D., 1996. Conductance and permeation of monovalent cations through depletion-activated Ca^{2+} channels (I_{CRAC}) in Jurkat T cells. Biophys. J. 71, 787–794.

[30] Kerschbaum, H.H., Cahalan, M.D., 1998. Monovalent permeability, rectification, and ionic block of store-operated calcium channels in Jurkat T lymphocytes. J. Gen. Physiol. 111, 521–537.

[31] Hille, B., 1992. Ionic channels of excitable membranes, 2nd ed. Sinauer Associates, Sunderland.

[32] Christian, E.P., Spence, K.T., Togo, J.A., Dargis, P.G., Patel, J., 1996. Calcium-dependent enhancement of depletion-activated calcium current in Jurkat T lymphocytes. J. Membr. Biol. 150, 63–71.

[33] Vaca, L., Kunze, D.L., 1994. Depletion of intracellular Ca^{2+} stores activates a Ca^{2+}-selective channel in vascular endothelium. Am. J. Physiol. 267, C920–C925.

[34] Delles, C., Haller, T., Dietl, P., 1995. A highly calcium-selective cation current activated by intracellular calcium release in MDCK cells. J. Physiol. 486, 557–569.

[35] Krause, E., Pfeiffer, F., Schmid, A., Schulz, I., 1996. Depletion of intracellular calcium stores activates a calcium conducting nonselective cation current in mouse pancreatic acinar cells. J. Biol. Chem. 271, 32523–32528.

[36] Kerschbaum, H.H., Cahalan, M.D., 1999. Single-channel recording of a store-operated Ca^{2+} channel in Jurkat T lymphocytes. Science 283, 836–839.

[37] Prakriya, M., Lewis, R.S., 2002. Separation and Characterization of Currents through Store-operated CRAC Channels and Mg^{2+}-inhibited Cation (MIC) Channels. J. Gen. Physiol. 119, 487–508.

[38] Hermosura, M.C., Monteilh-Zoller, M.K., Scharenberg, A.M., Penner, R., Fleig, A., 2002. Dissociation of the store-operated calcium current I_{CRAC} and the Mg-nucleotide-regulated metal ion current MagNuM. J. Physiol. 539, 445–458.

[39] Kozak, J.A., Kerschbaum, H.H., Cahalan, M.D., 2002. Distinct properties of CRAC and MIC channels in RBL cells. J. Gen. Physiol. 120, 221–235.

[40] Franzius, D., Hoth, M., Penner, R., 1994. Non-specific effects of calcium entry antagonists in mast cells. Pflügers Arch. 428, 433–438.

[41] Christian, E.P., Spence, K.T., Togo, J.A., Dargis, P.G., Warawa, E., 1996. Extracellular site for econazole-mediated block of Ca^{2+} release-activated Ca^{2+} current (I_{CRAC}) in T lymphocytes. Br. J. Pharmacol. 119, 647–654.

[42] Ross, P.E., Cahalan, M.D., 1995. Ca^{2+} influx pathways mediated by swelling or stores depletion in mouse thymocytes. J. Gen. Physiol. 106, 415–444.

[43] Aussel, C., Marhaba, R., Pelassy, C., Breittmayer, J.P., 1996. Submicromolar La^{3+} concentrations block the calcium release-activated channel, and impair CD69 and CD25 expression in CD3- or thapsigargin-activated Jurkat cells. Biochem. J. 313, 909–913.

[44] Mathes, C., Fleig, A., Penner, R., 1998. Calcium release-activated calcium current (I_{CRAC}) is a direct target for sphingosine. J. Biol. Chem. 273, 25020–25030.

[45] Lepple-Wienhues, A., Belka, C., Laun, T., Jekle, A., Walter, B., Wieland, U., Welz, M., Heil, L., Kun, J., Busch, G., Weller, M., Bamberg, M., Gulbins, E., Lang, F., 1999. Stimulation of CD95 (Fas) blocks T lymphocyte calcium channels through sphingomyelinase and sphingolipids. Proc. Natl Acad. Sci. USA 96, 13795–13800.

[46] Maruyama, T., Kanaji, T., Nakade, S., Kanno, T., Mikoshiba, K., 1997. 2APB, 2-aminoethoxydiphenyl borate, a membrane-penetrable modulator of Ins(1,4,5)P_3-induced Ca^{2+} release. J. Biochem. (Tokyo) 122, 498–505.

[47] Ma, H.T., Patterson, R.L., van Rossum, D.B., Birnbaumer, L., Mikoshiba, K., Gill, D.L., 2000. Requirement of the inositol trisphosphate receptor for activation of store-operated Ca^{2+} channels. Science 287, 1647–1651.

[48] Braun, F.J., Broad, L.M., Armstrong, D.L., Putney, J.W. Jr, 2001. Stable activation of single Ca^{2+} release-activated Ca^{2+} channels in divalent cation-free solutions. J. Biol. Chem. 276, 1063–1070.

[49] Prakriya, M., Lewis, R.S., 2001. Potentiation and inhibition of Ca^{2+} release-activated Ca^{2+} channels by 2-aminoethyldiphenyl borate (2-APB) occurs independently of IP_3 receptors. J. Physiol. 536, 3–19.

[50] Hoth, M., Fasolato, C., Penner, R., 1993. Ion channels and calcium signaling in mast cells. Ann. NY Acad. Sci. 707, 198–209.

[51] Zweifach, A., Lewis, R.S., 1995. Rapid inactivation of depletion-activated calcium current (I_{CRAC}) due to local calcium feedback. J. Gen. Physiol. 105, 209–226.

[52] Fierro, L., Parekh, A.B., 1999. Fast calcium-dependent inactivation of calcium release-activated calcium current (CRAC) in RBL-1 cells. J. Membr. Biol. 168, 9–17.

[53] Sugawara, H., Kurosaki, M., Takata, M., Kurosaki, T., 1997. Genetic evidence for involvement of type 1, type 2 and type 3 inositol 1,4,5-trisphosphate receptors in signal transduction through the B-cell antigen receptor. EMBO J. 16, 3078–3088.

[54] Zweifach, A., Lewis, R.S., 1995. Slow calcium-dependent inactivation of depletion-activated calcium current. Store-dependent and -independent mechanisms. J. Biol. Chem. 270, 14445–14451.

[55] Parekh, A.B., 1998. Slow feedback inhibition of calcium release-activated calcium current by calcium entry. J. Biol. Chem. 273, 14925–14932.

[56] Hoth, M., Fanger, C.M., Lewis, R.S., 1997. Mitochondrial regulation of store-operated calcium signaling in T lymphocytes. J. Cell Biol. 137, 633–648.

[57] Hoth, M., Button, D.C., Lewis, R.S., 2000. Mitochondrial control of calcium-channel gating: a mechanism for sustained signaling and transcriptional activation in T lymphocytes. Proc. Natl Acad. Sci. USA 97, 10607–10612.

[58] Gilabert, J.A., Parekh, A.B., 2000. Respiring mitochondria determine the pattern of activation and inactivation of the store-operated Ca^{2+} current I_{CRAC}. EMBO J. 19, 6401–6407.

[59] Zweifach, A., Lewis, R.S., 1996. Calcium-dependent potentiation of store-operated calcium channels in T lymphocytes. J. Gen. Physiol. 107, 597–610.

[60] Fewtrell, C., 1993. Ca^{2+} oscillations in non-excitable cells. Ann. Rev. Physiol. 55, 427–454.

[61] Beaven, M.A., Rogers, J., Moore, J.P., Hesketh, T.R., Smith, G.A., Metcalfe, J.C., 1984. The mechanism of the calcium signal and correlation with histamine release in 2H3 cells. J. Biol. Chem. 259, 7129–7136.

[62] Mohr, F.C., Fewtrell, C., 1987. Depolarization of rat basophilic leukemia cells inhibits calcium uptake and exocytosis. J. Cell Biol. 104, 783–792.

[63] Dolmetsch, R.E., Lewis, R.S., 1994. Signaling between intracellular Ca^{2+} stores and depletion-activated Ca^{2+} channels generates $[Ca^{2+}]_i$ oscillations in T lymphocytes. J. Gen. Physiol. 103, 365–388.

[64] Foskett, J.K., Roifman, C.M., Wong, D., 1991. Activation of calcium oscillations by thapsigargin in parotid acinar cells. J. Biol. Chem. 266, 2778–2782.

[65] Girard, S., Clapham, D., 1993. Acceleration of intracellular calcium waves in *Xenopus* oocytes by calcium influx. Science 260, 229–232.

[66] Cooper, D.M., Mons, N., Karpen, J.W., 1995. Adenylyl cyclases and the interaction between calcium and cAMP signalling. Nature 374, 421–424.

[67] Alvarez, J., Montero, M., Garcia-Sancho, J., 1991. Cytochrome P-450 may link intracellular Ca^{2+} stores with plasma membrane Ca^{2+} influx. Biochem. J. 274, 193–197.

[68] Serafini, A.T., Lewis, R.S., Clipstone, N.A., Bram, R.J., Fanger, C., Fiering, S., Herzenberg, L.A., Crabtree, G.R., 1995. Isolation of mutant T lymphocytes with defects in capacitative calcium entry. Immunity 3, 239–250.

[69] Fanger, C.M., Hoth, M., Crabtree, G.R., Lewis, R.S., 1995. Characterization of T cell mutants with defects in capacitative calcium entry: genetic evidence for the physiological roles of CRAC channels. J. Cell Biol. 131, 655–667.

[70] Timmerman, L.A., Clipstone, N.A., Ho, S.N., Northrop, J.P., Crabtree, G.R., 1996. Rapid shuttling of NF-AT in discrimination of Ca^{2+} signals and immunosuppression. Nature 383, 837–840.

[71] Feske, S., Giltnane, J., Dolmetsch, R., Staudt, L.M., Rao, A., 2001. Gene regulation mediated by calcium signals in T lymphocytes. Nat. Immunol. 2, 316–324.

[72] Chung, S.C., McDonald, T.V., Gardner, P., 1994. Inhibition by SK and F 96365 of Ca^{2+} current, IL-2 production and activation in T lymphocytes. Br. J. Pharmacol. 113, 861–868.

[73] Mills, G.B., Cheung, R.K., Grinstein, S., Gelfand, E.W., 1985. Increase in cytosolic free calcium concentration is an intracellular messenger for the production of interleukin 2 but not for expression of the interleukin 2 receptor. J. Immunol. 134, 1640–1643.

[74] Yoo, A.S., Cheng, I., Chung, S., Grenfell, T.Z., Lee, H., Pack-Chung, E., Handler, M., Shen, J., Xia, W., Tesco, G., Saunders, A.J., Ding, K., Frosch, M.P., Tanzi, R.E., Kim, T.W., 2000. Presenilin-mediated modulation of capacitative calcium entry. Neuron 27, 561–572.

[75] Leissring, M.A., Akbari, Y., Fanger, C.M., Cahalan, M.D., Mattson, M.P., LaFerla, F.M., 2000. Capacitative calcium entry deficits and elevated luminal calcium content in mutant presenilin-1 knockin mice. J. Cell Biol. 149, 793–798.

[76] Randriamampita, C., Tsien, R.Y., 1993. Emptying of intracellular Ca^{2+} stores releases a novel small messenger that stimulates Ca^{2+} influx. Nature 364, 809–814.

[77] Bird, G.S., Bian, X., Putney, J.W. Jr, 1995. Calcium entry signal? Nature 373, 481–482.

[78] Gilon, P., Bird, G.J., Bian, X., Yakel, J.L., Putney, J.W. Jr, 1995. The Ca^{2+}-mobilizing actions of a Jurkat cell extract on mammalian cells and *Xenopus laevis* oocytes. J. Biol. Chem. 270, 8050–8055.

[79] Csutora, P., Su, Z., Kim, H.Y., Bugrim, A., Cunningham, K.W., Nuccitelli, R., Keizer, J.E., Hanley, M.R., Blalock, J.E., Marchase, R.B., 1999. Calcium influx factor is synthesized by yeast and mammalian cells depleted of organellar calcium stores. Proc. Natl Acad. Sci. USA 96, 121–126.

[80] Trepakova, E.S., Csutora, P., Hunton, D.L., Marchase, R.B., Cohen, R.A., Bolotina, V.M., 2000. Calcium influx factor directly activates store-operated cation channels in vascular smooth muscle cells. J. Biol. Chem. 275, 26158–26163.

[81] Somasundaram, B., Norman, J.C., Mahaut-Smith, M.P., 1995. Primaquine, an inhibitor of vesicular transport, blocks the calcium-release-activated current in rat megakaryocytes. Biochem. J. 309, 725–729.

[82] Somasundaram, B., Mahaut-Smith, M.P., Floto, R.A., 1996. Temperature-dependent block of capacitative Ca^{2+} influx in the human leukemic cell line KU-812. J. Biol. Chem. 271, 26096–26104.

[83] Jaconi, M., Pyle, J., Bortolon, R., Ou, J., Clapham, D., 1997. Calcium release and influx colocalize to the endoplasmic reticulum. Curr. Biol. 7, 599–602.

[84] Petersen, C.C., Berridge, M.J., 1996. Capacitative calcium entry is colocalised with calcium release in *Xenopus* oocytes: evidence against a highly diffusible calcium influx factor. Pflügers Arch. 432, 286–292.

[85] Patterson, R.L., van Rossum, D.B., Gill, D.L., 1999. Store-operated Ca^{2+} entry: evidence for a secretion-like coupling model. Cell 98, 487–499.

[86] Kiselyov, K., Xu, X., Mozhayeva, G., Kuo, T., Pessah, I., Mignery, G., Zhu, X., Birnbaumer, L., Muallem, S., 1998. Functional interaction between $InsP_3$ receptors and store-operated Htrp3 channels. Nature 396, 478–482.

[87] Kiselyov, K., Mignery, G.A., Zhu, M.X., Muallem, S., 1999. The N-terminal domain of the IP_3 receptor gates store-operated hTrp3 channels. Mol. Cell 4, 423–429.

[88] Boulay, G., Brown, D.M., Qin, N., Jiang, M., Dietrich, A., Zhu, M.X., Chen, Z., Birnbaumer, M., Mikoshiba, K., Birnbaumer, L., 1999. Modulation of Ca^{2+} entry by polypeptides of the inositol 1,4,5-trisphosphate receptor (IP_3R) that bind transient receptor potential (TRP): evidence for roles of TRP and IP_3R in store depletion-activated Ca^{2+} entry. Proc. Natl Acad. Sci. USA 96, 14955–14960.

[89] Kiselyov, K.I., Shin, D.M., Wang, Y., Pessah, I.N., Allen, P.D., Muallem, S., 2000. Gating of store-operated channels by conformational coupling to ryanodine receptors. Mol. Cell 6, 421–431.

[90] Zitt, C., Obukhov, A.G., Strubing, C., Zobel, A., Kalkbrenner, F., Lückhoff, A., Schultz, G., 1997. Expression of TRPC3 in Chinese hamster ovary cells results in calcium-activated cation currents not related to store depletion. J. Cell Biol. 138, 1333–1341.

[91] Jayaraman, T., Ondriasova, E., Ondrias, K., Harnick, D.J., Marks, A.R., 1995. The inositol 1,4,5-trisphosphate receptor is essential for T-cell receptor signaling. Proc. Natl Acad. Sci. USA 92, 6007–6011.

[92] Ranganathan, R., Malicki, D.M., Zuker, C.S., 1995. Signal transduction in *Drosophila* photoreceptors. Annu. Rev. Neurosci. 18, 283–317.

[93] Hardie, R.C., Minke, B., 1993. Novel Ca^{2+} channels underlying transduction in *Drosophila* photoreceptors: implications for phosphoinositide-mediated Ca^{2+} mobilization. Trends Neurosci. 16, 371–376.

[94] Vaca, L., Sinkins, W.G., Hu, Y., Kunze, D.L., Schilling, W.P., 1994. Activation of recombinant *trp* by thapsigargin in Sf9 insect cells. Am. J. Physiol. 267, C1501–C1505.

[95] Xu, X.Z., Li, H.S., Guggino, W.B., Montell, C., 1997. Coassembly of TRP and TRPL produces a distinct store-operated conductance. Cell 89, 1155–1164.

[96] Ranganathan, R., Bacskai, B.J., Tsien, R.Y., Zuker, C.S., 1994. Cytosolic calcium transients: spatial localization and role in *Drosophila* photoreceptor cell function. Neuron 13, 837–848.

[97] Hardie, R.C., 1996. Excitation of *Drosophila* photoreceptors by BAPTA and ionomycin: evidence for capacitative Ca^{2+} entry? Cell Calcium 20, 315–327.

[98] Acharya, J.K., Jalink, K., Hardy, R.W., Hartenstein, V., Zuker, C.S., 1997. InsP$_3$ receptor is essential for growth and differentiation but not for vision in *Drosophila*. Neuron 18, 881–887.

[99] Raghu, P., Colley, N.J., Webel, R., James, T., Hasan, G., Danin, M., Selinger, Z., Hardie, R.C., 2000. Normal phototransduction in *Drosophila* photoreceptors lacking an InsP$_3$ receptor gene. Mol. Cell Neurosci. 15, 429–445.

[100] Chyb, S., Raghu, P., Hardie, R.C., 1999. Polyunsaturated fatty acids activate the *Drosophila* light-sensitive channels TRP and TRPL. Nature 397, 255–259.

[101] Raghu, P., Usher, K., Jonas, S., Chyb, S., Polyanovsky, A., Hardie, R.C., 2000. Constitutive activity of the light-sensitive channels TRP and TRPL in the *Drosophila* diacylglycerol kinase mutant, *rdgA*. Neuron 26, 169–179.

[102] Birnbaumer, L., Zhu, X., Jiang, M., Boulay, G., Peyton, M., Vannier, B., Brown, D., Platano, D., Sadeghi, H., Stefani, E., Birnbaumer, M., 1996. On the molecular basis and regulation of cellular capacitative calcium entry: roles for Trp proteins. Proc. Natl Acad. Sci. USA 93, 15195–15202.

[103] Montell, C., 1997. New light on TRP and TRPL. Mol. Pharmacol. 52, 755–763.

[104] Putney, J.W. Jr, 1999. "Kissin' cousins": intimate plasma membrane–ER interactions underlie capacitative calcium entry. Cell 99, 5–8.

[105] Harteneck, C., Plant, T.D., Schultz, G., 2000. From worm to man: three subfamilies of TRP channels. Trends Neurosci. 23, 159–166.

[106] Perraud, A.L., Fleig, A., Dunn, C.A., Bagley, L.A., Launay, P., Schmitz, C., Stokes, A.J., Zhu, Q., Bessman, M.J., Penner, R., Kinet, J.P., Scharenberg, A.M., 2001. ADP-ribose gating of the calcium-permeable LTRPC2 channel revealed by Nudix motif homology. Nature 411, 595–599.

[107] Sano, Y., Inamura, K., Miyake, A., Mochizuki, S., Yokoi, H., Matsushime, H., Brunichi, K., 2001. Immunocyte Ca^{2+} influx system mediated by LTRPC2. Science 293, 1327–1330.

[108] Nadler, M.J., Hermosura, M.C., Inabe, K., Perraud, A.L., Zhu, Q., Stokes, A.J., Kurosaki, T., Kinet, J.P., Penner, R., Scharenberg, A.M., Fleig, A., 2001. LTRPC7 is a MgATP-regulated divalent cation channel required for cell viability. Nature 411, 590–595.

[109] Runnels, L.W., Yue, L., Clapham, D.E., 2001. TRP-PLIK, a bifunctional protein with kinase and ion channel activities. Science 291, 1043–1047.

[110] Boulay, G., Zhu, X., Peyton, M., Jiang, M., Hurst, R., Stefani, E., Birnbaumer, L., 1997. Cloning and expression of a novel mammalian homolog of *Drosophila transient receptor potential* (Trp) involved in calcium entry secondary to activation of receptors coupled by the G$_q$ class of G protein. J. Biol. Chem. 272, 29672–29680.

[111] Okada, T., Shimizu, S., Wakamori, M., Maeda, A., Kurosaki, T., Takada, N., Imoto, K., Mori, Y, 1998. Molecular cloning and functional characterization of a novel receptor-activated TRP Ca^{2+} channel from mouse brain. J. Biol. Chem. 273, 10279–10287.

[112] Zitt, C., Zobel, A., Obukhov, A.G., Harteneck, C., Kalkbrenner, F., Lückhoff, A., Schultz, G., 1996. Cloning and functional expression of a human Ca^{2+}-permeable cation channel activated by calcium store depletion. Neuron 16, 1189–1196.

[113] Hofmann, T., Obukhov, A.G., Schaefer, M., Harteneck, C., Gudermann, T., Schultz, G., 1999. Direct activation of human TRPC6 and TRPC3 channels by diacylglycerol. Nature 397, 259–263.

[114] Schaefer, M., Plant, T.D., Obukhov, A.G., Hofmann, T., Gudermann, T., Schultz, G., 2000. Receptor-mediated regulation of the nonselective cation channels TRPC4 and TRPC5. J. Biol. Chem. 275, 17517–17526.

[115] Warnat, J., Philipp, S., Zimmer, S., Flockerzi, V., Cavalie, A., 1999. Phenotype of a recombinant store-operated channel: highly selective permeation of Ca^{2+}. J. Physiol. 518, 631–638.

[116] Freichel, M., Suh, S.H., Pfeifer, A., Schweig, U., Trost, C., Weissgerber, P., Biel, M., Philipp, S., Freise, D., Droogmans, G., Hofmann, F., Flockerzi, V., Nilius, B., 2001. Lack of an endothelial store-operated Ca^{2+} current impairs agonist-dependent vasorelaxation in TRP4$^{-/-}$ mice. Nat. Cell Biol. 3, 121–127.

[117] Peng, J.B., Chen, X.Z., Berger, U.V., Vassilev, P.M., Tsukaguchi, H., Brown, E.M., Hediger, M.A., 1999. Molecular cloning and characterization of a channel-like transporter mediating intestinal calcium absorption. J. Biol. Chem. 274, 22739–22746.

[118] Vennekens, R., Hoenderop, J.G., Prenen, J., Stuiver, M., Willems, P.H., Droogmans, G., Nilius, B., Bindels, R.J., 2000. Permeation and gating properties of the novel epithelial Ca^{2+} channel. J. Biol. Chem. 275, 3963–3969.

[119] Yue, L., Peng, J.B., Hediger, M.A., Clapham, D.E., 2001. CaT1 manifests the pore properties of the calcium-release-activated calcium channel. Nature 410, 705–709.

[120] Voets, T., Prenen, J., Fleig, A., Vennekens, R., Watanabe, H., Hoenderop, J.G., Bindels, R.J., Droogmans, G., Penner, R., Nilius, B. 2001. CaT1 and the calcium-release activated calcium channel manifest distinct pore properties. J. Biol. Chem. 30, 30.

[121] Zhu, X., Jiang, M., Peyton, M., Boulay, G., Hurst, R., Stefani, E., Birnbaumer, L., 1996. *trp*, a novel mammalian gene family essential for agonist-activated capacitative Ca^{2+} entry. Cell 85, 661–671.

[122] Philipp, S., Cavalie, A., Freichel, M., Wissenbach, U., Zimmer, S., Trost, C., Marquart, A., Murakami, M., Flockerzi, V., 1996. A mammalian capacitative calcium entry channel homologous to *Drosophila* TRP and TRPL. EMBO J. 15, 6166–6171.

[123] Philipp, S., Hambrecht, J., Braslavski, L., Schroth, G., Freichel, M., Murakami, M., Cavalié, A., Flockerzi, V., 1998. A novel capacitative calcium entry channel expressed in excitable cells. EMBO J. 17, 4274–4282.

[124] Lintschinger, B., Balzer-Geldsetzer, M., Baskaran, T., Graier, W.F., Romanin, C., Zhu, M.X., Groschner, K., 2000. Coassembly of Trp1 and Trp3 proteins generates diacylglycerol- and Ca^{2+}- sensitive cation channels. J. Biol. Chem. 275, 27799–27805.

[125] Sinkins, W.G., Estacion, M., Schilling, W.P., 1998. Functional expression of TrpC1: a human homologue of the *Drosophila* Trp channel. Biochem. J. 331, 331–339.

[126] Okada, T., Inoue, R., Yamazaki, K., Maeda, A., Kurosaki, T., Yamakuni, T., Tanaka, I., Shimizu, S., Ikenaka, K., Imoto, K., Mori, Y., 1999. Molecular and functional characterization of a novel mouse transient receptor potential protein homologue TRP7. Ca^{2+}-permeable cation channel that is constitutively activated and enhanced by stimulation of G protein-coupled receptor. J. Biol. Chem. 274, 27359–27370.

[127] Liu, X., Wang, W., Singh, B.B., Lockwich, T., Jadlowiec, J., O'Connell, B., Wellner, R., Zhu, M.X., Ambudkar, I.S., 2000. Trp1, a candidate protein for the store-operated Ca^{2+} influx mechanism in salivary gland cells. J. Biol. Chem. 275, 3403–3411.

[128] Groschner, K., Hingel, S., Lintschinger, B., Balzer, M., Romanin, C., Zhu, X., Schreibmayer, W., 1998. Trp proteins form store-operated cation channels in human vascular endothelial cells. FEBS Lett. 437, 101–106.

[129] Philipp, S., Trost, C., Warnat, J., Rautmann, J., Himmerkus, N., Schroth, G., Kretz, O., Nastainczk, W., Cavalié, A., Hoth, M., Flockerzi, V., 2000. TRP4 (CCE1) protein is part of native calcium release-activated Ca^{2+}-like channels in adrenal cells. J. Biol. Chem. 275, 23965–23972.

Neurodegenerative disease and the neuroimmunobiology of glutamate receptors

Lorise C. Gahring,[a,b] Noel G. Carlson,[a,b] Erin L. Meyer,[a,b]
Emily L. Days[a,b] and Scott W. Rogers[a,b,*]

[a]Geriatric Research, Education, and Clinical Center, Salt Lake City VA Medical Center, Salt Lake City,
UT 84132, USA
[b]The University of Utah School of Medicine, Salt Lake City, UT 84132, USA
*Correspondence address: Department of Neurobiology and Anatomy, University of Utah School of Medicine,
MREB 403, Salt Lake City, UT 84132
E-mail: scott.rogers@hsc.utah.edu

A hallmark of many neurodegenerative diseases is the extraordinary specificity through which they progress through the nervous system. Even in the severest forms of neurodegeneration, it is not uncommon to find that discrete regions of brain are destroyed while adjacent regions are spared. In this review, we present a brief introduction to the family of glutamate receptors [1,2], which provide the principal means of fast-excitatory neurotransmission in the brain, and their likely role in the progression of many neurodegenerative diseases. We will then focus on the role of the immune system in neurologic disease and the ways in which we might utilize the specificity of the immune system to explore further how the GluR family contributes to the etiology and regional specificity of neurodegenerative disease progression.

1. Glutamate-activated ligand-gated ion channels (GluRs)

Neurotransmission in the brain involves the balanced activity of excitatory and inhibitory neurotransmitter receptor systems acting through ligand-activated ion channel receptors. Glutamate-activated ligand-gated ion channels (GluRs) are the primary receptors responsible for fast excitatory neurotransmission in the brain, and their activation is central to physiological processes associated with learning and memory. However, aberrant activity of GluRs, particularly excessive activation, can lead to pathophysiological consequences such as epilepsy and neuronal death, including that associated with stroke [3–5] and neurodegenerative disorders such as Parkinson's disease [4,6] and Alzheimer's disease [5]. These diseases often exhibit a very selective process of neuronal death that may in fact be regulated by the localization of neurotransmitter subunit expression.

Advances in Molecular and Cell Biology, Vol. 32, pages 141–159
© 2004 Elsevier B.V. All rights of reproduction in any form reserved.
ISSN: 1569-2558 / DOI: 10.1016/S1569-2558(03)32007-7

Glutamate receptors are assembled from various combinations of at least 17 distinct subunits that are expressed in unique but overlapping regions of the mammalian brain (Fig. 1; for extensive review see Refs. [1,2]). These subunits are grouped into one of three major categories based upon their ability to associate into functional multimeric channels containing 4 or 5 subunits and their activation by the selective agonists n-methyl-*d*-aspartic acid (NMDA), kainic acid (KA), or α-amino-3-hydroxy-5-methyl-isoxazole-4-propionic acid (AMPA). Of the non-NMDA receptor subunits, GluR1–4 assemble with each other in various combinations to collectively form the AMPA receptor family, while GluR5–7 and KA1–2 assemble to form the KA receptors. Assembly of

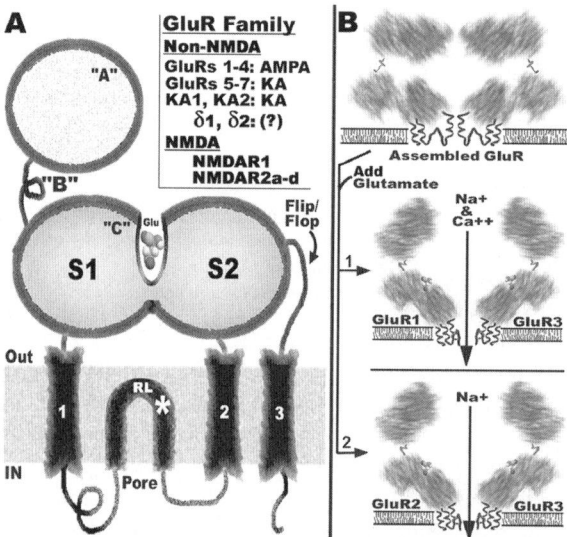

Fig. 1. The glutamate receptor family. In Panel A is a diagram of a typical GluR subunit (see Ref. [2] for detailed review). The structure is thought to be composed of two extracellular domains separated by three transmembrane domains (TM 1–3), and a "re-entry loop" (RL) that is believed to line the ion channel pore. The extracellular region prior to TM1 is divided roughly into two subdomains connected through a "hinge" region [51]. The subdomain, also termed "S1" folds with the second extracellular region termed "S2" domain to form the ligand-binding site where the major agonist, glutamate (Glu) binds. The site of Q/R editing in GluR2 is within the RL and is denoted by an asterisk. The location of the major Flip/Flop alternative splice variants on the extracellular face before TM3 are indicated. The approximate locations of epitopes designated "A", "B" and "C" are regions of poor sequence identity between different receptor subunits, and have been identified to react with autoantibodies in patients with various neurological diseases (see text). The insert lists the currently known GluR subunits. These can be divided into the non-NMDA class consisting of subunits that when assembled form predominantly GluRs that are activated by AMPA or KA (see text). The function of the delta subunits is currently unclear but they are known to participate in the proper function of neurotransmission in the cerebellum and other brain regions. NMDA receptors consist of five subunits of which NMDAR1 appears to be a subunit in common with possibly all functional NMDA receptors. In Panel B is a cross-section through a membrane revealing two subunits within a glutamate receptor complex. Upon binding of an agonist such as glutamate, the receptor undergoes a conformational change that opens the channel and allows cations to flow into the cell from the extracellular space. The cation selectivity through the channel varies with subunit composition, as illustrated in part B, where (1) a GluR composed of GluR1 and GluR3 subunits permits both the entry of sodium and calcium, and (2) a GluR assembled from GluR2 and GluR3 allows sodium to enter the cell but is impermeable to calcium. Similar scenarios are seen in all GluRs that vary in detail with subunit incorporation (see Ref. [2]).

different combinations of subunits within each of these receptor families imparts different pharmacological and electrophysiological properties. For instance, selective conductance of the ions Na^+, K^+, and Ca^{2+} by the activated AMPA receptor complex is altered by subunit composition (see Fig. 1b). Receptor complexes formed from combinations of GluR subunits lacking GluR2 are permeable to Ca^{2+} upon activation, while inclusion of the GluR2 subunit results in a receptor complex without appreciable calcium conductance [7,8].

Part of this functional diversity is also achieved by post-transcriptional modification of GluRs, including RNA editing and alternative splicing of transcripts (see Fig. 1a). For example, the regulation of Ca^{2+} permeability by GluR2 (as well as GluR5 and GluR6) is controlled by RNA editing [9–12], where the primary transcript for GluR2 undergoes an enzymatically catalyzed selective adenosine deamination. This modification results in a substitution from the glutamine (Q) codon (CAG) encoded by the genomic copy of the gene to an edited arginine (R) codon (CIG). The consequences of the substitution at this "Q/R site" in the M2 region of the protein (see Fig. 1) are low calcium permeability, lowered channel conductance, and altered rectification properties. Alternative RNA splicing also affects the functional properties of GluRs [13–15]. Transcripts of GluRs1–4 each exist as one of two splice variants, termed "Flip" or "Flop" depending on whether exon 14 or 15 is inserted just before the M4 region in the mature transcript. The Flip version (containing exon 14) differs from the Flop version by only five amino acids in this 38 amino acid long exon, yet desensitizes more slowly than the Flop form. This alternative splicing is regulated developmentally, with Flip variants predominating in the prenatal and early postnatal animal, and the Flop version eventually reaching levels similar to Flip in the adult animal. Therefore, because of the functional diversity resulting from regional expression of GluR subunits and their processed forms, alteration of the function of a specific GluR subunit not only directly affects neurons that express that receptor subtype, but also indirectly affects those regions of the brain dependent upon the fidelity of the neurotransmission by the affected GluR-expressing neurons. As a consequence, the dysfunction of one subunit in the receptor system could impart regional specificity to a disease process that would be closely related to the expression pattern for the subunit in question and reflect its unique role in neurotransmission.

2. Immune system specificity as a tool to examine glutamate receptor expression and function

Although an attractive hypothesis for specificity in the etiology of neurological disease can be attributed to the dysfunction of key GluR subunits, there remains the problem of explaining how specific subunits can become dysfunctional and how this contributes to disease etiology and progression. In an effort to understand the molecular mechanisms that contribute to neurodegeneration, we have proposed that in certain cases neuronal GluR subunits become targets of an autoimmune process that results in altered receptor function in a region-specific manner (see Fig. 2).

Autoimmune diseases afflict a significant and growing portion of the population of industrialized nations (e.g. Ref. [16]). Further, susceptibility to autoimmune disorders

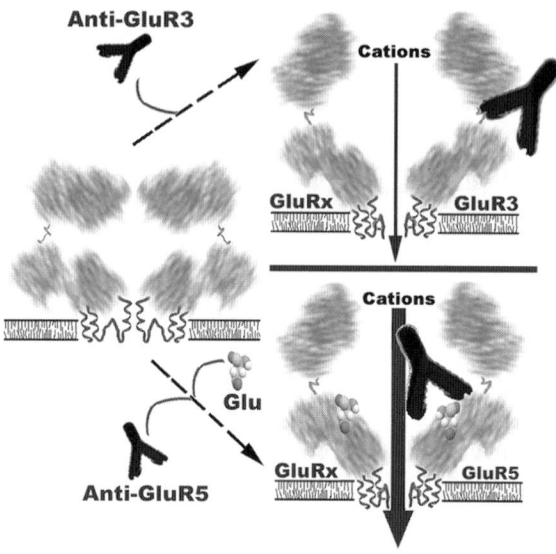

Fig. 2. Mechanisms of subunit specific activation or allosteric modulation of GluRs by autoantibodies. In this diagram GluRs are depicted similarly to those in Fig. 1b. In the upper diagram an anti-GluR3 agonist antibody binds to the glutamate receptor at the GluR3B sequence (hinge) and opens the channel through induction of conformational changes that mimic endogenous agonist binding. In the lower panel, glutamate binds to a site near the site of binding of anti-GluR5. In this case, there is enhanced current suggesting allosteric regulation by this autoreactivity. In this scenario, autoantibody alone does not alter detectably the receptor function. Although the antibody shown is bound to the approximate site of the "C" epitope (see Fig. 1a and text), its apparent location within the channel is a reflection of our pre-Renaissance artistic ability and is not meant to reflect the actual location of the antibody. (*For a colored version of this figure, see plate section, page 462.*)

increases with age [17] and the aged population is the most rapidly expanding segment of the population in the United States. Invoking autoimmune dysfunction as a mechanism to explain selectivity in neuronal loss reflects upon the basic tenet of the highly selective capabilities of the immune system for recognizing invading organisms and mounting a response that ultimately results in the preservation of the host. This recognition of self versus non-self includes a complex process in which self-reactive lymphocytes are selectively destroyed during maturation in the thymus [18,19]. Unfortunately, the immune system can fail in its ability to distinguish our own nucleic acids, sphingolipids, or proteins from those of the foreign agent. When this occurs, the immune system initiates an attack on self-tissues that express the target molecules through either cell-mediated processes involving T-lymphocytes and/or humoral processes where B-lymphocytes and antibody-mediated processes dominate. At present, what initiates the process that leads to the inability to distinguish between self- and non-self molecules is a topic of considerable research and debate. This topic will be returned to later in this discussion as it pertains to our examination of Rasmussen's encephalitis (RE) and other neurological disorders as models for understanding how GluRs become the target of the complex mechanisms that underlie and contribute to autoimmune-mediated neurological dysfunction and disease.

3. Rasmussen's encephalitis as a model for autoimmune neurodegeneration

RE was first described by Theodore Rasmussen over three decades ago (see Ref. [20] for review). It is a rare childhood disease of relentless and intractable focal seizures. The diagnosis of RE is histologically defined by the appearance of lymphocytic perivascular cuffing and microglial nodules that are generally restricted to one hemisphere of the brain. Included in these nodules are both microglial cells (inflammatory cells) and lymphocytes. The role of these microglial nodules in the pathogenic process of RE is unknown, and they have been used mostly as a diagnostic tool. However, valuable information is emerging from recent examination of these nodules. Quantitative analysis of the T-lymphocytes taken from the brain lesions of an RE patient indicates that these lymphocytes likely expanded from a few precursor T-cells responding to a discrete antigenic epitope [21]. Whether the microglial cells in the area serve as sources for cytokines and antigen presentation is not presently known.

Diseases grouped on the basis of similar clinical presentations rarely have a single causative agent in common. In fact, as the molecular mechanisms that contribute to diseases become apparent, we have come to expect that multiple pathways can lead to diseases of very similar clinical presentation. A particularly striking example of this has been the myasthenia syndromes, where despite a similarity in outcome of the disease, they may originate from an autoimmune attack on the muscle nicotinic acetylcholine receptor [22,23] or result from genetic defects within the receptor itself (see Ref. [24], and this volume). Therefore, proclamations of single pathways must be viewed with caution, particularly in diseases such as RE, where clinical diagnosis is often subjective and the incidence of disease is rare, occurring on the order of 30 proven cases in the United States per year [20,25].

Several hypotheses invoking the contribution by the immune system have been put forth to explain the origin of RE, and they have been based in part on the prominent inflammatory process in the brain (see Ref. [20] for reviews). These include the HLA haplotype as a genetic marker of disease susceptibility, or an underlying response to a viral component. For example, a viral etiology of RE has been widely explored, and there is evidence [26–29] for the occurrence of cytomegalovirus (CMV) or other Herpesvirus family members in RE patients. Notably, administration of ganciclovir [30] or the direct injection of interferon into the brain [31,32] has had efficacy in controlling seizures of some RE patients. Although the isolation of virus from brain tissue from RE patients or the identification of viral antigens has been largely unsuccessful, they cannot be readily dismissed as contributors to the disease, a point that will be returned to later in this discussion.

4. How an autoantibody contributes to the etiology of neurodegenerative disease: the autoimmune hypothesis for Rasmussen's encephalitis

For the past several years, we have examined the possibility that an autoimmune process underlies the etiology of disease in some cases of RE, and that it targets a glutamate receptor subunit in the brain termed GluR3. Our interest in RE as a model of neurological autoimmune disease stemmed from the observation that rabbits injected with a portion of the GluR3 subunit protein developed both symptoms and pathology similar to those observed in patients with RE [25]. Follow-up studies revealed that blood samples from RE patients harbored autoantibodies to GluR3 and that anti-GluR3 titers correlated with seizure frequency and disease severity, which diminished with autoimmune therapy such as plasmapheresis [25]. Finally, a key aspect of the anti-GluR3 antibodies in patients or rabbits was revealed when we demonstrated that these antibodies could directly activate GluRs expressed by cortical neurons maintained in culture for 3 weeks [33,34]. This observation led us to speculate [33–35] that the over-excitation of glutamatergic neurons by the constant exposure of susceptible neurons to antibody would create an electrical imbalance, consistent with both the clinical manifestation of intractable seizures in RE patients, and with the removal of the immune component resulting in clinical improvement.

The autoimmune hypothesis of RE has received additional support from another group [60,61] who reproduced our original findings using the mouse as the model organism and extended them to demonstrate that anti-GluR3 antibodies could cause neuronal cell death through a mechanism termed excitotoxicity. Excitotoxicity is the process whereby neuronal death results from over-excitation of GluRs [2,36,37]. The possibility that the dysfunction of a key receptor system can impart the recognized regional specificity and phenotypic outcome of a disease process is attractive, and as noted earlier has gained in popularity as an explanation for the selective neuronal loss and etiology of many neurological diseases. However, the origin of glutamate receptor dysfunction and the mechanism(s) contributing to neuronal death in discrete regions or neuronal groups in the brain remains to be determined. One possibility is the presence of autoantibodies that target neurotransmitter receptors and alter their normal function in the brain, leading to the excitotoxicity and other related mechanisms that contribute to neurodegenerative disorders.

5. Autoantibody modulation of receptor function and excitotoxicity

Antibodies have been used as highly specific modulators of receptor function by immunologists for many years. One need only examine the extensive lists found in bio-technology catalogs describing antibodies that specifically alter the function of cytokines, chemokines, or their respective receptors. But receptors expressed by the brain can also be targets of autoantibodies that alter the function of their targets. Examples of autoantibody modulation of CNS targets include voltage-gated calcium channels in Lambert–Eaton syndrome (and in some cases of sporadic amyotrophic lateral sclerosis), potassium channels as in Isaacs' syndrome (neuromyotonia), and sodium channels in some patients with acute inflammatory demyelinating polyneuropathy and Guillain–Barré syndrome (for examples see Refs. [38–43]). Autoantibodies that alter the function of glutamatergic metabotropic receptors have been found in the serum of patients with paraneoplastic disease [48] as well as autoantibodies that alter enzymes involved in neurotransmitter biosynthesis such as glutamic acid decarboxylase (GAD) [44]. Autoantibodies that modulate nicotinic acetylcholine receptor function [23,45,46] or thyroid hormone receptor signaling such as in Graves' disease [47] have been known for many years. In our studies, we have examined in detail autoantibodies to GluR3 that function as agonists of the receptor and an autoantibody to GluR5 that was discovered to have properties of an allosteric modulator (Fig. 2). The mechanism of autoantibody interaction with GluRs is described below. Notably, autoantibodies to GluRs and other neuronal proteins must first gain access to their targets through the blood–brain barrier. The added complication found in many autoimmune diseases of how autoantibodies gain access to intracellular proteins to disrupt their function is not a major obstacle since anti-GluR epitopes are extracellular. We have discussed previously several possibilities for how an autoantibody can access the brain through the blood–brain barrier [35,49], and consequently we will not repeat them here. However, it is important to remember that the permeability of the blood–brain barrier is closely associated with local factors, and can be increased dramatically by infection, fever, trauma or focal seizures, which are thought to contribute to the etiology of RE and other autoimmune neurological disorders.

6. Autoantibodies that act as direct agonists of GluRs

While our initial studies suggested a possible link between GluR3 autoimmunity and neuronal hyperexcitability, the mechanism remained unclear. This was solved when IgG from GluR3-injected rabbits or from active RE patients was demonstrated to directly activate GluRs expressed on a subset of cultured mouse cortical neuronal cells [33–35,49]. Notably, application of antibody to neurons induced rapid and reversible non-desensitizing inward currents in as many as 45% of the KA-responsive neurons. Further, antibody-evoked currents were essentially abolished by the competitive antagonist CNQX, indicating that antibodies were binding at sites similar to or nearby those that would bind typical glutamate receptor agonists (see Fig. 2).

In RE, a portion of the glutamate receptor subunit 3 (GluR3) is the antigenic determinant or epitope that is recognized by disease-associated antibodies. A protein has many epitopes (usually 9–22 amino acids) that can be linear or discontinuous as defined by conformations imparted by amino acid constituents of the sequence. The minimum GluR3 protein region required to produce an agonist-like antibody was identified as the region termed GluR3B (Fig. 3; [34]). Using a panel of GluR3B peptide carboxy-terminal deletion mutants as immunogens, the minimum GluR3B sequence that was alone sufficient to produce anti-GluR immunoreactivity was demonstrated to be: "NEYERFVPFSDQQISND" (AA 372–386; Fig. 3a). Antibody recognition of the GluR3B region was examined by the ability of different peptides to block anti-GluR3 agonist antibodies. Peptides that lacked the carboxyl ND residues still blocked while those lacking the C-terminal "ISND"-amino acids were inactive. Alanine substitution mutagenesis in GluR3B [34] revealed the most common epitope to be centered on the linear "VPF" tri-amino acid sequence, but key amino acids also included E375 and I385. Notably, this analysis also showed that binding by anti-GluR3 antibodies to F380 imparted both anti-GluR3 subunit specificity and relative agonist efficacy (Fig. 3a).

The GluR3B epitope resides immediately next to the N-terminus of a portion of the GluR protein originally proposed by molecular modeling and chimeric protein studies to form a portion of the ligand binding site [50–52] whose X-ray structure for the

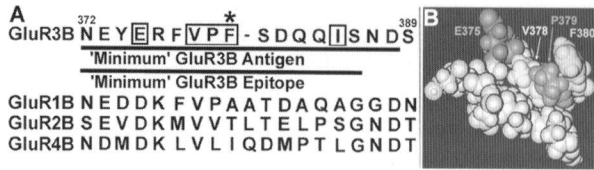

A
GluR3B N E Y E̲ R F V̲P̲F̲ - S D Q Q I̲ S N D S
                        *
'Minimum' GluR3B Antigen
'Minimum' GluR3B Epitope
GluR1B N E D D K F V P A A T D A Q A G G D N
GluR2B S E V D K M V V T L T E L P S G N D T
GluR4B N D M D K L V L I Q D M P T L G N D T

Fig. 3. Alignment of sequences homologous to GluR3B in closely related GluR subunits and the structural model of GluR3B. Panel A shows the regions homologous to GluR3B in GluRs1–4 (GluR1B, GluR2B and GluR4B, respectively). Key residues important for antibody immunoreactivity to GluR3B and subsequent GluR activation are boxed, and phenylalanine F380 which is important for agonist efficacy and subunit specificity of anti-GluR3 antibodies is denoted with an asterisk. The gap is placed in GluR3B to optimize the alignment with homologous regions of other GluR subunits. The minimum sequence required to generate an antigen capable of producing antibodies that are receptor agonists is noted. The minimum sequence required to block the agonist-like anti-GluR3B antibody is also noted (for details see the text and Ref. [34]). Panel B shows a space-filling structural model of GluR3B (residues 372–387; NEYERFVPFSDQQISN) described by McDonald et al. [55]. The key residues important for antibody binding to this sequence (which are boxed in panel A) are colored and numbered. Note that in three-dimensional space they align to form a "linear" epitope sequence. I385 is buried below V378 and is not visible in this view. Amino acids are identified by the commonly used single letter abbreviations. (*For a colored version of this figure, see plate section, page 463.*)

ligand-binding portion of GluR2 was recently solved [53]. Notably, deletion of the GluR3B sequence and all amino acids preceding it does not alter [^{3}H]ligand binding [51,54], but unfortunately this portion of the protein was not included in the final X-ray structure [53]. However, it was proposed to form a "link" or "hinge" between the ligand-binding region of the GluR and an additional extracellular domain [35,50,51] whose function is not yet defined (see Figs. 1 and 2 and below). To explore how receptor activation upon binding of antibody to GluR3B might occur, additional mutagenesis and molecular modeling were used to generate a model of the GluR3B peptide.

A structural model of GluR3B [55] offers suggestions as to how specific regions of GluR subunit sequence variability might translate into functional diversity upon binding by antibody (Fig. 3b). The tightly compact structure suggested for the GluR3B region was particularly intriguing. In this model, the GluR3B epitope is located in a reverse hairpin loop that places the key residues important for antibody recognition and receptor activation, E375, V378, P379 and F380, in a linear arrangement on the solvent-exposed surface. The C-terminal portion of the peptide forms a loop that is stabilized by a hydrophobic core centered on I385 that is critical for functional integrity of the epitope. Changing this residue to an alanine not only reduces immunoreactivity [34], but it substantially opens the structure, as suggested by energy minimization tests [55]. The proximity of the amino-terminal and carboxy-terminal residues also suggested that immunoreactivity to the GluR3B peptide would be unaffected when it is cyclized by replacement of these residues with cysteines to form a disulfide bond. This prediction was confirmed experimentally [55] since the cyclized peptide retained full immunoreactivity. Notably, cyclic peptides have been found to be useful in investigations examining the preferred conformations of constrained short peptides since they often have the ability to mimic surface loops on proteins. This has provided key structural insights into how critical amino acids within protein regions impart their function, since peptides have been used to model the active site of serine proteases, RGD-containing proteins, human growth hormone, adrenocorticotropic hormone, VCAM-1 and melanin-concentrating hormone (see Ref. [55] for discussion).

The model of GluR3B offers several possibilities for how this region could participate in transducing motion through receptor protein domains following ligand binding, and suggests that antibody binding would mimic these structural alterations [55]. As noted above, the compact configuration of GluR3B places the amino-terminal and Carboxy-terminal residues in close proximity, which is in part stabilized by a proline and a hydrophobic core. Consequently, motion between protein domains could be mediated through disruption of the hydrophobic core of this region or possibly through proline isomerization, as has been reported to be important in nicotinic acetylcholine receptor function [56]. Further, if the GluR3B region undergoes conformational changes during agonist-mediated activation, the antibody may no longer bind due to displacement of key contact residues, which could explain the rapid reversal of this antibody during receptor activation [33,34]. Although this model can be used to address these issues, this possibility awaits evaluation of the GluR3B region in the context of the entire receptor structure. However, this provides an important structural framework to begin to understand the binding of subunit-specific, agonist-like antibodies to GluRs, and to design novel therapeutic tools for intervening when such autoantibodies are present.

Immunoglobulin activation of receptors containing GluR2 was also found in a patient with olivopontocerbellar atrophy (OPCA; [57]). Notably, this autoimmune IgM bound GluR2 at the site homologous to GluR3B. The GluR agonist activity was blocked by CNQX, a competitive AMPA receptor antagonist, and by a GluR2 specific peptide (GluR2B). However, immunohistochemical studies with this antibody did not produce the extensive pattern of expression expected for GluR2, but rather only intense staining of cells in the cerebellum and pons, and some relatively weak staining of forebrain striatal and thalamic structures [57]. This suggests further complexity in interpreting how an autoantibody may produce disease, since their interaction with the target epitope may occur in only receptors of certain subunit composition or only in a specific cellular context. This has been observed for antibodies prepared to the extracellular domain of GluR1 where they preferentially label receptors that are extra synaptic despite the presence of concentrated GluR1 within the post-synaptic density (see Refs. [58,59]). The occurrence of autoantibodies to homologous but sequence diverse regions of different GluR subunits suggest a mechanism through which autoantibodies could impart region-specific alterations in neurotransmission. Hence, only neurons that express the appropriate subunit or the subunit in a particular context that is accessible to the antibody would be modulated in its function by the immunoglobulin.

7. The generation of autoantibodies to GluR3 in mice

In a recent series of papers, Levite and colleagues [60,61] presented a comprehensive set of experiments using various mouse strains and GluR3 peptide antigens to demonstrate the efficacy of these peptides toward generation of autoantibodies and pathogenesis associated with disease. GluR3 peptides used for immunization included GluR3A (residues 245–274) and GluR3B (residues 372–395), and a scrambled peptide of GluR3B.

Results demonstrate that mouse anti-GluR3B antisera (as well as purified antibody) was capable of binding neurons, evoking GluR ion channel activity, and killing neurons. Both neuronal activation and killing were blocked by either CNQX or the purified GluR3B peptide. Heat inactivation of complement proteins in mouse serum samples had no effect on neuronal toxicity, suggesting that complement fixation and activation was not playing a role in this experimental system. Mice immunized with GluR3B peptide also expressed a biased frequency of particular T-cell receptor Vβ chains (Vβ7, Vβ8, and Vβ11) compared to control mice. While mouse anti-GluR3A antisera and purified antibody were capable of binding neurons, neither was capable of killing neurons. Further, the antisera obtained by immunization with scrambled GluR3B peptide had no effect on neuronal activity or viability. These studies point to a complement-independent pathway of neuronal toxicity induced by autoantibodies to the region of GluR3 (GluR3B) previously mapped as the immunogenic region in RE [34,35]. While Levite et al. [60,61] did not observe seizures in any of the mouse strains employed (BALB/c, C3H/HeJ, SJL/J, and C57BL/6), they did note a change in brain pathology of the mice immunized with GluR3B. These changes included thickening of cerebral meninges with lymphocyte infiltrates, loss of Purkinje cells, occasional gliosis, and spongiform degeneration of white matter. Further, subclinical behavioral activity was noted, which in the mice immunized with the GLUR3B peptide included abnormally slow motor activity.

8. Autoantibodies as allosteric modulators of GluR function

Autoantibodies to other GluR subunits have also been identified from the serum of a subset of patients exhibiting features of paraneoplastic neurodegenerative syndrome [62]. While the occurrence of these autoantibodies is not likely to prove to be a reliable marker of this syndrome [72], when present these anti-GluR5 or anti-GluR6 autoantibodies have the interesting allosteric effect of enhancing receptor activation when co-applied with a traditional agonist such as glutamic acid to cultures of cortical neurons [62].

Epitope mapping of the autoreactivity to GluR5 [63] identified residues that span a region containing amino acids important for glutamic acid and KA interaction, but well away from the GluR3B epitope (Fig. 4). In this case, residues K497, N508, D509, K510 and E512 were found to be required for autoantibody binding (Fig. 4a). Residues N508 and E512 impart anti-GluR5 subunit specificity. As noted above, the binding site of the GluR agonists is formed by the interaction between the extracellular "S1" domain and a second extracellular domain termed "S2" [51,53]. The X-ray crystal structure of the KA-bound form of the S1S2 complex has been determined (Fig. 4b; [53]), and reveals a bi-lobed molecule with agonist bound in the cleft between the two lobes. Alignment of the GluR2 and GluR5 sequences identified the homologous residues in this structure to which anti-GluR5 modulating antibodies bind (Fig. 4b). In this model, residues N508, D509, K510 and E512 were located within a relatively exposed structure on the surface of the S1 domain termed "Loop2" [53]. The distal location of K497, which is found in the middle of the 4th beta strand 16 residues away from E512 (25Å), is well within the 40Å distance that can be spanned in the antigen-interaction residues of an immunoglobulin hypervariable region (see [63]).

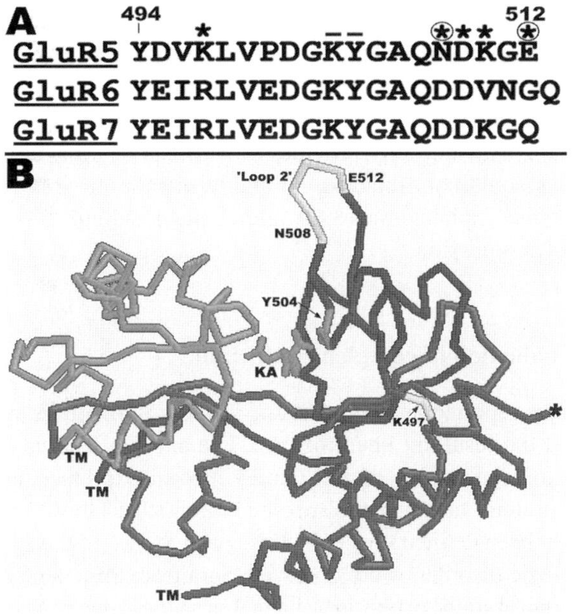

A

```
       494       *              ✪★★ ✪ 512
GluR5 YDVKLVPDGKYGAQNDKGE
GluR6 YEIRLVEDGKYGAQDDVNGQ
GluR7 YEIRLVEDGKYGAQDDKGQ
```

B

Fig. 4. Autoimmune reactivity towards homologous regions of GluR5, GluR6 and GluR7. Panel A shows the homologous sequences for GluR5 (residues 494–512), GluR6 (residues 478–497), and GluR7 (residues 481–499) termed the "C" autoimmune region (Fig. 1). Asterisks denote the residues important for binding or immune-recognition of the anti-GluR5 autoantibody with allosteric properties found in a patient with paraneoplastic disease. Circled asterisks are those residues defined through amino acid substitution mutagenesis to impart anti-GluR5 autoantibody subunit specificity. The residues overscored with single lines are reported [2] to interact directly with the agonist glutamate (Y504) or with KA (K503). Panel B shows a three-dimensional representation of the peptide backbone of the ligand-binding domain portion of GluR2 bound to kainic acid (KA, red) that was generated from the data of Ref. [2]; Protein Data base; 1GR2.PDB using the program RasMol. Highlighted features in the GluR2 X-ray structure homologous to GluR5 that are bound by autoantibody are shown. Segments are identified that are continuous with the transmembrane spanning residues (TM) and the region denoted by an asterisks is the last residue of the GluR3B region, and an initiating residue of the GluR-S1 domain (see Fig. 1a). Residues bound by the GluR-modulatory autoreactivity identified in a patient with paraneoplastic disease are highlighted in green, and residues key to immunoreactivity (K497) and subunit specificity (N508 and E512) are noted, as is the peptide region containing residue Y504 that functions in binding to glutamate. Note that autoantibody-binding residues span the residue(s) involved in ligand binding. The distance from the alpha carbons of residues K497 to E512 is approximately 25 Å. (*For a colored version of this figure, see plate section, page 464.*)

Of interest is that the autoantibody-binding residues span residue Y504 in GluR5 that is homologous to residue Y450 in the GluR2 X-ray structure [53]. This residue interacts directly with glutamate to secure the position of this agonist within the ligand-binding site. Because structural mobility around this residue contributes to agonist binding, the coincident location of autoantibody binding is of interest since it suggests how this auto-antibody may exert its allosteric effect on receptor action through modifying or distorting the structure surrounding residues that participate in ligand binding. This point warrants further study, especially since antibodies prepared to homologous sequences of other subunits could also serve as highly selective modulators of GluR function. In addition to autoantibodies providing a useful method to identify and develop subunit-specific agonists or potential allosteric regulators, the more important point is that autoantibodies to GluRs act to alter receptor function in at least two ways that depend upon both subunit-specificity and epitope selection. This not only provides a multitude of possibilities regarding how such autoreactivities could contribute to the highly specific progression of neurodegenerative disease, but also explains in part the often unique and puzzling variability in their etiology and rate of progression.

9. Autoantibody induction of complement fixation

He et al. [64] have reported another effect of antibodies to GluR3, namely the fixation of complement and the resulting neurotoxicity. The antisera used in these studies were generated against a large region of the GluR3 protein expressed as a glutathione-*S*-transferase fusion protein, however the specific region(s) of GluR3 recognized by these antisera have not yet been defined further by this group. In two of the five rabbits that were immunized, neurologic disorders were observed. Sera from these animals possessed anti-GluR3 antibodies, but did not evoke ion channel activity in neurons cultured for only 2 weeks, consistent with our findings that older cultures (at least 3 weeks old) are necessary to reveal this activity or with different epitope usage [33,34]. However, sera from these animals could induce neuronal toxicity if complement (C$'$) proteins were present. More recently this same group has reported [65] that while neurons were the major cell type recognized by these anti-GluR3 antibodies, glia were the primary targets of C$'$ mediated cell death induced by the anti-GluR3 antibodies. Neuronal protection against toxicity was afforded by high expression of CD59, which inhibits complement-mediated cell death. Glial cells express relatively low levels of CD59, presumably rendering them more susceptible to the toxicity of the antisera. The loss of glia as part of the pathogenesis of RE would certainly have an impact on the support capabilities of these cells, from the production of trophic factors to the buffering of extracellular glutamate. Inadequate glutamate buffering could ultimately result in the induction of seizure activity and might represent an additional mechanism by which autoreactive antibodies to GluR3 contribute to the different etiology and responses to therapy observed in individuals with this disease.

10. Loss or disruption of immunologic tolerance

In general, it is believed that during development we are tolerized to our own proteins. However, autoimmune processes are initiated when this tolerance either fails to be established, it is not maintained, or, in what is often termed as molecular mimicry [66], an invading organism expresses an antigen to which the immune system responds, but the epitope shares features in common with a self-protein. These topics will be considered for the origin of autoreactivity to GluR3B in the next sections.

11. Epitope mimicry and cross-reactivity

In an earlier study [67], we observed that the GluR3B sequence exhibits sequence identity (Fig. 5) with a region of both the Salmonella flagellar hook-associated protein-2 (FLID) fragment and the human interferon alpha receptor-1 (hIFNAR-1). Anti-GluR3B agonist-like antibodies prepared in rabbits exhibited immunoreactivity to the hIFNAR-1 protein fragment but not to FLID. A curious finding was that antibodies prepared to GluR3B immunogen exhibited greater reactivity towards hIFNAR-1 than GluR3B, but antibodies prepared to hIFNAR-1 lacked reactivity to GluR3B. These results demonstrate that antibodies prepared to a defined antigen can bind preferentially to a sequence in an unrelated protein with high, but not identical, sequence identity (Fig. 5). This leads us to suggest [35,67] that antibodies to the GluR3B sequence exhibit heteroclicity, which is defined as the ability of an antibody to react with equal or greater affinity towards an unrelated antigen. Therefore, hIFNAR-1 is a heteroclitic antigen of GluR3B. Of note is that the hIFNAR-1 sequence homologous to GluR3B is thought to create a "hinge" between two domains of the extracellular portion of hIFNAR-1 (see [67]), similar to the function proposed for GluR3B (Figs. 1 and 2). This observation raises the possibility that certain amino acid sequences are favored in structural motifs that allow flexible transitions between functional domains of proteins.

The interaction of anti-GluR3B antibodies with an unrelated protein suggests an important cautionary note: autoreactivity does not necessarily mean pathology. A feature common to antibody-mediated autoimmune diseases is the occurrence of autoantibodies to a protein peptide region in asymptomatic individuals. The absence of detectable pathology in these so-called "false positives" has often been attributed to the genetics of the individual or an incomplete autoimmune process in the individual harboring these autoantibodies. In the context of autoimmune disease, immunoreactivity to a region of a protein that contains the pathogenic epitope is alone not sufficient to determine whether pathology will occur. Rather, immunoreactivity directed towards particular amino acids within the epitope may be an important factor in determining whether or not disease develops as well as its pathologic course. This point is emphasized by two findings regarding antibodies that bind to GluR3B. First, as noted above, the interaction of antibodies with GluR3B–F380

Fig. 5. Amino acid sequence of GluR3B compared with two similar sequences in unrelated proteins. The human interferon alpha 1 receptor (IFNAR-1, amino acids 115–129) and a Salmonella flagellar hook-associated protein 2 (FLID, amino acids 25–39) are aligned with GluR3B (amino acids 372–386 are shown). Key residues important for agonist-like anti-GluR3B antibody binding are denoted by asterisks and boxed to highlight identical or similar residues in IFNAR or FLID. Residues in common between GluR3B and FLID are outlined. When GluR3B (amino acids 372–388, see Fig. 3) was used as immunogen in rabbits, anti-GluR3B antibodies bound to the human IFNAR sequence but not to FLID. Antibodies prepared to GluR3B peptide as immunogen exhibited greater immunoreactivity towards IFNAR than to the original immunogen. Notably, no immunoreactivity towards GluR3B was observed in rabbits immunized with the IFNAR-1 sequence suggesting that IFNAR-1 is a heteroclitic antigen to GluR3B (see text and Gahring et al., 1998 [35]).

produces a range of relative agonist efficacies [34,35] and consequently variable impact on neurotransmission. Second, while GluR3B exhibits relevant sequence identity to both the FLID and hIFNAR-1 sequences [67], the amino acids that compose the anti-GluR3B agonist antibody eptitope are most like those present in the hIFNAR-1 hetroclitic region, and are not common to the FLID sequence. This is consistent with the idea that anti-GluR3B antibodies react only with the hIFNAR-1 sequence, but it does not rule-out the possibility of antibody formation resulting from a molecular mimicry mechanism. Hence, although an antibody to FLID could hypothetically interact with GluR3B, this antibody would not likely have anti-GluR agonist properties nor would it contribute to an equivalent type of pathology. Finally, the enhanced reactivity of anti-GluR3B antibodies to the hIFNAR-1 protein could contribute to disease in a manner that does not involve the direct or immediate alteration of receptor function by antibody binding. For example, the likely autoimmune target based upon the antigen used for antibody production may not necessarily be the protein target it is thought to be, especially when the epitope target of the autoreactivity is not well defined. Further, exposure to GluR3B heteroclitic antibodies could have a long-term effect of altering the function of an unrelated receptor such as hIFNAR-1 to impair anti-viral host defenses. In this context, a relationship of GluR3B autoantibodies with viral infections and the accumulation of viral-like particles in the brain of RE patients (see above) is intriguing. Thus, depending upon the potential of a pathological autoantibody to react with heteroclitic autoantigens, a diverse range of distinct, but overlapping, pathophysiological effects could produce diseases with similar etiology but distinct clinical characteristics or progression.

12. Conditional proteolysis as a mechanism for the induction of an autoantigen

Although circumstances leading to the lack of tolerance and generation of autoimmunity are unknown, it is often associated with cell death resulting from DNA damage, growth factor deprivation, heat shock, oxidative stress, pro-immune intracellular infection by bacteria or viruses, and inflammation. Recently, Rosen and colleagues [68,69] recognized a mechanism that explains how novel protein fragments can be generated to which immune tolerance in the thymus has not been established. They discovered that proteolytic cleavage of intracellular proteins by granzyme B (GB) generated novel protein fragments to which immune tolerance was not established. GB is a serine protease that is released by activated T-lymphocytes and Natural Killer cells [70]. It enters the target cell upon completion of a pore produced by the co-released protein, perforin, and initiates caspase-mediated apoptosis cascades. Notably, GB can also directly cleave substrate proteins other than pro-caspases and when these protein fragments are released to the extracellular milieu upon apoptosis they can function as autoantigens.

The last four amino acids of the minimum GluR3B antigen fragment (residues 385–388 or ISND) defined in previous studies [34,35] is also a consensus GB proteolytic site (Fig. 6; see Ref. [71]). However, unlike intracellular GB sites, this GluR3B-GB site also harbors an internal N-linked glycosylation sequon (ISND*S) that when glycosylated inhibits cleavage by GB. Notably, this *N*-glycon sequon, while glycosylated normally, is inefficiently used and conversion of the S389 to the efficiently used N-glycosylation

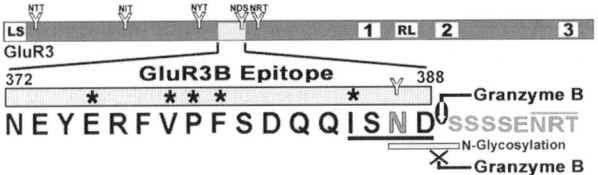

Fig. 6. GluR3 contains a granzyme B (GB) cleavage site at the autoreactive Glu3B epitope associated with autoimmune RE. Shown is a diagram of GluR3 that locates the proposed transmembrane domains (1, 2, 3), re-entry loop (RL) and leader sequence (LS). Other consensus N-glycosylation sequons are also noted ("Y") and their respective amino acid sequences indicated. The GluR3B autoantigen (light gray box; amino acids 372–388) and some adjacent sequence (gray) not required as immunogen or for antibody binding are expanded (see also Fig. 1a). Asterisks mark residues required for agonist autoantibody binding or epitope conformation. The black underline identifies a GB cleavage recognition sequence and cleavage-site (labeled by the out-lined arrow). Also underlined (out-lined) is the N-glycosylation sequon ("NDS"). Another N-glycosylation sequon (NRT) is also highlighted for orientation. The mouse GluR3B differs from the human GluR3 sequence at one residue where serine 390 is an alanine. When N387 is glycosylated, GB is unable to cleave this sequence. As noted in the text, this N-glycosylation sequon is inefficiently used.

sequon containing T389, renders this site unavailable to detectable GB cleavage. The occurrence of a relatively inefficient N-glycosylation sequon in GluR3B predicts that conditions that disrupt glycosylation, as might occur during inflammation and GB release, would increase the production of novel GB-generated fragments from the GluR3 surface protein. This would be particularly relevant during an infection when the immune system is poised to mount a response as may occur during a viral infection or in trauma. Both conditions, and especially viral infection, have long been viewed as candidates for initiating RE pathology, in part due to the close association with the onset of encephalitis, microglial nodules consisting of activated-T-cell infiltrates, and the subsequent appearance of clinical symptoms.

The relevance of GB sites in extracellular proteins to the generation of autoimmunity has not been fully explored, but a search of 140 unique human surface receptor proteins revealed the occurrence of a coincident GB/N glycosylation sequon site in only 11 other proteins [71]. Three of these proteins, the neuronal nicotinic acetylcholine receptor alpha7 subunit (nAChRα7), the voltage gated-calcium channel, and the voltage-gated potassium channel, are members of receptor families known to be autoimmune targets in myasthenia gravis, Lambert–Eaton syndrome, and Isaacs syndrome, respectively. In nAChRα7, the N-glycosylated-GB sequence is near the N-terminus and does not include the autoantigenic epitope commonly described as the main immunogenic region (MIR). However, autoantibodies towards both muscle and ganglionic nicotinic receptors that do not interact at the MIR have been described. Since nAChRα7 is expressed by ganglionic neurons (see Ref. [71]), and contains strong sequence identity with other neuronal and muscle nicotinic receptor subunits, this region of the molecule is a strong candidate for the location of the autoreactive epitope. In any case, GB is elevated in the plasma and sinovial fluid of individuals with inflammatory disease, providing an attractive hypothesis to explain both the origin of this autoantigen and the seemingly improbable convergence of these disease-initiation factors that make RE and similar neuro-autoimmune diseases relatively rare.

Collectively these studies present evidence that a contributor to neurological diseases may actually be found in the dysfunction of the immune system, where highly specific autoantibodies are generated that bind neuronal glutamate receptors and directly alter their normal function in neurotransmission. Among the implications of this are: first, autoantibody-mediated neurodegeneration offers testable mechanisms for rationalizing the remarkable specificity of many of these diseases. Second, identification of molecular mechanisms contributing to autoimmune targeting of GluRs will ultimately lead to more effective therapeutic approaches to these disorders. Finally, antibodies offer a novel strategy to approach subunit-specific modulation of GluR function. These advances already offer clinically applicable approaches to the treatment and diagnosis of these often sporadic, intractable, and frightening diseases, and suggest that better diagnosis and specific treatments are in the near future, as methods of functional genomics become common place and applicable to understanding the nature and etiology of these complex diseases.

Acknowledgements

This work was supported by NIH grant NS35181. Funding was also provided by Veterans Administration Merit grants (L.C.G. and S.W.R), the Pew Charitable Trust (S.W.R.) and the Val A. Browning Foundation of Utah.

References

[1] Hollmann, M., Heinemann, S., 1994. Cloned glutamate receptors. Annu. Rev. Neurosci. 17, 31–108.

[2] Dingledine, R., Borges, K., Bowie, D., Traynelis, S.F., 1999. The glutamate receptor ion channels. Pharmacol. Rev. 51(1), 7–61.

[3] Bruce, A.J., Boling, W., Kindy, M.S., Peschon, J., Kraemer, P.J., Carpenter, M.K., Holtsberg, F.W., Mattson, M.P., 1996. Altered neuronal and microglial responses to excitotoxic and ischemic brain injury in mice lacking TNF receptors. Nat. Med. 2(7), 788–794.

[4] Beal, M.F., 1992. Mechanisms of excitotoxicity in neurologic diseases. FASEB J. 6(15), 3338–3344.

[5] Arias, C., Becerra-Garcia, F., Tapia, R., 1998. Glutamic acid and Alzheimer's disease. Neurobiology 6(1), 33–43.

[6] Beal, M.F., 1998. Excitotoxicity and nitric oxide in Parkinson's disease pathogenesis. Ann. Neurol. 44(3 Suppl 1), S110–S114.

[7] Hume, R.I., Dingledine, R., Heinemann, S.F., 1991. Identification of a site in glutamate receptor subunits that controls calcium permeability. Science 253(5023), 1028–1031.

[8] Washburn, M.S., Numberger, M., Zhang, S., Dingledine, R., 1997. Differential dependence on GluR2 expression of three characteristic features of AMPA receptors. J. Neurosci. 17(24), 9393–9406.

[9] Lee, J.C., Greig, A., Ravindranathan, A., Parks, T.N., Rao, M.S., 1998. Molecular analysis of AMPA-specific receptors: subunit composition, editing, and calcium influx determination in small amounts of tissue. Brain Res. Brain Res. Protoc. 3(2), 142–154.

[10] Paschen, W., Schmitt, J., Uto, A., 1996. RNA editing of glutamate receptor subunits GluR2, GluR5 and GluR6 in transient cerebral ischemia in the rat. J. Cereb. Blood Flow Metab. 16(4), 548–556.

[11] Puchalski, R.B., Louis, J.C., Brose, N., Traynelis, S.F., Egebjerg, J., Kukekov, V., Wenthold, R.J., Rogers, S.W., Lin, F., Moran, T., Morrison, J.H., Heinemann, S.F., 1994. Selective RNA editing and subunit assembly of native glutamate receptors. Neuron 13(1), 131–147.

[12] Seeburg, P.H., Higuchi, M., Sprengel, R., 1998. RNA editing of brain glutamate receptor channels: mechanism and physiology. Brain Res. Brain Res. Rev. 26(2-3), 217–229.

[13] Sommer, B., Keinanen, K., Verdoorn, T.A., Wisden, W., Burnashev, N., Herb, A., Kohler, M., Takagi, T., Sakmann, B., Seeburg, P.H., 1990. Flip and flop: a cell-specific functional switch in glutamate-operated channels of the CNS. Science 249(4976), 1580–1585.

[14] Pollard, H., Heron, A., Moreau, J., Ben-Ari, Y., Khrestchatisky, M., 1993. Alterations of the GluR-B AMPA receptor subunit flip/flop expression in kainate-induced epilepsy and ischemia. Neuroscience 57(3), 545–554.

[15] Monyer, H., Seeburg, P.H., Wisden, W., 1991. Glutamate-operated channels: developmentally early and mature forms arise by alternative splicing. Neuron 6(5), 799–810.

[16] Schloot, N.C., Pozzilli, P., Mandrup-Poulsen, T., 1999. Immune markers for monitoring the progression of autoimmune disease. Diabetes Metab. Res. Rev. 15(2), 141–145.

[17] Burns, E.A., Goodwin, J.S., 1997. Immunodeficiency of aging. Drugs Aging 11(5), 374–397.

[18] Schonrich, G., Kalinke, U., Momburg, F., Malissen, M., Schmitt-Verhulst, A.M., Malissen, B., Hammerling, G.J., Arnold, B., 1991. Down-regulation of T-cell receptors on self-reactive T cells as a novel mechanism for extrathymic tolerance induction. Cell 65(2), 293–304.

[19] Bachmann, M.F., Rohrer, U.H., Steinhoff, U., Burki, K., Skuntz, S., Arnheiter, H., Hengartner, H., Zinkernagel, R.M., 1994. T helper cell unresponsiveness: rapid induction in antigen-transgenic and reversion in non-transgenic mice. Eur. J. Immunol. 24(12), 2966–2973.

[20] Andermann, F., 1991. Chronic Encephalitis and Epilepsy. Butterworth-Heinemann, Boston.

[21] Li, Y., Uccelli, A., Laxer, K.D., Jeong, M.C., Vinters, H.V., Tourtellotte, W.W., Hauser, S.L., Oksenberg, J.R., 1997. Local-clonal expansion of infiltrating T lymphocytes in chronic encephalitis of Rasmussen. J. Immunol. 158(3), 1428–1437.

[22] Conti-Tronconi, B.M., McLane, K.E., Raftery, M.A., Grando, S.A., Protti, M.P., 1994. The nicotinic acetylcholine receptor: structure and autoimmune pathology. Crit. Rev. Biochem. Mol. Biol. 29(2), 69–123.

[23] Lindstrom, J.M., 2000. Acetylcholine receptors and myasthenia. Muscle Nerve 23(4), 453–477.

[24] Engel, A.G., 1994. Congenital myasthenic syndromes. Neurol. Clin. 12(2), 401–437.

[25] Rogers, S.W., Andrews, P.I., Gahring, L.C., Whisenand, T., Cauley, K., Crain, B., Hughes, T.E., Heinemann, S.F., McNamara, J.O., 1994. Autoantibodies to glutamate receptor GluR3 in Rasmussen's encephalitis. Science 265(5172), 648–651.

[26] Topcu, M., Turanli, G., Aynaci, F.M., Yalnizoglu, D., Saatci, I., Yigit, A., Genc, D., Soylemezoglu, F., Bertan, V., Akalin, N., 1999. Rasmussen encephalitis in childhood. Childs Nerv. Syst. 15(8), 395–402 (discussion 403).

[27] Atkins, M.R., Terrell, W., Hulette, C.M., 1995. Rasmussen's syndrome: a study of potential viral etiology. Clin. Neuropathol. 14(1), 7–12.

[28] Farrell, M.A., Cheng, L., Cornford, M.E., Grody, W.W., Vinters, H.V., 1991. Cytomegalovirus and Rasmussen's encephalitis [letter; comment]. Lancet 337(8756), 1551–1552.

[29] Walter, G.F., Renella, R.R., 1989. Epstein–Barr virus in brain and Rasmussen's encephalitis. Lancet 1(8632), 279–280.

[30] McLachlan, R.S., Levin, S., Blume, W.T., 1996. Treatment of Rasmussen's syndrome with ganciclovir. Neurology 47(4), 925–928.

[31] Maria, B.L., Ringdahl, D.M., Mickle, J.P., Smith, L.J., Reuman, P.D., Gilmore, R.L., Drane, W.E., Quisling, R.G., 1993. Intraventricular alpha interferon therapy for Rasmussen's syndrome. Can. J. Neurol. Sci. 20(4), 333–336.

[32] Dabbagh, O., Gascon, G., Crowell, J., Bamoggadam, F., 1997. Intraventricular interferon-alpha stops seizures in Rasmussen's encephalitis: a case report. Epilepsia 38(9), 1045–1049.

[33] Twyman, R.E., Gahring, L.C., Spiess, J., Rogers, S.W., 1995. Glutamate receptor antibodies activate a subset of receptors and reveal an agonist binding site. Neuron 14(4), 755–762.

[34] Carlson, N.G., Gahring, L.C., Twyman, R.E., Rogers, S.W., 1997. Identification of amino acids in the glutamate receptor, GluR3, important for antibody-binding and receptor-specific activation. J. Biol. Chem. 272(17), 11295–11301.

[35] Gahring, L.C., Rogers, S.W., 1998. Autoimmunity to glutamate receptors in Rasmussen's encephalitis: a rare finding or the tip of an iceberg? Neuroscientist 4(5), 373–379.

[36] Rose, K., Goldberg, M.P., Choi, D.W., 1993. Cytotoxicity in Murine Neocortical Cell Culture. In: Tyson, C.A., Frazier, J.M., (Eds.), Methods in Toxicology. Academic Press, Inc., San Diego, New York, pp. 45–60.

[37] Choi, D.W., 1994. Calcium and excitotoxic neuronal injury. Ann. NY Acad. Sci. 747, 162–171.

[38] Lang, B., Vincent, A., 1996. Autoimmunity to ion-channels and other proteins in paraneoplastic disorders. Curr. Opin. Immunol. 8(6), 865–871.

[39] Smith, R.G., Siklos, L., Alexianu, M.E., Engelhardt, J.I., Mosier, D.R., Colom, L., Habib Mohamed, A., Appel, S.H., 1996. Autoimmunity and ALS. Neurology 47(4 Suppl 2), S40–S45 (discussion S45–6).

[40] Hart, I.K., Waters, C., Vincent, A., Newland, C., Beeson, D., Pongs, O., Morris, C., Newsome-Davis, J., 1997. Autoantibodies detected to expressed K^+ channels are implicated in neuromyotonia. Ann. Neurol. 41(2), 238–246.

[41] Lennon, V.A., Lambert, E.H., 1989. Autoantibodies bind solubilized calcium channel-omega-conotoxin complexes from small cell lung carcinoma: a diagnostic aid for Lambert–Eaton myasthenic syndrome. Mayo Clin. Proc. 64(12), 1498–1504.

[42] Takamori, M., Maruta, T., Komai, K., 2000. Lambert–Eaton myasthenic syndrome as an autoimmune calcium-channelopathy. Neurosci. Res. 36(3), 183–191.

[43] Archelos, J.J., Hartung, H.P., 2000. Pathogenetic role of autoantibodies in neurological diseases. Trends Neurosci. 23(7), 317–327.

[44] Solimena, M., De Camilli, P., 1991. Autoimmunity to glutamic acid decarboxylase (GAD) in Stiff-Man syndrome and insulin-dependent diabetes mellitus. Trends Neurosci. 14(10), 452–457.

[45] Patrick, J., Lindstrom, J., 1973. Autoimmune response to acetylcholine receptor. Science 180(88), 871–872.

[46] Jahn, K., Franke, C., Bufler, J., 2000. Mechanism of block of nicotinic acetylcholine receptor channels by purified IgG from seropositive patients with myasthenia gravis. Neurology 54(2), 474–479.

[47] Kendler, D.L., Rootman, J., Huber, G.K., Davies, T.F., 1991. A 64 kDa membrane antigen is a recurrent epitope for natural autoantibodies in patients with Graves' thyroid and ophthalmic diseases. Clin. Endocrinol. (Oxf) 35(6), 539–547.

[48] Sillevis Smitt, P., Kinoshita, A., De Leeuw, B., Moll, W., Coesmans, M., Jaarsma, D., Henzen-Logmans, S., Vecht, C., De Zeeuw, C., Sekiyama, N., Nakanishi, S., Shigemoto, R., 2000. Paraneoplastic cerebellar ataxia due to autoantibodies against a glutamate receptor. N. Engl. J. Med. 342(1), 21–27.

[49] Rogers, S.W., Twyman, R.E., Gahring, L.C., 1996. The role of autoimmunity to glutamate receptors in neurological disease. Mol. Med. Today 2(2), 76–81.

[50] O'Hara, P.J., Sheppard, P.O., Thogersen, H., Venezia, D., Haldeman, B.A., McGrane, V., Houamed, K.M., Thomsen, C., Gilbert, T.L., Mulvihill, E.R., 1993. The ligand-binding domain in metabotropic glutamate receptors is related to bacterial periplasmic binding proteins. Neuron 11(1), 41–52.

[51] Stern-Bach, Y., Bettler, B., Hartley, M., Sheppard, P.O., O'Hara, P.J., Heinemann, S.F., 1994. Agonist selectivity of glutamate receptors is specified by two domains structurally related to bacterial amino acid-binding proteins. Neuron 13(6), 1345–1357.

[52] Sutcliffe, M.J., Smeeton, A.H., Wo, Z.G., Oswald, R.E., 1998. Three-dimensional models of glutamate receptors. Biochem. Soc. Trans. 26(3), 450–458.

[53] Armstrong, N., Sun, Y., Chen, G.Q., Gouaux, E., 1998. Structure of a glutamate-receptor ligand-binding core in complex with kainate. Nature 395(6705), 913–917.

[54] Arvola, M., Keinanen, K., 1996. Characterization of the ligand-binding domains of glutamate receptor (GluR)-B and GluR-D subunits expressed in *Escherichia coli* as periplasmic proteins. J. Biol. Chem. 271(26), 15527–15532.

[55] McDonald, S., Carlson, N.G., Gahring, L.C., Ely, K.R., Rogers, S.W., 1999. A model for a glutamate receptor agonist antibody-binding site. J. Mol. Recognit. 12(4), 219–225.

[56] Helekar, S.A., Char, D., Neff, S., Patrick, J., 1994. Prolyl isomerase requirement for the expression of functional homo- oligomeric ligand-gated ion channels. Neuron 12(1), 179–189.

[57] Gahring, L.C., Rogers, S.W., Twyman, R.E., 1997. Autoantibodies to glutamate receptor subunit GluR2 in nonfamilial olivopontocerebellar degeneration. Neurology 48(2), 494–500.

[58] Rogers, S.W., Hughes, T.E., Hollmann, M., Gasic, G.P., Deneris, E.S., Heinemann, S., 1991. The characterization and localization of the glutamate receptor subunit GluR1 in the rat brain. J. Neurosci. 11(9), 2713–2724.

[59] Petralia, R.S., Wenthold, R.J., 1992. Light and electron immunocytochemical localization of AMPA-selective glutamate receptors in the rat brain. J. Comp. Neurol. 15, 15.

[60] Levite, M., Hermelin, A., 1999. Autoimmunity to the glutamate receptor in mice – a model for Rasmussen's encephalitis? J. Autoimmun. 13(1), 73–82.

[61] Levite, M., Fleidervish, I.A., Schwarz, A., Pelled, D., Futerman, A.H., 1999. Autoantibodies to the glutamate receptor kill neurons via activation of the receptor ion channel. J. Autoimmun. 13(1), 61–72.

[62] Gahring, L.C., Twyman, R.E., Greenlee, J.E., Rogers, S.W., 1995. Autoantibodies to neuronal glutamate receptors in patients with paraneoplastic neurodegenerative syndrome enhance receptor activation. Mol. Med. 1(3), 245–253.

[63] Carlson, N., Gahring, L.C., Rogers, S.W., 2003. Identification of the amino acids on a neuronal glutamate receptor recognized by an autoantibody from a patient with paraneoplastic syndrome. J. Neuro. Res. 63(6): 480–485.

[64] He, X.P., Patel, M., Whitney, K.D., Janumpalli, S., Tenner, A., McNamara, J.O., 1998. Glutamate receptor GluR3 antibodies and death of cortical cells. Neuron 20(1), 153–163.

[65] Whitney, K.D., McNamara, J.O., 2000. GluR3 autoantibodies destroy neural cells in a complement-dependent manner modulated by complement regulatory proteins. J. Neurosci. 20(19), 7307–7316.

[66] Albert, L.J., Inman, R.D., 1999. Molecular mimicry and autoimmunity. N. Engl. J. Med. 341(27), 2068–2074.

[67] Gahring, L.C., Carlson, N.G., Rogers, S.W., 1998. Antibodies prepared to neuronal glutamate receptor subunit3 bind IFNalpha-receptors: implications for an autoimmune process. Autoimmunity 28(4), 243–248.

[68] Casciola-Rosen, L., Andrade, F., Ulanet, D., Wong, W.B., Rosen, A., 1999. Cleavage by granzyme B is strongly predictive of autoantigen status: implications for initiation of autoimmunity. J. Exp. Med. 190(6), 815–826.

[69] Rosen, A., Casciola-Rosen, L., 1999. Autoantigens as substrates for apoptotic proteases: implications for the pathogenesis of systemic autoimmune disease. Cell Death Differ. 6(1), 6–12.

[70] Andrade, F., Roy, S., Nicholson, D., Thornberry, N., Rosen, A., Casciola-Rosen, L., 1998. Granzyme B directly and efficiently cleaves several downstream caspase substrates: implications for CTL-induced apoptosis. Immunity 8(4), 451–460.

[71] Gahring, L.C., Carlson, N.G., Meyer, E.L., Rogers, S.W., 2001. Granzyme B proteolysis of a neuronal glutamate receptor generates an autoantigen and is modulated by glycosylation. J. Immunol. 166, 1433–1438.

[72] Degenhardt, A., Duvoisin, R.M., Frennier, J., Gultekin, S.H., Rosenfeld, M.R., Posner, J.B., Dalmau, J., 1998. Absence of antibodies to non-NMDA glutamate-receptor subunits in paraneoplastic cerebellar degeneration. Neurology 50(5), 1392–1397.

Gap junction mutations in human disease

Matthew G. Hopperstad, Miduturu Srinivas, Alfredo Fort and David C. Spray*

Department of Neuroscience, Albert Einstein College of Medicine, 1300 Morris Park Avenue, Bronx, NY 10461, USA
Correspondence address: Tel.: +1-718-430-2537; fax: +1-718-430-8594
E-mail: spray@aecom.yu.edu

1. Introduction

Gap junctions are unique amongst membrane channels in that they are formed when adjacent cells each donate a half channel (hemichannel or connexon) across a gap (hence the name) between the two cells to form a full junctional channel, thereby providing an isolated aqueous pathway between the cellular interiors (Fig. 1A). The hemichannels are themselves oligomers of six subunit membrane proteins called connexins (the widely accepted nomenclature designates each subtype by its molecular weight, deduced from its cDNA clone, in kilodaltons, preceded by the prefix "Cx", e.g. Cx43). Topology is conserved among connexins, consisting of four transmembrane domains (TM1–TM4) joined together by a single cytoplasmic loop (CL) and two extracellular loops (EL1, EL2); the amino and carboxyl termini (NT, CT) are both located in the cytoplasm (Fig. 1B). Unger et al. [1] recently confirmed the tertiary and quaternary structure of a recombinant gap junction protein (Cx43, which lacked almost the entire CT domain) via electron crystallography with a resolution of 7.5 Å in the membrane plane and 21 Å in the vertical direction. Energy-minimized conformations of the first 15 amino acids of the NT of another connexin (Cx26) have recently been determined through ^{1}H-NMR [2]. Nevertheless, further high-resolution studies are necessary to determine function and structure for additional domains of these and other connexin molecules.

There are at least 20 members of the gap junction family of proteins, and most cells and tissues express more than one connexin subtype. The reasons for diversity in connexin subtype and expression are largely unknown, but individual subtypes clearly have unique distributions and distinct voltage and pH sensitivities. Furthermore, as considered below, different connexins are controlled by distinct gene regulatory mechanisms, and different connexins have differential abilities to interact with other subtypes in the formation of heterotypic and heteromeric channels (see Fig. 1A). Given that almost all cells express gap junction proteins (exceptions being mature spermatozoa, circulating erythrocytes,

Advances in Molecular and Cell Biology, Vol. 32, pages 161–187
ISSN: 1569-2558/DOI: 10.1016/S1569-2558(03)32008-9

M.G. Hopperstad et al.

A

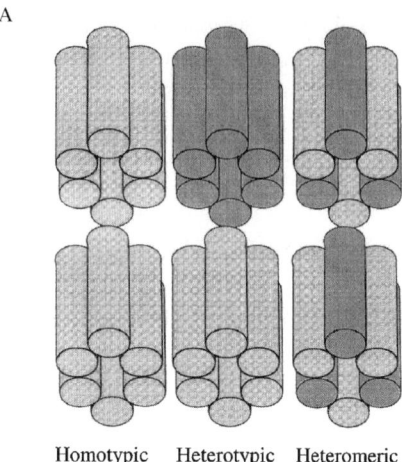

Homotypic   Heterotypic   Heteromeric

B

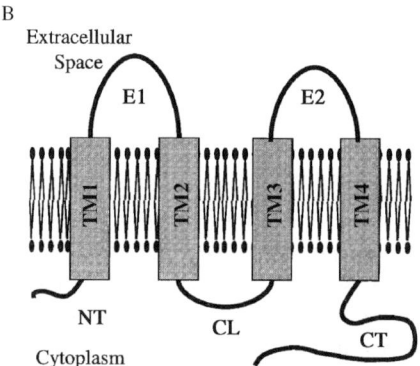

Fig. 1. Dodecameric arrangement of connexin subunits in the formation of cell–cell channels and connexin topography in relation to the cell membrane. (A) Gap junctions are formed by the oligomerization of six connexin subunits to form a connexon or hemichannel. Adjacent cells each contribute a hemichannel in the formation of a cell–cell channel. Channels may be formed of one connexin type (homotypic, left), two connexin types each one of which entirely composing a constituent hemichannel (heterotypic, middle) or, possibly, via heterosubtype interactions in the formation of one or both constituent hemichannels (heteromeric channels, right). (B) Connexin proteins are integral membrane proteins, with four transmembrane domains (TM1–TM4), two extracellular loops (E1, E2), and an intracellular loop (IL); both carboxyl and amino termini are located within the cytoplasm.

and differentiated skeletal muscle) it appears likely that the intercellular communication provided by these channels fulfills one or more fundamental tissue functions. In addition to the apparent homeostatic necessity of baseline intercellular communication, there appear to be multiple tissue-specific functional roles for gap junctions. For example, gap junctions provide a path for ionic current flow between excitable cells, leading to synchronized electrical activity among coupled neurons (where these channels form electrical synapses) and rhythmic activation of the heart. Gap junctions also allow the exchange of necessary nutrients and metabolites in the avascular lens core and the relay of intracellular second messenger signals between hepatocytes in the liver, glial cells in the nervous system, and among endocrine and exocrine glands.

Mutations in multiple connexin genes have been implicated in several genetic disorders including: X-linked Charcot-Marie-Tooth disease (CMTX; Cx32; [3]), senile cataract (Cx46, [4] and Cx50, [5]), non-syndromic hearing loss (NSHL) (Cx26, [6], and Cx30, [7]), erythrokeratodermia variabilis (Cx31, [8]), and palmoplantar keridermia (Cx26, [9]). Loss or alteration in gap junctional communication has been suggested as the underlying cause in each case, although discrete mechanisms of pathogenesis have generally not been determined. When investigated in vitro the majority of the mutations implicated result in the total loss of functional coupling, though a number of the mutant proteins appear to form functional channels, often with altered properties. A fundamental issue at present is to what extent these mutations produce diseases that involve altered function of gap junction channels rather than disrupting motifs recognized by trafficking signals or necessary for correct folding or membrane insertion of the connexin proteins. The possible molecular and functional determinants of the physiological changes induced by connexin mutants are discussed below, with an aim to reconcile current experimentally derived models of function with findings concerning the physiological basis of these diseases.

2. Regulation of connexin gating

2.1. Voltage sensitivity

The sensitivity of gap junction channels to transjunctional voltage (V_j) has long been recognized, and parameters of voltage dependence have been quantified for many of the connexins through exogenous expression in mammalian cell lines or *Xenopus* oocytes (see Ref. [10] for recent review). Although such studies have revealed interesting similarities and differences in the gating behavior of different connexins, and between wild-type connexins and disease-causing mutants, it should be recognized that (with the possible exception of a role in amphibian development; [11]) an adequate and compelling explanation of the significance of the sensitivity of gap junctions to transmembrane voltage has yet to be found. All types of gap junction channels have been shown to be sensitive to voltage gradients across the channel and relatively insensitive to voltage gradients between the inside and outside of the cell (V_{i-o} or transmembrane voltage, V_m) though such sensitivity has been reported for Cx26 [12] as well as other subtypes. Although sensitivity to V_m may have a significant role in regulating coupling strength, sensitivity to V_m appears to be slower and much less strong than to V_j. The importance, if any, of the relatively small degree of modulation over the range of physiological resting potentials under normal and pathological conditions is not known (see Ref. [13] for review).

Transjunctional voltage sensitivity (V_j) of junctional conductance is illustrated in Fig. 2. In response to moderate, sustained V_j gradients macroscopic junctional currents show slow (with time constants on the order of hundreds of milliseconds to seconds) transjunctional voltage-dependent relaxation to non-zero (residual) current levels. To a first approximation, the dominant component of this decline is monoexponential, implying a first-order gating process that is amenable to analysis using Boltzmann equations.

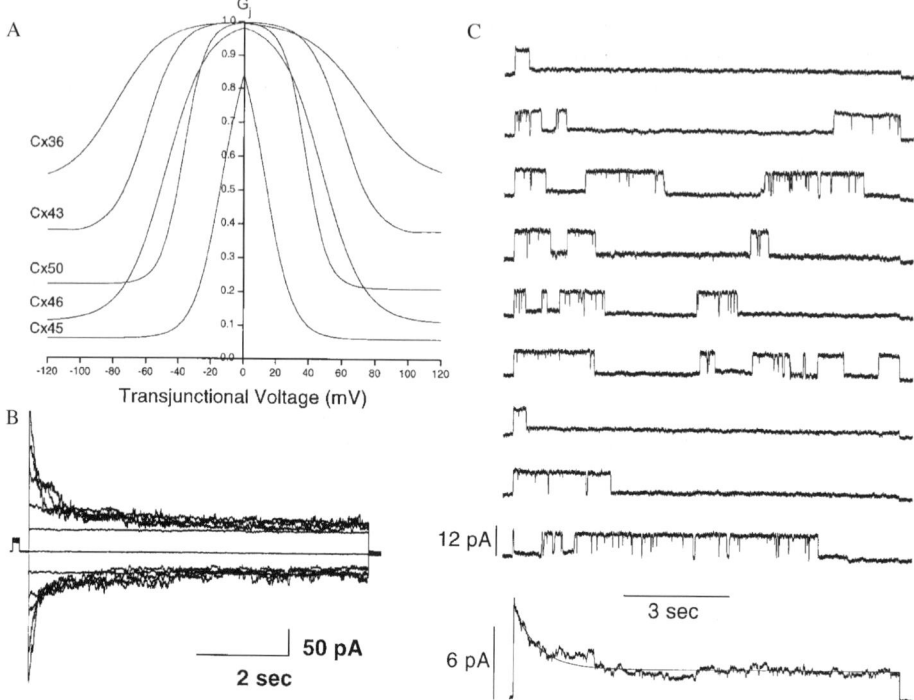

Fig. 2. Macroscopic voltage sensitivity to transjunctional voltage (V_j) of gap junction channels formed by various connexin subtypes and the relationship between single channel substates and macroscopic residual conductance. (A) Two-state Boltzmann equation fits (see Ref. 33 for details) to data derived from experiments using mammalian cells expressing specific connexin subtypes (Cx36 [14]; Cx43 [15]; Cx45 [16]; Cx46 [17]; Cx50 [18]). A broad range of voltage sensitivities are apparent, as are residual conductance levels, ranging from less than 10% of the maximal level observed at $V_j = 0$ mV for Cx45 to greater than 50% for Cx36. (B) A family of 11 current traces recorded from a Cx50 expressing neuroblastoma (N2A) cell pair using the dual whole-cell voltage clamp technique. The currents are in response to V_j step pulses that varied between -100 and 100 mV in 20 mV increments (inset). The Boltzmann fits shown in A were derived from currents measured at the end of pulses for similar data for each of the separate subtypes. (C) Single channel currents from N2A cells expressing Cx50 in response to a series of -50 mV pulses of 12 s duration. The bottom trace in C is an average of 35 such pulses. The solid line superimposed on the data is a monoexponential fit to the ensemble data with a time constant of 905 mS, a value that was shown to be similar to the decline observed at macroscopic levels ([18], compare to B). Careful inspection of the series of traces in C reveals that transitions occur predominantly between the main state and a dominant substate, suggesting that voltage independent substates observed at the single channel level are responsible for residual conductance at the macroscopic level.

Fig. 2A shows the fits of such two-state Boltzmann equations to the transjunctional voltage-induced current relaxation of a number of connexin sub-types. The traces shown in Fig. 2B are a representative family of transjunctional currents from a mammalian cell line endogenously expressing a single connexin subtype (Cx50). The data from similar experiments with cells expressing Cx50 and other subtypes were used to generate the fits in Fig. 2A. Fig. 2C indicates that macroscopic residual conductance is the result of the V_j insensitive substate observed at the single channel level [18].

Although the two-state (mainstate–substate) model has been broadly applied, and macroscopic records are apparently well fit by such a model, recent reports indicate much more complicated gating processes are present for many gap junction channel subtypes. For example, macroscopic currents from Cx46 are reasonably well fit to a two-state Boltzmann, though at the single channel level multiple substate conductance levels are apparent, as well as transitions to the fully closed state [17].

The single channel data presented in Fig. 3, recorded from a transfected mammalian cell line expressing Cx46 or Cx50, suggest that multiple mechanisms exist for V_j dependent gap junction closure. It should be noted that Fig. 3 also demonstrates that single channel gating behavior can be recorded from gap junction channels without the use of uncoupling agents (see Ref. [19] for review).

The bilateral symmetry of gap junction channels complicates the study of the voltage gating, indeed any type of channel closure, due to the contribution of gating mechanisms from both connexons. Thus voltage sensors and their gates are in series across the pore so

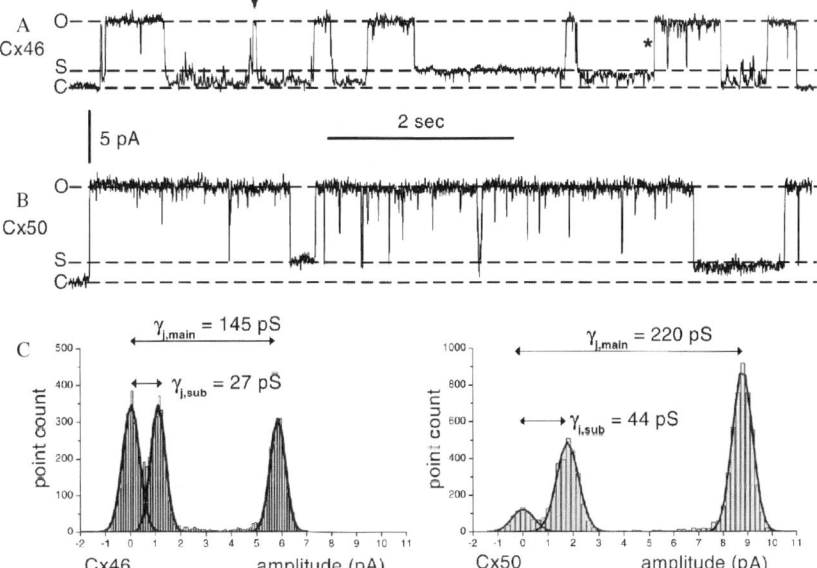

Fig. 3. Voltage-dependant gating and conductance of gap junction channels formed by Cx46 and Cx50. (A). Voltage-dependent gating of a single Cx46 channel in response to a -40 mV transjunctional voltage step applied for 8 s. The channel exhibits multiple voltage gating processes including fast (<2 mS) and slow (>5 mS) transitions to multiple substates. The arrowhead in (A) indicates a slow transition from a low conductance substate ($\gamma < 20$ pS) to the main state and back. The asterisk indicates a fast transition from a relatively stable substate to the main state. (B). Voltage-dependent gating by a single Cx50 channel under the same voltage gradient (-40 mV) for the same period as in (A). Cx50 channels show a preponderance of fast gating transitions (rise and decay times <2 mS) between the fully open state and a dominant substate. In contrast to Cx46 channels, Cx50 channels appear to have a single voltage-dependant mechanism for closure. (C) All-points current histograms with corresponding multi-peak Gaussian fits of the Cx46 trace (left) and the Cx50 trace (right). The left histogram shows peaks at 0, 1.08 and 5.82 pA corresponding to the closed state, a relatively stable substate of 27 pS and a main state of 145 pS. The histogram on the right has peaks at 0, 1.77, and 8.78 pA corresponding to the closed, substate (44 pS), and main state (220 pS) conductance levels.

that gating mechanisms in either connexon are capable of closing the channel. In homotypic combinations, where both connexons are composed of the same subtype (Fig. 1A), symmetrical gating is observed (Fig. 2) while heterotypic junctions have been shown to reflect the properties of the constituent connexons [20]. Such pairings have been instrumental for the isolation of connexin specific properties, such as gating polarity, selectivity, and the presence of multiple V_j gating mechanisms.

Isolation of functional hemichannels by expression of connexin mRNA in oocytes has added significantly to the understanding of full channel function by removing the inherent problems of studying membrane channels in series. Application of depolarizing potentials to oocytes expressing connexin mRNAs results in open hemichannel activity when oocytes are bathed in low calcium containing external solutions. Hemichannel expression in oocytes allows the study of macroscopic as well as isolated channels in excised patches, a technique which has proven useful for various investigations. Unfortunately only Cx46 [21] and a chimeric construct (Cx32 with the first extracellular loop of Cx43) [22] have been conclusively shown to form active hemichannels in oocytes so that the scope of functional studies has been somewhat limited. Nevertheless oocyte expression has resulted in partial mapping of regions implicated in selectivity and permeation [23,24] and voltage gating [25–27]. Whether the function of hemichannels and full, cell–cell channels, are directly comparable is not clear though several properties appear to be conserved (see Ref. [28]). In addition to the study of hemichannels in oocytes, various groups have recently suggested the presence of functional hemichannels in mammalian cells (e.g. Refs. [29–31]) under certain experimental conditions but definitive evidence is lacking, and certainly no clear evidence currently exists for functional hemichannels in mammalian cells in vivo.

The various molecular mechanisms involved in V_j dependent closure of gap junction channels continue to be the subject of ongoing investigations. Consequently, our understanding of the biophysics and molecular determinants of the voltage dependant closure of gap junctions is growing, but its relevance remains a mystery.

2.2. pH sensitivity

It has been known for some time that gap junctions are sensitive to changes in intracellular pH [32,33], though as with voltage sensitivity, the physiological implications in various tissue systems largely remain to be determined. However, because reductions in intercellular pH severe enough to close gap junction channels can occur under ischemic conditions, it is likely that acidification may compromise intercellular communication in infarcted heart tissue and the nervous system. The data in the field are not entirely consistent due to the inherent differences in speed of alteration and control of intracellular pH in the variety of assays utilized. Nevertheless, comparative studies of the pH sensitivity and molecular determinants for pH-induced closure of multiple wild-type and mutant connexin channels have been performed [34,35]. The comparative data are largely, though not entirely, restricted to exogenous expression systems, particularly paired Xenopus oocytes. The use of such systems is, of course, problematic if multiple and, perhaps non-direct, effects of pH are responsible because important cellular structure and/or molecular machinery may differ from system to system and from in vivo conditions. However, it is abundantly clear that as with voltage, a range of pH sensitivity (and possibly mechanisms for pH-dependent closure) exists for the family of connexin subtypes. Fig. 4A shows a clear difference in the pH sensitivity for two connexin subtypes, Cx50 and Cx43, expressed in mammalian cells and Fig. 4B illustrates the diversity of junctional conductance sensitivities to pH of a number of connexins expressed in oocytes [36].

Ek-Vitorin et al. [34] used the *Xenopus* oocyte expression system to show that the carboxyl terminal (CT) domain is necessary for the pH-induced channel closure of Cx43, a connexin highly expressed throughout mammalian tissue including, notably, the heart and brain. Expressions of various mutants with alterations in the CT and other regions

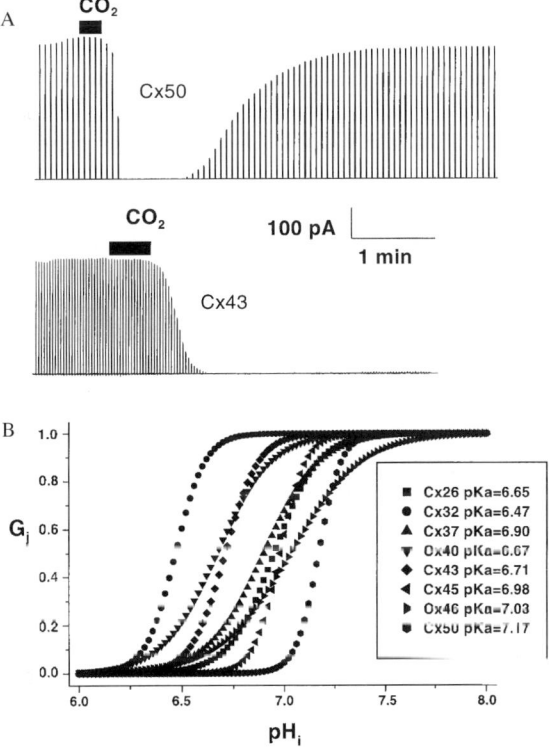

Fig. 4. pH Sensitivity differs among gap junction channels formed of different connexins A. Differences in sensitivity to CO_2 equilibrated external solutions for two connexin subtypes, Cx43 and Cx50, expressed in mammalian cells. (Top trace) Junctions formed of Cx50 are readily closed by exposure to CO_2. A 15 s exposure to CO_2 equilibrated solution closes a highly conductive (~ 40 nS) junctional membrane within the 15 s of exposure. The response is clearly reversible, with recovery occurring within a few minutes of washout. (Lower trace) Junctions formed of Cx43 are also sensitive to CO_2 equilibrated solutions, though longer exposures to CO_2 are required to entirely close all channels. Little or no recovery was observed for these junctions after such exposures. In both cases, the dual whole cell voltage clamp method was used; the junctional currents shown were in response to a series of 200 ms, -10 mV steps. Normal and 100% CO_2 equilibrated solutions were gravity perfused throughout both experiments at a rate of ~ 5 mL/min. (B) pH_i–conductance relations plotted for several connexins as determined in *Xenopus* oocytes following expression of corresponding cRNAs. Note the range of sensitivities of different connexins, as judged by apparent pK_as and slopes of the pH_i–g_j relations. Plotted from data presented in Table 2 in Ref. [36].

were studied from which came the hypothesis that pH-induced channel closure occurs via a particle–receptor mechanism analogous to the ball-and-chain model hypothesized for the voltage-dependent inactivation of K^+ channels. Later studies compared the pH sensitivity of connexin subtypes and their corresponding mutants [36] and found that Cx40 appeared to close via a similar intramolecular mechanism. The dominant and most compelling experimental result in these studies was the complete abolition of pH sensitivity in CT truncation mutants of Cx43 and Cx40, and the restoration of pH sensitivity with co-expression of the CT fragment. Interestingly, co-expression of Cx40 CT with the Cx43 CT-truncated mutant (and vice versa) appeared to restore pH sensitivity to the truncated channels, suggesting that functional domains are not subtype specific. In the hypothesized mechanism, the C-tail contains a region that functions as a particle that binds in a pH-dependent manner to the intracellular loop region of the molecule and thereby directly occludes the channel pore [37]. It is also conceivable that the hypothesized binding induces a conformational change in the channel. The role of the CT in the pH gating of other connexin subtypes is not clear, and there have been conflicting results concerning the alteration in pH sensitivity induced by truncation of Cx50 (see below). Consequently, it appears likely that multiple mechanisms exist for pH-dependent closure, some of which may be sensitive to experimental conditions or expression systems.

Whether protons bind directly to the connexin or their action is transduced through an intermediary regulatory molecule remains unclear in many cases. A long-lived series of investigations has suggested a role for calcium or calcium/calmodulin in pH-induced closure of Cx32, with changes in pH sensitivity induced by chelation of cytosolic Ca^{2+} or block of calmodulin [38]. The intracellular loop has been implicated as a site involved in $pH-Ca^{2+}$-calmodulin-dependent closure of Cx32 [40] and both the NT and CT have been shown to have calmodulin-binding domains [39]. In contrast to this implied Ca^{2+} dependence of closure, Trexler et al. [35] showed a direct effect of pH on Cx46 hemichannels excised from oocytes. A harmonious interpretation of these data sets is that pH-induced changes in junctional coupling can occur via distinct mechanisms in different cell types, depending on the connexins expressed.

For many years it has been recognized that cells are completely uncoupled by intracellular acidification, whereas residual conductance remains even under large V_j gradients. This difference, together with the lack of correlation of specific connexin sensitivities to these two stimuli, implies that gating mechanisms may be distinct. In support of this argument, Bukauskas and Peracchia [41] have shown that Cx43 endogenously expressed in fibroblasts (or exogenously expressed in HeLa cells), exhibits predominantly fast transitions (<2 ms) between the open and a prominent sub-state when exposed to a transjunctional voltage gradient, while reduction in intracellular pH caused by exposure to CO_2 entirely closed channels via slow gating transitions (>5 ms). Other studies have reported similar slow gating to the fully closed state under V_j conditions alone (Cx46 hemichannels, [26]; Cx46 channels [17]; Cx43 channels [42]). However, whether the "pH gate" and "slow V_j gate" are the same remains to be determined, although some evidence is provided by a recent study showing that truncations of the CT have different effects on the voltage sensitivity of Cx43 and Cx32 when expressed in oocytes [43]. Briefly, truncation of Cx43 resulted in significant changes in voltage sensitivity, while truncation of Cx32 produced only minor changes. Comparison of the pH sensitivities of the wild-type and mutant channels follows similar lines: Cx43 truncation abolished pH sensitivity while Cx32 truncation did not [44]. Therefore, the CT is implicated in V_j gating as well as pH-induced closure. Caution must be exercised in interpreting these results, however, as connexin-specific sequences and the highly variable length of the CT may account, at least in part, for differences observed in the sensitivity of connexin subtypes to intracellular pH.

An interesting physiological setting to consider the role of the CT and pH sensitivity is the avascular (and consequently reduced pH) environment of the lens interior, which is composed of cells expressing two highly pH-sensitive connexins, Cx46 and Cx50 (see Fig. 3B). These gap junctions provide intercellular pathways for the transport of nutrients and metabolites between the core and outer regions of the lens. It is now known that in the core region of the lens proteases cleave the carboxyl terminus of both Cx46 and Cx50 [45], thus possibly altering the pH sensitivity of the channels. Reports on the pH sensitivities of these truncation mutants vary in regards to the degree of alteration from wild-type. Stergiopoulus et al. [36] reported a reduction in the pKa of ~ one-tenth of a pH unit for the Cx50 C-tail truncation, while Eckert et al. [46] reported nearly complete loss of pH sensitivity (both studies used oocyte pairs and CO_2 equilibrated solutions to alter pH). Truncation of Cx46 appears to have no effect on its pH sensitivity [35]. Thus, the issue remains open as to whether the pH sensitivity of these channels is a determinant for their expression in the mammalian lens (see below for a discussion of Cx46 and Cx50 in lens homeostasis and disease).

2.3. Sensitivity to other agents

Lipophilic molecules, including heptanol, octanol, halothane, and oleic acid close gap junction channels by a mechanism that results in open–closed state transitions that may or may not be the same as involved in acidification-induced uncoupling. Unfortunately, there have been few studies performed under steady state conditions of partial blockade and none so far investigating possible interactions.

The hydrophobicity of halothane and long-chain alcohols and fatty acids suggests that their uncoupling action may be due to interaction with hydrophobic connexin domains or at the connexin–lipid interface [47]; more recently, heptanol blockade has been attributed to interaction with non-cholesterol lipids [48]. Site of action of these uncoupling compounds is a fundamental question that should be readily addressable using the new generation of methods for acute cholesterol exchange (e.g. treatment with cyclodextrins). Another category of uncoupling agents includes the licorice components α and β glyccerhetinic acid and the synthetic relative carbenoxolone [49]. Although these compounds are useful experimentally (and appear to be selective in closing gap junctions without grossly affecting other channels), their mechanism of action is almost completely unknown.

2.4. Sensitivity to phosphorylation

All connexins (with the exception of Cx26) are phosphoproteins, but the issue of how phosphorylation affects channel behavior remains to be clarified in detail. A number of studies have demonstrated alternate channel open states favored after phosphorylating or dephosphorylating treatments, yet variability in responses of cells expressing Cx43 treated with tumor promoting phorbol esters to activate protein kinase C has ranged from total uncoupling to enhanced intercellular communication (see review of phosphorylation induced changes [50]). Thus, correlating physiological consequences with post-translational processing remains a major obstacle in the field. Furthermore, the much larger number of kinases and phosphatases that are now known and the intricate overlapping and branching of kinase activation pathways demands more rigorous studies to quantify phosphate incorporation under conditions where functional changes are determined.

Connexin43 and closely related Group II connexins exhibit serine-rich domains close to the carboxyl terminus. Initial reports suggested that mutations within or adjacent to this region were associated with heterovisceral atriataxia [51–53], a syndrome involving profound axis asymmetry, consistent with the notion that this region was important, perhaps with regard to the phosphorylation of the channel. Although such mutations have not been discovered subsequently in much larger populations of patients suffering similar syndromes, the hypothesis stands that a gap junction "channelopathy" might arise from mutation of a phosphorylation site in the connexin molecule.

2.5. Overview of gating

The background presented above describes mechanisms of action of stimuli and pharmacological agents currently known to affect gap junction channels. It is clear that at least two of these gating mechanisms (mediated by transjunctional voltage (V_j) and by intracellular acidification) are different [54], and presumably involve distinct domains of the gap junction proteins; whether gating by other agents acts on independent or overlapping residues is unclear at present. As is indicated below, a number of connexin mutants implicated in hereditary disease are defective in gating by one or more of these stimuli, while channels from other mutants are relatively normal in their responses. Although defects in gating to these stimuli are generally interpreted as explaining the dysfunction, it should be emphasized that the role of such gating in normal tissue is not well understood and the real defect might lie in another role of connexins (see below).

2.6. Permeation

Permselectivity studies of multiple connexin types have shown that compared with other ion channels, gap junctions are relatively large, non-selective pores that allow the passage of molecules up to $\sim$ 1 kDa, with modest ionic selectivities that range from 8: 1 cation selective (Cx40, [55] and Cx43, [56]) to mildly anion selective (Cx32, [57]). A recent report implicates the EL1 domain as a specific molecular region in the determination of selectivity for gap junction channels. The identity of pore lining regions remains to be conclusively determined, though both the first and third transmembrane domains have been implicated. It has been hypothesized that differences in pore size and selectivity may have physiologically significant roles in intercellular signaling pathways by differentially allowing the passage of regulatory molecules from cell to cell in a tissue. Recent studies have confirmed connexin specific permeability to different intracellularly applied fluorescent tracers as well as to the second messenger molecules IP_3 [58] and cGMP and cAMP [59]. The physiological consequences of the differences in permselectivity are difficult to determine. Until a clear picture emerges about the distinct molecular determinants of permselectivity, it will be difficult to assess tissue specificity for connexin expression and signaling to understand how changes in permselectivity might result in dysfunction.

3. Expression patterns and connexin diversity

Given their widespread expression throughout vertebrate tissues, it is likely that gap junction-mediated intercellular diffusion is necessary for baseline tissue homeostasis. Most tissue types express multiple connexin sub-types; often they appear as dominantly expressed pairs (Cx32 and Cx26 in hepatocytes [60], Cx46 and Cx50 in lens fiber cells [21], and Cx40 and Cx43 in cardiac tissue [61]). The physiological consequences of such patterns are largely unknown. Expression of multiple connexin types in a given tissue may occur simply to fill the need for pathways of sufficiently high conductance or permeability between cells. Some degree of redundancy may also be a cause for the plurality of connexin subtypes. Studies of connexin "knock-out" animals provide evidence both for specific function and redundancy in tissues expressing multiple connexin subtypes (see discussion of diseases associated with specific subtypes). Alternatively distinct subtype (or mixed subtype) specific pathways may be required under different homeostatic and pathological conditions.

The picture becomes more complex due to the possibility of selective pairing between hemichannels of different connexin subtype composition (heterotypic channel formation) and interactions which occur between subunits of different connexin types which potentially lead to heteromeric channel formation (see Fig. 1A). These routes for channel formation can conceivably lead to the formation of thousands of heterologous forms with different functional properties. The voltage gating properties of heterotypic channels have been widely investigated; the properties are generally consistent with those expected from the individual hemichannels [17,20] and the second extacellular loop (E2) has been shown to be relevant in studies of compatibility between different connexin subtypes [62]. Investigations of heteromeric channel function have produced more ambiguous results (see below for discussion of heteromerization in the lens). Changes in other properties (i.e. pH sensitivity, permselectivity, and regulation by phosphorylation) due to heterologous channel formation have not been widely investigated.

4. Charcot-Marie-Tooth Syndrome and Cx32

Charcot-Marie-Tooth syndrome (CMT) is the most common inherited peripheral neuropathy, with a frequency of 1 in 2500. Patients suffer from a progressive wasting of foot and distal leg musculature. Because of the unopposed action of the long toe flexors and extensors, "clawed" foot deformities arise. Hand muscles are also affected, giving the "clawed" and unopposable thumb appearance. The phenotype of CMT syndrome is quite varied, with 10–20% of individuals with the mutation showing no observable symptoms except at the neuromyographic and/or neurological level. Very few patients are extremely incapacitated, although severity generally increases with age [3].

CMT is subgrouped into two major categories: CMT1 (associated with demyelination and reduced nerve conduction velocities), and CMT2 (a non-demyelinating neuropathy with no reduction in nerve conduction). Genes encoding compact peripheral myelin components are responsible for two CMT1 subtypes. CMT1A is the most common form and involves a gene on chromosome 17 encoding peripheral myelin protein 22 (PMP-22), while CMT1B has been localized to chromosome 1 and a gene encoding myelin protein zero (MPZ or P_0). Thus both CMT1A and 1B affect components of compact peripheral nervous system myelin [see Ref. 68 for review].

The inheritance of the CMT1 disease phenotype in another subgroup appeared to follow an X-linked pattern. Subsequent analysis of X-linked CMT (CMTX) identified Cx32 as the culprit gene [3], and additional studies of afflicted families have thus far revealed more than 160 coding region mutations responsible for this disease (see Table 1). Although frameshift, deletion, missense and non-sense mutations have been reported, the majority ($\sim 75\%$) are missense mutations. The mutational profile of Cx32 CMTX mutants does not reveal any particular genetic hot spots, although mutations are somewhat more common in sequences predicted to be transmembrane domains.

Although compelling electron microscopy is still lacking, fluorescence microscopy has localized Cx32 to paranodal regions and Schmidt–Lanterman incisures in myelinating Schwann cells [71]. The role of Cx32 has been hypothesized to be the formation of reflexive junctions interconnecting the external cytoplasm with the internal, adaxonal cytoplasm, thereby shortening the otherwise tortuous pathway caused by as many as 100 wrappings of the Schwann cell around the nerve. The strongest evidence in support of the hypothesis has come from dye injection studies showing connectivity between the outer (perinuclear) and adaxonal regions of Schwann cells [71]. Interestingly, the dye spread persisted in myelinating Schwann cells from mice lacking the Cx32 gene, suggesting the possible presence of other connexins in addition to Cx32 in wild-type Schwann cells or the upregulation of other connexins in Cx32 knockout animals [71].

Broadly speaking, the mutations associated with CMTX can be divided into four categories of functionality: 1) mutations in non-coding regions that abolish the initiation of transcription (nonsense mutations) [72], 2) mutations that result in premature termination of translation or that are quickly recognized as defective by intracellular copy editors [63], 3) mutations that cause alterations in connexin trafficking to the plasma membrane resulting in either protein accumulation within membranes in the cytoplasm or rapid degradation [63,69], and 4) mutations that appear to be correctly targeted to the plasma membrane, generally with altered gating as demonstrated in vivo [see Ref. 68].

Some of the numerous CMTX mutations that potentially fall into the second and third categories above have been studied in vitro, using oocyte and mammalian expression systems. The results of these studies are given in Table 1. Diverse alterations are clearly apparent within this group of mutations, which occur in all topological regions of the Cx32 protein. Studies using immunostaining techniques to determine localization of the mutant protein within the oocyte or cell have shown that a large number of these mutants are unable to localize to the plasma membrane (see Table 1 for references). A recent study by VanSlyke et al. [69] analyzed the specific intracellular trafficking patterns of

Table 1
Cx32 CMTX mutations that potentially form functional channels

Domain	Mutation	System	Immunostaining	Dye transfer	Coupling/V_j sensitivity	References
NT	W3Y	Oocytes/Cos-7	ER/golgi	No	No	[67]
NT	W3S	Oocytes/cCos-7	ER/golgi		No	[67,68]
NT	G12S	Oocytes/Cos-7	ER/golgi	No	No	[56,63,67]
NT	V13L	Oocytes, Cos-7	Plasma membrane	Yes		[67]
NT	R15Q	Oocyte	Plasma membrane			[63]
TM1	R22G,P,Q	Oocytes			No	[64]
TM1	S26L	Oocytes/N2A			Slightly shifted[a]	[57]
TM1	I30N	N2A			Very slightly shifted	[57]
TM1	M34T	Oocytes/N2A			Strongly shifted[b]	[57]
TM1	V35M	Oocytes			Strongly shifted	[57]
TM1	V38M	Oocytes			Moderately shifted	[57]
EL1	C53S	Glioma	ER/golgi	No		[65]
EL1	L56F	Oocytes			Slightly shifted	[64]
EL1	C60F, V63I	HeLa	Membrane, reduced	Reduced		[66]
TM2	P87A	Oocytes			Very slightly shifted	[57]
TM2, CL	L90H, V95M	Oocytes		No	No	[57]
CL	E102G	Oocytes			Normal[c]	[57,64]
CL	Del111-116	Oocytes, Cos-7	Membrane, reduced	Reduced	Very slightly shifted[c]	[57,64,67,68]
TM3	V139M	HeLa	Membrane, reduced	No		[63,66]
TM3	R142W	Oocytes/PC12	PM(oocytes) ER/golgi(cells) perinuclear (70)	No	No[d]	[63,64,68,69]
EL2	P172R	Glioma	ER/golgi	No		[65]
EL2	P172S	Oocytes			No	[64]
EL2	V175 frameshift	PC12	No detectable protein			[63]
EL2	E186K	PC12	ER/golgi or perinuclear (70)			[63,69]
CT	E208K	Oocytes/PC12/Cos-7	ER/golgi		No	[63,67,69,70]
CT	E208L	Oocytes			No	[64]
CT	R215W	HeLa	Membrane, reduced	No		[66]
CT	R217stop	Oocytes			Yes[e]	[70]
CT	R220stop	Oocytes/glioma	Plasma membrane	Yes	Very slightly shifted	[63,64,66]

(continued on next page)

Table 1 (*continued*)

Domain	Mutation	System	Immunostaining	Dye transfer	Coupling/V_j sensitivity	References
CT	Y211stop	Oocytes			No	[64,70]
CT	R238H	Oocytes			Yes[e]	[70]
CT	R265stop	Oocytes			Yes[e]	[70]
CT	C280G	Oocytes			Yes[e]	[70]
CT	S281stop	Oocytes			Yes[e]	[70]

[a]Reduced permeability to large solutes.
[b]Preferential substate residency.
[c]Altered recovery from CO_2 induced uncoupling.
[d]Dominant negative effect when expressed with WT C26.
[e]Reported to form functional hemichannels in oocytes.

three such mutants (R142W, E186K, and E208K). E208K appeared to be entirely retained within the ER, whereas R142W and E186K seemed to localize initially to perinuclear compartments before additional trafficking either to lysosomes (R142W) or back to the ER (E186K). The results of this study suggest that Cx32 CMTX mutations can have significant effects on the movement and localization of the protein within the cell. In any case, these mutants, like others that are not correctly localized, are unable to perform the role of intercellular communication presumably necessary for homeostasis in Schwann cells.

A number of other mutants are clearly capable of localizing at the plasma membrane and forming functional channels. Ressot et al. [64] described the properties of several functional CMTX mutants expressed in oocytes. These mutants (L56F, E102G, Del111-116, and R220stop) displayed minimal to moderate alterations in V_j sensitivity at the macroscopic level, and the Boltzmann parameters determined for all mutants were well within the range observed for other connexin subtypes. Fig. 5 compares the V_j sensitivity of one of these mutants (R220stop) to WTCx32. The data in Fig. 5 were generated in our lab by experiments with mutant and wild-type channels expressed in N2A cells. Junctional currents in response to 100 and -100 mV transjunctional voltages from wild-type and R220stop channels are exhibited in upper and lower traces. The Boltzmann fits to these data show only marginal differences in V_j sensitivity between the two channel types, largely consistent with the data from Ressot et al. [64].

Two of the mutants in Ref. [64] (Del111-116 and E102G) showed an increased sensitivity to CO_2 equilibrated external solutions, suggesting that increased pH sensitivity may contribute to the cause for the CMTX disease phenotype associated with these mutants. A report by Oh et al. [57] found additional alterations in channel properties that were construed as possible factors in the genesis of CMTX. The study investigated nine mutants associated with CMTX, eight of which were capable of forming functional channels in oocytes and N2A cells. The single channel data from one of the mutants, M34T, showed preferential residency in a subconductance state with very few and short-lived sojourns in the open state when compared to wild-type, while a second mutant, S26L, was shown to have greatly reduced permeability to large molecules. The permeability of the M34T was not evaluated but the authors hypothesized that both mutants may play a role in CMTX by reducing or abolishing permeation pathways important for the maintenance of homeostasis in Schwann cells. Thus at least three alterations in channel function (changes in V_j sensitivity, pH sensitivity, and permeation) have been hypothesized as possible reasons for the development of the CMTX disease phenotype when functional channels are likely to be present in the Schwann cell.

Although there is certainly evidence for multiple types of altered behavior of Cx32 CMTX channels, it remains possible that these mutations may also alter other connexin functions. The multiple effects of mutations, the lack of correlation between clinical phenotype with genotype, the variability in severity and onset of the disease, as well as the specificity of the disease to peripheral nervous system raises the possibility that there may be interaction between Cx32 and the normal maintenance of myelination. For example, Cx32 binds to the tight junction protein occludin in liver cells [73], Cx26 is associated with occludin in an intestinal line cell (T84) [74], and Cx43 binds to zonula occludens (ZO)-1 and to c- and v-src [75,76]. Thus, it is conceivable that one additional dysfunctional

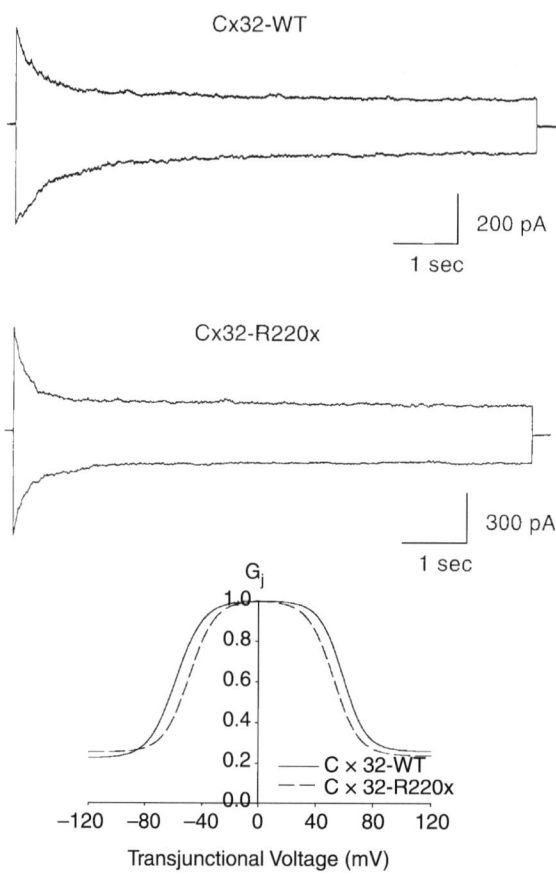

Fig. 5. Transjunctional voltage dependence of a CMTX truncation mutant (R220x) is not different from wild-type Cx32. (Top) Recordings of wild-type Cx32 and (Middle) R220x Cx32 gap junctional currents measured in one cell of a pair in response to 9 s step pulses that varied the transjunctional voltage from 0 to ± 100 mV (in 20 mV increments). Junctional currents were maximal at the beginning of the pulses and declined to steady-state values in a time- and voltage-dependent manner when V_j exceeded ± 40 mV. (Bottom) Relationship between V_j and steady state junctional conductance of wild-type Cx32 (solid line) and R220x Cx32 (dashed line) gap junctions. The lines are fits of the data to a two state Boltzmann equation. The Boltzmann parameters of Cx32–R220x channels were not significantly different from those of wild-type channels.

category of Cx32 CMTX mutations may involve altered binding affinity for other proteins, either decreasing functional linkage that is normally present or initiating binding that is normally absent.

5. Senile cataract and the involvement of Cx46 and Cx50

Inherited cataract is a heterogeneous lens disorder that most often presents itself as a congenital autosomal trait [77–79]. Autosomal dominant cataracts have been mapped to nine separate loci [80]. Recent studies have linked some types of congenital cataracts to mutations in Cx50 and Cx46 [4,5,81,82]. Both of these connexins are primarily expressed in the lens fiber cells, where they play important roles in maintaining tissue homeostasis [21,83]. Direct confirmation of the importance of gap junctions in the lens has been provided by studies of transgenic mice lacking expression of each of these lens connexins [84,85].

Mutations of the Cx50 gene have recently been found in families of diverse ethnic origin with zonular pulverulent cataracts. A missense mutation in the second transmembrane domain (P88S) was reported in an English family with zonular pulverulent cataract [5], whereas a second mutation has been identified in a Pakistani family with zonular nuclear pulverulent cataract (E48K) [81]. In addition, a missense mutation in mouse connexin 50 gene has been associated with the nuclear opacity 2 (No2) mouse cataract, a congenital hereditary bilateral cataract [82]. This missense mutation results in a substitution of aspartic acid to alanine at amino acid position 47 (D47A) in the first extracellular domain of Cx50. Similarly, in the case of Cx46, a missense mutation at postion 63 in the first extracellular loop, as well as a frameshift mutation resulting from a single base insertion at nucleotide position 1137, have been discovered in families with congenital cataracts [4]. The frameshift caused by this insertion (after codon 379) results in aberrant translation of the last 56 amino acids of the wild-type protein and the addition of 31 amino acids to the C-terminus of the mutant protein. This resulting sequence does not contain consensus phosphorylation sites that are present in wild-type Cx46.

The functional consequences of these mutations have been primarily investigated in the paired *Xenopus* oocyte system [86–88]. None of the mutations formed functional gap junction channels, and, with the exception of P88S, most mutations acted like "loss-of-function" mutations without dominant negative inhibition [86–88]. Co-expression of Cx46 mutations N63S and fs380 with wild-type Cx46 or Cx50 did not affect the junctional conductance induced by either wild-type Cx46 or wild-type Cx50 ([88]. Similarly, co-expression of D47A channels with wild-type Cx50 did not result in an inhibition of junctional conductance of wild-type channels [86]. In both cases, the authors speculated that cataracts form due to a reduction in the level of coupling below a critical value. The P88S mutant acted like a dominant negative inhibitor when co-expressed with wild-type Cx50 [86]. A 61% reduction in Cx50 gap junctional conductance was found when the mutant and wild-type Cx50 channels were injected in a ratio of 1:11, suggesting that a single mutant subunit per gap junctional channel is sufficient to abolish channel function. These characteristics may explain the autosomal dominant pattern of inheritance observed in this family. The effect of P88S Cx50 mutant channels on gap junctional conductance of Cx46 channels has not yet been investigated.

The existence of mutations in Cx46 and Cx50 that are linked to cataractogenesis highlights the importance of gap junctional communication in the lens. The lens is an avascular cyst, and therefore, maintenance of homeostasis is critically dependent on the presence of gap junctions [89–91]. The cells comprising the lens can be divided into three different layers. A single layer of epithelial cells lines its anterior surface. At the lens equator, these epithelial cells differentiate to form new fiber cells (also called differentiating fiber cells), which subsequently mature and become a part of the inner mass of fiber cells (mature fiber cells). Mature fiber cells do not contain intracellular organelles, and have a low level of metabolic activity. Thus, maintenance of precise intercellular ionic concentrations in the fiber cells is dependent on gap junctions, which provide low-resistance pathways for diffusion of metabolites between the various cells of the lens. Three different connexins are expressed in the lens. Epithelial cells express Cx43 whereas differentiating fiber cells [92] express Cx46 and Cx50 [21,83]. As the fiber cells mature, both Cx46 and Cx50 are modified by phosphorylation and are cleaved by endogenous proteases.

As mentioned earlier, transgenic mice lacking Cx46 or Cx50 develop cataracts, although the mice exhibit distinct phenotypes. Cx46 knockout mice develop a senile-type cataract, with lens growth and development being normal. Cx46 contributes as much as 71% of the conductance in the differentiating fiber cell zone [93]. The contribution of this connexin to mature fiber coupling is less certain. Cataracts were proposed to occur due to the uncoupling of cells in mature fiber cell zone from the surface cells [85,93]. In contrast, deletion of the Cx50 gene resulted in reduced growth of the lens and eye. In addition, these mice develop zonular pulverulent nuclear cataracts [84]. These studies on Cx46 and Cx50 knockout mice indicate that the presence of both connexins is required for lens homeostasis. In addition, the differences in phenotypes strongly suggest that the two connexins subserve different roles in the lens. For example, the retardation of growth in Cx50 knockout mice suggests that Cx50 channels allow the propagation of unknown growth signals more efficiently than Cx46 channels [84]. In addition, because both connexins have overlapping patterns of expression in the lens, these connexins may interact to form functional heterotypic and heteromeric channels. Such channels may play a significant role in the formation and determination of specific compartments in the lens.

Cx46 and Cx50 are known to form heterotypic junctions in both oocyte and mammalian expression systems where macroscopic and single channel properties have been well documented [17,18,94]. In contrast, electrophysiological evidence for the formation of heteromeric channels is less conclusive. Our investigation of electrical coupling between cells simultaneously expressing Cx46 and Cx50 was unable to provide unambiguous evidence for the existence of such channels, possibly due to the fact that Cx46 and Cx50 have rather similar properties. Macroscopic and single channel data were not strikingly different from homotypic and heterotypic recordings, though the data were not strictly counterindicative of heteromerization. Nevertheless, there is strong evidence for the heteromerization of these subtypes at the biochemical level, as both connexins co-immunoprecipitate from lens organ cultures [95,96]. Although the functional consequences of multiple connexin subtype expression in the lens are unknown, the interactions between the subtypes may be a determinant for their expression and heterologous forms may play specific roles in maintaining lens homeostasis.

6. The role of Cx26 (and other connexins?) in non-syndromic hearing loss (NSHL)

Mutations in the gene coding Cx26 have recently been implicated in the most common form of non-syndromic hereditary deafness [6,97,98], as currently more than 45 different mutations have been detected. A 4.76% reported frequency of carriers for these mutations predicts one deaf person per 1765 people in the Mediterranean population analyzed by Morell et al. [99]. Both autosomal dominant and recessive mutations have been described, though some controversy surrounds at least one of the reportedly dominant mutations. It appears there is little correlation between specific genotypes (homozygous recessive, heterozygous with mixed recessive mutant alleles, or heterozygous dominant) and severity of hearing loss. Mutations in the gene encoding Cx26 have recently been implicated in syndromes involving skin disorders in addition to hearing loss [100–103].

The vast majority of the Cx26 mutations implicated in hearing loss are autosomal recessive. A base deletion, 35delG, is the most common mutation [104] and results in a reading frame shift and premature cessation of translation at the 12th amino acid. The result is an obvious loss of channel function, though a range of hearing loss, from moderate to profound, has been reported in individuals homozygous for this mutation [98]. Various studies have reported numerous other recessive mutations implicated in NSHL, the majority of which lead to premature cessation of translation and/or likely loss of function. Unfortunately, unlike for CMTX mutations, very few studies have been conducted to assess alterations in channel properties induced by such missense mutations that could conceivably produce functional channels.

A study by Richard et al. [103] used the oocyte expression system to assess the properties of two mutations implicated in NSHL, a recessive W77R point mutation, and the dominant R75W point mutation. The results of the study were consistent with the inheritance patterns for hearing loss associated with these two mutants. The recessive mutant was unable to form functional channels when expressed in oocyte pairs, but did not appear to affect the formation of normal channels when simultaneously expressed with wild-type Cx26 mRNA. The purportedly dominant R75W mutation was similarly unable to form functional channels though, unlike W77R, it appeared to have a dominant negative effect when co-expressed with wild-type Cx26 (an effect similar to those observed in Cx50 mutants associated with cataractogenesis, see above). A recent study confirms loss of functionality for these two mutants (R75W, W77R) as well as another Cx26 mutant (M34T) when expressed in mammalian cells [105].

The M34T mutation has been a source of some controversy as it was initially associated with a dominant pattern of expression as described by Refs. [106,107] with apparent confirmation by the dominant negative result when simultaneously expressed with wild-type Cx26 in oocytes [107]. Other reports, describing individuals with normal hearing as heterozygous for this mutation have appeared to contradict these findings [108,109]. Recently, a novel Cx26 mutation (C202F) has also been associated with autosomal dominant hearing loss [11]. Further studies of the functional consequences of these and other mutations in exogenous expression systems are clearly warranted in order to determine the underlying relationship between expression, phenotype, and channel function. It would not be surprising if some of the Cx26 mutant proteins were capable of forming functional channels given the evidence for functional CMTX Cx32 mutants. Comparative studies of alterations in function may add to our understanding of physiological function of gap junctions in both the peripheral nervous system and the cochlea.

Currently, the specific role of intercellular communication in auditory function is not well understood, though gap junction expression in cochlear tissue is well documented [106,110–112]. There is no evidence for Cx26 expression between hair cells or between hair cells and supporting cells, but clear evidence is now available for the strong expression of Cx26 between a variety of supporting cells of the organ of Corti [106, 112]. The cochlea contains three fluid-filled chambers, the scala tympani, scala media, and scala vestibuli. The scala tympani and scala vesitibuli are filled with perilymph, which is similar in ionic composition to the cerebral spinal fluid, while the scala media is filled with high potassium containing endolymph. Auditory transduction occurs through mechanical stimulation of cochlear hair cells. The apical surfaces of hair cells are bathed in endolymph, and during auditory stimulation mechano-sensitive channels open and allow the influx of K^+. This results in the opening of Ca^{2+} channels and the generation of a receptor potential. Though the tissue specific role of gap junctional communication in the cochlea is not known, it is plausible that it plays a vital role in the maintenance of the cochlear ionic compartmentalization by recycling the K^+ known to accumulate in the perilymphatic environment surrounding the basolateral membranes of hair cells during auditory stimulation [113].

Recently Cx30 and, to a lesser extent, Cx43 have been shown to be expressed in the supporting cells of the organ of Corti in addition to Cx26 [114], indicating that Cx26 is not solely responsible for gap junctional intercellular communication in the inner ear. In this study Cx30 appeared to have an expression pattern that closely overlapped that of Cx26, suggesting the possibility of heterotypic and heteromeric channel formation and begging a comparison to the other disease phenotypes where loss of connexin expression is implicated in cells that nevertheless continue to be communication competent. A recent investigation of coupling between dissociated supporting cells (Hensen cells) showed electrophysiological properties consistent with multiple subtype expression and possibly the functional consequences of heterologous forms [115]. In addition, a previous study has shown that Cx30 and Cx26 form functional heterotypic channels using the oocyte expression system [116]. Recently, a mutation in Cx30 has also been associated with NSHL [7], following a dominant autosomal pattern of inheritance. This mutant also exhibited dominant negative effects when co-expressed with WT Cx30 in oocytes. Thus, further investigations of the functional interactions between Cx30, Cx26, and mutant proteins associated with hearing loss are clearly warranted in order to understand the interaction of these connexin types and the physiological role(s) of the intercellular communication they provide.

7. Summary and conclusions

Connexin mutations are the cause of genetic diseases with moderate frequency in man, especially those affecting the nervous system (non-syndromic hereditary deafness and X-linked Charcot-Marie-Tooth demyelinating neuropathy) and lens (senile and zonular pulverulent cataracts). Behavior of a number of these mutations have been characterized in exogenous expression systems, which has revealed a variety of diverse phenotypes, ranging from total disappearance of protein, to apparent mistargeting and to changes in channel properties that can be profoundly disruptive or seemingly benign. For some mutants it seems clear that the functional channel deficit should diminish intercellular coupling, whereas for others it remains possible that the relevant channel deficit has not yet been identified.

Because many of the connexin subtypes associated with human disease are expressed in multiple areas, the specificity of the afflicted tissues remains unexplained. It was initially surprising that peripheral neuropathy and no other symptomatology was associated with Cx32 mutations, because this connexin is expressed mainly in liver, as well as in alveolar cells of lactating mammary gland, pancreatic acinar cells, proximal kidney tubules, thyroid follicular cells, visceral yolk sac, Schwann cells, and oligodendrocytes [117,118]. This lack of associated dysfunction in other organs is also surprising because Cx32KO mice are reported to have higher incidence of hepatomas and reduced gluconeogenic responses compared to wild-type mice [119], whereas nerves do not seem to be as greatly affected as in CMTX patients. In at least some cases, phenotypic differences between knockout mice and patients with loss of function mutations may be due to physiological differences between the species. One such example is illustrated by the embryonic lethality of Cx26 knockout mice, whereas humans with severe Cx26 mutations suffer from hearing loss [120]. In this case, the explanation appears to be the species-specific organization of the placenta; whereas the human placenta possesses a single trophoblast layer, that of the mice is multi-layered, with cells connected by Cx26. Thus, understanding the physiological importance of connexin subtype expression and its relationship to specific pathologies will require comparison of experiments from several species and from several different tissues.

Acknowledgements

Support for studies of connexin mutants in our laboratory is derived from NIH DK, MH, NS and EY. In addition, a post-doctoral fellowship from the Heritage Chapter of the American Heart Association provided partial salary support for MS.

References

[1] Unger, V.M., Kumar, N.M., Gilula, N.B., Yeager, M., 1999. Three-dimensional structure of a recombinant gap junction membrane channel. Science 283, 1176–1180.

[2] Purnick, P.E., Benjamin, D.C., Verselis, V.K., Bargiello, T.A., Dowd, T.L., 2000. Structure of the amino terminus of a gap junction protein. Arch. Biochem. Biophys 381(2), 181–190.

[3] Bergoffen, J., Scherer, S.S., Wang, S., Scott, M.O., Bone, L.J., Paul, D.L., Chen, K., Lensch, M.N., Chance, P.F., Fischbeck, K.H., 1993. Connexin mutations in X-linked Charcot-Marie-Tooth disease. Science 262, 2039–2042.

[4] Mackay, D., Ionides, A., Kibar, Z., Rouleau, G., Berry, V., Moore, A., Shiels, A., Bhattacharya, S., 1999. Connexin46 mutations in autosomal dominant congenital cataract. Am. J. Hum. Genet. 64, 1357–1364.

[5] Shiels, A., Mackay, D., Ionides, A., Berry, V., Moore, A., Bhattacharya, S., 1998. A missense mutation in the human Connexin50 gene (GJA8) underlies autosomal dominant "Zonular Pulverulent" cataract, on chromosome 1q. Am. J. Hum. Genet. 62, 526–532.

[6] Kelsell, D.P., Dunlop, J., Stevens, H.P., Lench, N.J., Liang, J.N., Parry, G., Mueller, R.F., Leigh, I.M., 1997. Connexin26 mutations in hereditary non-sydnromic sensorineural deafness. Nature 38, 80–83.

[7] Grifa, A., Wagner, C.A., D'Ambrosio, L., Melchionda, S., Bernardi, F., Lopez-Bigas, N., Rabionet, R., Arbones, M., Monica, M.D., Estivill, X., Zelante, L., Lang, F., Gasparini, P., 1999. Mutations in GJB6 cause nonsyndromic autosomal dominant deafness at DFNA3 locus. Nat. Genet. 23(1), 16–18.

[8] Richard, G., Smith, L.E., Bailey, R.A., Itin, P., Hohl, D., Epstein, E.H., DiGiovanna, J.J., Compton, J.G., Bale, S.J., 1998. Mutations in the human connexin gene GJB3 cause erythrokeratodermia variabilis. Nat. Genet. 20(4), 366–369.

[9] Richard, G., White, T.W., Smith, L.E., Bailey, R.A., Compton, J.G., Paul, D.L., Bale, S.J., 1998. Functional defects of Cx26 resulting from a heterozygous missense mutation in a family with dominant deaf-mutism and palmoplantar keratoderma. Hum. Genet. 103(4), 393–399.

[10] Verselis, V.K., Veenstra, R.D., 2000. Gap junction channels: permeability and voltage gating. In: Bittar, E., Hertzberg, E. (Eds.), Advances in Molecular and Cell Biology, vol. 30. Jal Press, Standford CT pp. 129–192.

[11] Harris, A.L., Spray, D.C., Bennett, M.V., 1983. Control of intercellular communication by voltage dependence of gap junctional conductance. J. Neurosci. 3(1), 79–100.

[12] Barrio, L.C., Capel, J., Jarillo, J.A., Castro, C., Revilla, A., 1997. Species-specific voltage-gating properties of connexin-45 junctions expressed in Xenopus oocytes. Biophys. J. 73(2), 757–769.

[13] Barrio, L.C., Revilla, A., Gomez-Hernandez, J.M., De Miguel, M., Gonzalez, D., 2000. Membrane potential dependence of gap junction in vertebrates. In: Peracchia, C. (Ed.), Current Topics in Membranes (Gap Junctions: Molecular Basis of Cell Communication in Health and Disease), vol. 49. Academic Press, San Diego, pp. 175–188.

[14] Srinivas, M., Rozental, R., Kojima, T., Dermietzel, R., Mehler, M., Condorelli, D.F., Kessler, J.A., Spray, D.C., 1999. Functional properties of channels formed by the neuronal gap junction protein connexin36. J. Neurosci. 19(22), 9848–9855.

[15] Moreno, A.P., Rook, M.B., Fishman, G.I., Spray, D.C., 1994. Gap junction channels: distinct voltage-sensitive and -insensitive conductance states. Biophys. J. 67(1), 113–119.

[16] Moreno, A.P., Laing, J.G., Beyer, E.C., Spray, D.C., 1995. Properties of gap junction channels formed of connexin 45 endogenously expressed in human hepatoma (SKHep1) cells. Am. J. Physiol. 268(2 Pt 1), C356–C365.

[17] Hopperstad, M.G., Srinivas, M., Spray, D.C., 2000. Properties of gap junction channels formed by Cx46 alone and in combination with Cx50. Biophys. J. 79(4), 1954–1966.

[18] Srinivas, M., Costa, M., Gao, Y., Fort, A., Fishman, G.I., Spray, D.C., 1999. Voltage dependence of macroscopic and unitary currents of gap junction channels formed by mouse connexin50 expressed in rat neuroblastoma cells. J. Physiol. 517(Pt 3), 673–689.

[19] Rozental, R., Srinivas, M., Spray, D.C., 2001. How to close a gap junction channel. Efficacies and potencies of uncoupling agents. Methods Mol. Biol. 154, 447–476.

[20] Verselis, V.K., Ginter, C.S., Bargiello, T.A., 1994. Opposite voltage gating polarities of two closely related connexins. Nature 368(6469), 348–351.

[21] Paul, D.L., Ebihara, L., Takemoto, L.J., Swenson, K.I., Goodenough, D.A., 1991. Connexin46, a novel lens gap junction protein, induces voltage-gated currents in nonjunctional plasma membrane of Xenopus oocytes. J. Cell Biol. 115, 1077–1089.

[22] Pfahnl, A., Zhou, X.W., Werner, R., Dahl, G., 1997. A chimeric connexin forming gap junction hemichannels. Pflugers Arch. 433(6), 773–779.

[23] Nicholson, B.J., Weber, P.A., Cao, F., Chang, H., Lampe, P., Goldberg, G., 2000. The molecular basis of selective permeability of connexins is complex and includes both size and charge. Braz. J. Med. Biol. Res. 33(4), 369–378.

[24] Trexler, E.B., Bukauskas, F.F., Kronengold, J., Bargiello, T.A., Verselis, V.K., 2000. The first extracellular loop domain is a major determinant of charge selectivity in connexin46 channels. Biophys. J. 79(6), 3036–3051.

[25] Pfahnl, A., Dahl, G., 1998. Localization of a voltage gate in connexin46 gap junction hemichannels. Biophys. J. 75(5), 2323–2331.

[26] Trexler, E.B., Bennett, M.V., Bargiello, T.A., Verselis, V.K., 1996. Voltage gating and permeation in a gap junction hemichannel. Proc. Natl Acad. Sci. USA 93(12), 5836–5841.

[27] Purnick, P.E., Oh, S., Abrams, C.K., Verselis, V.K., Bargiello, T.A., 2000. Reversal of the gating polarity of gap junctions by negative charge substitutions in the N-terminus of connexin 32. Biophys. J. 79(5), 2403–2415.

[28] Verselis, V.K., Trexler, E.B., Bukauskas, F.F., 2000. Connexin hemichannels and cell–cell channels: comparison of properties. Braz. J. Med. Biol. Res. 33(4), 379–389.

[29] Bruzzone, S., Guida, L., Zocchi, E., Franco, L., De Flora, A., 2001. Connexin 43 hemi channels mediate Ca^{2+}-regulated transmembrane NAD+ fluxes in intact cells. FASEB J. 15(1), 10–12.

[30] Valiunas, V., Weingart, R., 2000. Electrical properties of gap junction hemichannels identified in transfected HeLa cells. Pflugers Arch. 440(3), 366–379.

[31] Quist, A.P., Rhee, S.K., Lin, H., Lal, R., 2000. Physiological role of gap-junctional hemichannels. Extracellular calcium-dependent isosmotic volume regulation. J. Cell Biol. 6;148(5), 1063–1074.

[32] Turin, L., Warner, A., 1977. Carbon dioxide reversibly abolishes ionic communication between cells of early amphibian embryo. Nature 270(5632), 56–57.

[33] Spray, D.C., Harris, A.L., Bennett, M.V., 1981. Gap junctional conductance is a simple and sensitive function of intracellular pH. Science 211(4483), 712–715.

[34] Ek-Vitorin, J.F., Calero, G., Morley, G.E., Coombs, W., Taffet, S.M., Delmar, M., 1996. PH regulation of connexin43: molecular analysis of the gating particle. Biophys. J. 71(3), 1273–1284.

[35] Trexler, E.B., Bukauskas, F.F., Bennett, M.V., Bargiello, T.A., Verselis, V.K., 1999. Rapid and direct effects of pH on connexins revealed by the connexin46 hemichannel preparation. J. Gen. Physiol. 113(5), 721–742.

[36] Stergiopoulos, K., Alvarado, J.L., Mastroianni, M., Ek-Vitorin, J.F., Taffet, S.M., Delmar, M., 1999. Hetero-domain interactions as a mechanism for the regulation of connexin channels. Circ. Res. 84(10), 1144–1155.

[37] Homma, N., Alvarado, J.L., Coombs, W., Stergiopoulos, K., Taffet, S.M., Lau, A.F., Delmar, M., 1998. A particle–receptor model for the insulin-induced closure of connexin43 channels. Circ. Res. 83(1), 27–32.

[38] Peracchia, C., Wang, X., Li, L., Peracchia, L.L., 1996. Inhibition of calmodulin expression prevents low-pH-induced gap junction uncoupling in Xenopus oocytes. Pflugers Arch. 431, 379–387.

[39] Torok, K., Stauffer, K., Evans, W.H., 1997. Connexin32 of gap junctions contains two cytoplasmic calmodulin-binding domains. Biochem. J. 326, 479–483.

[40] Peracchia, C., Wang, X.C., 1997. Connexin domains relevant to the chemical gating of gap junction channels. Braz. J. Med. Biol. Res. 30(5), 577–590 (Review).

[41] Bukauskas, F.F., Peracchia, C., 1997. Two distinct gating mechanisms in gap junction channels-CO_2-sensitive and voltage-sensitive. Biophys. J. 72, 2137–2142.

[42] Banach, K., Weingart, R., 2000. Voltage gating of Cx43 gap junction channels involves fast and slow current transitions. Pflugers Arch. 439(3), 248–250.

[43] Revilla, A., Castro, C., Barrio, L.C., 1999. Molecular dissection of transjunctional voltage dependence in the connexin-32 and connexin-43 junctions. Biophys. J. 77(3), 1374–1383.

[44] Werner, R., Levine, E., Rabadan-Diehl, C., Dahl, G., 1991. Gating properties of connexin32 cell–cell channels and their mutants expressed in Xenopus oocytes. Proc. R. Soc. Lond. B. Biol. Sci. 243(1306), 5–11.

[45] Lin, J.S., Fitzgerald, S., Dong, Y., Knight, C., Donaldson, P., Kistler, J., 1997. Processing of the gap junction protein connexin50 in the ocular lens is accomplished by calpain. Eur. J. Cell Biol. 73(2), 141–149.

[46] Eckert, R., Donaldson, P., Lin, J.S., Bond, J., Green, C., Merriman-Smith, R., Tunstall, M., Kistler, J., 2000 Gating of gap junction channels and hemichannels in lens: a role for cataract? In: Peracchia, C. (Ed.), Current Topics in Membranes (Gap Junctions: Molecular Basis of Cell Communication in Health and Disease, 49. Academic Press, San Diego, pp. 343–356.

[47] Spray, D.C., Burt, J.M., 1990. Structure–activity relations of the cardiac gap junction channel. Am. J. Physiol. 258(2 Pt 1), C195–C205.

[48] Bastiaanse, E.M., Jongsma, H.J., van der Laarse, A., Takens-Kwak, B.R., 1993. Heptanol-induced decrease in cardiac gap junctional conductance is mediated by a decrease in the fluidity of membranous cholesterol-rich domains. J. Membr. Biol. 136(2), 135–145.

[49] Davidson, J.S., Baumgarten, I.M., 1988. Glycyrrhetinic acid derivatives: a novel class of inhibitors of gap-junctional intercellular communication. Structure–activity relationships. J. Pharmacol. Exp. Ther. 246(3), 1104–1107.

[50] Lau, A.F., Warn-Cramer, B., Liu, R., 2000. Regulation of connexin43 by tyrosine protein kinases. In: Peracchia, C. (Ed.), Gap Junctions: Molecular Basis of Cell Communication in Health and Disease. Academic Press, San Diego, pp. 315–341.

[51] Britz-Cunningham, S.H., Shah, M.M., Zuppan, C.W., Fletcher, W.H., 1995. Mutations of the Connexin43 gap-junction gene in patients with heart malformations and defects of laterality. N. Engl. J. Med. 332(20), 1323–1329.

[52] Penman Splitt, M., Tsai, M.Y., Burn, J., Goodship, J.A., 1997. Absence of mutations in the regulatory domain of the gap junction protein connexin 43 in patients with visceroatrial heterotaxy. Heart 77(4), 369–370.

[53] Gebbia, M., Towbin, J.A., 1996. Casey B Failure to detect connexin43 mutations in 38 cases of sporadic and familial heterotaxy. Circulation 94(8), 1909–1912.

[54] Spray, D.C., Campos de Carvalho, A., Bennett, M.V., 1986. Sensitivity of gap junctional conductance to H ions in amphibian embryonic cells is independent of voltage sensitivity. Proc. Natl Acad. Sci. USA 83(10), 3533–3536.

[55] Beblo, D.A., Veenstra, R.D., 1997. Monovalent cation permeation through the connexin40 gap junction channel. Cs, Rb, K, Na, Li, TEA, TMA, TBA, and effects of anions Br, Cl, F, acetate, aspartate, glutamate, and NO_3. J. Gen. Physiol. 109(4), 509–522.

[56] Wang, H.Z., Veenstra, R.D., 1997. Monovalent ion selectivity sequences of the rat connexin43 gap junction channel. J. Gen. Physiol. 109(4), 491–507.

[57] Oh, S., Ri, Y., Bennett, M.V.L., Trexler, E.B., Verselis, V.K., Bargiello, T.A., 1997. Changes in permeability caused by connexin 32 mutations underlie X-linked Charcot-Marie-Tooth disease. Neuron 19, 927–938.

[58] Niessen, H., Harz, H., Bedner, P., Kramer, K., Willecke, K., 2000. Selective permeability of different connexin channels to the second messenger inositol 1,4,5-trisphosphate. J. Cell Sci. 113(Pt 8), 1365–1372.

[59] Bevans, C.G., Kordel, M., Rhee, S.K., Harris, A.L., 1998. Isoform composition of connexin channels determines selectivity among second messengers and uncharged molecules. J. Biol. Chem. 273(5), 2808–2816.

[60] Stutenkemper, R., Geisse, S., Schwarz, H.J., Look, J., Traub, O., Nicholson, B.J., Willecke, K., 1992. The hepatocyte-specific phenotype of murine liver cells correlates with high expression of connexin32 and connexin26 but very low expression of connexin43. Exp. Cell Res. 201(1), 43–54.

[61] Kanter, H.L., Laing, J.G., Beau, S.L., Beyer, E.C., Saffitz, J.E., 1993. Distinct patterns of connexin expression in canine Purkinje fibers and ventricular muscle. Circ. Res. 72(5), 1124–1131.

[62] White, T.W., Bruzzone, R., Wolfram, S., Paul, D.L., Goodenough, D.A., 1994. Selective interactions among the multiple connexin proteins expressed in the vertebrate lens: the second extracellular domain is a determinant of compatibility between connexins. J. Cell Biol. 125, 879–892.

[63] Deschenes, S.M., Walcott, J.L., Wexler, T.L., Scherer, S.S., Fischbeck, K.H., 1997. Altered trafficking of mutant connexin32. J. Neurosci. 17, 9077–9084.

[64] Ressot, C., Gomes, D., Dautigny, A., PhamDinh, D., Bruzzone, R., 1998. Connexin32 mutations associated with X-linked Charcot-Marie-Tooth disease show two distinct behaviors: loss of function and altered gating properties. J. Neurosci. 18, 4063–4075.

[65] Yoshimura, T., Satake, M., Ohnishi, A., Tsutsumi, Y., Fujikura, Y., 1998. Mutations of connexin32 in Charcot-Marie-Tooth disease type X interfere with cell-to-cell communication but not cell proliferation and myelin-specific gene expression. J. Neurosci. Res. 51(2), 154–161.

[66] Omori, Y., Mesnil, M., Yamasaki, H., 1996. Connexin 32 mutations from X-linked Charcot-Marie-Tooth disease patients: functional defects and dominant negative effects. Mol. Biol. Cell 7, 907–916.

[67] Martin, P.E., Mambetisaeva, E.T., Archer, D.A., George, C.H., Evans, W.H., 2000. Analysis of gap junction assembly using mutated connexins detected in Charcot-Marie-Tooth X-linked disease. J. Neurochem. 74(2), 711–720.

[68] Ressot, C., Bruzzone, R., 2000. Connexin channels in Schwann cells and the development of the X-linked form of Charcot-Marie-Tooth disease. Brain. Res. Brain. Res. Rev. 32(1), 192–202.

[69] VanSlyke, J.K., Deschenes, S.M., Musil, L.S., 2000. Intracellular transport, assembly, and degradation of wild-type and disease-linked mutant gap junction proteins. Mol. Biol. Cell 11(6), 1933–1946.

[70] Castro, C., Gomez-Hernandez, J.M., Silander, K., Barrio, L.C., 1999. Altered formation of hemichannels and gap junction channels caused by C-terminal connexin-32 mutations. J. Neurosci. 19(10), 3752–3760.

[71] Balice-Gordon, R.J., Bone, L.J., Scherer, S.S., 1998. Functional gap junctions in the schwann cell myelin sheath. J. Cell Biol. 142(4), 1095–1104.

[72] Ionasescu, V.V., Searby, C., Ionasescu, R., Neuhaus, I.M., Werner, R., 1996. Mutations of the noncoding region of the connexin32 gene in X-linked dominant Charcot-Marie-Tooth neuropathy. Neurology 47(2), 541–544.

[73] Kojima, T., Sawada, N., Chiba, H., Kokai, Y., Yamamoto, M., Urban, M., Lee, G.H., Hertzberg, E.L., Mochizuki, Y., Spray, D.C., 1999. Induction of tight junctions in human connexin 32 (hCx32)-transfected mouse hepatocytes: connexin 32 interacts with occludin. Biochem. Biophys. Res. Commun. 266(1), 222–229.

[74] Nusrat, A., Chen, J.A., Foley, C.S., Liang, T.W., Tom, J., Cromwell, M., Quan, C., Mrsny, R.J., 2000. The coiled-coil domain of occludin can act to organize structural and functional elements of the epithelial tight junction. J. Biol. Chem. 275(38), 29816–29822.

[75] Toyofuku, T., Yabuki, M., Otsu, K., Kuzuya, T., Hori, M., Tada, M., 1998. Direct association of the gap junction protein connexin-43 with ZO-1 in cardiac myocytes. J. Biol. Chem. 273(21), 12725–12731.

[76] Swenson, K.I., Piwnica-Worms, H., McNamee, H., Paul, D.L., 1990. Tyrosine phosphorylation of the gap junction protein connexin43 is required for the pp60v-src-induced inhibition of communication. Cell Regul. 1(13), 989–1002.

[77] Lund, A.M., Eiberg, H., Rosenberg, T., Warburg, M., 1992. Autosomal dominant congenital cataract: linkage relations: clinical and genetic heterogeneity. Clin. Genet. 41, 65–69.

[78] Scott, M.H., Hejtmancik, J.F., Wozencraft, L.A., Reuter, L.M., Parks, M.M., Kaiser-Kupfer, M.I., 1994. Autosomal dominant congenital cataract. Interocular pheontypic variability. Ophthalmology 101, 866–871.

[79] Nettleship, E., 1909. Seven new pedigrees of heredity cataract. Trans. Opthalmol. Soc. UK 29, 188–211.

[80] Hejtmancik, J.F., 1998. The genetics of cataract: our vision becomes clearer. Am. J. Hum. Genet. 62, 520–525.

[81] Berry, V., Mackay, D., Khaliq, S., Francis, P.J., Hameed, A., Anwar, K., Mehdi, S.Q., Newbold, R.J., Ionides, A., Shiels, A., Moore, T., Bhattacharya, S.S., 1999. Connexin50 mutation in a family with congenital "zonular nuclear" pulverulent cataract of Pakistani origin. Hum. Genet. 105, 168–170.

[82] Steele, E.C. Jr, Lyon, M.F., Glenister, P.H., Guillot, P., Church, R.L., 1997. Identification of a mutation in the connexin50 (Cx50) gene of the No2 cataractous mouse mutant. In: Werner, R. (Ed.), Gap Junctions. IOS press, Amsterdam, pp. 289–293.

[83] White, T.W., Bruzzone, R., Goodenough, D.A., Paul, D.L., 1992. Mouse Cx50, a functional member of the connexin family of gap junction proteins, is the lens fiber protein MP70. Mol. Biol. Cell 3, 711–720.

[84] White, T.W., Goodenough, D.A., Paul, D.L., 1998. Targeted ablation of connexin50 in mice results in microphthalmia and zonular pulverulent cataracts. J. Cell Biol., 815–825.

[85] Gong, X., Li, E., Klier, G., Huang, Q., Wu, Y., Lei, H., Kumar, N.M., Horwitz, J., Gilula, N.B., 1997. Disruption of alpha3 connexin gene leads to proteolysis and cataractogenesis in mice. Cell 91, 833–843.

[86] Xu, X., Ebihara, L., 1999. Characterization of a mouse Cx50 mutation associated with the No1 mouse cataract. Investig. Ophthalmol. Vis. Sci. 40, 1844–1850.

[87] Pal, J.D., Liu, X., Mackay, D., Shiels, A., Berthoud, V.M., Beyer, E.C., Ebihara, L., 1999. Connexin46 mutations linked to congenital cataract show loss of gap junction channel function. Am. J. Physiol. Cell Physiol. 279, C596–C602.

[88] Pal, J.D., Berthoud, V.M., Beyer, E.C., Mackay, D., Shiels, A., Ebihara, L., 1999. Molecular mechanism underlying a Cx50-linked congenital cataract. Am. J. Physiol. Cell Physiol. 276, C1443–C1446.

[89] Mathias, R.T., Rae, J.L., Baldo, G.J., 1997. Physiological properties of the normal lens. Physiol. Rev. 77, 21–50.

[90] Goodenough, D.A., 1992. The crystalline lens: a system networked by gap junctional intercellular communication. In: Gilula, N.B. (Ed.), Seminars in Cell Biology. Saunders Sci., London, pp. 49–58.

[91] Goodenough, D.A., Dick, J.S.B. II, Lyons, J.E., 1980. Lens metabolic cooperation: a study of mouse lens transport and permeability visualized with freeze-substitution autoradiography and electron microscopy. J. Cell Biol. 86, 576–589.

[92] Beyer, E.C., Kistler, J., Paul, D.L., Goodenough, D.A., 1989. Antisera directed against connexin43 peptides react with a 43-kD protein localized to gap junctions in myocardium and other tissues. J. Cell Biol. 108, 595–605.

[93] Gong, X., Baldo, G.J., Kumar, N.M., Gilula, N.B., Mathias, R.T., 1998. Gap junctional coupling in lenses lacking alpha3 connexin. Proc. Natl Acad. Sci. USA 95, 15303–15308.

[94] White, T.W., Bruzzone, R., Goodenough, D.A., Paul, D.L., 1994. Voltage gating of connexins. Nature 371, 208–209.

[95] Jiang, J.X., Goodenough, D.A., 1996. Heteromeric connexons in lens gap junction channels. Proc. Natl Acad. Sci. USA 93, 1287–1291.

[96] Konig, N., Zampighi, G., 1995. Purification of bovine lens cell-cell channels composed of connexin44 and connexin50. J. Cell Sci. 108, 3091–3098.

[97] Zelante, L., Gasparini, P., Estivill, X., Melchionda, S., D'Agruma, L., Gove, N., Mira, M., Moni, M.D., Lutti, J., Shoni Mansfield, E., Delg, K., Rappaport, E., Surrey, S., Fortir, S., 1997. Connexin26 mutations associated with the most common form of non-syndromic neurosensory autosomal recesseive deafness (DFNB1) in Mediterraneans. Hum. Mol. Genet. 6, 1605–1609.

[98] Denoyelle, F., Weil, D., Maw, M.A., Wilcox, S.A., Lench, N.J., Allen-Powell, D.R., Osborn, A.H., Dahl, H.H., Middleton, A., Houseman, M.J., Dode, C., Marlin, S., Boulila-ElGaied, A., Grati, M., Ayadi, H., BenArab, S., Bitoun, P., Lina-Granade, G., Godet, J., Mustapha, M., Loiselet, J., El-Zir, E., Aubois, A., Joannard, A., Petit, C., 1997. Prelingual deafness: high prevalence of a 30delG mutation in the connexin26 gene. Hum. Mol. Genet. 6, 2173–2177.

[99] Morell, R.J., Kim, H.J., Hood, L.J., Goforth, L., Friderici, K., Fisher, R., Van Camp, G., Berlin, C.I., Oddoux, C., Ostrer, H., Keats, B., Friedman, T.B., 1998. Mutations in the connexin 26 gene (GJB2) among Ashkenazi Jews with nonsyndromic recessive deafness. N. Engl. J. Med. 339(21), 1500–1505.

[100] Kelsell, D.P., Wilgoss, A.L., Richard, G., Stevens, H.P., Munro, C.S., Leigh, I.M., 2000. Connexin mutations associated with palmoplantar keratoderma and profound deafness in a single family. Eur. J. Hum. Genet. 8(6), 469–472.

[101] Heathcote, K., Syrris, P., Carter, N.D., Patton, M.A., 2000. A connexin 26 mutation causes a syndrome of sensorineural hearing loss and palmoplantar hyperkeratosis (MIM 148350). J. Med. Genet. 37(1), 50–51.

[102] Maestrini, E., Korge, B.P., Ocana-Sierra, J., Calzolari, E., Cambiaghi, S., Scudder, P.M., Hovnanian, A., Monaco, A.P., Munro, C.S., 1999. A missense mutation in connexin26, D66H, causes mutilating keratoderma with sensorineural deafness (Vohwinkel's syndrome) in three unrelated families. Hum. Mol. Genet. 8(7), 1237–1243.

[103] Richard, G., White, T.W., Smith, L.E., Bailey, R.A., Compton, J.G., Paul, D.L., Bale, S.J., 1998. Functional defects of Cx26 due to a heterozygous missense mutation in a family with dominant deaf-mutism and palmoplantar keratoderma. Hum. Genet. 103(4), 393–399.

[104] Green, G.E., Scott, D.A., McDonald, J.M., Woodworth, G.G., Sheffield, V.C., Smith, R.J., 1999. Carrier rates in the midwestern United States for GJB2 mutations causing inherited deafness. JAMA 281(23), 2211–2216.

[105] Martin, P.E., Coleman, S.L., Casalotti, S.O., Forge, A., Evans, W.H., 1999. Properties of connexin26 gap junctional proteins derived from mutations associated with non-syndromal heriditary deafness. Hum. Mol. Genet. 8(13), 2369–2376.

[106] Kelsell, D.P., Dunlop, J., Stevens, H.P., Lench, N.J., Liang, J.N., Parry, G., Mueller, R.F., Leigh, I.M., 1997. Connexin 26 mutations in hereditary non-syndromic sensorineural deafness. Nature 387(6628), 80–83.

[107] White, T.W., Deans, M.R., Kelsell, D.P., Paul, D.L., 1998. Connexin mutations in deafness. Nature 394, 630–631 (Sci. Correspond.).

[108] Kelley, P.M., Harris, D.J., Comer, B.C., Askew, J.W., Fowler, T., Smith, S.D., Kimberling, W.J., 1998. Novel mutations in the connexin26 gene (GJB2) that cause autosomal recessive (DFNB1) hearing loss. Am. J. Hum. Genet. 62, 792–799.

[109] Scott, D.A., Kraft, M.L., Stone, E.M., Sheffield, V.C., Smith, R.J., 1998. Connexin mutations and hearing loss. Nature 391, 32 (Lett.).

[110] Hama, K., Saito, K., 1977. Gap junctions between the supporting cells in some acoustico–vestibular receptors. J. Neurocytol. 6(1), 1–12.

[111] Forge, A., 1984. Gap junctions in the stria vascularis and effects of ethacrynic acid. Hear. Res. 13(2), 189–200.

[112] Kikuchi, T., Kimura, R.S., Paul, D.L., Adams, J.C., 1995. Gap junctions in the rat cochlea – immunohistochemical and ultrastructural analysis. Anat. Embryol. 191, 101–118.

[113] Johnstone, B.M., Patuzzi, R., Syka, J., Sykova, E., 1989. Stimulus-related potassium changes in the organ of Corti of guinea pig. J. Physiol. 408, 77–92.

[114] Lautermann, J., ten Cate, W.J., Altenhoff, P., Grummer, R., Traub, O., Frank, H., Jahnke, K., Winterhager, E., 1998. Expression of the gap-junction connexins 26 and 30 in the rat cochlea. Cell Tissue Res. 294(3), 415–420.

[115] Zhao, H.B., Santos-Sacchi, J., 2000. Voltage gating of gap junctions in cochlear supporting cells: evidence for nonhomotypic channels. J. Membr. Biol. 175(1), 17–24.

[116] Dahl, E., Manthey, D., Chen, Y., Schwarz, H.J., Chang, Y.S., Lalley, P.A., Nicholson, B.J., Willecke, K., 1996. Molecular cloning and functional expression of mouse connexin-30, a gap junction gene highly expressed in adult brain and skin. J. Biol. Chem. 271(30), 17903–17910.

[117] Bruzzone, R., White, T.W., Paul, D.L., 1996. Connections with connexins: the molecular basis of direct intercellular signaling. Eur. J. Biochem. 238(1), 1–27.

[118] Paul, D.L., 1986. Molecular cloning of cDNA for rat liver gap junction protein. J. Cell Biol. 103(1), 123–134.

[119] Moennikes, O., Buchmann, A., Willecke, K., Traub, O., Schwarz, M., 2000. Hepatocarcinogenesis in female mice with mosaic expression of connexin32. Hepatology 32(3), 501–506.

[120] Gabriel, H.D., Jung, D., Butzler, C., Temme, A., Traub, O., Winterhager, E., Willecke, K., 1998. Transplacental uptake of glucose is decreased in embryonic lethal connexin26-deficient mice. J. Cell Biol. 140(6), 1453–1461.

Molecular physiology and pathophysiology of ClC-type chloride channels

Christoph Fahlke[a,b,*]

[a]*RWTH Aachen, Institute of Physiology, Aachen, Germany*
[b]*Centro de Estudios Científicos (CECS), Avenida Prat 514,Valdivia, Chile*
[*]*Correspondence address: Institute of Physiology, RWTH Aachen, Pauwelsstr. 30, D-52057 Aachen, Germany.*
Tel.: +49-241-80-88-810, fax: +49-241-80-82-434
E-mail: chfahlke@physiology.rwth-aachen.de

1. Summary

13 years ago, Thomas Jentsch and co-workers cloned a voltage-gated chloride channel from the electric organ of *Torpedo marmorata*. This channel became the precursor of a large gene family of voltage-gated chloride channels, the ClC family. ClC channels can be found in virtually every living cell and fulfill various important cellular tasks. The dysfunction of four ClC isoforms is known to cause human inherited diseases. This chapter gives an overview of the molecular, physiological and pathophysiological features of this ion channel family with a focus on ClC channels in mammalian systems. Many aspects of these fascinating ion channels remain incompletely understood, despite the great progress made in the last few years.

2. Introduction

Anion channels exist in almost every living cell and fulfill a variety of physiological functions. These channels, often simply called chloride channels for the most abundant physiological anion, regulate excitability in nerve and muscle [1,2], they are involved in transepithelial solute transport [3,4] and in volume [5] and pH regulation in organelles [6]. In the past, anion channels have attracted much less attention than cation channels and their physiology and biophysics is still poorly understood.

As chloride channels are difficult to study in native tissues, a major breakthrough in their investigation was the molecular cloning and subsequent heterologous expression. In contrast to potassium channels that all exhibit an evolutionarily conserved pore region [7] there are several gene families coding for anion channels. One of those is the ClC family of voltage-gated chloride channel. It was identified by Thomas Jentsch and colleagues in 1990, when they cloned a chloride channel, ClC-0, from the electric organ of

Advances in Molecular and Cell Biology, Vol. 32, pages 189–217
© 2004 Elsevier B.V. All rights of reproduction in any form reserved.
ISSN: 1569-2558/DOI: 10.1016/S1569-2558(03)32009-0

Torpedo [8]. In the following years, a large number of homologs were identified in a variety of organisms [3]. The ClC family is in many respects distinct from other ion channel gene families. Prior to molecular identification little was known about these channels, and ClC channel biophysics and physiology, to a great degree, started with the availability of cDNAs and the possibility to study these channels in heterologous expression systems. For many ClC channels, the primary sequence was known before a native channel or a physiological role could be correlated.

These peculiarities of the ClC families have been taken into account in the organization of this chapter. It starts with a description of the molecular features of this channel family, their biophysical properties and ends with some physiological and pathophysiological considerations. This review focuses on mammalian ClC channels; readers specifically interested in other aspects of ClC channels are also directed to other recent review articles [9–11].

3. Molecular properties of ClC-type chloride channels

More than 20 years ago, a voltage-gated chloride channel from *T. california* electroplax was discovered during attempts to reconstitute a nicotinic acetylcholine receptor from the same source into planar lipid bilayers [12]. In the following years, this channel was studied in great detail [13–16]. It is highly anion-selective and exhibits a Cl > Br > I selectivity; it activates upon depolarization and inactivates at the same potentials on a much slower time scale. A hallmark of this channel is the occurrence of two equally spaced subconductance states that appear independently gated [17].

In 1990, Jentsch and colleagues used a novel hybrid-depletion expression cloning approach to identify a complementary DNA encoding a voltage-gated chloride channel from the electric organ of a related species, *T. marmorata* [8]. Heterologous expression of the so identified cDNA, dubbed ClC-0, in *Xenopus* ooctyes gave rise to an anion current with all functional properties of the native *T. california* channel [18], indicating that a single peptide is sufficient to produce these anion channels.

In the following years, several mammalian ClC isoforms were cloned. The first one was a chloride channel from rat skeletal muscle, ClC-1, cloned by homology screening [19], and later similar homology screening identified several other ClC channels: the almost ubiquitously expressed ClC-2 [20] and ClC-3 [21] originally cloned from rat, kidney specific channels from rats (ClC-K1 [22] and ClC-K2 [23]) and humans (ClC-Ka and ClC-Kb [24]) and the ubiquitously expressed ClC-6 and ClC-7 [25]. Two other ClC channels were found using genetic approaches. In attempts to map a region of the human X-chromosome, Xp22.3, Van Slegtenhorst and colleagues identified a gene encoding a protein with high sequence similarity to known members of the ClC family [26], expressed in heart, brain and skeletal muscle that were named ClC-4. ClC-5 is the disease gene for a familial nephrolithiasis, Dent's disease and was identified by positional cloning [27,28]. Fig. 1 shows a dendrogram of the currently known human ClC isoforms.

After the initial description of mammalian ClC channels, ClC isoforms were reported for almost every known creature [3], and thus the ClC family currently represents the

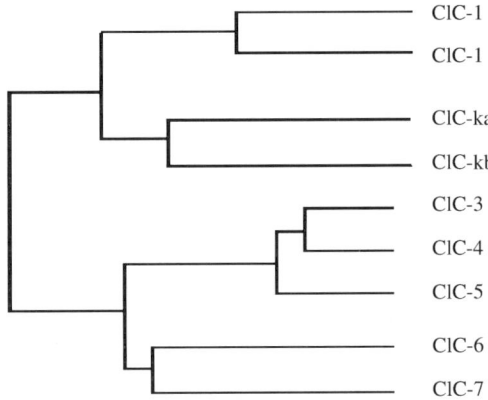

Fig. 1. Dendrogram of the known human ClC channels [3].

largest known anion channel gene family. The coding region of ClC channels is between 400 and 1000 amino acids long with the shortest isoforms found in prokaryotes [29]. All known ClC channels exhibit a similar hydropathy plot (Fig. 2).

The membrane topology of ClC channels was a matter of scientific debate for many years. The three-dimensional structures of two prokaryotic ClC channels (from *Salmonella typhi* and *Escherichia coli*) by X-ray crystallography finally settled this discussion [31]. ClC channels exhibit a total of 18 transmembrane helices per ClC subunit (Fig. 3), these are 6 more than expected from hydropathy plots. In eukaryotic ClC channels, the last transmembrane domain is followed by a cytoplasmic carboxy-terminus with two so-called CBS domains.

ClC channels are dimeric proteins [29,31–35]. Each subunit forms its own ion conduction pathway resulting in a unique architecture with two separate pores per channel. The "double-barreled" pore stoichiometry was first postulated by Christopher Miller to explain the outcome of single channel recordings on ClC-0 [17]. ClC-0 exhibits two equally spaced and independently gated subconductance states that can be individually inhibited by DIDS [36]. Final proof of this model was provided by the high resolution structure of bacterial ClC channels that directly showed the two separate pores [31].

4. Barttin, a ClC channel β-subunit

Recently, a ClC channel β-subunit, Barttin, was identified by Birkenjäger and colleagues as the disease gene of BSND, a special from of antenatal Bartter syndrome associated with sensorineural deafness and kidney failure [37]. This protein is around 320 AA long, and consists of two membrane-spanning α-helices with a cytoplasmic N- and C-terminus. It is expressed in the inner ear as well as in the kidney [37,38] and co-localizes with ClC-K and ClC-Kb in the basolateral membrane of renal tubules and of potassium-secreting epithelia of the inner ear [38]. Its specific role seems to be the proper membrane targeting of these two ClC-isoforms [38].

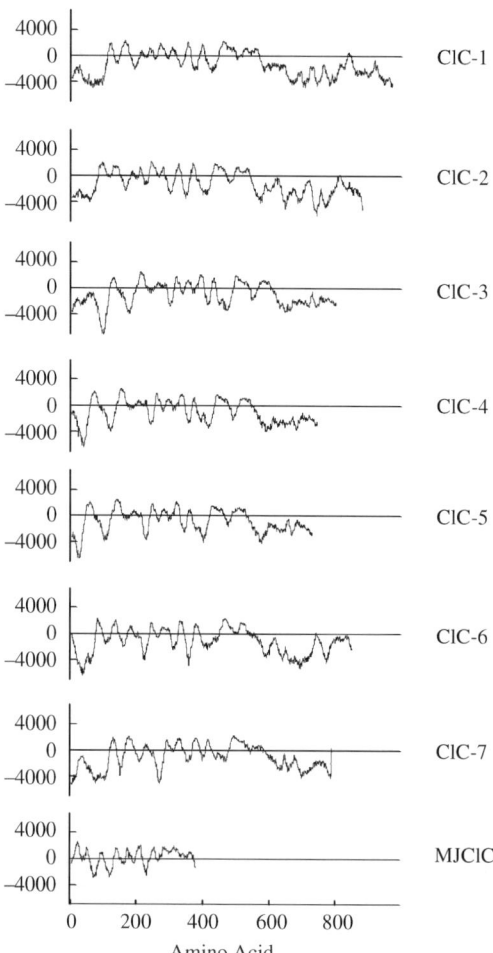

Fig. 2. Hydropathy profile of seven human isoforms (ClC-1 to ClC-7) and one prokaryotic ClC channel from *Methanococcus jannaschii*, computed according to the algorithm of Kyte and Doolttle [30].

5. Functional properties of ClC channels

Of the known mammalian ClC isoforms, functional forms of ClC-0, ClC-1, ClC-2, ClC-K1, ClC-4 and ClC-5 can be routinely expressed. They display a wide variety of gating and permeation properties (Table 1). Fig. 4 shows representative current recordings from four ClC isoforms, ClC-0 from the electric organ of *T. marmorata* (Fig. 4A), human ClC-1 (Fig. 4C), rat ClC-2 (Fig. 4E) and human ClC-4 (Fig. 4G). ClC-0 and ClC-1 are open at 0 mV, and currents deactivate upon membrane hyperpolarization (Fig. 4A and C). ClC-0 displays a linear instantaneous current–voltage relationship (Fig. 4B); hClC-1 is pronouncedly inwardly rectifying (Fig. 4D). ClC-2 is closed at positive potentials, and

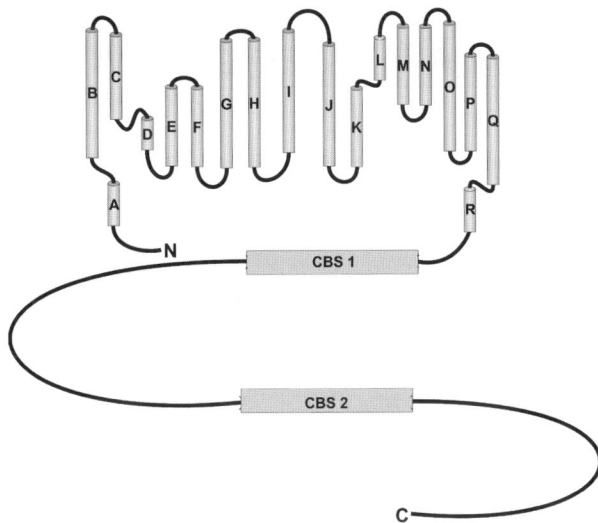

Fig. 3. Membrane topology of eukaryotic ClC channels based on the three-dimensional structure of StClC by Dutzler et al. [31].

activates upon membrane hyperpolarization (Fig. 4E), the I–V curve of the open channel is linear (Fig. 4F). ClC-4 outwardly rectifies, and activates upon membrane depolarization (Fig. 4G and H).

All ClC channels that have been studied so far are anion-selective and one can assume that they exhibit an evolutionary conserved anion-selective pore. The functional and structural features of this pore have been well characterized over the last years making ion permeation the best understood aspect of ClC channel function.

5.1. Selectivity in ClC-type chloride channels

In biological media, only two anions, Cl^- and HCO_3^-, exist in millimolar concentrations. Perfect discrimination between these two appears not to be crucial, as these anions are similarly distributed across the cell membrane. Instead, the physiological role of anion channels is more likely to select between anions and cations. This selectivity is thought to occur by electrostatic interactions between these ions and the channel protein. Analogous to cation channels where negative charges in pore vestibules attract cations and repel anions, anion channels likely employ a positive electrostatic potential to prevent the permeation of cations.

All known ClC channels are able to conduct multiple species of anions and can, to a certain extent, select between these. This anion selectivity is at least partially a side effect of the pore features necessary to maintain a high anion to cation selectivity [39] (see below). Although physiologically of minor importance, studying the mechanism by which ClC channels select between anions has provided many insights into the function of the ClC pore. Miller and White [15] were the first to probe a ClC channel with various

Table 1
Functional properties of several ClC-isoforms

ClC-Isoform	Permeability sequence	Open channel I–V relationship	Gating	Origin of tested channels
ClC-0	Cl > Br ≫ I, SCN	Linear	Fast gate activating upon depolarisation, slow gate upon hyperpolarization	Native channels, reconstituted in planar lipid bilayers [15]
ClC-0	Cl > Br > NO$_3$ > I	Linear	Fast gate activating upon depolarization, slow gate upon hyperpolarization	Expressed in *Xenopus* oocytes [8]
rClC-1	SCN > Cl > Br > NO3 > I ≫ CH$_3$SO$_3$	Inwardly rectifying	Activation upon depolarization	Expressed in Sf9 [42]
hClC-1	Cl > SCN > Br > NO3 > I	Inwardly rectifying	Activation upon depolarization	Expressed in tsA201 [40,41]
rClC-2	Cl > Br > I	Linear	Activation upon hyperpolarization	Expressed in *Xenopus* oocytes [20]
rClC-3	I > Cl > Br	Outwardly rectifying	Inactivation upon depolarization	Expressed in *Xenopus* oocytes [21]
rClC-3	I > Cl > Br	Outwardly rectifying	Activation upon depolarization	Expressed in CHOK1 [80]
gpClC-3	I > Cl	Outwardly rectifying	Inactivation upon depolarization	Expressed in NIH3t3 [81]
hClC-4	NO$_3$ > Cl > Br > I	Outwardly rectifying	Activation upon depolarization	Expressed in Xenopus oocytes and HEK293 [44]
hClC-4	SCN > NO$_3$ > Cl > F > Br	Outwardly rectifying	Activation upon depolarization	Expressed in HEK293 [47]
rClC-5	NO$_3$ > Cl > Br > HCO$_3$ > I	Outwardly rectifying	Activation upon depolarization	Expressed in *Xenopus* oocytes [96]
RClC-K1	Br > Cl > I	Outwardly rectifying	Activation upon depolarization	Expressed in *Xenopus* oocytes [22]

anions. They showed that ClC-0 is permeable only to Cl$^-$ and Br$^-$, and that chloride currents are blocked by other anions including I$^-$ and SCN$^-$. ClC-1 exhibits a similar anion permeability sequence of Cl$^-$ > Br$^-$ > NO$_3^-$ > I$^-$, and in a series of experiments with this isoform, it became clear that the blockade of anionic current by other anions is of crucial importance for anion selectivity in this channel [40–43]. All permeant anions, I$^-$, SCN$^-$, NO$_3^-$, or Br$^-$, block chloride current through ClC-1 [40–43] by binding to sites within the ion conduction pathway with higher affinity than Cl$^-$ [40,41]. These anions dwell longer at this site than Cl$^-$, and this feature reduces their permeability and accounts for the observed block of Cl$^-$ current [40–43]. The ion pore of hClC-1 exhibits at least two functionally distinct binding sites that can be distinguished by the unique kinetic effects, which the occupation of these sites exerts. The two binding sites differ in their relative anion selectivity, i.e. the extracellular site exhibits an affinity sequence of SCN$^-$ > I$^-$ > NO$_3^-$ > CH$_3$SO$_3^-$ > Br$^-$ > Cl$^-$ > F$^-$ and the intracellular one of I$^-$ > NO$_3^-$ >

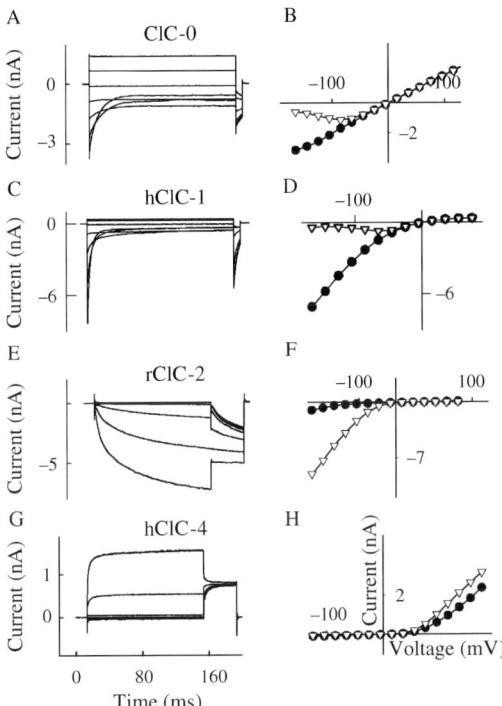

Fig. 4. (A,C,E,G) Representative tight whole-cell current recordings from tsA201 cells transiently expressing various ClC isoforms: ClC-0(A), hClC-1(C), rClC-2(E), and hClC-4(G). Cells were bathed in a standard external solution containing (in mM) 140 NaCl, 4 KCl, 2 CaCl$_2$, 1 MgCl$_2$, 5 HEPES, pH 7.4, and perfused with a standard internal solution (in mM: 130 NaCl, 2 MgCl$_2$, 5 EGTA, 10 HEPES, pH 7.4). The holding potential was chosen to be 0 mV and voltage steps between −165 and +75 mV were applied as shown in the pulse diagram. (B,D,F,H) Voltage dependence of the instantaneous (●) and the late current amplitude (∨) from recordings shown in (A,C,E,G).

SCN$^-$ > Cl$^-$ > F$^-$ (41). Using the characteristic kinetic changes induced by external and internal I$^-$, the K_D of the internal and external binding sites was independently determined. The different dissociation constants so obtained are in the millimolar range [41]. Because iodide binds more tightly than chloride, this value represents a lower limit for the chloride dissociation constant, and indicates that hClC-1 binds anions with affinities that are an order of magnitude smaller than for the cationic pores in potassium [44] and calcium channels [45].

The first ClC channel construct demonstrated to exhibit a larger permeability for I$^-$ than for Cl$^-$ was a mutant ClC-1 channel that carries a disease-causing G230E mutation [40]. G230E hClC-1 exhibits an anion permeability sequence of SCN$^-$ > NO$_3^-$ > I$^-$ > Br$^-$ > Cl$^-$. Experiments revealed that neither external nor internal I$^-$ blocks chloride currents, although its ability to change gating properties indicates that I$^-$ still binds better to the channel than Cl$^-$ [40]. This result together with other experimental data suggested the selectivity mechanism [39,41] outlined below.

ClC-type chloride channels select between ions by selective binding to several anion-selective binding sites. All these sites prefer large or polyatomic anions over small uniatomic ones; they are weak interacting sites in Eisenman terminology [1]. By binding to these sites, the anions dehydrate and the smaller radius of the dehydrated anions permits passage through a narrow pore constriction (Fig. 5). Some anions simply do not permeate, as they are too large to pass, for all others the binding to and unbinding from these sites is crucial in determining the anion passage through the channel. One can roughly divide ClC channels into two functionally distinct classes. For one class of ClC channels, the rate-limiting step in anion permeation is the dissociation of anions from one of these sites. Anions that bind with high affinity to this site are less permeant and cause block of current flow by anions, which bind with lower affinity. In these channels, the anion that binds with the lowest affinity is the most permeant, i.e. Cl^- exhibits a higher permeability and conductance than larger and polyatomic anions. Examples of this anion selectivity pattern are ClC-0 and ClC-1. In the other group of channels, the association and not the dissociation of permeant anions to these binding sites is the rate-limiting step. For this reason, the permeability sequence in these channels follows the affinity sequence of the binding sites, and larger and polyatomic anions are more permeant than smaller ones. This pattern is observed in hClC-4 [46]. ClC channels with I^- or $NO_3^- > Cl^-$ permeability sequence differ from those with the inverse permeability sequence in the absolute value of the interaction energy between binding site and anion, not in the selectivity of these binding sites.

ClC channels have multi-ion pores [39,41], and therefore it is possible that binding sites with different properties co-exist in a same pore, and that due to the distinct voltage dependence of binding to different sites, one is the rate-limiting site under certain conditions and another under other conditions. This can explain, why for example for K231A hClC-1 channels [39], external I^- blocks Cl^- currents at negative potentials, but is more permeant than Cl^- at positive potentials [39].

The selectivity mechanism of ClC channels between different anions is quite awkward and would not allow effective selection between different anions as it slows permeation of the more permeant anion [39]. However, the physiological role of ClC channels is not to

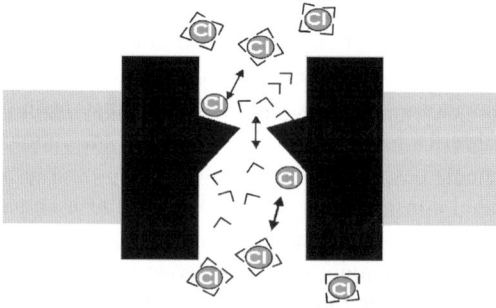

Fig. 5. Cartoon illustrating the postulated passage of anions through a ClC channel pore. The hydrated anion dehydrates by binding to a side within the pore, then passes a narrow part of the pore, binds to a distinct side and leaves the pore after rehydration.

select between various anions, but between anions and cations, and the selectivity mechanism described above indicates a likely physiological importance of the existence of anion channels with distinct anion selectivity sequences. As discussed above, the two types of anion channels differ in the absolute anion affinity of their ion conduction pathway. Channels with $Cl^- > Br^- > I^-$ sequence will be more selective for anions over cations than chloride channels with the inverse permeability sequence because of the higher electrostatic potentials of their ion conduction pathway. This feature is important in excitable tissues where small alterations of cation permeability cause dramatic alterations of excitability. On the other hand, anion channels with inverse selectivity sequences can exhibit larger unitary currents as the binding affinities are smaller, and this is advantageous under conditions in which high anion to cation selectivity is not crucial, for example in epithelia or in volume-regulation processes.

5.2. Gating of ClC channels

ClC channels exhibit broad variance in the nature of their voltage-dependent gating (Table 1, Fig. 4). Specific gating properties are crucial for the unique physiological roles of certain ClC channels. Small alterations of gating properties can cause organ dysfunction, as demonstrated by several naturally occurring point mutations that cause myotonia congenita by altering ClC-1 gating (see below).

Some channels activate upon depolarization, some upon hyperpolarization, and ClC-0 even exhibits two distinct gating mechanisms; a depolarization-induced fast activation, and a slow gating step causing channel closure at positive potentials and channel opening at negative potentials. The mechanisms underlying these processes are still very poorly understood.

To date, the characterization of gating in ClC channels has focussed primarily on two isoforms, ClC-0 from *Torpedo* electroplax [14,47–51] and the closely related ClC-1 from skeletal muscle [41,52–56]. For both isoforms, permeant anions have been shown to play a major role in determining the open probability of these two isoforms. Gating of ClC-0 is characterized by two gates that have opposite voltage dependencies and act on time scales which differ by orders of magnitude (Table 1). At positive potentials, a fast gate opens with time constants in the range of milliseconds, while membrane hyperpolarization activates a slow gate within seconds [14]. Using macroscopic current recordings from *Xenopus* oocytes expressing ClC-0 channels, Pusch et al. demonstrated tight coupling between the external anion concentration and the steady-state activation curve of the fast gate [48]. A subsequent analysis of the gating of single ClC-0 channels reconstituted in planar lipid bilayers demonstrated a clear dependence of opening rates, but not of the closing rates, on the extracellular Cl^- concentration [49].

The muscle ClC isoform, ClC-1, lacks the slow gate of the *Torpedo* isoform [19,55]. ClC-1 channels exhibit activation and deactivation with similar time and voltage dependence as the fast gate of ClC-0 [52]. Rychkov et al. demonstrated a dependence of the steady-state activation curve of ClC-1 on the extracellular Cl^- concentration [56] and proposed that external anions play a similar role in gating processes of ClC-1. Fahlke et al. [53,55] showed the influence of internal anions on ClC channel gating. In WT hClC-1, the minimum open probability at negative potentials depends on the internal chloride concentration [55]. In a mutant channel, D136G hClC-1, the voltage dependence of activation is determined by the internal chloride concentration, and this prevents the opening of the channel at potentials positive to the chloride equilibrium potential [53].

To account for the effect of anions on gating, Pusch and colleagues postulated that the apparent gating charge of ClC-0 is fully conferred by the permeating ion itself [48]. In this model, the observed depolarization-induced activation is caused by voltage-dependent binding of extracellular anions to a binding site within the pore. Occupation of this site causes opening of a cytoplasmic gate by a yet unexplained mechanism. This model was an interesting and novel hypothesis in clear contrast to the obvious independence of permeation and gating observed in voltage-gated cation channels. Nevertheless, the effects of permeant anions on ClC channel gating do not exclude the existence of an intrinsic ClC voltage sensor. In general, it is difficult to experimentally distinguish between gating models with linear state diagrams where the binding of a ligand precedes opening of a channel, and those with cyclic state diagrams, where transitions between open and closed states are mediated by intrinsic mechanisms, and ligands exert their effects only by stabilizing the open state. Recent examples, including calcium-activated potassium channels [57] and cGMP-gated channels [58], have established the role of intrinsic transitions in channels which were earlier thought to be strictly ligand-gated. There are a number of experimental findings that indicate the existence of endogenous voltage sensors that provide a part of the gating charge necessary to open ClC channels [40,46,53,55]. The mechanism of voltage-dependent gating of ClC channels is not settled and novel experimental approaches are needed to fully understand these processes.

6. Identification of molecular determinants of certain ClC channel functions

The three-dimensional structures of two prokaryotic isoforms defined the structure of the ion conduction pathway of ClC channels. This, together with site-directed mutagenesis experiments on eukaryotic isoforms, provided much information about the molecular determinants of ion conduction. On the other hand, we know almost nothing about the molecular process underlying channel opening and closing.

6.1. The high resolution structure of a prokaryotic ClC channel

Dutzler and colleagues [31] solved the X-ray structures of two prokaryotic ClC channels from *S. typhimurium* and *E. coli* at 3.0 and 3.5 Å, respectively. The two structures are quite similar; they reveal a dimeric protein with two identical pores, each pore formed by a separate subunit. Each individual subunit folds into two halves that span the membrane with opposite orientations in an antiparallel architecture (Fig. 6).

Similar to other ion channels, the ion conduction pathway of StClC exhibits only a short constriction displaying the major determinants of selectivity. Such an architecture reduces the passage of the ion through a low dielectric medium and thus significnatly diminishes the resistance of the ion pore. The helices D, F, and N (see Fig. 6) are oriented with the N-terminus pointing towards a crevice where an anion is bound in the crystal [31]. Due to

A

B

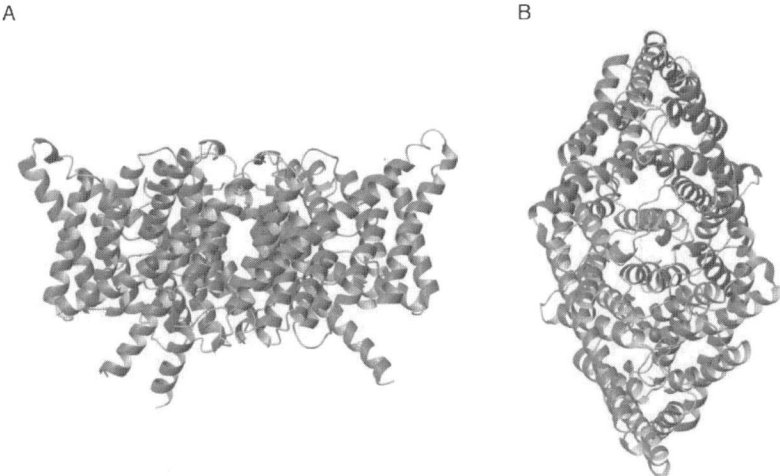

Fig. 6. Structure of the StClC channel [31]. View of the StClC dimer from within the membrane with the external solution above (A) or from the extracellular side (B). The two subunits are shown in red and blue. This figure was generated with the program MOLMOL [59]. (*For a colored version of this figure, see plate section, page 465.*)

the dipole moment of helices, this arrangement allows to create a positive electrostatic potentials attracting and binding negatively charged ions. The anion is coordinated by main-chain amide nitrogen atoms from amino acids Ile356 and Phe357 and by the side-chain oxygen atoms from Ser107 and Tyr445 [31]. The Cl⁻ ion is furthermore surrounded by a number of hydrophobic amino-acid side chains. There is no direct interaction of this anion with charged sidechains.

At present, it is unclear in which functional state the channel was crystallized. Next to the bound anion, there is a negatively charged sidechain of a highly conserved glutmate residue (position E148 in *S. typhi*). This would prevent the passage of permeant anion due to electrostatic repulsion and therefore the authors reasoned that the channel must be in its closed position. An attractive hypothesis might be that opening/closing transitions are mediated by a rotation of F helix.

6.2. Comparison with results of site-directed mutagenesis in human ClC channels

A combination of site-directed mutagenesis has provided data about sequence determinants of ion permeation and selectivity for eukaryotic ClC channels. The now available three-dimensional structure allows an evaluation and interpretation of these data.

Structure–function investigations identified the D and the F helix as pore-forming regions in the human muscle chloride channel, hClC-1. A disease-causing mutation [60], resulting in the substitution of a glycine by glutamic acid just before helix F (G230E), was the first mutation that inverted the permeability sequence of ClC-1 and increased the relative cation permeability [40]. In the framework of the proposed selectivity mechanism described above these results suggested that amino acid 230 is close to ion binding sites within the pore. Addition of a negative charge at this position would weaken the interaction with the permeant anion, and decrease the on-rate constant. This could explain the higher permeability of larger and polyatomic anions (see above). It would make the electrostatic potential within the pore less positive, thus accounting for the observed increase in cation permeability. In StClC, the corresponding G146 is located immediately before the F helix. A glutamic acid there could project into the selectivity filter and thus fully account for the observed changes of selectivity.

Extensive mutagenesis on the muscle ClC isoform ClC-1 identified the highly conserved N-terminal end of the D (P3 region) and F helix (P1 region) as pore-forming regions [61,62]. A combination of cellular electrophysiology, site-directed mutagenesis and chemical modification showed that these regions are critical determinants of ion selectivity, that there are side-chains that appear to be in contact with permeating anions and that the F helix contributes to an aqueous pore that connects the extra- and intracellular mediums and that the D helix is in contact with the intracellular aqueous medium. All these results are in good agreement with the high-resolution structure [31]. Similar experiments reveal a role of a region at the end of the H helix (P2 region) in forming the aqueous pore [61]. In contrast to the P1 and P3 regions, mutagenesis had minor effects on ion permeation. These experiments can be nicely explained by the three-dimensional structure that demonstrates that this region contributes to the wall of the aqueous pore, but is not very close to the selectivity filter.

The R helix of StClC is located at the inner entrance of the ion conduction pathway quite distant from the selectivity filter. Such a location was already suggested earlier by Middlethon et al. [33] who examined various side chain substitutions at the corresponding position K519 in ClC-0 and demonstrated that that this residue influences conduction mainly by an electrostatic interaction, whereby the positively charged amino group attracts Cl^- ions into the pore and thus increases the conductance. This residue is not conserved in every ClC isoform; there are certain prokaryotic isoforms that lack this region, and there are eukaryotic isoforms that have a neutral amino acid in this position.

There are nevertheless data that appear less compatible with the StClC structure. In ClC-1, K231 is of crucial importance for the electrostatic potential within the pore and thus for the anion selectivity. Distinct side chains at this position affect anion selectivity in a charge dependent way. Neutral and negatively charged residues exhibit a I > Cl selectivity and an absence of block. Adding a small cationic adduct to Cys 231(by application of MTSEA) restores selectivity to the wild-type pattern in a fast monoexponential time-course. For ClC-0, it was shown that ion permeation properties depend on the length of the positively charged side chain in position 165 that corresponds to position 231 in hClC-1, and this indicates that this side chain is in contact with the permeating anion and crucial for its binding properties [63]. The corresponding residue in StClC, R147, is projecting away from the co-ordinated anion. In contrast, mutations in E232 only little affect anion selectivity in hClC-1, although this resiude in the three-dimensional structure is just next to the bound anion.

6.3. Primary sequence regions affecting ClC channel gating

Preliminary information about the determinants of gating of ClC channels is now available. The above mentioned interdependence of ion permeation and gating complicates an identification of gating domains as many mutations affecting ion permeation also alter channel gating.

Lin and colleagues showed that a mutation from serine to alanine at position 212 in ClC-0 abolishes the action of the slow gate [64]. It is not clear how this mutation exerts this effect and moreover, in combination with another mutation, K165C, the slow gate is active again, indicating that this mutation impairs slow gating simply by allosteric interaction and is not involved in this gating process itself [63].

There are a number of mutations spread over the whole coding region of ClC-1 that shift the voltage dependence of activation to more positive potentials [54]. The mechanisms underlying these effects are not known and the role of these residues for ClC channel gating is currently unknown. However, Fahlke and colleagues reported that a D136G mutation in the muscle chloride channel (see below) inverts the voltage dependence of activation and likely couples channel opening to the binding of an intracellular anion [53]. To explain these results they proposed that this mutation abolishes the voltage sensor in ClC-1. This hypothesis has not been further evaluated yet, and the authors did not show subsequently that this residue moves upon voltage steps or that this movement is coupled to channel opening. Moreover, such an inversion of voltage-dependent gating has been also described as an effect of other mutations [32,65] although these mutations were not evaluated in detail and at the moment we do not know whether they are functionally similar. Further experiments are needed to judge whether this residue is indeed directly involved in voltage sensing or simply exerts allosteric effects on ClC-1 gating.

7. Properties of mammalian ClC isoforms in comparison with native chloride channels

For many mammalian ClC channels, primary sequence information was available prior to insights to their the function and cellular roles. The correlation between cloned ClC channels and native anion channels has been difficult for a variety of reasons: the incomplete biophysical characterization of native anion channels, the poorly developed pharmacology of this class of channels, and in many cases problems expressing cloned ClC channels in heterologous expression systems at levels high enough to distinguish them faithfully from endogenous channels.

7.1. ClC-1

Skeletal muscle fibers were one of the first tissues in which chloride channels were studied [66]. The determination of the muscle chloride conductance using a two-microelectrode cable technique provided first insights into gating, permeation as well as pharmacological properties of the underlying channels [67,68]. By showing that myotonia in the myotonic goat is due to a decreased sarcolemmal chloride conductance in affected muscle (see below), Bryant and colleagues demonstrated the physiological importance of this conductance component in mammalian muscle fibers [68]. The first mammalian ClC isoform, ClC-1, is exclusively expressed in adult skeletal muscle fibers, and exhibits the same developmental regulation and pharmacological properties as the chloride conductance in skeletal muscle [19]. The finding that mutations in the *CLCN1* gene cause myotonia congenita provided additional evidence that ClC-1 is either itself or a most important subunit of the muscle chloride channel (see below). The functional properties of native muscle chloride channels were difficult to study, and several groups failed in recording single channel currents. The use of the double-vaseline gap voltage clamp technique on manually dissected rat muscle fibers [69] finally provided a means for describing these important channels. These recordings show a high degree of similarity with recordings of heterologously expressed ClC-1 channels. There are nevertheless certain differences between ClC-1 in expression systems and these recordings from rat muscle: anion permeability and gating appear to be slightly different. This may be due to the expression of other ClC isoforms and/or the occurrence of heterodimeric channels in native muscle.

7.2. ClC-2

ClC-2 was the second identified mammalian chloride channel [20], and is almost ubiquitously expressed in several mammalian tissues. In *Xenopus* oocytes [20] as well as in mammalian cells [70], expression of ClC-2 causes the appearance of a voltage-dependent chloride current that is small at positive potential but exhibits pronounced activation upon hyperpolarization. Chloride currents with similar properties were reported for *Aplysia* neurones [71], parotid glands [70], certain neuronal cells [72], as well as in cardiac cells [73]. Gründer and colleagues [74] reported that ClC-2 channels in *Xenopus* oocytes are volume-sensitive, and proposed that this channel represents a ubiquitous anion pathway involved in regulatory volume changes. This report attracted much interest as volume changes of cells are key to many processes such as cell growth, division, migration, and are extremely important for many physiological and pathophysiological conditions.

Upon hypo-osmolar challenges, almost every cell tested so far exhibits outwardly rectifying, inactivating (at depolarized potentials) chloride channels termed VRAC [5] or VSOC [75] that are clearly distinct from ClC-2. For this reason, the physiological role of ClC-2 in regulatory volume decreases remains a topic of discussion. Bond et al [76] reported that while known blocker of VSOC had a significant impact on regulatory volume decrease (RVD) in T84 cells, a specific ClC-2 blocker did not have any effect, questioning any role of ClC-2 in volume regulation. In contrast, heterologous expression ClC-2 alters RVD in Sf9 cells [77] as well as in *Xenopus* oocytes [78]. Taken together these results suggest that ClC-2 is not the volume-sensitive chloride channel and not a major determinant of volume responses, but may regulate those in certain cell types where ClC-2 density is quite high.

7.3. ClC-3

ClC-3 is expressed in several organs [21] and represents another molecular candidate for the volume-sensitive anion channel. There is an ongoing debate about the physiological role and functional properties of this ClC isoform that is partly due to the fact that the functional properties of ClC-3 are not fully understood and heterologous expression studies have provided conflicting results. Injection of rClC-3 cRNA into *Xenopus* oocytes gave rise to the appearance of an outwardly rectifying chloride channel without obvious voltage-dependent gating at potentials negative to $+60\,mV$ [21]. A later study by the same group reported distinct functional features of rClC-3 channels when expressed in CHO-K1 cells, i.e. voltage-dependent activation at very positive potentials [79]. In 1997, Duan and colleagues hypothesized that ClC-3 is the volume-sensitive chloride channel mentioned above [80]. They showed that heterologous expression of guinea pig ClC-3 in NIH3T3 cells caused the appearance of a current component with many properties of VRAC or VSOC. This result could have been explained by proposing that ClC-3 expression simply activated endogenous volume-sensitive channels, but mutagenesis of the ClC-3 sequence changed functional properties of the observed current [80]. In contrast to these results, Li et al. [81] reported that expression of ClC-3 gave rise to chloride currents with properties clearly distinguishable from volume-sensitive currents. They argued that the results of the earlier mutagenesis study could have been due to the fact that Duan and colleagues observed a current carried by a mixture of different channels, including ClC-3 whose biophysical properties could be affected by mutagenesis and another volume sensitive one. Huang et al. [82] cloned the human ClC-3 isoform from T84 colonic cells and expressed them transiently and stably in tsA201 cells. They reported the appearance of an outwardly rectifying current activated by CAMKII kinase, that is insensitive to cell swelling and has functional features distinct from the currents described by Duan and colleagues [80] and those reported by Li et al. [81]. Mutagenesis in putative pore regions affected anion selectivity of this current. This discussion clearly requires additional data, most notably single channel recordings from cells expressing mutated ClC-3 channels.

Stobrawa and colleagues [83] demonstrated that disruption of ClC-3 in a knock-out mouse does not alter properties of swelling-activated chloride currents in hepatocytes and in pancreatic acinar cells. These data provide strong argument against the notion that ClC-3 forms VRAC or VSOC. Instead, these authors reported several experimental results that indicate that ClC-3 is a neuronal chloride channel responsible for acidification of endosomal compartments and synaptic vesicles [83]. There are nevertheless several lines of evidence supporting the claim that ClC-3 has something to do with volume-sensitive channels. Wang et al. [84] reported that volume-sensitive chloride currents were decreased in bovine non-pigmented ciliary epithelial cells in which endogeneous ClC-3 expression was reduced by using a ClC-3 antisense oligonucleotide. Duan and colleagues [85] demonstrated that antibodies against ClC-3 blocked native volume-sensitive chloride currents in guinea-pig cardiac cells, canine pulmonary arterial smooth muscle cells and *Xenopus laevis* oocytes.

7.4. ClC-4

ClC-4 is expressed in intestinal epithelium, heart, brain and skeletal muscle. In *Xenopus* oocytes as well as in mammalian cells, ClC-4 forms chloride channels that are functionally quite distinct from ClC-0, 1, and 2 (Fig. 1). They are outwardly rectifying and activate upon depolarization with a midpoint of activation of 80 mV [44,47]. The relative permeability and conductivity is highest for polyatomic and large anions, i.e. $SCN^- >$ $NO_3^- > Cl^- > Br^-$. Moreoever, the voltage dependence of the relative open probability did not significantly shift upon variation of external permeant anions [44,47].

ClC-4 was recently demonstrated to be expressed in the apical brush border membrane of the intestinal epithelium of mice, rats and humans [86]. This result suggests that ClC-4 operating in parallel to CFTR may participate in the chloride secretion in epithelium and could represent a possible therapeutic target in cystic fibrosis. In other tissues, little is currently known about potential physiological roles. There are two reports of native chloride channels that functionally resemble this isoform. Zachar and colleagues [87] described a macroscopic chloride current in human myoballs, a cultured human embryonic muscle preparation that outwardly rectified and displayed voltage-dependent activation upon depolarization. Such a current component disappears during muscle development as in adult fibres an anion current component with matching gating properties cannot be observed [69]. Later, another functionally similar chloride current was described in human astrocytoma cells [88] and a role in cell cycling was proposed for the underlying channels. While it is clear that more research is required in order to fully understand the physiological role of these ion channels, these two observations are consistent with the observed expression of hClC-4 in tissues [26] and suggest a possible role of ClC-4 in development of neurones and muscle.

7.5. ClC-5

ClC-5 was first identified by cloning of the disease gene responsible for an inherited human kidney disease, Dent's disease [27]. Later, the rat isoform was cloned [89] and expressed heterologously. Functional properties of ClC-5 are quite similar to those of ClC-4[44]. The currents are profoundly outwardly rectifying, with only very small inward currents at negative potentials. ClC-5 is located in endosomes in the proximal tubules, and for this reason it was proposed that ClC-5 forms a background anion channel that allows proper function of a H^+-ATPase and thus acidification of endosomes [90,91] (see below).

However, the properties of ClC-5 differ profoundly from reports of native chloride channels in endosomal vesicles. In endocytotic vesicles from rabbit proximal tubules [6], activation of a stilbene-sensitive chloride conductance by protein kinase A was shown to cause acidification of the vesicle lumen. In a subsequent study, Reenstra and colleagues [92] demonstrated that the molecular weight of the phosphorylated endosomal channel is 67 kDa, clearly distinct from the 86 kDa band that ClC-5 would produce. Using patch-clamp recordings on fused endocytotic vesicles from rat kidney cortex, Schmid and colleagues [93] described a chloride channel with a linear I–V relationship, a slope conductance of 73 pS under symmetrical chloride concentrations, a Cl = Br = I > SO$_4$ >

F permeability sequence, and sensitivity to DIDS and NPPB. Unfortunately, while all of the descriptions of native endosomal chloride channels are consistent, they differ profoundly from properties reported for ClC-5. ClC-5 in heterologous expression systems is profoundly outwardly rectifying, exhibits a $NO_3^- > Cl^- > Br^- > HCO_3^- > I^-$ conductivity sequence, is not regulated by protein kinase A and is insensitive to stilbenes [89,94].

7.6. ClC-6 and ClC-7

The most recently identified members of the ClC channel family are ClC-6 and 7 [25], and these two isoforms remain the ClC channels that are least understood. Northern blot analysis revealed that they are broadly expressed in almost every tissue tested, and that expression starts early in embryogenesis. When heterologously expressed in CHO cells, ClC-6 channels appear in intracellular membrane compartments. There are currently no functional data for these two isoforms [25].

Recently, a ClC-7 knock-out mouse was reported [95]. Disruption of ClC-7 causes severe osteopetrosis, as osteoclasts fail to resorb bones in the absence of ClC-7. Similar to ClC-5, ClC-7 seems to be important for proper function of a proton pump [95], and in the absence of ClC-7 this pump fails to properly acidify the extracellular space between the osteoclast and the bone and therefore resorption of the bone by osteoclasts cannot occur.

7.7. Kidney-specific ClC channels

Chloride channels play an important role in fluid reabsorption in the nephron. In 1992 and 1994, two ClC-isoforms from rat were cloned, ClC-K1 [22] and ClC-K2 [23] that were thought to be restricted to the kidney at the beginning. These two channels exhibit a distinct expression pattern along the nephron: ClC-K1 is restricted to the thin ascending limb of Henle's loop (tAL) and forms channels in both the apical and basolateral plasma membranes [96]. This pattern explains the large chloride permeability of this part of the nephron. ClC-K2 is more broadly expressed, with high expression in the distal convoluted tubules, connecting tubules, and cortical collecting ducts and moderate expression in the medullary thick ascending limb of Henle's loop (mTAL) [97]. ClC-K2 appears to be a basolateral chloride channel in the segments of the nephron that exhibit sodium dependent chloride transport, suggesting a role for this isoform in vectorial transepithelial chloride transport (reabsorption). Subsequently, Kieferle and colleagues reported the cloning of the human isoforms [24]. Surprisingly, the two human isoforms turned out to be more similar to each other than to each of the rat kidney ClC channels and for this reason a distinct nomenclature was chosen for human isoforms, ClC-Ka and ClC-Kb. The distinct roles and expression pattern now support the view that ClC-K1 is related to ClC-Ka and ClC-K2 to ClC-Kb.

The physiological role of these kidney ClC channels was identified with two distinct experimental approaches, the generation of mice lacking the ClC-K1 isoform and molecular genetic techniques. Mice lacking ClC-K1 by targeted gene disruption are unable to concentrate urine [98] and exhibit a nephrogenic diabetes insipidus. Moreover, mutations in the gene encoding ClC-Kb cause Bartter syndrome, an inherited human inability to concentrate urine [99] (see below). Although outwardly similar, the phenotypes of ClC-K1 and ClC-Kb dysfunction are quite distinct, indicating separate physiological roles of these isoforms. The observed changes of kidney function show the crucial role of ClC-K channels in urine concentration and make these isoforms interesting targets for antihypertensive drugs.

ClC-Ka and Kb are also expressed in the stria vascularis of the inner ear [37,38] where they fulfill an essential task for the K^+ secretion into the endolymph. In strial marginal cells, both isoforms colocalize with Barttin and NKCC1 at the basolateral membrane. While the Cl^- ions transported by the $Na^+/K^+/2Cl^-$ cotransporter recycle through the ClC channels back into the blood, K^+ accumulates inside the cell providing a driving force for the diffusion of K^+ through the apical KCNQ1/KCNE1 channel into the endolymph [37,38].

When heterologously expressed in oocytes, ClC-K1 forms an outwardly rectifying anion channel with a $Br^- > NO_3^- > Cl^- > I^-$ selectivity [96]. For human ClC isoforms, functional data were absent for a long time. In 2001, it was reported that the first ClC channel β-subunit, Barttin, is essential for the functional expression of the two human ClC-K isoforms [38]. Coexpression of Barttin with ClC-Ka and Kb results in the appearance of anion selective currents with linear instantaneous current–voltage relationship and a Cl > Br > I anion selectivity [38]. Gating properties somehow differ between the two used expression systems, *Xenopus* oocytes and tsA201 cells [38], a feature that is currently not understood.

All mammalian ClC channels known were cloned between 1990 and 1995. Since then no novel member has been cloned, and no new channels have been found in the eukaryotic genome projects completed to data, including *Caenorhabditis elegans* and *Drosophila melanogaster*. These findings suggest that the ClC family is complete, and that novel mammalian ClC isoforms are not to be expected. Unfortunately, some important native chloride channels do not belong to the ClC family, for example the Ca^{2+}-activated chloride channels, and it is currently unclear whether volume-sensitive chloride channels are ClC channels or not. Obviously, there are other physiologically important structural motifs that form anion channels that await discovery.

8. ClC-type chloride channels in inherited diseases

8.1. Myotonia congenita

Myotonia congenita is an inherited human muscle disease that is characterized by stiffness upon sudden forceful movement that improves when the same movement is repeated several times ("warm-up phenomenon"). Myotonia congenita is transmitted either dominantly (Thomson's myotonia) [100] or recessively (Becker-type myotonia) [101]. Its pathophysiology allowed important insights into the function of chloride channels. It was first studied in an animal model by Shirley Bryant who examined a strain of goats that exhibit bizarre attacks of stiffness and rigidity and were known as "falling goats" or "fainting goats". In a series of publications, Bryant and colleagues established that an alteration of the muscle membrane excitability [2,102,103] is the reason for the abnormal behavior of the goat. Muscle fibers from myotonic goats exhibit a drastically reduced chloride conductance (g_{Cl}), and normal muscle can be made myotonic by blocking muscle chloride channels pharmacologically [103]. These results firmly established chloride channel dysfunction as the cause of myotonia. Later a reduced g_{Cl} was identified also as a pathomechanism for human myotonias [104].

The detailed investigation of myotonic muscle fibers allowed an elucidation of the role chloride channels play in regulating membrane excitability. Although the muscle chloride conductance is the dominant conductance at resting potentials, it is nevertheless very small compared with voltage-activated sodium and potassium currents at more depolarized potentials. Muscle chloride channels do not play a role in the repolarization of the action potential. This was most clearly demonstrated by the finding that action potentials in normal and myotonic muscle fibers are very similar [105]. Myotonia is not due to impaired muscle repolarization [2] (Fig. 7), but to a sarcolemmal depolarization that is due to potassium accumulation in the t-tubule. In myotonic muscle, after a number of action potentials, the membrane potential fails to return to the resting potential and stays depolarized for a longer period of time ("afterdepolarization"). Depending on the magnitude of this afterdepolarization, the muscle fiber may start to fire action potentials in the absence of external stimuli, the electrical correlate to the myotonic muscle stiffness. The reason for this afterdepolarization is a potassium accumulation that occurs in the t-tubules of every muscle fiber because of the restricted diffusion in this muscle

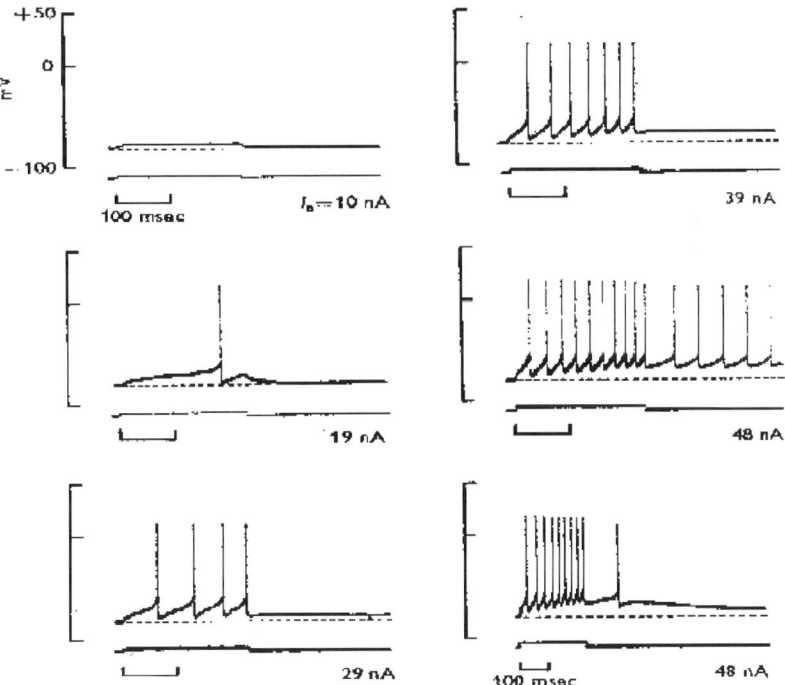

Fig. 7. Intracellular membrane potentials recorded from an intercental muscle fiber of a myotonic goat. Temperature 37°C, normal Ringer. For each record, a certain current amplitude was injected at a magnitude and duration given in the figure. This figure is reprinted from Ref. [2] with permission.

compartment. In normal muscle, the depolarization of the t-tubular membrane does not affect the surface membrane potential because of the small length constant of healthy muscle fibers. In myotonic muscle, the length constant is significantly increased and t-tubular depolarization also depolarizes the surface membrane. Muscle chloride channels simply reduce the length constant of muscle and thus allow muscle to electrically tolerate the t-tubule depolarization. Myotonia congenita thus demonstrates an important mechanism by which background conductances can regulate excitability. It appears very likely that a similar modulation also plays an important role in certain parts of the nervous system.

In 1992, Koch and colleagues [106] demonstrated linkage of both forms of myotonia congenita to *ClCN1* and reported the first point mutation found to cause recessive myotonia, F413C. Shortly thereafter, George and colleagues identified the first dominant myotonia mutation, G230E [60]. In the following years, a great number of disease-causing mutations were reported for Thomsen's as well as for Becker's myotonia, spreading over the whole coding region the ClC-1 [36,40,53,54,64,65,128,107,109–111]. Initially, the functional effects of these mutations were studied by a single group using two-microelectrode recordings in *Xenopus* oocytes, and all of them were reported as simply non-functional for recessive mutations and to be non-functional with an additional dominant negative effect in case of the dominant mutations [112]. Recently, it has become clear that the some of these results are simply due to technical problems of the expression system, and that many myotonia mutations originally thought to abolish channel function produce functional channels with altered functional properties [111].

The first myotonia mutation that was reported to be functional was the D136G mutation [53,108]. The D136G mutation was found in a patient with a very pronounced recessive generalized myotonia. D136G hClC-1 channels express as well as WT channels in HEK 293 cells, but exhibit dramatically different functional properties (Fig. 8A and B). In symmetric chloride concentrations, these channels activate upon hyperpolarization and thus exhibit an inverted voltage dependence of gating as compared to WT channels. The reason for this inverted voltage dependence and moreover the reason for the reduced chloride conductance of muscle fibers expressing these channels is in a distinct dependence of D136G gating on the intracellular chloride concentration. D136G hClC-1 channels open only at potentials negative to the chloride reversal potential [53]. For this reason muscle expressing mutant chloride channels exhibit only a negligible chloride conductance at the resting potential. Inversion of the voltage dependence of activation gating is also caused by other myotonia causing mutations [65]. Later, a second type of alteration of ClC-1 gating was reported. Some mutations with a dominant inheritance mode shift the activation curve to more positive potentials [54,109, 113–115]. This significantly reduces the open probability and thus the macroscopic chloride conductance at resting potentials (Fig. 8C). This phenotype is found in dominant as well as in recessive mutations. However, mutations causing recessive myotonia differ from those of dominant forms in the effect the mutation has on the activation curve of heterodimeric channels consisting of one WT and one mutant subunit [114,115]. G230E represents a third phenotype of myotonia causing mutations in hClC-1, resulting in alteration of pore properties [40] (Fig. 8D). Several other mutations do not alter

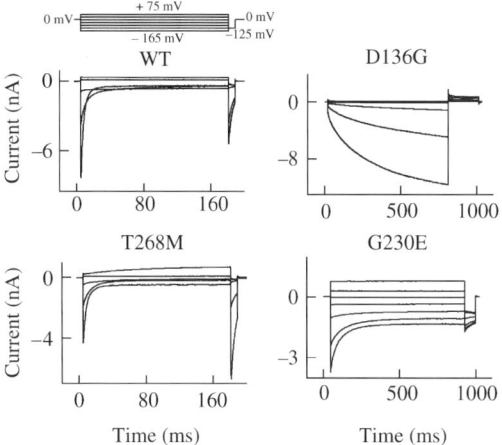

Fig. 8. Representative current recordings from WT(A) as well as from mutant hClC-1 channels carrying three distinct myotoria causing mutations. D136G (B) [51], T268M (C) [116], and G230E (D) [36]. Currents were recorded using tight whole-cell recordings from transiently transfected μA201 cells. Cells were bathed in a standard external and internal solution given in Fig. 4.

functional properties but cause reduced current density in heterologous expression systems [116].

Myotonia mutations have played an important role in the structure–function analysis of ClC chloride channels [32,36,40,107]. There are still a large number of mutations that have not been characterized thoroughly and it appears very likely that myotonia mutations will provide further insights into the molecular physiology of the ClC channel family.

8.2. Dent's disease

Dent's disease is an X-linked recessively inherited kidney disease that was defined only recently [27,28,117,118]. Between 1991 and 1993, four distinct clinical syndromes were described in different countries, with low-weight proteinuria, hypercalciuria and nephrolithiasis. As all these syndromes exhibit the same disease gene locus they are now thought to be one disease, Dent's disease [119,120]. Positional cloning of the disease locus by Rajesh Thakker's group in London [27,28] identified a gene (*CLCN5*) coding for an anion channel belonging to the ClC family of voltage-gated chloride channels (ClC-5). Later, several point mutations were identified and functional characterization in heterologous expression systems demonstrated that these mutations alter ClC-5 channel function [121]. These findings established *CLCN5* as the Dent's disease gene locus. In the initial publication reporting the cloning of the Dent's disease gene, Fisher et al. [27] reasoned that the affected chloride channel might play a role in endosomal acidification and might thus be important for absorptive endocytosis in the proximal tubule. In normal kidney function the glomerular barrier prevents the filtration of large proteins such as albumin, but smaller proteins may readily pass. These are then reabsorbed by proximal tubule cells through endocytosis [122,123] and later enzymatically degraded in lysosomes. All of the processes during protein internalization and degradation critically depend on an acidic environment in subcellular organelles [124]. The acidic pH inside endosomes is established by a proton pump, an H^+-ATPase that actively transports protons into these organelles. The transport is electrogenic and therefore capable of generating a positive membrane potential inside the organelle. In the absence of additional conductances, this potential would reach very high values and the transport of protons against the electrostatic field would become energetically prohibitive [125]. Chloride channels profoundly increase the conductance of this membrane and can therefore clamp the potential difference between vesicle inside and outside to zero. Chloride channel dysfunction results in hampered endosomal acidification, causing impaired reabsoptive protein endocytosis. ClC-5 is expressed in proximal tubule endosomes and is thought to provide this anion permeability in the endosomal membranes [90].

This hypothesis can explain many of the experimental results obtained so far. Nevertheless, its unique functional properties makes ClC-5 appear unsuited for this particular role. Moreover, the fact that the native channels that were shown to serve this function in proximal tubule endosome [6] prompted me to hypothesize that ClC-5 might serve an additional role in kidney endosomes [126]. Due to its pronounced rectification (see above), it would serve in endosomes as a sole efflux pathway and could thus cause a decrease of the endosomal HCO_3^- concentration. Proximal tubule cells exhibit high intracellular HCO_3^- concentrations and in endosomes this would cause a marked alkanization in the absence of carriers that decrease the amount of HCO_3^- in this cell compartment.

In the last several years a large number of Dent's disease causing mutations have been functionally studied in *Xenopus* oocytes, and surprisingly the alterations these mutations cause on the function of ClC-5 are quite uniform, in clear contrast to myotonia mutations. Every mutation associated with Dent's disease tested so far causes a decrease or a complete abolishment of chloride current [119,121,127–130].

8.3. Bartter syndrome

Bartter syndrome is characterized by hypotonia, hyponatriaemia and hypovolaemia due to an inability of the kidney to concentrate urine. Three subtypes of Bartter syndrome (I–III) can be clinically distinguished, and accordingly three distinct gene loci have been described: an inwardly rectifying K^+ channel ($K_{ir}1.1$ or ROMK1) [131], a $Na^+/K^+/Cl^-$-cotransporter [132] and a ClC-type chloride channel, ClC-Kb [99]. All these proteins are expressed in the thick ascending limb of Henle and are crucial for the establishment of a hyperosmotic kidney medulla. Patients with type III exhibit hypokalaemic alkalosis with salt-wasting and hypotonia. Urinary excretion of magnesium is normal, while hypercalciuria as well as normocalciuria can be observed.

Similar to disease-causing mutations in ClC-1 and ClC-5, Bartter mutations in ClC-Kb are spread throughout the whole coding region of this channel. The functional effects of these mutations were evaluated in *Xenopus* oocytes cotransfected with Barttin [38]. All of them significantly reduce macroscopic expression levels. This is in agreement with the notion that these mutations decrease the chloride conductance in the basolateral membrane of thick ascending limb of Henle's loop.

ClC-Kb is also expressed the stria vascularis, but mutations in *CLCKNB* do not cause deafness. The reason for that is that ClC-Ka and ClC-Kb coexpress in the basolateral membrane of strial marginal cells. As the non-affected isoform can still provide sufficient transport capacity, mutations in ClC-Kb do not prevent the basolateral Cl⁻ recycling. In contrast, mutations affecting Barttin, the common β-subunit of ClC-Ka and ClC-Kb, abolishes the anion pathway at the basolateral membrane and thus potassium secretion by the stria vascularis.

8.4. Infantile malignant osteopetrosis

Infantile malignant osteopetrosis is the most recently identified ClC channelopathy. Based on the phenotype of a ClC-7 knock-out mouse that exhibit severe osteopetrosis, Kornak and colleagues investigated patients with a human inherited form of osteopetrosis and found two mutations in the *CLCN7* gene, one nonsense (Q555X) and one missense mutation (R762Q) [95]. Osteopetrosis is caused by defective osteoclast function. Osteoclasts resorb bone material using a specialized cell compartment, the resorptive lacuna, in which an acidic pH is established by a proton pump present in the osteoclast ruffled membrane. The active transport is electrogenic and requires an anion channel in the same membrane to ensure electroneutrality. Mutations in ClC-7 impair the acidification of the lacuna and abolish the osteoclast function thus causing osteopetrosis [95].

9. Summary and Outlook

The ClC family of voltage-gated chloride channels has been known for 10 years now, and great progress in our understanding of these channels has been made. Our understanding of this fascinating ion channel family is still very limited, but recent work has provided us with many tools to address the remaining questions.

Acknowledgements

This work was supported by the DFG(Fa 301/4-1), and the Muscular Dystrophy Association. I would like to thank Dr. J.P. Johnson for helpful discussions and critical reading of the manuscript. I am very grateful to Dr T.J. Jentsch (Hamburg) and Dr. J.E. Melvin (Rochester) for kindly providing with the ClC-0 and rClC-2 clones that allowed the recordings shown in Fig. 4.

References

[1] Hille, B., 1992. Ionic Channels of Excitable Membranes, 2nd ed. Sinauer Associates Inc, Sunderland, MA.
[2] Adrian, R.H., Bryant, S.H., 1974. On the repetitive discharge in myotonic muscle fibres. J. Physiol. (Lond.) 240, 505–515.

[3] Jentsch, T.J., Günther, W., 1997. Chloride channels: an emerging molecular picture. Bioessays 19(2), 117–126.

[4] Uchida, S., 2000. In vivo role of CLC chloride channels in the kidney. Am. J. Physiol. Renal Fluid Electrolyte Physiol. 279, F802–F803.

[5] Nilius, B., Voets, T., Eggermont, J., Droogmans, G., 1999. VRAC: a multifunctional volume-regulated anion channel in vascuar endothelium. In: Koslowski, R.Z. (Ed.), Cl⁻ Channels. Isis Medical Ltd, Oxford, UK, pp. 47–63.

[6] Bae, H.R., Verkman, A.S., 1990. Protein kinase A regulates chloride conductance in endocytic vesicles from proximal tubule. Nature 348(6302), 637–639.

[7] Heginbotham, L., Lu, Z., Abramson, T., MacKinnon, R., 1994. Mutations in the K^+ channel signature sequence. Biophys. J. 66(4), 1061–1067.

[8] Jentsch, T.J., Steinmeyer, K., Schwarz, G., 1990. Primary structure of *T. marmorata* chloride channel isolated by expression cloning in *Xenopus* oocytes. Nature 348, 510–514.

[9] Jentsch, T.J., Friedrich, T., Schriever, A., Yamada, H., 1999. The CLC chloride channel family. Pflugers Arch. Eur. J. Physiol. 437(6), 783–795.

[10] George, A.L. Jr, Bianchi, L., Link, E.M., VanRaay, T.J., 2001. From stones to bones: the biology of ClC chloride channels. Curr. Biol. 11, R620–R628.

[11] Maduke, M., Miller, C., Mindell, J.A., 2000. A decade of CLC chloride channels: structure, mechanism, and many unsettled questions. Annu. Rev. Biophys. Biomol. Struct. 29, 411–438.

[12] White, M.M., Miller, C., 1979. A voltage-gated anion channel from the electric organ of *T. californica*. J. Biol. Chem. 254, 10161–10166.

[13] White, M.M., Miller, C., 1981. Probes of the conduction process of a voltage-gated Cl⁻ channel from *Torpedo*. electroplax. J. Gen. Physiol. 78, 1–18.

[14] Miller, C., Richards, E.A., 1990. The voltage-dependent chloride channel of *Torpedo* electroplax. In: Alvarex-Leefsman, F.J., Russel, J.M. (Eds.), Chloride Channels and Carriers in Nerve, Muscle, and Glial Cells. Plenum, New York, pp. 383–405.

[15] Miller, C., White, M.M., 1982. Probes of the conduction pathway of a voltage-gated Cl⁻ channel from *Torpedo*. J. Gen. Physiol. 78, 1–18.

[16] Miller, C., White, M.M., 1984. Dimeric structure of single chloride channels from *Torpedo* electroplax. Proc. Natl Acad. Sci. USA 81, 2772–2775.

[17] Miller, C., 1982. Open-state substructure of single chloride channels from *Torpedo* electroplax. Philos Trans. R. Soc. Lond. B. Biol. Sci. 299, 401–411.

[18] Bauer, C.K., Steinmeyer, K., Schwarz, J.R., Jentsch, T.J., 1991. Completely functional double-barreled chloride channel expressed from a single *Torpedo* cDNA. Proc. Natl Acad. Sci. USA 88, 11052–11056.

[19] Steinmeyer, K., Ortland, C., Jentsch, T.J., 1991. Primary structure and functional expression of a developmentally regulated skeletal muscle chloride channel. Nature 354, 301–304.

[20] Thiemann, A., Grunder, S., Pusch, M., Jentsch, T.J., 1992. A chloride channel widely expressed in epithelial and non-epithelial cells. Nature 356, 57–60.

[21] Kawasaki, M., Uchida, S., Monkawa, T., Miyawaki, A., Mikoshiba, K., Maruma, F., Sasaki, S., 1994. Cloning and expression of a protein kinase C-regulated chloride channel abundantly expressed in rat brain neuronal cells. Neuron 12, 597–604.

[22] Uchida, S., Sasaki, S., Furukawa, T., Hiraoka, M., Imai, T., Hirata, Y., Marumo, F., 1993. Molecular cloning of a chloride channel that is regulated by dehydration and expressed predominantly in kidney medulla. J. Biol. Chem. 268, 3821–3824.

[23] Adachi, S., Uchida, S., Ito, H., Hata, M., Hiroe, M., Marumo, F., Sasaki, S., 1994. Two isoforms of a chloride channel predominantly expressed in thick ascending limb of Henle's loop and collecting ducts of rat kidney. J. Biol. Chem. 269, 17677–17683.

[24] Kieferle, S., Fong, P., Bens, M., Vandewalle, A., Jentsch, T.J., 1994. Two highly homologous members of the ClC chloride channel family in both rat and human kidney. Proc. Natl Acad. Sci. USA 91, 6943–6947.

[25] Brandt, S., Jentsch, T.J., 1995. ClC-6 and ClC-7 are two novel broadly expressed members of the CLC chloride channel family. FEBS Lett. 377, 15–20.

[26] van Slegtenhorst, M.A., Bassi, M.T., Borsani, G., Wapenaar, M.C., Ferrero, G.B., de Conciliis, L., et al., 1994. A gene from the Xp22.3 region shares homology with voltage-gated chloride channels. Hum. Mol. Genet. 3(4), 547–552.

[27] Fisher, S.E., Black, G.C.M., Lloyd, S.E., Hatchwell, E., Wrong, O., Thakker, R.V., Craig, I.W., 1994. Isolation and partial characterization of a chloride channel gene which is expressed in kidney and is a candidate for Dent's disease (an X-linked hereditary nephrolithiasis). Hum. Mol. Genet. 3, 2053–2059.

[28] Fisher, S.E., Van Bakel, I., Lloyd, S.E., Pearce, S.H.S., Thakker, R.V., Craig, I.W., 1995. Cloning and characterization of CLCN5, the human kidney chloride channel gene implicated in Dent's disease (an X-linked hereditary nephrolithiasis). Genomics 29, 598–606.

[29] Maduke, M., Pheasant, D.J., Miller, C., 1999. High-level expression, functional reconstitution, and quaternary structure of a prokaryotic ClC-type chloride channel. J. Gen. Physiol. 114(5), 713–722.

[30] Kyte, J., Doolittle, R.F., 1982. A simple method for displaying the hydropathic character of a protein. J. Mol. Biol. 157(1), 105–132.

[31] Dutzler, R., Campbell, E.D., Cadene, M., Chait, M.B., MacKinnon, R., 2002. X-ray structure of a ClC chloride channel at 3.0 A reveals the molecular basis of anion selectivity. Nature 415, 287–294.

[32] Middleton, R.E., Pheasant, D.J., Miller, C., 1994. Purification, reconstitution, and subunit composition of a voltage-gated chloride channel from *Torpedo* electroplax. Biochemistry 33, 13189–13198.

[33] Middleton, R.E., Pheasant, D.J., Miller, C., 1996. Homodimeric architecture of a ClC-type chloride ion channel. Nature 383(6598), 337–340.

[34] Ludewig, U., Pusch, M., Jentsch, T.J., 1996. Two physically distinct pores in the dimeric ClC-0 chloride channel. Nature 383(6598), 340–343.

[35] Fahlke, Ch, Knittle, T.J., Gurnett, C.A., Campbell, K.P., George, A.L. Jr., 1997. Subunit stoichiometry of human muscle chloride channels. J. Gen. Physiol. 109, 93–104.

[36] Miller, C., White, M.M., 1984. Dimeric structure of single chloride channels from *Torpedo* electroplax. Proc. Natl Acad. Sci. USA 81, 2772–2775.

[37] Birkenjäger, R., Otto, E., Schurmann, M.J., Vollmer, M., Ruf, E.-M., Maier-Lutz, I., Beekmann, F., Fekete, A., Omran, H., Feldmann, D., Milford, D.V., Jeck, N., Konrad, M., Landau, D., Knoers, N.V.A.M., Antignac, C., Sudbrak, R., Kispert, A., Hildebrandt, F., 2001. Mutation of BSND causes Bartter syndrome with sensorineural deafness and kidney failure. Nat. Genet. 29, 310–314.

[38] Estévez, R., Boettger, T., Stein, V., Birkenjäger, R., Otto, E., Hildebrandt, F., Jentsch, T.J., 2001. Barttin is a Cl⁻ channel β-subunit crucial for renal Cl⁻ reabsorption and inner ear K⁺ secretion. Nature 414, 558–561.

[39] Fahlke, Ch., 2001. Ion permeation and selectivity in ClC-type chloride channels. Am. J. Physiol. Renal Fluid Electrolyte Physiol. 280(5), F748–F757.

[40] Fahlke, Ch., Beck, C.L., George, A.L. Jr, 1997. A mutation in autosomal dominant myotonia congenita affects pore properties of the muscle chloride channel. Proc. Natl Acad. Sci. USA 94, 2729–2734.

[41] Fahlke, Ch., Dürr, C., George, A.L. Jr, 1997. Mechanism of ion permeation in skeletal muscle chloride channels. J. Gen. Physiol. 110, 551–564.

[42] Rychkov, G.Y., Pusch, M., Roberts, M.L., Jentsch, T.J., Bretag, A.H., 1998. Permeation and block of the skeletal muscle chloride channel, ClC-1, by foreign anions. J. Gen. Physiol. 111(5), 653–665.

[43] Hutter, O.F., Mello, W.C., Warner, A.E., 1969. An application of the field strength theory. In: Testeson, D.C. (Ed.), The molecular basis of membrane function. Prentice-Hall, Inc, Englewood Cliffs, NJ, pp. 391–400.

[44] Vergara, C., Alvarez, O., Latorre, R., 1999. Localization of the K⁺ lock-in and the Ba²⁺ binding sites in a voltage-gated calcium-modulated channel. J. Gen. Physiol. 114, 365–376.

[45] Dang, T.X., McCleskey, E.W., 1999. Ion channel selectivity through stepwise changes in binding affinity. J. Gen. Physiol. 111, 185–193.

[46] Hebeisen, S., Heidtmann, H., Cosmelli, D., Gonzalez, C., Latorre, R., Alvarez, O., Fahlke, Ch. Anion permeation in human ClC-4 channels. Biophys. J. 84, 2306–2318.

[47] Hanke, W., Miller, C., 1983. Single chloride channels from *Torpedo* electroplax. Activation by protons. J. Gen. Physiol. 82, 25–45.

[48] Pusch, M., Ludewig, U., Rehfeldt, A., Jentsch, T.J., 1995. Gating of the voltage-dependent chloride channel ClC-0 by the permeant anion. Nature 373, 527–530.

[49] Chen, T.Y., Miller, C., 1996. Nonequilibrium gating and voltage dependence of the ClC-0 Cl⁻ channel. J. Gen. Physiol. 108(4), 237–250.

[50] Pusch, M., Ludewig, U., Jentsch, T.J., 1997. Temperature dependence of fast and slow gating relaxations of ClC-0 chloride channels. J. Gen. Physiol. 109(1), 105–116.

[51] Fong, P.Y., Rehfeldt, A., Jentsch, T.J., 1998. Determinants of slow gating in ClC-0, the voltage-gated chloride channel of *Torpedo marmorata*. Am. J. Physiol. Cell Physiol. 274(4), C966–C973.

[52] Pusch, M., Steinmeyer, K., Jentsch, T.J., 1994. Low single channel conductance of the major skeletal muscle chloride channel, ClC-1. Biophys. J. 66, 149–152.

[53] Fahlke, Ch., Rüdel, R., Mitrovic, N., Zhou, M., George, A.L. Jr, 1995. An aspartic acid residue important for voltage-dependent gating of human muscle chloride channels. Neuron 15, 463–472.

[54] Pusch, M., Steinmeyer, K., Koch, M.C., Jentsch, T.J., 1995. Mutations in dominant human myotonia congenita drastically alter the voltage-dependence of the ClC-1 chloride channel. Neuron 15, 1455–1463.

[55] Fahlke, Ch., Rosenbohm, A., Mitrovic, N., George, A.L. Jr, Rüdel, R., 1996. Mechanism of voltage-dependent gating in skeletal muscle chloride channels. Biophys. J. 71, 695–706.

[56] Rychkov, G.Y., Pusch, M., Astill, D.S.J., Roberts, M.L., Jentsch, T.J., Bretag, A.H., 1996. Concentration and pH dependence of skeletal muscle chloride channel ClC-1. J. Physiol. (Lond.) 497, 423–435.

[57] Cui, J., Cox, D.H., Aldrich, R.W., 1997. Intrinsic voltage dependence and Ca^{2+} regulation of mslo large conductance Ca-activated K^+ channels. J. Gen. Physiol. 109(5), 647–673.

[58] Tibbs, G.R., Goulding, E.H., Siegelbaum, S.A., 1997. Allosteric activation and tuning of ligand efficacy in cyclic-nucleotide-gated channels. Nature 386(6625), 612–615.

[59] Koradi, R., Billeter, M., Wüthrich, K., 1996. MOLMOL: a program for display and analysis of macromolecular structures. J. Mol. Graph. 14, 51–55.

[60] George, A.L., Crackower, M.A., Abdalla, J.A., Hudson, A.J., Ebers, G.C., 1993. Molecular basis of Thomsen's disease (autosomal dominant myotonia congenita). Nat. Genet. 3, 305–310.

[61] Fahlke, Ch., Yu, H.T., Beck, C.L., Rhodes, T.H., George, A.L. Jr, 1997. Pore-forming segments in voltage-gated chloride channels. Nature 390(6659), 529–532.

[62] Fahlke, Ch., Desai, R.R., Gillani, N., George, A.L. Jr, 2001. Residues lining the inner pore vestibule of human muscle chloride channels. J. Biol. Chem. 276, 1759–1765.

[63] Lin, C.W., Chen, T.Y., 2000. Cysteine modification of a putative pore residue in ClC-0: implication for the pore stoichiometry of ClC chloride channels. J. Gen. Physiol. 116, 535–546.

[64] Lin, Y.W., Lin, C.W., Chen, T.Y., 1999. Elimination of the slow gating of ClC-0 chloride channel by a point mutation. J. Gen. Physiol. 114(1), 1–12.

[65] Zhang, J., Sanguinetti, M.C., Kwiecinski, H., Ptacek, L.J., 2000. Mechanism of inverted activation of ClC-1 channels caused by a novel myotonia congenita mutation. J. Biol. Chem. 275(4), 2999–3005.

[66] Bryant, S.H., 1962. Muscle membrane of normal and myotonic goats in normal and low external chloride. Fed. Proc. 21, 312.

[67] Bryant, S.H., Morales-Aguilera, A., 1971. Chloride conductance in normal and myotonic muscle fibres and the action of monocarboxylic aromatic acids. J. Physiol. (Lond.) 219, 361–383.

[68] Warner, A.E., 1972. Kinetic properties of the chloride conductance of frog muscle. J. Physiol. (Lond.) 227, 291–312.

[69] Fahlke, Ch., Rüdel, R., 1995. Chloride currents across the membrane of mammalian skeletal muscle fibres. J. Physiol. (Lond.) 484, 355–368.

[70] Park, K., Arreola, J., Begenisich, T., Melvin, J.E., 1998. Comparison of voltage-activated Cl^- channels in rat parotid acinar cells with ClC-2 in a mammalian expression system. J. Membr. Biol. 163(2), 87–95.

[71] Chesnoy-Marchais, D., 1983. Characterization of a chloride conductance activated by hyperpolarization in Aplysia neurones. J. Physiol. (Lond.) 342, 277–308.

[72] Clark, S., Jordt, S.E., Jentsch, T.J., Mathie, A., 1998. Characterization of the hyperpolarization-activated chloride current in dissociated rat sympathetic neurons. J. Physiol. (Lond.) 506(Pt 3), 665–678.

[73] Duan, D., Ye, L., Britton, F., Horowitz, B., Hume, J.R., 2000. A novel anionic inward rectifier in native cardiac myocytes. Circ. Res. 86(4), E63–E71.

[74] Gründer, S., Thiemann, A., Pusch, M., Jentsch, T.J., 1992. Regions involved in the opening of the ClC-2 chloride channel by voltage and cell volume. Nature 360, 759–762.

[75] Kirk, K., Strange, K., 1998. Functional properties and physiological roles of organic solute channels. Ann. Rev. Physiol. 60, 719–739.

[76] Bond, T.D., Ambikapathy, S., Mohammad, S., Valverde, M.A., 1998. Osmosensitive Cl^- currents and their relevance to regulatory volume decrease in human intestinal T84 cells: outwardly vs. inwardly rectifying currents. J. Physiol. (Lond.) 511(Pt1), 45–54.

[77] Xiong, H., Li, C., Garami, E., Wang, Y., Ramjeesingh, M., Galley, K., et al., 1999. ClC-2 activation modulates regulatory volume decrease. J. Membr. Biol. 167(3), 215–221.

[78] Furukawa, T., Ogura, T., Katayama, Y., Hiraoka, M., 1998. Characteristics of rabbit ClC-2 current expressed in *Xenopus* oocytes and its contribution to volume regulation. Am. J. Physiol. Cell Physiol. 274(2), C500–C512.

[79] Kawasaki, M., Suzuki, M., Uchida, S., Sasaki, S., Marumo, F., 1995. Stable and functional expression of the ClC-3 chloride channel in somatic cell lines. Neuron 14, 1285–1291.

[80] Duan, D., Winter, C., Cowley, S., Hume, J.R., Horowitz, B., 1997. Molecular identification of a volume-regulated chloride channel. Nature 390(6658), 417–421.

[81] Li, X., Shimada, K., Showalter, L.A., Weinman, S.A., 2000. Biophysical properties of ClC-3 differentiate it from swelling-activated chloride channels in chinese hamster ovary-K1 Cells. J. Biol. Chem. 275, 35994–35998.

[82] Huang, P., Liu, J., Di, A., Robinson, N.C., Musch, M.W., Kaetzel, M.A., et al., 2001. Regulation of human CLC-3 channels by multifunctional Ca^{2+}/calmodulin-dependent protein kinase. J. Biol. Chem. 276(23), 20093–20100.

[83] Stobrawa, S., Breiderhoff, T., Takamori, S., Engel, D., Schweizer, M., Zdebik, A., et al., 2001. Disruption of ClC-3, a chloride channel expressed on synaptic vesicles, leads to a loss of the Hippocampus. Neuron 29, 185–196.

[84] Wang, L., Chen, L., Jacob, T.J., 2000. The role of ClC-3 in volume-activated chloride currents and volume regulation in bovine epithelial cells demonstrated by antisense inhibition. J. Physiol. (Lond.) 524(Pt 1), 63–75.

[85] Duan, D., Zhong, J., Hermoso, M., Satterwhite, C.M., Rossow, C.F., Hatton, W.J., et al., 2001. Functional inhibition of native volume-sensitive outwardly rectifying anion channels in muscle cells and *Xenopus* oocytes by anti-ClC-3 antibody. J. Physiol. (Lond.) 521(Pt 2), 437–444.

[86] Mohammad-Panah, R., Ackerley, C., Rommens, J.M., Choudhury, M., Bear, C.E., 2002. The chloride channel ClC-4 co-localizes with CFTR and may mediate chloride flux across the apical membrane of intestinal epithelia. J. Biol. Chem. 277(1), 566–574.

[87] Zachar, E., Fahlke, C., Rudel, R., 1992. Whole-cell recordings of chloride currents in cultured human skeletal muscle. Pflugers Arch. Eur. J. Physiol. 421(2-3), 101–107.

[88] Ullrich, N., Sontheimer, H., 1996. Biophysical and pharmacological characterization of chloride currents in human astrocytoma cells. Am. J. Physiol. Cell Physiol. 270(5), C1511–C1521.

[89] Steinmeyer, K., Schwappach, B., Bens, M., Vandewalle, A., Jentsch, T.J., 1995. Cloning and functional expression of rat CLC-5, a chloride channel related to kidney disease. J. Biol. Chem. 270, 31172–31177.

[90] Gunther, W., Luchow, A., Cluzeaud, F., Vandewalle, A., Jentsch, T.J., 1998. ClC-5, the chloride channel mutated in Dent's disease, colocalizes with the proton pump in endocytotically active kidney cells. Proc. Natl Acad. Sci. USA 95(14), 8075–8080.

[91] Devuyst, O., Christie, P.T., Courtoy, P.J., Beauwens, R., Thakker, R.V., 1999. Intra-renal and subcellular distribution of the human chloride channel, CLC-5, reveals a pathophysiological basis for Dent's disease. Hum. Mol. Genet. 8(2), 247–257.

[92] Reenstra, W.W., Sabolic, I., Bae, H.R., Verkman, A.S., 1992. Protein kinase A dependent membrane protein phosphorylation and chloride conductance in endosomal vesicles from kidney cortex. Biochemistry 31(1), 175–181.

[93] Schmid, A., Burckhardt, G., Gogelein, H., 1989. Single chloride channels in endosomal vesicle preparations from rat kidney cortex. J. Membr. Biol. 111(3), 265–275.

[94] Mo, L., Hellmich, H.L., Fong, P., Wood, T., Embesi, J., Wills, N.K., 1999. Comparison of amphibian and human ClC-5: similarity of functional properties and inhibition by external pH. J. Membr. Biol. 168, 253–264.

[95] Kornak, G., Kasper, D., Boesl, M.R., Kaiser, E., Schweizer, M., Schulz, A., et al., 2001. Loss of the ClC-7 chloride channel leads to osteopetrosis in mice and man. Cell 104, 205–215.

[96] Uchida, S., Sasaki, S., Nitta, K., Uchida, K., Horita, S., Nihei, H., Marumo, F., 1995. Localization and functional characterization of rat kidney-specific chloride channel, ClC-K1. J. Clin. Invest. 95, 104–113.

[97] Yoshikawa, M., Uchida, S., Yamauchi, A., Miyai, A., Tanaka, Y., Sasaki, S., Marumo, F., 1999. Localization of rat CLC-K2 chloride channel mRNA in the kidney. Am. J. Physiol. 276(4 Pt 2), F552–F558.

[98]   Matsumura, Y., Uchida, S., Kondo, Y., Miyazaki, H., Ko, S.B., Hayama, A., Morimotoooo, T., Liu, T., Liu, T., Liu, T., Liu, W.,Arisawa, M., Sasaki, S., Marumo, F., 1999. Overt nephrogenic diabetes insipidus in mice lacking the CLC-K1 chloride channel. Nat. Genet. 21(1), 95–98.

[99]   Simon, D.B., Bindra, R.S., Mansfield, T.A., Nilson-Williams, C., Mendonca, E., Stone, R., Schurman, S., Nayir, A., Alpay, H., Bakkaloglu, A., Rodriguez-Soriano, J., Morales, J.M., Sanjad, S.A., Taylor, C.M., Pilz, D., Brem, A., Trachtman, H., Griswold, W., Richard, G.A., John, E., Lifton, R.P., 1997. Mutations in the chloride channel gene, *CLCNKB*, cause Bartter's syndrome type III. Nat. Genet. 17, 171–178.

[100] Thomsen, J., 1876. Tonische Krämpfe in willkürlich beweglichen Muskeln in Folge von ererbter psychischer Disposition. Arch. Psychiat. Nervenkr. 6, 702–718.

[101] Becker, P.E., 1973. Generalized non-dystrophic myotonia. In: Desmedt, J.E. (Ed.), New Developments in Electromyography and Clinical Neurophysiology. S. Karger, Basal, pp. 407–412.

[102] Bryant, S.H., 1962. Muscle membrane of normal and myotonic goats in normal and low external chloride. Fed. Proc. 21, 312.

[103] Bryant, S.H., Morales-Aguilera, A., 1971. Chloride conductance in normal and myotonic muscle fibres and the action of monocarboxylic aromatic acids. J. Physiol. (Lond.) 219, 361–383.

[104] Rüdel, R., Lehmann-Horn, F., 1985. Membrane changes in cells from myotonia patients. Physiol. Rev. 65, 310–356.

[105] Bryant, S.H., 1982. Physical basis of myotonia. In: Schotland, D.L. (Ed.), Disorders of the Motor Unit. Wiley, New York, pp. 381–389.

[106] Koch, M.C., Steinmeyer, K., Lorenz, C., Ricker, K., Wolf, F., Otto, M., et al., 1992. The skeletal muscle chloride channel in dominant and recessive human myotonia. Science 257, 797–800.

[107] Steinmeyer, K., Lorenz, C., Pusch, M., Koch, M.C., Jentsch, T.J., 1994. Multimeric structure of ClC-1 chloride channel revealed by mutations in dominant myotonia congenita. EMBO J. 13, 737–743.

[108] Heine, R., George, A.L. Jr, Pika, U., Deymeer, F., Rüdel, R., Lehmann-Horn, F., 1994. Proof of a non-functional muscle chloride channel in recessive myotonia congenita (Becker) by detection of a four base-pair deletion. Hum. Mol. Genet. 3, 1123–1128.

[109] Wagner, S., Deymeer, F., Kürz, L.L., Benz, S., Schleithoff, L., Lehmann-Horn, F., et al., 1998. The dominant chloride channel mutant G200R causing fluctuating myotonia: Clinical findings, electrophysiology, and channel pathology. Muscle Nerve 21(9), 1122–1128.

[110] Lehmann-Horn, F., Mailander, V., Heine, R., George, A.L. Jr, 1995. Myotonia levior is a chloride channel disorder. Hum. Mol. Genet. 4, 1397–1402.

[111] Zhang, J., Bendahhou, S., Sanguinetti, M.C., Ptacek, L.J., 2000. Functional consequences of chloride channel gene (CLCN1) mutations causing myotonia congenita. Neurology 54(4), 937–942.

[112] Jentsch, T.J., Lorenz, C., Pusch, M., Steinmeyer, K., 1995. Myotonias due to ClC-1 chloride channel mutations. Soc. Gen. Physiologists Ser. 50, 149–159.

[113] Beck, C.L., Fahlke, Ch, George, A.L. Jr, 1996. Molecular basis for decreased muscle chloride conductance in the myotonic goat. Proc. Natl Acad Sci. USA 93, 11248–11252.

[114] Kubisch, C., Schmidt-Rose, T., Fontaine, B., Bretag, A.H., Jentsch, T.J., 1998. ClC-1 chloride channel mutations in myotonia congenita: variable penetrance of mutations shifting the voltage dependence. Hum. Mol. Genet. 7(11), 1753–1760.

[115] Rhodes, T.H., Vite, C.H., Giger, U., Patterson, D.F., Fahlke, Ch., George, A.L., 2000. A missense mutation in canine ClC-1 causes recessive myotonia congenita in the dog. FEBS Lett. 456, 54–58.

[116] Meyer-Kleine, C., Steinmeyer, K., Ricker, K., Jentsch, T.J., Koch, M.C., 1995. Spectrum of mutations in the major human skeletal muscle chloride channel gene (CLCN1) leading to myotonia. Am. J. Hum. Genet. 57(6), 1325–1334.

[117] Wrong, O.M., Norden, A.G., Feest, T.G., 1994. Dent's disease; a familial proximal renal tubular syndrome with low-molecular-weight proteinuria, hypercalciuria, nephrocalcinosis, metabolic bone disease, progressive renal failure and a marked male predominance. QJM 87(8), 473–493.

[118] Scheinman, S.J., 1998. X-linked hypercalciuric nephrolithiasis: Clinical syndromes and chloride channel mutations. Kidney Int. 53(1), 3–17.

[119] Lloyd, S.E., Günther, W., Pearce, S.H.S., Thomson, A., Bianchi, M.L., Bosio, M., Craig, I.W., Fisher, S.E., Scheinman, S.J., Wrong, O., Jentsch, T.J., Thakker, R.V., 1997. Characterisation of renal chloride channel, CLCN5, mutations in hypercalciuric nephrolithiasis (kidney stones) disorders. Hum. Mol. Genet. 6(8), 1233–1239.

[120] Akuta, N., Lloyd, S.E., Igarashi, T., Shiraga, H., Matsuyama, T., Yokoro, S., Cox, J.P., Thakker, R.V., 1997. Mutations of CLCN5 in Japanese children with idiopathic low molecular weight proteinuria, hypercalciuria and nephrocalcinosis. Kidney Int. 52(4), 911–916.

[121] Lloyd, S.E., Pearce, S.H.S., Fisher, S.E., Steinmeyer, K., Schwappach, B., Scheinman, S.J., Harding, B., Bolino, A., Devoto, M., Goodyer, P., Rigden, S.P., Wrong, O., Jentsch, T..J., Craig, I.W., Thakker, R.V., 1996. A common molecular basis for three inherited kidney stone diseases. Nature 379, 445–449.

[122] Wall, D.A., Maack, T., 1985. Endocytic uptake, transport, and catabolism of proteins by epithelial cells. Am. J. Physiol. 248, C12–C20.

[123] Christensen, E.I., Nielsen, S., 1991. Structural and functional features of protein handling in the kidney proximal tubule. Semin. Nephrol. 11, 414–439.

[124] Gekle, M., Mildenberger, S., Freudinger, R., Silbernagl, S., 1995. Endosomal alkalinization reduces Jmax and Km of albumin receptor-mediated endocytosis in OK cells. Am. J. Physiol. 268(5 Pt 2), F899–F906.

[125] Al-Awqati, Q., Barasch, J., Landry, D., 1992. Chloride channels of intracellular organelles and their potential role in cystic fibrosis. J. Exp. Biol. 172, 245–266.

[126] Fahlke, Ch., 2000. Dent's disease: an hereditary nephrolithiasis caused by dysfunction of a voltage-gated chloride channel. In: Lehmann-Horn, F., Jurkatt-Rott, K. (Eds.), Ion channelopathies. Elsevier Science, pp. 251–272.

[127] Igarashi, T., Gunther, W., Sekine, T., Inatomi, J., Shiraga, H., Takahashi, S., Suzuki, J., Tsuru, N., Yanagihara, T., Shimazu, M., Jentsch, T.J., Thakker, R.V., 1998. Functional characterization of renal chloride channel, CLCN5, mutations associated with Dent's Japan disease. Kidney Int. 54(6), 1850–1856.

[128] Lloyd, S.E., Pearce, S.H.S., Günther, W., Kawaguchi, H., Igarashi, T., Jentsch, T.J., Thakker, R.V., 1997. Idiopathic low molecular weight proteinuria associated with hypercalciuric nephrocalcinosis in Japanese children is due to mutations of the renal chloride channel (CLCN5). J. Clin. Invest. 99(5), 967–974.

[129] Morimoto, T., Uchida, S., Sakamoto, H., Kondo, Y., Hanamizu, H., Fukui, M., Tomino, Y., Nagano, N., Sasaki, S., Marumo, F., 1998. Mutations in CLCN5 chloride channel in Japanese patients with low molecular weight proteinuria. J. Am. Soc. Nephrol. 9(5), 811–818.

[130] Morimoto, T., Uchida, S., Ito, H., Hata, M., Hiroe, M., Marumo, F., 1997. A mutation of human CLCN5 chloride channel gene in patients with idiopathic low-molecular weight proteinuria. J. Am. Soc. Nephrol. 7, 1617–1617.

[131] Simon, D.B., Karet, F.E., Rodriguez-Soriano, J., Hamdan, J.H., DiPietro, A., Trachtman, H., et al., 1996. Genetic heterogeneity of Bartter's syndrome revealed by mutations in the K^+ channel, ROMK. Nat. Genet. 14(2), 152–156.

[132] Simon, D.B., Karet, F.E., Hamdan, J.M., DiPietro, A., Sanjad, S.A., Lifton, R.P., 1996. Bartter's syndrome, hypokalaemic alkalosis with hypercalciuria, is caused by mutations in the Na-K-2Cl cotransporter NKCC2. Nat. Genet. 13(2), 183–188.

Steroid modulation of GABA$_A$ receptors: from molecular mechanisms to CNS roles in reproduction, dysfunction and drug abuse

Leslie P. Henderson and Juan Carlos Jorge

Department of Anatomy, University of Puerto Rico, San Juan, PR 00936

1. Introduction

Gamma-aminobutyric acid type A (GABA$_A$) receptors are widely expressed throughout the vertebrate central nervous system (CNS) and subserve an essential role in mediating fast synaptic inhibitory transmission. Moreover, the GABA$_A$ receptor is the primary molecular target for a broad range of therapeutic and abused drugs, as well as a number of nervous system toxins. Because of their central importance in mediating so many aspects of CNS function and dysfunction, these neurotransmitter receptors have been the subject of many excellent recent reviews. Therefore, in this chapter, we will only highlight key aspects of GABA$_A$ receptor structure, function, and expression that have been reviewed more extensively elsewhere and instead focus primarily on how allosteric modulation of GABA$_A$ receptors by derivatives of gonadal steroids regulates normal CNS processing, as well as the role it may play in specific disorders and in drug abuse.

2. The GABA$_A$ receptor gene family

Gamma-aminobutyric acid (GABA) is the major inhibitory neurotransmitter in the adult mammalian CNS. This ubiquitous neurotransmitter binds to two structurally distinct classes of receptors: ion channel proteins, which include the GABA$_A$ and GABA$_C$ receptors, and the G protein-coupled GABA$_B$ receptor. GABA$_A$ receptors are members of the evolutionarily related superfamily of ligand-gated ion channels that includes the nicotinic acetylcholine receptor and the glycine receptor families [1–3], and the 5-HT$_3$ subclass of receptors gated by serotonin [4]. As with the other members of this superfamily, GABA$_A$ receptors are pentameric proteins [5] in which an ion selective pore and the binding sites for ligands are part of the same molecule [6]. The first GABA$_A$ receptor cDNAs were cloned in 1987 [7]. Since that time, the number of subunit genes identified in mammals has grown apace, with 19 different members (α_1–α_6, β_1–β_3, γ_1–γ_3, δ, ϵ, π, ρ_{1-3} and θ) now identified [8–10]. Designation in subunit classes has been

Advances in Molecular and Cell Biology, Vol. 32, pages 219–250
ISSN: 1569-2558/DOI: 10.1016/S1569-2558(03)32010-7

based on amino acid identity, which ranges from 30–50% between subunit families to 60–80% within a given family [10,11]. Structural heterogeneity is further enhanced by the existence of additional subunit genes for the β_4 [12] and the γ_4 [13] subunits in chick and by mRNA splice variants encoding alternative forms of the γ_2 [14–16], β_2 [17,18], β_3 [19], β_4 [12], and α_6 [20] subunits. It has been a point of contention as to whether or not receptors containing ρ subunits are truly GABA$_A$ receptors. With respect to sequence homology, there is no doubt that the ρ subunit class fits well within the GABA$_A$ receptor family [5]. However, the distinct pharmacology of ρ-containing receptors, as well as the fact that homo-oligomeric ρ-containing receptors form channels with properties that reflect those of native receptors in central visual pathways, have led some investigators to believe ρ-containing receptors would be better designated as a separate class of GABA-gated ion channels with the appellation of GABA$_C$ receptors [21]. In addition, it should be noted that π subunit mRNA has been detected in very limited regions of the CNS and only following a high number of cycles of PCR amplification [22]. Therefore, whether receptors containing the π subunit contribute significantly to CNS function remains to be determined.

A truly staggering array of pentameric combinations can be formed if the subunits described above are mixed and matched without restriction. However, nature has been somewhat more parsimonious, and while the exact stoichiometry of different subunit isoforms within native receptors has not been determined directly [10,23], the preponderance of biochemical and immunocytochemical data supports the assertion that native GABA$_A$ receptors are hetero-oligomeric, that most receptors contain two α, two β and a γ subunit, and that the most commonly expressed isoform consists of α_1, β_2 or β_3, and γ_2 subunits [23]. Less abundant subunits, such as γ_1, γ_3, δ, or ϵ may substitute for or be coexpressed with $\alpha\beta\gamma_2$ subunits [5], although specific restrictions for particular subunit combinations within a single receptor have been described. For example, while the splice variants of the γ_2 subunit (γ_{2Short} and γ_{2Long}) may coexist in the same receptor [24], γ_2 or γ_3 subunits are not coassembled with the γ_1 subunit [25]. Similarly, nonoverlapping patterns of regional expression suggest that δ subunits substitute for γ subunits rather than being coexpressed within the same receptor protein [26]. Finally, while the $\alpha_1\beta_{2/3}\gamma_2$-containing receptor is the most prevalent in the adult brain, it is by no means ubiquitous, and there is both tremendous developmental and regional diversity in the expression of less commonly expressed subunits which gives rise to a correspondingly high level of functional complexity (see below).

3. The GABA$_A$ receptor ion channel

The GABA$_A$ receptor is a transmembrane protein that forms a pore selective for the passage of anions, and physiologically relevant fluxes can be attributed to bicarbonate and chloride ions with a relative permeability ratio of bicarbonate to chloride of 0.2–0.3 to 1 [27]. While GABA$_A$-mediated synaptic potentials can be depolarizing or hyperpolarizing, depending on the chloride equilibrium potential, the generalization can be made that transmission mediated by the GABA$_A$ receptor is excitatory during early development, but provides the major mechanism of fast inhibition in the adult brain [8,27,28]. Although exact values will vary with experimental conditions and cell type, the majority of single channel analyses of GABA$_A$ receptors activated with low agonist concentrations under equilibrium conditions indicate that these channels have a main conductance state of ~25–30 pS (with less frequently observed openings corresponding to conductances of approximately 15–19 and 12 pS), apparent open duration distributions that are usually best described by two components (τ_1 and τ_2) of ~0.4–1.0 ms and 4–12 ms, respectively, and burst durations of 6–50 ms [29–45]. Recent reports suggest that openings corresponding to the lowest conductance state (12 pS) may arise from receptors comprising only α and β subunits (γ-less receptors), while transitions to the 15–19 pS state may reflect a substate of the predominant receptor containing α, β and γ subunits [44,45].

As with single channel analyses, the kinetics of macroscopic synaptic responses will vary depending upon biologically relevant parameters, including region-specific differences in subunit composition, age of the animal, and post-translational modifications of the receptor, as well as with experimental variables, such as temperature, holding potential, and ionic conditions. These variables impose limitations in making generalizations. Nonetheless, review of studies examining inhibitory postsynaptic currents (IPSCs) recorded using physiological salines and at room temperature from a multiplicity of brain regions including the cerebellum [46–50], hippocampus [51–55], thalamus [56], hypothalamus [57–60], cerebral cortex [61], spinal cord [62], septum [63], and pituitary [37], indicate that for most neurons, synaptic current decays are best described by two exponential components with time constants which range from 2 to 20 ms (τ_1) and 13–60 ms (τ_2) or, less frequently, by a monoexponential rate of decay described by a single time constant ranging from 10 to 21 ms.

In contrast to the classical synapse of the cholinergic neuromuscular junction, the duration and waveform of GABA$_A$-mediated synaptic currents in the CNS are not well described simply by summating single channel open durations. While the decay of synaptic currents is indeed influenced by the single channel open time, it has also been proposed that GABA$_A$ receptor channels enter a long-lived (but transmitter bound) desensitized state from which they can re-enter an open state long after the transmitter has been cleared from the synaptic cleft. The upshot of these late re-openings is thus to prolong the duration of the macroscopic current [48,49,64–67]. The ability of several classes of sedative/hypnotic and anesthetic allosteric modulators, including specific steroid modulators (see below), to enhance receptor desensitization and thus prolong the inhibitory synaptic current by promoting these late openings is believed to be a principal mechanism underlying the clinical efficacy of these drugs [64]. In addition, by altering desensitization, these modulators can also alter the integration of multiple synaptic responses in a postsynaptic cell. Specifically, enhanced desensitization may act to diminish amplitudes of sequential events elicited above a given frequency, and thus determine how faithfully postsynaptic cells can follow presynaptic inputs [65].

4. Allosteric modulation of GABA$_A$ receptors

As suggested above, in addition to their pivotal role in transducing inhibitory chemical transmission, GABA$_A$ receptors are the molecular targets of an extraordinarily diverse range of therapeutic drugs, toxins, and endogenous hormones that includes anxiolytic benzodiazepines (BZs) and related compounds, sedative/hypnotic barbiturates and neurosteroids, anticonvulsants, convulsants (including a number of insecticides), general anesthetics, ethanol, and zinc [5,8,68–70]. In addition, drugs long known to work via other molecular mechanisms, including the antipsychotic phenothiazines [71,55] and the anabolic androgenic steroids (AAS) [60,72–74], have also been shown recently to alter channel function via allosteric modulation of the GABA$_A$ receptor complex. While heterogeneity in the subunit composition of the GABA$_A$ receptor has repercussions for basic aspects of channel function, including receptor affinity, single channel conductance, and the kinetics of deactivation and desensitization [75], receptors comprising different subunit isoforms show the most striking variability in their responsiveness to these allosteric modulators. Without doubt, the structural basis underlying differential modulation is best understood for compounds acting at the BZ site [5,8,68,76]. In brief, both α and γ subunits are required for high affinity BZ site modulation, with specific isoforms of the α and the γ subunit classes determining both the degree of modulation and whether compounds enhance or diminish GABA-elicited responses. Beyond establishing how variability in subunit isoforms governs the effects mediated by BZ site compounds, critical amino acid residues within individual subunits that are required for BZ modulation have also been identified [5,8,68,76].

5. Steroid modulators of the GABA$_A$ receptor

Compounds classified as neurosteroids or neuroactive steroids [77] encompass both natural and synthetic derivatives of the gonadal steroids, progesterone and testosterone, and the adrenal steroid, deoxycorticosterone [78,79]. Naturally occurring steroid derivatives are synthesized from cholesterol in both peripheral organs and in the CNS by glial cells (and to a lesser extent by neurons). Biosynthetic pathways for these neurosteroids which have been well delineated and described in detail in recent reviews [78,79] are summarized in Fig. 1. Although there are still some ambiguities in categorically establishing the presence of specific requisite enzymes within the CNS, as well as questions concerning quantitative measurements of steroid synthesis and degradation in the brain [78,79], it is established that a host of steroid metabolites derived from endogenous sources are present in the brain and have significant physiological effects on neural function [68,69,78,79]. Moreover, levels of these compounds in the brain have been shown to vary significantly during different developmental epochs, between the sexes, amongst different brain regions, and as a function of hormonal state [79,80–84]. For example, the sulfate derivatives of pregnanolone (5β-pregnan-3α-ol-20-one) and dehydroepiandrosterone (3α-ol-5-andros-tene-17-one), which are pregnanolone sulfate (PS) and dehydroepiandrosterone sulfate (DHEAS), have been detected in human brain tissues at concentrations that are several fold higher than in plasma and are present at different levels in women than in men [85–87]. In addition, levels of the allosteric agonists, tetrahydro-progesterone (5α-pregnan-3α-ol-20-one; 3α-OH-DHP; 3α,5αTHP, or allopregnanolone)

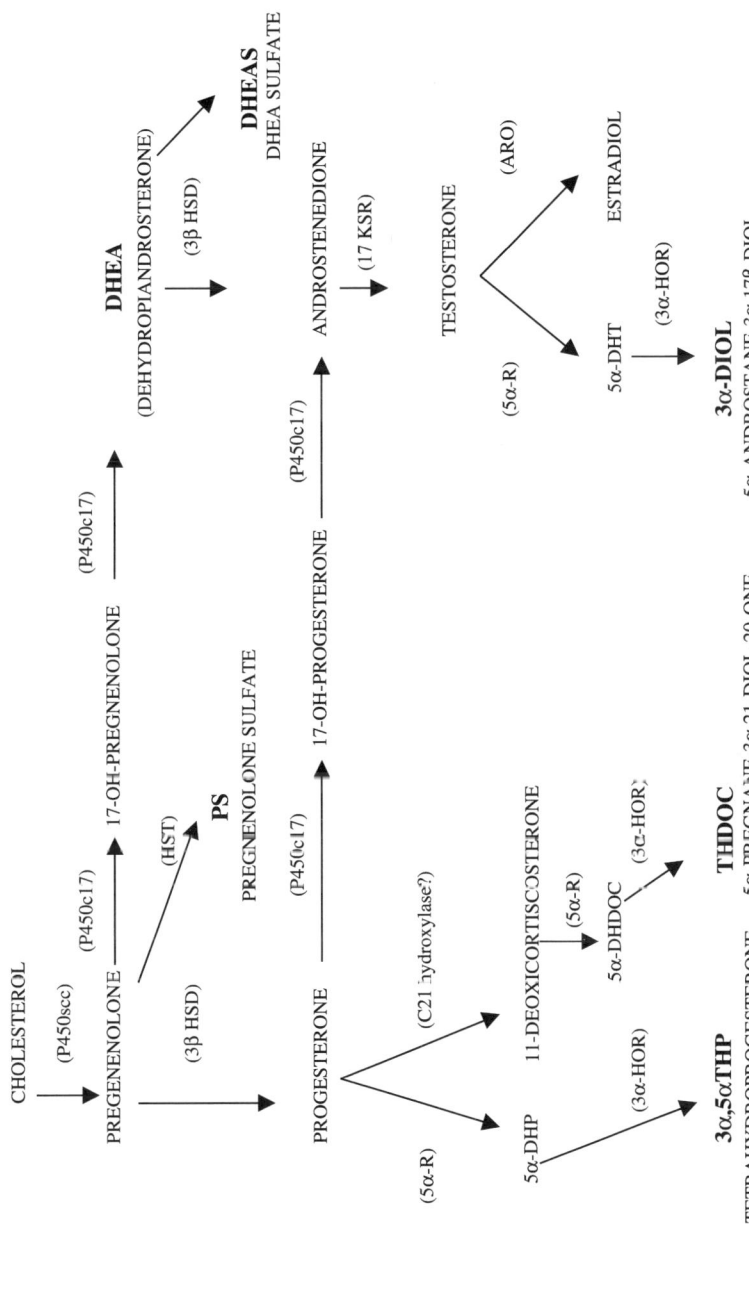

Fig. 1. Schematic summary of the biosynthetic pathways believed to generate the neurosteroids active in the vertebrate nervous system. Enyzmes (in parentheses) are given the following abbreviations: P450scc: side chain cleavage cytochrome P450; hydroxysteroid sulfotransferase (HST); 3β HSD: 3β hydroxysteroid dehydrogenase; P450c17: microsomal hydroxylase at C17; C21 hydroxylase ?: enzyme not definitively identified; 5α-R: 5α reductase; 3αHOR: 3α hydroxysteroid oxidoreductase; ARO: aromatase: converts C19 androgens to C18 estrogens; 17 KSR: 17-ketosteroid reductase. Neuroactive steroids discussed in the text are indicated in boldface.

and 5α-pregnane-3,20-dione (5α-DHP), show significant differences during the menstrual cycle and amongst brain regions [83]. Therefore, while the steroid modulators share therapeutic importance with the BZ compounds as anxiolytics and sedative/hypnotics, the endogenous variations in neurosteroid production imbues this class of modulators with an additional impact with respect to understanding both normal physiological function and pathophysiology within the CNS.

Beyond the *physiological* effects that may derive from transient fluctuations in neurosteroid modulation, recent studies indicating that the AAS can also allosterically modulate GABA$_A$ receptors [60,72–74] suggest that modulation of GABAergic transmission by these synthetic compounds may also contribute to the untoward psychological effects associated with supraphysiological levels of steroids that occur with steroid abuse. Thus, aberrant modulation of the GABA$_A$ receptor may be a common mechanism underlying the actions of a number of abused drugs that include the AAS, ethanol, and the barbiturates [8,68]. Although some steroid derivatives, in particular the AAS, can signal via classical nuclear hormone receptors [81,88], the primary mechanism of action of the majority of these compounds is allosteric modulation of GABA$_A$ and glutamate receptors [68,69,78–81]. Thus, defining the molecular mechanisms by which different steroid modulators alter GABAergic transmission, as well as identifying differences in the actions of these compounds as a function of brain region, age, sex, and hormonal state, has broad ramifications for understanding neural function and dysfunction.

6. Positive allosteric neurosteroid modulators of the GABA$_A$ receptor

Functionally, allosteric steroids can be divided into two general classes: those that mimic and enhance the effects of GABA, and those that antagonize GABA [68,69,80]. Of the plethora of neurosteroids known to act in the brain, the best studied positive modulators are the endogenous metabolites, 3α5αTHP, THDOC (5α-pregnane-3α,21-diol-20-one), 3α-diol (5α-androstane-3α,17β-diol), and androsterone (5α-androstan-3α-ol-17-one) (Fig. 1), as well as the synthetic steroid, alphaxalone (5α-pregnane-3α-ol-11,20-dione) that was the first steroid characterized as an allosteric modulator of the GABA$_A$ receptor [69]. Consistent with their role in augmenting GABAergic transmission, positive modulators act as sedative/hypnotics, anxiolytics and anticonvulsants in the adult brain [68,69,80]. These steroids potentiate both BZ and muscimol binding to the GABA$_A$ receptor, whereas they inhibit binding of *t*-butyl bicyclophosphorothionate (TBPS) [89–92], a caged convulsant that binds to the picrotoxin site of the receptor [68]. The weight of evidence supports the assertion that many neurosteroids induce their modulation via a specific binding site and interactions with selective residues of the GABA$_A$ receptor [92–96], while others may act to induce membrane perturbations that indirectly alter channel function [94–98]. A number of studies, however, support the hypothesis that there are multiple steroid recognition sites on the GABA$_A$ receptor protein [95,99–101] and that these sites are distinct from those of other well-characterized modulators of the GABA$_A$ receptor [68,69]. Support for the existence of specific binding sites has come from studies indicating that steroids must have specific structural entities in order to elicit a functional effect. Initial structure–activity studies indicated that a 5α or 5β-reduced pregnane or androstane skeleton with a 3α-OH and a ketone at C20 for pregnanes and C17 for androstanes was essential for positive modulatory effects [69,93]. However, more recent studies, especially those utilizing unnatural enantiomers and steroid-like benze[*e*]indenes [95,96,102], have provided numerous exceptions to this initial conclusion [69]. Finally, while it is evident that there are indeed specific interactions between selective steroids and the GABA$_A$ receptor, experimental limitations have made the exact identity of these binding sites difficult to pin down.

With respect to channel function, positive neurosteroid modulators, at nanomolar to low micromolar concentrations, act to prolong the duration of GABA$_A$ receptor-dependent macroscopic currents [35,43,54,60,66,102–107]. Under conditions when the concentration of GABA does not saturate available receptors, these modulators also enhance peak current amplitude [38,60,80,89,105,108]. In addition, while not related to modulation of postsynaptic GABA$_A$ receptors, neurosteroid modulators have also been reported to increase the frequency of presynaptic release ([58,109,110]; but see also Refs. [54,60,102]). While the end result of modulation by positive neurosteroid modulators is similar to that induced by the BZs and barbiturates, the single channel mechanisms by which these steroids induced their functional effects differ from that of these other widely used GABA$_A$ receptor modulators [11,68,69]. Single channel and noise analysis studies performed under steady state application of low concentrations of GABA indicate that positive steroid modulators augment channel burst durations [35,39, 90,108,111] by increasing the opening frequency (and therefore the relative proportion of long duration single channel events) without concomitant changes in the open duration time constants themselves [35,39]. Analysis of single channel currents elicited under nonequilibrium conditions by brief pulses of millimolar concentrations of GABA (conditions believed to mimic GABA kinetics in the synaptic cleft) indicate that positive neurosteroids increase the frequency of late channel openings [66]. This finding is consistent with the hypothesis that positive neurosteroid modulators act to slow the rate of recovery from desensitization [66], thus promoting late re-entry into an open state and prolongation of the GABA-elicited macroscopic responses [64–66]. One final facet that functionally distinguishes the neurosteroids from other positive allosteric modulators is their ability, when present at higher concentrations (>1 μM), to activate the receptor directly in the absence of GABA, a functional characteristic that is not shared by the BZs [68,69] or the AAS [60,74]. Finally, it should be noted that while allosteric modulation is believed to reflect a direct interaction of a specific compound with binding sites on the receptor, post-translational modifications of specific GABA$_A$ receptor subunits may influence the ability of neurosteroids to modulate receptor function. In particular, the GABA$_A$ receptor is known to be phosphorylated by a number of protein kinases [112], including protein kinase C [PKC; 113–115]. Recent studies have proposed that constitutive phosphorylation by PKC is required for binding and modulation induced by 3α,5αTHP [107], and that PKC may enhance the effects of positive, but not negative, neurosteroid modulators by altering receptor desensitization [116,117].

7. Negative allosteric neurosteroid modulators of the GABA$_A$ receptor

Negative allosteric neurosteroid modulators act as noncompetitive antagonists of the GABA$_A$ receptor, and, as would be expected with a diminution of GABA-mediated inhibition, these compounds enhance excitation and produce convulsant effects [68,69, 80]. For many of the positive modulators, addition of a sulfate moitey at C3 flips the effect of the steroid from agonist to antagonist [101]. The best characterized of these antagonists include PS and DHEAS (Fig. 1). Consistent with their role as noncompetitive antagonists, both PS and DHEAS interact with the TBPS/picrotoxin binding site of the receptor [118, 119] and inhibit ^{36}Cl$^-$ uptake into synaptoneurosomes [68,69,80]. As with positive steroid modulators, negative modulators also influence the binding of other GABA$_A$ receptor ligands, but do so in a more complex fashion than their agonist counterparts. Specifically, it has been shown that PS modulates ^{3}H-muscimol binding in a bimodal fashion, slightly potentiates BZ binding, and inhibits ^{35}S-TBPS binding [80,120,121]. Moreover, although both PS and DHEAS antagonize GABA effects, they differ in their capacity to displace ^{35}S-TBPS and to enhance ^{3}H-BZ binding [121,122].

With respect to mechanisms underlying their ability to modulate GABA$_A$ receptor currents, some negative modulators, like some positive modulators, appear to act directly via specific interactions with the channel protein and some indirectly via membrane perturbations to alter channel function [95]. Moreover, even for a single negative modulator, such as DHEAS, multiple binding sites have been inferred [123]. While addition of a sulfate group to pregnanolone or DHEA to form PS and DHEAS does indeed reverse the effect of modulation from positive to negative, the presence of a sulfate group per se does not dictate that a modulator will induce negative modulation. Rather the orientation of the sulfate moiety (α or β) on the A ring of the steroid determines whether the 3α-OH steroid sulfates potentiate or antagonize the action of GABA [124]. Finally, it appears that it is the negative charge at C3, rather than the sulfate group itself that is required for modulators to elicit inhibitory effects [101].

With respect to channel function, negative neurosteroid modulators act to reversibly inhibit peak GABA-elicited currents in a dose-dependent manner [95,101,105,109,120, 122,123,125–127]. Single channel analyses indicate that both PS [118] and DHEAS [125] decrease channel opening frequency without altering open or burst durations. Consistent with their effects on opening frequency and peak current amplitudes, both PS [127] and DHEAS [125] have been reported to increase the entry of receptors into the desensitized state, although arguments have also been made that the enhanced desensitization induced by DHEAS reflects activity arising from a separate and structurally distinct receptor population [125].

8. Structural correlates underlying neurosteroid modulation of the GABA$_A$ receptor

While the concept that steroids can induce modulation of GABA$_A$ receptor function via interactions with a specific binding site is well accepted as indicated above, identification of specific molecular structures required for this modulation has been more elusive. The prevailing tenet is that unlike the BZs and other receptor modulators, the actions of neurosteroids do not depend as absolutely on inclusion of specific subunit isoforms within the receptor for binding or functional effects. For example, potentiation of GABA$_A$ receptor-activated currents can be elicited from recombinant receptors expressed in human embryonic kidney (HEK293) cells by both 3α,5αTHP and THDOC, not only from receptors comprising α$_1$β$_1$γ$_2$ subunits, but also from those composed of only α$_1$β$_1$ or even β$_1$ subunits alone [35]. In this sense, neurosteroids can be regarded as "promiscuous" allosteric modulators of the GABA$_A$ receptor.

In spite of a lack of absolute requirements for a specific subunit to confer modulation by neurosteroids, differences in subunit composition have been reported to influence the degree of modulation induced by these compounds. Studies of recombinant receptors expressed in heterologous cells indicate that α subunit composition can influence the efficacy and potency of both positive modulators (including 3α,5αTHP, THDOC, and alphaxalone) and the negative modulators, PS and DHEAS [128–140]. However, the effects of α subunit substitutions are often subtle, and disparate results have been obtained from different laboratories (see references above). This lack of a clear consensus with respect to the role of the α subunits in determining the extent or sensitivity of modulation by neurosteroids has arisen, at least in part, from differences in experimental approaches. For example, both 3α,5αTHP and THDOC have been reported to enhance currents from recombinant α$_1$-containing receptors to a greater extent than from α$_6$-containing receptors [130,134,139]. However, the differences in neurosteroid potentiation for α$_1$- versus α$_6$-containing receptors has also been reported to vary with GABA concentration, such that potentiation by 3α,5αTHP is greater for α$_1$- versus α$_6$-containing receptors at micromolar concentrations of GABA, but the two receptor subtypes show equivalent potentiation at submicromolar concentrations of GABA [135]. Similarly, experimental preparations will limit interpretations of data. For example, while *Xenopus* oocytes are an excellent preparation for examining structure–function relationships and subunit specificities for neurosteroid modulation, the large size of the oocyte limits both voltage clamp control and rates of agonist application. These restrictions become particularly important in light of the fact that a major mechanism by which neurosteroids induce their effects is via changing rates of entry and recovery from desensitized states [66,127].

While fewer studies have been performed examining the role of different β isoforms, no apparent differences in modulation by either positive or negative neurosteroids have been uncovered for recombinant receptors differing in β subunit composition [100,141,142]. However, β subunit composition can regulate the extent of phosphorylation of the GABA$_A$ receptor [143], and phosphorylation, in turn, alters neurosteroid sensitivity [107,116,117, 137]. Thus, within the context of native neurons in their natural milieu, β subunit composition may influence the extent of modulation induced by neurosteroids. Gamma subunit composition has been shown to determine the efficacy of the positive modulator, 3α, 5αTHP [130,144], but to have no effect on modulation by the negative modulator, PS [136].

9. Region-specific differences in GABA$_A$ receptor subunit expression and neurosteroid modulation

A number of specific GABA$_A$ receptor subunits, including α$_6$, α$_4$, δ, γ$_1$ [145], ε [146–148], ρ$_{1-3}$ [149,150], π [22], and θ [147,151], show highly restricted patterns of expression in the brain that suggests that $GABA_A$ receptors expressed within particular regions may have correspondingly restricted or specific functional properties. While α subunit composition has marked effects for BZ modulation of the $GABA_A$ receptor [5,8,68], steroid modulation is most dramatically altered by differential expression of δ or ε subunits. For the δ subunit, high levels of expression are restricted to the cerebellum, thalamus and hippocampus [23,145]. Initial studies characterizing the effects of δ subunit composition on neurosteroid modulation suggested that inclusion of a δ subunit diminished the sensitivity of the $GABA_A$ receptor to the positive neurosteroid modulators. Specifically, electrophysiological analysis of currents elicited from recombinant receptors indicated that inclusion of the δ subunit resulted in reduced sensitivity to modulation by THDOC [134]. In addition, inactivation of the $α_6$ subunit gene in a mutant mouse line results in post-translational loss of δ subunits [152] and diminished capacity of 3α,5αTHP to modulate TBPS binding [153]. Finally in studies of both cerebellar [134] and hippocampal [54] granule cells, it was suggested that as δ subunit expression increases during development, there is a concomitant decrease in sensitivity to positive neurosteroid modulators. However, a substantive number of more recent studies have now convincingly shown that inclusion of the δ subunit actually confers *enhanced* sensitivity to positive neurosteroid modulators. First, behavioral studies of δ subunit-knockout mice demonstrated that these animals show attenuated responses to alphaxalone in behavioral sleep time, loss of righting reflex, and plus maze assays [154]. These results are consistent with a recent study demonstrating that $GABA_A$ receptor-mediated spontaneous IPSCs from granule cells of δ subunit-knockout mice show a diminished prologation of current decay with THDOC application when compared to wild-type mice [155]. Second, electrophysiological studies of recombinant receptors have convincingly demonstrated that substitution of the δ for the $γ_2$ subunit in recombinant receptors dramatically enhances neurosteroid potentiation [138,139,156,157]. Thus, there is now general agreement that inclusion of the δ subunit produces a receptor highly sensitive to the action of positive neurosteroid modulators.

Expression of the ε subunit also shows a highly restricted pattern in the forebrain [146–148,158,159]. As with the δ subunit, initial functional studies provided disparate findings with regard to the role of the ε subunit in steroid sensitivity. Recombinant receptors comprising $α_1$, $β_3$, and ε subunits were not modulated by 3α,5αTHP in one study [158], while currents from receptors containing $α_1$, $β_1$, and ε subunits were enhanced by 3α,5αTHP in a separate report [146]. Recent data [160] suggest that these differences in neurosteroid sensitivity may arise from overexpression of the ε construct from the pCDM8 vector relative to expression of $α_1$ and $β_1$ subunits that subsequently results in expression of receptors with aberrant stoichiometry and thus altered neurosteroid sensitivity. However, in preliminary experiments, Jones and Henderson have found that while 3α,5αTHP does induce positive modulation of currents induced by steady state applications of 1 μM GABA to recombinant $α_2β_3ε$ receptors, the magnitude of this potentiation is significantly less than that elicited by 3α,5αTHP for $α_2β_3γ_2$ receptors (Brian L. Jones and Leslie P. Henderson, unpublished data). Moreover, while the influence of ε subunit composition has not been tested directly for native $GABA_A$ receptors (e.g., by antisense experiments), the ability of the neurosteroid, 3α-diol, to modulate synaptic currents in the medial preoptic area (mPOA) of the forebrain has been shown to vary inversely with the steady state levels of ε mRNA over the course of the estrous cycle in mice [161], a result consistent with the finding from studies of recombinant receptors indicating that inclusion of the ε subunit leads to diminished, if still significant, modulation of the GABA$_A$ receptor by positive neurosteroid modulators.

The γ_1 subunit also exhibits a highly restricted pattern of expression throughout development that is limited to a handful of interconnected forebrain regions that are known to be important in the regulation of neuroendocrine function [59,145,162,163] (Fig. 2). This distribution of the γ_1 subunit has led to the suggestion that transmission mediated by γ_1-containing GABA$_A$ receptors may be of particular importance in the generation of sexual and reproductive behaviors [59,162,163]. Several studies support this assumption. Expression of γ_1 subunits in forebrain, but not hippocampal, neurons is regulated over the course of the estrous cycle [164] and by the hormonal milieu [163]. Interestingly, estrous cycle-dependent changes in the levels of $3\alpha,5\alpha$THP are also restricted to hypothalamic regions [165]. In addition to the dynamic changes in neuronal γ_1 expression in neuroendocrine regions, it is known that γ_1 subunits are highly expressed in glial cells [162] and that forebrain glial GABA$_A$ receptors are modulated by

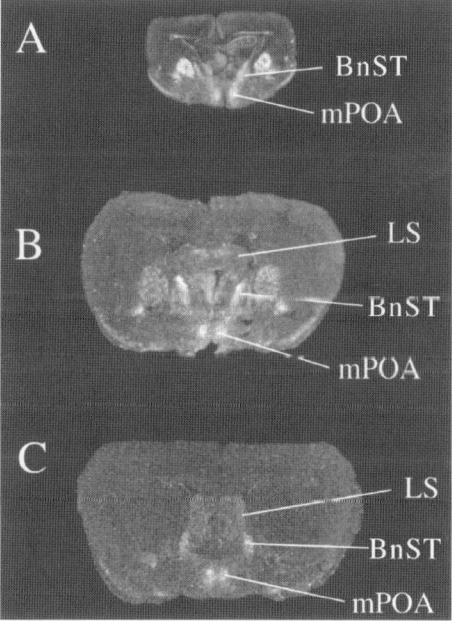

Fig. 2. Restricted expression of the GABA$_A$ receptor γ_1 subunit mRNA in the rat brain. Representative coronal sections from (A) postnatal day 0 (PN0), (B) PN14, and (C) adult rats demonstrating the distribution of γ_1 subunit mRNA as indicated by in situ hybridization with ^{35}S-labeled subunit specific oligonucleotides. Expression of the γ_1 mRNA is tightly regulated with high levels being restricted to the medial preoptic area (mPOA), bed nucleus of the stria terminalis (BnST), lateral septal nuclei (LS) and amygdala (not evident in this plane of section) at all developmental stages. At PN0, high levels of this subunit mRNA were also detected in the differentiating field of the globus pallidus; however this signal diminished to negligible levels by adulthood. Data are unpublished results from Henderson, L.P., Robinson, S., and Clark, A.S.

neurosteroids [32,105]. Since glial cells are the cellular sites for neurosteroid production in the CNS [78,79], expression of γ_1-containing receptors in these cells suggests the presence of a highly sensitive feedback mechanism in the hypothalamus and forebrain in which neurosteroids may govern their own production. Substitution of the γ_1 for the γ_2 subunit has been shown to double the efficacy (as assessed at 10 μM steroid), but has no appreciable effect on the potency (as assessed at 0.1 μM steroid) of 3α,5αTHP at $\alpha_1\beta_1\gamma_x$ recombinant receptors [130]. However, 10 μM concentrations of neurosteroids are well beyond the physiological range, and thus it is not clear if the activity of endogenous γ_1- versus γ_2-containing receptors would be differentially regulated under physiological conditions.

10. Anabolic-androgenic steroids: novel steroid modulators of the GABA$_A$ receptor

While modulation of GABA$_A$ receptors by both endogenous neurosteroids and their neuroactive synthetic derivatives has been studied extensively, it has only been in recent years that the actions of the AAS have been examined. AAS are synthetic derivatives of testosterone originally designed to provide enhanced anabolic potency with negligible androgenic effects [166,167]. Although first developed for clinical use, AAS administration is now predominantly one of abuse. Approximately 60 different AAS are available that vary in their chemical structure and thus in their metabolic fate and physiological effects [168–171]. Recent articles in the popular press have highlighted the rampant AAS abuse by professional athletes and the concomitant physiological, as well as psychological, effects that these drugs can produce. In particular, steroid abuse in both human subjects and in laboratory animals is associated with significant changes in female reproductive health, ranging from pubertal onset, fertility, and reproductive senescence to complex sexual behaviors [172–176]. AAS abuse also is reported to elicit detrimental changes in affect, including increased aggressive behavior, hypomania, hostility, dependency, acute psychosis and manic and/or depressive episodes in humans [175,176] and altered aggressive behaviors and anxiety in laboratory animals [173,177,178]. Moreover, recent data indicate that the more insidious abuse of these drugs is not among elite athletes, but among a growing number of adolescent boys and girls [167,175,179]. In spite of the increased use of these drugs and their detrimental effects on brain function, surprisingly little is known about how AAS act in the CNS. Recent data demonstrate that these steroids, much like other widely used psychoactive drugs such as the BZs and the neurosteroids, act to modulate signaling mediated by the GABA$_A$ receptor.

Acute exposure to AAS alters Cl$^-$ flux in synaptosomes, the binding of TBPS and BZs to the GABA$_A$ receptor [72,73], GABA$_A$ receptor-mediated synaptic currents in mammalian forebrain neurons [60], and currents elicited by direct application of GABA to recombinant receptors [74]. Modulation by AAS is distinct from that induced by the neurosteroids in that AAS at concentrations as high as 50 μM do not directly activate the receptor [60,74]. Moreover, binding studies indicate that in contrast to the neurosteroids which *enhance* BZ binding, the AAS, stanozolol and 17α-methyltestosterone, *decrease* BZ binding [72]. Finally, both the 17α-alkylated derivatives of testosterone, 17α-methyltestosterone and stanozolol, as well as the structurally distinct 19-nortestosterone derivative, nandrolone, were found to induce negative modulation of synaptic currents in the mPOA of the forebrain while the neurosteroids, 3α,5αTHP and 3α-diol, potentiated responses synaptic responses in this brain region [60,161].

Recent experiments suggest that the AAS act by a mechanism distinct from the neurosteroids and other allosteric modulators of the GABA$_A$ receptor. Specifically, in contrast to the neurosteroids which alter entry and exit from desensitized states [66], the AAS, 17α-methyltestosterone, was found to have no effect on either desensitization or recovery from desensitization for recombinant α$_1$β$_3$γ$_2$ receptors [74]. Kinetic modeling studies indicate that while effects of the AAS could be well described by altering receptor affinity, a facet of this modulation that might suggest a shared biophysical mechanism with the BZs, simultaneous fits for multiple data entries indicated that the effects of this AAS on α$_1$β$_3$γ$_2$ receptors were best described by enhancement of entry of singly liganded receptors into the open state [74]. Potentiation of entry into the singly liganded open state is also consistent with data from this study demonstrating that the AAS potentiate currents elicited by subsaturating concentrations of GABA, but had no appreciable effect on mIPSCs recorded from cerebellar Purkinje neurons where GABA concentrations are believed to reach near saturating levels [74]. Finally, in this study it was found that while the benzodiazepine site agonist, flumazenil, did not diminish the potentiation induced by 17α-methyltestosterone, preexposure to μM concentrations of zolpidem blunted the subsequent potentiation of currents elicited by low concentrations of GABA. This suggests that modulators, such as the BZs, that will promote rapid passage along the activation pathway, will "bypass" the ability of the AAS to promote singly liganded opening [74].

11. Steroid modulation of GABA$_A$ receptors in neuroendocrine control regions and the effect on reproductive behaviors

The expression of sexual and reproductive behaviors is regulated not only by steroid hormones, but also by a rich array of neurotransmitters and neuromodulators in the hypothalamus and forebrain, including GABA [180]. GABAergic transmission in selective hypothalamic/forebrain regions plays a central role in regulating the activity of gonadotropin releasing hormone (GnRH) neurons and the pulsatile gonadotropin secretion that is required for the onset of puberty and regular estrous cyclicity [181]. Quantitative ultrastructural analysis has shown that nearly half of all synaptic boutons within these neuroendocrine control regions are immunoreactive for GABA [182]. While many effects of the parent gonadal steroids on the GABA$_A$ receptor system most likely occur by nuclear hormone receptor-mediated regulation of GABA$_A$ receptor subunit genes and GABA synthetic enzymes [163,164,183,184], nongenomic actions of their neurosteroid derivatives, including allosteric modulation of GABA$_A$ receptors, are also implicated in playing a pivotal role in regulating these neuroendocrine behaviors [185–190]. For example, direct infusion of GABA$_A$ receptor agonists or antagonists into specific neuroendocrine control regions indicates that modulation of GABAergic transmission can elicit rapid and region-specific effects on the expression of sexual behaviors. Specifically, while infusion of GABA$_A$ agonists into either the VMN of the hypothalamus or the midbrain central gray facilitates sexual receptivity in female rats, infusion of these agonists into the mPOA inhibits receptivity [189]. Acute infusion of positive steroid modulators, including 3α,5αTHP, into these regions elicits modulation of sexual receptivity comparable to that induced by GABA$_A$ receptor agonists [188–190]. This acute modulation by neurosteroids is rapid (in minutes) [191] and can be antagonized by bicuculline [192]. Moreover, modulation of sexual behaviors is also elicited by BSA-conjugated neurosteroids (a modification that precludes genomic actions of this compound via nuclear hormone receptors) [191,193]. Infusion of progesterone has been shown to induce rapid facilitation of sexual receptivity in progesterone receptor-knockout mice [194], suggesting that this gonadal steroid evokes its primary effects on sexual receptivity via nongenomic actions of its neurosteroid metabolites. Similarly, 3α-diol (which regulates estrus termination [195]), has a low affinity for the androgen receptor [196,197] and has been suggested to regulate sexual behaviors via an androgen receptor-independent and anesthetic action ([198]; although see also Ref. [88] for discussion of opposing findings), which could reflect allosteric modulation of the GABA$_A$ receptor. Taken in conjunction with studies demonstrating that neurosteroid levels [80–84] and the ability of neurosteroids to modulate GABA$_A$ receptors [161] varies with hormonal state, these results suggest that endogenous changes in neurosteroids may modulate sexual behaviors via membrane-delimited actions at the GABA$_A$ receptor.

Steroid modulation of GABAergic circuits has also been shown to induce significant effects on parturition and lactation by regulating oxytocin release from magnocellular neurons in the hypothalamus. Magnocellular neurons release oxytocin directly into the blood stream, thus promoting uterine contractility during parturition [199]. Furthermore, these neurons are subject to positive feedback in which oxytocin facilitates their firing by decreasing peak current amplitudes of inhibitory synaptic responses mediated by GABA$_A$ receptors [200]. The pattern of electrical activity of these neurons changes from minimal firing during pregnancy to high frequency, synchronous bursting around parturition, and then back to minimal firing during lactation [201,202]. Recently, it has been shown that this dramatic change in the electrical activity at parturition is correlated with a decrease in the sensitivity of magnocellular neurons to the facilitating actions of 3α,5αTHP. Specifically, while 3α,5αTHP prolonged synaptic current decays ~2.5-fold in rats during late pregnancy, the majority of neurons tested were unresponsive to this neurosteroid at the time of parturition [183]. Thus, while high levels of 3α,5αTHP are evident during pregnancy, the fall in the 3α,5αTHP levels at the time of parturition coupled with the decreased sensitivity of GABA$_A$ receptors to modulation by this neurosteroid is hypothesized to lead to disinhibition of magnocellular neurons and thus enhanced firing and oxytocin release at this time [183]. Finally, the ability of oxytocin to antagonize GABA$_A$ receptor-mediated currents, and thus promote magnocellular neuron firing, requires that the GABA$_A$ receptor be phosphorylated by protein kinse C (PKC) [106]. 3α,5αTHP, in turn, blocks the ability of oxytocin to decrease GABA$_A$ receptor peak current amplitudes by blocking PKC-dependent phosphorylation of the receptor [106]. Therefore, the decrease in levels of 3α,5αTHP at parturition is postulated to augment firing and oxytocin release in magnocellular neurons both by removing the direct allosteric potentiation of the receptor and by permitting oxytocin/PKC-dependent decreases in GABA$_A$ receptor peak currents [106].

12. Clinical correlations of GABA$_A$ receptor modulation by steroids

In addition to the well-established role for steroid actions at the GABA$_A$ receptor in the expression of sexual behaviors, an emerging scenario is that steroid control of the GABA$_A$ receptor system has widespread repercussions with respect to neurological and psychiatric dysfunction. In particular, allosteric modulation of GABA$_A$ receptors by neurosteroids is believed to play a critical role in premenstrual syndrome (PMS), certain types of epilepsy, affective disorders (including depression and anxiety, aggressive behaviors, and stress reactions), and in affective components of sexual behaviors. Moreover, interactions between steroids and the GABA$_A$ receptor have been implicated as contributing to the etiology underlying abuse of specific substances.

13. Drugs of abuse and steroid modulation of the GABA$_A$ receptor: ethanol

Significant changes in GABA$_A$ receptor expression and function have been demonstrated to occur with chronic use and upon withdrawal of BZs, barbiturates, and ethanol [8]. While data from a number of studies strongly indicate that many of ethanol's CNS effects are mediated by changes in GABA$_A$ receptor function, defining the molecular mechanisms underlying the actions of ethanol at the GABA$_A$ receptor remains somewhat of an elusive pursuit [203–205]. Interestingly, a number of recent studies in rodents provide evidence that neurosteroids may be one of the mediators of ethanol's effects and, conversely, that chronic ethanol exposure alters the sensitivity of the brain to neurosteroids. Both processes are presumed to occur by convergent actions at the GABA$_A$ receptor. Specifically, the presence of endogenous neurosteroids has been shown to be permissive for ethanol enhancement of GABA$_A$ receptor-mediated inhibition [206]. In addition, behavioral studies suggest that endogenous 3α,5αTHP may mediate some of the reinforcing effects of ethanol [207,208]. More direct evidence has been provided by VanDoren et al. [209] who have shown that systemic administration of ethanol increases levels of cortical 3α,5αTHP to pharmacologically relevant levels and that blockade of 3α,5αTHP synthesis prohibits both the cellular and behavioral effects of ethanol. Finally, modulation induced by 3α,5αTHP has been shown to be altered both with chronic ethanol exposure and upon withdrawal after prolonged ethanol administration [204,210,211].

Results in rodents have been supported by studies in humans in which levels of 3α,5αTHP and THDOC were reported to be lower in alcoholic subjects during early phases of ethanol withdrawal, and these diminished levels were positively correlated with the anxiety and depression that occurs at that time [212]. These studies in human subjects suggest that chronic ethanol exposure may decrease normal levels of GABA$_A$ receptor-mediated inhibition that are maintained by endogenous neurosteroids. Finally, chronic ethanol exposure and subsequent ethanol withdrawal induce significant changes in GABA$_A$ receptor subunit expression [8]. In particular, the enhanced levels of γ$_1$ subunit expression that occur upon ethanol withdrawal [211] would be predicted to increase the efficacy of neurosteroid potentiation of the GABA$_A$ receptor [130,144]; a result consistent with the increased sensitivity to neurosteroids in ethanol-withdrawing rats [211]. Interestingly, chronic AAS exposure also diminishes neurosteroid production via inhibition of 5α-reductase and 3α-hydroxysteroid dehydrogenase (3α-HSD) [213] and

GABA$_A$ receptor expression [159]. Specifically, in this study [159], chronic (4 week) exposure of both male and female mice to the AAS, 17α-methyltestosterone, was found to decrease levels of GABA$_A$ receptor subunit mRNA in regions of the mouse forebrain that mediate both neuroendocrine behaviors and aggression. Interestingly, pubertal animals were found to be more sensitive than were adults to the AAS treatment with respect to eliciting changes in steady state receptor subunit mRNA levels, and pubertal females were more sensitive than pubertal males. The significance of AAS effects on reproductive health is highlighted by recent data indicating that the largest increases in AAS use is among adolescents, especially adolescent girls [176,179]. Moreover, these data suggest that common mechanisms may contribute to the psychological changes and mood disorders associated with both ethanol and AAS abuse [167].

14. Affective disorders and steroid modulation of the GABA$_A$ receptor

It is estimated that 10–15% of prescriptions written in the United States are intended to change mental processes: to sedate or stimulate, or otherwise to change mood, thinking, or behavior [214]. The clinical manifestation and the severity of disorders of mood, also known as affective disorders, vary extraordinarily. It has been noted that levels of 3α-reduced neurosteroids in plasma and cerebrospinal fluid (CSF) are diminished in clinically depressed patients, but are surprisingly normalized by treatment with selective serotonin reuptake inhibitors (SSRIs) and other antidepressants [215–217]. The effect of antidepressant treatments on 3α,5αTHP levels is selective for 3α-reduced neurosteroids, as neither progesterone, pregnanolone, nor DHEA levels are changed in human [215,216] or in animal models [218] with treatment. Moreover, experiments in vitro have shown that pretreatment with SSRIs leads to increased accumulation of 3α,5αTHP [218]. It has been suggested that antidepressants shift the activity of 3α-hydroxysteroid oxidoreductase (3α-HOR; Fig. 1) towards reduction and thus augment conversion of 5α-DHP to 3α,5αTHP levels [215]. These data suggest a putative role for neurosteroids, and as such their modulation of GABA$_A$ receptors, in the manifestation of clinical depression, as well as a potential avenue of antidepressant therapy [215,216,218].

Neurosteroids have also been clearly implicated in the etiology of anxiety. In animal studies, progesterone and 3α,5αTHP [219–222] have anxiolytic properties in the absence of sedation. Reduction to 3α,5αTHP is required for progesterone to elicit anxiolytic effects [220]. Moreover, as has been shown for potentiation of GABA$_A$ receptor function, the 3α-hydroxy group is a structural requirement for anxiolytic action in whole animals as 3β isomers are ineffective [219]. Finally, the anxiolytic effects of progesterone are blocked by inhibitors of GABA$_A$ receptors, but not by inhibitors of the progesterone receptor [220].

Perhaps the best evidence that neurosteroid modulation of GABA$_A$ receptors plays an important role in the mediation and modulation of anxiety comes from studies of PMS. The menstrual cycle in women is regulated by a complex neuroendocrine cascade that is regulated by estrogen and progesterone. PMS is a constellation of disorders primarily characterized by affective, behavioral, and somatic symptoms that occur during the luteal phase of the menstrual cycle. In particular, PMS is associated with enhanced anxiety, dysphoria and susceptibility to seizures [80,223,224]. Plasma progesterone levels decrease during the luteal phase in women during PMS [225], and premenstrual symptoms and susceptibility to seizures have been reported to be associated with precipitous decreases in progesterone (and concomitant decreases in $3\alpha,5\alpha$THP) at this time [80,223]. Measurements of the velocity of saccadic eye movements as an indicator of GABAergic tone [226–229] show that GABA$_A$ receptor sensitivity to neurosteroids is decreased in PMS patients in the late luteal phase [227]. These studies suggest that decreases in neurosteroid levels in association with a loss of modulation of GABA$_A$ receptors is involved in the anxiety and dysphoria associated with PMS.

In animal models that are readily tractable for manipulating dose and duration of progesterone so as to mimic changes in hormone levels associated with PMS, it has been directly demonstrated that progesterone withdrawal is anxiogenic [230] and proconvulsant [185,186]. Thus, progesterone withdrawal in the rat replicates the enhanced anxiety that characterizes PMS, as well as the decreased seizure threshold seen in catamenial epilepsy (see below). In the rat model, the anxiogenic effects of progesterone withdrawal are directly correlated with an increase in the expression of the BZ-insensitive α_4 subunit of the GABA$_A$ receptor (albeit in the hippocampus, which, as the authors of this study point out, is not a structure readily implicated in the expression of anxiety) and expression of receptors with enhanced rates of desensitization [185,186]. These effects on GABA$_A$ receptor expression and function were shown to be due to $3\alpha,5\alpha$THP, not to progesterone itself [185,186].

Consistent with the observation in rats that progesterone withdrawal induces increased expression of BZ-insensitive GABA$_A$ receptors, human studies have also demonstrated that patients with PMS have reduced sensitivity to BZs compared to control subjects [228,229]. These studies point to a mechanism in which decreases in endogenous positive neurosteroid modulators not only lead to less allosteric potentiation of GABA$_A$ receptor-mediated currents, but also to enhanced expression of receptor subtypes that may be less sensitive to regulation by anxiolytic compounds. In spite of the general agreement between human and animal studies, it should be noted that differences between neurosteroid levels in PMS versus control subjects are not consistently reported and that correlations between neurosteroid levels and dysphoric symptoms are not always evident in clinical reports [231–233]. This lack of reproducible findings in human studies may stem, in part, from the difficulties associated with measuring GABAergic function in living human subjects, as well as the variability inherent in self-reporting of subjective mood. However, it also must be considered that steroid-dependent changes in neurotransmission and associated changes in affect in humans may reflect a level of complexity that cannot be accurately modeled by progesterone withdrawal in animal models.

15. Epilepsy and steroid modulation of the GABA$_A$ receptor

Epilepsy is a common chronic neurological disorder that affects at least 1% of the general population. In women, certain types of epilepsy are associated with critical landmarks in the reproductive timeline, such as the onset of puberty, menstrual cycles, pregnancy, and menopause [224]. In particular, perimenstrual progesterone withdrawal with concomitant changes in neurosteroid metabolites has been implicated in the exacerbation of seizures that characterizes catamenial epilepsy, a disorder that occurs in 33–50% of women with epilepsy [224,234,235]. Because $GABA_A$ receptors play such a pivotal role in regulating neuronal excitability, there has been an inherent interest in determining how natural changes in steroid levels may trigger seizure activity and whether exogenous steroid modulation of GABAergic transmission can be used as a therapeutic approach.

It is well accepted that $3\alpha,5\alpha THP$ can act as an anticonvulsant and PS as a convulsant via their opposing actions at $GABA_A$ receptors [68,69]. Moreover, while progesterone withdrawal in animal models results in an increased susceptibility to seizures [185,186], administration of 5α-reductase inhibitors (which block the conversion of progesterone to $3\alpha,5\alpha THP$) results in a reduction in the threshold for seizures induced by the convulsant, pentylenetetrazole (PTZ) [235]. Moreover, while exogenous progesterone treatment protects against PTZ-induced seizures in mice, this protection is abrogated by coadministration of 5α-reductase inhibitors [236,237]. These studies strongly suggest that the enhanced seizure susceptibility in catamenial epilepsy may arise from diminished levels of endogenous $3\alpha,5\alpha THP$ and the concomitant decrease in the loss of $GABA_A$ receptor potentiation that would follow the precipitous decline in this endogenous neurosteroid. While experiments explicitly designed to determine if changes in $GABA_A$ receptor composition over the course of the menstrual cycle also contribute to cycle-dependent changes in seizure threshold have not been performed, studies in animal models indicate that expression of $GABA_A$ receptor subunit genes and the sensitivity of these receptors to neurosteroids change over the course of the estrous cycle [161,164]. These changes in $GABA_A$ receptor expression and function, in conjunction with cycle-dependent changes in endogenous neurosteroid levels, may lead to expression of $GABA_A$ receptors that provide diminished inhibition and thus increased susceptibility to seizures [185,186].

While progesterone therapy is thus an appealing avenue to consider in treatment of epilepsy (in particular, with respect to the enhanced seizure activity associated with changes in hormonal state), the lack of control over metabolic fate coupled with untoward side effects has limited the efficacy of the naturally occurring progesterone derivatives [69,238]. However, a synthetic analog of $3\alpha,5\alpha THP$, ganaxolone (3α-hydroxy-3β-methyl-5α-pregnan-20-one) [235,239,240], is a positive allosteric modulator of the $GABA_A$ receptor with enhanced bioavailability compared to $3\alpha,5\alpha THP$ and no progestational activity [94,235,241,242]. This synthetic neurosteroid has been shown to block seizures in acute seizure models and in chemical kindling models [238–240]. In addition, it has been shown that ganaxolone has an enhanced anticonvulsant potency during the period of heightened seizure susceptibility associated with progesterone withdrawal, thus making it a particularly attractive drug for treatment of catamenial epilepsy [235]. An initial study in human subjects shows that this drug is well tolerated with minimal adverse side effects [243]. Interestingly, ganaxolone has also been shown to be effective in the management of convulsions due to cocaine poisoning [240] and may also be beneficial in controlling anxiety, mood changes, and other behavioral alterations associated with preseizure activity [242].

There is clinical and anecdotal evidence that patients suffering from PMS or catamenial epilepsia also suffer from sexual dysfunction, most notably diminished fertility and a decrease in sexual libido [224]. As discussed earlier, AAS abusers have also reported changes in fertility, sexual performance and libido [176]. Hormonal imbalance, associated with disease states, drug abuse or therapeutic management of disease may all cause disruption of normal endocrine function that is ultimately reflected in compromised reproductive health and sexual performance. While many of the effects of steroids on reproduction and libido may be mediated via both peripheral and central actions at nuclear hormone receptors, the importance of GABA$_A$ receptors in controlling these key behaviors indicates that abnormal allosteric modulation of these receptors by AAS and neurosteroids must also be considered in understanding the sexual and reproductive dysfunction that is often associated with drug abuse and disease.

16. Concluding remarks

The net cast by steroid modulators encompasses an impressively wide range of CNS functions. The preponderance of data discussed here highlights allosteric modulation of the GABA$_A$ receptor as a critical, if not the key, mechanism by which steroids induce their myriad of effects. However, it is overly simplistic to suggest that this is the sole mechanism of steroid action in the nervous system. In addition to GABA$_A$ receptors, neurosteroids can act as allosteric modulators at nicotinic acetylcholine, 5HT-3, glutamate, oxytocin, and sigma type 1 opioid receptors [217], and acute effects of steroids at these other neurotransmitter receptors play a significant role in the production of psychotrophic effects. Moreover, the actions of steroids in the CNS are extended further by the fact that allosteric modulation of these receptors alters neuronal activity, which in turn can change receptor subunit expression. This can then come full circle to alter the sensitivity of the receptors to allosteric modulators (Fig. 3). Finally, steroids, including the parent gonadal steroids and the AAS, bind to nuclear hormone receptors to alter the expression not only of the neurotransmitter receptors themselves, but also of an array of genes that govern synaptic efficacy and neuronal structure. It is precisely the ability of steroids to provide this complexity of cross-talk between such a wide range of membrane and nuclear receptors that may be at the heart of the observed broad spectrum of CNS effects. In addition, in order to fully understand the behavioral consequences produced by steroid modulation of the GABA$_A$ receptor, it will be critical to reduce the gap between the established knowledge of how these steroids act at the molecular and cellular levels and the presently scant knowledge as to how this modulation alters neuronal circuits underlying specific behaviors. For example, the interactions of steroids with glial cells in modulating neural transmission has received little attention. Glial cells express GABA$_A$ receptors that are sensitive to steroid modulation [32,105], synthesize neurosteroids [78,79], and can modulate neuronal signaling [244]. Finally, in response to changes in the hormonal milieu, glial cells can induce significant synaptic remodeling [245,246]. Thus, modulation of GABA$_A$ receptors expressed in glial cells must be considered in determining both long-term and short-term effects of steroids. In addition, the efficacy of steroid modulation, with respect to potential therapeutic uses, susceptibility to specific mood disorders, and to steroid abuse, needs to be re-assessed within the framework of dynamic changes in GABA$_A$ receptor expression and function that occurs throughout the mammalian lifespan. In particular, a full understanding of ontological changes in the

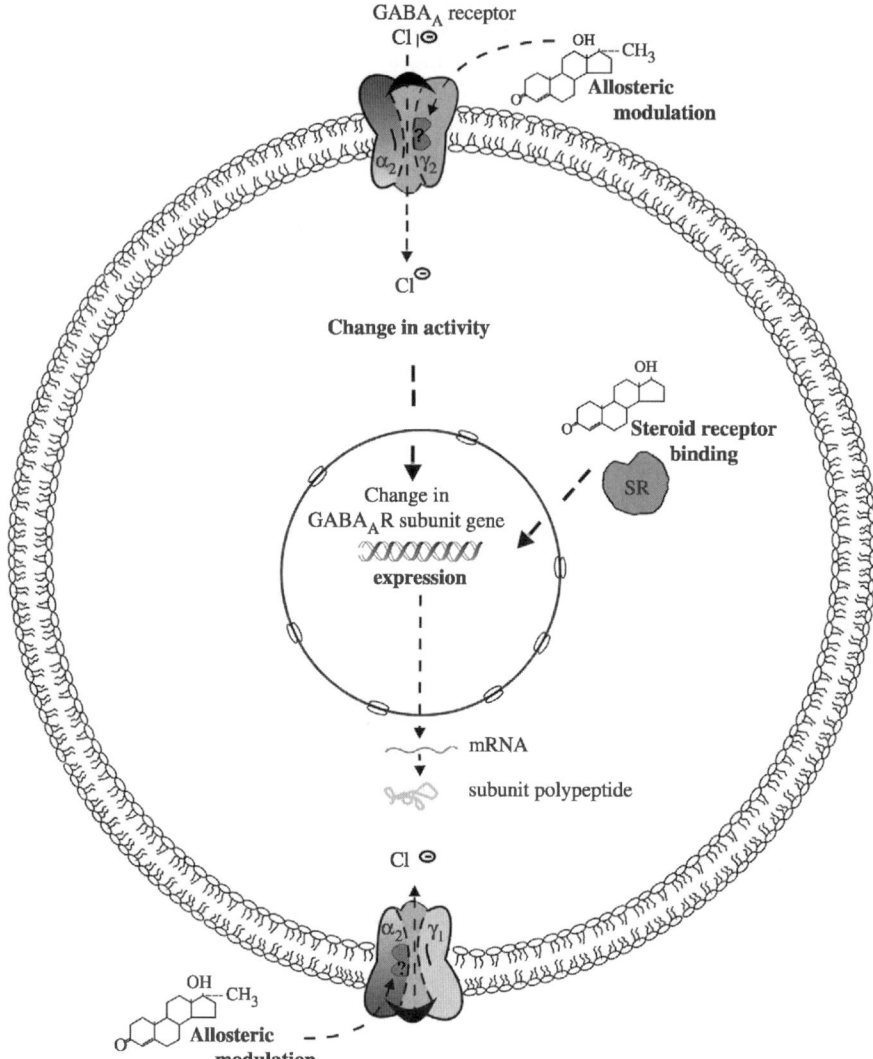

Fig. 3. Genomic and nongenomic regulation of GABA$_A$ receptor expression. Steroids can have a multiplicity of actions in regulating the expression and function of the GABA$_A$ receptor. Gonadal steroids, as well as the synthetic AAS, can induce significant changes in GABA$_A$ receptor subunit gene expression via binding to cognate nuclear hormone receptors and altering gene expression. Neurosteroid metabolites and the AAS, however, can also induce significant changes in GABA$_A$ receptor subunit gene expression by their ability to allosterically modulate currents carried by the receptor and thus alter the electrical activity in the cell. By either pathway, changes in subunit expression can, in turn, result in the assembly of receptors with altered subunit composition and therefore altered sensitivity to allosteric modulators. In summary, an individual neuron comprises a highly sensitive system in which steroids and the GABA$_A$ receptor can bidirectionally regulate signalling mediated by each system. (*For a colored version of this figure, see plate section, page 466.*)

GABA$_A$ receptor system, both with respect to changes in subunit gene expression and to receptor function, should be made against the background of determining comparable ontological changes in the expression and function of steroids in the CNS. At present, changes in receptor subunit gene expression and function during early development have been well characterized. However, the ability of the steroid milieu to govern these changes is not well understood, and very little is known as to changes in expression and function of these receptors at other critical endocrine junctures, such as puberty and reproductive senescence. An understanding of the interplay between the endogenous steroid and GABA$_A$ receptor systems, as well as a thorough assessment of the dynamic changes that occur in each, will be imperative in delineating how steroids govern this crucial receptor, as well as in designing successful therapeutic strategies in which steriods can be used efficaciously in both sexes and at all ages.

Acknowledgements

We thank Ann Clark, Brian Jones, Paul Yang and Kerry McIntyre for comments on the manuscript. We also thank Ann Clark and Siobhan Robinson for giving us permission to reproduce unpublished data in this chapter.

References

[1] Betz, H., 1990. Ligand-gated ion channels in the brain: the amino acid receptor superfamily. Neuron 5, 383–392.

[2] Boyd, R.T., 1997. The molecular biology of neuronal nicotinic acetylcholine receptors. Crit. Rev. Toxicol. 27(3), 299–318.

[3] Betz, H., Kuhse, J., Schmieden, V., Laube, B., Kirsch, J., Harvey, R.J., 1999. Structure and functions of inhibitory and excitatory glycine receptors. Ann. NY Acad. Sci. 868, 667–676.

[4] Jackson, M.B., Yakel, J.L., 1995. The 5-HT3 receptor channel. Annu. Rev. Physiol. 57, 447–468.

[5] Barnard, E.A., Skolnick, P., Olsen, R.W., Möhler, H., Sieghart, G., Biggio, C., Braestrup, A.N., Langer, S.Z., 1998. International Union of Pharmacology. XV. Subtypes of γ-aminobutyric acid$_A$ receptors: classification on the basis of subunit structure and receptor function. Pharmacol. Rev. 50(2), 291–313.

[6] Smith, G.B., Olsen, R.W., 1995. Functional domains of GABA$_A$ receptors. Trends Pharmacol. Sci. 16(5), 162–168.

[7] Schofield, P.R., Darlison, M.G., Fujita, N., Burt, D.R., Stephenson, F.A., Rodriquez, H., Rhee, L.M., Ramanchandran, J., Reale, V., Glencorse, T.A., Seeburg, P.H., Barnard, E.A., 1987. Sequence and functional expression of the GABA$_A$ receptor shows a ligand-gated receptor super-family. Nature 328, 221–227.

[8] Mehta, A.K., Ticku, M.K., 1999. An update on GABA$_A$ receptors. Brain Res. Revs. 29(2–3), 196–217.

[9] Whiting, P.J., 1999. The GABA-A receptor gene family: new targets for therapeutic intervention. Neurochem. Int. 34(5), 387–390.

[10] Whiting, P., Bonnert, T.P., McKernan, R.M., Farrar, S., Le Bourdellés, B., Heavens, R.P., Smith, D.W., Hewson, L., Rigby, M.R., Sirinathsinghji, D.J.S., Thompson, S.A., Wafford, K.A., 1999. Molecular and functional diversity of the expanding GABA-A receptor gene family. Ann. NY Acad. Sci. 868, 645–653.

[11] Macdonald, R.L., Olsen, R.W., 1994. GABA$_A$ receptor channels. Annu. Rev. Neurosci. 17, 569–602.

[12] Bateson, A.N., Lasham, A., Darlison, M.G., 1991. γ-Aminobutyric acid$_A$ receptor heterogeneity is increased by alternative splicing of a novel β-subunit gene transcript. J. Neurochem. 56(4), 1437–1440.

[13] Harvey, R.J., Kim, H.C., Darlison, M.G., 1993. Molecular cloning reveals the existence of a fourth γ-subunit of the vertebrate brain GABA$_A$ receptor. FEBS Lett. 331(3), 211–216.

[14] Whiting, P., McKernan, R.M., Iversen, L.L., 1990. Another mechanism for creating diversity in γ-aminobutyrate type A receptors: RNA splicing directs expression of two forms of γ_2 subunit, one of which contains a protein kinase C phosphorylation site. Proc. Natl Acad. Sci. USA 87, 9966–9970.

[15] Kofuji, P., Wang, J.B., Moss, S.J., Huganir, R.L., Burt, D.R., 1991. Generation of two forms of the γ-aminobutyric acid$_A$ receptor γ_2-subunit in mice by alternative splicing. J. Neurochem. 56(2), 713–715.

[16] Moss, S.J., Doherty, C.A., Huganir, R.L., 1992. Identification of the cAMP-dependent protein kinase and protein kinase C phosphorylation sites within the major intracellular domains of the β_1, γ_2S and γ_2L subunits of the γ-aminobutyric type$_A$ receptor. J. Biol. Chem. 267, 14470–14476.

[17] Harvey, R.J., Chinchetru, M.A., Darlison, M.G., 1994. Alternative splicing of a 51-nucleotide exon that encodes a putative protein kinase C phosphorylation site generates two forms of the chicken γ-aminobutyric acid$_A$ receptor β_2 subunit. J. Neurochem. 62(1), 10–16.

[18] McKinley, D.D., Lennon, D.J., Carter, D.B., 1995. Cloning, sequence analysis and expression of two forms of mRNA coding for the human β_2 subunit of the GABA$_A$ receptor. Mol. Brain Res. 28, 175–179.

[19] Kirkness, E.F., Fraser, C.M., 1993. A strong promoter element is located between alternative exons of a gene encoding the human γ-aminobutyric acid type A receptor β_3 subunit (GABRB3). J. Biol. Chem. 268(6), 4420–4428.

[20] Korpi, E.R., Kuner, T., Kristo, P., Kohler, M., Herb, A., Lüddens, H., Seeburg, P.H., 1994. Small N-terminal deletion by splicing in cerebellar α_6 subunit abolishes GABA$_A$ receptor function. J. Neurochem. 63(3), 1167–1170.

[21] Johnston, G.A., 1996. GABA$_C$ receptors: relatively simple transmitter-gated ion channels? Trends Pharmacol. Sci. 17(9), 319–323.

[22] Hedblom, E., Kirkness, E.F., 1997. A novel class of GABA$_A$ receptor subunit in tissues of the reproductive system. J. Biol. Chem. 272(24), 15346–15350.

[23] Fritschy, J.M., Möhler, H., 1995. GABA$_A$-receptor heterogeneity in the adult rat brain: differential regional and cellular distribution of seven major subunits. J. Comp. Neurol. 359(1), 154–194.

[24] Khan, Z.U., Gutiérrez, A., De Blas, A.L., 1994. Short and long form γ_2-subunits of the GABA$_A$/benzodiazepine receptors. J. Neurochem. 63(4), 1466–1476.

[25] Quirk, K., Gillard, N.P., Ragan, C.I., Whiting, P.J., McKernan, R.M., 1994. γ-Aminobutyric acid type A receptors in the rat brain can contain both γ_2 and γ_3 subunits, but γ_1 does not exist in combination with another γ-subunit. Mol. Pharmacol. 45, 1061–1070.

[26] Quirk, K., Whiting, P.J., Ragan, C.I., McKernan, R.M., 1995. Characterisation of δ-subunit containing GABA$_A$ receptors from rat brain. Eur. J. Pharmacol. 290(3), 175–181.

[27] Kaila, K., 1994. Ionic basis of GABA$_A$ receptor channel function in the nervous system. Prog. Neurobiol. 42(4), 489–537.

[28] Cherubini, E., Gaiarsa, J.L., Ben-Ari, Y., 1991. GABA: an excitatory transmitter in early postnatal life. Trends Neurosci. 14(12), 515–519.

[29] Sakmann, B., Bormann, J., Hamill, O.P., 1983. Ion transport by single receptor channels. Cold Spring Harb. Symp. Quant. Biol. 48(1), 247–257.

[30] Bormann, J., Hamill, O.P., Sakmann, B., 1987. Mechanism of anion permeation through channels gated by glycine and γ-aminobutyric acid in mouse cultured spinal neurones. J. Physiol. (Lond.) 385, 243–286.

[31] Bormann, J., Clapham, D.E., 1985. γ-Aminobutyric acid receptor channels in adrenal chromaffin cells: a patch-clamp study. Proc. Natl Acad. Sci. USA 82(7), 2168–2172.

[32] Bormann, J., Kettenmann, H., 1988. Patch-clamp study of γ-aminobutyric acid receptor Cl$^-$ channels in cultured astrocytes. Proc. Natl Acad. Sci. USA 85(23), 9336–9340.

[33] Macdonald, R.L., Rogers, C.J., Twyman, R.E., 1989. Kinetic properties of the GABA$_A$ receptor main conductance state of mouse spinal cord neurones in culture. J. Physiol. (Lond.) 410, 479–499.

[34] Weiss, D.S., Magleby, K.L., 1989. Gating scheme for single GABA-activated Cl$^-$ channels determined from stability plots, dwell-time distributions, and adjacent-interval durations. J. Neurosci. 9(4), 1314–1324.

[35] Puia, G., Santi, M.R., Vicini, S., Pritchett, D.B., Purdy, R.H., Paul, S.M., Seeburg, P.H., Costa, E., 1990. Neurosteroids act on recombinant human GABA$_A$ receptors. Neuron 4(5), 759–765.

[36] Newland, C.F., Colquhoun, D., Cull-Candy, S.G., 1991. Single channels activated by high concentrations of GABA in superior cervical ganglion neurones of the rat. J. Physiol. (Lond.) 432, 203–233.

[37] Schneggenburger, R., Konnerth, A., 1992. GABA-mediated synaptic transmission in neuroendocrine cells: a patch-clamp study in pituitary slice preparation. Pflügers Arch. 421(4), 364–373.

[38] Smart, T.G., 1992. A novel modulatory binding site for zinc on the GABA$_A$ receptor complex in cultured rat neurones. J. Physiol. (Lond.) 447, 587–625.

[39] Twyman, R.E., Macdonald, R.L., 1992. Neurosteroid regulation of GABA$_A$ receptor single-channel kinetic properties of mouse spinal cord neurons in culture. J. Physiol. (Lond.) 456, 215–245.

[40] Angelotti, T.P., Macdonald, R.L., 1993. Assembly of GABA$_A$ receptor subunits: $\alpha_1\beta_1$ and $\alpha_1\beta_1\gamma_{2S}$ subunits produce unique ion channels with dissimilar single-channel properties. J. Neurosci. 13(4), 1429–1440.

[41] Maconochie, D.J., Zempel, J.M., Steinbach, J.H., 1994. How quickly can GABA$_A$ receptors open? Neuron 12(1), 61–71.

[42] Rogers, C.J., Twyman, R.E., Macdonald, R.L., 1994. Benzodiazepine and β-carboline regulation of single GABA$_A$ receptor channels of mouse spinal cord neurones in culture. J. Physiol. (Lond.) 475.1, 69–82.

[43] Zhang, S.J., Jackson, M.B., 1995. Properties of the GABA$_A$ receptor of rat posterior pituitary nerve terminals. J. Neurophysiol. 73(3), 1135–1144.

[44] Brickley, S.G., Cull-Candy, S.G., Farrant, M., 1999. Single-channel properties of synaptic and extrasynaptic GABA$_A$ receptors suggest differential targeting of receptor subtypes. J. Neurosci. 19(8), 2960–2973.

[45] Lorez, M., Benke, D., Lüscher, B., Möhler, H., Benson, J.A., 2000. Single-channel properties of neuronal GABA$_A$ receptors from mice lacking the γ2 subunit. J. Physiol. (Lond.) 527.1, 11–31.

[46] Konnerth, A., Llano, I., Armstrong, C.M., 1990. Synaptic currents in cerebellar Purkinje cells. Proc. Natl Acad. Sci. USA 87(7), 2662–2665.

[47] Vincent, P., Armstrong, C.M., Marty, A., 1992. Inhibitory synaptic currents in rat cerebellar Purkinje cells: modulation by postsynaptic depolarization. J. Physiol. (Lond.) 456, 453–471.

[48] Tia, S., Wang, J.F., Kotchabhakdi, N., Vicini, S., 1996. Distinct deactivation and desensitization kinetics of recombinant GABA$_A$ receptors. Neuropharmacology 35(9/10), 1375–1382.

[49] Mellor, J.R., Randall, A.D., 1997. Frequency-dependent actions of benzodiazepines on GABA$_A$ receptors in cultured murine cerebellar granule cells. J. Physiol. (Lond.) 503.2, 353–369.

[50] Rossi, D.J., Hamann, M., 1998. Spillover-mediated transmission at inhibitory synapses promoted by high affinity α$_6$ subunit GABA$_A$ receptors and glomerular geometry. Neuron 20, 783–795.

[51] Collingridge, G.L., Gage, P.W., Robertson, B., 1984. Inhibitory post-synaptic currents in rat hippocampal CA1 neurones. J. Physiol. (Lond.) 356, 551–564.

[52] Edwards, F.A., Konnerth, A., Sakmann, B., 1990. Quantal analysis of inhibitory synaptic transmission in the dentate gyrus of rat hippocampal slices: a patch-clamp study. J. Physiol. (Lond.) 430, 213–249.

[53] Banks, M.I., Li, T.B., Pearce, R.A., 1998. The synaptic basis of GABA$_{A,slow}$. J. Neurosci. 18(4), 1305–1317.

[54] Cooper, E.J., Johnston, G.R., Edward, F.A., 1999. Effects of a naturally occurring neurosteroid on GABA$_A$ IPSCs during development in rat hippocampal or cerebellar slices. J. Physiol. (Lond.) 521.2, 437–449.

[55] Mozrzymas, J.W., Barberis, A., Michalak, K., Cherubini, E., 1999. Chlorpromazine inhibits miniature GABAergic currents by reducing the binding and by increasing the unbinding rate of GABA$_A$ receptors. J. Neurosci. 19(7), 2474–2488.

[56] Le Feuvre, Y., Fricker, D., Leresche, N., 1997. GABA$_A$ receptor-mediated IPSCs in rat thalamic sensory nuclei: patterns of discharge and tonic modulation by GABA$_B$ autoreceptors. J. Physiol. (Lond.) 502.1, 91–104.

[57] Smith, S.T., Brennan, C., Clark, A.S., Henderson, L.P., 1996. GABA$_A$ receptor-mediated responses in the ventromedial nucleus of the hypothalamus of female and male neonatal rats. Neuroendocrinology 64(2), 103–113.

[58] Haage, D., Johansson, S., 1999. Neurosteroid modulation of synaptic and GABA-evoked currents in neurons from the rat medial preoptic nucleus. J. Neurophysiol. 82(1), 143–151.

[59] Nett, S.T., Jorge-Rivera, J-C., Myers, M., Clark, A.S., Henderson, L.P., 1999. Properties and sex-specific differences of GABA$_A$ receptors in neurons expressing γ1 subunit mRNA in the preoptic area of the rat. J. Neurophysiol. 81(1), 192–203.

[60] Jorge-Rivera, J.C., McIntyre, K.L., Henderson, L.P., 2000. Anabolic steroids induce region- and subunit-specific rapid modulation of GABA$_A$ receptor-mediated currents in the rat forebrain. J. Neurophysiol. 83, 3299–3309.

[61] Galarreta, M., Hestrin, S., 1997. Properties of GABA$_A$ receptors underlying inhibitory synaptic currents in neocortical pyramidal neurons. J. Neurosci. 17(19), 7220–7227.

[62] Chéry, N., de Koninck, Y., 1999. Junctional versus extrajunctional glycine and GABA$_A$ receptor-mediated IPSCs in identified lamina I neurons of the adult rat spinal cord. J. Neurosci. 19(17), 7342–7355.

[63] Schneggenburger, R., López-Barneo, J., Konnerth, A., 1992. Excitatory and inhibitory synaptic currents and receptors in rat medial septal neurones. J. Physiol. (Lond.) 445, 261–276.

[64] Jones, M.V., Westbrook, G.L., 1995. Desensitized states prolong GABA$_A$ channel responses to brief agonist pulses. Neuron 15(1), 181–191.

[65] Jones, M.V., Westbrook, G.L., 1996. The impact of receptor desensitization on fast synaptic transmission. Trends Neurosci. 19(3), 96–101.

[66] Zhu, W.J., Vicini, S., 1997. Neurosteroid prolongs GABA$_A$ channel deactivation by altering kinetics of desensitized states. J. Neurosci. 17(11), 4022–4031.

[67] Jones, M.V., Sahara, Y., Dzubay, J.A., Westbrook, G.L., 1998. Defining affinity with the GABA$_A$ receptor. J. Neurosci. 18(21), 8590–8604.

[68] Sieghart, W., 1995. Structure and pharmacology of γ-aminobutyric acid$_A$ receptor subtypes. Pharmacol. Rev. 47(2), 181–234.

[69] Lambert, J.J., Belelli, D., Hill-Venning, C., Peters, J.A., 1995. Neurosteroids and GABA$_A$ receptor function. Trends Pharmacol. Sci. 16(9), 295–303.

[70] Narahashi, T., Ginsburg, K.S., Nagata, K., Song, J.H., Tatebayashi, H., 1998. Ion channels as targets for insecticides. Neurotoxicology 19(4–5), 581–590.

[71] Zorumski, C.F., Yang, J., 1988. Non-competitive inhibition of GABA currents by phenothiazines in cultured chick spinal cord and rat hippocampal neurons. Neurosci. Lett. 92(1), 86–91.

[72] Masonis, A.E.T., McCarthy, M.P., 1995. Direct effects of the anabolic/androgenic steroids, stanozolol and 17α-methyltestosterone, on benzodiazepine binding to the GABA$_A$ receptor. Neurosci. Lett. 189(1), 35–38.

[73] Masonis, A.E.T., McCarthy, M.P., 1996. Effects of the androgenic/anabolic steroids stanozolol on GABA$_A$ receptor function: GABA-stimulated of $^{36}Cl^-$ influx and $[^{35}S]$TBPS binding. J. Pharmacol. Exp. Ther. 279(1), 186–193.

[74] Yang, P., Jones, B.L., Henderson, L.P., 2002. Mechanisms of anabolic androgenic steroid modulation of α$_1$β$_3$γ$_{2L}$ GABA$_A$ receptors. Neuropharmacology 43(4), 619–633.

[75] Vicini, S., 1999. New perspectives in the functional role of GABA$_A$ channel heterogeneity. Mol. Neurobiol. 19(2), 97–110.

[76] Vafa, B., Schofield, P.R., 1998. Heritable mutations in the glycine, GABA$_A$, and nicotinic acetylcholine receptors provide new insights into the ligand-gated ion channel receptor superfamily. Int. Rev. Neurobiol. 42, 285–332.

[77] Revelli, A., Tesarik, J., Massobrio, M., 1998. Nongenomic effects of neurosteroids. Gynecol. Endocrinol. 12(1), 61–67.

[78] Baulieu, E.E., 1998. Neurosteroids: a novel function of the brain. Psychoneuroendocrinology; 23(8), 963–987.

[79] Compagnone, N.A., Mellon, S.H., 2000. Neurosteroids: biosynthesis and function of these novel neuromodulators. Front Neuroendocrinol. 21(1), 1–56.

[80] Majewska, M.D., 1992. Neurosteroids: endogenous bimodal modulators of the GABA$_A$ receptor: mechanisms of action and physiological significance. Prog. Neurobiol. 38(4), 379–395.

[81] Rupprecht, R., Hauser, C.A., Trapp, T., Holsboer, F., 1996. Neurosteroids: molecular mechanisms of action and psychopharmacological significance. J. Steroid Biochem. Mol. Biol. 56(1–6), 163–168.

[82] Wilson, M.A., 1996. GABA physiology: modulation by benzodiazepines and hormones. Crit. Rev. Neurobiol. 10(1), 1–37.

[83] Bixo, M., Andersson, A., Winblad, B., Purdy, R.H., Bäckström, T., 1997. Progesterone, 5α-pregnane-3, 20-dione and 3α-hydroxy-5α-pregnane-20-one in specific regions of the human female brain in different endocrine states. Brain Res. 764(1–2), 173–178.

[84] Sundström, I., Bäckström, T., Wang, M., Olsson, T., Seippel, L., Bixo, M., 1999. Premenstrual syndrome, neuroactive steroids and the brain. Gynecol. Endocrinol. 13(3), 206–220.

[85] Lacroix, C., Fiet, J., Benais, J.P., Gueux, B., Bonete, R., Villette, J.M., Gourmel, B., Dreux, C., 1987. Simultaneous radioimmunoassay of progesterone, androst-4-enedione, pregnenolone, dehydroepiandrosterone and 17-hydroxyprogesterone in specific regions of human brain. J. Steroid Biochem. 28(3), 317–325.

[86] Lanthier, A., Patwardhan, V.V., 1986. Sex steroids and 5-en-3 beta-hydroxysteroids in specific regions of the human brain and cranial nerves. J. Steroid Biochem. 25(3), 445–449.

[87] Herrington, D.M., 1998. DHEA: a biological conundrum. J. Lab. Clin. Med. 131(4), 292–294.

[88] Blasberg, M.E., Robinson, S., Henderson, L.P., Clark, A.S., 1998. Inhibition of estrogen-induced sexual receptivity by androgens: role of the androgen receptor. Horm. Behav. 34(3), 283–293.

[89] Cottrell, G.A., Lambert, J.J., Peters, J.A., 1987. Modulation of GABA$_A$ receptor activity by alphaxalone. Br. J. Pharmacol. 90(3), 491–500.

[90] Callachan, H., Cottrell, G.A., Hather, N.Y., Lambert, J.J., Nooney, J.M., Peters, J.A., 1987. Modulation of GABA$_A$ receptor by progesterone metabolites. Proc. R Soc. Lond. B 231, 359–369.

[91] Peters, J.A., Kirkness, E.F., Callachan, H., Lambert, J.J., Turner, A.J., 1988. Modulation of the GABA$_A$ receptor by depressant barbiturates and pregnane steroids. Br. J. Pharmacol. 94(4), 1257–1269.

[92] Korpi, E.R., Lüddens, H., 1993. Regional γ-aminobutyric acid sensitivity of t-butylbicyclophosphoro[^{35}S]thionate binding depends on γ-aminobutyric acid$_A$ receptor α subunit. Mol. Pharmacol. 44(1), 87–92.

[93] Harrison, N.L., Majewska, M.D., Harrington, J.W., Barker, J.L., 1987. Structure–activity relationships for steroid interaction with the γ-aminobutyric acid$_A$ receptor complex. J. Pharmacol. Exp. Ther. 241(1), 346–353.

[94] Gee, K.W., Bolger, M.B., Brinton, R.E., Coirini, H., McEwen, B.S., 1988. Steroid modulation of the chloride ionophore in rat brain: structure–activity requirements, regional dependence and mechanism of action. J. Pharmacol. Exp. Ther. 246(2), 803–812.

[95] Nilsson, K.R., Zorumski, C.F., Covey, D.F., 1998. Neurosteroid analogues. 6. The synthesis and GABA$_A$ receptor pharmacology of enantiomers of dehydroepiandrosterone sulfate, pregnenolone sulfate, and (3α,5β)-3hydroxypregnan-20-one sulfate. J. Med. Chem. 41(14), 2604–2613.

[96] Covey, D.F., Nathan, D., Kalkbrenner, M., Nilsson, K.R., Hu, Y., Zorumski, C.F., Evers, A.S., 2000. Enantioselectivity of pregnanolone-induced γ-aminobutyric acid$_A$ receptor modulation and anesthesia. J. Pharmacol. Exp. Ther. 293(3), 1009–1016.

[97] Gee, K.W., 1988. Steroid modulation of the GABA/benzodiazepine receptor-linked chloride ionophore. Mol. Neurobiol. 2, 291–317.

[98] Turner, D.M., Ransom, R.W., Yang, J.S.J., Olsen, R.W., 1989. Steroid anesthetics and naturally occurring analogs modulate the γ-aminobutyric acid receptor complex at a site distinct from barbiturates. J. Pharmacol. Exp. Ther. 248, 960–966.

[99] Morrow, A.L., Pace, J.R., Purdy, R.H., Paul, S.M., 1990. Characterization of steroid interactions with γ-aminobutyric acid receptor-gated chloride ion channels: evidence for multiple steroid recognition sites. Mol. Pharmacol. 37(2), 263–270.

[100] Maitra, R., Reynolds, J.N., 1998. Modulation of GABA$_A$ receptor function by neuroactive steroids: evidence for heterogeneity of steroid sensitivity of recombinant GABA$_A$ receptor isoforms. Can. J. Physiol. (Lond.) Pharmacol. 76(9), 909–920.

[101] Park-Chung, M., Malayev, A., Purdy, R.H., Gibbs, T.T., Farb, D.H., 1999. Sulfated and unsulfated steroids modulate gamma-aminobutyric acid$_A$ receptor function through distinct sites. Brain Res. 830(1), 72–87.

[102] Zorumski, C.F., Mennerick, S.J., Covey, D.F., 1998. Enantioselective modulation of GABAergic synaptic transmission by steroids and benz[e]indenes in hippocampal microcultures. Synapse 29(2), 162–171.

[103] Majewska, M.D., Harrison, N.L., Schwartz, R.D., Barker, J.L., Paul, S.M., 1986. Steroid hormone metabolites are barbiturate-like modulators of the GABA$_A$ receptor. Science 232, 1004–1007.

[104] Harrison, N.L., Vicini, S., Barker, J.L., 1987. A steroid anesthetic prolongs inhibitory postsynaptic currents in cultured rat hippocampal neurons. J. Neurosci. 7, 604–609.

[105] Chvátal, A., Kettenmann, H., 1991. Effect of steroids on γ-aminobutyrate-induced currents in cultured rat astrocytes. Pflügers Arch. 419(3–4), 263–266.

[106] Brussaard, A.B., Wossink, J., Lodder, J.C., Kits, K.S., 2000. Progesterone-metabolite prevents protein kinase C-dependent modulation of γ-aminobutyric acid type A receptors in oxytocin neurons. Proc. Natl Acad. Sci. USA 97(7), 3625–3630.

[107] Fáncsik, A., Linn, D.M., Tasker, J.G., 2000. Neurosteroid modulation of GABA IPSCs is phosphorylation dependent. J. Neurosci. 20(9), 3067–3075.

[108] Mistry, D.K., Cottrell, G.A., 1990. Actions of steroids and bemegride on the GABA_A receptor of mouse spinal neurones in culture. Exp. Physiol. 75(2), 199–209.

[109] Poisbeau, P., Feltz, P., Schlichter, R., 1997. Modulation of GABA_A receptor-mediated IPSCs by neuroactive steroids in a rat hypothalamo-hypophyseal coculture model. J. Physiol. (Lond.) 500(2), 475–485.

[110] Tsutsui, K., Ukena, K., 1999. Neurosteroids in the cerebellar Purkinje neuron and their actions (Review). Int. J. Mol. Med. 4(1), 49–56.

[111] Barker, J.L., Harrison, N.L., Lange, G.D., Owen, D.G., 1987. Potentiation of γ-aminobutyric-acid-activated chloride conductance by a steroid anaesthetic in cultured rat spinal neurones. J. Physiol. (Lond.) 386, 485–501.

[112] Moss, S.J., Smart, T.G., 1996. Modulation of amino acid-gated ion channels by protein phosphorylation. Int. Rev. Neurobiol. 39, 1–52.

[113] Kellenberger, S., Malherbe, P., Sigel, E., 1992. Function of α1β2γ2S γ-aminobutyric acid type A receptor is modulated by protein kinase C via multiple phosphorylation sites. J. Biol. Chem. 267(36), 25660–25663.

[114] Browning, M.D., Endo, S., Smith, G.B., Dudek, E.M., Olsen, R.W., 1993. Phosphorylation of the GABA_A receptor by cAMP-dependent protein kinase and by protein kinase C: analysis of the substrate domain. Neurochem. Res. 18(1), 95–100.

[115] Krishek, B.J., Xie, X., Blackstone, C., Huganir, R.L., Moss, S.J., Smart, T.G., 1994. Regulation of GABA_A receptor function by protein kinase C phosphorylation. Neuron 12(5), 1081–1095.

[116] Leidenheimer, N.J., McQuilkin, S.J., Hahner, L.D., Whiting, P.J., Harris, R.A., 1992. Activation of protein kinase C selectively inhibits the γ-aminobutyric acid_A receptor: role of desensitization. Mol. Pharmacol. 41(6), 1116–1123.

[117] Leidenheimer, N.J., Chapell, R., 1997. Effects of PKC activation and receptor desensitization on neurosteroid modulation of GABA_A receptors. Mol. Brain Res. 52, 173–181.

[118] Mienville, J.M., Vicini, S., 1989. Pregnenolone sulfate antagonizes GABA_A receptor-mediated currents via a reduction of channel opening frequency. Brain Res. 489(1), 190–194.

[119] Sousa, A., Ticku, M.K., 1997. Interactions of the neurosteroid dehydroepiandrosterone sulfate with the GABA_A receptor complex reveals that it may act via the picrotoxin site. J. Pharmacol. Exp. Ther. 282(2), 827–833.

[120] Demirgören, S., Majewska, M.D., Spivak, C.E., London, E.D., 1991. Receptor binding and electrophysiological effects of dehydroepiandrosterone sulfate, an antagonist of the GABA_A receptor. Neuroscience 45(1), 127–135.

[121] Majewska, M.D., Demirgören, S., Spival, C.E., London, E.D., 1990. The neurosteroid dehydroepiandrosterone sulfate is an allosteric antagonist of the GABA_A receptor. Brain Res. 526(1), 143–146.

[122] Majewska, M.D., Schwartz, R.D., 1987. Pregnenolone-sulfate: an endogenous antagonist of γ-aminobutyric acid receptor complex in brain? Brain Res. 404(1–2), 355–360.

[123] Hansen, S.L., Fjalland, B., Jackson, M.B., 1999. Differential blockade of γ-aminobutyric acid type A receptors by the neuroactive steroid dehydroepiandrosterone sulfate in posterior and intermediate pituitary. Mol. Pharmacol. 55(3), 489–496.

[124] El-Etr, M., Akwa, Y., Robel, P., Baulieu, E.E., 1998. Opposing effects of different steroid sulfates on GABA_A receptor-mediated chloride uptake. Brain Res. 790(1–2), 334–338.

[125] Spivak, C.E., 1994. Desensitization and noncompetitive blockade of GABA_A receptors in ventral midbrain neurons by a neurosteroid dehydroepiandrosterone sulfate. Synapse 16, 113–122.

[126] Majewska, M.D., Mienville, J.M., Vicini, S., 1988. Neurosteroid pregnenolone sulfate antagonizes electrophysiological responses to GABA in neurons. Neurosci. Lett. 90(3), 279–284.

[127] Shen, W., Mennerick, S., Covey, D.F., Zorumski, C.F., 2000. Pregnenolone sulfate modulates inhibitory synaptic transmission by enhancing GABA_A receptor desensitization. J. Neurosci. 20(10), 3571–3579.

[128] Lan, N.C., Gee, K.W., Bolger, M.B., Chen, J.S., 1991. Differential responses of expressed recombinant human γ-aminobutyric acidA receptors to neurosteroids. J. Neurochem. 57(5), 1818–1821.

[129] Shingai, R., Sutherland, M., Barnard, E.A., 1991. Effects of subunit type of the cloned GABA$_A$ receptor on the response to a neurosteroid. Eur. J. Pharmacol. 206(1), 77–80.

[130] Puia, G., Ducic, I., Vicini, S., Costa, E., 1993. Does neurosteroid modulatory efficacy depend on GABA$_A$ receptor subunit composition? Receptors Channels 1(2), 135–142.

[131] Belelli, D., Lambert, J.J., Peters, J.A., Gee, K.W., Lan, N.C., 1996. Modulation of human recombinant GABA$_A$ receptors by pregnanediols. Neuropharmacology 35(9/10), 1223–1231.

[132] Hadingham, K.L., Garrett, E.M., Wafford, K.A., Bain, C., Heavens, R.P., Sirinathsinghji, D.J., Whiting, P.J., 1996. Cloning of cDNAs encoding the human γ-aminobutyric acid type A receptor α6 subunit and characterization of the pharmacology of α6-containing receptors. Mol. Pharmacol. 49(2), 253–259.

[133] Hauser, C.A., Wetzel, C.H., Rupprecht, R., Holsboer, F., 1996. Allopregnanolone acts as an inhibitory modulator of α$_1$- and α$_6$-containing GABA$_A$ receptors. Biochem. Biophys. Res. Commun. 219(2), 531–536.

[134] Zhu, W.J., Wang, J.F., Krueger, K.E., Vicini, S., 1996. δ subunit inhibits neurosteroid modulation of GABA$_A$ receptors. J. Neurosci. 16(21), 6648–6656.

[135] Zorumski, C.F., Wittmer, L.L., Isenberg, K.E., Hu, Y., Covey, D.F., 1996. Effects of neurosteroid and benz[e]indene enantiomers on GABA$_A$ receptors in cultured hippocampal neurons and transfected HEK-293 cells. Neuropharmacology 35(9/10), 1161–1168.

[136] Maitra, R., Reynolds, J.N., 1999. Subunit dependent modulation of GABA$_A$ receptor function by neuroactive steroids. Brain Res. 819(1–2), 75–82.

[137] Lambert, J.J., Harney, S.C., Belelli, D., Peters, J.A., 2001. Neurosteroid modulation of recombinant and synaptic GABA$_A$ receptors. Int. Rev. Neurobiol. 46, 177–205.

[138] Wohlfarth, K.M., Bianchi, M.T., Macdonald, R.L., 2002. Enhanced neurosteroid potentiation of ternary GABA$_A$ receptors containing the delta subunit. J. Neurosci. 22(5), 1541–1549.

[139] Bianchi, M.T., Haas, K.F., Macdonald, R.L., 2002. α1 and α6 subunits specify distinct desensitization, deactivation and neurosteroid modulation of GABA$_A$ receptors containing the δ subunit. Neuropharmacology 43(4), 492–502.

[140] Belelli, D., Casula, A., Ling, A., Lambert, J.J., 2002. The influence of subunit composition on the interaction of neurosteroids with GABA$_A$ receptors. Neuropharmacology 43(4), 651–661.

[141] Hadingham, K.L., Wingrove, P.R., Wafford, K.A., Bain, C., Kemp, J.A., Palmer, K.J., Wilson, A.W., Wilcox, A.S., Sikela, J.M., Ragan, C.I., Whiting, P.J., 1993. Role of the β subunit in determining the pharmacology of human γ-aminobutyric acid type A receptors. Mol. Pharmacol. 44(6), 1211–1218.

[142] Belelli, D., Pistis, M., Peters, J.A., Lambert, J.J., 1999. The interaction of general anaesthetics and neurosteroids with GABA$_A$ and glycine receptors. Neurochem. Int. 34(5), 447–452.

[143] Swope, S.L., Moss, S.I., Raymond, L.A., Huganir, R.L., 1999. Regulation of ligand-gated ion channels by protein phosphorylation. Adv. Second Messenger Phosphoprotein Res. 33, 49–78.

[144] Puia, G., Vicini, S., Seeburg, P.H., Costa, E., 1991. Influence of recombinant γ-aminobutyric acid$_A$ receptor subunit composition on the action of allosteric modulators of γ-aminobutyric acid-gated Cl$^-$ currents. Mol. Pharmacol. 39(6), 691–696.

[145] Wisden, W., Laurie, D.J., Monyer, H., Seeburg, P.H., 1992. The distribution of 13 GABA$_A$ receptor subunit mRNAs in the rat brain. I. Telencephalon, diencephalon, mesencephalon. J. Neurosci. 12(3), 1040–1062.

[146] Whiting, P.J., McAllister, G., Vassilatis, D., Bonnert, T.P., Heavens, R.P., Smith, D.W., Hewson, L., O'Donnell, R., Rigby, M.R., Sirinathsinghji, D.J., Marshall, G., Thompson, S.A., Wafford, K.A., 1997. Neuronally restricted RNA splicing regulates the expression of a novel GABA$_A$ receptor subunit conferring atypical functional properties. J. Neurosci. 17(13), 5027–5037.

[147] Sinkkonen, S.T., Hanna, M.C., Kirkness, E.F., Korpi, E.R., 2000. GABA$_A$ receptor ε and θ subunits display unusual structural variation between species and are enriched in the rat locus ceruleus. J. Neurosci. 20(10), 3588–3595.

[148] Moragues, N., Ciofi, P., Lafon, P., Odessa, M.-F., Tramu, G., Garret, M., 2000. cDNA cloning and expression of a γ-aminobutyric acidA receptor ε-subunit in rat brain. Eur. J. Neurosci. 12, 4318–4330.

[149] Boue-Grabot, E., Roudbaraki, M., Bascles, G., Tramu, G., Bloch, B., Garret, M., 1998. Expression of GABA receptor ρ subunits in rat brain. J. Neurochem. 70(3), 899–907.

[150] Wegelius, K., Pasternack, M., Hiltunen, J.O., Rivera, C., Kaila, K., Saarma, M., Reeben, M., 1998. Distribution of GABA receptor ρ subunit transcripts in the rat brain. Eur. J. Neurosci. 10(1), 350–357.

[151] Bonnert, T.P., McKernan, R.M., Farrar, S., le Bourdellés, B., Heavens, R.P., Smith, D.W., Hewson, L., Rigby, M.R., Sirinathsinghji, D.J.S., Brown, N., Wafford, K.A., Whiting, P.J., 1999. θ, a novel γ-aminobutyric acid type A receptor subunit. Proc. Natl Acad. Sci. USA 96(17), 9891–9896.

[152] Jones, A., Korpi, E.R., McKernan, R.M., Pelz, R., Nusser, Z., Mäkelä, R., Mellor, J.R., Pollard, S., Bahn, S., Stephenson, F.A., Randall, A.D., Sieghart, W., Somogi, P., Smith, A.J., Wisden, W., 1997. Ligand-gated ion channel subunit partnerships: GABA$_A$ receptor α_6 subunit gene inactivation inhibits δ subunit expression. J. Neurosci. 17(4), 1350–1362.

[153] Makela, R., Uusi-Oukari, M., Homanics, G.E., Quinlan, J.J., Firestone, L.L., Wisden, W., Korpi, E.R., 1997. Cerebellar γ-aminobutyric acid type A receptors: pharmacological subtypes revealed by mutant mouse lines. Mol. Pharmacol. 52(3), 380–388.

[154] Mihalek, R.M., Banerjee, P.K., Korpi, E.R., Quinlan, J.J., Firestone, L.L., Mi, Z.P., Lagenaur, C., Tretter, V., Sieghart, W., Anagnostaras, S.G., Sage, J.R., Fanselow, M.S., Guidotti, A., Spigelman, I., Li, Z., DeLorey, T.M., Olsen, R.W., Homanics, G.E., 1999. Attenuated sensitivity to neuroactive steroids in γ-aminobutyrate type A receptor delta subunit knockout mice. Proc. Natl Acad. Sci. USA 96(22), 12905–12910.

[155] Vicini, S., Losi, G., Homanicws, G.E., 2002. GABA$_A$ receptor δ subunit deletion prevents neurosteroid modulation of inhibitory synaptic currents in cerebellar neurons. Neuropharmacology 43(4), 646–650.

[156] Adkins, C.E., Pillai, G.V., Kerby, J., Bonnert, T.P., Haldon, C., McKernan, R.M., Gonzalez, J.E., Oades, K., Whiting, P.J., Simpson, P.B., 2001. α4β3δ GABA$_A$ receptors characterized by fluorescence resonance energy transfer-derived measurements of membrane potential. J. Biol. Chem. 276, 38934–38939.

[157] Brown, N., Kerby, J., Bonnert, T.P., Whiting, P.J., Wafford, K.A., 2002. Pharmacological characterization of a novel cell line expressing human α4β3δ GABA$_A$ receptors. Br. J. Pharmacol. 136(7), 965–974.

[158] Davies, P.A., Hanna, M.C., Hales, T.G., Kirkness, E.F., 1997. Insensitivity to anaesthetic agents conferred by a class of GABA$_A$ receptor subunit. Nature 385, 820–823.

[159] McIntyre, K.L., Porter, D.M., Henderson, L.P., 2002. Anabolic androgenic steroids induce age-, sex-, and dose-dependent changes in GABA$_A$ receptor subunit mRNAs in the mouse forebrain. Neuropharmacology 43(4), 634–645.

[160] Thompson, S.A., Bonnert, T.P., Cagetti, E., Whiting, P.J., Wafford, K.A., 2002. Overexpression of the GABA$_A$ receptor ε subunit results in insensitivity to anaesthetics. Neuropharmacology 43(4), 662–668.

[161] Jorge, J.C., Mcintyre, K.L., Henderson, L.P., 2002. The function and the expression of forebrain GABA$_A$ receptors change with hormonal state in the adult mouse. J. Neurobiol. 50, 137–149.

[162] Ymer, S., Draguhn, A., Wisden, W., Werner, P., Keinänen, K., Schofield, P.R., Sprengel, R., Pritchett, D.B., Seeburg, P.H., 1990. Structural and functional characterization of the γ_1 subunit of GABA$_A$/benzodiazepine receptors. EMBO J. 9(10), 3261–3267.

[163] Herbison, A.E., Fénelon, V.S., 1995. Estrogen regulation of GABA$_A$ receptor subunit expression in preoptic area and bed nucleus of the stria terminalis of female rat brain. J. Neurosci. 15(3), 2328–2337.

[164] Clark, A.S., Myers, M., Robinson, S., Chang, P., Henderson, L.P., 1998. Hormone-dependent regulation of GABA$_A$ receptor γ subunit mRNAs in sexually dimorphic regions of the rat brain. Proc. R Soc. Lond. B 265, 1853–1859.

[165] Genazzani, A.R., Palumbo, M.A., de Micheroux, A.A., Artini, P.G., Criscuolo, M., Ficarra, G., Guo, A.L., Benelli, A., Bertolini, A., Petraglia, F., 1995. Evidence for a role for the neurosteroid allopregnanolone in the modulation of reproductive function in female rats. Eur. J. Endocrinol. 133(3), 375–380.

[166] Kochakian, C.D., 1993. Anabolic-androgenic steroids: a historical perspective. In: Yesalis, C.E. (Ed.), Anabolic Steroids in Sports and Exercise. Human Kinetics Publishers, Champaign IL, pp. 3–33.

[167] Lukas, S.E., 1996. CNS effects and abuse liability of anabolic-androgenic steroids. Annu. Rev. Pharmacol. Toxicol. 36, 333–357.

[168] Quincey, R.V., Gray, C.H., 1967. The metabolism of [1,2-^{3}H] 17α-methyl-testosterone in human subjects. J. Endocrinol. 37(1), 37–55.

[169] Winters, S.J., 1990. Androgens: endocrine physiology and pharmacology. NIDA Res. Monogr. 102, 113–130.

[170] Schänzer, W., Opfermann, G., Donike, M., 1990. Metabolism of stanozolol: identification and synthesis of urinary metabolites. J. Steroid Biochem. 36(1–2), 153–174.

[171] Kammerer, R.C., 1993. Drug testing and anabolic steroids. In: Yesalis, C.E. (Ed.), Anabolic Steroids in Sports and Exercise. Human Kinetics Publishers, Champaign, IL, pp. 283–308.

[172] Bronson, F.H., 1996. Effects of prolonged exposure to anabolic steroids on the behavior of male and female mice. Pharmacol. Biochem. Behav. 53(2), 329–334.

[173] Bronson, F.H., Nguyen, K.Q., De La Rosa, J., 1996. Effect of anabolic steroids on behavior and physiological characteristics of female mice. Physiol. Behav. 59(1), 49–55.

[174] Blasberg, M.E., Langan, C.J., Clark, A.S., 1997. The effects of 17α-methyltestosterone, methandrostenolone, and nandrolone decanoate on the rat estrous cycle. Physiol. Behav. 61(2), 265–272.

[175] Johnson, M.D., 1990. Anabolic steroid use in adolescent athletes. Pediatr. Clin. N. Am. 37(5), 1111–1123.

[176] Franke, W.W., Berendonk, B., 1997. Hormonal doping and androgenization of athletes: a secret program of the German Democratic Republic government. Clin. Chem. 43(7), 1262–1279.

[177] Melloni, R.H. Jr, Connor, D.F., Hang, P.T., Harrison, R.J., Ferris, C.F., 1997. Anabolic-androgenic steroid exposure during adolescence and aggressive behavior in golden hamsters. Physiol. Behav. 61(3), 359–364.

[178] Ågren, G., Thiblin, I., Tirassa, P., Lundeberg, T., Stenfors, C., 1999. Behavioural anxiolytic effects of low-dose anabolic androgenic steroid treatment in rats. Physiol. Behav. 66(3), 503–509.

[179] Yesalis, C.E., Barsukiewicz, C.K., Kopstein, A.N., Bahrke, M.S., 1997. Trends in anabolic-androgenic steroid use among adolescents. Arch. Pediatr. Adolesc. Med. 151(12), 1197–1206.

[180] Kow, L.M., Mobbs, C.V., Pfaff, D.W., 1994. Roles of second-messenger systems and neuronal activity in the regulation of lordosis by neurotransmitters, neuropeptides, and estrogen: a review. Neurosci. Biobehav. Rev. 18(2), 251–268.

[181] Herbison, A.E., Chapman, C., Dyer, R.G., 1991. Role of medial preoptic GABA neurones in regulating luteinising hormone secretion in the ovariectomised rat. Exp. Brain Res. 87(2), 345–352.

[182] Decavel, C., Van den Pol, A.N., 1990. GABA: a dominant neurotransmitter in the hypothalamus. J. Comp. Neurol. 302, 1019–1037.

[183] Brussaard, A.B., Kits, K.S., Baker, R.E., Willems, W.P., Leyting-Vermeulen, J.W., Voorn, P., Smit, A.B., Bicknell, R.J., Herbison, A.E., 1997. Plasticity in fast synaptic inhibition of adult oxytocin neurons caused by switch in GABA$_A$ receptor subunit expression. Neuron 19, 1103–1114.

[184] Herbison, A.E., 1997. Estrogen regulation of GABA transmission in rat preoptic area. Brain Res. Bull. 44(4), 321–326.

[185] Smith, S.S., Gong, G.H., Hsu, F.-C., Markowitz, R.S., ffrench-Mullen, J.M., Li, X., 1998. GABA$_A$ receptor α4 subunit suppression prevents withdrawal properties of an endogenous steroid. Nature 392, 926–930.

[186] Smith, S.S., Gong, Q.H., Li, X., Moran, M.H., Bitran, D., Frye, C.A., Hsu, F.-C., 1998. Withdrawal from 3α-OH-5α-pregnan-20-one using a pseudopregnancy model alters the kinetics of hippocampal GABA$_A$-gated current and increases the GABA$_A$ receptor α$_4$ subunit in association with increased anxiety. J. Neurosci. 18(14), 5275–5284.

[187] Delville, Y., 1991. Progesterone-facilitated sexual receptivity: a review of arguments supporting a nongenomic mechanism. Neurosci. Biobehav. Rev. 15, 407–414.

[188] Frye, C.A., Bayon, L.E., Pursnani, N.K., Purdy, R.H., 1998. The neurosteroids, progesterone and 3α,5α-THP, enhance sexual motivation, receptivity, and proceptivity in female rats. Brain Res. 808, 72–83.

[189] McCarthy, M.M., Beach Award, F.A., 1995. Functional significance of steroid modulation of GABAergic neurotransmission: analysis at the behavioral, cellular, and molecular levels. Horm. Behav. 29, 131–140.

[190] Frye, C.A., Vongher, J.M., 1999. Progestins' rapid facilitation of lordosis when applied to the ventral tegmentum corresponds to efficacy at enhancing GABA$_A$ receptor activity. J. Neuroendocrinol. 11(11), 829–837.

[191] Frye, C.A., Van Keuren, K.R., Rao, P.N., Erskine, M.S., 1996. Progesterone and 3α-androstanediol conjugated to bovine serum albumin affects estrous behavior when applied to the MBH and POA. Behav. Neurosci. 110(3), 603–612.

[192] Frye, C.A., Duncan, J.E., Basham, M., Erskine, M.S., 1996. Behavioral effects of 3α-androstanediol II: Hypothalamic and preoptic area actions via a GABAergic mechanism. Behav. Brain Res. 79, 119–130.

[193] Frye, C.A., Gardiner, S.G., 1996. Progestins can have a membrane-mediated action in rat midbrain for facilitation of sexual receptivity. Horm. Behav. 30(4), 682–691.

[194] Frye, C.A., Vongher, J.M., 1999. Progesterone has rapid and membrane effects in the facilitation of female mouse sexual behavior. Brain Res. 815, 259–269.

[195] Erskine, M.S., 1987. Serum 5α-androstane-3α,17β-diol increases in response to paced coital stimulation in cycling female rats. Biol. Reprod. 37, 1139–1148.

[196] Verhoeven, G., Heyns, W., De Moor, P., 1975. Ammonium sulfate precipitation as a tool for the study of androgen receptor proteins in rat prostrate and mouse kidney. Steroids 6, 149–167.

[197] Roselli, C.E., Horton, L.E., Resko, J.A., 1987. Time-course and steroid specificity of aromatase induction in rat hypothalamus-preoptic area. Biol. Reprod. 37, 628–633.

[198] Erskine, M.S., 1989. Effect of 5α-dihydrotestosterone and flutamide on the facilitation of lordosis by LHRH and naloxone in estrogen-primed female rats. Physiol. Behav. 45, 753–759.

[199] Poulain, D.A., Wakerley, J.B., 1982. Electrophysiology of hypothalamic magnocellular neurones secreting oxytocin and vasopressin. Neuroscience 7, 773–808.

[200] Brussaard, A.B., Kits, K.S., de Vlieger, T.A., 1996. Postsynaptic mechanism of depression of GABAergic synapses by oxytocin in the supraoptic nucleus of immature rat. J. Physiol. (Lond.) 497, 495–507.

[201] Summerlee, A.J., 1981. Extracellular recordings from oxytocin neurones during the expulsive phase of birth in unanaesthesized rats. J. Physiol. (Lond.) 321, 1–9.

[202] Leng, G., Brown, C.H., Russell, J.A., 1999. Physiological pathways regulating the activity of magnocellular neurosecretory cells. Prog. Neurobiol. 57, 625–655.

[203] Korpi, E.R., 1994. Role of GABA_A receptors in the actions of alcohol and in alcoholism: recent advances. Alcohol Alcohol. 29(2), 115–129.

[204] Mehta, A.K., Ticku, M.K., 1998. Chronic ethanol administration alters the modulatory effects of 5α-pregnan-3α-ol-20-one on the binding characteristics of various radioligands of GABA_A receptors. Brain Res. 805, 88–94.

[205] Grobin, A.C., Matthews, D.B., Devaud, L.L., Morrow, A.L., 1998. The role of GABA_A receptors in the acute and chronic effects of ethanol. Psychopharmacology 139(1–2), 2–19.

[206] Criswell, H.E., McCown, T.J., Ming, Z., Mueller, R.A., Breese, G.R., 1999. Interactive role for neurosteroids in ethanol enhancement of γ-aminobutyric acid-gated currents from dissociated substantia nigra reticulata neurons. J. Pharmacol. Exp. Ther. 291(3), 1054–1059.

[207] Janak, P.H., Redfern, J.E., Samson, H.H., 1998. The reinforcing effects of ethanol are altered by the endogenous neurosteroid, allopregnanolone. Alcohol Clin. Exp. Res. 22(5), 1106–1112.

[208] Bienkowski, P., Kotowski, W., 1997. Discriminative stimulus properties of ethanol in the rat: effects of neurosteroids and picrotoxin. Brain Res. 753(2), 348–352.

[209] VanDoren, M.J., Matthews, D.B., Janis, G.C., Grobin, A.C., Devaud, L.L., Morrow, A.L., 2000. Neuroactive steroid 3α-hydroxy-5α-pregnan-20-one modulates electrophysiological and behavioral actions of ethanol. J. Neurosci. 20(5), 1982–1989.

[210] Devaud, L.L., Purdy, R.H., Morrow, A.L., 1995. The neurosteroid 3α-hydroxy-5α-pregnan-20-one protects against bicuculline-induced seizures during ethanol withdrawal in rats. Alcohol Clin. Exp. Res. 19, 350–355.

[211] Devaud, L.L., Purdy, R.H., Finn, D.A., Morrow, A.L., 1996. Sensitization of γ-aminobutyric acid_A receptors to neuroactive steroids in rats during ethanol withdrawal. J. Pharmacol. Exp. Ther. 278, 510–517.

[212] Romeo E. Brancati, A., De Lorenzo, A., Fucci, P., Furnari, C., Pompili, E., Sasso, G.F., Spalleta, G., Troisi, A., Pasini, A., 1996. Marked decrease of plasma neuroactive steroids during alcohol withdrawal. Clin. Neuropharmacol. 19(4), 366–369.

[213] Sturenburg, H.J., Fries, U., Kunze, K., 1997. Glucocorticoids and anabolic/androgenic steroids inhibit the synthesis of GABAergic steroids in rat cortex. Neuropsychobiology 35(3), 143–146.

[214] Baldessarini, R.J., 1996. Drugs and the treatment of psychiatric disorders. In: Hardman, J.G., Limbird, L.E. (Eds.), The Pharmacological Basis of Therapeutics. 9th ed. McGraw-Hill, New York, pp. 399–430.

[215] Romeo, E., Ströhle, A., Spalletta, G., di Michele, F., Hermann, B., Holsboer, F., Pasini, A., Rupprecht, R., 1998. Effects of antidepressant treatment on neuroactive steroids in major depression. Am. J. Psychiatry 155(7), 910–913.

[216] Uzunova, V., Sheline, Y., Davis, J.M., Rasmusson, A., Uzunov, D.P., Costa, E., Guidottii, A., 1998. Increase in the cerebrospinal fluid content of neurosteroids in patients with unipolar major depression who are receiving fluoxetine or fluvoxamine. Proc. Natl Acad. Sci. USA 95, 3239–3244.

[217] Rupprecht, R., Holsboer, F., 1999. Neuroactive steroids: mechanism of action and neuropsychopharmacological perspectives. Trends Neurosci. 22, 410–416.

[218] Uzunov, D.P., Cooper, T.B., Costa, E., Guidotti, A., 1996. Fluoxetine-elicited changes in brain neurosteroid content measured by negative ion mass fragmentography. Proc. Natl Acad. Sci. USA 93, 12599–12604.

[219] Bitran, D., Hilvers, R.J., Kellogg, C.K., 1991. Anxiolytic effects of 3α-hydroxy-5α[β]-pregnan-20-one: endogenous metabolites of progesterone that are active at the GABA$_A$ receptor. Brain Res. 561(1), 157–161.

[220] Bitran, D., Shiekh, M., McLeod, M., 1995. Anxiolytic effect of progesterone is mediated by the neurosteroid allopregnanolone at brain GABA$_A$ receptors. J. Neuroendocrinol. 7, 171–177.

[221] Wieland, S., Lan, N.C., Mirasedeghi, S., Gee, K.W., 1991. Anxiolytic activity of the progesterone metabolite 5α-pregnan-3α-ol-20-one. Brain Res. 565, 263–268.

[222] Wieland, S., Belluzi, J.D., Stein, L., Lan, N.C., 1995. Comparative behavioral characterization of the neuroactive steroids 3α-OH, 5α-pregnan-20-one and 3α-OH, 5β-pregnan-20-one in rodents. Psychopharmacology 118, 65–71.

[223] Paul, S.M., Purdy, R.H., 1992. Neuroactive steroids. FASEB J. 6, 2311–2322.

[224] Morrell, M.J., 1999. Epilepsy in women: the science of why it is special. Neurology 53(4), S42–S48.

[225] Bäckström, T., Cartensen, A., 1974. Estrogen and progesterone in plasma in relation to premenstrual tension. J. Steroid Biochem. 5, 257–260.

[226] Ball, D.M., Glue, P., Wilson, S., Nutt, D.J., 1991. Pharmacology of saccadic eye movements in man. 1. Effects of the benzodiazepine receptor ligands midazolam and flumazenil. Psychopharmacology 105, 361–367.

[227] Sundström, I., Andersson, A., Nyberg, S., Ashbrook, D., Purdy, R.H., Bäckström, T., 1998. Patients with premenstrual syndrome have a different sensitivity to a neuroactive steroid during the menstrual cycle compared to control subjects. Neuroendocrinology 67(2), 126–138.

[228] Sundström, I., Nyberg, S., Bäckström, T., 1997. Patients with premenstrual syndrome have reduced sensitivity to midazolam compared to control subjects. Neuropsychopharmacology 17(6), 370–381.

[229] Sundström, I., Ashbrook, D., Bäckström, T., 1997. Reduced benzodiazepine sensitivity in patients with premenstrual syndrome: a pilot study. Psychoneuroendocrinology 22(1), 25–38.

[230] Gallo, M.A., Smith, S.S., 1993. Progesterone withdrawal decreases latency to and increases duration of electrified prod burial: a possible rat model of PMS anxiety. Pharmacol. Biochem. Behav. 46, 897–904.

[231] Schmidt, P.J., Purdy, R.H., Moore, P.H. Jr, Paul, S.M., Rubinow, D.R., 1994. Circulating levels of anxiolytic steroids in the luteal phase in women with premenstrual syndrome and in control subjects. J. Clin. Endocrinol. Metab. 79, 1256–1260.

[232] Wang, M., Seippel, L., Purdy, R.H., Bäckström, T., 1996. Relationship between symptom severity and steroid variation in women with premenstrual syndrome: study on serum pregnenolone, pregnenolone sulfate, 5α-pregnane-3,20-dione and 3α-hydroxy-5α-pregnan-20-one. J. Clin. Endocrinol. Metab. 81(3), 1076–1082.

[233] Rapkin, A.J., Morgan, M., Goldman, L., Brann, D.W., Simone, D., Mahesh, V.B., 1997. Progesterone metabolite allopregnanolone in women with premenstrual syndrome. Obstet. Gynecol. 90(5), 709–714.

[234] Herzog, A.G., Klein, P., Rensil, E.J., 1997. Three patterns of catamenial epilepsy. Epilepsia 38, 1082–1088.

[235] Reddy, D.S., Rogawski, M.A., 2000. Enhanced anticonvulsant activity of ganaxolone after neurosteroid withdrawal in a rat model of catamenial epilepsy. J. Pharmacol. Exp. Ther. 294(3), 909–915.

[236] Moran, M.H., Smith, S.S., 1998. Progesterone withdrawal I: pro-convulsant effects. Brain Res. 807(1–2), 84–90.

[237] Kokate, T.G., Banks, M.K., Magee, T., Yamaguchi, S., Rogawski, M.A., 1999. Finasteride, a 5α-reductase inhibitor, blocks the anticonvulsant activity of progesterone in mice. J. Pharmacol. Exp. Ther. 288(2), 679–684.

[238] Gasior, M., Carter, R.B., Goldberg, S.R., Witkin, J.M., 1997. Anticonvulsant and behavioral effects of neuroactive steroids alone and in conjunction with diazepam. J. Pharmacol. Exp. Ther. 282, 543–553.

[239] Carter, R.B., Wood, P.L., Wieland, S., Hawkinson, J.E., Belelli, D., Lambert, J.J., White, H.S., Wolf, H.H., Mirsadeghi, S., Tahir, H., Bolger, M.B., Lan, N.C., Gee, K.W., 1997. Characterization of the

anticonvulsant properties of ganaxolone (CCD 1042; 3α-hydroxy-3β-methyl-5α-pregnan-20-one), a selective high-affinity steroid modulator of the GABA$_A$ receptor. J. Pharmacol. Exp. Ther. 280, 1284–1295.

[240] Gasior, M., Carter, R.B., Witkin, J.M., 1999. Neuroactive steroids: potential therapeutic use in neurological and psychiatric disorders. Trends Pharmacol. Sci. 20, 107–112.

[241] Belelli, D., Bolger, M.B., Gee, K.W., 1989. Anticonvulsant profile of the progesterone metabolite 5α-pregnan-3α-ol-20-one. Eur. J. Pharmacol. 166, 325–329.

[242] Beekman, M., Ungard, J.T., Gasior, M., Carter, R.B., Dijkstra, D., Goldberg, S.R., Witkin, J.M., 1998. Reversal of behavioral effects of pentylenetetrazol by the neuroactive steroid ganaloxone. J. Pharmacol. Exp. Ther. 284, 868–877.

[243] Monaghan, E.P., Navalta, L.A., Shum, L., Ashbrook, D.W., Lee, D.A., 1997. Initial human experience with ganaxolone, a neuroactive steroid with antiepileptic activity. Epilepsia 38(9), 1026–1031.

[244] Araque, A., Parpura, R.P., Sanzgiri, R.P., Haydon, P.G., 1999. Tripartite synapses: glia, the unacknowledged partner. Trends Neurosci. 22, 208–215.

[245] Theodosis, D.T., MacVicar, B., 1996. Neurone–glial interactions in the hypothalamus and pituitary. Trends Neurosci. 19, 363–367.

[246] Jordan, C.L., 1999. Glia as mediators of steroid hormone action on the nervous system: an overview. J. Neurobiol. 40, 434–445.

Cyclic nucleotide-gated channels: multiple isoforms, multiple roles

Marie-Christine Broillet[a],[*] and Stuart Firestein[b]

[a]*Institute of Pharmacology and Toxicology, University of Lausanne, CH-1005 Lausanne, Switzerland*
[b]*Department of Biological Sciences, Columbia University, New York, NY 10027, USA*
[*]*Correspondence address: Tel.: +41-21-692-5370; fax: +41-21-692-5355*
E-mail: mbroille@ipharm.unil.ch

Cyclic nucleotide-gated (CNG) channels are a family of ligand-gated channels that are activated by the binding of at least two molecules of cAMP or cGMP at intracellular sites on the channel protein. They are non-selective cation channels conducting both mono- and divalent cations and they belong to the superfamily of cation channels with six transmembrane segments (Fig. 1A). This superfamily includes voltage-gated K^+, Na^+, and Ca^{2+} channels, hyperpolarization-activated cyclic nucleotide gated (HCN) channels, transient receptor potential (TRP) channels and the polycystins [1].

CNG channels were originally identified in vertebrate photoreceptor cells [2] and olfactory receptor neurons (ORNs) [3] where they mediate calcium entry, providing an intracellular calcium signal that is important for both excitation and adaptation [4–7]. Because of their strong calcium permeability, activation of CNG channels by the ubiquitous cyclic nucleotide second messengers lead not only to membrane depolarization but also to a significant calcium influx into the cells [8].

1. CNG channels: structure and nomenclature

The CNG channels are constructed from different but highly homologous subunits (Fig. 1A) that assemble in a yet unknown stoichiometry, probably as a heterotetramer [9]. The olfactory CNG channel is, for example, constructed of at least three different subunits: the CNGA2 subunit also called α, α3 or CNG2 (original designation OCNC1; [10]), the CNGA4 subunit also called β, α4 or CNG5 (original designation OCNC2; [11,12]), and the CNGB1b subunit also called β1b [13] or CNG4.3 [14]. In this review, we will refer to the recently adopted nomenclature [15].

The different channel subunits are similar in structure to voltage-gated K^+ channels except that they possess a cyclic nucleotide-binding site on the intracellular C-terminal tail, and have no apparent voltage-sensitivity [6]. Indeed, CNG channels possess

Advances in Molecular and Cell Biology, Vol. 32, pages 251–267
ISSN: 1569-2558 / DOI: 10.1016/S1569-2558(03)32011-9

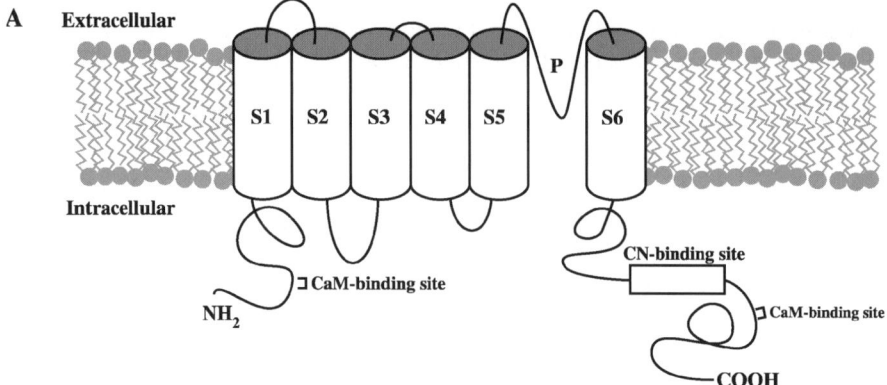

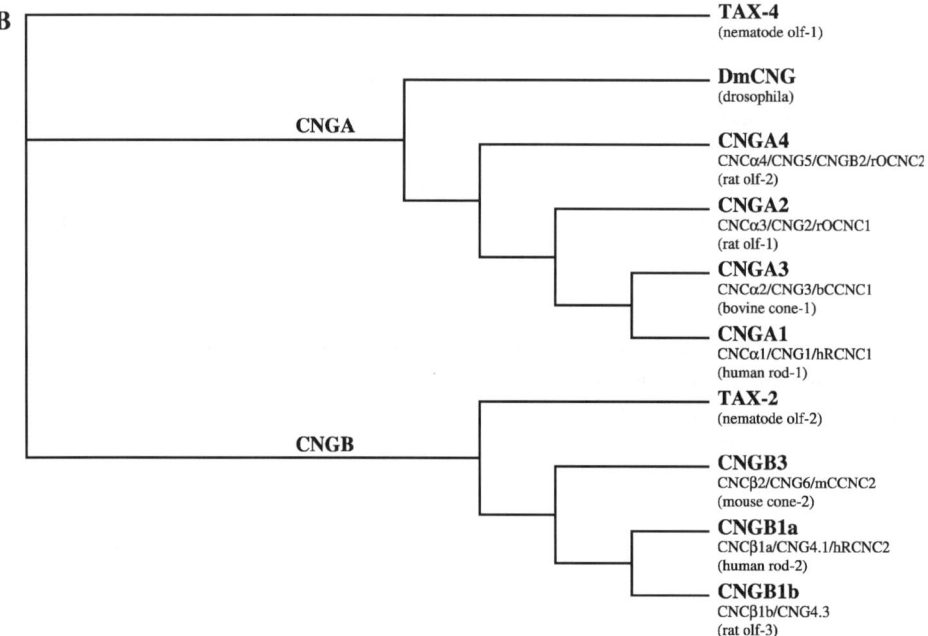

Fig. 1. The cyclic nucleotide-gated channels. (A) Hypothetical model of the two-dimensional architecture of a cyclic nucleotide-gated channel subunit. S1–S6 are the putative transmembrane domains, P is the putative pore region. The cyclic nucleotide (CN) binding site is defined by homology to the sequences of cAMP and cGMP binding proteins. (B) Phylogenetic tree of olfactory and retinal CNG channels subunits. The tree was calculated on the basis of sequence alignments with the transmembrane domains and the cyclic nucleotide binding site of the respective subunits. The different names used in the literature are indicated.

a voltage-sensor motif in S4 with a reduced number of positively charged amino acids. This domain might be the ancestral S4 segment that has evolved into the voltage sensor of voltage-activated cation channels [16].

A series of elegant experiments with cloned channels and chimeric constructs has revealed significant information regarding the binding and gating reactions that lead to CNG channel activation [6,17–27]. These studies have identified several regions as well as specific residues distributed throughout the approximately 500 amino acids of the rod or the olfactory proteins that play a key role in channel regulation [28].

The different CNG channel subunits can be grouped in two main types, called CNGA and CNGB (Fig. 1B). The CNGA subunits (or *principal* subunits) can form functional homomeric channels activable by cyclic nucleotides when expressed in heterologous systems (like HEK293 cells or *Xenopus laevis* oocytes). Three types of different CNGA subunits have been identified. The CNGA1 subunit was first identified in the rod photoreceptor cells [29], the CNGA2 is the corresponding subunit in ORNs [10] and the CNGA3 subunit has been cloned in the testis and cone photoreceptor cells [30,31]. These CNGA subunits share common structural features (60–70% homology) such as 6 transmembrane domains (S1–6), an S4-like voltage sensor motif, a pore region and a cyclic nucleotide (CN) binding-site in the intracellular N-terminal region (Fig. 1A).

Additional subunits have been discovered, sharing the same structural properties with the CNGA subunits but with some added diversity; they are called CNGB subunits. They are considered to be *modulatory* subunits, as they cannot form functional cyclic nucleotide-gated channels on their own in expression systems. The CNGB1a is the rod modulatory subunit [32]. It has an unusual bipartite structure consisting of a core region and a long N-terminal sequence, which is characterized by a large number of glutamic acid- and proline residues. This N-terminal domain is also expressed as two soluble protein variants called glutamic-acid-rich proteins (GARPs). In rod photoreceptors, GARPs are multivalent proteins that interact with the key players of cGMP signalling, phosphodi-esterase and guanylate cyclase, and with a retina-specific ATP-binding cassette transporter, through four, short, repetitive sequences [33]. Recently, it has been demonstrated that the rod CNG channel associates with the Na/Ca–K exchanger in the rod outer segment plasma membrane. This channel-exchanger complex and the soluble GARP proteins also interact with peripherin-2 oligomers in the rim region of outer segment disc membranes. These results suggest that channel/peripherin protein interactions mediated by the GARP part of the CNGB1a subunit play a role in connecting the rim region of discs to the plasma membrane and in anchoring the channel-exchanger complex in the rod outer segment plasma membrane [34].

The CNGA4 subunit is found in ORNs [11,12]; while it shares more sequence homology with the CNGA subunits (Fig. 1 B), it cannot form homomeric cyclic nucleotide-activated ion channels, and so from a functional standpoint is classified with the modulatory CNGB subunits. The CNGB1b subunit, a splice variant of the rod modulatory subunit (CNGB1a), is also a component of the olfactory channel [13,14]. Recently, the modulatory subunit of the cone CNG channel (CNGB3) has been cloned [35].

The subunit composition of the native retinal or olfactory CNG channels has not been determined yet, but experimental evidence suggests that they are most likely composed of a mix of different CNGA and CNGB subunits. For example, while the CNGA1 (retinal) or the CNGA2 (olfactory) subunits can form functional homomeric channels in heterologous expression systems, their biophysical properties are different from native channels [10,29] (Table 1). Furthermore, homomeric channels formed by the CNGB1a (retinal) and the CNGA4 (olfactory) subunits expressed in these cells cannot be activated by cyclic nucleotides [11,12].

The co-expression of CNGA and CNGB subunits in heterologous systems results in heteromeric channels with different properties than those observed for homomeric channels consisting of CNGA subunits. For example, each channel subunit seems to confer specific affinity for the ligands of the channel. When expressed in the same cell, CNGA3 and CNGB3 cone subunits have properties very similar to the native channel, including ligand affinity [35] implying the mixed nature of the native channel. The co-expression of CNGA2/CNGA4/CNGB1b olfactory subunits in a heterologous system leads to the formation of a channel whose properties resemble the native olfactory channel and differ from those observed for homomeric CNGA2 channels. The native olfactory channel is sensitive to both cGMP and cAMP (with a higher affinity for cGMP; $K_{1/2}$ for cAMP $= 4$ μM, $K_{1/2}$ for cGMP $= 1.8$ μM, (Table 1) [13]). The native channel subtypes also differ in their relative permeability to physiological concentrations of calcium such that the fractional current carried by calcium in the olfactory channel is greater than in the rod channel [44]. In heterologous expression systems, it has been shown that calcium permeation is determined by the subunit composition of the channel [45].

In summary, each homomeric or heteromeric channel combination has its own pharmacological characteristics (ligand affinity and efficacy) and electrophysiological properties (ion selectivity and calcium permeation, conductance, and gating kinetics). Thus, the subunit composition of CNG channels clearly determines both the ligand sensitivity and the electrophysiological properties, and therefore calcium entry in the cells.

2. CNG channels: multiple ligand sensitivity

It has been shown that in addition to their activation by cyclic nucleotides, nitric oxide (NO)-generating compounds can directly open the olfactory CNG channels through a redox reaction that results in the S-nitrosylation of a free SH group on a cysteine residue [46]. This post-translational modification, comparable to phosphorylation, has been shown to modulate the activity of other proteins, including caspases and NMDA receptor channels [47].

The NO target site on the CNG channel has been identified by mutating the four candidate intracellular cysteine residues Cys-460, Cys-484, Cys-520, and Cys-552 of the rat olfactory CNGA2 channel into serine residues. All mutant channels continue to be activated by cyclic nucleotides, but only one of them, the C460S mutant channel, exhibits a total loss of NO sensitivity [48]. This result is consistent with the lack of NO sensitivity of the CNG channel expressed in *Drosophila melanogaster* (DmCNG), which does not have this specific cysteine residue [48,49]. Cys-460 is located in the C-linker region of the channel known to be important in channel gating. Kinetic analyses suggest that at least two of these Cys-460 residues on different channel subunits are involved in

Table 1
Properties of cyclic nucleotide-gated channels of various species

Species	Composition	Conductance (pS)[a]	$K_{1/2}$ (µM)		n (Hill Coefficient)		References
			cAMP	cGMP	cAMP	cGMP	
Rod channels							
Human	CNGA1	25–30	–	60	–	2	[36]
	CNGA1CNGB1	Flickery	–	60	–	2	[36]
Bovine	Native	6	–	165	–	2	[37]
	CNGA1	20	2000	60	1.2	2	[18,29]
Salamander	Native		846	9.8	1.8	2.2	[38]
Cone channels							
Frog	Native	45–50	–	42	–	2.5	[39]
Olfactory channels							
Rat	Native	12–27	4	1.8	1.8	1.3	[13,40]
	CNGA2	35	55	2	2.6	2.5	[10]
	CNGA2CNGB1	19	14	4	1.5	1.4	[11,12]
	CNGA2CNGA4	34	10.3	–	1.8	–	[13]
	CNGA2CNGA4CNGB1	27	4.8	–	2.2	–	[13]
Catfish	Native	44	3.4	2.5	1.4	1.3	[40,41]
	CNGA2	55	64	75	1.5	1.2	[18]
Salamander	Native	45	20	2.1	4	2.1	[42]
Newt	Native	28	22	–	1.2	–	[43]
Toad	Native	–	3.4	1.5	1.5	–	[3]
Fro	Native	12–15	4	1.6	1.5	1.4	[40]

[a]Single channel conductance in the absence of divalent cations.

the activation by NO. These results show that one single cysteine residue is responsible for NO sensitivity but that several channel subunits need to be activated for channel opening by NO [50].

The corresponding cysteine on the olfactory CNGA4 subunit is the cysteine 352. This subunit cannot form homomeric channels that can be activated by cyclic nucleotides, but they can be activated by NO. There are striking differences between the properties of these NO-gated channels and either native CNG channels in rat olfactory neurons or homomeric CNGA2 channels expressed in heterologous systems, including the amount of open channel noise and the calcium permeability [50]. Thus, these homomeric CNGA4 channels might be expressed at different cellular locations and fulfill different physiological roles than the heteromeric CNG channels.

3. CNG channels: specific inhibitors

Although there is a formidable array of specific blockers for sodium, calcium and potassium channels, specific blockers for CNG channels are scarce. A number of pharmacological agents including L-*cis*-diltiazem [51], tetracaine [52], pimozide [53] and LY-58558 [54] have been reported to block the current through CNG channels, but these agents have significant drawbacks. First, they exert their effect on the cytoplasmic face of the channel, making their utility questionable in more intact preparations. Second, these blockers have multiple targets, making results difficult to interpret.

Recently, Brown et al. [55] have found that a peptide toxin purified from the venom of the Australian King Brown snake, that they have named pseudechetoxin (PsTx), when applied to the extracellular face of membrane patches containing the rat CNGA2 olfactory CNG channel, blocked the cGMP-dependent current with a K_i of 5 nM. The block was independent of voltage and required only a single molecule of the toxin. PsTx also blocked CNG channels containing the bovine rod CNGA1 with high affinity (100 nM), but it was less effective on the heteromeric version of the rod channel (K_i approximately 3 μM). PsTx promises to be a valuable pharmacological tool for the analyses of the structure and physiology of CNG channels.

Membrane-permeant phosphorothioate derivatives of cAMP and cGMP can also be used to elucidate the physiological roles of CNG channels in intact cells [56]. Based on the different structural features of the ligand, phosphorothioate derivatives can be either agonists or competitive antagonists of CNG channels. The photoreceptor channel uses the nature of the purine ring (adenine vs. guanine), whereas the olfactory channel uses the isomeric position of the thiophosphate S atom (Rp vs. Sp). Interestingly, the same ligand, Rp-cGMPS, has opposite effects on the two channels, activating the photoreceptor channel and antagonizing the olfactory channel. Because Rp-cGMPS binds to both channels but activates only one, the channels must differ in a protein region that couples binding to gating [56].

4. CNG channels: modulators

CNG channels are modulated by a plethora of cellular factors, including phosphorylation enzymes [57–59], Ca^{2+}/calmodulin (CaM) [20,60,61], endogenous Ca^{2+} binding proteins [20,62], transition metal divalent cations [19,63,64] and lipids, including diacylglycerol [65–67].

A characteristic of CNG channels is their Ca^{2+} permeability. At physiological extracellular Ca^{2+} concentrations, Ca^{2+} represents a significant fraction of the current passing through CNG channels. Ca^{2+} is very important for both adaptation and excitation of photoreceptors and olfactory neurons as it controls the activity of several signaling enzymes including the CNG channels themselves (for reviews see Refs. [68,69]). CaM attenuates the activity of rod and olfactory CNG channels by increasing their apparent $K_{1/2}$ for cGMP and cAMP (for review see Ref. [70]). This modulation is believed to serve as one of the several Ca^{2+}-mediated feedback mechanisms that terminate the electrical response and set the sensitivity of the photoreceptor cells and olfactory neurons (for reviews see Refs. [5,70]).

Weitz et al. [71] and Grunwald et al. [72] have examined the mechanism of CaM modulation using electrophysiological, biochemical techniques and surface plasmon resonance spectroscopy. They have identified a domain on the CNGB1a subunit of the rod CNG channel that mediates inhibition by CaM. Using heteromeric channels consisting of retinal CNGA1 and CNGB1a subunits that display a high CaM sensitivity (EC50 ≤ 5 nM) similar to the native channel, they identified two unconventional CaM-binding sites (CaM1 and CaM2), one in each of the N- and the C-terminal regions of the CNGB1a subunit. Ca^{2+} co-operatively stimulates binding of CaM to these sites exactly within the range of Ca^{2+} concentrations occurring during a light response. Deletion of the N-terminal CaM1 site results in channels that are no longer CaM-sensitive, whereas deletion of CaM2 has only minor effects. These results indicate that CaM controls the activity of the rod CNG channel by decreasing the apparent cGMP affinity through an unconventional binding site in the N-terminus of the CNGB1a subunit.

Müller et al. [59] showed that phosphorylation of a serine residue near the CaM-binding site sensitized the olfactory CNG channel to cAMP and thereby extended the range of CaM modulation. Indeed, in olfactory neurons, negative modulation by Ca^{2+} through calmodulin (CaM) on CNG channels underlies the adaptation of ORNs to odorants. Bardley et al. [73] have recently shown that this feedback mechanism requires the two modulatory subunits of the native olfactory channel, CNGA4 and CNGB1b, even though the machinery for CaM binding and modulation is present in the principal subunit CNGA2. This provides a rationale for the presence of three distinct subunits in the native olfactory channel and underscores the subtle link between the molecular make-up of an ion channel and its physiological function.

5. CNG channels: roles in the olfactory system

5.1. Olfactory sensory transduction

One important model system to study the different functions of CNG channels is the olfactory system [74]. The remarkable capacity to discriminate among a wide range of odor molecules begins at the level of the ORNs. These particular neurons perform the complex task of converting the chemical information contained in the odor molecules into changes in membrane potential [68]. In most vertebrates, the ORNs form a sensory epithelium within the nasal cavity. They are true neurons, sending an axon to the central nervous system. They have a bipolar morphology and a single dendrite extended to the epithelial surface bearing 10–12 cilia. The sensory transduction occurs at the level of the ciliary membrane. Odorant recognition involves membrane protein receptors and transduction components analogous to those that mediate the specific responses to hormones, growth factors and neurotransmitters [75]. Every molecular element of the olfactory transduction cascade has been isolated, cloned and expressed allowing the establishment of the scheme presented in Fig. 2A. The different steps of the transduction cascade can be summarized as follows: when a receptor molecule is occupied by an odorant, it activates a specific GTP-binding protein (G_{olf}), which modulates the activity of an adenylyl cyclase (AC type III), an enzyme producing the second messenger cAMP. cAMP directly activates a CNG channel representing the final step in the biochemical cascade and the first step in the generation of the electrical response. An additional, unique membrane conductance, a Ca^{2+}-activated chloride current, is also involved in the electrical response to odors [76].

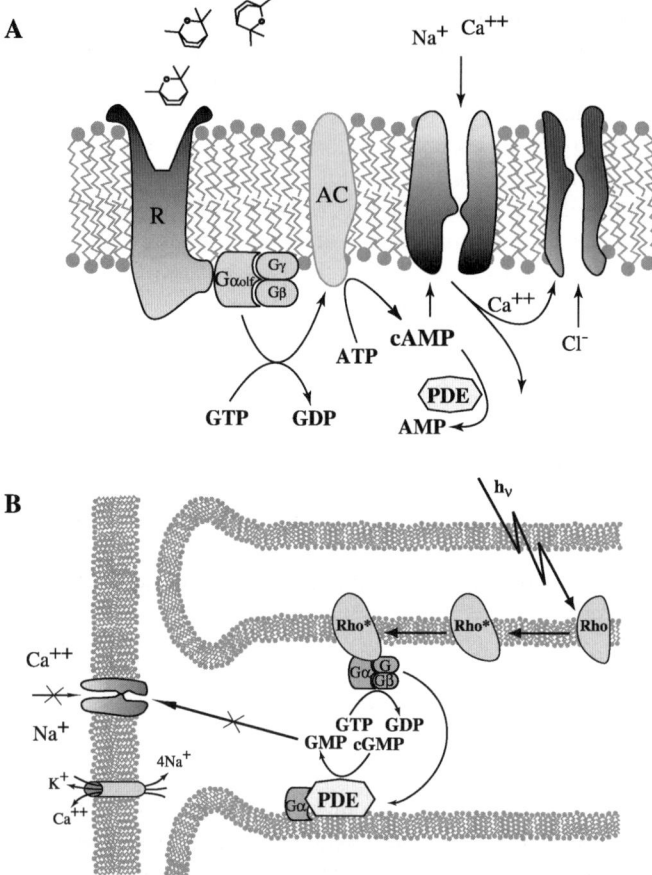

Fig. 2. (A) The olfactory signal transduction cascade. In this pathway, the binding of an odorant molecule (cineole is represented here) carried through the mucus layer via an odour binding protein (OBP) to the odorant receptor leads to the interaction of the receptor to a GTP-binding protein (G protein). This interaction in turn leads to the release of the GTP-coupled α subunit of the G protein, which then stimulates the adenylyl cyclase (AC) to produce elevated levels of cAMP. The increase of cAMP opens cyclic nucleotide-gated channels (CNG) causing an alteration of the membrane potential. (B) The visual signal transduction cascade. In the dark, the CNG channel is co-operatively kept open on binding of 4 cGMP molecules to its CNGB subunits. This causes an exchange of cations (Ca^{2+}, Na$^+$, K$^+$) between the cytoplasm and the surrounding inter photoreceptor space. To keep the ion gradients active the cations are actively pumped across the plasma membrane by Na$^+$/Ca^{2+}/K$^+$-exchanger. In the light cGMP is hydrolysed to 5′GMP by the phosphodiesterases (PDE). With decreased cGMP concentration, cGMP is removed from the CNG channel subunits and the channel is closed. This blocks the flow of Ca^{2+} and Na$^+$ inside the outer segment of the rod (ROS). In darkness, the inward flow of charges (Ca^{2+}, Na$^+$, K$^+$) is equal to the outward flow. This is obtained by a Na$^+$/Ca^{2+}/K$^+$-exchanger in the ROS membrane. At illumination the Na$^+$/Ca^{2+}/K$^+$-exchanger is still active but the inward Ca^{2+} and Na$^+$ flow through the CNG is blocked and the plasma membrane is hyperpolarized because the charge flow rates have become unequal.

This cascade of events results in the cell membrane shifting the resting potential from -65 to -45 mV. This depolarization spreads by passive current flow through the dendrite to the soma where it activates voltage-gated Na^+ channels, initiating impulse generation. The combination of Na^+ currents, voltage-dependent K^+ currents and small Ca^{2+} currents acts to produce one or more action potentials that can propagate via the axon to the olfactory bulb of the brain.

In olfactory neurons, CNG channels can be activated by either cAMP or cGMP (Table 1), although it is generally believed that under normal physiological conditions it is a rise in intracellular cAMP that is responsible for channel activation [68,74]. In summary, a high density of CNG channels is present on the ciliary membrane of ORNs. These channels are selective for cations and their activation plays a key role in cell membrane depolarization due to the olfactory stimulus.

5.2. Olfactory neuronal regeneration and development

Axonal pathfinding and target recognition are critical steps in the formation of specific axonal connections in the developing nervous system. These processes could be controlled by local cues, either molecular or structural that develop in the region of the axon tract. Recently, cyclic nucleotides have been receiving much attention as important regulators in determining whether environmental cues are functioning as attractive or repulsive for growth cone behavior [77–79]. The effects of cyclic nucleotides on growth cone behavior may be partially mediated by CNG channels. CNG channel subunits have been identified in chemosensory neurons in *C. elegans* (TAX-2, TAX-4) and shown to be important in the growth and targeting of axons [80]. It is possible that they play an important function in the regulation of Ca^{2+} in the growth cone. Active growth cones have been shown to exhibit Ca^{2+} waves which may spread into the neurite [81,82]. These effects are dependent on extracellular calcium, but none of the voltage-gated Ca^{2+} channels expressed in growth cones [83] seem to be involved [84]. In mammalian sperm cells, which also exhibit Ca^{2+} waves, CNG channels have been shown to control calcium entry and in this way possibly regulate flagellar motility [85]. CNG channels may also be involved in calcium entry in axonal growth cones of ORNs.

Besides the effects of cyclic nucleotides on growth cone behavior, nitric oxide has also been shown to significantly affect growth cone behavior. Cheung et al. [86] showed that NO can stabilize growth cone filopodia, similar to previous findings with 8-Br-cAMP. Van Wagenen and Rehder [87] have extended these findings by showing that the response of the growth cone to NO is mediated via a cascade implying cGMP and entry of external Ca^{2+}. These effects of NO on growth cone behavior may also be mediated by CNG channels. Thus, Kafitz et al. [88] have recently shown that cyclic nucleotides and NO activate a calcium entry mechanism in olfactory receptor growth cones with properties similar to the CNG channels. Since CNG channels, especially the NO-gated homomeric CNGA4 channels, are highly permeable to Ca^{2+} [50] this may provide a direct mechanism through which NO can regulate calcium levels and behavior of growth cones. In summary, CNG channels present at the membrane of the growth cones might play a key role in neural development and plasticity, particularly of ORNs.

6. CNG channels: role in the visual transduction cascade

Phototransduction is mediated by an enzymatic cascade that ultimately leads to the hydrolysis of cGMP. The photoreceptor cells, rods and cones, integrate and respond to cGMP hydrolysis via a CNG channel in the plasma membrane of the outer segment. The last step in phototransduction is the creation of a change in membrane potential which is mediated by CNG channels. In the dark, the CNG channel is co-operatively kept open by the binding of four cGMP molecules to its CNGB subunits [89]. This causes an exchange of cations (Ca^{2+}, Na^+, K^+) between the cytoplasm and the surrounding inter-photoreceptor space. To keep the ion gradients active the cations are actively pumped out across the plasma membrane by a $Na^+/Ca^{2+}/K^+$-exchanger [90]. In the light cGMP is hydrolyzed to $5'GMP$ by the phosphodiesterases (PDEA-PDEB). With decreased cGMP concentration, cGMP is released from the CNG channel subunits and the channel is closed (Fig. 2B). This blocks the flow of Ca^{2+} and Na^+ into the outer segment of the rod (ROS).

In darkness, the inward flow of charges (Ca^{2+}, Na^+, K^+) is equal to the outward flow. This is obtained by a $Na^+/Ca^{2+}/K^+$-exchanger in the ROS membrane. On illumination the $Na^+/Ca^{2+}/K^+$-exchanger is still active but the inward Ca^{2+} and Na^+ flow through the CNG is blocked and the plasma membrane is hyperpolarized because the charge flow rates have become unequal. This hyperpolarization leads to a change in synaptic activity and ultimately alters the nerve impulse pattern that is sent to the brain [91]. The CNG channel activity is also liable to high Ca^{2+}-calmodulin that leads to closure of the channel to reduce the Ca^{2+} influx [32,92].

7. CNG channels: roles in other systems

Over the past several years, CNG channels have been found in a variety of other tissues including kidney, testis and heart (for review see Refs. [93,94]) where they might fulfill various physiological functions.

More recently, these channels have been found in the central nervous system [7,95,96] and have been implicated in processes as diverse as synaptic modulation, central communication, plasticity and axon outgrowth in animals ranging from the nematode to mammals [80,97]. CNG channel subunits, in particular the olfactory CNG channel subunits, have been identified in the brain [95,96,98]. Specific subsets of neurons like the CA1 and CA3 neurons of the hippocampus express CNG channel subunits suggesting that these channels have a particular function in the central nervous system that is related specifically to certain cell types, rather than being of a general housekeeping nature [97].

In the heart, the I_h channel in the sinoatrial node controls the pacemaking activity and is regulated by the binding of cAMP [99]. Another CNG channel, similar to the olfactory CNGA2, is expressed throughout the heart of the mouse but its function remains unclear [100]. As sensors of cyclic nucleotide concentrations and conduits for Ca^{2+} entry, these CNG channels may play a role in regulating the heart rate and contraction.

Recently, the cloning and first functional characterization of a plant CNG channel occurred from the *Arabidopsis* cDNA [101]. This channel could, like its animal counterparts, translate stimulus-induced changes in cytosolic cyclic nucleotides into altered cell membrane potential and/or cation flux as part of signal cascade pathway(s) that remain to be identified.

8. CNG channels: clinical relevance in channelopathies

8.1. Achromatopsia

Complete achromatopsia is a rare, autosomal recessive disorder characterized by photophobia, low visual acuity, nystagmus and a total inability to distinguish colors [102]. Electroretinographic recordings show that rod photoreceptor function is normal, whereas cone photoreceptor responses are absent. In this disease, cone photoreceptors, the retinal sensory neurons mediating color vision, seem viable but fail to generate an electrical response to light. Achromatopsia, or rod monochromatism, was first mapped to 2p11–2q12 (MIM 216900; [103]), where it is associated with missense mutations in the CNGA3 gene which encodes the CNGA subunit of the cone CNG channel [102]. This CNG channel generates the light-evoked electrical responses of cone photoreceptors. A second locus at 8q21–q22 has been identified among the Pingelapese islanders of Micronesia, who have a high incidence of recessive achromatopsia (MIM 262300; [104]). The Pingelapese achromatopsia segregates with a missense mutation at a highly conserved site in CNGB3, a new gene that encodes the CNGB subunit of the cone CNG channel [105]. In summary, total color-blindness is caused by mutations in the genes encoding both the CNGA and the CNGB subunits of the cone photoreceptor CNG channels, which are essential for phototransduction in all three classes of cones.

8.2. Retinitis pigmentosa

Retinitis pigmentosa (RP) is a blinding retinal disease in which the photoreceptor cells degenerate. Dryja et al. [106] noted that RP is demonstrably a genetically heterogeneous set of diseases. This was indicated by the fact that dominant, recessive, X-linked, and digenic patterns of inheritance had been found in RP families and that more than 15 separate loci have been implicated by linkage studies. Five of the 15 genes had been identified before their relation to RP was established: two of them encode proteins functioning in the phototransduction pathway, rhodopsin (180380) and PDEB; two are photoreceptor-specific proteins of unknown function, peripherin/RDS (179605) and ROM1 (180721); and one is an unconventional myosin (276903), which is mutant in a form of Usher syndrome (RP and deafness). Dryja et al. [106] screened 94 unrelated patients with autosomal dominant RP and 173 unrelated patients with autosomal recessive RP for mutations in the CNGA1 subunit gene of the rod CNG channel. Five mutant sequences co-segregated with the disease among four unrelated families with autosomal recessive RP. Two of these were nonsense mutations early in the reading frame (glu76-to-ter, 123825.0001; and lys139-to-ter, 123825.0002) and one was a deletion encompassing most if not all of the transcriptional unit, these three alleles would not be expected to encode a functional channel. The remaining two mutations were missense mutations (ser316-to-phe; 123825.0003) and a frameshift caused by a 1-bp deletion (123825.0004) that truncated the last 32 amino acids of the C terminus. The latter two mutations were expressed in vitro and found to encode proteins that were retained predominantly inside the cell instead of being targeted to the plasma membrane.

In one family with recessive RP, the same authors found a nonsense mutation, glu76 to tyr, in a heterozygous state in two affected sisters, and in their unaffected father. No mutation was detected in the CNGA1 gene in the maternally derived allele in the two affected sisters. In fact, analysis with the CNGA1 polymorphism showed that they received different alleles at this locus from their mother. It was suggested that there was a pathogenic mutation in a gene encoding another subunit of the channel protein, or some other protein that interacts with the CNGA subunit, and that the combination of the two defects is the cause of RP. An alternative possibility is that this mutation is not the cause of RP in this family. In 1 of 5 families with autosomal recessive RP and mutations in the CNGA1 gene, they found that the affected individuals in one family were homozygous for a 1-bp deletion in the codon for arg654, resulting in frameshift and truncation of the last 32 amino acids in the C terminus.

In summary, although mutations of the CNGA1 subunit of the rod CNG channel are involved in RP, approximately 80% of the cases still remained without identification of the specific mutant gene in the disorder. The rhodopsin locus only accounts for approximately 10% of all cases, and the remaining 5 loci together also account for roughly 10% [106].

9. CNG channels: the future

CNG channels first identified in rods and olfactory receptors are now known to be distributed throughout many different cells of the body and may have important roles in processes such as cell motility, secretion, and development. Future work is likely to reveal new functions for CNG channels as well as more information about the functional significance of the particular structural features and the modulation of these channels.

Important physiological roles for CNG channels might be found in different tissues as, for example in the heart where they might be involved in regulating heart rate and contraction as sensors of cyclic nucleotide concentrations and conduits for Ca^{2+} entry. In the hippocampus or other parts of the central nervous system, they might play a role in neuronal pathfinding, synapse formation and plasticity.

Significant advances in our understanding of the chemosensory systems have occurred at a rapid pace over the last several years. While this knowledge has provided some very critical insights, many fundamental questions remain to be answered. For example, taste and pheromone transduction cascades have not been identified yet. Vertebrate taste transduction has recently received a lot of attention with the cloning of two new multigene families of G-protein coupled receptors [107–109]. The screening of the newly published human genome library allowed the discovery of one gene sharing strong homology with the mouse and rat vomeronasal genes [110] reopening the debate as whether humans can detect pheromones. The presence of CNG channels in both systems as well as the presence of nNOS may indicate a significant importance of these channels and NO for these two sensory systems.

A facet of CNG channels that also needs further exploration is the possibility that different combinations of subunits are differentially expressed in microdomains of the same cell. Evidence for this has already been described in spermatozoa, where CNGA subunits are expressed alone on the distal and midpart of the flagellum, while both CNGA and CNGB subunits are expressed on the proximal part [85]. Therefore, it is possible that not only are there different physiological functions for CNG channels in different tissues but also in different parts of the same cell.

Acknowledgements

We gratefully thank Olivier Randin for assistance in producing the Figures. M.-C.B. and S.F. are supported by grants from the NIH, ONR and Fonds National Suisse de la Recherche.

References

[1] Chen, X.Z., Vassilev, P.M., Basora, N., Peng, J.B., Nomura, H., Segal, Y., Brown, E.M., Reeders, S.T., Hediger, M.A., Zhou, J., 1999. Polycystin-1 is a calcium-regulated cation channel permeable to calcium ions. Nature 401, 383–386.

[2] Fesenko, E.E., Kolesnikov, S.S., Lyubarsky, A.L., 1985. Induction by cyclic GMP of cationic conductance in plasma membrane of retinal rod outer segment. Nature 313, 310–313.

[3] Nakamura, T., Gold, G.H., 1987. A cyclic nucleotide-gated conductance in olfactory receptor cilia. Nature 325, 442–444.

[4] Kaupp, U.B., 1995. Family of cyclic nucleotide-gated ion channels. Curr. Opin. Neurobiol. 5, 434–442.

[5] Finn, J.T., Yau, K.W., 1996. Cyclic nucleotide-gated channels: an extended family with diverse functions. Annu. Rev. Physiol. 58, 395–426.

[6] Zagotta, W.N., Siegelbaum, S.A., 1996. Structure and function of cyclic nucleotide-gated channels. Annu. Rev. Neurosci. 19, 235–263.

[7] Wei, J.Y., Roy, D.S., Leconte, L., Barnstable, C.J., 1998. Molecular and pharmacological analysis of cyclic nucleotide-gated channel function in the central nervous system. Prog. Neurobiol. 56, 37–64.

[8] Leinders-Zufall, T., Rand, M.N., Shepherd, G.M., Greer, C.A., Zufall, F., 1997. Calcium entry through cyclic nucleotide-gated channels in individual cilia of olfactory receptor cells: Spatiotemporal dynamics. J. Neurosci. 17, 4136–4148.

[9] Liu, D.T., Tibbs, G.R., Siegelbaum, S.A., 1996. Subunit stoichiometry of cyclic nucleotide-gated channels and effects of subunit order on channel function. Neuron 16, 983–990.

[10] Dhallan, R.S., Yau, K.W., Schrader, K.A., Reed, R.R., 1990. Primary structure and functional expression of a cyclic nucleotide-activated channel from olfactory neurons. Nature 347, 184–187.

[11] Bradley, J., Li, J., Davidson, N., Lester, H.A., Zinn, K., 1994. Heteromeric olfactory cyclic nucleotide-gated channels: A subunit that confers increased sensitivity to cAMP. PNAS 91, 8890–8894.

[12] Liman, E.R., Buck, L.B., 1994. A second subunit of the olfactory cyclic nucleotide-gated channel confers high sensitivity to cAMP. Neuron 13, 611–621.

[13] Bonigk, W., Bradley, J., Muller, F., Sesti, F., Boekhoff, I., Ronnett, G.V., Kaupp, U.B., Frings, S., 1999. The native rat olfactory cyclic nucleotide-gated channel is composed of three distinct subunits. J. Neurosci. 19, 5332–5347.

[14] Sautter, A., Zong, X., Hofmann, F., Biel, M., 1998. An isoform of the rod photoreceptor cyclic nucleotide-gated channel β subunit expressed in olfactory neurons. Proc. Natl Acad. Sci. USA 95, 4696–4701.

[15] Bradley, J., Frings, S., Yau, K.W., Reed, R., 2001. Nomenclature for ion channel subunits. Science 294, 2095–2096.

[16] Jan, L.Y., Jan, Y.N., 1990. A superfamily of ion channels. Nature 345, 672.

[17] Goulding, E.H., Tibbs, G.R., Liu, D., Siegelbaum, S.A., 1993. Role of H5 domain in determining pore diameter and ion permeation through cyclic nucleotide-gated channels. Nature, 364.

[18] Goulding, E.H., Tibbs, G.R., Siegelbaum, S.A., 1994. Molecular mechanism of cyclic nucleotide-gated channel activation. Nature 372, 369–374.

[19] Gordon, S.E., Zagotta, W.N., 1995. A histidine residue associated with the gate of the cyclic nucleotide-activated channels in rod photoreceptors. Neuron 14, 177–183.

[20] Gordon, S.E., Zagotta, W.N., 1995. Localization of regions affecting an allosteric transition in cyclic nucleotide-activated channels. Neuron 14, 857–864.

[21] Root, M., MacKinnon, R., 1993. Identification of an external divalent cation-binding site in the pore of a cGMP-activated channel. Neuron 11, 459–466.

[22] Root, M.J., MacKinnon, R., 1994. Two identical noninteracting sites on an ion channel revealed by proton transfer. Science 265, 1852–1856.

[23] Park, C.S., MacKinnon, R., 1995. Divalent cation selectivity in a cyclic nucleotide-gated ion channel. Biochemistry 34, 13328–13333.

[24] Sun, Z., Akabas, M.H., Goulding, E.H., Karlin, A., Siegelbaum, S.A., 1996. Exposure of residues in the cyclic nucleotide-gated channel pore: P-region structure and function in gating. Neuron 16, 141–149.

[25] Liu, D.T., Tibbs, G.R., Paoletti, P., Siegelbaum, S.A., 1998. Constraining ligand-binding site stoichiometry suggests that a cyclic nucleotide-gated channel is composed of two functional dimers. Neuron 21, 235–248.

[26] Brown, L.A., Snow, S.D., Haley, T.L., 1998. Movement of gating machinery during the activation of rod cyclic nucleotide-gated channels. Biophys. J. 75, 825–833.

[27] Gavazzo, P., Picco, C., Eismann, E., Kaupp, U.B., Menini, A., 2000. A point mutation in the pore region alters gating, Ca^{2+} blockage, and permeation of olfactory cyclic nucleotide-gated channels. J. Gen. Physiol. 116, 311–326.

[28] Broillet, M.-C., Firestein, S., 1999. Cyclic nucleotide-gated channel: Molecular mechanisms of activation. In: Rudy, B., Seeburg, P. (Eds.), Molecular and functional diversity of ion channels and receptors. The New York Academy of Sciences, New York, pp. 730–740.

[29] Kaupp, U.B., Niidome, T., Tanabe, T., Terada, S., Bonigk, W., Stuhmer, W., Cook, N.J., Kangawa, K., Matsuo, H., Hirose, T., Miyata, T., Numa, S., 1989. Primary structure and functional expression from complementary DNA of the rod photoreceptor cGMP-gated channel. Nature 342, 762–766.

[30] Bonigk, W., Altenhofen, W., Muller, F., Dose, A., Illing, M., Molday, R.S., Kaupp, U.B., 1993. Rod and cone photoreceptor cells express distinct genes for cGMP-gated channels. Neuron 10, 865–877.

[31] Weyand, I., Godde, M., Frings, S., Weiner, J., Müller, F., Altenhofen, W., Hatt, H., Kaupp, U.B., 1994. Cloning and functional expression of a cyclic-nucleotide-gated channel from mammalian sperm. Nature 368, 859–863.

[32] Chen, T.-Y., Illing, M., Molday, L.L., Hsu, Y.-T., Yau, K.-W., Molday, R.S., 1994. Subunit 2 (or beta) of retinal rod cGMP-gated cation channel is a component of the 240 kDa channel-associated protein and mediates Ca^{2+}-calmodulin modulation. PNAS 91, 11757–11761.

[33] Korschen, H.G., Beyermann, M., Muller, F., Heck, M., Vantler, M., Koch, K.W., Kellner, R., Wolfrum, U., Bode, C., Hofmann, K.P., Kaupp, U.B., 1999. Interaction of glutamic-acid-rich proteins with the cGMP signalling pathway in rod photoreceptors. Nature 400, 761–766.

[34] Poetsch, A., Molday, L.L., Molday, R.S., 2001. The cGMP-gated channel and related glutamic acid-rich proteins interact with peripherin-2 at the rim region of rod photoreceptor disc membranes. J. Biol. Chem. 276, 48009–48016.

[35] Gerstner, A., Zong, X., Hofmann, F., Biel, M., 2000. Molecular cloning and functional characterization of a new modulatory cyclic nucleotide-gated subunit from mouse retina. J. Neurosci. 20, 1324–1332.

[36] Chen, T.-Y., Peng, Y.-W., Dhallan, R.S., Ahamed, B., Reed, R.R., Yau, K.-W., 1993. A new subunit of the cyclic nucleotide-gated cation channel in retinal rods. Nature 362, 764–767.

[37] Quandt, F.N., Nicol, G.D., Schnetkamp, P.P.M., 1991. Voltage-dependent gating and block of the cyclic-GMP-dependent current in bovine rod outer segments. Neuroscience 42, 629–638.

[38] Menini, A., 1995. Cyclic nucleotide-gated channels in visual and olfactory transduction. Biophys. Chem. 55, 185–196.

[39] Picones, A., Korenbrot, J.I., 1992. Permeation and interaction of monovalent cations with the cGMP-gated channel of cone photoreceptors. J. Gen. Physiol. 100, 647–673.

[40] Frings, S., Lynch, J.W., Lindemann, B., 1992. Properties of cyclic nucleotide-gated channels mediating olfactory transduction. J. Gen. Physiol. 100, 45–67.

[41] Goulding, E., Ngai, J., Kramer, R., Colicos, S., Axel, R., Siegelbaum, S., Chess, A., 1992. Molecular cloning and single-channel properties of the cyclic nucleotide-gated channel from catfish olfactory neurons. Neuron 8, 45–58.

[42] Zufall, F., Firestein, S., Shepherd, G.M., 1991. Single channel recordings of a cyclic nucleotide-gated, odour sensitive channel in isolated salamander olfactory receptor neurones. J. Physiol. 438, 223P.

[43] Kurahashi, T., Kaneko, A., 1991. High density cAMP-gated channels at the ciliary membrane in the olfactory receptor cell. NeuroReport 2, 5–8.

[44] Frings, S., Seifert, R., Godde, M., Kaupp, U.B., 1995. Profoundly different calcium permeation and blockage determine the specific function of distinct cyclic nucleotide-gated channels. Neuron 15, 169–179.

[45] Dzeja, C., Hagen, V., Kaupp, U.B., Frings, S., 1999. Ca^{2+} permeation in cyclic nucleotide-gated channels. EMBO J. 18, 131–144.

[46] Broillet, M.-C., Firestein, S., 1996. Direct activation of the olfactory cyclic nucleotide-gated channel through modification of sulfhydryl groups by NO compounds. Neuron 16, 377–385.

[47] Broillet, M.-C., 1999. S-nitrosylation of proteins. Cell. Mol. Life Sci. 55, 1036–1042.

[48] Broillet, M.-C., 2000. A single intracellular cysteine residue is responsible for the activation of the olfactory cyclic nucleotide-gated channel by NO. J. Biol. Chem. 275, 15135–15141.

[49] Baumann, A., Frings, S., Godde, M., Seifert, R., Kaupp, U.B., 1994. Primary structure and functional expression of a drosophila cyclic nucleotide-gated channel present in eyes and antennae. EMBO J. 13, 5040–5050.

[50] Broillet, M.-C., Firestein, S., 1997. β subunits of the olfactory cyclic nucleotide-gated channel form a nitric oxide activated Ca^{2+} channel. Neuron 18, 951–958.

[51] Haynes, L.W., 1992. Block of the cyclic GMP-gated channel of vertebrate rod and cone photoreceptors by L-*cis*-diltiazem. J. Gen. Physiol. 100, 783–801.

[52] Fodor, A.A., Black, K.D., Zagotta, W.N., 1997. Tetracaine reports a conformational change in the pore of cyclic nucleotide-gated channels. J. Gen. Physiol. 110, 591–600.

[53] Nicol, G.D., 1993. The calcium channel antagonist, pimozide, blocks the cyclic GMP-activated current in rod photoreceptors. J. Pharmacol. Exp. Therapeutics 265, 626–632.

[54] Leinders-Zufall, T., Zufall, F., 1995. Block of cyclic nucleotide-gated channels in salamander olfactory receptor neurons by the guanylyl cyclase inhibitor LY-83583. J. Neurophysiol. 74, 2759–2762.

[55] Brown, R.L., Halay, T.L., West, K.A., Crabb, J.W., 1999. Pseudochetoxin: A peptide blocker of cyclic nucleotide-gated ion channels. Proc. Natl Acad. Sci. USA 96, 754–759.

[56] Kramer, R.H., Tibbs, G.R., 1996. Antagonists of cyclic nucleotide-gated channels and molecular mapping of their site of action. J. Neurosci. 16, 1285–1293.

[57] Gordon, S.E., Brautigan, D.L., Zimmerman, A.L., 1992. Protein phosphatases modulate the apparent agonist affinity of the light-regulated ion channel in retinal rods. Neuron 9, 739–748.

[58] Molokanova, E., Trivedi, B., Savchenko, A., Kramer, R.H., 1997. Modulation of rod photoreceptor cyclic nucleotide-gated channels by tyrosine phosphorylation. J. Neurosci. 17, 9068–9076.

[59] Muller, F., Bonigk, W., Sesti, F., Frings, S., 1998. Phosphorylation of mammalian olfactory cyclic nucleotide-gated channels increases ligand sensitivity. J. Neurosci. 18(1), 164–173.

[60] Hsu, Y.-T., Molday, R.S., 1993. Modulation of the cGMP-gated channel of rod photoreceptor cells by calmodulin. Nature 361, 76–79.

[61] Chen, T.-Y., Yau, K.-W., 1994. Direct modulation by Ca^{2+}-calmodulin of cyclic nucleotide-activated channel of rat olfactory receptor neurons. Nature 368, 545–548.

[62] Balasubramanian, S., Lynch, J.W., Barry, P.H., 1996. Calcium-dependent modulation of the agonist affinity of the mammalian olfactory cyclic nucleotide-gated channel by calmodulin and a novel endogenous factor. Membr. Biol. 152, 13–23.

[63] Ildefonse, M., Crouzy, S., Bennett, N., 1992. Gating of retinal rod cation channel by different nucleotides: Comparative study of unitary currents. J. Membr. Biol. 130, 91–104.

[64] Karpen, J.W., Brown, R.L., Stryer, L., Baylor, D.A., 1993. Interactions between divalent cations and the gating machinery of cyclic GMP-activated channels in salamander retinal rods. J. Gen. Physiol. 101, 1–25.

[65] Gordon, S.E., Downing-Park, J., Tam, B., Zimmerman, A.L., 1995. Diacylglycerol analogs inhibit the rod cGMP-gated channel by a phosphorylation-independent mechanism. Biophys. J. 69, 409–417.

[66] Crary, J.I., Dean, D.M., Maroof, F., Zimmerman, A.L., 2000. Mutation of a single residue in the S2–S3 loop of CNG channels alters the gating properties and sensitivity to inhibitors. J. Gen. Physiol. 116, 769–780.

[67] Crary, J.I., Dean, D.M., Nguitragool, W., Kurshan, P.T., Zimmerman, A.L., 2000. Mechanism of inhibition of cyclic nucleotide-gated ion channels by diacylglycerol. J. Gen. Physiol. 116, 755–768.

[68] Shepherd, G.M., 1994. Discrimination of molecular signals by the olfactory receptor neuron. Neuron 13, 771–790.

[69] Zufall, F., Firestein, S., Shepherd, G.M., 1994. Cyclic nucleotide-gated ion channels and sensory transduction in olfactory receptor neurons. Annu. Rev. Biophys. Biomol. Struct. 23, 577–607.

[70] Molday, R.S., 1996. Calmodulin regulation of cyclic nucleotide-gated channels. Curr. Opin. Neurobiol. 6(4), 445–452.

[71] Weitz, D., Zoche, M., Muller, F., Beyermann, M., Korschen, H.G., Kaupp, U.B., Koch, K.W., 1998. Calmodulin controls the rod photoreceptor CNG channel through an unconventional binding site in the N-terminus of the beta-subunit. EMBO J. 17, 2273–2284.

[72] Grunwald, M.E., Yu, W.P., Yu, H.H., Yau, K.W., 1998. Identification of a domain on the beta-subunit of the rod cGMP-gated cation channel that mediates inhibition by calcium-calmodulin. J. Biol. Chem. 273(15), 9148–9157.

[73] Bradley, J., Reuter, D., Frings, S., 2001. Facilitation of calmodulin-mediated odor adaptation by cAMP-gated channel subunits. Science 294, 2176–2178.

[74] Firestein, S., Zufall, F., 1994. The cyclic nucleotide-gated channel of olfactory receptor neurons. Semin. Cell Biol. 5, 39–46.

[75] Hibert, M.F., Trump-Kallmeyer, S., Bruinvels, A., Hoflack, J., 1991. Three-dimensional models of neurotransmitter G-binding protein coupled receptors. Mol. Pharmacol. 40, 8–15.

[76] Lowe, G., Gold, G.H., 1993. Nonlinear amplification by calcium-dependent chloride channels in olfactory receptor cells. Nature 366, 283–286.

[77] Hopker, V.H., Shewan, D., Tessier-Lavigne, M., Poo, M., Holt, C., 1999. Growth-cone attraction to netrin-1 is converted to repulsion by laminin-1. Nature 401, 69–73.

[78] Song, H.J., Poo, M.M., 1999. Signal transduction underlying growth cone guidance by diffusible factors. Curr. Opin. Neurobiol. 9, 355–363.

[79] Song, H.J., Ming, G.L., Poo, M.M., 1997. cAMP-induced switching in turning direction of nerve growth cones [published erratum appears in nature sep 25. Nature 388, 275–279, 389(6649) (1997) 412.

[80] Coburn, C.M., Bargmann, C.I., 1996. A putative cyclic nucleotide-gated channel is required for sensory development and function in *C. elegans*. Neuron 17, 695–706.

[81] Gu, X., Spitzer, N.C., 1995. Distinct aspects of neuronal differentiation encoded by frequency of spontaneous Ca^{2+} transients. Nature 375, 784–787.

[82] Spitzer, N.C., Olson, E., Gu, X., 1995. Spontaneous calcium transients regulate neuronal plasticity in developing neurons. J. Neurobiol. 26, 316–324.

[83] Gottmann, K., Lux, H.D., 1995. Growth cone calcium ion channels: Properties, clustering, and functional roles. Perspect. Dev. Neurobiol. 2, 371–377.

[84] Gomez, T.M., Snow, D.M., Letourneau, P.C., 1995. Characterization of spontaneous calcium transients in nerve growth cones and their effect on growth cone migration. Neuron 14, 1233–1246.

[85] Wiesner, B., Weiner, J., Middendorff, R., Hagen, V., Kaupp, U.B., Weyand, I., 1998. Cyclic nucleotide-gated channels on the flagellum control Ca^{2+} entry into sperm. J. Cell Biol. 142, 473–484.

[86] Cheung, W.S., Bhan, I., Lipton, S.A., 2000. Nitric oxide (NO.) stabilizes whereas nitrosonium (NO^+) enhances filopodial outgrowth by rat retinal ganglion cells in vitro. Brain Res. 868, 1–13.

[87] Van Wagenen, S., Rehder, V., 1999. Regulation of neuronal growth cone filopodia by nitric oxide. J. Neurobiol. 39, 168–185.

[88] Kafitz, K.W., Leinders-Zufall, T., Greer, C.A., 2000. Cyclic GMP evoked calcium transients in olfactory receptor cell growth cones. Chem. Neurosci. 11, 677–681.

[89] Brown, R.L., Gramling, R., Bert, R.J., Karpen, J.W., 1994. Identification by photoaffinity labeling of peptide regions within retinal rod cGMP-activated channel subunits involved in cGMP binding. Invest. Ophthalmol. Vis. Sci. 35, 1473.

[90] Tucker, J.E., Winkfein, R.J., Cooper, C.B., Schnetkamp, P.P., 1998. cDNA cloning of the human retinal rod Na-Ca-K exchanger: Comparison with a revised bovine sequence. Invest. Ophthalmol. Vis. Sci. 39, 435–440.

[91] Yau, K.W., 1994. Phototransduction mechanism in retinal rods and cones. The friedenwald lecture. Invest. Ophthalmol. Vis. Sci. 35, 9–32.

[92] Hsu, Y.T., Molday, R.S., 1994. Interaction of calmodulin with the cyclic GMP-gated channel of rod photoreceptor cells. Modulation of activity, affinity purification, and localization. J. Biol. Chem. 269, 29765–29770.

[93] Kaupp, U.B., 1991. The cyclic nucleotide-gated channels of vertebrate photoreceptors and olfactory epithelium. TINS 14, 150–157.

[94] Yau, K.-W., 1994. Cyclic nucleotide-gated channels: An expanding new family of ion channels. PNAS 91, 3481–3483.

[95] Bradley, J., Zhang, Y., Bakin, R., Lester, H.A., Ronnett, G.V., Zinn, K., 1997. Functional expression of the heteromeric "olfactory" cyclic nucleotide-gated channel in the hippocampus: A potential effector of synaptic plasticity in brain neurons. J. Neurosci. 17, 1993–2005.

[96] Kingston, P.A., Zufall, F., Barnstable, C.J., 1996. Rat hippocampal neurons express genes for both rod retinal and olfactory cyclic nucleotide-gated channels: Novel targets for cAMP/cGMP function. Proc. Natl Acad. Sci. USA 93, 10440–10445.

[97] Zufall, F., Shepherd, G.M., Barnstable, C.J., 1997. Cyclic nucleotide gated channels as regulators of CNS development and plasticity. Curr. Opin. Neurobiol. 7, 404–412.

[98] Strijbos, P.J., Pratt, G.D., Khan, S., Charles, I.G., Garthwaite, J., 1999. Molecular characterization and in situ localization of a full-length cyclic nucleotide-gated channel in rat brain. Eur. J. Neurosci. 11, 4463–4467.

[99] DiFrancesco, D., Tortora, P., 1991. Direct activation of cardiac pacemaker channels by intracellular cyclic AMP. Nature 351, 145–147.

[100] Ruiz, M.L., London, B., Nadal-Ginard, B., 1996. Cloning and characterization of an olfactory cyclic nucleotide-gated channel expressed in mouse heart. J. Mol. Cell Cardiol. 28, 1453–1461.

[101] Leng, Q., Mercier, R.W., Yao, W., Berkowitz, G.A., 1999. Cloning and first functional characterization of a plant cyclic nucleotide-gated cation channel. Plant Physiol. 121, 753–761.

[102] Kohl, S., Marx, T., Giddings, I., Jagle, H., Jacobson, S.G., Apfelstedt-Sylla, E., Sharpe, E., Wissinger, B., 1998. Total colourblindness is caused by mutations in the gene encoding the alpha-subunit of the cone photoreceptor cGMP-gated cation channel. Nat. Genet. 19, 257–259.

[103] Wissinger, B., Jagle, H., Kohl, S., Broghammer, M., Baumann, B., Hanna, D.B., Hedels, C., Apfelstedt-Sylla, E., Randazzo, G., Jacobson, S.G., Zrenner, E., Sharpe, L.T., 1998. Human rod monochromacy: Linkage analysis and mapping of a cone photoreceptor expressed candidate gene on chromosome 2q11. Genomics 51, 325–331.

[104] Kohl, S., Baumann, B., Broghammer, M., Jagle, H., Sieving, P., Kellner, U., Spegal, R., Anastasi, M., Zrenner, E., Sharpe, L.T., Wissinger, B., 2000. Mutations in the CNGB3 gene encoding the beta-subunit of the cone photoreceptor cGMP-gated channel are responsible for achromatopsia (achm3) linked to chromosome 8q21. Hum. Mol. Genet. 9, 2107–2116.

[105] Sundin, O.H., Yang, J.M., Li, Y., Zhu, D., Hurd, J.N., Mitchell, T.N., Silva, E.D., Maumenee, I.H., 2000. Genetic basis of total colorblindness among the pingelapese islanders. Nat. Genet. 25, 289–293.

[106] Dryja, T.P., Finn, J.T., Peng, Y.W., McGee, T.L., Berson, E.L., Yau, K.W., 1995. Mutations in the gene encoding the alpha subunit of the rod cGMP-gated channel in autosomal recessive retinitis pigmentosa. Proc. Natl Acad. Sci. USA 92, 10177–10181.

[107] Hoon, M.A., Adler, E., Lindemeier, J., Battey, J.F., Ryba, N.J.P., Zuker, C.S., 1999. Putative mammalian taste receptors: A class of taste-specific GPCRs with distinct topographic selectivity. Cell 96, 541–551.

[108] Adler, E., Hoon, M.A., Mueller, K.L., Chandrashekar, J., Ryba, N.J.P., Zuker, C.S., 2000. A novel family of mammalian taste receptors. Cell 100, 693–702.

[109] Chandrashekar, J., Mueller, K.L., Hoon, M.A., Adler, E., Feng, L., Guo, W., Zuker, C.S., Ryba, J.P., 2000. T2Rs function as bitter taste receptor. Cell 100, 703–711.

[110] Rodriguez, I., Greer, C.A., Mok, M.Y., Mombaerts, P., 2000. A putative pheromone receptor gene expressed in human olfactory mucosa. Nat. Genet. 26, 18–19.

Nicotinic acetylcholine receptors in the nervous system

Joseph P. Margiotta* and Phyllis C. Pugh

Department of Anatomy and Neurobiology, Medical College of Ohio, Toledo, OH 43614, USA
*Correspondence address: Tel.: +1-419-383-4119; fax: +1-419-383-3008
E-mail: jmargiotta@mco.edu

1. Introduction

Nicotinic acetylcholine receptor channels (AChRs) are widely expressed in the nervous system where they are involved in signaling, neurological disease, and motivational/addictive behaviors. Like their relatives in muscle and electric organ, neuronal AChRs are pentameric transmembrane proteins belonging to a superfamily of ligand-gated ion channels that also includes $GABA_A$, glycine and serotonin (5-HT_3) receptors [1]. In each case, occupation of the receptor's ligand binding domain by agonist induces an allosteric interaction between its constituent subunits causing the channel domain to make a transition from closed to open state. Despite these similarities, there are important differences between individual members of the ligand-gated ion channel superfamily, specifically between AChRs on neurons and those on muscle fibers. In particular, when compared with muscle AChRs, neuronal AChRs are assembled from a broader palette of homologous yet distinct subunits, resulting in a much wider array of receptor subtypes. Given this diversity it is not surprising that when compared to muscle AChRs, neuronal AChRs have broader functional roles, and respond differently to both extracellular ligands and intracellular regulatory signals. While such findings illustrate a clear correlation between the diverse composition and functional roles of neuronal AChR subtypes in vivo, they also indicate that a coherent picture of such receptors is complex and still not well understood. The goal of this chapter is to provide readers interested in studying neuronal AChRs an introduction to their basic structural and functional features, and an appreciation of some of their recently discovered roles. Thus in the first part we describe the structural elements, and functions associated with neuronal AChRs by summarizing and updating information on these topics from previous treatments [1–6]. The second part of the chapter will address more recent findings, selecting those that have either not been reviewed recently, or that implicate AChRs in novel functional processes. This part will focus on native neuronal AChRs and work relevant to (1) receptor structure–function

Advances in Molecular and Cell Biology, Vol. 32, pages 269–302
ISSN: 1569-2558 / DOI: 10.1016/S1569-2558(03)32012-0

relationships, (2) cellular mechanisms regulating receptor function, and (3) possible receptor roles in neuronal survival and development. Readers interested in the involvement of neuronal AChRs in other important processes such as transmitter release, neurological disease, cognition, pain, and drug addiction are directed to recent reviews of these topics [6–14].

2. Neuronal nicotinic receptors and synapses

2.1. General features of AChRs

The current view of neuronal AChR structure and function is based largely on related AChRs from muscle. We therefore first provide a snapshot of muscle AChRs, then overlay the fuzzier, inferred structural elements of neuronal AChRs for comparison. Muscle AChRs were the first ligand-gated channel proteins whose constituent subunits were biochemically purified, the first whose cDNAs were isolated, and the only ones where information about tertiary structure is available. This characterization was aided by the serendipitous availability of two tools. First, *Torpedo* electric organ (a muscle-derived tissue) was found to be a rich source of AChRs, permitting their isolation in milligram quantities. Second, α-bungarotoxin (αBgt) was found to bind nearly irreversibly to an extracellular site on α subunits of *Torpedo*, and many other muscle AChRs, potently blocking their function [15]. These tools provided a means of biochemically quantifying AChRs, allowing them to be isolated and their component subunits separated and purified [16]. Such approaches were eventually combined with N-terminal amino acid sequencing [17] revealing that *Torpedo* AChRs are composed of four homologous subunits ($\alpha\beta\gamma\delta$) each having estimated masses of 40–65 kDa. Assembled receptors contain two principal ligand-binding (α) subunits, and one of each of the other subunits, indicating that the *Torpedo* AChR is a heteropentamer having stoichiometry of $\alpha_2\beta\delta\gamma$ (Fig. 1) and a predicted molecular mass of about 255 kDa (somewhat larger after glycosylation and other post-translational modifications). Mammalian muscle AChRs are believed to have a similar stoichiometry with γ- and ε-subunit containing subtypes predominating in embryonic and mature muscle fibers, respectively [18,19].

Based on the partial sequence information, Numa and colleagues [20–22] as well as other investigators [23–25] applied molecular approaches to isolate cDNA clones encoding *Torpedo* and muscle AChR subunits and deduce their primary amino acid structure. Amino acid hydrophobicity analyses [26] indicated that each AChR subunit has a characteristic secondary structure featuring extracellular, hydrophilic amino- (N-) and carboxyl- (-COOH) terminal regions that bracket four hydrophobic membrane-spanning segments (M1–M4), and a long cytoplasmic region between M3 and M4 (Fig. 1A) [22,25]. In the assembled AChR, individual subunits are folded such that their membrane-spanning regions (M1–M4) form compact, rod-like conformations encircling the channel pore (Seen en-face in Fig. 1B). X-ray analysis and electron microscopy of paracrystalline arrays of the receptor in the membrane [27,28] suggested that the native *Torpedo* receptor resembles a transmembrane chalice having its long axis perpendicular to the lipid bilayer (Fig. 1C). The large extracellular N-terminal domains and the M3–M4 cytoplasmic region

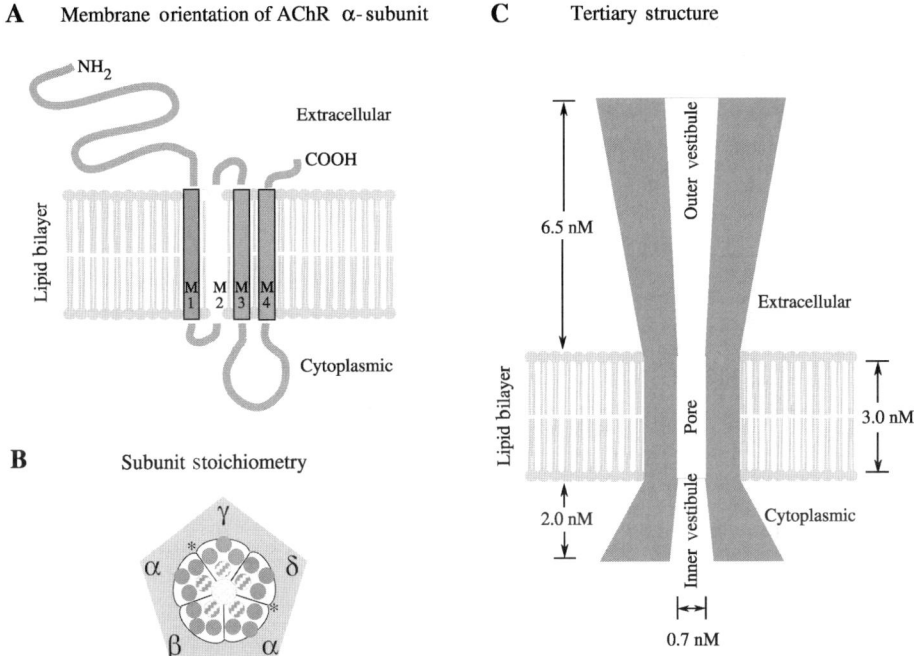

A   Membrane orientation of AChR α-subunit   **C**   Tertiary structure

B   Subunit stoichiometry

Fig. 1. Current structural model of the muscle nicotinic receptor. A. Transverse view depicting features of a single AChR subunit (after Ref. [22]). Each subunit has a long, extracellular N-terminus, and four transmembrane segments (M1–M4). Of these M1, M3 and M4 are believed to be arranged as beta sheets, and M2 (striped) as an alpha helix [27,28]. A long intracellular loop connects M3 and M4; a short carboxy terminus is extracellular. The α subunit shown displays two disulfide bridges (between residues 126–141 and 192–193), the latter being associated with the ACh binding site (See text for details). B. En-face extracellular, membrane level view showing the pentameric arrangement of individual subunits with M2 for each depicted as lining the channel pore. Ligand binding interfaces are indicated by asterisks. The position of the β and γ subunits [272,273] may be reversed [28]. C. Tertiary AChR structure depicting relationship between extracellular, cytoplasmic and transmembrane receptor domains. Neuronal AChRs are presumed to be similar in their overall pentameric structure.

appear to form funnel-like vestibules that protrude out from the plasma membrane and into the cytoplasm. The large vestibular openings narrow in the transmembrane region where the effective pore diameter is estimated to be 0.7 nm.

Much progress has been made in determining the structural elements of AChRs that underlie specific functions. The principal binding site for agonist [e.g. ACh, nicotine, carbamylcholine (CCh)] and competitive antagonist [e.g. αBgt, d-tubocurarine (d-TC)] ligands, for example, is in the vestibular, N-terminal region of each α subunit. The precise ligand binding site was originally presumed to be near disulfide bonds associated with cysteines at residues 128, 142, 192 and 193 [29]. Studies with radioactive alkylating agents indicated that uniquely paired cysteines at positions 192 and 193 form a disulfide bond [30,31] in what is now known as the C loop of the principal ligand binding site. More detailed photoaffinity labeling experiments demonstrated that the ligand binding site is actually formed at interfaces between the principal A, B, and C loops on α subunits, and complementary D, E, and F loops on adjacent δ and γ (or ε) subunits (references in [32]). Very recently, a soluble protein released by molluscan glial cells was described that binds ACh (AChBP), and modulates cholinergic transmission [33]. High resolution X-ray diffraction analyses indicate that AChBP protomers resemble the N-terminal domains of nicotinic receptor α subunits, and confirm earlier predictions that ligand binding sites occur at interfaces between A, B, and C loops of one subunit forming a binding pocket with D and E loops of adjacent subunits [32].

To test whether AChR subunit genes actually encode functional receptors, they are expressed in a permissive heterologous system such as *Xenopus* oocytes [34] or, with somewhat more difficulty, in mammalian cell lines [35]. The presence of AChR channels can be then be assessed by binding αBgt or other biochemical ligands, and their ability to function verified with electrophysiological or ion flux assays. In the case of *Torpedo* AChRs, for example, injection of mRNAs encoding α, β, δ, and γAChR subunits in *Xenopus* oocytes resulted in specific, saturable [^{125}I]αBgt binding within three days, and application of ACh induced currents in voltage-clamped oocytes that were reversibly blocked by d-TC [36]. All four subunits seem to be required for maximally functional AChRs since omitting any one of the subunit mRNAs from the injection mixture greatly reduced or eliminated subsequent ACh-induced responses [36–38]. When combined with site-directed mutations, heterologous expression studies have yielded powerful insights concerning the structural determinants of specific, muscle AChR channel functions [39]. For example, α subunit mutations at cysteine residues 128, 142, 192 or 193 result in receptors that fail to respond to ACh [39]. Since mutations elsewhere had little or no effect, these findings confirmed the parallel biochemical studies together indicating that residues within a rather small region of the first extracellular domain contain the ACh binding site, with cysteine residues at 192 and 193 being particularly important [30,31,39]. Similar mutagenesis experiments combined with the use of open channel blockers indicate that M2 regions of each subunit line the channel pore [40] while other studies suggest that rings of negatively charged amino acids bordering M2 determine the rate of ion transport through the channel and thereby confer selectivity for cations over anions [41,42]. Thus the combination of molecular, biochemical and electrophysiological approaches have provided powerful, albeit indirect, insights into AChR structural elements that subserve specific functions such as agonist binding, channel gating, permeation, and selectivity.

A long cytoplasmic domain of each AChR subunit connects the transmembrane regions M3 and M4. This domain contains consensus sites for phosphorylation by protein kinases, resulting in the modulation of AChR function (see below) [43–45]. The M3–M4 domain of the muscle AChR α subunit also appears to interact with rapsyn (receptor associated protein at synapses) [46]. Rapsyn is a 43 kDa cytoplasmic protein that colocalizes with AChRs at the neuromuscular junction (NMJ) [47] where AChRs are packed at densities of ≈ $10^4 \, \mu m^{-2}$ [48] to ensure reliable synaptic transmission. Rapsyn clusters AChRs in heterologous expression systems [49–51] and is a critical component in a series of molecular interactions that lead to the development of high density AChR clusters at the NMJ in vivo [52,53].

2.2. *Neuronal AChRs: subunits, structure and function*

Because no source comparable to *Torpedo* electric organ exists for neuronal AChRs, the approaches taken to isolate them and determine their structure/function motifs have been, by necessity, more indirect. Moreover, since muscle and neuronal AChRs were known to display somewhat different channel properties and be preferentially activated and blocked by different agonists and antagonists, they were suspected to be assembled from distinct subunits. Reflecting this presumed molecular distinction was the consistently obtained finding that αBgt failed to block neuronal AChRs, such as those present on Renshaw cells [54], autonomic ganglion neurons [55,56], neural crest-derived chromaffin cells [57] or PC12 cells [58]. The failure of αBgt to block neuronal AChR function was nonetheless puzzling since many of the same cells displayed abundant [^{125}I]-αBgt binding which could be competed with cholinergic ligands (d-TC, ACh, CCh) indicating that αBgt binds to a widely distributed cholinergic receptor (αBgt-AChR). Consistent with these results were the observations that functionally detectable AChRs on chick ciliary ganglion (CG) neurons were regulated independently from αBgt-AChRs and blocked by neuronal-bungarotoxin (N-Bgt), a second, distinct toxin from *Bungarus multicinctus* venom [56,59]. N-Bgt, alternatively termed Bgt 3.1 [56], κ-Bgt [60] or toxin F [61], blocks nicotinic responses and transmission in autonomic ganglion preparations [56,60–62] as well as nicotinic responses in some but not all neuronal populations expressing AChRs in the central nervous system (reviewed in Ref. [2]). The toxin and regulation studies prompted speculation that neuronal cholinergic receptors recognized by αBgt represented inactive AChRs or degradation products [63]. This view prevailed through the 1980s, suggesting that the αBgt approach, which had proven so valuable for the eventual cloning of muscle AChRs, could not be used to isolate functional AChRs in the nervous system. In spite of their presumed and later-confirmed structural differences, neuronal and muscle AChRs were expected to be homologous proteins since antibodies raised against the latter would block the former [58,64].

Within this historical framework, progress towards identifying neuronal AChR genes was first achieved using a molecular cloning approach based on their presumed homology with muscle AChRs. Briefly, a cDNA fragment encoding the N-terminal region and putative ACh-binding site of the muscle AChR α subunit was used to probe a PC12 cDNA library [65] at low stringency. A single hybridizing cDNA clone was isolated and sequenced, which revealed a 1497 base-pair open reading frame, predicting a protein of 499 amino acids having 47% amino acid homology overall with the mouse α1 AChR subunit. The deduced protein displayed a secondary structure reminiscent of the muscle AChR α subunit, featuring an extracellular N-terminal domain, four transmembrane regions (M1–M4) highly homologous to those of the muscle α subunit (72–87% identity), a highly divergent long M3–M4 cytoplasmic loop, and an extracellular C-terminus (e.g. Fig. 1). Critical to identifying the deduced neuronal protein as an AChR α subunit (now known as α3) was the presence of the cysteine loop between residues 128 and 142 and paired cysteines at residues 192 and 193, both hallmarks of the ACh-binding site. Using similar approaches and criteria, other neuronal "α1-like" subunits (α2, α4, α5, α6) and "α1-unlike" subunits (β2, β3, β4) were isolated and identified from chicken [66,67] and rodent [68–73] sources (Table 1). All of the neuronal α and β subunits proteins are

Table 1
Vertebrate AChR subunit genes and products

Name	Alternate name(s)	Original source	References
α1, β1, γ, δ, ε	α, β	*Torpedo* electroplax, muscle	For review, see Ref. [74]
α2		Genomic library	[66,70,75]
α3		PC12 cDNA library	[65–67,76]
α4		Brain and PC12 cDNA libraries	[66,69,77–79]
α5		Brain and PC12 cDNA libraries	[67,68,80]
α6		Brain and neuroretina cDNA libraries	[75,81,82]
α7	αBgtBP1	Brain cDNA library	[83–87]
α8	αBgtBP2	Brain cDNA library	[84]
α9		Genomic library	[88–90]
α10		Cochlea cDNA library	[91,92]
β2	nα, nα1	Brain and PC12 cDNA libraries	[66,71,93,94]
β3	nα2	Brain cDNA library	[72,95,96]
β4	nα3	PC12 cDNA library	[67,68,73,97]

between 49 and 67 kDa in mass, share between 40 and 55% homology with muscle AChR subunits from the same species, and are believed from hydrophobicity analyses to display a secondary structure similar to that depicted in Fig. 1. Only the α subunits have paired cysteines at residues corresponding to 192 and 193 of *Torpedo* α1 and are presumed by analogy with α1 to be the subunits that bind ACh.

A nomenclature for AChR subunits has recently been adopted [98]. An alternative nomenclature is provided for convenience of those consulting older references (Table 1). Receptor subunits from muscle and electric organ (muscle-type AChRs) are identified by the suffix "1" as in α1 and β1, with γ, δ, and ε currently receiving no additional designation. Neuronal AChR subunits are identified as α subunits if they display adjacent cysteines at residues corresponding to 192, 193 on the *Torpedo* α1 subunit. All non-α neuronal AChR subunits are identified as βs. Both neuronal α and β subunits have been given numerical suffixes corresponding roughly to the order of their discovery. (Adapted from Ref. [2].)

Contrasting with the muscle AChR composition requiring four different subunits, functional neuronal AChRs are detected in oocytes after expression of genes encoding one ACh-binding (α) subunit (α2, α3 or α4) and one "structural" (β) subunit (β2 or β4) [67,70,71,77,99]. The neuronal AChRs expressed from these αβ combinations had different biophysical properties and pharmacological profiles [100,101], but as expected from the inability of αBgt to block ACh-induced responses in neurons, none was blocked by αBgt, even when the toxin was applied at concentrations sufficient to saturate muscle AChRs. In an ingenious experiment to determine the subunit stoichiometry of neuronal AChRs, α4 and β2 subunits, mutated at sites flanking M2 to change the rate of ion transport, were expressed in oocytes together with wild-type subunits. Subsequent single-channel measurements revealed the expression of normal and hybrid receptors that displayed correspondingly different conductances that were consistent with receptor assembly in $(\alpha4)_2(\beta2)_3$ stoichiometry [102]. Based largely on this experiment, it has been generally concluded that neuronal AChRs assemble as heteropentamers having two α subunits, at least in oocytes. By extension, the 300 kDa size of neuronal AChRs immunopurified from rodent, bovine, and chick brain is also consistent with native neuronal AChRs being assembled from five 40–70 kDa subunits. Other studies using subunit specific antibodies to immunopurify and probe AChRs from chick [103], rat [104], and bovine brain [105] have demonstrated that the purified neuronal receptors contain only two types of subunits (α and β) where each receptor contains two ligand-binding α subunits. These findings indicate that heteromeric neuronal AChRs from brain are expressed as pentamers having $(\alpha)_2(\beta)_3$ stoichiometry. As for subunit genes other than α2, α3, α4, β2, and β4, the β3 subunit did not participate in a functional receptor when its mRNA was coinjected with that for α2, α3, or α4 [72], and similar results were obtained with rodent or chicken α5 or α6 when expressed in combination with a comparable β2 or β4 ([3,106] but see also Ref. [82]). More recent experiments indicate that α5, α6, and β3 can contribute to functional AChRs, but only efficiently do so when coexpressed with at least two other subunits [82,107,108]. These and other findings have led to the suggestion that the structurally similar α5 and β3 subunits occupy a position like that of β1 in muscle AChRs where their presence influences agonist binding and channel properties [109]. More recent studies involving expression of human α6 with other subunits have revealed a similar yet somewhat more promiscuous picture with this subunit able to form functional AChRs when coassembled with β4 alone, or in combination with other α and β subunits [110]. In fact immunoprecipitation studies using subunit-specific antibodies indicate that many native neuronal AChRs are composed of more than just two subunit types [111,112]. These results indicate that while many native neuronal AChRs in brain are α4β2 heteropentamers, others are likely built from more complex α and β subunit combinations. Determining the composition of native neuronal AChRs and how the presence and arrangement of subunits determines receptor function (see below) represents a major challenge to both basic and clinical research involving neuronal AChRs.

In 1990 two groups using different approaches finally isolated cDNAs corresponding to αBgt-AChRs. In one case, a cDNA encoding the α7 AChR subunit was isolated from a chicken brain library using an α3 N-terminal probe [85]. In the second case, previous N-terminal sequencing of αBgt-purified receptors [113] led to the isolation of both α7 and α8 cDNAs, also from chick brain [84]. The deduced α7 protein was composed of 479 amino acids, had 35–40% homology with other neuronal and muscle AChR subunits, and the features expected of an α subunit. Similarly, the α8 protein was composed of 511 amino acids with 62% identity to α7 and less than 50% homology to other AChR α subunits, while sharing the features expected for an α subunit. These features included four transmembrane motifs homologous to those of α1 and the characteristic cysteines in the N-terminal region at positions corresponding to 128, 142, 192 and 193 (for α7). Several features of α7, however, made it unique to the other AChR subunits that had been previously discovered. *First,* unlike other neuronal AChRs which require at least 2 subunits (α + β) for function, α7 or α8 formed a functional AChR channel when expressed alone in *Xenopus* oocytes [85,114]. *Second,* α7-AChRs also have unusual electrophysiological properties, displaying large conductances [115,116] and much higher Ca^{2+} permeabilities ($P_{Ca}/P_{Na} \approx 20$) when compared to muscle AChRs (P_{Ca}/P_{Na} or $P_{Ca}/P_{Cs} \approx 0.2$) or heteromeric neuronal AChR channels formed from α3β4 subunits ($P_{Ca}/P_{Cs} \approx 1.5$) [83,117]. *Third,* unlike heteromeric AChRs displaying comparable sensitivity for nicotine and ACh [118], the homomeric $\alpha 7$-AChRs displayed a 5-fold higher sensitivity for nicotine than for ACh [85]. *Fourth*, the N-terminal sequence of the $\alpha 7$ protein was nearly identical to that of the αBgt binding protein previously isolated from chicken optic tectum [113]. *Fifth*, unlike $\alpha\beta$ heteromeric receptors, $\alpha 7$-AChRs were blocked by low concentrations of αBgt (IC$_{50}$ = 0.7 nM) [85], thus the terms $\alpha 7$-AChRs and αBgt-AChRs are often used synonymously. These results were later accompanied by the identification in cochlear hair cells of $\alpha 9$ and $\alpha 10$ subunits which share significant homology with each other (55% amino acid identity) and moderate homology with other neuronal AChR α subunits (35–42% amino acid identity in human). The $\alpha 9$ subunit forms functional AChRs when expressed alone or in combination with $\alpha 10$, and these receptors are blocked by αBgt [88,91].

The discovery of functional $\alpha 7$-containing αBgt-AChRs might at least partly reconcile the historical disparity between neuronal αBgt binding sites and functional receptors on neurons, while $\alpha 9$ and $\alpha 10$ could account for αBgt binding in some specialized tissues such as cochlear hair cells. (While $\alpha 8$ is a major subunit in chick retina and binds αBgt, no mammalian homologue of this gene has yet been identified.) It was soon demonstrated that $\alpha 7$ subunits are widely expressed in central and autonomic neurons populations of birds and mammals, with studies using subunit specific antibodies showing that most αBgt-AChRs are likely to be $\alpha 7$ homopentamers in vivo [112,119–121]. Moreover, parallel functional experiments indicate that currents attributable to αBgt-AChRs could be induced in these same populations using fast perfusion of nicotinic agonists [122,123]. It now appears that previous failures to detect currents attributable αBgt-AChRs were due to both their rapid rate of desensitization and their higher affinity for nicotine over ACh [122,124]. These revelations make it clear that αBgt-sensitive and insensitive neuronal AChRs represent distinct receptor classes and have launched a series of ongoing studies seeking to identify and assess the roles that subtypes within each class play in vivo.

2.3. Neuronal AChRs in synaptic transmission

2.3.1. Ganglionic synapses

Nicotinic AChRs were first implicated in ganglionic neurotransmission over a century ago [125]. Modern electrophysiological recording methods have revealed that pre-ganglionic stimulation evokes quantal, d-TC-sensitive nicotinic excitatory post-synaptic potentials (EPSPs) in autonomic ganglion neurons (e.g. Ref. [126]) that are normally suprathreshold for eliciting action potentials. In recent years, studies have focused on the importance of contributions made by two prominent AChR types on autonomic neurons [111,112]: αBgt-AChRs that contain $\alpha 7$ subunits; and αBgt-insensitive AChRs that contain $\alpha 3$ and other subunits ($\alpha 3^*$-AChRs) but lack $\alpha 7$ subunits. Morphological studies using αBgt and subunit-specific antibodies, in conjunction with electron or confocal microscopy, indicate that some of both AChR types are clustered in association with defined post-synaptic specializations. Similar to the arrangement of muscle AChRs at the NMJ, ganglionic $\alpha 3^*$-AChRs are present in post-synaptic densities adjacent to pre-synaptic terminals [127,128], although in ciliary ganglia a sizable proportion are also peripheral to the immediate post-synaptic membrane [129]. In contrast, clustered αBgt-AChRs are excluded from the post-synaptic membrane and are localized instead to nearby perisynaptic somatic spines [127,129,130]. Earlier extracellular recordings had suggested that αBgt-AChRs were unnecessary for autonomic transmission because αBgt failed to block evoked compound action potentials in post-ganglionic nerve [55,131,132]. More recent whole-cell current recordings have shown that both αBgt- and α3*-AChRs contribute to evoked EPSCs, suggesting receptors of either type normally cause sufficient depolarization to generate post-synaptic action potentials that follow 1:1 with the pre-synaptic stimulus [133,134]. EPSCs generated by αBgt-AChRs are large and desensitize rapidly, while those generated by α3*-AChRs are smaller, slower to desensitize, and insensitive to αBgt but blocked by α-Conotoxin MII (αCTx-MII), a snail toxin that recognizes the mammalian α3β2 subunit interface [135]. By virtue of their post-synaptic cellular localization and excitatory function, α3*-AChRs are well suited to serve a primary role in synaptic transmission, analogous to the role of muscle AChRs at the NMJ. This conclusion is further supported by studies in mice bearing a null mutation of the α3 gene where homozygous (-/-), "knockouts" display autonomic abnormalities associated with deficits in fast nicotinic transmission, including urinary retention, bladder enlargement and widely dilated pupils [136]. In contrast, αBgt-AChRs localized to somatic spines assume a more regulatory role. Depolarization carried by these AChRs increases the reliability of post-ganglionic action potentials following high frequency pre-ganglionic stimulation, and this effect is more apparent in ganglia from embryonic than adult animals [137]. Since αBgt-AChRs are highly permeable to Ca^{2+} [83,138], their synaptic activation [139] may influence regulatory or early developmental processes, such as survival, growth, and synapse formation, many of which are sensitive to changes in intracellular Ca^{2+} [6,140–142].

How α3*- and αBgt-AChRs become clustered and localized to different cell-surface domains on autonomic neurons are still open questions. At the NMJ, pre-synaptic axons release agrin, a protein that triggers a series of events critical to synaptic differentiation [143,144]. Agrin associates with a muscle-specific tyrosine kinase receptor, initiating a cascade of signals that recruit rapsyn, which in turn then recruits AChRs causing them to cluster at high density in the junctional membrane [52,53]. Despite the homology of muscle and neuronal AChRs, it has not been possible to demonstrate a corollary to the agrin → rapsyn → AChR clustering hypothesis relevant to neuronal nicotinic synapses. For example, while agrin mRNA can be detected by in situ hybridization in the CNS, particularly in cranial somatic and visceral efferent neurons [144,145], its role in the differentiation of neuronal nicotinic synapses has not been established. Also, while α3β2, α4β2 and α7-AChRs form clusters when AChR subunits and muscle rapsyn are coexpressed in QT-6 cells, the aggregates are retained intracellularly [128,146] rather than on the cell surface, as would be required for them to participate in synaptic transmission. Lastly, although muscle rapsyn RNA transcripts are expressed in chick ciliary ganglion neurons and in mouse sympathetic ganglia, rapsyn protein is undetectable in either tissue, and rapsyn -/- knockout mice display normally clustered AChRs [128,147,148]. These findings indicate that *muscle* rapsyn is not required for the formation of neuronal nicotinic synapses. It remains possible that an as yet unidentified neuronal rapsyn isoform performs the obligatory role of clustering AChRs at synapses in the nervous system, however, recent studies show that PDZ-containing proteins (i.e. PSD-95 and PSD-93a) co-localize with

AChRs on ciliary ganglion neurons and that their disruption prevents normal synaptic firing [274]. While further experiments are required to identify the nature of the PDZ-AChR interactions, this study indicates that PDZ-containing proteins form an organizing scaffold that maintains functional alignment of pre- and postsynaptic components at nicotinic synapses.

Results from other experiments suggest subunit requirements for the ultimate targeting of different AChR types into specific cellular domains. In particular, experiments with chimeric AChRs indicate that residues in the M3–M4 cytoplasmic loop of the $\alpha3$ subunit are important for appropriate receptor targeting. When $\alpha7$ subunits containing the M3–M4 cytoplasmic loop of $\alpha3$ (but not $\alpha5$) are expressed in ciliary ganglion neurons using retroviral mediated gene transfer, the resulting $\alpha7/\alpha3$ chimeric homomers are targeted to the post-synaptic membrane [149]. It is not known if the $\alpha3$ M3–M4 cytoplasmic loop is required for the post-synaptic targeting of native, heteromeric $\alpha3$*-AChRs in vivo, or if the $\alpha7$ cytoplasmic loop has a parallel role in the targeting of native $\alpha7$ homomers to somatic spines. Nevertheless these studies suggest that the M3–M4 cytoplasmic loops of constituent subunits contain information important for the ultimate targeting of different AChR types to appropriate cell surface domains. Of relevance here is the recent observation that in heterologous cells PSD-93a will associate with chimeric AChRs containing the $\alpha3$ but not the $\alpha7$ cytoplasmic loop [274] suggesting that PSD-protein and M3–M4 loop are critical to forming an intracellular scaffold, potentially one that appropriately organizes AChRs at nicotinic synapses.

2.3.2. Central synapses

Until recently, central synapses where transmission is mediated by post-synaptic nicotinic AChRs were thought rare, and examples were confined to spinal cord Renshaw cells [54,150] and nucleus ambiguus [151]. This list has now grown with AChR-mediated synapses recently reported in cortex [152], lateral spiriform nucleus [153], retina [154], and CA1 hippocampal interneurons [155,156]. Nicotinic synapses may also be present in cranial nuclei and substantia nigra (reviewed in Refs. [6,155]). The functional significance of nicotinic synapses in the CNS, where the bulk of excitatory transmission is mediated by glutamate receptors, remains to be identified. For CA1 hippocampal interneurons, however, it is known that AChR activation leads to both inhibition and disinhibition of pyramidal neurons, suggesting that nicotinic synapses on the neurons can serve to fine-tune circuits mediated by other neurotransmitters [157]. In addition to mediating transmission at some CNS synapses, AChRs assume prominent pre-synaptic and non-synaptic functions in regulating transmitter release and modulating neuronal excitability, as has been stressed in previous reviews [6,8–10]. The most direct evidence for a pre-synaptic function comes primarily from experiments where nicotinic agonists (nicotine, ACh, CCh) are shown to increase the frequency of spontaneous synaptic currents at glutamatergic and GABAergic central synapses while nicotinic antagonists block the facilitation [158–161]. In many cases, the pre-synaptic effects are mediated by αBgt-AChRs since they are blocked by αBgt, but other subtypes, notably those containing $\alpha4\beta2$ subunits, also can participate [6]. Although places in the CNS where AChRs are known to participate in post-synaptic signaling still appear to be quite rare, it is evident that AChRs play a role in CNS signaling, since nicotine can significantly improve attention in non-smoking subjects with no pre-existing attentional deficits [13]. Further, nicotine treatment has been shown to improve attention, memory, and learning in patients with Alzheimer's disease (AD) (reviewed in Ref. [13]). Although the mechanisms underlying the actions of nicotine in improving cognition are not well understood, it has been hypothesized that many of these effects are likely mediated by pre-synaptic AChRs modulating the release of neurotransmitters.

3. Recent findings and trends

3.1. Structure–function

Neuronal AChRs are ubiquitous, diverse and assume multiple roles relevant to signaling, development and disease. Thus an identification of native, functional AChR channels with their corresponding molecular subtypes is critical from both basic research and therapeutic perspectives. While functional expression of α and β subunits in *Xenopus* oocytes or mammalian cell lines indicates that at least two subunits (e.g. $\alpha3\beta2$, $\alpha4\beta2$, $\alpha2\beta4$, $\alpha3\beta4$, $\alpha4\beta4$) are required for AChR function [3,162], the properties of receptors expressed in heterologous cells do not match those of native neuronal AChRs known to contain the same subunits [163,164]. This shortfall is likely to partly reflect a greater subunit complexity of native AChRs. In ciliary ganglion neurons, for example, two $\alpha3$*-AChR subtypes containing $\alpha3$, $\alpha5$, and $\beta4$ subunits or $\alpha3$, $\alpha5$, $\beta4$, and $\beta2$ subunits have been identified by coprecipitation using subunit-specific monoclonal antibodies (Fig. 2) ([111,112], see also Ref. [165]). It was also recently shown that $\alpha5$ coassembles with AChRs containing $\alpha4\beta2$ [107,166,167], and that $\alpha7$ may contribute to as many as three heteromeric complexes containing other α and β subunits in sympathetic neurons [168]. Additional potential for native AChR complexity is possible if one further considers that additional subunits may also await discovery [112] or identification [121].

How can we relate the functional signature of a native AChR to its potentially complex subunit composition? A general approach to this question is to assess the contribution of candidate subunits by functionally deleting them. One relevant method involves manipulating the expression of individual subunit genes. For example, many of the known neuronal AChR subunit genes ($\alpha3$, $\alpha4$, $\beta2$, $\beta3$, $\beta4$, $\alpha7$) have been deleted in knockout mice that were subsequently examined for abnormalities and behavioral deficits, as well as alterations in receptor expression and function. These studies provide important information about the functional roles for hetero- and homomeric receptors that are presumed to contain the deleted subunits in vivo (reviewed extensively in Refs. [169,170]). In particular, they confirmed that $\alpha7$- and $\alpha4\beta2$-AChRs are the principal receptor subtypes in brain, that $\alpha7$ subunits are responsible for high-affinity αBgt binding and fast αBgt-sensitive currents, and that $\beta2$ subunits underlie the high-affinity nicotine binding associated with $\alpha4\beta2$ receptors.

A similar method involves treating neuronal cultures with antisense subunit oligonucleotides to "knock-down" expression of native AChR subunits. Working with cultured sympathetic neurons, Role and colleagues found that $\alpha3$ but not $\alpha4$ antisense oligonucleotides decreased AChR openings and altered channel properties, implicating

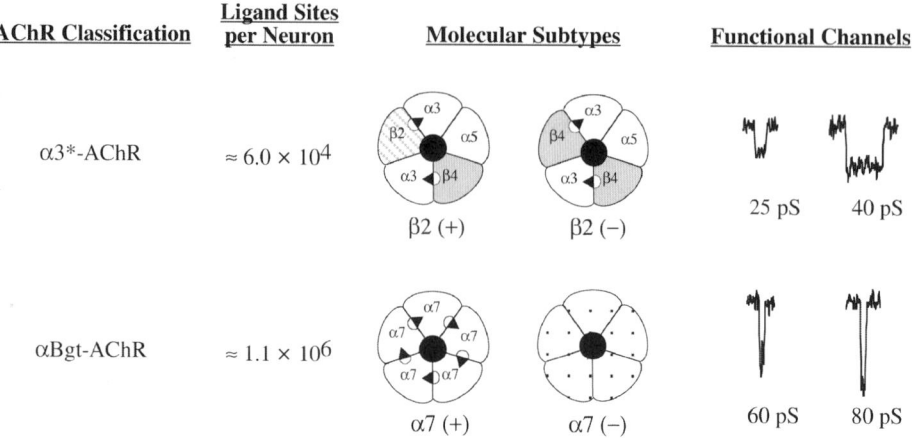

Fig. 2. Heterogeneity of AChRs on chick CG neurons. Ligand binding sites were assessed in surface labeling assays using [^{125}I]-mAb35 or [^{125}I]-αBgt to detect α3*- or αBgt-AChRs, respectively [124,193]. Hetero- and homomeric subtypes were assessed using subunit specific antibodies and toxins in pull-down or solid phase assays [111,112,121]. All detectable α3*-AChRs are heteromeric, containing α3, α5, and β4 subunits and lacking α7 subunits. Most α3*-AChRs (80%) lack β2 subunits (β2−), but a sizeable fraction (20%) do contain β2 (β2+). By contrast most αBgt-AChRs (95%) are believed to be homomeric, containing only α7 subunits and lacking α3, α5, β4, or β2 subunits. A tiny fraction (5%) of αBgt-AChRs is of unknown composition. Functional AChRs are identified in single channel recordings from excised membrane patches [124,181]. Brief-open duration AChR channels (those having conductances of 60 and 80 pS are shown, but see [181]) represent αBgt-AChRs since they are blocked by the toxin. Long-open duration AChR channels having conductances of 25 and 40 pS are insensitive to αBgt and thus considered α3*-AChRs. Long-duration 40 pS channel openings are attributed to β2(+) AChRs since they were blocked by αCTx-MII [181], an α3/β2 specific toxin [135].

α3 as a major component of the normal AChR repertoire present on these neurons [171]. More recent studies using α7 and α5 antisense oligonucleotides suggest that a subset of α7-containing AChRs may be heteromeric and that α5 contributes to at least two AChR subtypes [167,168]. Despite the power of these methods for identifying the participation of individual subunits, a nagging problem with both is that compensatory changes can occur [172] resulting in the expression of altered channels [136,171] that may not reflect the role of the subunit in vivo. A second approach avoids this problem by using subunit specific reagents to acutely perturb receptor function. αBgt, for example, can be used to identify AChR subtypes containing α7 subunits since it binds with high affinity to many native and heterologously expressed α7-containing receptors and efficiently blocks them [85,122, 123,173,174]. Exceptions are the small fraction (5%) of αBgt-sensitive αT/35-AChRs on ciliary ganglion neurons that appear to lack α7 (Fig. 2) [121] and some AChRs on sympathetic neurons, which can be deleted by α7 antisense treatment, but appear insensitive to αBgt [168]. αBgt-AChRs are highly abundant on ciliary ganglion neurons, and αBgt blocks all brief-duration (*200*s) AChR openings, including those having conductances of 25 and 40 pS [181], as well as 60 and 80 pS openings (Fig. 2) [124,181]. Despite this conductance heterogeneity, recent studies suggest that the brief-duration events can all be attributed to homomeric α7-AChRs [181]. In addition to the traditionally used αBgt, α-Conotoxins from snail venom have recently been used in identifying neuronal AChR subtypes [175]. α-Conotoxin ImI, like αBgt, blocks homomeric α7-AChRs expressed in oocytes [176] and inhibits presumed α7-AChR mediated currents in the rat hippocampus [177]. Two other *Conus* toxins have generated particular interest recently because of their specificity for subunit combinations. For example, α-Conotoxin AuIB (αCTx-AuIB) blocks rodent α3β4 AChRs expressed in *Xenopus* oocytes and putative α3β4 AChRs in rat medial habenula but not α3β2 AChRs in rat habenula or muscle type AChRs [178,179]. In contrast, αCTx-MII blocks rodent α3β2 AChRs expressed in *Xenopus* oocytes with nanomolar affinity and is 1000-fold more selective for α3β2 versus muscle AChRs [135]. αCTx-MII also blocks the slow component of the EPSC attributable to α3*-AChRs in ciliary ganglion neurons [133,180]. In addition, αCTx-AuIB and -MII both block nearly all of the whole-cell current attributable to α3*-AChRs in ciliary ganglion neurons [181]. In single channel experiments, αCTx-AuIB blocked the αBgt-insensitive, long-duration (∗ 200 ∗s) 40 pS AChR channel openings that dominate such recordings, as well as the less abundant long-duration 25 pS openings; αCTx-MII by contrast selectively blocked long-duration 40 pS openings [181]. The findings suggest that the predominant long-duration 40 pS AChR channel openings can be identified with α3*-AChRs having α3β2α3β4α5 composition and the less abundant long-duration 25 pS openings with α3α3β4α5 composition (β2+ and β2− in Fig. 2) [181]. These studies illustrate the utility of employing toxins that recognize individual AChR subunits (αBgt, αCTx-ImI) and subunit combinations (αCTx-AuIB and αCTx-MII) in conjunction with single channel recording as a means of dissecting the complex contribution of multiple subunits to native receptor subtypes. In addition to toxins, subunit-specific antibodies have been successfully employed for assessing AChR function in intracardiac ganglion neurons [182]. Here, results show that α7-specific, but not α3, α5, or α1-specific, antibodies applied intracellularly will block currents attributed to αBgt-AChRs. How the antibody blocks receptor function is unknown, but may involve interaction with the M3−M4 intracellular loop. An obvious caveat of both the toxin and the antibody approaches is the necessity to demonstrate the specificity of the reagent in the particular system under study. In this regard, specific reagents recognizing prevalent subunit combinations (e.g. α3β2, α3β4, α4β2) will be most useful since single neurons can express more than one heteromeric receptor subtype. Thus future studies will likely employ a combination of heterologous expression, gene knockout, and acute deletion approaches using subunit specific reagents to identify the composition of functionally relevant AChR subtypes expressed by neurons.

3.2. Post-translational receptor regulation

As with many voltage- and ligand-gated ion channels, the functional status of neuronal AChRs can be altered in response to short-term changes in synaptic inputs, growth factors and neuropeptides. Changes in synaptic activity, for example, can influence levels of extracellular Ca^{2+} [183], and elevated external Ca^{2+} is known to rapidly potentiate AChR responses in central and autonomic neurons by increasing channel opening probability [117,184,185]. Similarly, recent results indicate that certain albumins alter single AChR channel kinetics, raising the possibility that they mimic endogenous compounds capable of directly modulating AChR function in-vivo [275]. Neuronal AChRs are also potently and rapidly regulated by cytoplasmic signaling cascades, notably those initiated by the effector enzymes adenylate cyclase (AC) and phosphoinositide-specific phospholipase (PLC) (e.g. Fig. 3). AC activity increases cAMP levels and subsequent recruitment of cAMP-dependent protein kinase (PKA), while PLC activity increases phosphoinositol turnover, thereby elevating IP_3, $[Ca^{2+}]_i$, and phospholipid/calcium-dependent protein kinase (PKC).

The downstream second messengers and kinases mobilized by AC and PLC cascades can differentially regulate both voltage- and ligand-gated ion channels (reviewed in Refs. [44,45]). Activation of PLC or AC signal pathways, either directly or in response to neuropeptides, usually alters AChR function in different ways. For example, when phorbol esters are applied to chick sympathetic neurons to activate PKC, the rate of rapid AChR desensitization is enhanced without affecting the peak amplitude of the ACh response [186]. Substance P (SP) is a neuropeptide known to activate the PLC pathway and similarly increases the rate of rapid AChR desensitization in sympathetic [187,188] and ciliary ganglion neurons [187,189], presumably by activating PKC. SP may function as an endogenous AChR modulator since SP-like immunoreactivity is present in both sympathetic [190] and ciliary [191] ganglia. In contrast, cAMP elevation, produced either by bath-application of membrane permeant analogs or by intracellular injection of cAMP, enhanced AChR-induced currents about 2-fold in ciliary ganglion neurons without changing AChR desensitization or requiring that new receptors be inserted into the cell membrane [192,193]. Pituitary adenylate cyclase activating polypeptide (PACAP) is endogenous to the ciliary ganglion, and the neurons abundantly express high-affinity PACAP type 1 receptors (PAC_1Rs) that efficiently raise cAMP levels

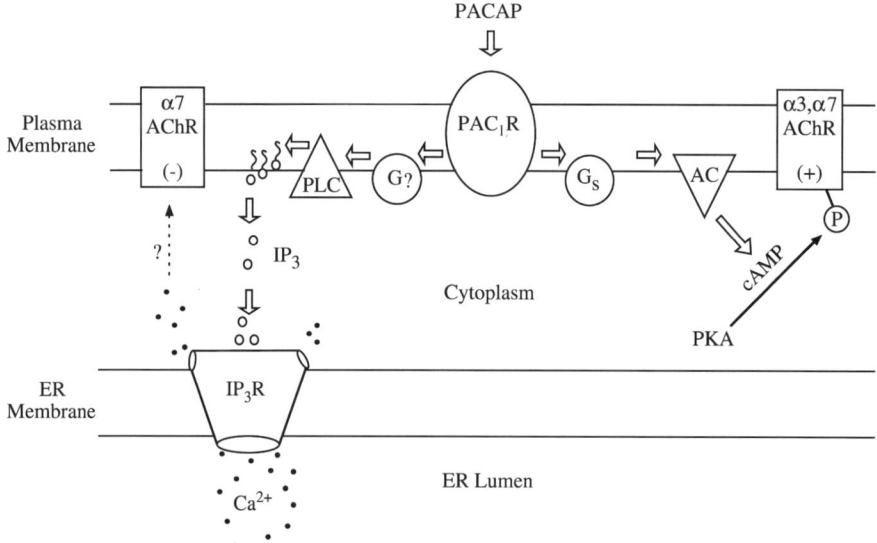

Fig. 3. Summary of PLC- and AC-dependent intracellular signal cascades mobilized by PAC_1Rs that reciprocally modulate neuronal AChR function. See text for details. Adapted from Ref. [195].

($EC_{50} \approx 0.4$ nM; [194]). As expected, PACAP application also rapidly potentiated currents mediated by both αBgt- and α3*-AChRs about 2-fold [195]. Since the ability of PACAP to enhance the currents was blocked by inhibiting PKA, this up-regulation may involve phosphorylation of αBgt- and α3*-AChRs, both of which are PKA substrates [196,197]. As in other systems [198,199], however, PAC$_1$Rs on ciliary ganglion neurons also couple to PLC, increasing IP turnover ($EC_{50} \approx 0.2$ nM) and producing IP$_3$-dependent release of $[Ca^{2+}]_i$ (Fig. 3). After blocking AC, subsequent PACAP application leads to reduced αBgt-AChR currents, an effect that requires PLC, Ca^{2+} and IP$_3$ and does not influence α3*-AChRs. Presumably PACAP induces a selective PLC-dependent inhibition of αBgt-AChRs that is normally overridden by tonic AC activation and unmasked after its blockade [195]. These results illustrate the ability of physiological interactions that change Ca^{2+} gradients and levels of neuropeptides to produce potent and differential regulation of several neuronal AChR classes.

Activating even one intracellular signaling pathway, however, can have diverse effects on ligand-gated channel currents. Like neuronal AChRs, currents mediated by GABA$_A$, glycine, and glutamate receptors [200–202] can be enhanced by cAMP/PKA-dependent signaling as are ACh-activated and synaptic currents in *Xenopus* muscle [203,204]. In other cases, however, things clearly go the other way with cAMP/PKA-dependent signaling decreasing GABA$_A$-mediated currents [205] and accelerating desensitization of mammalian muscle and *Torpedo* AChRs (reviewed in Refs. [44,45]). Direct phosphorylation of *Torpedo* AChRs by PKA and PKC indicate a requirement for phosphorylation of γ and δ subunits on specific serine residues to accelerate desensitization [206]. Studies with recombinant GABA$_A$ receptors provide a molecular perspective for the disparate outcomes of cAMP/PKA-dependent modulation. GABA$_A$ receptors containing $\alpha_1\beta_3\gamma_{2S}$ subunits expressed in HEK cells are potentiated by PKA phosphorylation of β_3 at S408 and S409, and a β_3 serine to alanine mutation at 408 converted the enhancement to an inhibition [207]. These findings suggest that the availability and position of specific residues within the intracellular domains of individual receptor subunits govern outcomes of cAMP/PKA-dependent phosphorylation. Expression of subunits containing residues phosphorylated by specific kinases would thus represent a simple mechanism for controlling the forms of modulation available to different receptor subtypes.

The ability of cAMP/PKA-dependent signaling to change ACh-induced currents has been ascribed to regulation of AChR functional properties; principally desensitization rate and/or channel open probability. A novel mechanism, involving recruitment of silent AChRs, may better explain the regulation seen in chick ciliary ganglion neurons. Using [^{125}I]-mAb35 and -αBgt, we determined the total numbers (N_T) of surface α3*- and αBgt-AChRs per ciliary ganglion neuron and compared these values with the numbers of functional receptors (N_F) obtained from whole-cell and single-channel recordings. The studies revealed that $N_T \gg N_F$ for both α3*- and αBgt-AChRs, suggesting that surface AChRs of both receptor types reside in functional and much larger "silent" pools [124,193, 208]. Interestingly, a large silent surface receptor pool is also implicated for α4β2 AChRs expressed in oocytes where the number of ligand binding sites greatly exceeds the estimated number of functional channels [209]. The possibility that cAMP-dependent signaling can recruit these sleeping receptors was suggested by previous studies showing that cAMP elevation rapidly increases the number of available Ca^{2+} channels on cardiac myocytes [210,211]. To explore this idea, we modeled the peak whole-cell current response of ciliary ganglion neurons attributable to α3*-AChRs (I_p) using $I_p = N_F p_o i$, where i is the single α3*-AChR channel current and p_o is the probability a fully-liganded AChR channel will enter the open state [212]. Since elevation of cAMP increased I_p 2-fold without changing N_T, single channel current amplitude (i) or channel opening probability (p_o), we inferred that cAMP-dependent signaling increases N_F [192] and a similar conclusion may also hold for αBgt-AChRs [124,195]. These findings suggest that AChRs residing in silent and functional pools exist in a dynamic equilibrium that can be influenced by intracellular signal cascades. The ability of such signals to alter both the entry of AChRs into a desensitized state, and the equilibrium between silent and functional receptors may, in fact, be related. Numerous studies indicate the surface AChRs cycle between a single functional state, and short- and long-lived desensitized states [213–215]. In addition, while PKA- and PKC-dependent phosphorylation accelerate the rapid phase of AChR desensitization [44,45], PKA phosphorylation speeds the recovery of recombinant muscle AChRs from a more long-lived, deeply desensitized state [216]. Interestingly, PKC phosphorylation also accelerated the recovery of neuronal α4β2 AChRs expressed in *Xenopus* oocytes from slow desensitization, possibly by reducing receptor accumulation in a deeply desensitized state [217]. Based on these considerations, it is tempting to speculate that the abundant silent AChRs on ciliary ganglion neurons occupy a deep, long-lived desensitized state and that intracellular signals, particularly those triggering PKA phosphorylation, can help awaken them.

3.3. AChRs: beyond synaptic transmission

3.3.1. AChRs are present early in development

Missing from the above accounts of receptor actions, however, is the observation that AChRs are evident in many neuronal precursors and neurons well before synapse formation, and thus prior to any role in synaptic transmission. For example, the α7 AChR subunit promoter is even active in cultures of undifferentiated mesoderm stem cells [218]. Similarly, transcripts for the α3, α5, α7, β2, and β4 subunits were detected in neural crest cells, coinciding both with immunoreactivity for the subunits and functional AChR expression on a population of those cells [219]. Both, α3 and α8 AChR subunits are expressed in the chick retina by E4.5 and E5.5, respectively, before choline acetyltransferase (ChAT)-positive neurons can be detected and well before synaptogenesis [220]. Further, in the chick CG, ACh-induced currents are present by E6, prior to innervation that occurs at E7 [221]. In the prenatal human CNS, both subunit transcripts and assembled AChRs have been detected as early as 4 weeks of gestational age [222], certainly prior to the vast majority of synapse formation (reviewed in Ref. [223]). What is the function of AChRs expressed early in the development of the nervous system? One possibility is based on the observation that neurotransmitters frequently influence neuronal survival and morphology (reviewed in Refs. [224,225]). Thus, neuronal AChRs might participate in such processes by virtue of their high permeability to Ca^{2+}, since Ca^{2+} influxes have been shown to have important influences on many aspects of neuronal development (reviewed in Refs. [140,141]).

3.3.2. AChRs and neuronal survival

The activity of neuronal AChRs has been implicated in neuronal survival in a number of in vivo and in vitro studies. Interestingly, though, the application of nicotine or nicotinic agonists has been shown to have both neurotrophic and neurotoxic effects. In general, AChR activation appears to support the survival of post-mitotic neurons, while similar activation seems to decrease the survival of developing neurons or neuroblasts. The remainder of this subsection will outline examples of both types of influence as elucidated by in vivo and in vitro studies.

Perhaps one of the more intriguing observations regarding AChRs and survival has been the presumptive neuroprotective effects of nicotine self-administration in disease states involving post-mitotic neurons (e.g. AD and Parkinson's disease (PD); reviewed in Ref. [226]). Work done in vitro has suggested that β-amyloid, a component of the neurotoxic machinery in AD, binds directly to AChRs, especially αBgt-AChRs, and induces toxicity, and that nicotine can prevent both this binding and toxicity. In lesion studies, both the α7-specific agonist 2,4-dimethoxybenzylidene anabaseine (DMXB) and the acetylcholinesterase (AChE) inhibitor tetrahydroaminoacridine (THA) prevented neocortical atrophy and degeneration following nucleus basalis lesions [227]. Thus, AChR binding or activation may directly promote neuronal survival. Alternatively, nicotinic agonist administration has been shown to alter both neurotrophin and neurotrophin receptor levels in the rodent CNS [228,229], suggesting that AChR activation can also play an initiatory role in other survival cascades. While these systems continue to be pursued, they strengthen the suggestion of a role for AChRs in the survival of neurons.

Assessing the effects of nicotinic receptor activation on mammalian development is complicated somewhat by the possibility that the drugs administered alter the maternal system. It has been reported, however, that pre-natal nicotine exposure in developing rats evokes cell death during neurulation [230] and persistent *c-fos* expression [231]. Further studies have been done in early neonatal rat pups, particularly during the period of rapid neonatal brain growth, where nicotine treatment can induce lasting behavioral deficits, including hyperactivity [232] which may be a model for similar observations in human smoking mothers (reviewed in Ref. [233]). Finally, using a rat model of adolescent nicotine exposure, Slotkin and colleagues report that nicotine administration alters behavior and neurotransmitter turnover, cell number and size, and AChR and *p53* expression (reviewed in Ref. [234]). Another way to address the role of AChR activation in development, without the complications of maternal exposure, would be the generation of a constituitively active AChR model, and recently a transgenic mouse which expresses α7-AChRs with a "gain of function" mutation (L250T) has been generated. Interestingly, this mouse exhibits increased apoptosis throughout the somatosensory cortex and dies shortly after birth [235]. Together these results suggest that exogenous activation of AChRs has profound consequences for the developing mammalian nervous system, especially with regards to neuronal survival. Whether survival or connectivity is altered in the adolescent nicotine exposure model remains to be seen.

Avian systems have been attractive for developmental studies because the potential for maternal complications is absent. Numerous studies in ovo have utilized AChR blockers to elucidate a role for receptor activity in neuronal survival. Previous work has shown that administration of ganglionic AChR channel blockers to embryos during the period of naturally-occurring cell death in the CG can potentiate cell death [236–238]. In vivo application of more general AChR blockers (d-TC and αBgt) has produced mixed effects on ganglionic survival, with some blockers preferentially promoting survival of one subpopulation of CG neurons (ciliary neurons) over another (choroid neurons), presumably through interactions which include the peripheral targets [237]. Notably absent in these studies are in ovo applications of nicotinic agonists, likely because such compounds are known teratogens for the developing chicken embryo [239]. In general, the in vivo avian experiments have pointed to the complex interactions between the activity of AChRs on neurons and their targets. Given the recent progress in defining central nicotinic synapses, perhaps new studies of CNS neuronal survival after AChR activity alteration in avians are warranted.

In vitro application of nicotinic ligands to AChR-expressing cells in culture has similarly produced mixed effects. Delbono and colleagues, for example, showed that activation of AChRs could increase the survival of chick spinal motorneurons [240], and nicotine has been shown to rescue PC12 cells from NGF-deprivation [241,242]. Additionally, nicotine has been shown to promote both proliferation and survival of mouse cerebellar granule neuroblasts maintained in culture [243]. In the CG system, application of the nicotinic ligands nicotine and CCh also produced a survival effect comparable to that seen with depolarization in high K^+. Both αBgt and d-TC prevented the ability of AChR activation to promote survival, implicating αBgt-AChRs in this process. The agonists apparently achieved this survival promotion through additional activation of voltage-dependent Ca^{2+} channels (VDCCs) and by preventing apoptotic death in the cultures [244]. In both primary and immortalized hippocampal progenitor cells, however, nicotine treatment induced apoptosis, while having no effect on the same cells when they have differentiated [245]. Interestingly, there are several other reports of AChR activation having mixed effects depending on context [242,246]. For instance, the α7-specific agonist DMXB exerts both neuroprotective and toxic effects on PC12 cells in a dose-dependent manner. In this case, use of blockers of intracellular signaling cascades suggests that PKC is involved in the neuroprotective effects, while an unspecified protein tyrosine kinase in implicated in the toxicity [242]. In primary rat cortical neuron cultures, the nicotinic receptor-mediated protection against β-amyloid neurotoxicity is thought to utilize the phosphatidylinositol 3-kinase to pathway to *bcl* gene expression (reviewed in Ref. [247]). Interestingly, none of the reports to date have implicated PKA in the AChR survival pathway since it is thought to be a likely player in K^+-depolarization-induced survival [248].

These sometimes contradictory findings have led to an interest in possible interactions between AChR activation and growth factors. For example, while nicotine- and K^+-induced depolarization were very similar in their support of CG neuronal survival, they were non-equivalent in at least one aspect-nicotine failed to support additional survival in the presence of FGF, unlike the synergism seen with high K^+ [249], although both were additive with PACAP [244]. Interactions between AChR and growth factor receptor activation do not always lead to survival of the affected neurons. Hory-Lee and Frank [250] demonstrated that blocking AChRs potentiated the effect of muscle extract on the survival of spinal motorneurons in vitro. In cortical neuron cultures under NMDA assault, either nicotine or tumor necrosis factor-α (TNF-α) can partially support the survival of the neurons, while simultaneous treatment with both agents fails to protect from the NMDA-induced toxicity [246]. Thus, a clear picture of AChR-mediated survival both in vivo and in vitro is still emerging. Fully understanding the effects of AChR activation on neuronal survival will likely require further elucidation of the pathways utilized by such activation and the interactions of those pathways with those of other neurotrophic substances.

3.3.3. AChRs and neurite outgrowth

Together with their early developmental appearance, the ability of AChRs, especially αBgt-AChRs, to elevate Ca^{2+} efficiently paired with their primarily non-synaptic [127,129,251], often pre-synaptic, locations has led to much speculation about the roles these receptors might play in neurite outgrowth and synapse formation. That these receptors often first appear at the time of synapse formation, or shortly before, also supports these ideas. Indeed, growth cones of *Xenopus* spinal neurons have been shown to contain functional αBgt-AChRs [252]. Additionally, cholinesterase is often another early marker of neuronal populations, and many axon tracts in the periphery contain cholinesterase [253].

Are neuronal AChRs being activated as an axon navigates through the periphery? What is the source of agonist (ACh) for that activation? While we still know little about the former question, evidence has emerged regarding the latter. One plausible source of agonist is the growth cones themselves, since growth cones of cholinergic neurons can release ACh in vitro [254,255]. This observation suggests that ACh released from the growth cone might activate receptors on that same growth cone or its neighbors. Since many axon tracts in the periphery contain cholinesterase, any released ACh would normally be quickly catalyzed and growth could continue. If, however, the growth cone turned away from the axon tract and its endogenous cholinesterase released, ACh might persist long enough to inhibit further growth in that direction. This effect could thus contribute to keeping growth cones confined to appropriate peripheral pathways. In studies of developing amphibian NMJs in culture, Poo and collaborators demonstrated that ACh is released from spinal neurons at the earliest moments of neuron–muscle contact [255]. Fischbach and colleagues have also shown that such ACh release from the growing axon tips occurs in chick neurons without the necessity of contact [254]. Thus, it is possible that the ACh released from outgrowing axons might be sufficient to activate AChRs on those same axons or their immediate neighbors and thus guide pathfinding.

There is also evidence that embryonic nerve pathways express cholinesterases. In the anterior sclerotome of the chicken, where motor axons first enter the periphery, both butyrylcholinesterase (BChE) and AChE expression precede myotome differentiation, motor axon outgrowth and first establishment of synaptic contacts [253]. Since motorneurons express AChE early in development, the motor axons themselves might be the source for AChE detected in the axon tracts [253,256]. Layer [257] also proposes that if neurite outgrowth could be inhibited by ACh, then those cholinesterases could create ACh-free spaces and thereby direct neurite extension. Furthermore, blockade of either BChE or AChE activity in retinal explant cultures strongly decreases neurite growth [258]. The authors, however, argue that cholinesterases may be acting as cell adhesion molecules (via their HNK-1 epitopes), independent of their enzymatic activities, because blocking both cholinesterase activities (using a different reagent) does not change neuritic growth in vitro [258,259]. Therefore, cholinesterases lining the pathways through which the outgrowing axons grow could provide an "ACh-free zone" which would prevent activation of AChRs and allow axonal growth to progress as long as the growth cone remained within the appropriate pathway.

Few studies directly address the contribution of such AChR mechanisms to axonal pathfinding in vivo. Transgenic mice carrying either the loss or the gain of function $\alpha 7$ mutations, for example, manifest no gross morphological deficits in pathfinding [235,260]. Mice lacking functional $\beta 2$ subunits are missing early retinal wave patterns, have delayed stratification of retinal ganglion cell dendrites, and are lacking eye-specific layers within the retinogeniculate projections [276]. Certainly these results and others [277] suggest a role for AChR-mediated activity in the refinement of these projections, but these deficits are unlikely to result directly from pathfinding errors [261]. AChRs, however, have been implicated in pathfinding in vivo in the regenerating retinotectal projections of goldfish. Schmidt [262] has demonstrated that eliminating the modulatory cholinergic systems, using the cholinergic neurotoxin AF64A prior to the regeneration period, prevented activity-dependent retinotopic sharpening of the regenerated projection. Interestingly, while blocking the AChRs with αBgt, N-Bgt, or pancuronium had little effect on the recorded field potentials, each of those drugs also prevented the sharpening of the retinal projections. Additionally, in the developing retinotectal system of *Xenopus* tadpoles, pre-synaptic AChRs have been shown to elevate $[Ca^{2+}]_i$ in growing retinal ganglion cell axon arbors [263]. This calcium increase has been implicated in branch addition to the arbors, and its induction by a light stimulus can be blocked by αBgt. Both of these studies strongly suggest that αBgt-AChRs play a role in activity-dependent sharpening. Thus, it appears that AChRs may be capable of influencing axonal outgrowth in vivo, although further studies will certainly be necessary to understand the differences between the results from the knockout mice where compensatory changes may mask subtle effects and those from the pharmacological manipulations in non-mammalian vertebrates.

In culture, cholinergic drugs have been demonstrated to influence neurite outgrowth directly. Depending on the neuronal model system used, activation of AChRs has been reported to promote, to inhibit, and to alter the direction of neurite outgrowth. Nicotinic receptor antagonists have been shown to enhance the number and mean length of neurites from retinal ganglion cells in culture, implicating AChRs in the regulation of neurite outgrowth [264]. In mouse, the outgrowth of neurites emanating from spinal cord explants was halted upon exposure to ACh, although the receptor subtype specificity was not assessed [265]. In the case of differentiated PC12 cells, long-term nicotine treatment has been shown to result in lessened neuritic outgrowth, an effect prevented by coincubation with αBgt [266]. Additionally in undifferentiated PC12 cells, αBgt appeared to promote neurite outgrowth on its own, although at a higher concentration than would be expected given its affinity for the AChRs [267]. While blocking αBgt-AChRs in chick CG neuronal cultures did not promote neurite outgrowth, short-term, repetitive activation of those receptors did elicit neurite retraction over the period of an hour. In this case, the retraction required the additional activation of VDCCs, much like that seen in the survival studies [268]. Contrasting with these inhibitory effects, AChR activation appears to be neurotropic and enhance neurite outgrowth from *Xenopus* spinal neurons. Poo and colleagues have shown that pulsatile ACh application can change the directionality of growth cones, with them turning towards the source of ACh over a 2-h period. This effect can be blocked by d-TC [269]. In these same cells, the galvinotropism (the turning of a neurite in response to an applied magnetic field) of the growing neurite appears to be mediated by AChRs as the resulting alterations in growth speed, branching direction, and growth cone orientation are all prevented by d-TC blockade of AChRs [270]. Thus, while we are beginning to understand that AChRs are capable of influencing neurite outgrowth, and likely cell migration, the ways in which such things change appears to be specific to each cell type, and additional studies will be necessary to elucidate the mechanisms underlying such specificity. Thus, the early and ubiquitous expression of AChRs may indeed be indicative developmental roles for these receptors. The studies outlined above implicate AChRs in neuronal survival, both in mature and developing nervous systems. The possibility that AChRs may play a role in axonal pathfinding is intriguing and remains to be fully explored. Contributions of AChRs to the activity-dependent refinement of projections suggest that AChRs may also play a role in synapse formation and remodeling. Certainly, our understanding of the ramifications of early nicotine exposure will continue to grow in parallel with our understanding of the mechanisms underlying developmental events.

4. Conclusions and future directions

In the 40 years since nicotinic AChRs were first identified on neurons, our understanding of their importance and ability to study them have both grown steadily. We now know that neuronal AChRs are ligand-gated ion channel proteins, similar to their relatives on skeletal muscle fibers in both pentameric subunit organization and basic functional properties. And yet there are important differences, notably in channel properties, regulation, synaptic localization, and pharmacology that may reflect the greater flexibility required in a neuronal setting. Equally revealing has been the discovery that unlike muscle AChRs, neuronal AChRs are constructed from a different, more diverse, palette of subunits. This discovery is consistent with the functional heterogeneity of neuronal AChRs, but also implicates them in a wide range of functions. Indeed, the past 10 years have yielded an impressive harvest of novel functions attributable to neuronal AChRs. We have mentioned their ability to mediate and modulate synaptic transmission, and participate in nicotine addiction, cognitive behavior, and neurological disease. A particular focus of this chapter has been a discussion of neuronal AChR involvement in processes relevant to development, such as neuronal survival and neurite pathfinding. The range functions attributable to neuronal AChRs are somewhat of an embarrassment of riches at times because a detailed understanding of how, or if, the post-, pre- and non-synaptic functions are inter-related is still incomplete. It already seems apparent, however, that efforts to unravel how diverse AChR functions and channel subtypes might be coordinated to achieve complex integrative functions are underway [271].

More fundamental issues about neuronal AChRs also persist. We still have an incomplete picture of the composition and function of many native neuronal AChRs. Three basic questions seem most relevant here. First, have we identified all of the genes encoding native neuronal AChR subunits? Second, which subunits and in which arrangement are neuronal AChRs assembled to form the diverse subtypes present on neurons? Third, how are the formation, function, and localization of individual subtypes regulated by development, and by cell interactions within and between neuronal populations? Answers to these questions will likely result in further discoveries of new receptor subtypes and new functional roles. They will also present a directed approach for drug design to target individual subtypes and functions. A second fundamental issue is that we need to better understand the roles neuronal AChRs play in the complex cell interactions that occur during neuronal development. Certainly, the early expression of neuronal AChRs, and their ability to influence survival and neuronal pathfinding are all consistent with AChRs contributing to developmental processes, but the critical details of this picture are still missing. Here several questions are noteworthy. Are AChRs necessary for any of these processes in vivo? Is neuronal development influenced by an interaction between AChR-generated signals and those generated by other cell interactions such as migratory route and neurotrophic factors? Does the influence of AChRs extend to regulating synapse formation, and if so how does this occur? With our expanding repertoire of tools, these exciting questions are all now approachable and it seems likely that some answers and, of course, new questions will be forthcoming.

Acknowledgements

Supported by National Institutes of Health (NS24417) and National Science Foundation (IBN-9514560) awards to JFM.

References

[1] Karlin, A., Akabas, M., 1995. Toward a structural basis for the function of nicotinic acetylcholine receptors and their cousins. Neuron 15, 1231–1244.
[2] Sargent, P.B., 1993. The diversity of neuronal nicotinic acetylcholine receptors. Annu. Rev. Neurosci. 16, 403–443.
[3] Role, L.W., 1992. Diversity in primary structure and function of neuronal nicotinic acetylcholine receptor channels. Curr. Opin. Neurobiol. 2, 254–262.
[4] McGehee, D.S., Role, L.W., 1995. Physiological diversity of nicotinic acetylcholine receptors expressed by vertebrate neurons. Annu. Rev. Physiol. 57, 521–546.
[5] Lindstrom, J., 1997. Nicotinic acetylcholine receptors in health and disease. Mol. Neurobiol. 15(2), 193–222.
[6] Role, L.W., Berg, D.K., 1996. Nicotinic receptors in the development and modulation of CNS synapses. Neuron 16, 1077–1085.
[7] MacDermott, A.B., Role, L.W., Siegelbaum, S.A., 1999. Presynaptic ionotropic receptors and the control of transmitter release. Annu. Rev. Neurosci. 22, 443–485.
[8] Dani, J.A., 2001. Nicotine mechanisms in Alzheimer's disease. Overview of nicotinic receptors and their roles in the central nervous system. Biol. Psychiatry 49, 166–174.
[9] Wonnacott, S., 1997. Presynaptic nicotinic ACh receptors. Trends Neurosci. 20, 92–98.
[10] Vizi, E.S., Lendvai, B., 1999. Modulatory role of presynaptic nicotinic receptors in synaptic and non-synaptic communication in the central nervous system. Brain Res. Brain Res. Rev. 30, 219–235.
[11] Weiland, S., Bertrand, D., Leonard, S., 2000. Neuronal nicotinic acetylcholine receptors: from the gene to the disease. Behav. Brain Res. 113(1-2), 43–56.

[12] Mansvelder, H.D., McGehee, D.S., 2002. Cellular and synaptic mechanisms of nicotine addition. J. Neurobiol. 53, 606–617.

[13] Levin, E.D., 2002. Nicotinic receptor subtypes and cognitive function. J. Neurobiol. 53, 633–640.

[14] Flores, C.M., 2000. The promise and pitfalls of a nicotinic cholinergic approach to pain management. Pain 88(1), 1–6.

[15] Lee, C., Tseng, L., Chiu, T., 1967. Influence of denervation on localization of neurotoxins from elapid venoms in rat diaphragm. Nature 215, 1177–1178.

[16] Changeux, J.-P., Kasai, M., Lee, C.Y., 1970. The use of a snake venom to characterize the cholinergic receptor protein. Proc. Natl Acad. Sci. USA 67, 1241–1247.

[17] Raftery, M.A., Hunkapiller, M.W., Strader, C.D., Hood, L.E., 1980. Acetylcholine receptor: complex of homologous subunits. Science 208, 1454–1457.

[18] Brehm, P., Henderson, L., 1988. Regulation of acetylcholine receptor channel function during development of skeletal muscle. Dev. Biol. 129, 1–11.

[19] Mishina, M., Takai, T., Imoto, K., Noda, M., Takahashi, T., Numa, S., 1986. Molecular distinction between fetal and adult forms of muscle acetylcholine receptor. Nature 321, 406–411.

[20] Noda, M., Takahashi, H., Tanabe, T., Toyosato, M., Furutani, Y., Hirose, T., et al., 1982. Primary structure of α-subunit precursor of *Torpedo californica* acetylcholine receptor deduced from cDNA sequence. Nature 299(5886), 793–797.

[21] Noda, M., Takahashi, H., Tanabe, T., Toyosato, M., Kikyotani, S., Furutani, Y., et al., 1983. Structural homology of *Torpedo californica* acetylcholine receptor subunits. Nature 302(5908), 528–532.

[22] Noda, M., Takahashi, H., Tanabe, T., Toyosato, M., Kikyotani, S., Hirose, T., et al., 1983. Primary structures. of β- and δ-subunit precursors of *Torpedo californica* acetylcholine receptor deduced from cDNA sequences. Nature 301(5897), 251–255.

[23] Sumikawa, K., Houghton, M., Smith, J.C., Bell, L., Richards, B.M., Barnard, E.A., 1982. The molecular cloning and characterisation of cDNA coding for the a subunit of the acetylcholine receptor. Nucleic Acids Res. 10(19), 5809–5822.

[24] Devillers-Thiéry, A., Giraudat, J., Bentaboulet, M., Changeux, J.P., 1983. Complete mRNA coding sequence of the acetylcholine binding alpha-subunit of *Torpedo marmorata* acetylcholine receptor: a model for the transmembrane organization of the polypeptide chain. Proc. Natl Acad. Sci. USA 80, 2067–2071.

[25] Claudio, T., Ballivet, M., Patrick, J., Heinemann, S., 1983. *Torpedo californica* acetylcholine receptor 60,000 dalton subunit: nucleotide sequence of cloned cDNA deduced amino acid sequence, subunit structural predictions. Proc. Natl Acad. Sci. USA 80, 1111–1115.

[26] Kyte, J., Doolittle, R.F., 1982. A simple method for displaying the hydropathic character of a protein. J. Mol. Biol. 157, 105–132.

[27] Unwin, N., 1995. Acetylcholine receptor channel imaged in the open state. Nature 373, 37–43.

[28] Unwin, N., 1993. Nicotinic acetylcholine receptor at 9 Å resolution. J. Mol. Biol. 229, 1101–1124.

[29] Karlin, A., Cowburn, D., 1973. The affinity-labeling of partially purified acetylcholine receptor from electric tissue of Electrophorus. Proc. Natl Acad. Sci. USA 70, 3636–3640.

[30] Kao, P.N., Dwork, A.J., Kaldany, R.R., Silver, M.L., Wideman, J., Stein, S., et al., 1984. Identification of the alpha subunit half-cystine specifically labeled by an affinity reagent for the acetylcholine receptor binding site. J. Biol. Chem. 25, 11662–11665.

[31] Kao, P.N., Karlin, A., 1986. Acetylcholine receptor binding site contains a disulfide cross-link between adjacent half-cystinyl residues. J. Biol. Chem. 261, 8085–8088.

[32] Brejc, K., van Dijk, W.J., Klaassen, R.V., Schuurmans, M., van der Oost, J., Smit, A.B., et al., 2001. Crystal structure of an ACh-binding protein reveals the ligand-binding domain of nicotinic receptors. Nature 411, 269–276.

[33] Smit, A.B., Syed, N.I., Schaap, D., van Minnen, J., Klumperman, J., Kits, K.S., et al., 2001. A glia-derived acetylcholine-binding protein that modulates synaptic transmission. Nature 411, 261–268.

[34] Colman, A., 1984. Translation of eukaryotic messenger RNA in Xenopus oocytes. In: Hanes, B.D., Higgins, S.J. (Eds.), Transcription and Translation: a Practical Approach. IRL Press, Oxford.

[35] Claudio, T., Green, W.N., Hartman, D.S., Hayden, D., Paulson, H.L., Sigworth, F.J., et al., 1987. Genetic reconstitution of functional acetylcholine receptor channels in mouse fibroblasts. Science 238, 1688–1694.

[36] Mishina, M., Kurosaki, T., Tobimatsu, T., Morimoto, Y., Noda, M., Yamamoto, T., et al., 1984. Expression of functional acetylcholine receptor from cloned cDNAs. Nature 307, 604–608.

[37] White, M.M., Mayne, K.M., Lester, H.A., Davidson, N., 1985. Mouse-Torpedo hybrid acetylcholine receptors: functional homology does not equal sequence homology. Proc. Natl Acad. Sci. USA 82, 4852–4856.

[38] Kurosaki, T., Fukuda, K., Konno, T., Mori, Y., Tanaka, K., Mishina, M., et al., 1987. Functional properties of nicotinic acetylcholine receptor subunits expressed in various combinations. FEBS Lett. 3214, 253–258.

[39] Mishina, M., Tobimatsu, T., Imoto, K., Tanaka, K., Fujita, Y., Fukuda, K., et al., 1985. Location of functional regions of acetylcholine receptor α-subunit by site-directed mutagenesis. Nature 313, 364–369.

[40] Charnet, P., Labarca, C., Leonard, R.J., Vogelaar, N.J., Czyzyk, L., Gouin, A., et al., 1990. An open-channel blocker interacts with adjacent turns of α-helices in the nicotinic acetylcholine receptor. Neuron 4, 87–95.

[41] Imoto, K., Methfessel, C., Sakmann, B., Mishina, M., Mori, Y., Konno, T., et al., 1986. Location of a δ-subunit region determining ion transport through the acetylcholine receptor channel. Nature 324, 670–674.

[42] Imoto, K., Busch, C., Sakmann, B., Mishina, M., Konno, T., Nakai, J., et al., 1988. Rings of negatively charged amino acids determine the acetylcholine receptor channel conductance. Nature 335, 645–648.

[43] Swope, S.L., Qu, Z., Huganir, R.L., 1995. Phosphorylation of the nicotinic acetylcholine receptor by tyrosine kinases. Ann. NY Acad. Sci. 757, 197–214.

[44] Swope, S.L., Moss, S.I., Raymond, L.A., Huganir, R.L., 1999. Regulation of ligand-gated ion channels by protein phosphorylation. Adv. Second Messenger Phosphoprot. Res. 33, 49–78.

[45] Swope, S.L., Moss, S.J., Blackstone, C.D., Huganir, R.L., 1993. Phosphorylation of ligand-gated ion channels: a possible mode of synaptic plasticity. FASEB J. 6, 2514–2523.

[46] Maimone, M.M., Enigk, R.E., 1999. The intracellular domain of the nicotinic receptor alpha subunit mediates its coclustering with rapsyn. Mol. Cell Neurosci. 14, 350–354.

[47] Sealock, R., Wray, B.E., Froehner, S.C., 1984. Ultrastructural localization of the Mr 43,000 protein and the acetylcholine receptor in *Torpedo* postsynaptic membranes using monoclonal antibodies. J. Cell Biol. 98, 2239–2244.

[48] Fertuck, H.C., Salpeter, M.M., 1974. Localization of acetylcholine receptor by ^{125}I-labeled alpha bungarotoxin binding at mouse motor endplates. Proc. Natl Acad. Sci. USA 71, 1376–1378.

[49] Froehner, S.C., Luetje, C.W., Scotland, P.B., Patrick, J., 1990. The postsynaptic 43K protein clusters muscle nicotinic acetylcholine receptors in *Xenopus* oocytes. Neuron 5, 403–410.

[50] Phillips, W.D., Kopta, C., Blount, P., Gardner, P.D., Steinbach, J.H., Merlie, J.P., 1991. ACh receptor-rich membrane domains organized fibroblasts by recombinant 43-kilodalton protein. Science 251, 568–570.

[51] Yu, X.-M., Hall, Z.W., 1994. The role of the cytoplasmic domains of individual subunits of the acetylcholine receptor in 43 kDa protein-induced clustering in COS cells. J. Neurosci. 14, 785–795.

[52] Apel, E.D., Merlie, J.P., 1995. Assembly of the postsynaptic apparatus. Curr. Opin. Neurobiol. 5, 62–67.

[53] Colledge, M., Froehner, S.C., 1998. Signals mediating ion channel clustering at the neuromuscular junction. Curr. Opin. Neurobiol. 8(3), 357–363.

[54] Duggan, A.W., Hall, J.G., Lee, C.Y., 1976. Alpha-bungarotoxin, cobra neurotoxin and excitation of Renshaw cells by acetylcholine. Brain Res. 107(1), 166–170.

[55] Carbonetto, S.T., Fambrough, D.M., Muller, K.J., 1978. Nonequivalence of alpha-bungarotoxin receptors and acetylcholine receptors in chick sympathetic neurons. Proc. Natl Acad. Sci. USA 75, 1016–1020.

[56] Ravdin, P.M., Berg, D.K., 1979. Inhibition of neuronal acetylcholine sensitivity by α-toxins from *Bungarus multicinctus* venom. Proc. Natl Acad. Sci. USA 76, 2072–2076.

[57] Wilson, S.P., Kirshner, N., 1977. The acetylcholine receptor of the adrenal medulla. J. Neurochem. 28, 687–695.

[58] Patrick, J., Stallcup, B., 1977. α-Bungarotoxin binding and cholinergic receptor function on a rat sympathetic nerve line. J. Biol. Chem. 252, 8629–8633.

[59] Oswald, R.E., Sutcliffe, M.J., Bamburger, M., Loring, R.H., Braswell, E., Dobson, C.M., 1991. Solution structure of neuronal bungarotoxin determined by two-dimensional NMR spectroscopy: sequence specific assignments, secondary structure and dimer formation. Biochemistry 30, 4901–4909.

[60] Chiappinelli, V.A., 1983. Kappa-bungarotoxin: a probe for the neuronal nicotinic receptor in the avian ciliary ganglion. Brain Res. 277, 9–22.

[61] Loring, R.H., Chiappinelli, V.A., Zigmond, R.E., Cohen, J.B., 1984. Characterization of a snake venom neurotoxin which blocks nicotinic transmission in the avian ciliary ganglion. Neuroscience 11, 989–999.

[62] Sargent, P.B., Bryan, G.K., Streichert, L.C., Garrett, E.N., 1991. Denervation does not alter the number of neuronal bungarotoxin binding sites on autonomic neurons in the frog cardiac ganglion. J. Neurosci. 11, 3610–3623.

[63] Smith, M.A., Margiotta, J.F., Berg, D.K., 1983. Differential regulation of acetylcholine sensitivity and α-bungarotoxin-binding sites on ciliary ganglion neurons in cell culture. J. Neurosci. 3, 2395–2402.

[64] Patrick, J., Stallcup, W.B., 1977. Immunological distinction between acetylcholine receptor and the α-bungarotoxin binding component on sympathetic neurons. Proc. Natl Acad. Sci. USA 74, 4689–4692.

[65] Boulter, J., Evans, K., Goldman, D., Martin, G., Treco, D., Heinemann, S., et al., 1986. Isolation of a cDNA clone coding for a possible neural nicotinic acetylcholine receptor alpha-subunit. Nature 319, 368–374.

[66] Nef, P., Oneyser, C., Alliod, C., Couturier, S., Ballivet, M., 1988. Genes expressed in the brain define three distinct neuronal nicotinic acetylcholine receptors. EMBO J. 7, 595–601.

[67] Couturier, S., Erkman, L., Valera, S., Rungger, D., Bertrand, S., Boulter, J., et al., 1990. α5, α3, and non-α3: three clustered avian genes encoding nicotinic acetylcholine receptor-related subunits. J. Biol. Chem. 265, 17560–17567.

[68] Boulter, J., O'Shea-Greenfield, A., Duvoisin, R., Connolly, J., Wada, E., Jensen, A., et al., 1990. α3, α5 and β4: three members of the rat neuronal nicotinic acetylcholine receptor-related gene family form a gene cluster. J. Biol. Chem. 265, 4472–4482.

[69] Goldman, D., Deneris, E., Luyten, W., Kochhar, A., Patrick, J., Heinemann, S., 1987. Members of a nicotinic acetylcholine receptor gene family are expressed in different regions of the mammalian central nervous system. Cell 48, 965–973.

[70] Wada, K., Ballivet, M., Boulter, J., Connolly, J., Wada, E., Deneris, E., et al., 1988. Functional expression of a new pharmacological subtype of Brain nicotinic acetylcholine receptor. Science 240, 330–334.

[71] Deneris, E.S., Connolly, J., Boulter, J., Wada, E., Wada, K., Swanson, L.W., et al., 1988. Primary structure and expression of β2: a novel subunit of neuronal nicotinic acetylcholine receptors. Neuron 1, 45–54.

[72] Deneris, E.S., Boulter, J., Swanson, L., Patrick, J., Heinemann, J., 1989. β3: a new member of nicotinic acetylcholine receptor gene family is expressed in brain. J. Biol. Chem. 264, 6268–6272.

[73] Duvoisin, R.M., Deneris, E.S., Patrick, J., Heinemann, S., 1989. The functional diversity of nicotinic acetylcholine receptors is increased by a novel subunit: β4. Neuron 3, 487–496.

[74] Lindstrom, J., Schoepfer, R., Conroy, W., Whiting, P., Das, M., Saedi, M., et al., 1991. The nicotinic acetylcholine receptor gene family: structure of nicotinic receptors from muscle and neurons and neuronal α-bungarotoxin-binding proteins. Adv. Exp. Med. Biol. 287, 255–278.

[75] Elliott, K.J., Ellis, S.B., Berckhan, K.J., Urrutia, A., Chavez-Noriega, L.E., Johnson, E.C., et al., 1996. Comparative structure of human neuronal α2–α7 and β2–β4 nicotinic acetylcholine receptor subunits and functional expression of the α2, α3, α4, α7, β2, and β4 subunits. J. Mol. Neurosci. 7(3), 217–228.

[76] Fornasari, D., Chini, B., Tarroni, P., Clementi, F., 1990. Molecular cloning of human neuronal nicotinic receptor α3-subunit. Neurosci. Lett. 111, 351–356.

[77] Boulter, J., Connolly, J., Deneris, E., Goldman, D., Heinemann, S., Patrick, J., 1987. Functional expression of two neuronal nicotinic acetylcholine receptors from cDNA clones identifies a gene family. Proc. Natl Acad. Sci. USA 84, 7763–7767.

[78] Steinlein, O., Smigrodzki, R., Lindstrom, J., Anand, R., Kohler, M., Tocharoentanaphol, C., et al., 1994. Refinement of the localization of the gene for neuronal nicotinic acetylcholine receptor α4 subunit (CHRNA4) to human chromosome 20q13.2-q13.3. Genomics 22(2), 493–495.

[79] Monteggia, L.M., Gopalakrishnan, M., Touma, E., Idler, K.B., Nash, N., Arneric, S.P., et al., 1995. Cloning and transient expression of genes encoding the human α4 and β2 neuronal nicotinic acetylcholine receptor (nAChR) subunits. Gene 155(2), 189–193.

[80] Chini, B., Clementi, F., Hukovic, N., Sher, E., 1992. Neuronal type a-bungarotoxin receptors and the α5 nicotinic receptor subunit gene are expressed in neuronal and non-neuronal human cell lines. Proc. Natl Acad. Sci. USA 89, 1572–1576.

[81] Le Novère, N., Zoli, M., Changeux, J., 1996. Neuronal nicotinic α6 subunit mRNA is selectively concentrated in catecholaminergic nuclei of the rat brain. Eur. J. Neurosci. 8, 2428–2439.

[82] Fucile, S., Matter, J-M., Erkman, L., Ragozzino, D., Barbaino, B., Grassi, F., et al., 1998. The neuronal α_6 subunit forms functional heteromeric acetylcholine receptors in human transfected cells. Eur. J. Neurosci. 10, 172–178.

[83] Séguéla, P., Wadiche, J., Dineley-Miller, K., Dani, J.A., Patrick, J.W., 1993. Molecular cloning, functional properties, and distribution of rat Brain α7: a nicotinic cation channel highly permeable to calcium. J. Neurosci. 13, 596–604.

[84] Schoepfer, R., Conroy, W.G., Whiting, P., Gore, M., Lindstrom, J., 1990. Brain α-bungarotoxin binding protein cDNAs and MAbs reveal subtypes of this Brain of the ligand-gated ion channel gene superfamily. Neuron 5, 35–48.

[85] Couturier, S., Bertrand, D., Matter, J-M., Hernandez, M-C., Bertrand, S., Millar, N., et al., 1990. A neuronal nicotinic acetylcholine receptor subunit (α7) is developmentally regulated and forms a homo-oligomeric channel blocked by α-Btx. Neuron 5, 847–856.

[86] Chini, B., Raimond, E., Elgoyhen, A., Moralli, D., Balzaretti, M., Heinemann, S., 1994. Molecular cloning and chromosomal localization of the human α7-nicotinic receptor subunit gene (CHRNA7). Genomics 19, 379–381.

[87] Peng, X., Katz, M., Gerzanich, V., Anand, R., Lindstrom, J., 1994. Human α7 acetylcholine receptor: cloning of the α7 subunit from SH-SY5Y cell line and determination of pharmacological properties of native receptors and function α7 homomers expressed in *Xenopus* oocytes. Mol. Pharmacol. 45, 546–554.

[88] Elgoyhen, A.B., Johnson, D.S., Boulter, J., Vetter, D.E., Heinemann, S., 1994. α9 an acetylcholine receptor with novel pharmacolgical properties expressed in rat cochlear hair cells. Cell 79, 705–715.

[89] Hiel, H., Luebke, A., Fuchs, P., 2000. Cloning and expression of the α9 nicotinic acetylcholine receptor subunit in cochlear hair cells of the chick. Brain Res. 858(1), 215–225.

[90] Nguyen, V., Ndoye, A., Grando, S., 2000. Novel human α9 acetylcholine receptor regulating keratinocyte adhesion is targeted by *Pemphigus vulgaris* autoimmunity. Am. J. Pathol. 157(4), 1377–1391.

[91] Elgoyhen, A.B., Vetter, D.E., Katz, E., Rothlin, C.V., Heinemann, S.F., Boulter, J., 2001. α10: a determinant of nicotinic cholinergic receptor function in mammalian vestibular and cochlear mechanosensory hair cells. Proc. Natl Acad. Sci. USA 98(6), 3051–3056.

[92] Lustig, L.H.P., Hiel, H., Yamamoto, T., Fuchs, P., 2001. Molecular cloning and mapping of the human nicotinic acetylcholine receptor α10 (CHRNA10). Genomics 73(3), 272–283.

[93] Schoepfer, R., Whiting, P., Esch, F., Blacher, R., Shimasaki, S., Lindstrom, J., 1988. cDNA clones coding for the structural subunit of a chicken brain nicotinic acetylcholine receptor. Neuron 1, 241–248.

[94] Anand, R., Lindstron, J., 1990. Nucleotide sequence of the human nicotinic acetylcholine receptor β2 subunit gene. Nucleic Acids Res. 18, 4272.

[95] Hernandez, M.C., Erkman, L., Matter-Sadzinski, L., Roztocil, T., Ballivet, M., Matter, J.M., 1995. Characterization of the nicotinic acetylcholine receptor β3 gene. Its regulation within the avian nervous system is effected by a promoter 143 base pairs in length. J. Biol. Chem. 270, 3224–3233.

[96] Willoughby, J., Ninkina, N., Beech, M., Latchman, D., Wood, J., 1993. Molecular cloning of a human neuronal nicotinic acetylcholine receptor β3-like subunit. Neurosci. Lett. 155, 136–139.

[97] Tarroni, P., Rubboli, F., Chini, B., Zwart, R., Oortgiesen, M., Sher, E., et al., 1992. Neuronal-type nicotinic receptors in human neuroblastoma and small-cell lung carcinoma cell lines. FEBS Lett. 312, 66–70.

[98] Lukas, R.J., Changeux, J.-P., Le Novère, N., Albuquerque, E.X., Balfour, D.J., Berg, D.K., Bertrand, D., et al., 1999. International union of pharmacology. X.X. Current status of the nomenclature for nicotinic acetylcholine receptors and their subunits. Pharmacol. Rev. 51, 397–401.

[99] Ballivet, M., Nef, P., Couturier, S., Rungger, D., Bader, C.R., Bertrand, D., et al., 1988. Electrophysiology of a chick neuronal nicotinic acetylcholine receptor expressed in *Xenopus* oocytes after cDNA injection. Neuron 1, 847–852.

[100] Luetje, C.W., Patrick, J., 1991. Both α- and β-subunits contribute to the agonist sensitivity of neuronal nicotinic acetylcholine receptors. J. Neurosci. 11, 837–845.

[101] Luetje, C.W., Wada, K., Rogers, S., Abramson, S.N., Tsuji, K., Heinemann, S., et al., 1990. Neurotoxins distinguish between different neuronal nicotinic acetylcholine receptor subunit combinations. J. Neurochem. 55, 632–640.

[102] Cooper, E., Couturier, S., Ballivet, M., 1991. Pentameric structure and subunit stoichiometry of a neuronal nicotinic acetylcholine receptor. Nature 350, 235–238.

[103] Whiting, P.J., Lindstrom, J.M., 1986. Purification and characterization of a nicotinic acetylcholine receptor from chick brain. Biochemistry 25, 2082–2093.

[104] Whiting, P., Lindstrom, J., 1987. Purification and characterization of nicotinic acetylcholine receptor from rat brain. Proc. Natl Acad. Sci. USA 84, 595–599.

[105] Whiting, P.J., Lindstrom, J.M., 1988. Characterization of bovine and human neuronal nicotinic acetylcholine receptors using monoclonal antibodies. J. Neurosci. 8, 3395–3404.

[106] Gerzanich, V., Kuryatov, A., Anand, R., Lindstrom, J., 1997. "Orphan" α6 nicotinic AChR subunit can form a functional heteromeric acetylcholine receptor. Mol. Pharmacol. 51, 320–327.

[107] Ramirez-Latorre, J., Yu, C.R., Qu, X., Perin, F., Karlin, A., Role, L., 1996. Functional contributions of α5 subunit to neuronal acetylcholine receptor channels. Nature 380, 347–351.

[108] Groot-Kormelink, P.J., Luyten, W.H., Colquhoun, D., Sivilotti, L.G., 1998. A reporter mutant approach shows incorporation of the "orphan" subunit β3 into a functional nicotinic receptor. J. Biol. Chem. 273, 15317–15320.

[109] Lindstrom, J., 1999. Purification and cloning of nicotinic acetylcholine receptors. In: Arneric, S.P., Brioni, J.D. (Eds.), Pharmacology and Therapeutic Opportunities. Wiley-Liss, New York, pp. 3–23.

[110] Kuryatov, A., Olale, F., Cooper, J., Choi, C., Lindstrom, J., 2000. Human α6 AChR subtypes: subunit composition, assembly and pharmacological responses. Neuropharmacology 39, 2570–2590.

[111] Conroy, W.G., Berg, D.K., 1995. Neurons can maintain multiple classes of nicotinic acetylcholine receptors distinguished by different subunit compositions. J. Biol. Chem. 270, 4424–4431.

[112] Vernallis, A.B., Conroy, W.G., Berg, D.K., 1993. Neurons assemble AChRs with as many as 3 kinds of subunits while maintaining subunit segregation among subtypes. Neuron 10, 451–464.

[113] Conti-Tronconi, B., Dunn, S., Barnard, E., Dolly, J., Lai, F., Ray, N., et al., 1985. Brain and muscle nicotinic acetylcholine receptors are different but homologous proteins. Proc. Natl Acad. Sci. USA 82, 5208–5212.

[114] Gerzanich, V., Anand, R., Lindstrom, J., 1994. Homomers of α8 subunits of nicotinic receptors functionally expressed in Xenopus oocytes exhibit similar channel but contrasting binding site properties compared to α7 homomers. Mol. Pharmacol. 45, 212–220.

[115] Galzi, J.-L., Devillers-Thiery, A., Hussy, N., Bertrand, S., Changeux, J.-P., Bertrand, D., 1992. Mutations in the channel domain of a neuronal nicotinic receptor convert ion selectivity from cationic to anionic. Nature 359, 500–505.

[116] Revah, F., Bertrand, D., Galzi, J.-L., Devillers-Thiéry, A., Mulle, C., Hussy, N., et al., 1991. Mutations in the channel domain alter desensitization of a neuronal nicotinic receptor. Nature 353, 846–849.

[117] Verino, S., Amador, M., Luetje, C.W., Patrick, J., Dani, J.A., 1992. Calcium modulation and high calcium permeability of neuronal nicotinic acetylcholine receptors. Neuron 8, 127–134.

[118] Bertrand, D., Ballivet, M., Rungger, D., 1990. Activation and blocking of neuronal nicotinic acetylcholine receptor reconstituted in *Xenopus* oocytes. Proc. Natl Acad. Sci. USA 87, 1993–1997.

[119] Chen, D., Patrick, J.W., 1997. The α-bungarotoxin-binding nicotinic acetylcholine receptor from rat brain contains only the α7 subunit. J. Biol. Chem. 272, 24024–24029.

[120] Drisdel, R.C., Green, W.N., 2000. Neuronal α-bungarotoxin receptors are α7 subunit homomers. J. Neurosci. 20, 133–139.

[121] Pugh, P.C., Corriveau, R.A., Conroy, W.G., Berg, D.K., 1995. Novel subpopulation of neuronal acetylcholine receptors among those binding α-bungarotoxin. Mol. Pharmacol. 47, 717–725.

[122] Zhang, Z-W., Vijayaraghavan, S., Berg, D.K., 1994. Neuronal acetylcholine receptors that bind α-bungarotoxin with high affinity function as ligand-gated ion channels. Neuron 12, 167–177.

[123] Alkondon, M., Albuquerque, E.X., 1993. Diversity of nicotinic acetylcholine receptors in rat hippocampal neurons. I. Pharmacological and functional evidence for distinct structural subtypes. J. Pharmacol. Exp. Ther. 265, 1455–1473.

[124] McNerney, M.E., Pardi, D., Pugh, P.C., Nai, Q., Margiotta, J.F., 2000. Expression and channel properties of α-bungarotoxin-sensitive acetylcholine receptors on chick ciliary and choroid neurons. J. Neurophysiol. 84, 1314–1329.

[125] Langley, J.N., Anderson, H.K., 1892. The action of nicotine on the ciliary ganglion and on the endings of the third cranial nerve. J. Physiol. 13, 460–468.

[126] Martin, R.A., Pilar, G., 1963. Dual mode of synaptic transmission in the avian ciliary ganglion. J. Physiol. 168, 443–463.

[127] Jacob, M.H., Berg, D.K., 1983. The ultrastructural localization of α-bungarotoxin binding sites in relation to synapses on chick ciliary ganglion neurons. J. Neurosci. 3, 260–271.

[128] Feng, G., Steinbach, J.H., Sanes, J.R., 1998. Rapsyn clusters neuronal acetylcholine receptors but is inessential for formation of an interneuronal cholinergic synapse. J. Neurosci. 18, 4166–4176.

[129] Wilson Horch, H.L., Sargent, P.B., 1995. Perisynaptic surface distribution of multiple classes of nicotinic acetylcholine receptors on neurons in the chicken ciliary ganglion. J. Neurosci. 15, 7778–7795.

[130] Shoop, R.D., Martone, M.E., Yamada, N., Ellisman, M.H., Berg, D.K., 1999. Neuronal acetylcholine receptors with α7 subunits are concentrated on somatic spines for synaptic signaling in embryonic chick ciliary ganglia. J. Neurosci. 19, 692–704.

[131] Brown, D.A., Fumagalli, L., 1977. Dissociation of α-bungarotoxin binding and receptor block in the rat superior cervical ganglion. Brain Res. 129, 165–168.

[132] Chiappinelli, V.A., Dryer, S.E., 1984. Nicotinic transmission in sympathetic ganglia: blockade by the snake venom neurotoxin kappa-bungarotoxin. Neurosci. Lett. 50, 239–244.

[133] Ullian, E.M., McIntosh, J.M., Sargent, P.B., 1997. Rapid synaptic transmission in the avian ciliary ganglion is mediated by two distinct classes of nicotinic receptors. J. Neurosci. 17, 7210–7219.

[134] Zhang, Z-w, Coggan, J.S., Berg, D.K., 1996. Synaptic currents generated by neuronal acetylcholine receptors sensitive to α-bungarotoxin. Neuron 17, 1231–1240.

[135] Cartier, G.E., Yoshikami, D., Gray, W.G., Luo, S., Olivera, B.M., McIntosh, J.M., 1996. A new α-conotoxin which targets α3β2 nicotinic acetylcholine receptors. J. Biochem. 271, 7522–7528.

[136] Wu, W., Gelber, S., Orr-Urtreger, A., Armstrong, D., Lewis, R.L., Ou, C-N., et al., 1999. Megacystis, mydriasis, and ion channel defect in mice lacking the α3 neuronal nicotinic acetylcholine receptor. Proc. Natl Acad. Sci. USA 96, 5746–5751.

[137] Chang, K.T., Berg, D.K., 1999. Nicotinic acetylcholine receptors containing α7 subunits are required for reliable synaptic transmission in situ. J. Neurosci. 19, 3701–3710.

[138] Bertrand, D., Galzi, J.L., Devillers-Thiéry, A., Bertrand, S., Changeux, J.-P., 1993. Mutations at two distinct sites within the channel domain M2 alter calcium permeability of neuronal α7 nicotinic receptor. Proc. Natl Acad. Sci. USA 90, 6971–6975.

[139] Shoop, R.D., Chang, K.T., Ellisman, M.H., Berg, D.K., 2001. Synaptically driven calcium transients via nicotinic receptors on somatic spines. J. Neurosci. 21(3), 771–781.

[140] Spitzer, N., Lautermilch, N., Smith, R., Gomez, T., 2000. Coding of neuronal differentiation by calcium transients. BioEssays 22, 811–817.

[141] Toescu, E.C., 1998. Apoptosis and cell death in neuronal cells: where does Ca^{2+} fit in? Cell Calcium 24, 387–403.

[142] Berridge, M., 1998. Neuronal calcium signaling. Neuron 21, 13–26.

[143] McMahan, U.J., Wallace, B.G., 1989. Molecules in basal lamina that direct the formation of synaptic specializations at neuromuscular junctions. Dev. Neurosci. 11, 227–247.

[144] Bowe, M.A., Fallon, J.F., 1995. The role of agrin in synapse formation. Annu. Rev. Neurosci. 18, 443–462.

[145] McAvoy, M., Smith, M.A., Fujii, J.T., 1996. Agrin mRNA expression in the developing chick Edinger–Westphal nucleus. Vis. Neurosci. 13, 293–301.

[146] Kassner, P.D., Conroy, W.G., Berg, D.K., 1998. Organizing effects of rapsyn on neuronal nicotinic acetylcholine receptors. Mol. Cell Neurosci. 10, 258–270.

[147] Burns, A.L., Benson, D., Howard, M.J., Margiotta, J.F., 1997. Chick ciliary ganglion neurons contain transcripts coding for acetylcholine receptor-associated protein at synapses (Rapsyn). J. Neurosci. 17, 5016–5026.

[148] Conroy, W.G., Berg, D.K., 1999. Rapsyn variants in ciliary ganglia and their possible effects on clustering of nicotinic receptors. J. Neurochem. 73, 1399–1408.

[149] Williams, B.M., Temburni, M.K., Levey, M.S., Bertrand, S., Bertrand, D., Jacob, M., 1998. The long internal loop of the α3 subunit targets nAChRs to subdomains within individual synapses on neurons in vivo. Nat. Neurosci. 1, 557–562.

[150] Curtis, D.R., Ryall, R.W., 1966. The synaptic excitation of Renshaw cells. Exp. Brain Res. 107, 166–170.

[151] Zhang, M., Wang, Y.T., Vyas, D.M., Neuman, R.S., Bieger, D., 1993. Nicotinic cholinoceptor-mediated excitatory postsynaptic potentials in the rat nucleus ambiguous. Exp. Brain Res. 96, 83–88.

[152] Roerig, B., Nelson, D.A., Katz, L.C., 1997. Fast synaptic signaling by nicotinic acetylcholine and serotonin 5-HT$_3$ receptors in developing visual cortex. J. Neurosci. 17, 8353–8362.

[153] Nong, Y., Sorenson, E.M., Chiappinelli, V.A., 1999. Fast excitatory nicotinic transmission in the chick lateral spiriform nucleus. J. Neurosci. 19, 7804–7811.

[154] Feller, M.B., Wellis, D.P., Stellwagen, D., Werblin, F.S., Shatz, C.J., 1996. Requirement for cholinergic synaptic transmission in the propogation of spontaneous retinal waves. Science 272, 1182–1186.

[155] Frazier, C.J., Buhler, A.V., Weiner, J.L., Dunwiddie, T.V., 1998. Synaptic potentials mediated via α-bungarotoxin-sensitive nicotinic acetylcholine receptors in rat hippocampal interneurons. J. Neurosci. 18, 8228–8235.

[156] Alkondon, M., Pereira, E.F., Albuquerque, E.X., 1998. α-bungarotoxin and methyllycaconitine-sensitive nicotinic receptors mediate fast synaptic transmission in interneurons of rat hippocampal slices. Brain Res. 810, 257–263.

[157] Ji, D., Dani, J.A., 2000. Inhibition and disinhibition of pyramidal neurons by activation of nicotinic receptors on hippocampal interneurons. J. Neurophysiol. 83, 2682–2690.

[158] Gray, R., Rajan, A., Radcliffe, K., Yakehiro, M., Dani, J., 1996. Hippocampal synaptic transmission enhanced by low concentrations of nicotine. Nature 383, 713–716.

[159] Guo, J.-Z., Tredway, T.L., Chiappinelli, V.A., Glutamate, A.B.A., 1998. release are enhanced by different subtypes of presynaptic nicotinic receptors in the lateral geniculate nucleus. J. Neurosci. 18, 1963–1969.

[160] McGehee, D., Heath, M., Gelber, S., Devay, P., Role, L.W., 1995. Nicotinic enhancement of fast excitatory synaptic transmission in CNS by presynaptic receptors. Science 269, 1692–1697.

[161] McMahon, L.L., Yoon, K.-W., Chiappinelli, V.A., 1994. Nicotinic receptor activation facilitates GABAergic neurotransmission in the avian lateral spiriform nucleus. Neuroscience 59, 689–698.

[162] Whiting, P., Schoepfer, R., Lindstrom, J., Priestley, T., 1991. Structural and pharmacological characterization of the major brain nicotinic acetylcholine receptor subtype stably expressed in mouse fibroblasts. Mol. Pharmacol. 40, 463–472.

[163] Covernton, P.J.O., Kojima, H., Sivilotti, L.G., Gibb, A.J., Colquhoun, D., 1994. Comparison of neuronal nicotinic receptors in rat sympathetic neurons with subunit pairs expressed in *Xenopus* oocytes. J. Physiol. 481, 27–34.

[164] Sivilotti, L.G., McNeil, D.K., Lewis, T.M., Nassar, M.A., Schoepfer, R., Colquhoun, D., 1997. Recombinant nicotinic receptors, expressed in Xenopus oocytes, do not resemble native rat sympathetic ganglion receptors in single-channel behaviour. J. Physiol. 500, 123–138.

[165] Sheffield, E.B., Quick, M.W., Lester, R.A.J., 2000. Nicotinic acetylcholine receptor subunit mRNA expression and channel function in medial habenula neurons. Neuropharmacology 39, 2591–2603.

[166] Conroy, W.G., Berg, D.K., 1998. Nicotinic receptor subtypes in the developing chick brain: appearance of a species containing the α4, β2, and α5 gene products. Mol. Pharmacol.(1998), 392–401.

[167] Yu, C.R., Role, L.W., 1998. Functional contribution of the α5 subunit to neuronal nicotinic channels expressed by chick sympathetic neurons. J. Physiol. 509, 667–681.

[168] Yu, C.R., Role, L.W., 1998. Functional contribution of the α7 subunit to multiple subtypes of nicotinic receptors in embryonic chick sympathetic neurons.. J. Physiol. 509, 651–665.

[169] Marubio, L.M., Changeux, J.-P., 2000. Nicotinic acetylcholine receptor knockout mice as animal models for studying receptor function. Eur. J. Pharmacol. 393, 113–121.

[170] Cordero-Erausquin, M., Marubio, L.M., Klink, R., Changeux, J.-P., 2000. Nicotinic receptor function: new perspectives from knockout mice. Trends Pharmacol. Sci. 21, 211–217.

[171] Listerud, M., Brussaard, A.J., Devay, P., Colman, D.R., Role, L.W., 1991. Functional contribution of neuronal AChR subunits revealed by antisense oligonucleotides. Science 254, 1518–1521.

[172] Gingrich, J.A., Hen, R., 2000. The broken mouse: the role of development, plasticity, and environment in the interpretation of phenotypic changes in knockout mice. Curr. Opin. Neurobiol. 10, 146–152.

[173] Zorumski, C.F., Thio, L.L., Isenberg, K.E., Clifford, D.B., 1992. Nicotinic acetylcholine currents in cultured postnatal rat hippocampal neurons. Mol. Pharmacol. 41, 931–936.

[174] Anand, R., Peng, X., Lindstrom, J., 1993. Homomeric and native α7 acetylcholine receptors exhibit remarkably similar but non-identical pharmacological properties, suggesting that the native receptor is a heteromeric protein complex. FEBS Lett. 327, 241–246.

[175] McIntosh, J.M., Santos, A.D., Olivera, B., 1999. *Conus* peptides targeted to specific nicotinic acetylcholine receptor subtypes. Annu. Rev. Biochem. 68, 59–88.

[176] Johnson, D., Wu, S.M.-S., 1995. Foundations of cellular neurophysiology. The MIT Press, Cambridge.

[177] Pereira, E.F., Alkondon, M., McIntosh, J.M., Albuquerque, E.X., 1996. α-conotoxin-ImI: a competitive antagonist at α-bungarotoxin-sensitive neuronal nicotinic receptors in hippocampal neurons. J. Pharmacol. Exp. Ther. 278, 1472–1483.

[178] Quick, M.W., Ceballos, R.M., Kasten, M., McIntosh, J.M., Lester, R.A.J., 1999. α3β4 subunit-containing nicotinic receptors dominate function in rat medial habenula neurons. Neuropharmacology 38, 769–783.

[179] Luo, S., Kulak, J.M., Cartier, G.E., Jacobsen, R.B., Yoshikami, D., Olivera, B.M., et al., 1998. α-Conotoxin AuIB selectively blocks α3β4 nicotinic acetylcholine receptors and nicotine-evoked norepinepherine release. J. Neurosci. 18, 8571–8579.

[180] Chen, M., Pugh, P.C., Margiotta, J.F., 2001. Nicotinic synapses formed between chick ciliary ganglion neurons in culture resemble those present on the neurons in vivo. J. Neurobiol. 47, 265–279.

[181] Nai, Q., McIntosh, J.M., Margiotta, J.F., 2003. Relating neuronal nicotinic acetylcholine receptor subtypes defined by subunit composition and channel function. Molecular Pharmacology 63, 311–324.

[182] Cuevas, J., Berg, D.K., 1998. Mammalian nicotinic receptors with α7 subunits that slowly desensitize and rapidly recover from α-bungarotoxin blockade. J. Neurosci. 18, 10335–10344.

[183] Pumain, R., Heinemann, U., 1985. Stimulus- and amino acid- induced calcium and potassium changes in the rat neocortex. J. Neurophysiol. 53, 1–16.

[184] Mulle, C., Lena, C., Changeux, J.-P., 1992. Potentiation of nicotinic receptor response by external calcium in rat central neurons. Neuron 8, 937–945.

[185] Amador, M., Dani, J.A., 1995. Mechanism for modulation of nicotinic acetylcholine receptors that can influence synaptic transmission. J. Neurosci. 15, 4525–4532.

[186] Downing, J.E.G., Role, L.W., 1987. Activators of protein kinase C enhance acetylcholine receptor desensitization in sympathetic ganglion neurons. Proc. Natl Acad. Sci. USA 84, 7739–7743.

[187] Role, L.W., 1984. Substance P modulation of acetylcholine-induced currents in embryonic chicken sympathetic and ciliary ganglion neurons. Proc. Natl Acad. Sci. USA 81, 2924–2928.

[188] Simmons, L.K., Schuetze, S.M., Role, L.W., 1990. Substance P modulates single-channel properties of neuronal nicotinic acetylcholine receptors. Neuron 4, 393–403.

[189] Margiotta, J.F., Berg, D.K., 1986. Enkephalin and substance P modulate synaptic properties of chick ciliary ganglion neurons in cell culture. Neuroscience 18, 175–782.

[190] Hayashi, M.D., Edgar, D., Thoenen, H., 1983. The development of substance P., somatostatin, and vasoactive intestinal peptide in sympathetic and spinal sensory ganglia of the chick embryo. Neuroscience 10, 31–39.

[191] Reiner, A., 1987. A VIP-like peptide co-occurs with substance P and enkephalin in cholinergic preganglionic terminals of the avian ciliary ganglion. Neurosci. Lett. 78, 22–28.

[192] Margiotta, J.F., Berg, D.K., Dionne, V.E., 1987. Cyclic AMP regulates the proportion of functional acetylcholine receptors on chick ciliary ganglion neurons. Proc. Natl Acad. Sci. USA 84, 8155–8159.

[193] Margiotta, J.F., Gurantz, D., 1989. Changes in the number, function and regulation of nicotinic acetylcholine receptors during neuronal development. Dev. Biol. 135, 326–399.

[194] Margiotta, J.F., Pardi, D., 1995. Pituitary adenylate cyclase-activating polypeptide type I receptors mediate cyclic AMP-dependent enhancement of neuronal acetylcholine sensitivity. Mol. Pharmacol. 48, 63–71.

[195] Pardi, D., Margiotta, J.F., 1999. Pituitary adenylate cyclase-activating polypeptide activates a phospholipase C-dependent signal pathway in chick ciliary ganglion neurons that selectively inhibits α7-containing nicotinic receptors. J. Neurosci. 19, 6327–6337.

[196] Vijayaraghavan, D., Schmid, H.A., Halvorsen, S.W., Berg, D.K., 1990. Cyclic AMP-dependent phosphorylation of a neuronal acetylcholine receptor α-type subunit. J. Neurosci. 10, 3255–3262.

[197] Moss, S.J., McDonald, B.J., Rudhard, Y., Schoepfer, R., 1996. Phosphorylation of the predicted major intracellular domains of the rat and chick nicotinic α7 subunit by cAMP-dependent protein kinase. Neuropharmacology 35, 1023–1028.

[198] Deutsch, P.J., Sun, Y., 1992. The 38-amino acid form of pituitary adenylate cyclase-activating polypeptide stimulates dual signaling cascades in PC12 cells and promotes neurite outgrowth. J. Biol. Chem. 267, 5108–5113.

[199] Spengler, D., Waeber, C., Pantaloni, C., Holsboer, F., Bockaert, J., Seeburg, P., et al., 1993. Differential signal transduction by five splice variants of the PACAP receptor. Nature 365, 170–175.

[200] Kapur, J., Macdonald, R.L., 1996. Cyclic AMP-dependent protein kinase enhances hippocampal dentate granule cell GABA$_A$ receptor currents. J. Neurophysiol. 76, 2626–2634.

[201] Song, Y.M., Huang, L.Y., 1990. Modulation of glycine receptor chloride channels by cAMP-dependent protein kinase in spinal trigeminal neurons. Nature 348, 242–245.

[202] Greengard, P., Jen, J., Nairn, A.C., Stevens, C.F., 1991. Enhancement of the glutamate response by cAMP-dependent protein kinase in hippocampal neurons. Science 253, 1135–1137.

[203] Fu, W.M., 1993. Potentiation of acetylcholine responses in *Xenopus* embryonic muscle cells by dibutyryl cAMP. Pflugers Arch. 425, 439–445.

[204] Lu, B., Fu, W.M., Greengard, P., Poo, M.-m., 1993. Calcitonin gene-related peptide potentiates synaptic responses at developing neuromuscular junctions. Nature 363, 76–79.

[205] Porter, N.M., Twyman, R.E., Uhler, M.D., Macdonald, R.L., 1990. Cyclic AMP-dependent protein kinase decreases GABA$_A$ receptor current in mouse spinal neurons. Neuron 5, 789–796.

[206] Huganir, R.L., Delcour, A.H., Greengard, P., Hess, G.P., 1986. Phosphorylation of the nicotinic acetylcholine receptor regulates its rate of desensitization. Nature 321, 774–776.

[207] McDonald, B.J., Amato, A., Connolly, C.N., Benke, D., Moss, S.J., Smart, T.G., 1998. Adjacent phosphorylation sites on GABA$_A$ receptor β subunits determine regulation by cAMP-dependent protein kinase. Nat. Neurosci. 1, 23–28.

[208] Margiotta, J.F., Berg, D.K., Dionne, V.E., 1987. The properties and regulation of functional acetylcholine receptors on chick ciliary ganglion neurons. J. Neurosci. 7, 3612–3622.

[209] Fenster, C.P., Whitworth, T., Quick, M.W., Lester, R.A.J., 1999. Upregulation of surface α4β2 nicotinic receptors is initiated by receptor desensitization following chronic exposure to nicotine. J. Neurosci. 19, 4804–4818.

[210] Bean, B.P., Nowycky, M.C., Tsien, R.W., 1984. Beta-adrenergic modulation of calcium channels in frog ventricular heart cells. Nature 307, 371–375.

[211] Tsien, R.W., Bean, B.P., Hess, P., Lansman, J.B., Nilius, B., Nowycky, M.C., 1986. Mechanisms of calcium channel modulation by beta-adrenergic agents and dihydropyridine calcium agonists. J. Mol. Cell Cardiol. 18, 691–710.

[212] Hille, B., 1992. Ionic channels of excitable membranes, 2nd ed. Sinauer Associates, Sunderland, MA.

[213] Katz, B., Thesleff, S., 1957. A study of the "desensitization" produced by acetylcholine at the motor end-plate. J. Physiol. 138, 63–80.

[214] Feltz, A., Trautmann, A., 1982. Desensitization at the frog neuromuscular junction: a biphasic process. J. Physiol. 322, 257–272.

[215] Boyd, N.D., 1987. Two distinct kinetic phases of desensitization of acetylcholine receptors of clonal rat PC12 cells. J. Physiol. 389, 45–67.

[216] Paradiso, K., Brehm, P., 1998. Long-term desensitization of nicotinic acetylcholine receptors is regulated via protein kinase A-mediated phosphorylation. J. Neurosci. 18, 9227–9237.

[217] Fenster, C.P., Beckman, M.L., Parker, J.C., Sheffield, E.B., Whitworth, T.L., Quick, M.W., et al., 1999. Regulation of α4β2 nicotinic receptor desensitization by calcium and protein kinase C. Mol. Pharmacol. 55, 432–443.

[218] Matter-Sadzinski, L., Hernandez, M.-C., Rotzocil, T., Ballivet, M., Matter, J.M., 1992. Neuronal specificity of the α7 nicotinic promoter develops during morphogenesis of the central nervous system. EMBO J. 11, 4529–4538.

[219] Howard, M.J., Gershon, M.D., Margiotta, J.F., 1995. Expression of nicotinic acetylcholine receptors and subunit mRNA transcripts in cultures of neural crest cells. Dev. Biol. 170, 479–495.

[220] Hamassaki-Britto, D.E., Gardino, P.F., Hokoc, J.N., Keyser, K.T., Karten, H.J., Lindstrom, J.M., et al., 1994. Differential development of α-bungarotoxin-sensitive and α-bungarotoxin-insensitive nicotinic acetylcholine receptors in the chick retina. J. Comp. Neurol. 347, 161–170.

[221] Blumenthal, E.M., Shoop, R.D., Berg, D.K., 1999. Developmental changes in the nicotinic responses of ciliary ganglion neurons. J. Neurophysiol. 81, 111–120.

[222] Hellström-Lindahl, E., Gorbounova, O., Seiger, A., Mousavi, M., Nordberg, A., 1998. Regional distribution of nicotinic receptors during prenatal development of human brain and spinal cord. Brain Res. Dev. Brain Res. 108, 147–160.

[223] Levitt, P., 1998. Prenatal effects of drugs of abuse on brain development. Drug Alcohol Depend. 51, 109–125.

[224] Lipton, S.A., Kater, S.B., 1989. Neurotransmitter regulation of neuronal outgrowth, plasticity and survival. Trends Neurosci. 12, 265–270.

[225] Lauder, J.M., Schambra, U.B., 1999. Morphogenetic roles of acetylcholine. Environ. Health Perspect. 107, 65–69.

[226] Fratiglioni, L., Wang, H.-X., 2000. Smoking and Parkinson's and Alzheimer's disease: review of the epidemiological studies. Behav. Brain Res. 113, 117–120.

[227] Meyer, E.M., King, M.A., Meyers, C., 1998. Neuroprotective effects of 2,4-dimethoxybenzylidene anabaseine (DMXB) and tetrahydroaminoacridine (THA) in neocortices of nucleus basalis lesioned rats. Brain Res. 786, 252–254.

[228] Belluardo, N., Mudo, G., Blum, M., Fuxe, K., 2000. Central nicotinic receptors, neurotrophic factors and neuroprotection. Behav. Brain Res. 113, 21–34.

[229] French, S.J., Humby, T., Horner, C.H., Sofroniew, M.V., Rattray, M., 1999. Hippocampal neurotrophin and trk receptor mRNA levels are altered by local administration of nicotine, carbachol and pilocarpine. Brain Res. Mol. Brain Res. 67, 124–136.

[230] Roy, T.S., Andrews, J.E., Seidler, F.J., Slotkin, T.A., 1998. Nicotine evokes cell death in embryonic rat brain during neurulation. J. Pharmacol. Exp. Ther. 287, 1136–1144.

[231] Trauth, J.A., Seidler, F.J., McCook, E.C., Slotkin, T.A., 1999. Persistent c-fos induction by nicotine in developing rat brain regions: interaction with hypoxia. Pediatr. Res. 45, 38–45.

[232] Thomas, J.D., Garrison, M.E., Slaweki, C.J., Ehlers, C.L., Riley, E.P., 2000. Nicotine exposure during the neonatal brain growth spurt produces hyperactivity in preweanling rats. Neurotoxicol. Teratol. 22, 695–701.

[233] Ferriero, D.M., Dempsey, D.A., 1999. Impact of addictive and harmful substances on fetal brain development. Curr. Opin. Neurol. 12, 161–166.

[234] Slotkin, T.A., 2002. Nicotine and the adolescent brain: insights from an animal model. Neurotoxicol. Teratol. 24, 369–394.

[235] Orr-Urtreger, A., Broide, R., Kasten, M., Dang, H., Dani, J., Beaudet, A., et al., 2000. Mice homozygous for the L250T mutation in the α7 nicotinic acetylcholine receptor show increased neuronal apoptosis and die within 1 day of birth. J. Neurochem. 74, 2154–2166.

[236] Wright, L., 1981. Cell survival in chick embryo ciliary ganglion is reduced by chronic ganglionic blockade. Dev. Brain Res. 1, 283–286.

[237] Meriney, S.D., Pilar, G., Ogawa, M., Nuñez, R., 1987. Differential survival in the avian ciliary ganglion after chronic acetylcholine receptor blockade. J. Neurosci. 7, 1840–1849.

[238] Maderut, J.L., Oppenheim, R.W., Prevette, D., 1988. Enhancement of naturally occurring cell death in the sympathetic and parasympathetic ganglia of the chicken embryo following blockade of ganglionic transmission. Brain Res. 444, 189–194.

[239] Landauer, W., 1975. Cholinomimetic teratogens: studies with chicken embryos. Teratology 12, 125–146.

[240] Messi, M.L., Renganathan, M., Grigorenko, E., Delbono, O., 1997. Activation of α7 nicotinic acetylcholine receptor promotes survival of spinal cord motoneurons. FEBS Lett. 411(1), 32–38.

[241] Yamashita, H., Nakamura, S., 1996. Nicotine rescues PC12 cells from death induced by nerve growth factor deprivation. Neurosci. Lett. 213, 145–147.

[242] Li, Y., Papke, R., He, Y-J., Millard, W.J., Meyer, E.M., 1999. Characterization of the neuroprotective and toxic effects of α7 nicotinic receptor activation in PC12 cells. Brain Res. 830, 218–225.

[243] Opanashuk, L.A., Pauly, J.R., Hauser, K.F., 2001. Effect of nicotine on cerebellar granule neuron development. Eur. J. Neurosci. 13, 48–56.

[244] Pugh, P.C., Margiotta, J.F., 2000. Nicotinic acetylcholine receptor agonists promote survival and reduce apoptosis of chick ciliary ganglion neurons. Mol. Cell Neurosci. 15, 113–122.

[245] Berger, F., Gage, F.H., Vijayaraghavan, S., 1998. Nicotinic receptor-induced apoptotic cell death of hippocampal progenitor cells. J. Neurosci. 18, 6871–6881.

[246] Carlson, N.G., Bacchi, A., Rogers, S.W., Gahring, L.C., 1998. Nicotine blocks TNF-α-mediated neuroprotection to NMDA by an α-bungarotoxin-sensitive pathway. J. Neurobiol. 35, 29–36.

[247] Shimohama, S., Kihara, T., 2001. Nicotinic receptor-mediated protection against β-amyloid neurotoxicity. Biol. Psychiatry 49, 233–239.

[248] Grewal, S.S., Horgan, A.M., York, R.D., Withers, G.S., Banker, G.A., Stork, P.J.S., 2000. Neuronal calcium activates a Rap1 and B-Raf signaling pathway via the cyclic adenosine monophosphate-dependent protein kinase. J. Biol. Chem. 275, 3722–3728.

[249] Schmidt, M.F., Kater, S.B., 1993. Fibroblast growth factors, depolarization, and substratum interact in a combinatorial way to promote neuronal survival. Dev. Biol. 158, 228–237.

[250] Hory-Lee, F., Frank, E., 1995. The nicotinic blocking agents d-tubocurare and α-bungarotoxin save motoneurons from naturally occurring death in the absence of neuromuscular blockade. J. Neurosci. 15, 6453–6460.

[251] Loring, R.H., Dahm, L.M., Zigmond, R.E., 1985. Localization of α-bungarotoxin binding sites in the ciliary ganglion of the embryonic chick: an autoradiographic study at the light and electron microscopic level. Neuroscience 14, 645–660.

[252] Fu, W.M., Liu, J.J., 1997. Regulation of acetylcholine release by presynaptic nicotinic receptors at developing neuromuscular synapses. Mol. Pharmacol. 51, 390–398.

[253] Layer, P.G., Alber, R., Rathjen, F.G., 1988. Sequential activation of butyrylcholinesterase in rostral half somites and acetylcholinesterase in motoneurons and myotomes preceding growth of motor axons. Development 102, 396–398.

[254] Hume, R.I., Role, L.W., Fischbach, G.D., 1983. Acetylcholine release from growth cones detected with patches of acetylcholine receptor-rich membranes. Nature 305, 632–634.

[255] Young, S.H., Poo, M.-m., 1983. Spontaneous release of transmitter from growth cones of embryonic neurons. Nature 305, 634–637.

[256] Rotundo, R.L., Carbonetto, S.T., 1987. Neurons segregate clusters of membrane-bound acetylcholinesterase along their neurites. Proc. Natl Acad. Sci. USA 84, 2063–2067.

[257] Layer, P.G., 1992. Expression and possible functions of cholinesterases during chicken neurogenesis. In: Jea, M. (Ed.), Cholinesterases: Structure, Function, Mechanism, Genetics, and Cell Biology. American Chemical Society, Washington, pp. 350–357.

[258] Layer, P.G., Weikert, T., Alber, R., 1993. Cholinesterases regulate neurite outgrowth from chick nerve cells in vitro by means of a non-enzymatic mechanism. Cell Tissue Res. 273, 219–226.

[259] Layer, P.G., Willbold, E., 1994. Cholinesterases in avian neurogenesis. Int. Rev. Cytol. 151, 139–181.

[260] Orr-Urtreger, A., Goldner, F.M., Saeki, M., Lorenzo, I., Goldberg, L., De Biasi, M., et al., 1997. Mice deficient in the α7 neuronal nicotinic acetylcholine receptor lack α-bungarotoxin binding sites and hippocampal fast nicotinic currents. J. Neurosci. 17, 9165–9171.

[261] Rossi, F.M., Pizzorusso, T., Porciatti, V., Marubio, L.M., Maffei, L., Changeux, J.P., 2001. Requirement of the nicotinic acetylcholine receptor β2 subunit for the anatomical and functional development of the visual system. Proc. Natl Acad. Sci. USA 98, 6453–6458.

[262] Schmidt, J.T., 1995. The modulatory cholinergic system in goldfish tectum may be necessary for retinotopic sharpening. Vis Neurosci. 12, 1093–1103.

[263] Edwards, J.A., Cline, H.T., 1999. Light-induced calcium influx into retinal axons is regulated by presynaptic nicotinic acetylcholine receptor activity in vivo. J. Neurophysiol. 81, 895–907.

[264] Lipton, S.A., Frosch, M.P., Phillips, M.D., Tauck, D.L., Aizenman, E., 1988. Nicotinic antagonists enhance process outgrowth by rat retinal ganglion cells in culture. Science 239, 1293–1296.

[265] Owen, A., Bird, M., 1995. Acetylcholine as a regulator of neurite outgrowth and motility in cultured embryonic mouse spinal cord. NeuroReport 6, 2269–2272.

[266] Chan, J., Quik, M., 1993. A role for the nicotinic α-bungarotoxin receptor in neurite outgrowth in PC12 cells. Neuroscience 56, 441–451.

[267] Quik, M., Cohen, R., Audhya, T., Goldstein, G., 1990. Thymopoietin interacts at the α-bungarotoxin site of and induces process formation in PC12 pheochromocytoma cells. Neuroscience 39, 139–150.

[268] Pugh, P.C., Berg, D.K., 1994. Neuronal acetylcholine receptors that bind α-bungarotoxin mediate neurite retraction in a calcium-dependent manner. J. Neurosci. 14, 889–896.

[269] Zheng, J.Q., Felder, M., Connor, J.A., Poo, M.-m., 1994. Turning of nerve growth cones induced by neurotransmitters. Nature 38, 140–144.

[270] Erskine, L., McCaig, C.D., 1995. Growth cone neurotransmitter receptor activation modulates electric field-guided nerve growth. Dev. Biol. 171, 330–339.

[271] Ji, D., Lape, R., Dani, J.A., 2001. Timing and location of nicotinic activity enhances or depresses hippocampal synaptic plasticity. Neuron 31, 131–141.

[272] Toyoshima, C., Unwin, N., 1988. Ion channel of acetylcholine receptor reconstituted from images of postsynaptic membranes. Nature 336, 247–250.

[273] Stroud, R.M., Finer-Moore, J., 1985. Acetylcholine receptor structure, function and evolution. Annu. Rev. Cell Biol. 1, 317–351.

[274] Conroy, W.G., Liu, Z., Nai, Q., Coggan J.S., Berg, D.K., 2003. PDZ-containing proteins provide a functional postsynaptic scaffold for nicotinic receptors in neurons. Neuron 38, 759–771.

[275] Conroy, W.G., Liu, Q.S., Nai, Q., Margiotta J.F., Berg, D.K., 2003. Potentiation of α7-containing nicotinic acetylcholine receptors by select albumins. Molecular Pharmacology 63, 419–428.

[276] Feller, M.A., 2002. The role of nAChR-mediated spontaneous retinal activity in visual system development. J. Neurobiol. 53, 556–567.

[277] Penn, A.A., Riquelme, P.A., Feller, M.B., Shatz, C.J., 1998. Competition in retinogeniculate patterning driven by spontaneous activity. Science 279, 2108–2112.

The ENaC/Deg family of cation channels

Sylvie Coscoy[a] and Pascal Barbry[b,*]

[a]Laboratoire de Physico-Chimie, Institut Curie, CNRS UMR 168, 26 rue d'Ulm, 75005 Paris, France
[b]Institut de Pharmacologie Moléculaire et Cellulaire, CNRS UPR 411, 660 route des Lucioles 06560 Sophia Antipolis, France
*Correspondence address: E-mail: barbry@ipmc.cnrs.fr

The ENaC/Deg family comprises more than 60 homologous ion channels, expressed from epithelia to nervous system. The members of the family are usually selective for sodium over potassium and sensitive to the diuretic amiloride. These channels are involved in a diversity of physiological roles. Some of them, such as the epithelial sodium channel ENaC, are constitutively active channels. Other members, such as the degenerins from the nematode *Caenorhabditis elegans*, might be activated by mechanical stress. Still others, such as the acid-sensing ion channels (ASICs) (activated by protons), are ligand gated.

Numerous studies have contributed to the current structural model of these channels and their interaction with pharmacological agents. Independent experiments on different members of the family show that ENaC/Deg channels are organized into tetrameric complexes of subunits. Individual ENaC/Deg subunits contain two transmembrane alpha-helices, a large extracellular domain, and two short NH_2- and COOH-terminal cytoplasmic domains. Structure–function studies suggest that the ion pore is lined by the transmembrane domain TM2, and by the short pre-TM2 and pre-TM1 domains, which might form re-entrant loops. Amiloride interacts with the entrance of the ion pore, but might also interact with additional extracellular sites.

Evidence suggests these channels play important roles in human physiology and pathology. ASICs, which are expressed in the nervous system, may be involved in the perception of pain following tissular acidosis. The epithelial sodium channel plays a crucial physiological role in Na^+ homeostasis. It can be activated by aldosterone and vasopressin via transcriptional and/or post-translational mechanisms. Some mutations on ENaC subunits are associated with human pathologies, such as pseudohypoaldosteronism type I or Liddle's syndrome. These mutations have highlighted the role of interacting proteins such as Nedd4.

Advances in Molecular and Cell Biology, Vol. 32, pages 303–329
ISSN: 1569-2558 / DOI: 10.1016/S1569-2558(03)32013-2

1. Introduction

The epithelial sodium channel ENaC is the molecular entity responsible for the apical electrodiffusion of sodium across tight epithelium. It is found in distal kidney and distal colon, in proximal and distal respiratory tract, as well as in many other reabsorping epithelia. This channel is characterized by a low unitary conductance (~ 5 pS in 150 M Na^+), and by its "exquisite" sensitivity to the diuretic amiloride. Its cloning in 1991 revealed that it was structurally different from the classical voltage-gated sodium channel, and has led to the identification of a new gene superfamily of ion channels, comprising more than 60 homologous proteins. The members of this super-family are found in many animals (about 10 different members in mammals, two members in snail, more than 20 different members in *Drosophila*, and more than 25 different members in nematode) and are involved in a diversity of physiological roles, from sodium vectorial absorption to perception.

The first part of this review will present the main characteristics of the ENaC/Deg family of channels: primary structure, pharmacological and electrophysiological properties, localization and potential physiological roles.

The second part will focus on their tetrameric organization, ion pore, amiloride binding site, and gating properties.

While the channels of the family share important functional domains, the mode of regulation is usually unique to each member of the family. The regulatory properties of the epithelial channel have been extensively studied and are reviewed in the third part. ENaC activity is regulated, directly or indirectly, by a large number of proteins, including cytoskeleton proteins, proteins possibly involved in endocytosis (Nedd4), and protein kinases (PKA and PKC). In kidney, colon and lung, ENaC activity is primarily controlled by aldosterone and vasopressin, with tissue-specific effects explained by the variety of short-term, as well as long-term (transcriptional), regulatory mechanisms. Truncation of the transcripts and altered interaction of one ENaC subunit with associated proteins can both lead directly to the development of rare human diseases, such as pseudohypoaldosteronism type I or Liddle's syndrome.

2. Description of the family members

The ENaC/Deg superfamily of cation channels comprises at least 60 different members, showing a relatively weak homology (about 20%) (Fig. 1). All proteins are 500–700 amino acids long and are inserted into the lipid bilayer through two transmembrane domains (Fig. 2). Three modes of activity (constitutive, gating by mechanical stress, gating by a ligand) have been described for the members of this family, and this subdivision into three subfamilies is to some extent confirmed by sequence analysis.

2.1. Constitutively activated channels

2.1.1. The epithelial sodium channel ENaC
The first physiological description of the epithelial sodium channel was achieved in 1958 by Kœfœd-Johnsen and Ussing, who provided the general model for sodium

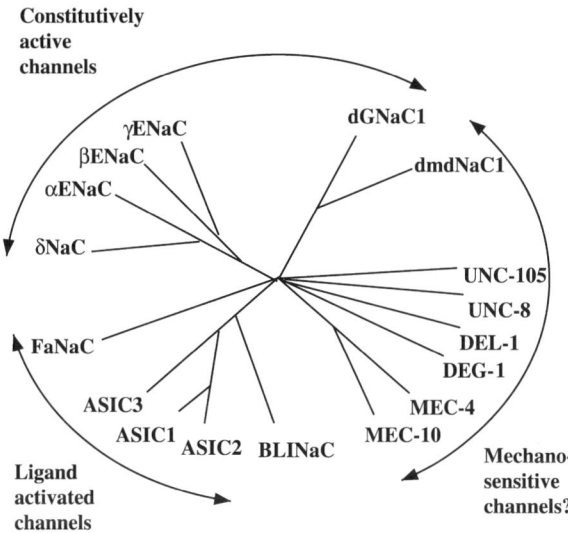

Fig. 1. Phylogenic tree of some members of the ENaC/Deg family.

reabsorption across frog skin [1]. In this model, coupling between passive electrodiffusion of sodium through the apical membrane and active extrusion of intracellular sodium by basolateral Na^+/K^+/ATPase generates a vectorial transcellular sodium transport. This sodium channel was to be identified in many tight junction epithelia, in amphibian tissues (toad urinary bladder, A6 cells) as well as in mammalian tissues (cortical collecting duct, distal colon and airway epithelium, etc., – reviewed in Refs. [2,3]). The epithelial sodium channel is characterized by a small unitary conductance (4–6 pS in 140 mM NaCl), a strong selectivity for sodium and lithium over potassium, long opening and closing times. It is efficiently blocked by the diuretic amiloride ((3,5-diamino-*N*(aminoiminomethyl)-6-chloropyrazine-carboxamide) and some of its derivatives like benzamil or phenamil; its pharmacological profile (benzamil > amiloride > ethylisopropylamiloride (EIPA))

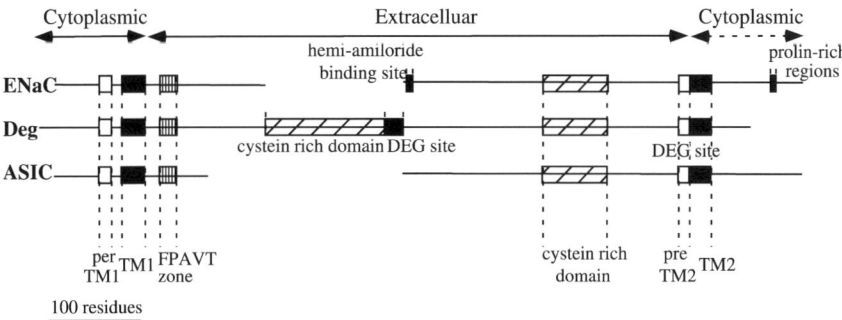

Fig. 2. Conserved motifs of ENaC/Deg proteins. An alignment between the three main subfamilies of ENaC/Deg channels (ENaC, degenerins, ASIC) is presented. TM: transmembrane domain. DEG sites: "degeneration" sites.

allows it to be distinguished from other amiloride targets like the Na^+/H^+ exchanger, which is efficiently blocked by EIPA.

The first subunit of the channel, αENaC (or RCNaCh), was cloned by expression in *Xenopus laevis* oocytes from rat colon [4,5]. αENaC encodes a 698 amino-acid protein. The subunits βENaC and γENaC, sharing 35% homology with αENaC, were subsequently identified by functional complementation or homology cloning [6,7]. These subunits do not alone generate current in *Xenopus* oocytes, but potentiate αENaC function. Co-expression of the three subunits in oocytes increases the amplitude of the channel by two orders of magnitude, with electrophysiological and pharmacological properties identical to those of the native channel [6].

α, β and γhENaC have been cloned from human lung [8,9] and kidney [10,11]; the human channel has the same electrophysiological properties as the rat channel, with a unitary conductance of 5–6 pS in 140 mM NaCl. The human αENaC gene maps to chromosome 12p13 [8], while βENaC and γENaC genes are co-localized within a common 400 kilobase fragment on chromosome 16p12-p13 [9]. Transcripts encoding the subunits are expressed at a high level in epithelial tissues, such as renal cortex and medulla, distal colon, urinary bladder, lung, placenta, and salivary glands [6,8,12], at a low level in proximal colon, in uterus, in thyroid, and in intestine [12]. Transcripts and proteins are also detected in isolated taste buds, in particular in the anterior tongue, where salt taste is transduced by apical amiloride-sensitive channels [13].

2.1.2. The δNaC subunit

A subunit showing a high similitude with αENaC, δNaC, has been cloned from a expressed sequence tag (EST) from human testis [14]. Human δNaC gene maps to chromosome 1p36.3–p36.2 [14]. δNaC encodes a protein of 638 amino-acid, with 37% identity with αENaC, the closest family member. Expressed in *Xenopus* oocyte, δNaC generates a weak amiloride-sensitive sodium current, potentiated after co-expression of β and γENaC. Despite its similarity with αENaC, δNaC is not a fourth subunit for the epithelial sodium channel, and differs from it by its biophysical parameters. The role of δENaC subunit, which has been detected in ovaries, testis, brain, lung and pancreas, as well as its partners are still to be defined.

2.1.3. The Drosophila channel dGNaC1

An homologue of the epithelial sodium channel has been identified in *Drosophila melanogaster*. *Drosophila* gonad Na^+ channel (dGNaC1) or *ripped pocket* (dGNaC1) was cloned from a *Drosophila* EST [15,16]. This 562 amino-acid protein shares less than 20% homology with the other family members, the closest member being another *Drosophila* member of the ENaC/Deg family, dmdNaC1 (38%). Expressed in *Xenopus* oocytes, dGNaC1 generates a constitutive sodium amiloride-sensitive current, with a pharmaco-logical profile different from ENaC (equal sensitivity to amiloride and EIPA). dGNaC1 mRNA is strongly expressed in early embryogenesis (0–4 h), before the initiation of zygotic transcription, as well as in ovaries and testis. In gametes, transcripts are maximally expressed in late ovariogenesis, during the constitution of the reserves in nutrients of the oocytes, as well in nurse cells and follicle cells as in the oocyte itself. In testis, dGNaC1 mRNAs are detected in the cysts that contain primary spermatocysts and in seminal vesicles. In embryos, the mRNAs are uniformly distributed in the syncitium, then in cellular blastoderm, and disappear at the beginning of gastrulation.

The physiological role of dGNaC1 can only be hypothesized, and might be linked to the regulation of water movements in gametogenesis or early embryogenesis.

2.2. Degenerins

2.2.1. Degenerins

Degenerins are proteins homologous to ENaC that have been first identified in the nematode *C. elegans*. Gain of function mutations in degenerins induce the degeneration of the cells expressing them. Degenerins are expressed in various sensory neurons, interneurons, motor neurons and in muscles. MEC-4 and MEC-10 (for *MEChanosensitive*) participate to light touch perception, and they are both expressed in the six sensory neurons that mediate light perception. MEC-10 is also present in two additional neurons involved in the perception of brutal touch and mechanical stimuli. UNC-8 (for *UNCoordinated*) and perhaps DEL-1 (for *DEgenerin-like*) participate to locomotion: loss of function mutations induce a modification of sinusoidal traces generated during *C. elegans* locomotion [17]. *Fluoride resistant* (FLR1), a degenerin-related protein present in the intestine, is involved in the control of the defecation rythm in *C. elegans* [18]. The *C. elegans* sequencing project has revealed the existence of ~ 29 degenerin related proteins, but additional work is needed to characterize the function of these genes.

An extensive genetic dissection of the interactions existing between MEC-4 and MEC-10 has been made by Chalfie, Driscoll and co-workers [19,20]. It has led to a model where MEC-4 and MEC-10 form a complex linked to the cytoskeleton by MEC-2 (a protein sharing homology with stomatin [21]) and to the extracellular matrix by MEC-5 (a collagen) and MEC-9 (a secreted protein rich in EGF and Kunitz domains [22]). The tension transmitted by these proteins would directly open the channel and lead to the generation of an action potential. An additional protein whose molecular nature is not known, MEC-6, is required for MEC-4/MEC-10 function. Similarly, an interaction between UNC-105 and collagen has been identified and may serve as a mechanism of stretch-activated muscle contraction [23].

The existence of mutations leading to cell swelling and lysis led Chalfie and colleagues to consider the possibility that there were abnormal ion fluxes. Together with sequence homologies with ENaC subunits, and a functional expression of αENaC/MEC-4 chimeras, this suggests that degenerins can indeed form channels, which might be activated by mechanical stress. However, functional expression of a degenerin in an heterologous system has not yet been reported, except for the muscle protein UNC-105 in its constitutively active form [24]. In that case, two gain-of-function UNC-105 mutants P134S and A692V were successfully expressed in HEK-293 cells and in *Xenopus* oocytes, and were associated with the development of a non-selective cationic current that could be blocked by amiloride [24].

2.2.2. The Drosophila channel dmdNaC1

Degenerins have only been reported in the nematode *C. elegans*. However the *Drosophila* homologue, *Drosophila* multidendritic neuron Na$^+$ channel (dmdNaC1); also called *pickpocket* might be functionally related to degenerins. In insects, multidendritic neurons mediate the sensations of touch and stress, and are involved in the perception of many mechanical stimuli, including muscular tension, intestinal motility, and the position of limbs and wings. More precisely, dmdNaC1 is expressed, from late embryogenesis (14–17 h), in a subset of multidendritic neurons with arborescent dendrites (da) in abdominal and thoracic segments, and subcellularly in discrete varicosities[16]; such varicosities are mechanotransduction sites in the butterfly *Pieris rapae crucivora*. dmdNaC1 has been identified from *Drosophila* EST [16,25]. It encodes a 606 amino-acid protein with the hallmarks of ENaC/Deg family, but lacking the extracellular region specific for degenerin. Like the degenerins, dmdNaC1 does not generate any current when expressed into an heterologous system. The *Drosophila* sequencing project has revealed the existence of more than 20 different ENaC/Deg related proteins.

2.3. Ligand-gated channels

2.3.1. The snail FMRFamide-gated channel FaNaCh

In *Helix aspersa* neurons, as well as in *Aplysia californica* bursting and motor neurons, the molluscan neuropeptide PheMetArgPheNH$_2$ (FMRFamide) and structurally related peptides induce a fast excitatory depolarizing response. This response is due to direct activation of an amiloridesensitive Na$^+$ channel [26,27].

A cDNA, FMRFamide-activated Na$^+$ channel (FaNaCh), has been isolated by homology with ENaC/degenerins from *H. aspersa* nervous tissue [28]. It encodes a 625 amino-acid protein with the structural organization characteristic of this channel family. FaNaCh mRNA is principally located in excitable tissues: nervous system and pedious muscle. Expressed in *Xenopus* oocytes, it generates a FMRFamide-activated current, forming a channel that has a unitary conductance of 13.1 pS, very strong Na$^+$/K$^+$ selectivity, and can be blocked by amiloride.

FaNaCh represents the first example of an ionotropic receptor for a peptide. Interestingly, FMRFamide and neuropeptide FF (a related mammalian peptide) potentiate the current of other ENaC/Deg family channels, the ASIC channels [29]. However, true FaNaCh mammalian homologues have not yet been identified.

2.3.2. Acid-Sensing Ion Channels (ASICs)

ASIC subunits were first identified from ESTs, or by homology cloning from rat brain. They form a subfamily of six members showing about 67% homology between them. They have been named ASIC1a (or ASIC, or BNaC2) [30,31], ASIC2a (or ASIC2, MDEG, BNaC1 or BNC1) [32,33], ASIC3 [34] and ASIC4 [35]; the two splice variants ASIC1b (or ASICβ) [36] and ASIC2b [37] differ respectively from ASIC1a and ASIC2a by their NH$_2$ ends. ASIC subunits assemble in homomeric or heteromultimeric complexes to form proton-activated channels, with different characteristics of activation and kinetics. The channels are located in brain, where their physiological role is currently unknown, or in dorsal root ganglia (DRG), where ASIC channels could be involved in the perception of the pain accompanying tissular acidosis (review [38]). Some ASIC channels have a more extented distribution (for instance, human ASIC3 is also abundantly present in testis and lung) [39–41].

The channels ASIC1a, 1b, 2a and 3, when expressed in *Xenopus* oocytes, generate proton-activated currents of unitary conductances 10–14 pS, blocked by amiloride with a weak efficiency (IC_{50} about 10 μM), and selective for sodium, lithium and protons over potassium. ASIC1a is also selective for Ca^{2+} ($pNa^+/pCa^{2+} = 2.5$), a unique characteristic among ENaC/Deg family channels [30]. ASIC channels differ mainly by their kinetics and pH of half activation. In *Xenopus* oocytes, ASIC1a, 1b and ASIC2a generate transitory Na^+ currents activated by protons, while ASIC3 current is biphasic (transient component + sustained component) [34]. ASIC2b is not functional by itself but can associate with other channels to modulate their activity [37]; ASIC4 does not generate proton-gated channels [35]. Heteromultimeric associations ASIC1/ASIC2a [42], ASIC2a/ASIC2b [37], ASIC3/ASIC2b [37] and ASIC3/ASIC2a [43] have been described. The channels ASIC3 and ASIC2a/ASIC3 are inhibited by lanthanides including gadolinium, which have been reported to block some types of stretch-activated responses in neurons ($IC_{50} = 50$ μM) [43]. A more specific and potent inhibitor has been identified for ASIC1a, PcTX1. PcTX1 is a 40-amino acid peptide isolated from tarantula venom. Its IC_{50} for ASIC1a is 0.9 nM. The toxin is only active on the homomultimeric form of the channel [44].

ASIC1a, ASIC2a, ASIC2b, and ASIC3 are transcribed in the central nervous system in cerebellum, hippocampus, cortex and olfactive bulb [30,32,34,45]. ASIC2a and ASIC3 are also co-localized in the cerebellar cortex [43]. ASIC4 is mainly located in pituitary gland, in thalamus, in substantia nigra, in putamen, in temporal and occipital lobe [35,46], but is also present in inner ear, at the level of the vestibular system in the organ of Corti [46]. ASIC1a and ASIC2a are co-localized and can associate in heterologous systems [42]. The resulting current observed after acidification is similar to some currents observed in cortex cells and granular layer of hippocampus after acidification, but the pH dependence of its activation is shifted to the left (more acidic). ASIC1a seems also to be present in its homomultimeric form in cerebellar granule cells, where the current generated after an acidification is completely blocked by 10 nM PcTXT1 [44].

ASIC1a, 1b, 2b, ASIC3 and ASIC4 are also transcribed in the peripheral nervous system, and more precisely in small diameter neurons of the DRG [30,34,36,45]. This population of afferent neurons is characterized by its sensitivity to capsaicin and is responsible for thermic or pain perceptions. A putative role for proton-activated channels in these structures is the perception of pain accompanying tissular acidosis; acidosis is generated by pathologies such as muscle and cardiac ischemia, corneal injury, bone resorption, inflammation or local infection [47–49]. Since the pain sensation persists while the pH is low, their perception is more probably associated to a sustained current. Such a current, with a biphasic behavior (a transitory current selective for sodium, followed by a sustained, cationic non-selective current) has been recorded in rat native DRG. Association between ASIC3 and ASIC2b, which indeed co-localize in DRG, generates in a heterologous expression system a current very close to the native current [37]. ASIC1a is also expressed in DRG and in trigeminal ganglia, where a calcium-permeable current can be blocked by the ASIC1a inhibitor PcTX1 [44]. While the variety of transitory currents detected in DRG after acidification [50] can probably partly be explained by different associations between ASIC proteins, the exact contribution and interactions between ASIC channels and VR1 vanilloid receptors [51, 52] have yet to be carefully evaluated. VR1 are not only activated by heat or capsaicin in the DRG, but also by protons, and play a major role in pain perception [53]. Therefore, analysis of possible interactions would contribute to the important task of determining the physiological role of transitory proton-gated currents in DRG.

The human gene for ASIC1a maps to chromosome 12q12 [31]. The human gene for ASIC2a maps to chromosome 17q11.212 [33,54]; a human form of ASIC2b has not been detected so far. The human gene for ASIC3 (or hTNaC1 human Testis Na + channel) maps to chromosome 7q35 [41].

2.3.3. The brain–liver–intestine channel BLINaC

A channel belonging phylogenetically to the group of ligand-gated channels, but whose activator has not yet been identified, has been cloned by homology from rat brain [55]. As the name suggests, brain–liver–intestine Na^+ channel (BLINaC) mRNA is detected in brain, liver (including hepatocytes) and small intestine. It is not functional when expressed in *Xenopus* oocytes, and not gated by known activators of ENaC/Deg family channels, like protons. Like some degenerins and ASIC channels, however, it is constitutively active when specific residues known to induce degeneration are mutated. The resulting channel displays low conductance (9–10 pS), high selectivity for sodium over potassium and is blocked by amiloride ($IC_{50} = 1.3$ μM). The human homologue, human intestine Na+ channel (hINaC) is principally present in small intestine, where its role and activator have to be elucidated. hINaC has been mapped on the 4q313.3–q32 region of the human genome [56].

3. Membrane topology and stoichiometry

The ENaC/Deg family subunits are 500–700 amino-acid long proteins, with two hydrophobic 30 amino-acid long segments near N- and C-terminus, and without signal peptide. This profile suggests proteins composed of two short cytoplasmic moities, two transmembrane domains, and a large extracellular loop; this transmembrane organisation has been experimentally confirmed for α, β, γENaC and MEC-4 [57–60]. Other non-related channels share a similar membrane topology: the purinergic receptors have a large extracellular loop and cysteine-rich domains [61]; inward-rectifier potassium channels have two transmembrane domains surrounding a short extracellular loop.

The requirement of three different ENaC subunits, α, β and γ, to form efficient channels, as well as genetic evidence showing that MEC-4/MEC-10 complexes contained at least two MEC-4 subunits, suggested that ENaC/Deg family channels were composed of at least three subunits. In order to establish this stoichiometry, biochemical, electrophysiological and electron microscopy experiments have been performed with FaNaCh and ENaC.

FaNaCh has been used as a model in biochemical experiments because, unlike ENaC, it functions as a homomultimer, and because it is very efficiently matured and expressed at the cell surface when expressed in an heterologous system (HEK-293 cells). Covalent bonding by bifunctional cross-linkers results in the formation of covalent multimers that contain up to four subunits. The combination of sucrose sedimentation and gel filtration experiments on the FaNaCh complex solubilized with Triton X-100 also indicates a stoichiometry of four subunits per complex; similar profiles are obtained after solubilization of FaNaCh with CHAPS [62].

This stoichiometry clearly differs from Snyder et al., who used sucrose gradient sedimentation on ENaC complexes solubilized in digitonin after expression in COS cells [63,64]. Two peaks were identified, a "light" one (containing at least two subunits), and a "heavy" one at 25 S. Since this experiment was performed with digitonin, which partial specific volume is similar to those of proteins, it was not possible to evaluate the contribution of digitonin to the size of the complex. It is therefore possible that the heavy peak reported by Snyder et al. corresponded indeed to a tetramer.

A distinct biochemical approach has been developed by Firsov et al., who have quantified the relative number of ENaC subunits present at the cell surface of *Xenopus* oocytes. It has been shown that the three ENaC subunits assemble according to a fixed stoichiometry, αENaC being more abundant than βENaC and γENaC [65].

Several independant electrophysiological experiments have been performed on ENaC channels, principally by co-expressing wild-type and mutant subunits in *Xenopus* oocytes and studying the properties of the expressed channels. Firsov et al. have used mutants in the pre-TM2 domain with modified sensitivities to Zn^{2+} (rat α S583C) or amiloride (βS525C and γ S537C) to conclude that the channel has a tetrameric structure with the stoichiometry α2βγ. In these studies, the authors not only confirmed there was free association between wild-type and mutant subunits, but also further reinforced their conclusion through the use of concateners, which allowed them to establish the order (αβαγ) of the subunits [65]. Kosari et al. used the modified sensitivity of the same mutants to MTSEA (α) and amiloride (β, γ) to also conclude there is a tetrameric stoichiometry of α2βγ [66]. On the other hand, Snyder et al., using the modified sensitivity of the corresponding human γ mutant (γ S536C) to MTSET, found that the complex contains at least three γ; the reason for this discrepancy is unknown. Additional experiments with other mutants (α S549C, β S520C, γ S529C) and MTSEA led them to suggest a nonameric stoichiometry of α3β3γ3 [63]. Finally, Berdiev et al. have taken advantage of the modified sensitivity of an α subunit mutated in the extracellular loop (αDWYRFHY(278-283)) to amiloride; after in vitro translation of a 1:1 wild-type /mutant mixture and reconstitution in planar bilayers, five channel subtypes with distinct sensitivities to amiloride are observed. Their relative abundance is consistant with a tetrameric organization [67].

A freeze-fracture electron microscopy study has also been performed on the ENaC complex expressed in *Xenopus* oocytes [68]. Particles with a near-square geometry are observed, strongly suggesting that the stoichiometry is four or a multiple of four. Making the assumption that the transmembrane domains correspond solely to the two hydrophobic alpha-helices, the cross-sectional area of 24 nm² suggests the presence of 17 ± 2 transmembrane α helices, favoring an octameric, or even nonameric, organization of channel subunits. However, this conclusion relies on precise assumptions about the size occupied by the transmembrane parts of a subunit, which is probably wrong in the case of these proteins due to the existence of pre-TM1 and pre-TM2 segments, which might also be buried in the membrane.

Overall, despite some conflicting data, a tetrameric organization seems to emerge for the ENaC family of channels. FaNaCh is a homotetramer, ENaC appears to assemble in the order $\alpha\beta\alpha\gamma$, and it might be hypothesized from genetic studies in the nematode that the MEC4/MEC-10 complex exists in the form $(MEC-4)_2(MEC-10)_2$. A tetrameric architecture is also observed for inward-rectifier K^+ channels, whose membrane topology is the same as ENaC/Deg family.

4. Structure–function

All ENaC/Deg family subunits, though having quite a weak similarity (about 20%), have some very conserved domains, as illustrated in Fig. 2.

4.1. Ion pore and sodium selectivity

4.1.1. The TM2, pre-TM2 domains and the "degeneration" position

The TM2 domain, which shows 30% homology between ENaC/Deg family members, is implicated in the conductance, selectivity and opening/closing properties of these channels. Some mutations in the TM2 domain of MEC-4 activate the channel. Other mutations in the same domain induce a loss of function [19,20]. Replacing αENaC TM2 with MEC-4 TM2 changes sodium conductance, P_{Li+}/P_{Na+}, and amiloride sensitivity, with the main amino-acids implicated being αENaC S589 and 593 [69]. In the FaNaCh mutant corresponding to ENaC S589I, permeability to potassium and lithium increases, inactivation kinetics slow down, and amiloride sensitivity is modified [28].

Located immediately before the TM2 domain is a very conserved region known as the pre-TM2 domain, which has the consensus sequence GGQXG. This motif belongs to the ENaC ion pore, since some residues inside (for α, β and γENaC subunits), when replaced by cysteine, are accessible to zinc, MTSEA and MTSET [70]. Zinc block slightly increases with hyperpolarization, suggesting that its binding site is located inside the electric field. Channels containing mutant subunits βG525C or γG537C show a 40% decrease in unitary currents; but the gating properties and the selectivity properties are not modified. That suggests that the pre-TM2 domain belongs to the entrance of ENaC ion pore but is not a major determinant for selectivity.

A position adjacent to this pre-TM2 domain, which we refer to here as the "degeneration position", plays a key role in the opening/closing properties of these channels. Replacement of A713 in MEC-4 with large side chain amino acids leads to the non-apoptotic degeneration of sensory neurons expressing it, which could be due to a constitutive activity of the associated channel [71]. Identical results have been reported in MEC-10 [19], DEG-1 [72], and UNC-105 [24]. The death phenotype can be suppressed in MEC-4 [20] or MEC-10 [19] by some mutations in TM2 or pre-TM2 domains known to plug the channels [73]. Degeneration mutations have various effects on the different members of this family of channels. They constitutively activate degenerins and ASIC2a [32], they increase current in dGNaC1 [16], they have no effect on the epithelial sodium channel [63], and do not constitutively activate ASIC1a [30], ASIC2b [37] or dmdNaC1 [25]. Finally, they abolish FaNaCh activity (Champigny, unpublished results).

About 20 amino-acids above this domain, several other mutations perturb opening/closing properties: the semi-dominant mutation in UNC-105 (E677K) leads to muscle hypercontraction, mutations of some lysines (K504E, K515E) in bovine αENaC reduce the periods of inactivity of the channel and decrease Na^+/K^+ selectivity by a factor of 3–5 [74].

4.1.2. The TM1 and pre-TM1 domains

Replacing αEnaC TM1 with the TM1 from MEC-4 slightly increases unitary conductances for sodium and lithium, modifies the sensitivity to amiloride by a factor of 3, and decreases mean open time by three orders of magnitude [69]. This suggests some participation of this domain in the formation of the ion pore.

The pre-TM1 domain is composed by a conserved domain of 12 amino-acids, located in the N-terminal putative intracellular region, about 13 amino-acids before the TM1 domain; it ends with an "HG" doublet, which is present in all ENaC/Deg family members. Studies of loss-of-function ENaC mutants associated with pseudohypoaldosteronism type I, a genetic disease characterized by severe salt wasting, have shed light on the importance of the conserved glycine: one of the variants found, βG37S, is mutated at this residue. When the α, β or γENaC mutated at that position (G into S) are co-expressed in *Xenopus* oocytes with the two other wild-type subunits, a diminution of the current (40% for β and γ, 85% for α) is observed. These mutations do not affect cell surface expression or unitary conductance; but (for α), the open probability is diminished by a factor 10, and, contrary to the wild-type channel, only short openings are recorded [75]. This proves that the pre-TM1 domain is involved in opening/closing properties of ENaC. In other family members (FaNaCh, dGNaC1), corresponding mutations lead to a diminution or a suppression of the current (Champigny et al., unpublished results). Two amino-acids before the glycine, mutations of a serine also lead to MEC-4 or MEC-10 loss-of-function [19].

A recent work on ASIC channels strongly suggests the participation of the pre-TM1 domain in the ion pore [76]. ASIC2b is a splice variant of ASIC2a, which differs only by its first third (intracellular part, TM1 domain, beginning of the extracellular loop), but has very different properties. ASIC2a encodes a functional acid-sensing channel, selective for sodium over potassium. ASIC2b is not active of its own, but, when associated with ASIC3, confers to this channel a loss of selectivity for sodium over potassium. Construction of chimeras between ASIC2a and ASIC2b has shown that a nine amino-acid region in the pre-TM1 domain, just before the HG doublet, is crucial for selectivity. In particular, point mutations in the pre-TM1 region of ASIC2a (I19P, F20S, T25K) nearly abolish Na^+/K^+ selectivity. These mutations lead to a loss of function in the other ASICs or in FaNaCh, except for one ASIC3 mutant where a loss of selectivity is also observed [76].

4.1.3. Ion pore and sodium selectivity

A putative model for the ion pore of ENaC/Deg family members can be proposed from the structure–function data (Fig. 3). The ion pore is lined by the TM2 domain, the pre-TM2 domain (probably as a reentrant loop), which constitutes the entrance of the pore, and the pre-TM1 domain, which contribute to the selectivity filter and could also form a reentrant loop. This model is reminiscent to the ion pore of Shaker K^+ channels, lined by the transmembrane domain S6, the re-entrant loop H5 (corresponding to the pre-TM2 domain) and the S4-5 domain [77] (corresponding to the pre-TM1 domain).

ENaC/Deg family channels are characterized by their very strong selectivity for sodium over potassium. The only other ions to permeate ENaC are small ions, H^+ and Li^+ [78], with a resulting Eisenman type XI selectivity sequence, $Li^+ > Na^+ > K^+ > Rb^+ > Cs^+$ [5] (Li^+/Na^+ permeability is inverted for the other ENaC/Deg family channels). ENaC is impermeant to small organic cations (hydroxyammonium, hydrazinium) [79], which imposes a restriction on the size of the pore (smaller than 3 Å X 5 Å), and makes

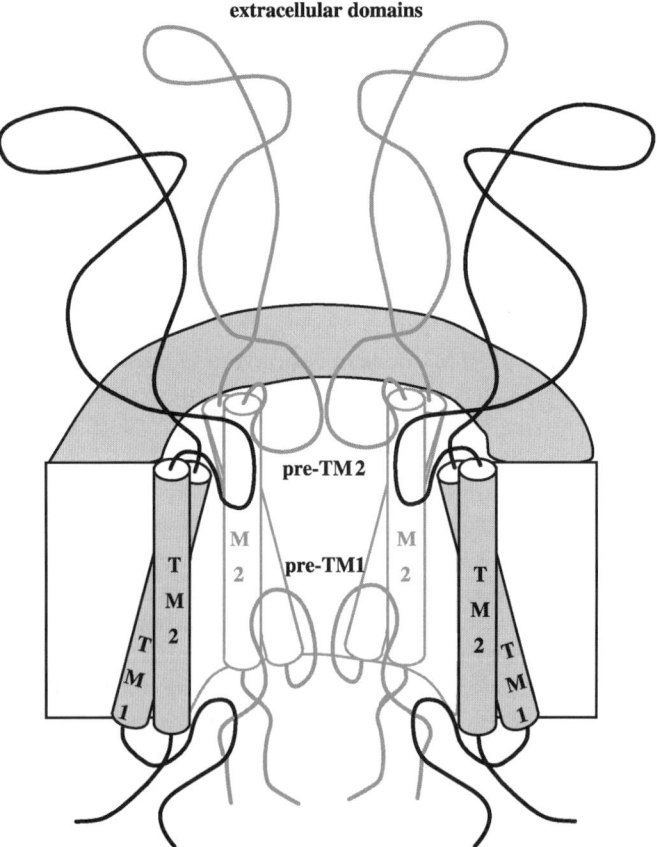

Fig. 3. A working model for the organization of the ENaC/Deg ion pore. The pore is lined by the 4 TM2 domains from 4 distinct subunits, and the corresponding pre-TM1 and pre-TM2 domains, forming re-entrant loops.

this type of channel more selective than voltage-gated sodium channels. Water permeability, which has been observed for the ENaC reconstituted in lipid bilayers [80], but not in native systems like toad urinary bladder [79], would indicate a pore diameter at least equal to 3 Å.

Future studies will have to establish experimentally the secondary structure and dimensions of the ion pore, to identify residues interacting directly with sodium, and to explain the selectivity differences of some family members, like ASIC1, which is permeable to divalent cations [30].

4.2. Amiloride binding

ENaC/Deg family members are blocked by amiloride and by its derivatives like benzamil or phenamil. Amiloride induces rapid opening/closing of the channel, and decreases the mean open time of ENaC [81]. A stoichiometry of one amiloride molecule per channel has been proposed [82]. An increase of the membrane potential increases the inhibition by amiloride, suggesting that part of amiloride binding site belongs to the ion pore [78]. Molecules having hydrophobic groups on the guanidinium terminal amine, like phenamil or benzamil, have a better affinity for ENaC than amiloride; on the contrary, molecules substituted on the amine moiety of pyrazine C5 have a reduced affinity for ENaC. Although amiloride blocks all characterized ENaC members, differences of sensitivity are high. Amiloride blocks very efficiently ENaC ($IC_{50} =$ 100 nM) but is a poor blocker of ASIC family ($IC_{50} = 10 \, \mu m$ for ASIC1 [30]) and even of a member closer to ENaC, dGNaC1 ($IC_{50} = 2.6 \, \mu m$) [15]. The sequence of inhibition also differs: benzamil > amiloride > EIPA for ENaC subunits, equal sensitivity to benzamil, amiloride and EIPA for ASIC1, and equal sensitivity to amiloride and EIPA for dGNaC1.

Mutants of the TM2 or pre-TM2 domains, in either the α, β or γ ENaC subunits (S589 in the α subunit), show a strong alteration in amiloride sensitivity [69,70], suggesting that all three subunits participate in amiloride binding. The WYRFHY sequence in the extracellular site of ENaC also interacts with amiloride. This region has been identified during production of anti-amiloride antibodies: the histidine in the interaction motif (YYGHY) of the antibody interacts directly with the halide of pyrazine group. Similar motifs have subsequently been found on αENaC, γENaC (WYKLHY) and δNaC (WYHFHY). Deletion of this sequence in αENaC strongly decreases amiloride affinity (by a factor of 150) [83].

The first part of the putative amiloride binding site, in domains pre-TM2 and TM2, is well conserved (although not identical) among family members. The second region around the histidine is not conserved. If the pyrazine group indeed links to this motif, it could explain the difference between amiloride and EIPA binding. It is interesting, albeit speculative, to note that δNaC and FaNaCh, which possess motifs close to WYRFHY, have a better affinity for amiloride than for EIPA.

The inhibition by amiloride is a common characteristic for ENaC/Deg family members. Some channels also have activators: FaNaCh is gated by the neuropeptide FMRFamide, ASIC channels by protons. However, nothing is known about the binding sites, nor the mechanism(s) by which these activators work. Mutations modifying the $pH_{0.5}$ of ASIC2a have been found in degeneration position, shifting the $pH_{0.5}$ two units higher [32]. In the pre-TM1 domain, mutations shift the $pH_{0.5}$ one unit higher [76], but also have a profound effect on the ion pore properties.

4.3. The extracellular region

This region is characterized by the presence of cysteine-rich domains (approximately 70 amino acids long). One extra cysteine-rich region is found in degenerins. These cysteine-rich domains, distinct from those found in EGF domains, might confer a rigid structure to the domain after formation of disulfide bridges. A mutation in one of these domains in UNC-8 leads to neurodegeneration, and to a lack of coordination phenotype [17].

Another motif extremely well conserved in the extracellular loop is the "FPAVT" motif, near the first transmembrane domain. Its role is currently unknown, but point mutations affecting Pro^{134} (P134T, or P134S) of *C. elegans* UNC105, two amino-acids before this motif, lead to hypercontraction and paralysis [23].

Recessive mutations in a 22 amino-acid region specific to degenerins lead to the same degeneration phenotype as described above (MEC-4, DEG-1) [73]. In the same region, a mutation of UNC-8 leads to a swelling of motor neurons and to a lack of correlation in movements [17]. Since these mutations are recessive (contrary to mutations in degeneration position), it has been proposed that this degenerin domain may serve as an inactivation ball, with one unique ball being sufficient to close the pore.

4.4. Intracellular domain: ENaC binding proteins

The C⊃OH-terminal extremity of α, β and γENaC contains two proline-rich regions. They have been implicated in the binding of intracellular proteins, which could be involved in ENaC trafficking.

The first proline-rich region is a binding motif for SH3 domains, and in αENaC is $_{663}$PPLALTAPPPA. It interacts with the SH3 domain of α-spectrin after immobilization on filters and in vivo, in MDCK cells [84]. In contrast, corresponding proline-rich domains (PXXP, in PXXPPPXY) in β and γENaC do not seem to bind to SH3 domains [85].

The second proline-rich regions are PY domains in α, β and γENaC, of consensus sequence PPPXY. These regions are altered in Liddle's syndrome, an inherited form of hypertension, which leads to an increase in amiloride-sensitive current. The observed increase in ENaC surface expression might be linked to a reduced endocytosis in Liddle's syndrome, as ENaC is normally endocytosed by clathrin-rich vesicles. Consistent with this, α and γENaC half-lives are doubled by chloroquine treatment (which inhibits lysosomal degradation) or after expression of a dominant mutant of dynamin. This is not observed for the mutants observed in Liddle's disease [86]. Tyrosine in the PPPXY motif can act as an internalization signal, both on β and γ subunits. A double-hybrid approach has suggested that Nedd4 interacts with this PY motif [85], consistent with the finding that Nedd4 interacts with rat αENaC in transfected MDCK cells. Furthermore, Nedd4 contains

3 WW domains, 38 amino-acid sequences that bind PY domains. Nedd4 also contains a calcium-binding site and a ligase ubiquitin domain [85]. Thus, Nedd4 may act as a trigger, causing ubiquitination of subunits that then undergo endocytosis [86]. This ENaC regulation by Nedd4 would be organ dependent, since Nedd4 is co-localized with ENaC in kidney and lungs, but not in colon [87].

5. Regulation of ENaC

5.1. Cell biology and regulation by the cytoskeleton

After its synthesis, ENaC is rapidly oligomerized in the endoplasmic reticulum [63,88]. Multiple early contacts are generated between ENaC subunits: for example, the amino terminus of γENaC is sufficient for interaction with αENaC, but not necessary [89]. Similarly, the glysosylation of ENaC does not seem to be required for efficient biosynthesis. αENaC possesses 6 potential glycosylation sites, and this subunit is indeed glycosylated when expressed in *Xenopus* oocytes. However, mutation of these sites does not perturb channel expression or properties [58,59].

ENaC subunits have a short life-time, estimated to be between 1 and 4 h in different systems [86,90,91]; this contrasts with another channel of the family, FaNaCh, which appears very stable in heterologous systems [62]. One of the determinants of this short half-life is probably degradation by proteosomes following ubiquitination, as α and γENaC are ubiquitinated on N-terminal lysines, and their half-lives are doubled after treatment with lactacystin, a proteasome inhibitor [91].

Syntaxins 1A and 3, two vesicle traffic regulatory proteins associated with target membranes (t-SNAREs) [92,93], co-immunoprecipitate with ENaC subunits in *Xenopus* oocytes. They have opposite functional effects on ENaC: in the oocyte, syntaxin 1A inhibits ENaC activity by a factor of two, while syntaxin 3 stimulates it by a factor of two.

Near the apical cell surface, ENaC probably belongs to a cytoskeleton-linked complex containing α-spectrin [84,94]. A protein co-sedimenting with spectrin/ENaC complexes has been identified in amphibian tissues: Apx (apical protein *Xenopus*) is a 160 kDa protein that could represent an ENaC regulatory protein (ENaC current is strongly reduced after injection of Apx antisense oligonucleotides [94]) or have by itself a sodium conductance [95]. The role of spectrin/Apx association with ENaC is currently unknown; it might cluster ENaC in the apical domain, or link ENaC-containing vesicles to the spectrin-based terminal web, therefore regulating its trafficking.

ENaC seems also directly regulated by actin at the cell surface. Short actin filaments (tri- or tetramers) stimulate or induce ENaC activity in lipid bilayers, and modify its unitary parameters. Moreover, the presence of short actin filaments seems necessary for the activation of ENaC by PKA [96]. The substrate for phosphorylation seems to be actin itself; Cantiello has shown that phosphorylated actin becomes a poor substrate for sustained polymerization, so that PKA phosphorylation would favor the formation of short actin filaments, and therefore activate ENaC [97].

5.2. Regulation by protein kinases

5.2.1. Regulation by cyclic AMP dependent protein kinase (PKA)

Cyclic AMP stimulates ENaC activity in rat cortical collecting tubules [98] but not in rat colon [99]. In toad urinary bladder, a cyclic AMP stimulation is observed, but is lost after disruption of the cells [100]. In frog skin, frog colon, toad urinary bladder and A6 cells, cyclic AMP increases the number of conductive Na^+ channels without affecting the open probability [101–103]. These observations are consistent with an indirect stimulatory mechanism rather than direct modification of ENaC biophysical properties by PKA.

A variety of indirect mechanisms could underlie this stimulation of ENaC activity. First, through its actions on actin, PKA could activate a population of silent channels located at or near the cell surface. Second, as proposed for aquaporin channels, PKA could stimulate the exocytosis of vesicles containing functional ENaC complexes to the apical membrane. Third, in lung, where there is an increase in amiloride-sensitive fluid clearance observed after stimulation with β-adrenergic agonists [104], the stimulation could act, not at the level of ENaC, but at the level of a chloride conductance, which is also necessary for NaCl absorption [105]. This has been recently confirmed by Fang et al., who have demonstrated that the activation of fluid clearance by cAMP was abolished in mice harboring a deficient CFTR, the cAMP-activated Cl^- channel mutated during Cystic Fibrosis [106].

5.2.2. Regulation by PKC

PKC has been shown to inhibit ENaC activity equally well in cultured cell lines (A6 cells) and mammalian tissues. The effects are generaly slow (20–30 min on sheep trachea [107] or rabbit CCD [108]), and mediated by different PKC isoforms: calcium sensitive (in rat CCD [109]) or insensitive (frog skin [110], rabbit CCD [108]). In contrast, ENaC activity can be stimulated by PKC in frog skin [110].

6. Physiology and pathophysiology of the epithelial sodium channel ENaC

6.1. Kidney

The three ENaC subunit mRNAs are co-expressed in the rat renal distal convoluted tubules, connecting tubules, cortical collecting ducts, and outer medullary collecting ducts [111]. Overall, ENaC activity in the kidney is upregulated by aldosterone and vasopressin [112].

The regulation by aldosterone is exemplified by the ~ 100-fold increase in the amiloride-sensitive whole-cell current in rat cortical collecting tubule cells in response to a low Na^+ diet [113]. This stimulation is hardly explained by a direct transcriptional effect on ENaC subunits, since ENaC mRNA levels are not [114] or only slightly [115,116] altered by steroid treatment. In kidney (contrary to A6 cells [117]), the number of active Na^+ channels is increased, without modification in the open probability [118]. The serine–threonine kinase SGK, whose transcription is increased by aldosterone or glucocorticoids, leads to a 7-fold increase in the amiloride-sensitive current in Xenopus oocyte [119]; it might correspond to one of the early activators of the Na^+ current.

Changes in internal pH [120], methylation ([121] but see also Ref. [98]), changes in internal calcium [122], or mobilization of ENaC subunits stored in cytoplasmic vesicles to the apical membrane [123] have also been proposed as mechanisms underlying the early response to aldosterone.

Vasopressin acts primarily in the distal part of the nephron by stimulating water permeability and sodium reabsorption. Consistent with this, vasopressin elicits a 2 to 5-fold increase in the amiloride-sensitive signal in the collecting tubules within minutes [112]. This is not due to new protein synthesis, but to the increase of cAMP after vasopressin binding to the basolateral V2 receptor. While this rapid activation of Na^+ channels by vasopressin is consistent with the existence of a pool a silent channels, this hormone also does have a slight transcriptional effect on ENaC subunit gene expression [124].

Due to the role of ENaC in precise sodium and ion reabsorption, small perturbations in the activity of the channel will lead to subsequent changes in the quantity of fluid and salt reabsorbed. A slight (2–3 times) increase in activity will cause excessive reabsorption and hypertension, as in Liddle's syndrome, while a partial loss of function will lead to salt waste and dehydration observed in pseudohypoaldosteronism type I.

Liddle's syndrome is a rare autosomal dominant form of essential hypertension, where an excessive sodium reabsorption is accompanied by hypokaliemy and hypoaldostero-nemy. It is efficiently cured by weak doses of amiloride. Genetic analyses have led to identification of mutations in βENaC and γEnaC that are associated with this disorder. Some mutations affect a 95 bp segment of $3'$ terminal extremity of βENaC, leading to a translation stop or a frameshift [125]. Another mutation, X574stop, has been identified on γENaC [126]. Finally, a point mutation in βENaC (P616L) has been found in the conserved proline-rich PY motif [127]. All these mutations, reproduced in heterologous systems, increase amiloride-sensitive current (2 to 4-fold for the β and γ truncated form, 7.5-fold for the γW574stop, 8.8-fold for βP616L [126–130]), showing that the cause of Liddle's syndrome is the loss of negative regulation. As described earlier, these mutants lead to an inhibition of endocytosis, possibly due to the loss of the interaction of the PY domains with the cellular factor Ncdd4.

Several genetic screens have been aimed at identifying a putative role for ENaC in various forms of hypertension. A genetic linkage between β and γENaC and systolic blood pressure has been identified [131], and a number of variants of β and γENaC have been identified in humans, especially in populations of African, African-American, and West Indian origins [132]. An increased Na^+ channel activity was detected in some of these patients, with a loss of PKC inhibition associated with a variant (T594M) of the β subunit [133,134]. However, expression of these variants in *Xenopus* oocytes was not associated with altered levels of current, although tissue-specific mechanisms of regulation may underlie this disparity.

Pseudohypoaldosteronism type I (PHA1) is a rare inherited disease, characterized by severe neonatal salt wasting, hyperkaliemy, metabolic acidosis and unresponsiveness to mineralocorticoid hormones. It is often fatal in the neonatal period in the absence of treatment. Transmission is autosomal, and may be recessive or dominant. To date, only mutations corresponding to recessive forms have been identified on ENaC subunits [135–137]. All of these mutations lead to a loss of function, with some resulting in the introduction of a stop codon before the TM2 domain (αR508stop, α168(frameshift), nucleotidic mutations γG318A, α1449delC, α729delA), while others involve extremely conserved residues: γKYS106-108N (another splice variant caused by γG318A mutation), αS562L, or βG37S in the conserved dinucleotide HG of the pre-TM1 domain.

6.2. Colon

The epithelial sodium channel mediates electrogenic reabsorption of sodium in distal colon. The three subunits are expressed in the surface epithelial cells, but not in the crypt cells [111,114]. This electrogenic sodium reabsorption is stimulated by aldosterone [138], whose action is mediated by type I mineralocorticoid receptors [139]. This stimulation is partly due to a long-term, transcriptional effect. αENaC is constitutively expressed in colon, but does not reach the apical surface in the absence of β and γ subunits [114,140]. Aldosterone, dexamethasone or low-salt diet stimulate the transcription of βENaC and γENaC genes by a factor of 5–10, so that the expression level of the three subunits becomes equivalent [7,114,141]. In addition to this long-term effect, a short-term stimulation by aldosterone is observed: the short-circuit current is increased within 3 h of aldosterone infusion, before any transcriptional effects on βENaC or γENaC genes [141].

6.3. Lung

ENaC plays a key role in the transition from a fluidfilled to an airfilled lung at the time of birth [142,143]; in adult, it controls the quantity and composition of the respiratory tract fluid. The three subunits are detected in distal airways, at the level of Clara cells and of ciliated cells [114]; human β and γENaC have also been detected in large airways, in ciliated cells, in cells lining glandular ducts and in the serous gland cells [144]. More recently, Johnson et al. have demonstrated the presence of α, β and γENaC into alveolar epithelial type I cells, as well as into alveolar type II cells, suggesting that type I cells may contribute significantly in maintaining alveolar fluid balance [145].

At birth, an increase in Na^+ channel transcription and expression results in a switch of the ionic transport in lung from active Cl secretion to active Na^+ reabsorption [8,10,146]. The ENaC activity is controlled in lungs by corticosteroids, which stimulate the transcription of the three subunits genes [147]. The increase in Na^+ channel activity observed in the lung around birth might be related to the raise of corticosteroids, although other triggers, such as change in P_{O_2} [148], could also intervene.

For most ENaC variants associated with pathologies, the pulmonary function does not seem strongly affected, although inactivation of murine αENaC causes respiratory distress of neonates and death with 40 h of birth [149]. This might be due to the existence of alternative sodium channels in human lung [150,151]. One case of respiratory distress syndrome at birth, probably associated with PHA1, has been reported [152]. Pulmonary symptoms to various degrees have been found for four Swedish patients with PHA1, with bacterial profiles reminiscent of Cystic Fibrosis, but without chronic infection or progressive decline of lung function [137].

In Cystic Fibrosis, ENaC is not mutated but there are reports suggesting that Na^+ absorption may be increased in some respiratory tissues [153]. When the human chloride channel cystic fibrosis transmembrane conductance regulator (CFTR) is co-expressed with rat ENaC in Xenopus oocytes or MDCK cells, cAMP inhibits ENaC, while it stimulates it in the absence of CFTR [154–156]. This negative regulation is not observed with human CFTR co-expressed with human ENaC (Barbry, unpublished data) it is not observed in sweat gland where ENaCs and CFTR are also coexpressed [157], and some strains of transgenic mice, obtained by knockout of the CFTR gene, do not develop airway $hyperNa^+$ reabsorption [158]. Many possible mechanisms have been proposed; a direct interaction between ENaC and CFTR[159], an interaction through actin cytoskeleton [160], a perturbation of membrane recycling in the absence of CFTR (which would affect ENaC vesicle stimulation by PKA [161]), an indirect control of ENaC by intracellular anions, which in submandibulary salivary glands involves specific heterotrimeric G proteins [162,163].

7. Conclusions and perspectives for the future

Cation channels of ENaC family are found in a variety of organisms, from nematode to human, and implicated in various functions, from sodium reabsorption to mechano-transduction. However, the physiological roles of many of these channels remain to be elucidated. Identifying the role of ASIC channels in pain perception following tissue acidosis and in central nervous system remain to be elucidated. The role of ENaC-related cation channels during early development and functional expression of stretch-activated channels are other objectives of the active research. Another active area of research centers is around the cell biology of ENaC proteins. It will be particularly important to precisely understand the mechanisms of short-term regulation by PKA and PKC, and to characterize in vivo the interactions of ENaC with intracellular components.

Acknowledgements

This work was supported by the Centre National de la Recherche Scientifique, the Association de Recherche contre le Cancer, the association Vaincre la Mucoviscidose and the Fondation pour la Recherche Médicale (20000374/2). We are grateful to Franck Aguila for artwork.

References

[1] Kœfœd-Johnsen, V., Ussing, H.H., 1958. The nature of frog skin potential. Acta Physiol. Scand. 42, 298–308.
[2] Barbry, P., Lazdunski, M., 1996. Structure and regulation of the amiloride-sensitive epithelial sodium channel. Ion Channels 4, 115–167.
[3] Palmer, L.G., 1992. Epithelial Na channels: function and diversity. Annu. Rev. Physiol. 54, 51–66.
[4] Lingueglia, E., Voilley, N., Waldmann, R., Lazdunski, M., Barbry, P., 1993. Expression cloning of an epithelial amiloride-sensitive Na+ channel. A new channel type with homologies to *Caenorhabditis elegans* degenerins. FEBS Lett. 318, 95–99.

[5] Canessa, C.M., Horisberger, J.D., Rossier, B.C., 1993. Epithelial sodium channel related to proteins involved in neurodegeneration. Nature 361, 467–470 (see comments).

[6] Canessa, C.M., Schild, L., Buell, G., Thorens, B., Gautschi, I., Horisberger, J.D., Rossier, B.C., 1994. Amiloride-sensitive epithelial Na+ channel is made of three homologous subunits. Nature 367, 463–467 (see comments).

[7] Lingueglia, E., Renard, S., Waldmann, R., Voilley, N., Champigny, G., Plass, H., Lazdunski, M., Barbry, P., 1994. Different homologous subunits of the amiloride-sensitive Na+ channel are differently regulated by aldosterone. J. Biol. Chem. 269, 13736–13739.

[8] Voilley, N., Lingueglia, E., Champigny, G., Mattei, M.G., Waldmann, R., Lazdunski, M., Barbry, P., 1994. The lung amiloride-sensitive Na+ channel: biophysical properties, pharmacology, ontogenesis, and molecular cloning. Proc. Natl Acad. Sci. USA 91, 247–251.

[9] Voilley, N., Bassilana, F., Mignon, C., Merscher, S., Mattei, M.G., Carle, G.F., Lazdunski, M., Barbry, P., 1995. Cloning, chromosomal localization, and physical linkage of the beta and gamma subunits (SCNN1B and SCNN1G) of the human epithelial amiloride-sensitive sodium channel. Genomics 28, 560–565.

[10] McDonald, F.J., Snyder, P.M., McCray, P.J., Welsh, M.J., 1994. Cloning, expression, and tissue distribution of a human amiloride- sensitive Na+ channel. Am. J. Physiol. 266, L728–L734.

[11] McDonald, F.J., Price, M.P., Snyder, P.M., Welsh, M.J., 1995. Cloning and expression of the beta- and gamma-subunits of the human epithelial sodium channel. Am. J. Physiol. 268, C1157–C1163.

[12] Voilley, N., Galibert, A., Bassilana, F., Renard, S., Lingueglia, E., Coscoy, S., Champigny, G., Hofman, P., Lazdunski, M., Barbry, P., 1997. The amiloride-sensitive Na+ channel: from primary structure to function. Comp. Biochem. Physiol. A Physiol. 118, 193–200.

[13] Kretz, O., Barbry, P., Bock, R., Lindemann, B., 1999. Differential expression of RNA and protein of the three pore-forming subunits of the amiloride-sensitive epithelial sodium channel in taste buds of the rat. J. Histochem. Cytochem. 47, 51–64.

[14] Waldmann, R., Champigny, G., Bassilana, F., Voilley, N., Lazdunski, M., 1995. Molecular cloning and functional expression of a novel amiloride- sensitive Na+ channel. J. Biol. Chem. 270, 27411–27414.

[15] Darboux, I., Lingueglia, E., Champigny, G., Coscoy, S., Barbry, P., Lazdunski, M., 1998. dGNaC1, a gonad-specific amiloride-sensitive Na+ channel [In Process Citation]. J. Biol. Chem. 273, 9424–9429.

[16] Adams, C.M., Anderson, M.G., Motto, D.G., Price, M.P., Johnson, W.A., Welsh, M.J., 1998. Ripped pocket and pickpocket, novel Drosophila DEG/ENaC subunits expressed in early development and in mechanosensory neurons. J. Cell Biol. 140, 143–152.

[17] Tavernarakis, N., Shreffler, W., Wang, S., Driscoll, M., 1997. unc-8, a DEG/ENaC family member, encodes a subunit of a candidate mechanically gated channel that modulates *C. elegans* locomotion. Neuron 18, 107–119.

[18] Take-Uchi, M., Kawakami, M., Ishihara, T., Amano, T., Kondo, K., Katsura, I., 1998. An ion channel of the degenerin/epithelial sodium channel superfamily controls the defecation rhythm in *Caenorhabditis elegans*. Proc. Natl Acad. Sci. USA 95, 11775–11780.

[19] Huang, M., Chalfie, M., 1994. Gene interactions affecting mechanosensory transduction in *Caenorhabditis elegans*. Nature 367, 467–470 (see comments).

[20] Hong, K., Driscoll, M., 1994. A transmembrane domain of the putative channel subunit MEC-4 influences mechanotransduction and neurodegeneration in *C. elegans*. Nature 367, 470–473 (see comments).

[21] Huang, M., Gu, G., Ferguson, E.L., Chalfie, M., 1995. A stomatin-like protein necessary for mechanosensation in *C. elegans*. Nature 378, 292–295.

[22] Du, H., Gu, G., William, C.M., Chalfie, M., 1996. Extracellular proteins needed for *C. elegans* mechanosensation. Neuron 16, 183–194.

[23] Liu, J., Schrank, B., Waterston, R.H., 1996. Interaction between a putative mechanosensory membrane channel and a collagen. Science 273, 361–364 (see comments).

[24] Garcia, A.J., Garcia, J.A., Liu, J.D., Corey, D.P., 1998. The nematode degenerin UNC-105 forms ion channels that are activated by degeneration- or hypercontraction-causing mutations [In Process Citation]. Neuron 20, 1231–1241.

[25] Darboux, I., Lingueglia, E., Pauron, D., Barbry, P., Lazdunski, M., 1998. A new member of the amiloride-sensitive sodium channel family in *Drosophila melanogaster* peripheral nervous system [In Process Citation]. Biochem. Biophys. Res. Commun. 246, 210–216.

[26] Belkin, K.J., Abrams, T.W., 1993. FMRFamide produces biphasic modulation of the LFS motor neurons in the neural circuit of the siphon withdrawal reflex of Aplysia by activating Na+ and K+ currents. J. Neurosci. 13, 5139–5152.

[27] Green, K.A., Falconer, S.W., Cottrell, G.A., 1994. The neuropeptide Phe-Met-Arg-Phe-NH2 (FMRFamide) directly gates two ion channels in an identified Helix neurone. Pflugers Arch. 428, 232–240.

[28] Lingueglia, E., Champigny, G., Lazdunski, M., Barbry, P., 1995. Cloning of the amiloride-sensitive FMRFamide peptide-gated sodium channel. Nature 378, 730–733.

[29] Askwith, C.C., Cheng, C., Ikuma, M., Benson, C., Price, M.P., Welsh, M.J., 2000. Neuropeptide FF and FMRFamide potentiate acid-evoked currents from sensory neurons and proton-gated DEG/ENaC channels. Neuron 26, 133–141.

[30] Waldmann, R., Champigny, G., Bassilana, F., Heurteaux, C., Lazdunski, M., 1997. A proton-gated cation channel involved in acid-sensing. Nature 386, 173–177.

[31] Garcia, A.J., Derfler, B., Neville, G.J., Hyman, B.T., Corey, D.P., 1997. BNaC1 and BNaC2 constitute a new family of human neuronal sodium channels related to degenerins and epithelial sodium channels. Proc. Natl Acad. Sci. USA 94, 1459–1464.

[32] Waldmann, R., Champigny, G., Voilley, N., Lauritzen, I., Lazdunski, M., 1996. The mammalian degenerin MDEG, an amiloride-sensitive cation channel activated by mutations causing neurodegeneration in *Caenorhabditis elegans*. J. Biol. Chem. 271, 10433–10436.

[33] Price, M.P., Snyder, P.M., Welsh, M.J., 1996. Cloning and expression of a novel human brain Na+ channel. J. Biol. Chem. 271, 7879–7882.

[34] Waldmann, R., Bassilana, F., de, W.J., Champigny, G., Heurteaux, C., Lazdunski, M., 1997. Molecular cloning of a non-inactivating proton-gated Na+ channel specific for sensory neurons. J. Biol. Chem. 272, 20975–20978.

[35] Akopian, A.N., Chen, C.C., Ding, Y., Cesare, P., Wood, J.N., 2000. A new member of the acid-sensing ion channel family [In Process Citation]. Neuroreport 11, 2217–2222.

[36] Chen, C.C., England, S., Akopian, A.N., Wood, J.N., 1998. A sensory neuron-specific, proton-gated ion channel. Proc. Natl Acad. Sci. USA 95, 10240–10245.

[37] Lingueglia, E., de, W.J., Bassilana, F., Heurteaux, C., Sakai, H., Waldmann, R., Lazdunski, M., 1997. A modulatory subunit of acid sensing ion channels in brain and dorsal root ganglion cells. J. Biol. Chem. 272, 29778–29783.

[38] Waldmann, R., Lazdunski, M., 1998. Proton-gated cation channels- neuronal acid sensors in the amiloride-sensitive Na+ channel/degenerin family of ion channels. Curr. Opin. Neurobiol. 8, 418–424.

[39] Babinski, K., Le, K.T., Seguela, P., 1999. Molecular cloning and regional distribution of a human proton receptor subunit with biphasic functional properties. J. Neurochem. 72, 51–57.

[40] Ishibashi, K., Marumo, F., 1998. Molecular cloning of a DEG/ENaC sodium channel cDNA from human testis [In Process Citation]. Biochem. Biophys. Res. Commun. 245, 589–593.

[41] de, W.J., Bassilana, F., Lazdunski, M., Waldmann, R., 1998. Identification, functional expression and chromosomal localisation of a sustained human proton-gated cation channel [In Process Citation]. FEBS Lett. 433, 257–260.

[42] Bassilana, F., Champigny, G., Waldmann, R., de, W.J., Heurteaux, C., Lazdunski, M., 1997. The acid-sensitive ionic channel subunit ASIC and the mammalian degenerin MDEG form a heteromultimeric H+-gated Na+ channel with novel properties. J. Biol. Chem. 272, 28819–28822.

[43] Babinski, K., Catarsi, S., Biagini, G., Seguela, P., 2000. Mammalian ASIC2a and ASIC3 subunits co-assemble into heteromeric proton- gated channels Sensitive to Gd3+. J. Biol. Chem. 275, 28519–28525.

[44] Escoubas, P., De Weille, J.R., Lecoq, A., Diochot, S., Waldmann, R., Champigny, G., Moinier, D., Menez, A., Lazdunski, M., 2000. Isolation of a tarantula toxin specific for a class of proton-gated Na+ channels [In Process Citation]. J. Biol. Chem. 275, 25116–25121.

[45] Olson, T.H., Riedl, M.S., Vulchanova, L., Ortiz, G.X., Elde, R., 1998. An acid sensing ion channel (ASIC) localizes to small primary afferent neurons in rats [In Process Citation]. Neuroreport 9, 1109–1113.

[46] Grunder, S., Geissler, H.S., Bassler, E.L., Ruppersberg, J.P., 2000. A new member of acid-sensing ion channels from pituitary gland. Neuroreport 11, 1607–1611.

[47] Pan, H.L., Longhurst, J.C., Eisenach, J.C., Chen, S.R., 1999. Role of protons in activation of cardiac sympathetic C-fibre afferents during ischaemia in cats. J. Physiol. (Lond.) 518, 857–866.

[48] Chen, X., Gallar, J., Belmonte, C., 1997. Reduction by antiinflammatory drugs of the response of corneal sensory nerve fibers to chemical irritation. Investig. Ophthalmol. Vis. Sci. 38, 1944–1953.

[49] Kress, M., Zeilhofer, H.U., 1999. Capsaicin, protons and heat: new excitement about nociceptors. Trends Pharmacol. Sci. 20, 112–118.

[50] Akaike, N., Ueno, S., 1994. Proton-induced current in neuronal cells. Prog. Neurobiol. 43, 73–83.

[51] Caterina, M.J., Schumacher, M.A., Tominaga, M., Rosen, T.A., Levine, J.D., Julius, D., 1997. The capsaicin receptor: a heat-activated ion channel in the pain pathway. Nature 389, 816–824 (see comments).

[52] Tominaga, M., Caterina, M.J., Malmberg, A.B., Rosen, T.A., Gilbert, H., Skinner, K., Raumann, B.E., Basbaum, A.I., Julius, D., 1998. The cloned capsaicin receptor integrates multiple pain-producing stimuli. Neuron 21, 531–543 (see comments).

[53] Caterina, M.J., Leffler, A., Malmberg, A.B., Martin, W.J., Trafton, J., Petersen-Zeitz, K.R., Koltzenburg, M., Basbaum, A.I., Julius, D., 2000. Impaired nociception and pain sensation in mice lacking the capsaicin receptor. Science 288, 306–313 (see comments).

[54] Waldmann, R., Voilley, N., Mattei, M.G., Lazdunski, M., 1996. The human degenerin MDEG, an amiloride-sensitive neuronal cation channel, is localized on chromosome 17q11.2–17q12 close to the microsatellite D17S798. Genomics 37, 269–270.

[55] Sakai, H., Lingueglia, E., Champigny, G., Mattei, M.G., Lazdunski, M., 1999. Cloning and functional expression of a novel degenerin-like Na+ channel gene in mammals. J. Physiol. (Lond.) 519(Pt 2), 323–333.

[56] Schaefer, L., Sakai, H., Mattei, M., Lazdunski, M., Lingueglia, E., 2000. Molecular cloning, functional expression and chromosomal localization of an amiloride-sensitive Na(+) channel from human small intestine. FEBS Lett. 471, 205–210.

[57] Renard, S., Lingueglia, E., Voilley, N., Lazdunski, M., Barbry, P., 1994. Biochemical analysis of the membrane topology of the amiloride- sensitive Na+ channel. J. Biol. Chem. 269, 12981–12986.

[58] Canessa, C.M., Merillat, A.M., Rossier, B.C., 1994. Membrane topology of the epithelial sodium channel in intact cells. Am. J. Physiol. 267, C1682–1690.

[59] Snyder, P.M., McDonald, F.J., Stokes, J.B., Welsh, M.J., 1994. Membrane topology of the amiloride-sensitive epithelial sodium channel. J. Biol. Chem. 269, 24379–24383.

[60] Lai, C.C., Hong, K., Kinnell, M., Chalfie, M., Driscoll, M., 1996. Sequence and transmembrane topology of MEC-4, an ion channel subunit required for mechanotransduction in *Caenorhabditis elegans*. J. Cell Biol. 133, 1071–1081.

[61] Torres, G.E., Egan, T.M., Voigt, M.M., 1998. Topological analysis of the ATP-gated ionotrophic P2X2 receptor subunit. FEBS Lett. 425, 19–23.

[62] Coscoy, S., Lingueglia, E., Lazdunski, M., Barbry, P., 1998. The phe-met-arg-phe-amide-activated sodium channel is a tetramer [In Process Citation]. J. Biol. Chem. 273, 8317–8322.

[63] Snyder, P.M., Cheng, C., Prince, L.S., Rogers, J.C., Welsh, M.J., 1998. Electrophysiological and biochemical evidence that DEG/ENaC cation channels are composed of nine subunits. J. Biol. Chem. 273, 681–684.

[64] Cheng, C., Prince, L.S., Snyder, P.M., Welsh, M.J., 1998. Assembly of the epithelial Na+ channel evaluated using sucrose gradient sedimentation analysis. J. Biol. Chem. 273, 22693–22700.

[65] Firsov, D., Gautschi, I., Merillat, A.M., Rossier, B.C., Schild, L., 1998. The heterotetrameric architecture of the epithelial sodium channel (ENaC). EMBO J. 17, 344–352.

[66] Kosari, F., Sheng, S., Li, J., Mak, D., Foskett, J.K., Kleyman, T.R., 1998. Subunit stoichiometry of the epithelial sodium channel. J. Biol. Chem. 273, 13469–13474.

[67] Berdiev, B.K., Karlson, K.H., Kozari, F., Stanton, B.A., Kleyman, T.R., Ismailov, I.I., 1998. Subunit stoichiometry of a core conduction element in a cloned epithelial amiloride-sensitive Na+ channel. Biophys. J. 74, A402.

[68] Eskandari, S., Snyder, P.M., Kreman, M., Zampighi, G.A., Welsh, M.J., Wright, E.M., 1999. Number of subunits comprising the epithelial sodium channel. J. Biol. Chem. 274, 27281–27286.

[69] Waldmann, R., Champigny, G., Lazdunski, M., 1995. Functional degenerin-containing chimeras identify residues essential for amiloride-sensitive Na+ channel function. J. Biol. Chem. 270, 11735–11737.

[70] Schild, L., Schneeberger, E., Gautschi, I., Firsov, D., 1997. Identification of amino acid residues in the alpha, beta, and gamma subunits of the epithelial sodium channel (ENaC) involved in amiloride block and ion permeation. J. Gen. Physiol. 109, 15–26.

[71] Driscoll, M., Chalfie, M., 1991. The mec-4 gene is a member of a family of *Caenorhabditis elegans* genes that can mutate to induce neuronal degeneration. Nature 349, 588–593 (see comments).

[72] Chalfie, M., Wolinsky, E., 1990. The identification and suppression of inherited neurodegeneration in *Caenorhabditis elegans*. Nature 345, 410–416.

[73] Garcia, A.J., Ma, C., Chalfie, M., 1995. Regulation of *Caenorhabditis elegans* degenerin proteins by a putative extracellular domain. Curr. Biol. 5, 441–448 (published erratum appears in Curr. Biol. 5(6), 686).

[74] Fuller, C.M., Berdiev, B.K., Shlyonsky, V.G., Ismailov, I.I., Benos, D.J., 1997. Point mutations in alpha bENaC regulate channel gating, ion selectivity, and sensitivity to amiloride. Biophys. J. 72, 1622–1632.

[75] Grunder, S., Firsov, D., Chang, S.S., Jaeger, N.F., Gautschi, I., Schild, L., Lifton, R.P., Rossier, B.C., 1997. A mutation causing pseudohypoaldosteronism type 1 identifies a conserved glycine that is involved in the gating of the epithelial sodium channel. EMBO J. 16, 899–907.

[76] Coscoy, S., de Weille, J.R., Lingueglia, E., Lazdunski, M., 1999. The pre-transmembrane 1 domain of acid-sensing ion channels participates in the ion pore. J. Biol. Chem. 274, 10129–10132.

[77] Slesinger, P.A., Jan, Y.N., Jan, L.Y., 1993. The S4–S5 loop contributes to the ion-selective pore of potassium channels. Neuron 11, 739–749.

[78] Palmer, L.G., 1984. Voltage-dependent block by amiloride and other monovalent cations of apical Na channels in the toad urinary bladder. J. Membr. Biol. 80, 153–165.

[79] Palmer, L.G., 1987. Ion selectivity of epithelial Na channels. J. Membr. Biol. 96, 97–106.

[80] Ismailov, I.I., Shlyonsky, V.G., Benos, D.J., 1997. Streaming potential measurements in alphabetagamma-rat epithelial Na+ channel in planar lipid bilayers. Proc. Natl Acad. Sci. USA 94, 7651–7654.

[81] Palmer, L.G., Frindt, G., 1986. Amiloride-sensitive Na channels from the apical membrane of the rat cortical collecting tubule. Proc. Natl Acad. Sci. USA 83, 2767–2770.

[82] Bindslev, N., Cuthbert, A.W., Edwardson, J.M., Skadhauge, E., 1982. Kinetics of amiloride action in the hen coprodaeum in vitro. Pflügers Arch. 392, 340–346.

[83] Ismailov, I.I., Kieber, E.T., Lin, C., Berdiev, B.K., Shlyonsky, V.G., Patton, H.K., Fuller, C.M., Worrell, R., Zuckerman, J.B., Sun, W., Eaton, D.C., Benos, D.J., Kleyman, T.R., 1997. Identification of an amiloride binding domain within the alpha-subunit of the epithelial Na+ channel. J. Biol. Chem. 272, 21075–21083.

[84] Rotin, D., Bar, S.D., O'Brodovich, H., Merilainen, J., Lehto, V.P., Canessa, C.M., Rossier, B.C., Downey, G.P., 1994. An SH3 binding region in the epithelial Na+ channel (alpha rENaC) mediates its localization at the apical membrane. EMBO J. 13, 4440–4450.

[85] Staub, O., Dho, S., Henry, P., Correa, J., Ishikawa, T., McGlade, J., Rotin, D., 1996. WW domains of Nedd4 bind to the proline rich PY motifs in the epithelial Na+ channel deleted in Liddle's syndrome EMBO J. 15, 2371–2380.

[86] Shimkets, R.A., Lifton, R.P., Canessa, C.M., 1997. The activity of the epithelial sodium channel is regulated by clathrin- mediated endocytosis. J. Biol. Chem. 272, 25537–25541.

[87] Staub, O., Yeger, H., Plant, P.J., Kim, H., Ernst, S.A., Rotin, D., 1997. Immunolocalization of the ubiquitin-protein ligase Nedd4 in tissues expressing the epithelial Na+ channel (ENaC). Am. J. Physiol. 272, C1871–C1880.

[88] Awayda, M.S., Ismailov, I.I., Berdiev, B.K., Benos, D.J., 1995. A cloned renal epithelial Na+ channel protein displays stretch activation in planar lipid bilayers. Am. J. Physiol. 268, C1450–C1459.

[89] Adams, C.M., Snyder, P.M., Welsh, M.J., 1997. Interactions between subunits of the human epithelial sodium channel. J. Biol. Chem. 272, 27295–27300.

[90] May, A., Puoti, A., Gaeggeler, H.P., Horisberger, J.D., Rossier, B.C., 1997. Early effect of aldosterone on the rate of synthesis of the epithelial sodium channel alpha subunit in A6 renal cells. J. Am. Soc. Nephrol. 8, 1813–1822.

[91] Staub, O., Gautschi, I., Ishikawa, T., Breitschopf, K., Ciechanover, A., Schild, L., Rotin, D., 1997. Regulation of stability and function of the epithelial Na+ channel (ENaC) by ubiquitination. EMBO J. 16, 6325–6336.

[92] Qi, J., Peters, K.W., Liu, C., Wang, J.M., Edinger, R.S., Johnson, J.P., Watkins, S.C., Frizzell, R.A., 1999. Regulation of the amiloride-sensitive epithelial sodium channel by syntaxin 1A. J. Biol. Chem. 274, 30345–30348.

[93] Saxena, S., Quick, M.W., Tousson, A., Oh, Y., Warnock, D.G., 1999. Interaction of syntaxins with the amiloride-sensitive epithelial sodium channel. J. Biol. Chem. 274, 20812–20817.

[94] Zuckerman, J.B., Chen, X., Jacobs, J.D., Hu, B., Kleyman, T.R., Smith, P.R., 1999. Association of the epithelial sodium channel with Apx and alpha- spectrin in A6 renal epithelial cells. J. Biol. Chem. 274, 23286–23295.

[95] Prat, A.G., Holtzman, E.J., Brown, D., Cunningham, C.C., Reisin, I.L., Kleyman, T.R., McLaughlin, M., Jackson, G.R. Jr, Lydon, J., Cantiello, H.F., 1996. Renal epithelial protein (Apx) is an actin cytoskeleton-regulated Na + channel. J. Biol. Chem. 271, 18045–18053.

[96] Berdiev, B.K., Prat, A.G., Cantiello, H.F., Ausiello, D.A., Fuller, C.M., Jovov, B., Benos, D.J., Ismailov, I.I., 1996. Regulation of epithelial sodium channels by short actin filaments. J. Biol. Chem. 271, 17704–17710.

[97] Cantiello, H.F., 1995. Role of the actin cytoskeleton on epithelial Na+ channel regulation. Kidney Int. 48, 970–984.

[98] Frindt, G., Palmer, L.G., 1996. Regulation of Na channels in the rat cortical collecting tubule: effects of cAMP and methyl donors. Am. J. Physiol. 271, F1086–F1092.

[99] Bridges, R.J., Rummel, W., Wollenberg, P., 1984. Effects of vasopressin on electrolyte transport across isolated colon from normal and dexamethasone-treated rats. J. Physiol. (Lond.) 355, 11–23.

[100] Lesage, F., Reyes, R., Fink, M., Duprat, F., Guillemare, E., Lazdunski, M., 1996. Dimerization of TWIK-1 K+ channel subunits via a disulfide bridge. EMBO J. 15, 6400–6407.

[101] Helman, S.I., Cox, T.C., Van Driessche, W., 1983. Hormonal control of apical membrane Na transport in epithelia. Studies with fluctuation analysis. J. Gen. Physiol. 82, 201–220.

[102] Krattenmacher, R., Fischer, H., van Driessche, W., Clauss, W., 1988. Noise analysis of cAMP-stimulated Na current in frog colon. Pflugers Arch. 412, 568–573.

[103] Marunaka, Y., Eaton, D.C., 1991. Effects of vasopressin and cAMP on single amiloride-blockable Na channels. Am. J. Physiol. 260, C1071–C1084.

[104] Berthiaume, Y., Staub, N.C., Matthay, M.A., 1987. Beta-adrenergic agonists increase lung liquid clearance in anesthetized sheep. J. Clin. Investig. 79, 335–343.

[105] Jiang, X., Ingbar, D.H., O'Grady, S.M., 1998. Adrenergic stimulation of Na+ transport across alveolar epithelial cells involves activation of apical Cl-channels. Am. J. Physiol. 275, C1610–C1620.

[106] Fang, X., Fukuda, N., Barbry, P., Sarton, C., Verkman, A.S., Matthay, M., 2003. Fluid absorption from the Distal Airspaces of the lung. J. Gen. Phys. (in press).

[107] Graham, A., Steel, D.M., Alton, E.W.F.W., Geddes, D.M., 1992. Second-messenger regulation of sodium transport in mammalian airway epithelia. J. Physiol. 453, 475–491.

[108] DeCoy, D.L., Snapper, J.R., Breyer, M.D., 1995. Anti sense DNA down-regulates proteins kinase C-epsilon and enhances vasopressin-stimulated Na+ absorption in rabbit cortical collecting duct. J. Clin. Investig. 95, 2749–2756.

[109] Silver, R.B., Frindt, G., Windhager, E.E., Palmer, L.G., 1993. Feedback regulation of Na channels in rat CCT. I. Effects of inhibition of Na pump. Am. J. Physiol. 264, F557–F564.

[110] Civan, M.M., Oler, A., Peterson, Y.K., George, K., O'Brien, T.G., 1991. Ca(2+)-independent form of protein kinase C may regulate Na+ transport across frog skin. J. Membr. Biol. 121, 37–50.

[111] Duc, C., Farman, N., Canessa, C.M., Bonvalet, J.P., Rossier, B.C., 1994. Cell-specific expression of epithelial sodium channel alpha, beta, and gamma subunits in aldosterone-responsive epithelia from the rat: localization by in situ hybridization and immunocytochemistry. J. Cell Biol. 127, 1907–1921.

[112] Reif, M.C., Troutman, S.L., Schafer, J.A., 1986. Sodium transport by rat cortical collecting tubule. Effects of vasopressin and desoxycorticosterone. J. Clin. Investig. 77, 1291–1298.

[113] Frindt, G., Sackin, H., Palmer, L.G., 1990. Whole-cell currents in rat cortical collecting tubule: low-Na diet increases amiloride-sensitive conductance. Am. J. Physiol. 258, F562–F567.

[114] Renard, S., Voilley, N., Bassilana, F., Lazdunski, M., Barbry, P., 1995. Localization and regulation by steroids of the alpha, beta and gamma subunits of the amiloride-sensitive Na+ channel in colon, lung and kidney. Pflugers Arch. 430, 299–307.

[115] Volk, K.A., Sigmund, R.D., Snyder, P.M., McDonald, F.J., Welsh, M.J., Stokes, J.B., 1995. rENaC is the predominant Na+ channel in the apical membrane of the rat renal inner medullary collecting duct. J. Clin. Investig. 96, 2748–2757.

[116] Denault, D.L., Fejes-Toth, G., Naray-Fejes-Toth, A., 1996. Aldosterone regulation of sodium channel gamma-subunit mRNA in cortical collecting duct cells. Am. J. Physiol. 271, C423–C428.

[117] Kemendy, A.E., Kleyman, T.R., Eaton, D.C., 1992. Aldosterone alters the open probability of amiloride-blockable sodium channels in A6 epithelia. Am. J. Physiol. 263, C825–C837.

[118] Pacha, J., Frindt, G., Antonian, L., Silver, R.B., Palmer, L.G., 1993. Regulation of Na channels of the rat cortical collecting tubule by aldosterone. J. Gen. Physiol. 102, 25–42.

[119] Chen, S.Y., Bhargava, A., Mastroberardino, L., Meijer, O.C., Wang, J., Buse, P., Firestone, G.L., Verrey, F., Pearce, D., 1999. Epithelial sodium channel regulated by aldosterone-induced protein sgk. Proc. Natl Acad. Sci. USA 96, 2514–2519.

[120] Harvey, B.J., Thomas, S.R., Ehrenfeld, J., 1988. Intracellular pH controls cell membrane Na+ and K+ conductances and transport in frog skin epithelium. J. Gen. Physiol. 92, 767–791.

[121] Sariban, S.S., Burg, M., Wiesmann, W.P., Chiang, P.K., Johnson, J.P., 1984. Methylation increases sodium transport into A6 apical membrane vesicles: possible mode of aldosterone action. Science 225, 745–746.

[122] Petzel, D., Ganz, M.B., Nestler, E.J., Lewis, J.J., Goldenring, J., Akcicek, F., Hayslett, J.P., 1992. Correlates of aldosterone-induced increases in Ca_i^{2+} and Isc suggest that Ca_i^{2+} is the second messenger for stimulation of apical membrane conductance. J. Clin. Investig. 89, 150–156.

[123] Palmer, L.G., Li, J.H.-Y., Lindemann, B., Edelman, I.S., 1982. Aldosterone control of the density of sodium channels in the toad urinary bladder. J. Membr. Biol. 64, 91–102.

[124] Djelidi, S., Fay, M., Cluzeaud, F., Escoubet, B., Eugene, E., Capurro, C., Bonvalet, J.P., Farman, N., Blot-Chabaud, M., 1997. Transcriptional regulation of sodium transport by vasopressin in renal cells. J. Biol. Chem. 272, 32919–32924.

[125] Shimkets, R.A., Warnock, D.G., Bositis, C.M., Nelson-Williams, C., Hansson, J.H., Schambelan, M., Gill, J.R.J., Ulick, S., Milora, R.V., Finding, J.W., Canessa, C.M., Rossier, B.C., Lifton, R.P., 1994. Liddle's syndrome: heritable human hypertension caused by mutations in the beta chain of the epithelial sodium channel. Cell 79, 407–414.

[126] Hansson, J.H., Schild, L., Lu, Y., Wilson, T.A., Gautschi, I., Shimkets, R., Nelson, W.C., Rossier, B.C., Lifton, R.P., 1995. A de novo missense mutation of the beta subunit of the epithelial sodium channel causes hypertension and Liddle syndrome, identifying a proline-rich segment critical for regulation of channel activity. Proc. Natl Acad. Sci. USA 92, 11495–11499.

[127] Hansson, J.H., Nelson, W.C., Suzuki, H., Schild, L., Shimkets, R., Lu, Y., Canessa, C., Iwasaki, T., Rossier, B., Lifton, R.P., 1995. Hypertension caused by a truncated epithelial sodium channel gamma subunit: genetic heterogeneity of Liddle syndrome. Nat. Genet. 11, 76–82.

[128] Snyder, P.M., Price, M.P., McDonald, F.J., Adams, C.M., Volk, K.A., Zeiher, B.G., Stokes, J.B., Welsh, M J , 1995. Mechanism by which Liddle's syndrome mutations increase activity of a human epithelial Na+ channel. Cell 83, 969–978.

[129] Schild, L., Canessa, C.M., Shimkets, R.A., Gautschi, I., Lifton, R.P., Rossier, B.C., 1995. A mutation in the epithelial sodium channel causing Liddle disease increases channel activity in the *Xenopus laevis* oocyte expression system. Proc. Natl Acad. Sci. USA 92, 5699–5703.

[130] Schild, L., Lu, Y., Gautschi, I., Schneeberger, E., Lifton, R.P., Rossier, B.C., 1996. Identification of a PY motif in the epithelial Na channel subunits as a target sequence for mutations causing channel activation found in Liddle syndrome. EMBO J. 15, 2381–2387.

[131] Wong, Z.Y., Stebbing, M., Ellis, J.A., Lamantia, A., Harrap, S.B., 1999. Genetic linkage of beta and gamma subunits of epithelial sodium channel to systolic blood pressure. Lancet 353, 1222–1225 (see comments).

[132] Persu, A., Barbry, P., Bassilana, F., Houot, A.M., Mengual, R., Lazdunski, M., Corvol, P., Jeunemaitre, X., 1998. Genetic analysis of the beta subunit of the epithelial Na+ channel in essential hypertension. Hypertension 32, 129–137.

[133] Baker, E.H., Dong, Y.B., Sagnella, G.A., Rothwell, M., Onipinla, A.K., Markandu, N.D., Cappuccio, F.P., Cook, D.G., Persu, A., Corvol, P., Jeunemaitre, X., Carter, N.D., MacGregor, G.A., 1998. Association of hypertension with T594M mutation in beta subunit of epithelial sodium channels in black people resident in London. Lancet 351, 1388–1392.

[134] Cui, Y.S.Y., Rutkowski, M., Reif, M., Menon, A.G., Pun, R.Y., 1997. Loss of protein kinase C inhibition in the beta-T594M variant of the amiloride-sensitive Na+ channel. Proc. Natl Acad. Sci. USA 94, 9962–9966.

[135] Chang, S.S., Grunder, S., Hanukoglu, A., Rosler, A., Mathew, P.M., Hanukoglu, I., Schild, L., Lu, Y., Shimkets, R.A., Nelson, W.C., Rossier, B.C., Lifton, R.P., 1996. Mutations in subunits of the epithelial

sodium channel cause salt wasting with hyperkalaemic acidosis, pseudohypoaldosteronism type 1. Nat. Genet. 12, 248–253.

[136] Strautnieks, S.S., Thompson, R.J., Gardiner, R.M., Chung, E., 1996. A novel splice-site mutation in the gamma subunit of the epithelial sodium channel gene in three pseudohypoaldosteronism type 1 families. Nat. Genet. 13, 248–250.

[137] Schaedel, C., Marthinsen, L., Kristoffersson, A.C., Kornfalt, R., Nilsson, K.O., Orlenius, B., Holmberg, L., 1999. Lung symptoms in pseudohypoaldosteronism type 1 are associated with deficiency of the alpha-subunit of the epithelial sodium channel. J. Pediatr. 135, 739–745.

[138] Sandle, G.I., Binder, H.J., 1987. Corticosteroids and intestinal ion transport. Gastroenterology 93, 188–196.

[139] Bastl, C.P., Hayslett, J.P., 1992. The cellular action of aldosterone in target epithelia. Kidney Int. 42, 250–264 (editorial).

[140] Firsov, D., Schild, L., Gautschi, I., Merillat, A.M., Schneeberger, E., Rossier, B.C., 1996. Cell surface expression of the epithelial Na channel and a mutant causing Liddle syndrome: a quantitative approach. Proc. Natl Acad. Sci. USA 93, 15370–15375.

[141] Asher, C., Garty, H., 1988. Aldosterone increases the apical Na+ permeability of toad bladder by two different mechanisms. Proc. Natl Acad. Sci. USA 85, 7413–7417.

[142] O'Brodovich, H., 1991. Epithelial ion transport in the fetal and perinatal lung. Am. J. Physiol. 261, C555–C564.

[143] Strang, L.B., 1991. Fetal lung liquid: secretion and reabsorption. Physiol. Rev. 71, 991–1016.

[144] Gaillard, D., Hinnrasky, J., Coscoy, S., Hofman, P., Matthay, M.A., Puchelle, E., Barbry, P., 2000. Early expression of beta- and gamma-subunits of epithelial sodium channel during human airway development. Am. J. Physiol. Lung Cell Mol. Physiol. 278, L177–L184.

[145] Johnson, M.D., Widdicombe, J.H., Allen, L., Barbry, P., Dobbs, L.G., 2003. Alveolar epithelial type I cells contain transport proteins and transport sodium, supporting an active role for type I cells in regulation of lung liquid homeostasis. Proc. Natl Acad. Sci. USA 99(4), 1966–1971.

[146] O'Brodovich, H., Canessa, C., Ueda, J., Rafii, B., Rossier, B.C., Edelson, J., 1993. Expression of the epithelial Na+ channel in the developing rat lung. Am. J. Physiol. 265, C491–C496.

[147] Champigny, G., Voilley, N., Lingueglia, E., Friend, V., Barbry, P., Lazdunski, M., 1994. Regulation of expression of the lung amiloride-sensitive Na+ channel by steroid hormones. EMBO J. 13, 2177–2181.

[148] Tchepichev, S., Ueda, J., Canessa, C., Rossier, B.C., O'Brodovich, H., 1995. Lung epithelial Na channel subunits are differentially regulated during development and by steroids. Am. J. Physiol. 269, C805–C812.

[149] Hummler, E., Barker, P., Gatzy, J., Beermann, F., Verdumo, C., Schmidt, A., Boucher, R., Rossier, B.C., 1996. Early death due to defective neonatal lung liquid clearance in alpha- ENaC-deficient mice. Nat. Genet. 12, 325–328.

[150] Xu, W., Leung, S., Wright, J., Guggino, S.E., 1999. Expression of cyclic nucleotide-gated cation channels in airway epithelial cells. J. Membr. Biol. 171, 117–126.

[151] Junor, R.W., Benjamin, A.R., Alexandrou, D., Guggino, S.E., Walters, D.V., 1999. A novel role for cyclic nucleotide-gated cation channels in lung liquid homeostasis in sheep. J. Physiol. (Lond.) 520(Pt 1), 255–260.

[152] Malagon-Rogers, M., 1999. A patient with pseudohypoaldosteronism type 1 and respiratory distress syndrome. Pediatr. Nephrol. 13, 484–486.

[153] Boucher, R.C., Stutts, M.J., Knowles, M.R., Cantley, L., Gatzy, J.T., 1986. Na+ transport in cystic fibrosis respiratory epithelia. Abnormal basal rate and response to adenylate cyclase activation. J. Clin. Investig. 78, 1245–1252.

[154] Stutts, M.J., Canessa, C.M., Olsen, J.C., Hamrick, M., Cohn, J.A., Rossier, B.C., Boucher, R.C., 1995. CFTR as a cAMP-dependent regulator of sodium channels. Science 269, 847–850 (see comments).

[155] Mall, M., Hipper, A., Greger, R., Kunzelmann, K., 1996. Wild type but not deltaF508 CFTR inhibits Na + conductance when coexpressed in *Xenopus* oocytes. FEBS Lett. 381, 47–52.

[156] Mall, M., Kunzelmann, K., Hipper, A., Busch, A.E., Greger, R., 1996. cAMP stimulation of CFTR-expressing *Xenopus* oocytes activates a chromanol-inhibitable K+ conductance. Pflugers Arch. 432, 516–522.

[157] Quinton, P.M., 1990. Cystic fibrosis: a disease in electrolyte transport. FASEB J. 4, 2709–2717.

[158] Barbry, P., Lazdunski, M., 1996. Structure and regulation of the amiloride-sensitive epithelial Na+ channel. In: Narahashi, T. (Ed.), Ion Channels. Plenum Press, New York, pp. 115–167.

[159] Kunzelmann, K., Mall, M., Briel, M., Hipper, A., Nitschke, R., Ricken, S., Greger, R., 1997. The cystic fibrosis transmembrane conductance regulator attenuates the endogenous Ca2+ activated Cl- conductance of *Xenopus* oocytes. Pflugers Arch. 435, 178–181.

[160] Ismailov, I.I., Berdiev, B.K., Shlyonsky, V.G., Fuller, C.M., Prat, A.G., Jovov, B., Cantiello, H.F., Ausiello, D.A., Benos, D.J., 1997. Role of actin in regulation of epithelial sodium channels by CFTR. Am. J. Physiol. 272, C1077–C1086.

[161] Bradbury, N.A., Jilling, T., Berta, G., Sorscher, E.J., Bridges, R.J., Kirk, K.L., 1992. Regulation of plasma membrane recycling by CFTR. Science 256, 530–532 (see comments).

[162] Dinudom, A., Komwatana, P., Young, J.A., Cook, D.I., 1995. A forskolin-activated Cl⁻ current in mouse mandibular duct cells. Am. J. Physiol. 268, G806–G812.

[163] Komwatana, P., Dinudom, A., Young, J.A., Cook, D.I., 1996. Cytosolic Na+ controls and epithelial Na+ channel via the Go guanine nucleotide-binding regulatory protein. Proc. Natl Acad. Sci. USA 93, 8107–8111.

Vanilloid receptor 1 (VR1): an integrator of noxious and inflammatory stimuli

Stuart Bevan* and Alison J. Reeve

Novartis Institute for Medical Sciences, 5 Gower Place, London WC1E 6BN, UK
*Correspondence address: Tel.: +44-207-333-2103; fax: +44-207-383-5109
E-mail: stuart.bevan@pharma.novartis.com

1. Introduction

The pungent ingredient of chili peppers, capsaicin, has long been used as a tool in sensory neuron biology. This compound activates a subset of primary afferent neurons, the polymodal nociceptors, which can respond to noxious heat, noxious mechanical pressures and to noxious chemicals (i.e. capsaicin). Capsaicin-sensitive neurons innervate all major organs and tissues and are thought to play a role in both normal and patho-physiological functions.

The relationships between the various modalities sensed by polymodal nociceptors and the underlying transduction mechanisms were unknown until the cloning of a capsaicin sensitive receptor (vanilloid receptor 1, VR1) by Caterina et al. [1]. It is now clear that the expression of VR1 confers on cells the property of both heat and capsaicin sensitivity as well as the ability to respond to some mediators produced during inflammation. Emerging evidence also indicates that other membrane receptors can be involved in the transduction of noxious thermal stimuli. However, the link between noxious mechanical transduction and responsiveness to heat and noxious chemicals remains unclear. VR1 has a pivotal role in nociception as it integrates the response to thermal stimuli and a range of inflammatory mediators.

2. The cloning and initial characterization of VR1

The receptor conferring sensitivity to capsaicin was discovered by expression cloning from a rat dorsal root ganglia (DRG) cDNA library, using the property that capsaicin-activated channels in DRG neurons have a high permeability to calcium ions (see, e.g. Wood et al., [2]). The resultant cDNA from rat encodes a protein with 838 amino acids with a predicted molecular mass (M_r) of 95 kDa. Hydrophobicity analysis suggests a

The literature covered in this chapter is up to and including April 2001.

Advances in Molecular and Cell Biology, Vol. 32, pages 331–350
ISSN: 1569-2558 / DOI: 10.1016/S1569-2558(03)32014-4

topology of six transmembrane domains with a putative pore region between transmembrane domains 5 and 6. The long (432 amino acids) sequence of the cytoplasmic N-terminus region contains three ankyrin repeat domains. Three consensus protein kinase A (PKA) and up to 20 consensus protein kinase C (PKC) phosphorylation sites are predicted on the intracellular domains. The predicted topology of VR1 is shown in Fig. 1.

The human homologue of VR1 has now been cloned and characterized [3,4]. This shows 85–86% identity and 92% similarity to rat VR1 at the amino acid level, with the ankyrin repeat and transmembrane domains conserved. The N-terminus region shows the lowest homology with increasing divergence towards the N-terminus, while the C-terminus shows a greater degree of conservation. Differences in the amino acid sequence have been shown to result in divergent pharmacological properties of rat and human VR1 (Ref. [4], see below), although the residues responsible for these changes remain unknown.

VR1 expressed in either *Xenopus* oocytes or in mammalian cells such as HEK293 or CHO cells confers sensitivity to capsaicin and other structurally related molecules, which had been postulated, on the basis of their pharmacological properties, to act via the same (or similar) receptors in sensory neurons. Thus VR1 can be activated by resiniferatoxin (RTX) [1] and structurally related phorbol ester derivatives including phorbol 12,13-didecanoate 20-homovanillate and phorbol 12-phenylacetate 13 acetate 20-homovanillate [4,5] as well as the natural and synthetic capsaicin congeners, zingerone (from ginger, [6]) and olvanil [5]. Furthermore the responses of VR1 to capsaicin can be inhibited by capsazepine, a competitive antagonist of capsaicin in DRG neurons [7], and ruthenium red, a non-competitive antagonist [8].

Capsaicin acting at the VR1 receptor channel opens a cation-selective channel that is permeable to both monovalent (e.g. Na^+, K^+) and divalent (Ca^{2+}, Mg^{2+}) cations. Although there is some variation in the calculated relative permeabilities, the majority of studies indicate that the permeability of divalent cations is greater than for monovalent

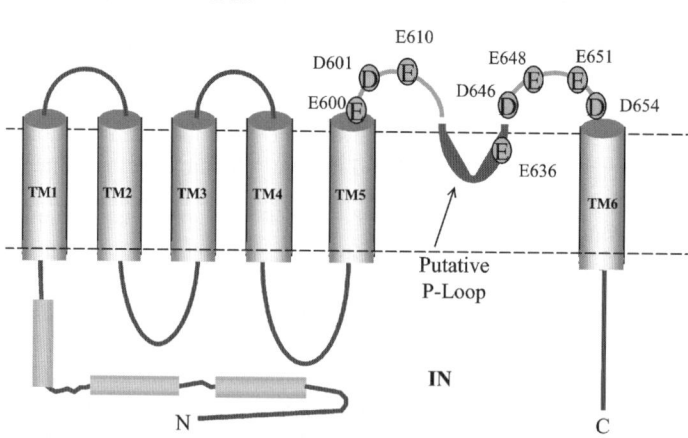

Fig. 1. Schematic diagram illustrating the predicted topology of VR1. The three ankyrin repeat segments are shown as cylinders and the putative transmembrane domains numbered TM1–TM6. Some amino acids in the P-loop targeted in site-directed mutagenesis studies are indicated. (*For a colored version of this figure, see plate section, page 467.*)

cations, with $P_{Ca}/P_{Na} = 9.6$ and $P_{Mg}/P_{Na} = 4.99$ and a permeability sequence $Ca^{2+} > Mg^{2+} > Na^+ \approx K^+ \approx Cs^+$[1]. This permeability sequence is essentially identical to that seen for native capsaicin receptors in DRG neurons [9–11].

The composition of the native channel is unknown although four or more "VR1-like" subunits are predicted by analogy to other ion channels of similar transmembrane topology (e.g. K^+ channels) where the stoichiometry has been investigated. This estimate fits reasonably well with values of about 270 kDa for the receptor/channel complex derived from irradiation inactivation studies [12]. Examination of the log(concentration)–response curves for capsaicin activation of cloned VR1 yields EC_{50} values that range from about 30 nM to more than 1 µM [3–5,13,14] with the estimates higher in *Xenopus* oocytes than in mammalian cell expression systems. The reason for these differences is unclear but may relate to variation in channel regulation in the different cell types (see below). In general, EC_{50} values obtained for capsaicin in DRG neurons (about 200–300 nM [9,15]) are close to the estimates for VR1 in mammalian cells, although in a comparative study Shin et al. [16] found that the EC_{50} value for capasicin activation of single channels was 3.5-fold less for cloned VR1 than for native channels. The slope factor of the log (concentration)–response curves for both cloned and native receptors is about two [1,10,15,17], which suggests that activation of VR1 requires the binding of at least two molecules of capsaicin.

3. VR-1 related channels

The structural and functional features of VR1 place it in a family together with members of the Ca^{2+} permeable transient receptor potential protein (TRP) channels. These TRP channels have been subdivided into three sub-families [18]. VR1 has been grouped in one sub-family with some TRP-like channels that can be activated by physical and chemical stimuli. Other members include the osmosensitive channels OSM-9 from *Caenorhabditis elegans* [19], and OTPRC4 [20], VR-OAC [21] and Trp12 [22] from mammalian tissues as well as the high temperature, heat-activated channel VRL-1 [23].

Two splice variants of VR1 have been identified. One is an N-terminal splice variant (VR.5'sv) which varies from VR1 in the N-terminal region and has only one ankyrin repeat domain, but has an identical amino acid sequence to VR1 in the transmembrane and C-terminal domains [24]. Despite the identity with VR1 in these regions, heterologously expressed VR.5'sv is unresponsive to capsaicin, RTX, low pH solutions or noxious thermal stimuli (50 °C). The second splice variant has the same start site as VR.5'sv and shares the transmembrane sequence with VR1, but contains a divergent C-terminal region. This channel shows mechanosensitivity and is inhibited by stretch, and is not activated by either noxious heat or capsaicin. Unlike VR1, this stretch inactivated channel (SIC) has a relatively low relative permeability to calcium ions ($P_{Ca}/P_{Na} = 0.24$, [25]).

A further VR1 homologue cloned from rat brain, VRL-1, shows 49% identity and 66% similarity to rat VR1 at the amino acid level. VRL-1 is insensitive to either capsaicin or RTX but is thermosensitive with a higher threshold temperature for activation (52 °C) than VR1 [23]. Another homologue, VRL-2, has also been identified but unlike VR1 and VRL-1, is not located in the cell bodies of DRG or sensory nerve fibres but is found in epithelia in the airways, distal tubules of the kidney and sympathetic and parasympathetic neurons [26].

4. The expression of VR-1 in nervous tissues

In situ hydridisation, Northern blot studies and immunostaining with VR1 specific antibodies have shown that VR1 mRNA and protein are expressed in small and medium diameter primary afferent neurons of the dorsal root, trigeminal and nodose ganglia [1,27–30]. VR1 is co-expressed in both peptide (substance P and CGRP) and non-peptide containing sensory neurons that terminate in laminae I and II of the dorsal horn of the spinal cord. In contrast, VR1 expression has not been found in sympathetic neurons. The pattern of VR1 expression in this sub-population of cells corresponds well with the hypothesis that VR1 occurs exclusively in nociceptive afferent neurons.

Studies with more sensitive PCR methods have described low levels of VR1 mRNA in regions of the central nervous system, with the highest levels detected in the hypothalamus and cerebellum [31]. Further studies with VR-1 specific antibodies have supported these results and shown that VR-1 immunostaining can be found throughout the whole neuroaxis including layers 3 and 5 of the cerebral cortex, hippocampus, central amygdala, medial and lateral habenula, striatum, hypothalamus, thalamic nuclei, substantia nigra, locus coeruleus, cerebellum and inferior olive [32]. Expression of VR1 in the hypothalamus may correlate with the ability of hypothalamic injections of capsaicin to influence body temperature [33]. At present the relevance of VR-1 immunostaining throughout the central nervous system is unclear, although it raises questions about roles for this receptor in functions other than pain perception. In this regard, the discovery of VR-1 mRNA and immunostaining in kidney [34] and functional responses to capsaicin and RTX in mast cells [35] suggests that VR-1 or closely related channels may have roles outside the nervous system.

5. Control of VR-1 expression and inflammation

VR-1 expression is controlled by the provision of growth factors, notably by the neurotrophins, nerve growth factor (NGF), glial derived growth factor (GDNF) and brain derived neurotrophic factor (BDNF). The regulation depends on the presence of the appropriate neurotrophin receptor in populations of sensory neurons. DRG neurons express receptors for NGF (trkA) and GDNF (ret), while nodose ganglion neurons fail to express these receptors but do express the receptor for BDNF (trkB). Capsaicin sensitivity, RTX binding and VR-1 immunoreactivity are lost reversibly in almost all adult rat DRG neurons when they are depleted of NGF and GDNF in culture [36–39]. The changes occur with a half time of about 2–3 days, which is consistent with transcriptional control of VR1 expression. Capsaicin sensitivity of DRG neurons is also lost in vivo following administration of a trkA–IgG fusion molecule which neutralizes NGF [40]. Similarly, the capsaicin sensitivity of nodose ganglion neurons depends on the presence of BDNF [41].

The regulation of VR-1 expression by changes in neurotrophin levels are likely to be significant in inflammatory conditions as the production of NGF is elevated in experimental inflammatory conditions [42,43] and in the synovial fluid of patients with arthritis [44,45]. This idea is strengthened by the finding that the increases in the capsaicin sensitivity of DRG neurons seen after peripheral inflammation can be blocked by antibodies to NGF [46].

In contrast to the upregulation of VR1 expression seen with neurotrophins, axotomy or spinal nerve section leads to a decrease in VR1 expression in dorsal root ganglia [30] possibly because the cell bodies are deprived of the retrograde supply of neurotrophins from the target tissues.

6. Heat activation is mediated by VR1 and VRL-1

Noxious heat stimuli ($>40\ ^\circ$C) activate two types of responses in mammalian primary afferent neurons characterized by different threshold temperatures. This can be seen with recordings from intact nerve preparations (e.g. Ref. [47]) and with isolated DRG neurons [48]. One response has a threshold of about $42-45\ ^\circ$C (Fig. 2B) and the other about $49-51\ ^\circ$C.

6.1. VR1 encodes a low threshold temperature response

A strong correlation between capsaicin sensitivity and activation at the lower threshold temperature in both intact sensory fibres [47] and isolated DRG neurons [49,50], suggested that responses to the two modalities were linked. The molecular relationship between heat and capsaicin sensitivity could not, however, be explored until the properties of VR1 were studied in heterologous expression systems and in VR1 "knockout" mice. One remarkable

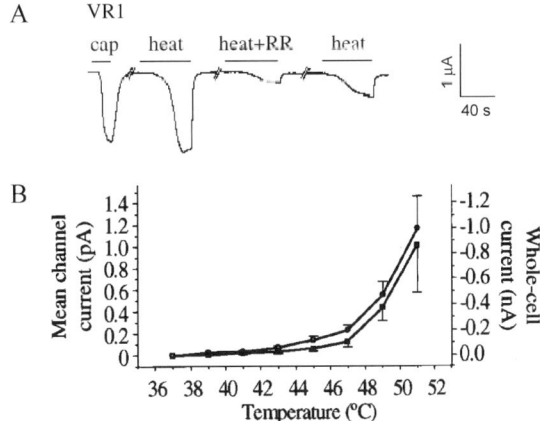

Fig. 2. A. Noxious heat-evoked currents in a VR1 expressing cell blocked by ruthenium red (RR) (Reproduced with permission from Ref. [1]) B. Temperature–response relationships for whole cell and single channel currents in DRG neurons (Reproduced with permission from Ref. [11]).

finding of Caterina et al. [1] was that expression of VR1 in *Xenopus* oocytes or mammalian cells conferred sensitivity to noxious temperatures (Fig. 2A). Untransfected cells showed little or no changes in evoked membrane currents or intracellular calcium concentration when the temperature was raised above 40 °C. In contrast, a heat challenge evoked a robust response in rat or human VR1 expressing cells, with a threshold temperature of about 42–43 °C [4,27,51]. Studies in VR1-null mice yielded results consistent with a strong role for VR1 in thermoreception [52,53]. DRG neurons from VR1-null mice show a loss of the lower threshold heat response and VR1-null animals also display a deficit in noxious thermal responses, although they remain thermosensitive.

The properties of heat- and capsaicin-evoked responses are remarkably similar for native channels in DRG neurons and heterologously expressed VR1. The whole cell currents show outward rectification due to time- and voltage-dependent increases in open channel probability at positive membrane potentials ([17,54] and the single channel properties are very similar [11]). Both heat and capsaicin responses can be blocked by the capsaicin receptor antagonist ruthenium red (see Fig. 2A). Another capsaicin receptor antagonist, capsazepine [7], shows interesting properties. Although equally active at both rat and human VR1 as a capsaicin antagonist, it shows a species dependent block of thermal responses. Capsazepine blocks the heat responses of human VR1 at low concentrations, but has only a weak effect on heat responses of rat VR1 [4].

Ion substitution experiments indicate that like capsaicin-activated channels in DRG neurons or VR1 expressing cells, the conductance underlying the heat-evoked current discriminates relatively poorly between sodium, caesium and calcium ions [11,49,55]. However, studies on DRG neurons [11] and VR1 expressing cells [27] showed that the relative calcium permeability of the heat-evoked conductance was about half that of the capsaicin response. This may indicate a variation in the conformation of VR1 with different modes of activation.

The precise correspondence between capsaicin-sensitive channels and noxious heat-activated channels has been challenged by several findings. Although Nagy and Rang [48] found that all capsaicin-sensitive DRG neurons studied had a low noxious heat threshold (43 °C), occasional capsaicin-sensitive, heat-insensitive DRG neurons have been described [49,50]. Furthermore, Nagy and Rang [11] found a divergence in the capsaicin- and heat-evoked responses at the single channel level in membrane patches from rat DRG neurons. A relatively low percentage (14%) of responsive channels were dually sensitive to the two stimuli (Fig. 3A,B), while 36% responded only to capsaicin and 50% were thermoresponsive but not activated by capsaicin. One possibility is that distinct channels are responsible for the currents evoked by either heat or capsaicin. Alternatively, both channel types may be VR1 gene products differentiated by nucleic acid splicing, post-translational modification or association with other membrane proteins [11]. In this respect it is perhaps noteworthy that Kirschstein et al., [49] reported that some heat-insensitive neurons gained thermal sensitivity after a response to capsaicin.

One interesting additional divergence between capsaicin and thermal sensitivity is found in chicken DRG neurons. These are thermally sensitive, with a similar temperature threshold to VR1 but do not show capsaicin sensitivity [2,56,57]. It seems likely that VR1 is responsible for the currents evoked by low noxious temperatures as they are blocked by

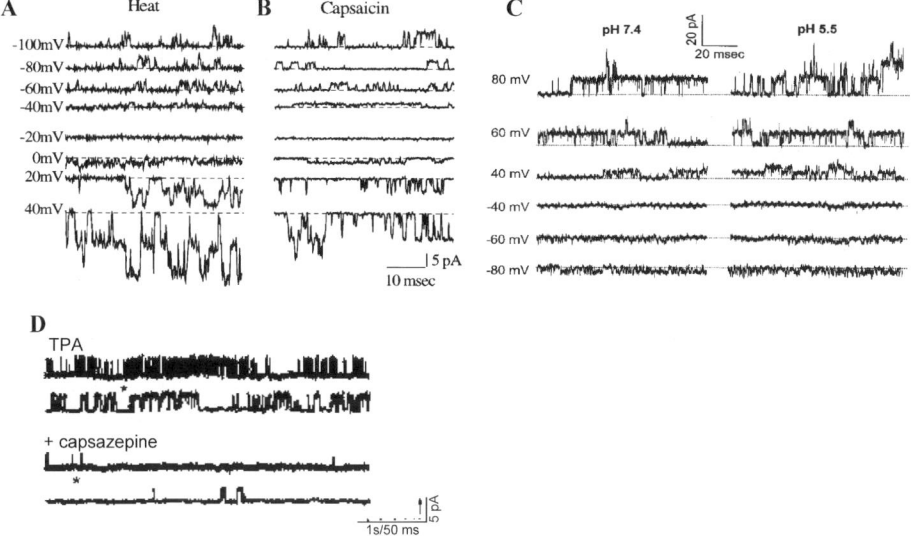

Fig. 3. Single channel currents evoked by A. Heat and B. Capsaicin in a membrane patch from a DRG neuron (Reproduced with permission from Ref. [11]). C. Single channel currents evoked by low pH and capsaicin in a DRG membrane patch (Reproduced with permission from Ref. [84]). D. TPA-evoked currents in VR1 transfected cells (Reproduced with permission from Ref. [72]).

capsazepine [56]. This result suggests that the chicken VR1 homologue lacks the sequence responsible for capsaicin sensitivity.

6.2. Lessons from mice lacking VR1

Studies with VR1 null mice [52,53] have indicated that VR-1 is not the only mechanism responsible for thermoreception of moderate heat stimuli. Although low threshold noxious heat responses are absent in DRG neurons from VR1 null mice, recordings from primary afferent fibres in a skin nerve preparation revealed responses with the same temperature threshold (40–41 °C) as wild-type mice [52]. Both studies also showed that the null mice display an unexpectedly small deficit in their response to a noxious heat stimulus in the temperature range where VR1 would be expected to play a role. Paradoxically the response to high noxious heat stimuli (>52 °C) was compromised in VR1 null mice for reasons that are not understood.

In contrast to the results with thermal stimuli in naive animals, the role of VR1 in thermal hyperalgesia was demonstrated in both VR1 null mice studies. Thermal thresholds normally decrease after an inflammatory insult and this reduction was completely lost in the VR1 knockout animals.

6.3. VRL-1 and high threshold heat responses

Medium or large diameter DRG neurons can also respond to a noxious temperature challenge. Typically these cells are capsaicin insensitive and show a very high temperature threshold (51 °C) [11]. High temperatures also activate larger diameter, capsaicin-insensitive primary afferent nerves (see, e.g. Ref. [47]). It seems likely that a molecule, such as VRL-1, is responsible for such high threshold responses, since VRL-1 shows similar temperature characteristics and is expressed in a subset of medium and large diameter DRG neurons [23]. There are, however, no direct data to support this assumption.

7. Inflammatory mediators and VR1

Although the actions of capsaicin have been known for a long time, the identity of any endogenous chemical agonist at VR1 has remained elusive. The finding that hyperalgesic neuronal responses produced by inflammation were attenuated by capsazepine, suggested that one or more mediators were activating VR1 [58]. Now it is clear that the endogenous cannabinoid, anandamide and several products of the lipoxygenase pathway can activate VR1. Furthermore other important inflammatory mediators such as bradykinin may exert an excitatory action by activating VR1 via PKC.

7.1. Lipid agents activators of VR1

The first indication of lipid activators of VR1 was the discovery that the endogenous cannabinoid (CB) receptor agonist, anandamide, could evoke the release of CGRP from isolated vascular preparations in a capsazepine-sensitive manner [59]. In addition, anandamide activated native and cloned VR1 channels [59–61] whereas the synthetic CB1 and CB2 receptor agonists, HU 210, WIN 55,212-2 and CP 55,940 failed to mimic the action of anandamide [59]. Another natural CB agonist, 2-arachidonylglycerol, was also weakly active as an agonist at rat VR1, while related endogenous fatty acid compounds (2-arachidonic acid, palmitoylethanolamide) proved inactive in the same study [59]. Subsequently anandamide, methanandamide and palmitoylethanolamine were shown to act as agonists at human VR1 channels [60,62]. The responses of VR1 to anandamide at both rat and human VR1 were blocked by capsazepine, although it has not been shown whether or not the inhibition is competitive.

Studies with an ionizable analogue of capsaicin indicated that the binding site for capsaicin is intracellular [63]. Such a location is consistent with the "membrane delimited" binding site for fatty acids reported for the structurally related TRPC3 and TRPC6 channels [64] and suggested that the search for endogenous VR1 activators should include intracellular mediators. Subsequently Hwang and colleagues [65] discovered that certain lipoxygenase products had the ability to open capaicin-activated channels. 12- and 15-(s) Hydroperoxyeicosatetraenoic acids (HPETE), 5- and 15-(s)-hydroxyeicosatetraenoic acids and leukotriene B_4 evoked currents in isolated inside-out membrane patches from DRG neurones and VR1 transfected cells. These currents were blocked by 10 μM capsazepine, and the capsaicin- and 12(s) HPETE-evoked responses showed similar current–voltage relationships. Recent studies have also demonstrated that compounds that inhibit the transport of anandamide across the cell membrane can block the effects of externally applied anandamide, but not of capsaicin, on VR1. These results suggest a translocation across the plasma membrane and intracellular site of action for anandamide, but a passive membrane diffusion for capsaicin.

Despite the evidence for VR1 activation by the above lipid mediators, the (patho) physiological relevance of the findings has to be established, This is particularly true for anandamide, where there is considerable debate about the amounts produced under normal and inflammatory conditions and the concentrations required for VR1 and CB1 receptor activation (see Refs. [66–69]). This question is of physiological relevance as the effects of anandamide at CB1 receptor will antagonize the neuronal responses to VR1 activation.

7.2. Bradykinin, PKC and VR1

Bradykinin is a potent inflammatory mediator that is able to both activate and sensitise nociceptive sensory neurons, in part through activation of PKC [55,70,71]. Phorbol esters that activate PKC induce an excitatory current in DRG neurons and a recent report showed that PKC activation by 12-*O*-tetradecanoylphorbol-13-acetate (TPA) induced a current in *Xenopus* oocytes expressing VR1 but not in control oocytes. This response was blocked by the VR1 antagonists, ruthenium red and capsazepine, and also by the selective PKC inhibitor, bisindolylmaleimide (BIM). TPA also evoked capsazepine-sensitive responses in DRG neurons (Fig. 3D, [72]). The potential link between bradykinin and VR1 was shown when bradykinin was applied to the outside surface of DRG membrane patches. This evoked single channel currents with current–voltage relationships that were identical to those for both capsaicin and TPA-evoked currents. The bradykinin-induced currents were greatly enhanced by the application of PKC-ε to the intracellular surface [72]. Overall, the characteristics of the bradykinin-evoked currents are consistent with activation of VR1, although surprisingly the authors did not report on the capsazepine sensitivity of the response.

8. Low pH solutions activate VR1

A reduction in extracellular pH is seen in ischaemic and inflammatory conditions and can lead to primary afferent excitation [73–75]. Reducing the pH evokes two types of proton-activated currents in primary afferent nerves. The first of these is a transient current, which activates rapidly when exposed to low pH solutions and then typically inactivates within a few seconds. This response is found in many neuronal cell types including brain, spinal cord and sympathetic neurons, as well as nociceptive and non-nociceptive afferent neurons (see Ref. [73]). Such transient responses are unlikely to underlie the continuous pain measured in psychophysical experiments in response to prolonged application of acid solutions to the skin [76]. These transient currents are probably due to activation of ASIC-like channels found in DRG neurons [77] and have been proposed to play a role in the sensation of cardiac pain in ischaemic conditions [78].

The second type of low-pH activated current is probably responsible for prolonged nociceptor activation by acidic solutions in most conditions. It is activated at room temperature when the external pH falls below about pH 6.5 and is sustained for minutes in the continued presence of low pH solutions. This current shares many characteristics with the current evoked by capsaicin in DRG neurons [79]. These similarities include co-expression in a sub-population of DRG neurons, co-regulation of expression by NGF, similar relative permeabilities to monovalent cations and similarities in single channel properties (Fig. 3C, [80]). One difference is that the relative permeability of calcium ions appears to be lower with low pH as the agonist than with capsaicin stimulation [81]. This slow response to a low pH challenge is completely lost in DRG neurons from VR1 null mice [52,53], consistent with a role of VR1 in the response to acidification.

Rat and human VR1 channels expressed in both mammalian cells and *Xenopus* oocytes respond to challenges with acidic solutions, although the pH required for half maximum activation is lower in *Xenopus* oocytes (pH 4.8–4.9, [13,34]) than in mammalian expression systems (EC_{50} pH 5.4–6.4, [4,13,27]). The values in mammalian cells compare well with the EC_{50} value estimated for DRG neurons (pH 5.8, [79]). The slope factors estimated from pH–response curves are close to 2 for both cloned [13,14] and native [79] channels, which suggests that the binding of 2 protons is needed to activate VR1.

9. Interactions between stimuli

In addition to their ability to directly gate channels, low pH solutions also facilitate the response of DRG neurons to capsaicin [82–84] such that low concentrations of capsaicin can have a large agonist effect (Fig. 4B). The augmentation results from an increase in the probability of channel opening at these capsaicin concentrations with little or no change in single channel conductance [84,85]. This is due to a shift in the capsaicin log (concentration)–response relationship to lower concentrations without any increase in the maximally attainable response (Fig. 4A). The relationship between pH and the EC_{50} concentrations for capsaicin is sigmoidal with a mid-point at pH7.4 [84]. This means that the chemosensitivity is finely tuned for small shifts in pH around the normal physiological values. A similar facilitation of capsaicin responses by low pH solutions is also found for

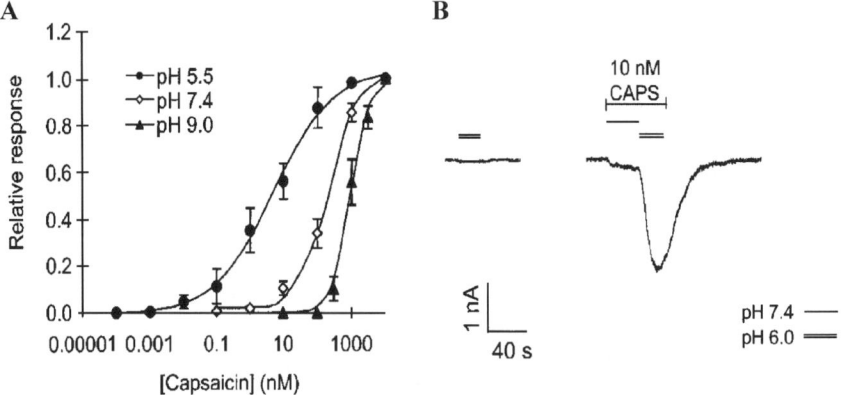

Fig. 4. A. Log(concentration)–response curve for capsaicin-evoked whole cell currents in DRG neurons. B. Augmentation of current evoked by 10 nM capsaicin with low pH. (Reproduced with permission from Ref. [84]).

the rat [1,5,27] and human [3,34,60] cloned VR1 channels. Here, in contrast to the findings with native channels in rat DRG neurons, the augmented capsaicin response was associated with an increase in the maximal response [3]. Lowering extracellular pH also increases the response of VR1 to noxious heat by reducing the threshold temperature [27] but, perhaps surprisingly, no such interaction was reported between low pH and anadamide [60].

This interplay between temperature, pH and, potentially, lipid mediators is significant as it may lower the threshold for neuronal excitation to body temperature in inflammatory conditions.

10. Molecular determinants for VR1 function

The cloning of VR1 has enabled the search for the molecular determinants responsible for its functional properties using site-directed mutagenesis. Although in its infancy, these studies have already revealed important features of VR1, especially for residues in the extracellular domain between the fifth and sixth membrane domains containing the putative pore forming loop (P-loop). The P-loop and the flanking residues connecting it to the transmembrane domains 5 and 6 contain a number of acidic glutamate (E) and aspartate (D) residues including, E^{600}, E^{636}, D^{646} and E^{648} that probably have a role in determining channel function and the pore properties [13,14,86].

In the flanking regions, neutralising E^{600} (E^{600}Q) increases the sensitivity to capsaicin, lowers the threshold to heat, and abolishes the potentiation of capsaicin responses by protons [13]. It is highly likely that this residue is a key site for the acid potentiation of the VR1. Jordt et al. [13] also proposed that E^{648} influences the ability of low pH solutions to activate VR1, but not to affect capsaicin or heat-evoked activity. Replacement of this glutamate residue with either glutamine (E^{648}Q) or alanine (E^{648}A) led to a decrease in the currents produced by protons, but normal responses to capsaicin. The alanine substitution is of particular interest as alanine is found in the comparable position in VRL-1, which is heat activated but proton insensitive. In contrast to the results of this study, Welch et al. [14] showed no change in acid sensitivity with the E^{618}Q mutation. The location of these acidic residues fit with the studies of Jung et al. [63] and McLatchie and Bevan [84], who have demonstrated that lowering the intracellular pH does not activate a current or potentiate the capsaicin response, suggesting that the protonation site is on an extracellularly exposed portion of the receptor/channel complex.

In the P-loop, E^{636}, D^{646} and E^{648} may influence capsaicin binding as, in one study [14], the EC_{50} concentrations for capsaicin activation was reduced about 3-fold with the substitutions E^{636}Q, D^{646}N and E^{648}Q. In single channel experiments, the mean channel open probabilities for capsaicin were greater for E^{636}Q and E^{648}Q compared to wild-type VR1, consistent with a decrease in EC_{50} concentration. Furthermore, the log(concentration)−response curve for the E^{636} mutant, E^{636}Q, showed a loss of positive co-operativity with a slope factor of 1 compared to the slope of 2 seen for wild-type VR1. In contrast to the change in capsaicin sensitivity, the responses of these point mutations to lowered pH or to temperature was not different to wild-type VR1 [14]. It should be noted, however, that

Garcia-Martinez et al. [86] did not find any change in the efficacy of capsaicin with the same P-loop mutations.

In addition to the identification of sites on the P-loop region, two modulatory sites have been discovered on the intracellular cytoplasmic tails [87]. These are putative Walker type nucleotide-binding domains on the N-terminus (D^{178}) and the C-terminus (lysine, K^{735}). The intracellular application of ATP or non-hydrolysable ATP analogues increased capsaicin-activated VR1 channel activity without involvement of phosphorylation. Single point mutations within these domains ($D^{178}N$, $K^{735}R$) abolished the ability of ATP to modulate VR1 activity.

11. Regulation of VR1 activity by phosphorylation

11.1. Sensitisation

Treatments that promote protein phosphorylation can potentiate the activity of VR1 and there is evidence that this can occur via PKA and PKC pathways. Increasing the level of phosphorylation by introduction of ATPγS into *Xenopus* oocytes activated the expressed rat VR1 channels indicating that the effects of phosphorylation may be manifested not only as sensitisation but also as agonism [72].

11.1.1. PKA-mediated effects

Prostaglandin E_2 (PGE_2), which mediates its effects in many cells via a PKA-induced protein phosphorylation, evokes a time dependent increase in the responses of DRG neurons to a sub-maximally effective concentration (100 nM) of capsaicin [88]. This effect was mimicked by raising the cAMP levels with forskolin or by application of membrane permeant cAMP analogues and was completely inhibited when a peptide inhibitor of PKA was introduced into the cell [89]. The location of the phosphorylation site responsible for this sensitisation is unclear as the sensitising effects of cAMP on the capasicin response of VR1 were not seen when VR1 is expressed in *Xenopus* oocytes or *Aplysia* neurons [90] or in CHO cells (Reeve and Bevan, unpublished observations). In these latter cell types, raising the cAMP level reduced the amplitude of the capsaicin responses.

11.1.2. PKC

In addition to exciting sensory neurons, bradykinin is also able to sensitise them to noxious heat by shifting the activation to lower temperatures [55]. This sensitisation can be mimicked by phorbol esters, consistent with an action via PKC. Five PKC isoforms are found in DRG neurons and using the translocation of PKC from the cytoplasm to the membrane as an index of activation, Cesare et al. [91] found that bradykinin caused the translocation of only one isoform (PKC-ε). The functional significance of this result was shown by the discovery that intracellular application of constitutively active PKC-ε increased the thermal response in DRG neurons, while a specific peptide PKC-ε inhibitor attenuated the bradykinin-induced sensitisation. Bradykinin-induced sensitisation of VR1 would also explain the finding that bradykinin increased the proportions of DRG neurons that responded to either capsaicin or low pH solutions as well as the amplitudes of the responses [92].

11.1.3. Other mechanisms of sensitisation

Some findings indicate that there are other pathways of sensitisation. NGF has been shown to have an acute sensitising effect on capsaicin sensitivity in DRG neurons in addition to its effects on VR1 expression levels [93]. It is unclear how this effect is mediated. A rise in intracellular calcium levels has also been implicated in the sensitisation of the heat- and proton-induced responses in afferent fibres and DRG neurons [94]. Again the underlying mechanism is unclear.

11.2. Desensitisation

Desensitisation of VR1 responses with capsaicin activation is, in part, dependent on calcium influx through the activated ion channel and the consequent rise in intracellular calcium concentration. Little desensitisation is seen either when calcium is omitted from the external medium [95–97] or when cells are loaded with the fast calcium chelator BAPTA [95,97]. Similarly the low pH-evoked responses of VR1 show a calcium-dependent desensitisation [27] whereas studies with heat as the activator have shown little effect of extracellular calcium and intracellular BAPTA on desensitisation [50,98]. At present, the reason for this disparity is unknown but could be related to the lower calcium permeability of heat-activated VR1.

Desensitisation of VR1 to capsaicin is associated with development of an outward rectification in the VR1 current–voltage relationship, which can also be inhibited by omission of extracellular calcium or the presence of BAPTA inside the cell [54]. It is likely that the rise in intracellular calcium concentration promotes desensitisation by stimulating a phosphatase, calcincurin (phosphatase 2B), since desensitisation in calcium containing medium is reduced when calcineurin is inhibited either by a complex of cyclosporin A and cyclophilin or by FK506 [97]. Intracellular application of ATP or the non-hydrolysable analogue ATPγS reduces the degree of desensitisation evoked by repeated applications of capsaicin [95] presumably by promoting or maintaining the levels of phosphorylation. Perhaps surprisingly, intracellular application of an autoinhibitory peptide of calcineurin failed to attenuate the acute desensitisation to capsaicin although it did inhibit the development of rectification [54]. A degree of desensitisation is still observed in the presence of calcineurin inhibition, which suggests the existence of other mechanisms.

Further clues to the mechanisms regulating VR1 sensitivity were found in the studies of PGE$_2$ sensitisation of capsaicin currents [88]. Here the duration of PGE$_2$-induced sensitisation was greatly prolonged when extracellular calcium was removed. Application of the membrane permeant cGMP analogue, 8-Br-cGMP, rapidly reversed the sensitisation and this effect could be mimicked by the nitric oxide donors, SNAP and SIN. Conversely the duration of the PGE$_2$-induced sensitisation of VR1 was lengthened by the nitric oxide synthase inhibitor, L-NAME as well as a protein kinase G antagonist, KT-5823. These results complement the earlier finding that capasicin-induced calcium entry into DRG neurons raised the cGMP levels [99]. One possible scheme is that that elevation of intracellular calcium levels leads to an enhanced production of cGMP via a nitric oxide pathway. This ultimately activates GMP-dependent protein kinase and so terminates the cAMP-mediated sensitisation process by some unidentified mechanism [88].

The available evidence indicates that the activity and properties of VR1 are controlled by the degree of phosphorylation of the receptor and/or associated molecules and that inflammatory mediators can influence VR1 activity in sensory neurons. The details of the precise molecular mechanisms are, however, currently unclear.

12. Therapeutic uses of VR1 ligands

Although capsaicin initially stimulates nociceptive neurons, there is a subsequent period of analgesia when these neurons become unresponsive to a range of stimuli [100]. This analgesia, which can often be long lasting, is probably due to a variety of effects triggered by calcium entry. These include receptor desensitisation, neuronal damage and inhibition of voltage-gated calcium channels [80]. Topical application of capsaicin has been used in the treatment of a variety of intractable pain conditions including arthritis and painful diabetic neuropathy [101]. The major problem is that the excitatory side effects of agonists, which include significant, and often complex, changes in blood pressure, hypothermia and bronchoconstriction, preclude the use of systemic agonists. There are, however, vanilloid ligands such as SDZ249-665 that show the analgesic properties but minimise or eliminate the excitatory effects [102]. Such ligands suggest that analgesic therapies based on VR1 agonists are feasible. However, in the absence of clinically approved synthetic compounds with such properties, various strategies are being considered to reduce the excitatory effects associated with agonism. These include local epidural application of RTX to produce a segmental analgesia [103] or application of capsaicin together with regional anaesthesia [104].

Capsaicin and RTX have also been used to treat bladder dysfunctions with local instillation which minimises the side effect problems. This treatment gives useful and long-lasting improvement in bladder control in, for example, multiple sclerosis patients [105].

The utility of VR1 antagonists is unclear. The analgesic/anti-hyperalgesic effects of capsazepine have been investigated by several groups. Doses of capsazepine that inhibited capsaicin-induced analgesia (100 μmols/kg s.c.) had no analgesic effects in acute thermal pain models or mechanical anti-hyperalgesic effects after inflammation [100,102]. In contrast, other groups [58,106] have reported that intradermal injections of capsazepine inhibit the behavioural responses, but not the inflammation, induced by an intradermal injection of formalin. Intradermal administration of capsazepine also reversed the thermal hyperalgesia in carrageenan-induced inflammation [58]. These latter results are consistent with an action of capsazepine to antagonize endogenous inflammatory mediators acting at VR1. It should be noted, however, that capsazepine acts at other sites at higher concentrations and can block voltage-gated calcium channels [107]. Furthermore high concentrations of capsazepine inhibit the response to noxious heat [49] which could explain the inhibitory effects on thermal hyperalgesia.

13. Conclusion

The knowledge gained since the cloning of VR1 has illustrated the importance of this ion channel in the transduction of noxious signals. VR1 is responsive to thermal, pH and some chemical stimuli, including inflammatory mediators derived from arachidonic acid breakdown. The interplay between thermal stimuli and extracellular pH illustrates the way in which VR1 is finely tuned to respond to shifts in the environment. VR1 clearly plays a key role in thermal hyperalgesia as illustrated by the studies in VR1 null mice, but other roles for VR1 remain to be discovered. The availability of these mice and specific pharmacological tools will allow us to explore in more detail the ways in which VR1 contributes to the physiology of sensory nerves in both health and disease.

References

[1] Caterina, M.J., Schumacher, M.A., Tominaga, M., Rosen, T.A., Levine, J.D., Julius, D., 1997. The capsaicin receptor: a heat-activated ion channel in the pain pathway. Nature 389, 816–824.

[2] Wood, J.N., Winter, J., James, I.F., Rang, H.P., Yeats, J., Bevan, S., 1988. Capsaicin-induced ion fluxes in dorsal root ganglion cells in culture. J. Neurosci. 8, 3208–3220.

[3] Hayes, P., Meadows, H.J., Gunthorpe, M.J., Harries, M.H., Duckworth, D.M., Cairns, W., Harrison, D.C., Clarke, C.E., Ellington, K., Prinjha, R.K., Barton, A.J., Medhurst, A.D., Smith, G.D., Topp, S., Murdock, P., Sanger, G.J., Terrett, J., Jenkins, O., Benham, C.D., Randall, A.D., Gloger, I.S., Davis, J.B., 2000. Cloning and functional expression of a human orthologue of rat vanilloid receptor-1. Pain 88, 205–215.

[4] McIntyre, P., McLatchie, L.M., Chambers, A., Phillips, E., Clarke, M., Savidge, J., Toms, C., Peacock, M., Shah, K., Winter, J., Weerasakera, N., Webb, M., Rang, H.P., Bevan, S., James, I.F., 2001. Pharmacological differences between the human and rat vanilloid receptor 1 VR1. Br. J. Pharmacol. 132, 1084–1094.

[5] Jerman, J.C., Brough, S.J., Prinjha, R., Harries, M.H., Davis, J.B., Smart, D., 2000. Characterization using FLIPR of rat vanilloid receptor (rVR1) pharmacology. Br. J. Pharmacol. 130, 916–922.

[6] Liu, L., Welch, J.M., Erickson, R.P., Reinhart, P.H., Simon, S.A., 2000. Different responses to repeated applications of zingerone in behavioral studies, recordings from intact and cultured TG neurons, and from VR1 receptors. Physiol. Behav. 69, 177–186.

[7] Bevan, S., Hothi, S., Hughes, G., James, I.F., Rang, H.P., Shah, K., Walpole, C.S., Yeats, J.C., 1992. Capsazepine: a competitive antagonist of the sensory neurone excitant capsaicin. Br. J. Pharmacol. 107, 544–552.

[8] Amann, R., Maggi, C.A., 1991. Ruthenium red as a capsaicin antagonist. Life Sci. 49, 849–856.

[9] Bevan, S., Docherty, R., 1996. The ionic basis of capsaicin-evoked responses. In: Geppetti, P., Holzer, P. (Eds.), Neurogenic Inflammation. CRC Press, Boca Raton, pp. 53–67.

[10] Oh, U., Hwang, S.W., Kim, D., 1996. Capsaicin activates a nonselective cation channel in cultured neonatal rat dorsal root ganglion neurons. J. Neurosci. 16, 1659–1667.

[11] Nagy, I., Rang, H.P., 1999. Similarities and differences between the responses of rat sensory neurons to noxious heat and capsaicin. J. Neurosci. 19, 10647–10655.

[12] Szallasi, A., Blumberg, P.M., 1991. Molecular target size of the vanilloid (capsaicin) receptor in pig dorsal root ganglia. Life Sci. 48, 1863–1869.

[13] Jordt, S.E., Tominaga, M., Julius, D., 2000. Acid potentiation of the capsaicin receptor determined by a key extracellular site. Proc. Natl Acad. Sci. USA 97, 8134–8139.

[14] Welch, J.M., Simon, S.A., Reinhart, P.H., 2000. The activation mechanism of rat vanilloid receptor 1 by capsaicin involves the pore domain and differs from the activation by either acid or heat. Proc. Natl Acad. Sci. USA 97, 13889–13894.

[15] Vlachová, V., Vyklický, L., 1993. Capsaicin-induced membrane currents in cultured sensory neurons of the rat. Physiol. Res. 42, 301–311.

[16] Shin, J.S., Wang, M., Hwang, S.W., Cho, H., Cho, S.Y., Kwon, M.J., Lee, S., Oh, U., 2001. Differences in sensitivity of vanilloid receptor 1 transfected to human embryonic kidney cells and capsaicin-activated channels in cultured rat dorsal root ganglion neurons to capsaicin receptor agonists. Neurosci. Lett. 299, 135–139.

[17] Gunthorpe, M.J., Harries, M.H., Prinjha, R.K., Davis, J.B., Randall, A., 2000. Voltage- and time-dependent properties of the recombinant rat vanilloid receptor (rVR1). J. Physiol. 525, 747–759.

[18] Harteneck, C., Plant, T.D., Schultz, G., 2000. From worm to man: three subfamilies of TRP channels. Trends Neurosci. 23, 159–166.

[19] Colbert, H.A., Smith, T.L., Bargmann, C.I., 1997. OSM-9, a novel protein with structural similarity to channels, is required for olfaction, mechanosensation, and olfactory adaptation in *Caenorhabditis elegans*. J. Neurosci. 17, 8259–8269.

[20] Strotmann, R., Harteneck, C., Nunnenmacher, K., Schultz, G., Plant, T.D., 2000. OTRPC4, a nonselective cation channel that confers sensitivity to extracellular osmolarity. Nat. Cell Biol. 2, 695–702.

[21] Liedtke, W., Choe, Y., Marti-Renom, M.A., Bell, A.M., Denis, C.S., Sali, A., Hudspeth, J., Friedman, J.M., Heller, S., 2000. Vanilloid receptor-related osmotically activated channel (VR-OAC), a candidate vertebrate osmoreceptor. Cell 103, 525–535.

[22] Wissenbach, U., Bodding, M., Freichel, M., Flockerzi, V., 2000. Trp12, a novel Trp related protein from kidney. FEBS Lett. 485, 127–134.

[23] Caterina, M.J., Rosen, T.A., Tominaga, M., Brake, A.J., Julius, D., 1999. A capsaicin-receptor homologue with a high threshold for noxious heat. Nature 398, 436–441.

[24] Schumacher, M.A., Moff, I., Sudangunta, S.P., Levine, J.D., 2000. Molecular cloning of an N-terminal splice variant of the capsaicin receptor. Loss of N-terminal domain suggests functional divergence among capsaicin receptor subtypes. J. Biol. Chem. 275, 2756–2762.

[25] Suzuki, M., Sato, J., Kutsuwada, K., Ooki, G., Imai, M., 1999. Cloning of a stretch-inhibitable nonselective cation channel. J. Biol. Chem. 274, 6330–6335.

[26] Delany, N.S., Hurle, M., Facer, P., Alnadaf, T., Plumpton, C., Kinghorn, I., See, C.G., Costigan, M., Anand, P., Woolf, C.J., Crowther, D., Sanseau, P., Tate, S.N., 2001. Identification and characterization of a novel human vanilloid receptor-like protein, VRL-2. Physiol. Genomics 4, 165–174.

[27] Tominaga, M., Caterina, M.J., Malmberg, A.B., Rosen, T.A., Gilbert, H., Skinner, K., Raumann, B.E., Basbaum, A.I., Julius, D., 1998. The cloned capsaicin receptor integrates multiple pain-producing stimuli. Neuron 21, 531–543.

[28] Guo, A., Vulchanova, L., Wang, J., Li, X., Elde, R., 1999. Immunocytochemical localization of the vanilloid receptor 1 (VR1): relationship to neuropeptides, the P2X3 purinoceptor and IB4 binding sites. Eur. J. Neurosci. 11, 946–958.

[29] Helliwell, R.J., McLatchie, L.M., Clarke, M., Winter, J., Bevan, S., McIntyre, P., 1998. Capsaicin sensitivity is associated with the expression of the vanilloid (capsaicin) receptor (VR1) mRNA in adult rat sensory ganglia. Neurosci. Lett. 250, 177–180.

[30] Michael, G.J., Priestley, J.V., 1999. Differential expression of the mRNA for the vanilloid receptor subtype 1 in cells of the adult rat dorsal root and nodose ganglia and its downregulation by axotomy. J. Neurosci. 19, 1844–1854.

[31] Sasamura, T., Sasaki, M., Tohda, C., Kuraishi, Y., 1998. Existence of capsaicin-sensitive glutamatergic terminals in rat hypothalamus. Neuroreport 9, 2045–2048.

[32] Mezey, E., Toth, Z.E., Cortright, D.N., Arzubi, M.K., Krause, J.E., Elde, R., Guo, A., Blumberg, P.M., Szallasi, A., 2000. Distribution of mRNA for vanilloid receptor subtype 1 (VR1), and VR1-like immunoreactivity, in the central nervous system of the rat and human. Proc. Natl Acad. Sci. USA 97, 3655–3660.

[33] Jancso-Gabor, A., Szolcsanyi, J., Jancso, N., 1970. Stimulation and desensitization of the hypothalamic heat-sensitive structures by capsaicin in rats. J. Physiol. 208, 449–459.

[34] Cortright, D.N., Crandall, M., Sanchez, J.F., Zou, T., Krause, J.E., White, G., 2001. The tissue distribution and functional characterization of human VR1. Biochem. Biophys. Res. Commun. 281, 1183–1189.

[35] Biro, T., Maurer, M., Modarres, S., Lewin, N.E., Brodie, C., Acs, G., Acs, P., Paus, R., Blumberg, P.M., 1998. Characterization of functional vanilloid receptors expressed by mast cells. Blood 91, 1332–1340.

[36] Winter, J., Walpole, C.S., Bevan, S., James, I.F., 1993. Characterization of resiniferatoxin binding sites on sensory neurons: co-regulation of resiniferatoxin binding and capsaicin sensitivity in adult rat dorsal root ganglia. Neuroscience 57, 747–757.

[37] Bevan, S., Winter, J., 1995. Nerve growth factor (NGF) differentially regulates the chemosensitivity of adult rat cultured sensory neurons. J. Neurosci. 15, 4918–4926.

[38] Ogun-Muyiwa, P., Helliwell, R., McIntyre, P., Winter, J., 1999. Glial cell line derived neurotrophic factor (GDNF) regulates VR1 and substance P in cultured sensory neurons. Neuroreport 10, 2107–2111.

[39] Winston, J., Toma, H., Shenoy, M., Pasricha, P.J., 2001. Nerve growth factor regulates VR-1 mRNA levels in cultures of adult dorsal root ganglion neurons. Pain 89, 181–186.

[40] McMahon, S.B., Bennett, D.L., Priestley, J.V., Shelton, D.L., 1995. The biological effects of endogenous nerve growth factor on adult sensory neurons revealed by a trkA–IgG fusion molecule. Nat. Med. 1, 774–780.

[41] Winter, J., 1998. Brain derived neurotrophic factor, but not nerve growth factor, regulates capsaicin sensitivity of rat vagal ganglion neurones. Neurosci. Lett. 241, 21–24.

[42] Donnerer, J., Schuligoi, R., Stein, C., 1992. Increased content and transport of substance P and calcitonin gene-related peptide in sensory nerves innervating inflamed tissue: evidence or a regulatory function of nerve growth factor in vivo. Neuroscience 49, 693–698.

[43] Woolf, C.J., Safieh-Garabedian, B., Ma, Q.P., Crilly, P., Winter, J., 1994. Nerve growth factor contributes to the generation of inflammatory sensory hypersensitivity. Neuroscience 62, 327–331.

[44] Aloe, L., Tuveri, M.A., Carcassi, U., Levi-Montalcini, R., 1992. Nerve growth factor in the synovial fluid of patients with chronic arthritis. Arthritis Rheum. 35, 351–355.

[45] Halliday, D.A., Zettler, C., Rush, R.A., Scicchitano, R., McNeil, J.D., 1998. Elevated nerve growth factor levels in the synovial fluid of patients with inflammatory joint disease. Neurochem. Res. 23, 919–922.

[46] Nicholas, R.S., Winter, J., Wren, P., Bergmann, R., Woolf, C.J., 1999. Peripheral inflammation increases the capsaicin sensitivity of dorsal root ganglion neurons in a nerve growth factor-dependent manner. Neuroscience 91, 1425–1433.

[47] Treede, R.D., Meyer, R.A., Raja, S.N., Campbell, J.N., 1995. Evidence for two different heat transduction mechanisms in nociceptive primary afferents innervating monkey skin. J. Physiol. 483, 747–758.

[48] Nagy, I., Rang, H., 1999. Noxious heat activates all capsaicin-sensitive and also a sub-population of capsaicin-insensitive dorsal root ganglion neurons. Neuroscience 88, 995–997.

[49] Kirschstein, T., Greffrath, W., Busselberg, D., Treede, R.D., 1999. Inhibition of rapid heat responses in nociceptive primary sensory neurons of rats by vanilloid receptor antagonists. J. Neurophysiol. 82, 2853–2860.

[50] Vyklický, L., Vlachová, V., Vitásková, Z., Dittert, I., Kabat, M., Orkand, R.K., 1999. Temperature coefficient of membrane currents induced by noxious heat in sensory neurones in the rat. J. Physiol. 517, 181–192.

[51] Savidge, J.R., Ranasinghe, S.P., Rang, H.P., 2001. Comparison of intracellular calcium signals evoked by heat and capsaicin in cultured rat dorsal root ganglion neurons and in a cell line expressing the rat vanilloid receptor, VR1. Neuroscience 102, 177–184.

[52] Caterina, M.J., Leffler, A., Malmberg, A.B., Martin, W.J., Trafton, J., Petersen-Zeitz, K.R., Koltzenburg, M., Basbaum, A.I., Julius, D., 2000. Impaired nociception and pain sensation in mice lacking the capsaicin receptor. Science 288, 306–313.

[53] Davis, J.B., Gray, J., Gunthorpe, M.J., Hatcher, J.P., Davey, P.T., Overend, P., Harries, M.H., Latcham, J., Clapham, C., Atkinson, K., Hughes, S.A., Rance, K., Grau, E., Harper, A.J., Pugh, P.L., Rogers, D.C., Bingham, S., Randall, A., Sheardown, S.A., 2000. Vanilloid receptor-1 is essential for inflammatory thermal hyperalgesia. Nature 405, 183–187.

[54] Piper, A.S., Yeats, J.C., Bevan, S., Docherty, R.J., 1999. A study of the voltage dependence of capsaicin-activated membrane currents in rat sensory neurones before and after acute desensitization. J. Physiol. 518, 721–733.

[55] Cesare, P., McNaughton, P., 1996. A novel heat-activated current in nociceptive neurons and its sensitization by bradykinin. Proc. Natl Acad. Sci. USA 93, 15435–15439.

[56] Marin-Burgin, A., Reppenhagen, S., Klusch, A., Wendland, J.R., Petersen, M., 2000. Low-threshold heat response antagonized by capsazepine in chick sensory neurons, which are capsaicin-insensitive. Eur. J. Neurosci. 12, 3560–3566.

[57] Nagy, I., Rang, H., 2000. Comparison of currents activated by noxious heat in rat and chicken primary sensory neurons. Regul. Pept. 96, 3–6.

[58] Kwak, J.Y., Jung, J.Y., Hwang, S.W., Lee, W.T., Oh, U., 1998. A capsaicin-receptor antagonist, capsazepine, reduces inflammation-induced hyperalgesic responses in the rat: evidence for an endogenous capsaicin-like substance. Neuroscience 86, 619–626.

[59] Zygmunt, P.M., Petersson, J., Andersson, D.A., Chuang, H., Sorgard, M., Di Marzo, V., Julius, D., Hogestatt, E.D., 1999. Vanilloid receptors on sensory nerves mediate the vasodilator action of anandamide. Nature 400, 452–457.

[60] Smart, D., Gunthorpe, M.J., Jerman, J.C., Nasir, S., Gray, J., Muir, A.I., Chambers, J.K., Randall, A.D., Davis, J.B., 2000. The endogenous lipid anandamide is a full agonist at the human vanilloid receptor (hVR1). Br. J. Pharmacol. 129, 227–230.

[61] Tognetto, M., Amadesi, S., Harrison, S., Creminon, C., Trevisani, M., Carreras, M., Matera, M., Geppetti, P., Bianchi, A., 2001. Anandamide excites central terminals of dorsal root ganglion neurons via vanilloid receptor-1 activation. J. Neurosci. 21, 1104–1109.

[62] De Petrocellis, L., Bisogno, T., Davis, J.B., Pertwee, R.G., Di Marzo, V., 2000. Overlap between the ligand recognition properties of the anandamide transporter and the VR1 vanilloid receptor: inhibitors of anandamide uptake with negligible capsaicin-like activity. FEBS Lett. 483, 52–56.

[63] Jung, J., Hwang, S.W., Kwak, J., Lee, S.Y., Kang, C.J., Kim, W.B., Kim, D., Oh, U., 1999. Capsaicin binds to the intracellular domain of the capsaicin-activated ion channel. J. Neurosci. 19, 529–538.

[64] Hofmann, T., Obukhov, A.G., Schaefer, M., Harteneck, C., Gudermann, T., Schultz, G., 1999. Direct activation of human TRPC6 and TRPC3 channels by diacylglycerol. Nature 397, 259–263.

[65] Hwang, S.W., Cho, H., Kwak, J., Lee, S.Y., Kang, C.J., Jung, J., Cho, S., Min, K.H., Suh, Y.G., Kim, D., Oh, U., 2000. Direct activation of capsaicin receptors by products of lipoxygenases: endogenous capsaicin-like substances. Proc. Natl Acad. Sci. USA 97, 6155–6160.

[66] Szolcsanyi, J., 2000. Are cannabinoids endogenous ligands for the VR1 capsaicin receptor? Trends Pharmacol. Sci. 21, 41–42.

[67] Szolcsanyi, J., 2000. Anandamide and the question of its functional role for activation of capsaicin receptors. Trends Pharmacol. Sci., 203–204.

[68] Zygmunt, P.M., Julius, I., Di Marzo, I., Hogestatt, E.D., 2000. Anandamide – the other side of the coin. Trends Pharmacol. Sci. 21, 43–44.

[69] Smart, D., Jerman, J.C., 2000. Anandamide: an endogenous activator of the vanilloid receptor. Trends Pharmacol. Sci. 21, 134.

[70] Burgess, G.M., Mullaney, I., McNeill, M., Dunn, P.M., Rang, H.P., 1989. Second messengers involved in the mechanism of action of bradykinin in sensory neurons in culture. J. Neurosci. 9, 3314–3325.

[71] McGehee, D.S., Goy, M.F., Oxford, G.S., 1992. Involvement of the nitric oxide-cyclic GMP pathway in the desensitization of bradykinin responses of cultured rat sensory neurons. Neuron 9, 315–324.

[72] Premkumar, L.S., Ahern, G.P., 2000. Induction of vanilloid receptor channel activity by protein kinase C. Nature 408, 985–990.

[73] Bevan, S., 1998. Proton-gated ion channels in neurons. In: Kaila, K., Ransom, B.R. (Eds.), pH and Brain Function. Wiley-Liss, New York, pp. 447–475.

[74] Steen, K.H., Reeh, P.W., Anton, F., Handwerker, H.O., 1992. Protons selectively induce lasting excitation and sensitization to mechanical stimulation of nociceptors in rat skin, in vitro. J. Neurosci. 12, 86–95.

[75] Pan, H.L., Longhurst, J.C., Eisenach, J.C., Chen, S.R., 1999. Role of protons in activation of cardiac sympathetic C-fibre afferents during ischaemia in cats. J. Physiol. 518, 857–866.

[76] Steen, K.H., Issberner, U., Reeh, P.W., 1995. Pain due to experimental acidosis in human skin: evidence for non-adapting nociceptor excitation. Neurosci. Lett. 199, 29–32.

[77] Waldmann, R., Champigny, G., Lingueglia, E., De Weille, J.R., Heurteaux, C., Lazdunski, M., 1999. H^+ gated cation channels. Ann. N Y Acad. Sci. 868, 67–76.

[78] Sutherland, S.P., Benson, C.J., Adelman, J.P., McCleskey, E.W., 2001. Acid-sensing ion channel 3 matches the acid-gated current in cardiac ischemia-sensing neurons. Proc. Natl Acad. Sci. USA 98, 711–716.

[79] Bevan, S., Yeats, J., 1991. Protons activate a cation conductance in a sub-population of rat dorsal root ganglion neurones. J. Physiol. 433, 145–161.

[80] Bevan, S., 1999. Capsaicin and pain mechanisms. In: Brain, S.D., Moore, P.K. (Eds.), Pain and Neurogenic Inflammation. Birkhauser Verlag, Basel, pp. 61–80.

[81] Zeilhofer, H.U., Kress, M., Swandulla, D., 1997. Fractional Ca^{2+} currents through capsaicin- and proton-activated ion channels in rat dorsal root ganglion neurones. J. Physiol. 503, 67–78.

[82] Petersen, M., LaMotte, R.H., 1993. Effect of protons on the inward current evoked by capsaicin in isolated dorsal root ganglion cells. Pain 54, 37–42.

[83] Kress, M., Fetzer, S., Reeh, P.W., Vyklicky, L., 1996. Low pH facilitates capsaicin responses in isolated sensory neurons of the rat. Neurosci. Lett. 211, 5–8.

[84] McLatchie, L., Bevan, S., 2001. The effects of pH on the interaction between capsaicin and the vanilloid receptor in rat dorsal root ganglia neurons. Br. J. Pharmacol. 132, 899–908.

[85] Baumann, T.K., Martenson, M.E., 2000. Extracellular protons both increase the activity and reduce the conductance of capsaicin-gated channels. J. Neurosci. 20, RC80.

[86] Garcia-Martinez, C., Morenilla-Palao, C., Planells-Cases, R., Merino, J.M., Ferrer-Montiel, A., 2000. Identification of an aspartic residue in the P-loop of the vanilloid receptor that modulates pore properties. J. Biol. Chem. 275, 32552–32558.

[87] Kwak, J., Wang, M.H., Hwang, S.W., Kim, T.Y., Lee, S.Y., Oh, U., 2000. Intracellular ATP increases capsaicin-activated channel activity by interacting with nucleotide-binding domains. J. Neurosci. 20, 8298–8304.

[88] Lopshire, J.C., Nicol, G.D., 1997. Activation and recovery of the PGE2-mediated sensitization of the capsaicin response in rat sensory neurons. J. Neurophysiol. 78, 3154–3164.

[89] Lopshire, J.C., Nicol, G.D., 1998. The cAMP transduction cascade mediates the prostaglandin E2 enhancement of the capsaicin-elicited current in rat sensory neurons: whole-cell and single-channel studies. J. Neurosci. 18, 6081–6092.

[90] Lee, Y.S., Lee, J.A., Jung, J., Oh, U., Kaang, B.K., 2000. The cAMP-dependent kinase pathway does not sensitize the cloned vanilloid receptor type 1 expressed in xenopus oocytes or Aplysia neurons. Neurosci. Lett. 288, 57–60.

[91] Cesare, P., Dekker, L.V., Sardini, A., Parker, P.J., McNaughton, P.A., 1999. Specific involvement of PKC-epsilon in sensitization of the neuronal response to painful heat. Neuron 23, 617–624.

[92] Stucky, C.L., Abrahams, L.G., Seybold, V.S., 1998. Bradykinin increases the proportion of neonatal rat dorsal root ganglion neurons that respond to capsaicin and protons. Neuroscience 84, 1257–1265.

[93] Shu, X., Mendell, L.M., 1999. Nerve growth factor acutely sensitizes the response of adult rat sensory neurons to capsaicin. Neurosci. Lett. 274, 159–162.

[94] Guenther, S., Reeh, P.W., Kress, M., 1999. Rises in $[Ca^{2+}]_i$ mediate capsaicin- and proton-induced heat sensitization of rat primary nociceptive neurons. Eur. J. Neurosci. 11, 3143–3150.

[95] Koplas, P.A., Rosenberg, R.L., Oxford, G.S., 1997. The role of calcium in the desensitization of capsaicin responses in rat dorsal root ganglion neurons. J. Neurosci. 17, 3525–3537.

[96] Cholewinski, A., Burgess, G.M., Bevan, S., 1993. The role of calcium in capsaicin-induced desensitization in rat cultured dorsal root ganglion neurons. Neuroscience 55, 1015–1023.

[97] Docherty, R.J., Yeats, J.C., Bevan, S., Boddeke, H.W., 1996. Inhibition of calcineurin inhibits the desensitization of capsaicin-evoked currents in cultured dorsal root ganglion neurones from adult rats. Pflugers Arch. 431, 828–837.

[98] Schwarz, S., Greffrath, W., Busselberg, D., Treede, R.D., 2000. Inactivation and tachyphylaxis of heat-evoked inward currents in nociceptive primary sensory neurones of rats. J. Physiol. 528, 539–549.

[99] Burgess, G.M., Mullaney, I., McNeill, M., Coote, P.R., Minhas, A., Wood, J.N., 1989. Activation of guanylate cyclase by bradykinin in rat sensory neurones is mediated by calcium influx: possible role of the increase in cyclic GMP. J. Neurochem. 53, 1212–1218.

[100] Perkins, M.N., Campbell, E.A., 1992. Capsazepine reversal of the antinociceptive action of capsaicin in vivo. Br. J. Pharmacol. 107, 329–333.

[101] Robbins, W., 2000. Clinical applications of capsaicinoids. Clin. J. Pain 16(suppl), S86–S89.

[102] Urban, L., Campbell, E.A., Panesar, M., Patel, S., Chaudhry, N., Kane, S., Buchheit, K., Sandells, B., James, I.F., 2000. In vivo pharmacology of SDZ 249-665, a novel, non-pungent capsaicin analogue. Pain 89, 65–74.

[103] Szabo, T., Olah, Z., Iadarola, M.J., Blumberg, P.M., 1999. Epidural resiniferatoxin induced prolonged regional analgesia to pain. Brain Res. 840, 92–98.

[104] Robbins, W.R., Staats, P.S., Levine, J., Fields, H.L., Allen, R.W., Campbell, J.N., Pappagallo, M., 1998. Treatment of intractable pain with topical large-dose capsaicin: preliminary report. Anesth. Analg. 86, 579–583.

[105] De Ridder, D., Chandiramani, V., Dasgupta, P., Van Poppel, H., Baert, L., Fowler, C.J., 1997. Intravesical capsaicin as a treatment for refractory detrusor hyperreflexia: a dual center study with long-term followup. J. Urol. 158, 2087–2092.

[106] Santos, A.R., Calixto, J.B., 1997. Ruthenium red and capsazepine antinociceptive effect in formalin and capsaicin models of pain in mice. Neurosci. Lett. 235, 73–76.

[107] Docherty, R.J., Yeats, J.C., Piper, A.S., 1997. Capsazepine block of voltage-activated calcium channels in adult rat dorsal root ganglion neurones in culture. Br. J. Pharmacol. 121, 1461–1467.

Novel roles for aquaporins as gated ion channels

Andrea J. Yool[a,b,*] and W. Daniel Stamer[b,c]

[a]Department of Physiology, University of Arizona College of Medicine, Tucson, AZ 85724-5051, USA
[b]Department of Pharmacology, University of Arizona College of Medicine, Tucson, AZ 85724-5051, USA
[c]Department of Ophthalmology, University of Arizona College of Medicine, Tucson, AZ 85724-5051, USA
[*]Correspondence address: Department of Physiology, PO Box 245051, University of Arizona College of Medicine, Tucson, AZ 85724-5051, Tel.: +1-520-626-2198; fax: +1-520-626-2383
E-mail: ayool@u.arizona.edu

Aquaporins (AQPs) are members of the major intrinsic protein (MIP) gene family, a diverse group of water and solute channels found throughout the phyla of plants, animals, and bacteria. Membrane lipid bilayers have an inherently low permeability to water, an attribute that benefits life in aqueous and terrestrial environments. In specialized water-absorbing and secreting cells of mammals, membrane permeability to water is greatly enhanced by the expression of AQPs, a family of membrane proteins which provide channels for osmotically driven water fluxes [1]. An expanding role for mammalian AQPs as pathways for regulated solute transport is becoming evident, and these channels are now proposed to mediate transmembrane fluxes for a growing list of other substrates, including ions, glycerol, and carbon dioxide.

Osmotic water and glycerol permeabilities of MIP channels have been described previously in numerous reviews. This review considers a new perspective, focusing on the structural and functional basis of regulated ion channel function in a subset of MIP channels. Even if only a small proportion of water channels function at any given time as ion channels, their contribution may be significant to the physiological function of tissues in which the channels are expressed. Given the fundamental coupling of water and salt fluxes in transport epithelia, and the influence of membrane potential on transport processes, the role of AQPs as regulated ion channels will be of great interest for understanding the physiological and pathophysiological processes of transmembrane transport and signaling in a variety of tissues. Sequence comparisons and electrophysiological analyses of these MIP family members reveal structural features that correlate with the properties of activation, block and ionic permeability, particularly for AQP1. Ongoing research promises more intriguing links between structural and functional properties of AQPs.

Advances in Molecular and Cell Biology, Vol. 32, pages 351–379
© 2004 Elsevier B.V. All rights of reproduction in any form reserved.
ISSN: 1569-2558/DOI: 10.1016/S1569-2558(03)32015-6

1. Structural features of major intrinsic protein (MIP) channels

1.1. General structure

The existence of selective water channels in biological membranes was postulated based on physiological evidence [2] before the first molecular identification of AQPs. The first member of the MIP family to be sequenced and characterized was the MIP of bovine lens [3], now known as AQP0. AQP1 (first called CHIP28) was cloned and characterized [4,5], and launched new interest in proteins involved in transmembrane water transport. The family of mammalian AQPs has since expanded to include at least ten members, as well as a great number of related proteins in bacteria, plants and other animals.

AQP1 channels associate as tetramers of subunits, each with six predicted transmembrane domains and internal amino and carboxy terminals [1]. Other members of the MIP family show a comparable transmembrane topology. A schematic diagram of AQP1 is shown in Fig. 1. This topology also is a hallmark of the ion channel gene family including potassium channels and cyclic nucleotide-gated (CNG) channels [6]. Within each AQP subunit, the first and second halves of the six transmembrane domain core show sequence similarities, suggested to have arisen by tandem duplication of an ancient prokaryotic genetic element and since diversified for a myriad of specialized functions, particularly in the carboxy half [7]. Notably, loop B and loop E show a conserved asparagine–proline–alanine (NPA) sequence that is a hallmark of the diverse family of MIP channels. From three-dimensional structure analyses of AQP1 and an *E. coli* glycerol facilitator (GlpF) in the MIP family, the NPA regions have been suggested to form an interface between the two repeat sets of transmembrane helical domains [8–10]. GlpF crystallizes as a symmetric tetramer of channels, each holding three glycerol molecules. Each channel has an outer vestibule approximately 15 Å wide, and a constricted selectivity filter ~ 3.8 Å by 3.4 Å located 8 Å above the quasi-twofold axis of symmetry where proline rings of the conserved NPA sequences interact. The central cavity at the fourfold axis of symmetry has dimensions similar to those of a K^+ channel, and appears to have two sites for coordinating ion binding, suggesting that this pathway should be considered as a candidate ion channel in regulated AQPs [10].

1.2. Sequence homology between carboxy tail domains for aquaporin (AQP) and cyclic-nucleotide-gated (CNG) channels

In Fig. 2, MIP and CNG channels are grouped based on amino acid sequence patterns in the carboxy tail regions. Most notably in Group I, AQP1 shows a limited pattern of amino acid sequence similarity with CNG channels. In mutagenesis studies of CNG channels, this region of the carboxy tail has been shown to be involved in cyclic nucleotide binding and channel activation [11,12]. Alignment of the carboxyl terminal domains provides support for the existence of a cyclic-nucleotide binding domain in the AQP1 channel, which is consistent with binding studies with AQP1-expressing SF9 cells [13], and the effect of carboxy tail deletion in removing cGMP-gated channel activity in AQP1 proteins reconstituted in bilayers [14].

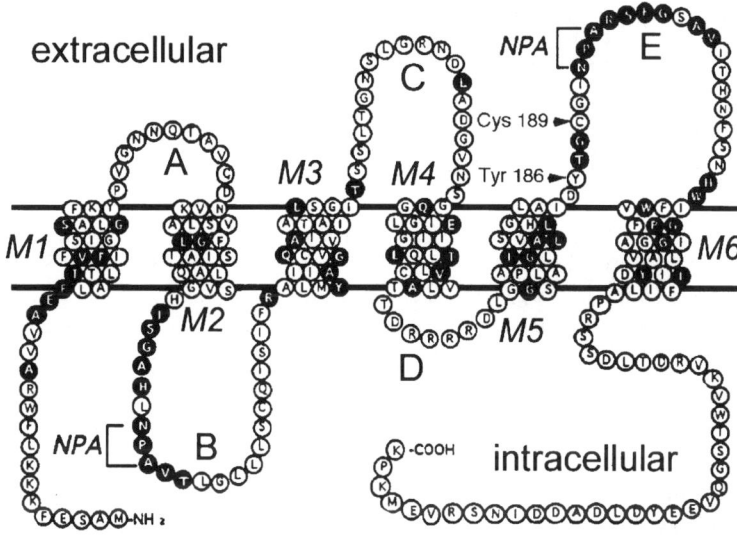

Fig. 1. Predicted transmembrane topology of Aquaporin 1, showing six transmembrane domains connected by loops A through E. Loops B and E carry the characteristic NPA motif that is a hallmark of the MIP family. Filled circles indicate amino acids that are conserved among AQPs. Cys189 is the site mediating the block of osmotic water permeability by Hg^{2+}. Tyr186 influences the block of water permeability by external TEA.

The sequence alignments shown in Fig. 2 include the complete AQP1 carboxy tail from the end of the sixth membrane-spanning (M6) domain; however, only a portion of the CNG channel carboxy terminal is shown (from the middle of the putative β7, through β8, Bα and Cα domains). For the CNG channels, the secondary structure proposed for the C-terminal is an alpha helix (Aα), followed by an eight stranded β barrel (β1-8), and two additional alpha helices (Bα and Cα) [15]. Crystal structure analysis of a cAMP binding protein (catabolite gene activator protein) has shown that residues that participate in binding are clustered in β7 and Cα[16]. Interestingly, the positions of these clusters frame the limits of the complete AQP1 carboxy domain. Other CNG domains that are not represented in the AQP1 carboxy terminal could conceivably be contributed by intracellular loops, or may not be necessary to cGMP-gated function in AQP1. The existence of a cGMP binding site (K_D 0.2 μM) was confirmed by expression of AQP1 in SF9 cells [13]. Red blood cells are known to express AQP1, and show an abundant protein that binds cGMP (K_D 0.16 μM) and is antagonized by cAMP [17]. Both properties are consistent with those proposed for the AQP1 channel.

In the CNG channel binding domain, a number of residues have been proposed to interact with substituent groups, coordinately binding the cyclic-nucleotide and dictating relative sensitivities to cAMP and cGMP. For example, mutations of arginine 559 and threonine 560 in β7 decreased the $K_{0.5}$ value for cGMP [18,19]. Mutations of lysine 596 and aspartate 604 in Cα influenced cGMP and cAMP activation of bovine retinal CNG channels [12,20]. Other residues also influence cyclic-nucleotide binding and channel

```
Group I       β7- - - -   β8- - - - - -   B α - - - - - - -       C α - - - - - - ★ - - - - - - -
bRet          N I K S I G Y S D L F C L S K D D L M E A L T E Y P D A K G M L E E K G K Q I L M K +
rOlf          N I R S L G Y S D L F C L S K D D L M E A V T E Y P D A K K V L E E R G R E I L M K +
rOlf          N I K S L G Y S D L F C L S K E D L R E V L S E Y P Q A Q A V M E E K G R E I L L K +
fOlf          N I R S I G Y S D L·F C L S K D D L M E A V A E Y P D A Q K V L E E R G R E I L R K +
fEAG          N V R A L T Y C D L H A I K R D D K L L E V L I F Y S A F A N S F A R N L V L T Y N +
hAQP1         F I L A P R S S D L T D R V K V W T S G Q V E E Y D L D A D D I N S R V E M K P K
rAQP1         F I L A P R S S D F T D R M K V W T S G Q V E E Y D L D A D D I N S R V E M K P K

Group II
hAQP0         F L L F P R L K S I S E R L S V L K G A K . P D V S N G Q P E V T G E . P V E L N T Q A L
mAQP0         F L L F P R L K S V S E R L S I L K G A R . P S D S N G Q P E G T G E . P V E L K T Q A L
bAQP0         F L L F P R L K S V S E R L S I L K G S R . P S E S N G Q P E V T G E . P V E L K T Q A L
hAQP2         Y V L F P P A K S L S E R L A V L K G L E . P D T D W E E R E V R R R Q S V E L H S P Q S +
mAQP2         Y L L F P S T K S L Q E R L A V L K G L E . P D T D W E E R E V R R R Q S V E L H S P Q S +
mAQP5         Y L L F P S S L S L H D R V A V V K G T Y E P E E D W E D H R E E R K K T I E L T A H
rAQP5         Y L L F P S S L S L H D R V A V V K G T Y E P E E D W E D H R E E R K K T I E L T A H
rAQP6         F I L F P D T K T V A Q R L A I L V G T T K V E K V V D L E P Q K K E S Q T N S E D T E V +
fBIB          L V Y E Y I F N S R N R N L R H N K G S I D N D S S S I H S E D E L N Y D M D M E K P N K +
mAQP7         G L I H P S I P Q D P Q R L E N F T A R D Q K V T A S Y K N A A S A N I S G S V P L E H F
nod26         G A W V Y N I V R Y T D K P L S E T T K S A S F L K G R A A S K

Group III
rAQP3         S P L L G S I G G V F V Y Q L M I G C H L E Q P P P S T E A E N V K L A H M K H H E Q I
mAQP3         S P L L G S I A G V F V Y Q L M I G C H L E Q P P P S T E E E N V K L A H M K H H E Q I
xAQP3         S P L L G S F A G V L V Y Q L M I G C H I E P A P Q S T Q Q E N I K L A D V K N K D R L
hAQP9         G P L V G A V I G G L I Y V L V I E I H H P E P D S V F K A E Q S E D K P E K Y E L S V +

Group IV
bAQP4         V E D N R S Q V E T D D L I L K P G V V H V I D I D R G E E K K G K D P S G E V L S S V
mAQP4         V E D N R S Q V E T E D L I L K P G V V H V I D I D R G E E K K G K D S S G E V L S S V
rAQP8         Y W D F H W I Y W L G P L L A G L F V G L L I R L F I G D E K T R L I L K S R
hAQP8         H W N F H W I Y W L G P L L A G L L V G L L I R C F I G D G K T R L I L K A R
mAQP8         Y W D F H W I Y W L G P L L A G L F V G L L I R L L I G D E K T R L I L K S R
```

Fig. 2. Alignment of the carboxy terminal domains of AQP1, other MIP family channels, and cyclic-nucleotide-activated ion channels suggests four groups (I–IV) based on sequence similarities. Bold type indicates conserved amino acid residues within a group. Underlined segments are similar sequences that may be displaced in position. The asterisk in Group I indicates a positively charged amino acid residue that is important for channel gating in retinal CNG channels. β7, β8, Bα and Cα represent putative structural elements of CNG channels within the binding domain. (+) indicates additional sequence not shown. Sources of sequences: human AQP0 – Ohtaka-Maruyama et al., 1998 [118]; mouse AQP0 – Shiels and Bassnett, 1996 [105]; bovine AQP0 – Gorin et al., 1984 [3]; human AQP1 – Preston and Agre, 1991 [4]; rat AQP1 – Li et al., 1994 [119]; human AQP2 – Deen et al., 1994 [120]; mouse AQP2 – Yang et al., 1999 [121]; rat AQP3 – Ishibashi et al., 1994 [122]; mouse AQP3 – Ma et al., 2000 [123]; *Xenopus* AQP3; Schreiber R et al. (unpubl.,1998 GenBank # AJ131847); rat AQP4 – Jung et al, 1994 [124]; bovine AQP4 – Sobue et al., 1999 [125]; mouse AQP4 – Turtzo et al., 1997 [126]; mouse AQP5 – Krane et al., 1999 [127]; rat AQP5 – Raina et al., 1995 [57]; rat AQP6 – Yasui et al., 1999 [66]; mouse AQP7; Ishibashi K (unpubl., 1998 GenBank # AB010100); rat AQP7 – Ishibashi et al., 1997 [128]; rat AQP8 – Ishibashi et al., 1997 [129]; Koyama et al., 1997 [130]; human AQP8 – Koyama et al., 1998 [131]; mouse AQP8 – Calamita et al., 1999 [132]; human AQP9 – Ishibashi et al., 1998 [133]; rat olfactory Olf(1) – Dhallan et al., 1990 [134]; rat olfactory Olf(2) – Bradley et al., 1994 [135]; fish olfactory Olf – Goulding et al., 1992 [136]; bovine retinal Ret – Kaupp et al., 1989 [137]; nodulin (soybean) nod26 – Fortin et al., 1987 [138]; fly (*Drosophila*) bib– Rao et al., 1990 [112]; fly (*Drosophila*) eag – Warmke et al., 1991 [139].

activation. Lysine 596 (marked by an asterisk in Fig. 2, Group I) markedly influences channel gating [20]. An equivalent positively charged residue is also present in AQP1. The amino acids that are conserved in nature and positioned between AQP1 and the CNG channels offer clues to the sites that may be important for ligand binding and channel activation. Future studies using chimeric substitutions and site-directed mutagenesis of AQPs might lend insight into the functional significance of the apparent sequence similarity with CNG channels.

The CNG-like pattern in the carboxyl terminus is not immediately apparent for the other members of the MIP family. Instead, there appear to be different patterns shared among the other MIP channels. Fig. 2 shows the segregation of MIP members into groups by patterns of amino acid sequence similarity in the carboxy tail, without reference to possible functional connotations. Group II includes AQP0, 2, 5 and 6 that share sequence elements at conserved positions along the length of the carboxy tail. This group may include Big Brain, AQP7 and Nodulin 26, although the similarities there are less robust. Groups III and IV are also sorted by patterns of similarity without reference to possible functional implications. Systematic scanning of a variety of second messenger cascades for effects on various AQPs could provide clues for linking structure with function in the channels of Groups II, III and IV, and might compel future revision of the groupings based on experimental evidence.

1.3. MIP family sequence homologies in loops B and E

The ion conducting channels identified thus far in the MIP family (AQP0, AQP1, AQP6, Nod26 and BIB) exhibit regions of amino acid sequence identities within loops B and E that can be compared with other members of the MIP family (Figs. 3 and 4). These and other conserved sequences may offer clues to the identities of the amino acids that are important for ion and water channel function.

The conservation of loop B sequences (Fig. 3) implicates this domain in an essential function, yet it is interesting to note that AQP1 channels are more tolerant of equivalent site-directed mutations in loop B than in loop E [21]. Mutation of N76 to D (NPA in loop B) produced channels that allow water flux while the equivalent mutation in loop E (N192D) resulted in a non-functional phenotype. Similarly, the mutant channel A78V showed virtually normal water transport, whereas A194V was non-functional. Nonetheless, other loop B mutations such as A73M do disrupt function [21]. From structural imaging studies, an integral involvement of loop B appears to be important for MIP channel structure and functionality [8–10]. The AQP loop B sequences show two highly conserved domains. The first domain (box 1) differs in the presence of either a glycine or an alanine residue in the third position, but the residues are otherwise identical. The boxed homology domains do not correspond to known differences in solute permeability, and thus do not distinguish between the water-selective and the glycerol-transporting AQPs. The second conserved domain (box 2) also shows high sequence identity. The differences are subtle and retain the character of the amino acid R group (hydrophobic V or I; and polar S or T). The lack of substantial differences in conserved sequences might imply that the B loop does not govern solute selectivity. However, mutation of the lysine residue of AQP6 (near box 1) to glutamate increased the relative permeability of the channel to cations [22]. Molecular imaging has shown that loop B contributes directly to part of the pore lining region that contains

```
                    1        2
hAQP0    Q S V G H I|S G A H V|N P A V T|F A F L V G S Q M S L L R A F C Y
mAQP0    Q T V G H I|S G A H V|N P A V T|F A F L V G S Q M S L L R A F C Y
bAQP0    Q A V G H I|S G A H V|N P A V T|F A F L V G S Q M S L L R A I C Y
hAQP1    Q S V G H I|S G A H L|N P A V T|L G L L L S C Q I S I F R A L M Y
rAQP1    Q S V G H I|S G A H S|N P A V T|L G L L L S C Q I S I L R A V M Y
hAQP2    Q A L G H I|S G A H I|N P A V T|V A C L V G C H V S V L R A A F Y
mAQP2    Q A L G H V|S G A H I|N P A V T|V A C L V G C H V S F L R A A F Y
rAQP3    L V A G Q V|S G A H L|N P A V T|F A M C F L A R E P W I K L P I Y
mAQP3    L V A G Q V|S G A H L|N P A V T|F A M C F L A R E P W I K L P I Y
rAQP6    Q I S W K T|S G A H A|N P A V T|L A Y L V G S H I S L P R A V A Y
mAQP7    H V A G G I|S G A H M|N P A V T|F T N C A L G R M T W K K F P V Y
rAQP7    H V A G G I|S G A H M|N P A V T|F T N C A L G R M A G R K F P I Y
fBIB     Q C F L H I|S G A H I|N P A V T|L A L C V V R S I S P I R A A M Y

rAQP4    Q C F G H I|S G G H I|N P A V T|V A M V C T R K I S I A K S V F Y
bAQP4    Q C F G H I|S G G H I|N P A V T|V A M V C T R R I S I A K S V F Y
mAQP4    Q C F G H I|S G G H I|N P A V T|V A M V C T R K I S I A K S V F Y
nod26    Y T V G H I|S G G H F|N P A V T|I A F A S T R R F P L I Q V P A Y
mAQP5    Q A L G P V|S G G H I|N P A I T|L A L L I G N Q I S L L R A I F Y
rAQP5    Q A L G P V|S G G H I|N P A I T|L A L L I G N Q I S L L R A V F Y
rAQP8    A T L G N I|S G G H F|N P A V S|L A V T L V G G L K T M L L I P Y
hAQP8    A T L G N I|S G G H F|N P A V S|L A A M L I G G L N L V M L L P Y
mAQP8    A T L G N I|S G G H F|N P A V S|L A V T V I G G L K T M L L I P Y
hAQP9    Y V A G G V|S G G H I|N P A V S|L A M C L F G R M K W F K L P F Y
```

Loop B

Fig. 3. Alignment of the Loop B domains of AQPs and related channels shows a highly conserved pattern in two segments indicated by boxes (1 and 2). The underlined lysine residue in AQP6 influences ionic selectivity. Sources of sequences: As in Fig. 2.

the selectivity filter for glycerol or water, and indirectly by linking structural domains in the channel [10,23].

In contrast to loop B, the patterns of sequence similarity in loop E do correspond to the permeability profiles known thus far for the various AQPs. The sequences for the last half of M5 (the 5th transmembrane domain) and all of loop E are sorted into three groups, based on the nature of the residue in the last position of box 5. The NPARS group contains serine in the fifth position, and includes the MIP channels that carry ionic conductances when reconstituted in lipid bilayers or expressed in *Xenopus* oocytes, including AQP0, AQP1, AQP6, Nod26 and BIB. The members of the NPARS group are thought to assemble and function as tetramers of subunits, as are the members of the voltage-gated ion channel superfamily. The serine in box 5 that distinguishes this group has been postulated to play an important role in channel assembly [24] or tetrameric stability. A valine in M5 (box 1) is suggested to be required for maximum water channel activity [25]. Notably, this residue (as well as others) differs for the NPARS channels shown to have low

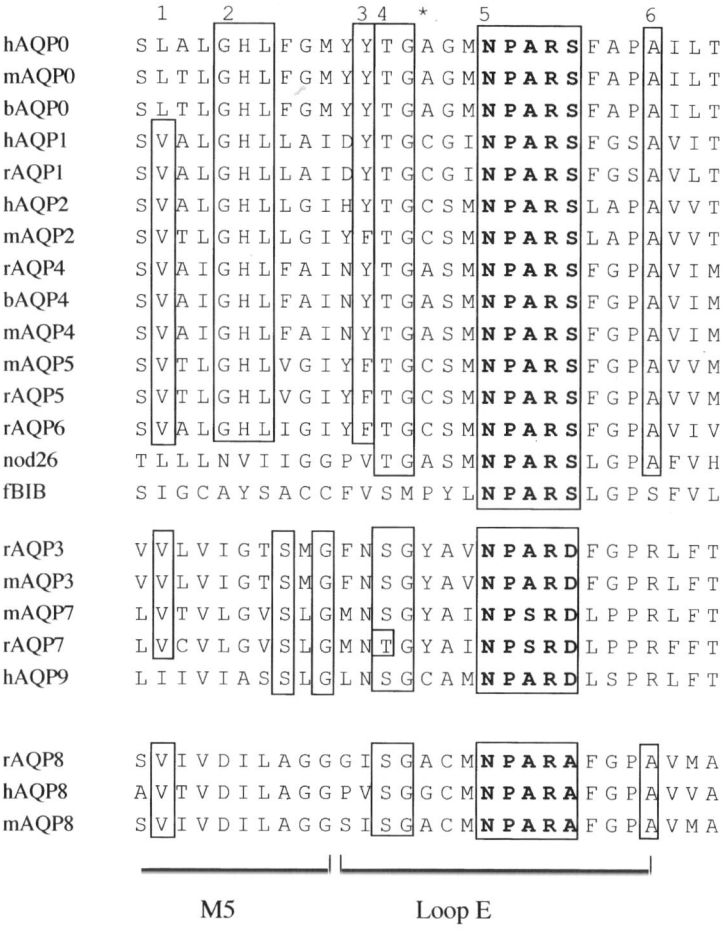

Fig. 4. Alignment of the Loop E domains of AQPs and related channels shows sequence similarities that distinguish the water selective (NPARS, box 5) and the glycerol transporting (NPARD, box 5) members of the aquaporin family. The sequence alignments also include the 2nd half of the 5th transmembrane domain (M5). Patterns of conserved sequences within each group are indicated by boxes (1–6). The asterisk indicates the location of the mercury-sensitive cysteine residue.

water permeability, including AQP0 [26] and *Drosophila* Big Brain [27]. Within the NPARS group, ion channel function has not been studied for AQP2, AQP4 and AQP5, so possible activation mechanisms have not been identified. AQP5 does not carry an ionic current in response to cGMP [13], but could be regulated by a different mechanism. The likelihood of alternative activation mechanisms is supported by the diversity of patterns of amino acid sequence seen in the carboxy terminal domains, where only AQP1 shows a discernable CNG-like pattern (Fig. 2).

The NPAR*D* group (Fig. 4) contains aspartate in the 5th position, and includes the glycerol-facilitator-related members of the AQP family. Prokaryotic glycerol facilitator protein (GlpF) in the MIP family carries the NAPRD motif in loop E, and appears to separate during protein isolation into monomeric subunits [24], although oligomeric complexes can be obtained depending on the isolation method [28]. The NPARA group comprises AQP8, but may expand as other members of the MIP family are cloned.

In AQP1, the residue Y186 (box 3, Fig. 4) determines the sensitivity of osmotic water flux to block by external tetraethylammonium (TEA) [29]. Water flux through AQP1 channels is inhibited approximately 30% by TEA, classically known as a blocker of K^+ channels. This sensitivity to TEA is removed by the mutation of Y186 to F [29]. The effect of TEA on water permeability provides new insight into AQP1 pore-blocking compounds other than mercury, and although not clinically favored may define a "lead" compound for the pharmaceutical development of related derivatives with potential relevance to the treatment of glaucoma, hydrocephaly, and other diseases involving AQP1-mediated water flux across membranes. By comparing sequences, one might speculate that water permeability of other AQPs could be sensitive to inhibition by external TEA. A role for tyrosine in the pore-lining sequence between M5 and M6 as a TEA-binding site has precedent in studies of voltage-gated K^+ channels that showed that the presence of tyrosine increased the affinity for TEA [30]. The critical tyrosine in K^+ channels is separated by two residues from the GYG sequence that contributes to the ionic selectivity filter and constitutes a "signature sequence" for K^+ channels [31]. Interestingly, in AQP1 the critical tyrosine (Y186) is separated by one residue from the sequence GCG that contains a cysteine residue (C189, Fig. 4, asterisk) that enables the block of water flux by Hg^{2+} [32], and this may similarly contribute to one of the conduction-limiting sites within the water permeation pathway. These similarities hint at a conservation of general strategies for matching designs with functions across AQPs and voltage-gated ion channels.

1.4. The five-pore hypothesis for aquaporin channels

Structural models based on imaging indicate that MIP channels have independent pores for water or glycerol in each subunit. The individual subunit pores are lined by loops B and E, folded inward to meet at the center and to create a transmembrane pathway for solute movement. The 5th and 6th transmembrane helices (M5 and M6) and loop E are critical determinants of water permeability properties in AQPs. In chimeras made from AQP0 (low water permeability) and AQP2 (high water permeability), transfer of the AQP2 region containing the second half of M5 and loop E into AQP0 conferred the high water permeability properties of AQP2, indicating that the water conduction property was strongly influenced by loop E and its flanking regions [25]. Site-directed mutagenesis of an insect AQP channel (AQPcic) to convert serine S205 in NPARS to the corresponding aspartate of GlpF disrupted permeability to water [24], perhaps by destabilizing subunit interactions. Mutation of alanine (box 6 of the NPARS group) to lysine (as found in GlpF) also disrupted water permeability, but did not compromise tetrameric stability. Mutation of AQPcic residues (Y, W) in transmembrane domain M6 to the corresponding amino acids of GlpF (P, L) switched the selectivity of the channel; water permeability was lost but a new glycerol permeability similar to that of GlpF was created [33].

Well before structural images were available for AQPs, functional analyses suggested that individual water pores existed in each subunit, rather than at a central site of interaction between subunits [34], although these studies were not definitive. When Hg^{2+}-insensitive C186S and wild-type AQP1 were co-expressed in oocytes, the magnitude of the Hg^{2+}-sensitive osmotic water permeability was equal to that expected from the ratio of wild-type subunits, suggesting the contribution of each type of subunit was simply additive and thus that the subunits were independent [32]. However, this interpretation presumed that a single Hg^{2+} binding site in a central pore would reduce water permeability completely. If instead each Hg^{2+}-bound cysteine contributed proportionally to the magnitude of block, then the same data could have been consistent with a central pore hypothesis. In another study, tandem fusion constructs of AQP1 wild-type with non-functional mutant channels (C189W) showed that the osmotic water permeability of the wild-type/mutant dimer was approximately half that of the wild-type/wild-type dimer [35]. This supported the idea of subunit independence with wild-type being unaffected by the presence of non-functional components in the same tetramer; however, the investigators acknowledged the possibility that the non-functional subunits might not have contributed to the channel core. This caveat was worth considering particularly since trimer constructs also showed some unexpected functionality. In theory, a functional tetrameric channel should not have been able to assemble if all subunits in the trimer contributed to the pore (unless monomers were present to fill the 4th position). A 3rd line of evidence for the independent subunit hypothesis came from radiation-inactivation studies that suggested the functional water channel has a target size of 30 kD, which is about the mass of a single subunit (28 kD) [36]. The independent subunit hypothesis was favored, but not conclusively demonstrated until the crystal structure of the glycerol facilitator GlpF showed glycerol molecules lined up in transport pathways within each individual subunit [10]. A similar strategy for water permeation through individual subunits of AQPs is a reasonable hypothesis.

The hydrophobic nature of the independent pores for water or glycerol within each subunit is not consistent with ionic permeation, and suggests that the regulated ionic conductance pathway is located elsewhere in the AQP protein complex. Insight may be obtained by considering the voltage-gated channels, which have a single pore located in the center of the tetramer [37]. The voltage-gated ion channel pore is lined by P domains (the loops between the 5th and 6th transmembrane helices) contributed by each subunit [38], and also is influenced by the 6th transmembrane domains [39,40]. Similar structural themes in related ion channels include the two-pore-domain potassium channels made of dimers of subunits with four transmembrane segments and two P domains in each, and the inward rectifier class of channels made of tetramers of subunits with two membrane spanning domains and one P domain in each [38]. The ion channel family, despite their different strategies, are envisioned as operating with a single central pore formed from the association of four subunits or pseudosubunits. Tetrameric subunit organization similarly is required for AQP function [1]. It is conceivable that AQP-mediated ion conductances may also rely on a central regulated pore located in the center of the tetramer of subunits.

An attractive hypothesis is that ion-channel AQPs have a five-pore structure, with four constitutively active water pores located in individual subunits, and a fifth regulated ion-selective pore in the center of the tetramer [41]. The structure of AQP1 determined by electron crystallography shows a distinct central pore-like passageway in the middle of the tetramer [42]. By analogy with voltage-gated ion channels, Loop E at first might appear to be a good candidate for the functional equivalent of the P region. However, high resolution analyses of AQP1 indicate that the central cavity is bounded by helical domains M2 and M5, not by Loop E directly. Thus, the interesting correlation between the sequence of Loop E and the ion conduction properties of MIP channels (see Fig. 4) suggests that although Loop E may not line the central region, it could influence regions of the protein (such as M5) that are in a position to form a central ionic pore.

The transmembrane passageway in AQP1 at the central 4-fold axis of symmetry is formed by M2 and M5 and narrowest on the extracellular side. At that site, a 4-fold symmetry of glutamine residues contributed by each of the M2 domains creates the narrowest constriction, with a diameter estimated at less than approximately 3 Å [23]. One could postulate that such a structure, perhaps in combination with other limiting sites, could contribute to channel gating. Ion selectivity studies demonstrated that the activated AQP1 channel is permeant to K^+, Na^+, Cs^+ and to a lesser degree, TEA [43]. Thus, if ion flow is mediated by the central pathway, these data would suggest that channel opening must be accompanied by a conformational change that effectively increases the limiting size of the pore.

Despite evidence for independent pores for water or glycerol in subunits, it is clear that the tetrameric organization of the channel is essential. An interesting set of complementation studies showed that non-functional mutations of AQP1 channels could be rescued by co-expression, demonstrating functional complementation [21] that is consistent with necessary interactions between subunits. Perhaps the most compelling evidence for a necessary interaction between subunits is the demonstration of a dominant negative suppression of function by the co-expression of wild-type AQP1 with a non-functional AQP1 mutant A73M [34]. These results were not attributed to gross misfolding, but instead suggested that the mutant subunit must somehow interfere with the function of the wild-type subunit. The dominant negative approach has been utilized by many researchers in the K^+ channel field, taking advantage of the ability of certain mutations in the pore region to prevent ion channel conductance even if only one subunit in the tetramer is defective. Thus it appears that the "independent" individual pores for water in AQPs are not truly independent of one another, but rely on interactions in the tetrameric complex for water channel functionality.

2. Ion channel function in MIP proteins

AQPs 0, 1, 2, 4, 5 and 8 have been classified as "water selective" channels, whereas AQPs 3, 7 and 9 have been classified as a "small molecule" transporter group for substrates including glycerol, urea, and other molecules [44]. The simple segregation of these channels into "orthodox" AQPs (strictly water permeant) and less selective aquaglyceroporins [45] may be premature, as increasing evidence suggests that members of the water-selective channel group may also mediate fluxes of other molecules, including ions. Along with the similarities in transmembrane topology, tetrameric organization, and structure–function relationships, these findings would place AQPs within the larger group of ion channel proteins.

2.1. Aquaporin-0 (lens MIP) ion channel activity in bilayers

AQP0 is abundant in lens and shows a relatively low permeability to water (less than 3% of that of AQP1 [46]). Reconstitution of bovine and chicken lens AQP0 proteins into lipid bilayers yields large conductance ion channels [47–49]. Bovine single channel amplitudes in 100 mM KCl saline are bimodal, with unitary conductances of 180 and 380 pS. These channels display long open times, with symmetrical activation over a range of voltages. Conductance increases with ionic concentration; at 1 M KCl, the unitary channel conductances are 3700 and 3100 pS for the high conductance state, and 1600 and 1200 pS for the low conductance state. The voltage dependence of AQP0 ion channel activity is modulated by phosphorylation [50]. Similarly, chicken lens AQP0 reconstituted in lipid bilayers yields channels with unitary conductances of 60 and 290 pS, with the higher conductance state favored at potentials within $\pm$ 10 mV [49].

Although ion channel function was not observed for bovine AQP0 expressed in oocytes during osmotically induced swelling [26], in this study only the signaling effects that might accompany a volume change were tested. Frog AQP0 expressed in *Xenopus* oocytes increased water permeability and glycerol permeability over control levels, and caused a depolarizing shift in the reversal potential of the total membrane current. A change in reversal potential typically indicates an alteration in ionic selectivity of the membrane currents, but the effect in this case was not attributed to ion channels because of the lack of change in total current amplitude [51].

2.2. Aquaporin-1 ion channel activity gated by cyclic GMP

The activation of ionic permeability in AQP1-expressing oocytes requires intracellular signaling, whereas water permeability is a feature of both the stimulated and unstimulated states [43]. AQP1 expressed in oocytes also has been reported to mediate transport of other solutes such as CO_2 [52,53] and glycerol [54], although not in all cases [55,56,140]. While controversies surround AQP-mediated transport, it is important to recognize that intracellular signaling pathways may influence relative solute permeabilities, and that the various experimental findings are not necessarily mutually exclusive.

The ability of AQP1 to form functional ion channels [43] was initially disputed, but has since been confirmed [13,14]. Without stimulation, *Xenopus* oocytes injected with AQP1 cRNA show no difference in ionic permeability as compared with control oocytes [5,43,57]. However, after stimulation with cGMP, large-conductance non-selective cation channels are observed [13], as shown in Fig. 5. The properties of AQP1 channels show interesting similarities to those of AQP0 and Nodulin26, in that they have a large unitary conductance and long open times; however, AQP0 and Nod26 show a weak selectivity for anions whereas AQP1 is cation-selective ($K^+ = Cs^+ \geq Na^+ > TEA^+$). The conductance of AQP1 channels in 100 mM KCl saline when expressed in oocytes is $\sim$ 150 pS [13]. The unitary channel conductance for reconstituted AQP1 in bilayers is 2.4, 5.9 and 9.8 pS [14].

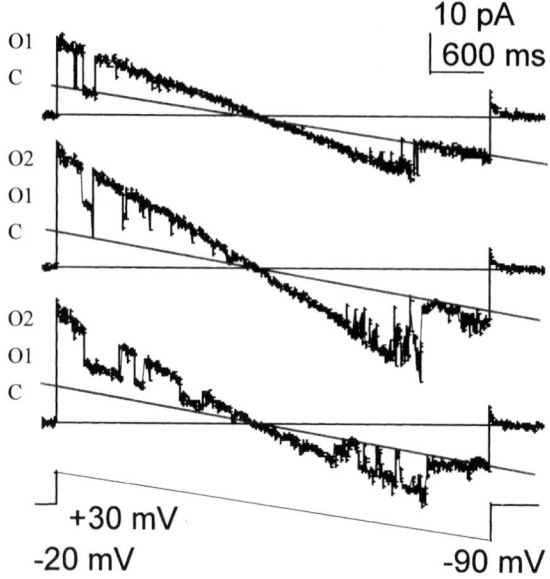

Fig. 5. Activation by cGMP of large-conductance long-opening channels in inside–out patches from AQP1-expressing oocytes. Voltage ramps from +30 to −90 mV show cGMP-induced activity in 120 mM KCl saline (with 5 mM MgCl$_2$ and 10 mM HEPES, pH 7.3) with long open times and approximately linear current-voltage relationships. Traces are unsubtracted; the baseline is marked by a slanted line. C is closed; O1 is one channel open; O2 is two channels open.

This difference may provide clues to additional contributions of cytoskeletal associations, regulatory proteins, kinases, or second messengers that modulate channel properties.

Initial uncertainty with regard to AQP1 as an ion channel may have stemmed from the indirect nature of forskolin activation, reflecting a variability inherent in the oocyte system. The lack of reliability in the response to forskolin was first evident in a set of technical comments published in *Science* (1997, 275, pp. 1490–1492). Seasonal and genetic variability [58] as well as variability in cAMP signaling pathways in *Xenopus* oocytes have been known for more than a decade [59]. The effect of forskolin is mediated by the surrounding follicular cells via gap junctions [60,61], and in routine preparations of oocytes, the removal of follicular cells from the inner vitelline membrane varies between 20 and 100% [60]. The variability in cAMP signaling has been suggested post hoc to result in part from differential retention of follicular cells [61]. Patil and colleagues subsequently confirmed increased water flux in AQP1-expressing oocytes after direct injection of 8Br-cAMP or stimulation with arginine vasopressin, thought to act via PKA [62]. The ion channel function of AQP1 was subsequently confirmed in studies demonstrating that the direct activating agent was cGMP [13,14].

A schematic diagram (Fig. 6) illustrates a possible hierarchical pathway for ionic current activation of AQP1 channels in *Xenopus* oocytes, and incorporates the indirect nature of the mechanism of action of forskolin. As illustrated, AQP1 ion channels are activated by direct binding of cGMP [13,14], although more slowly than are the diffusion-limited retinal

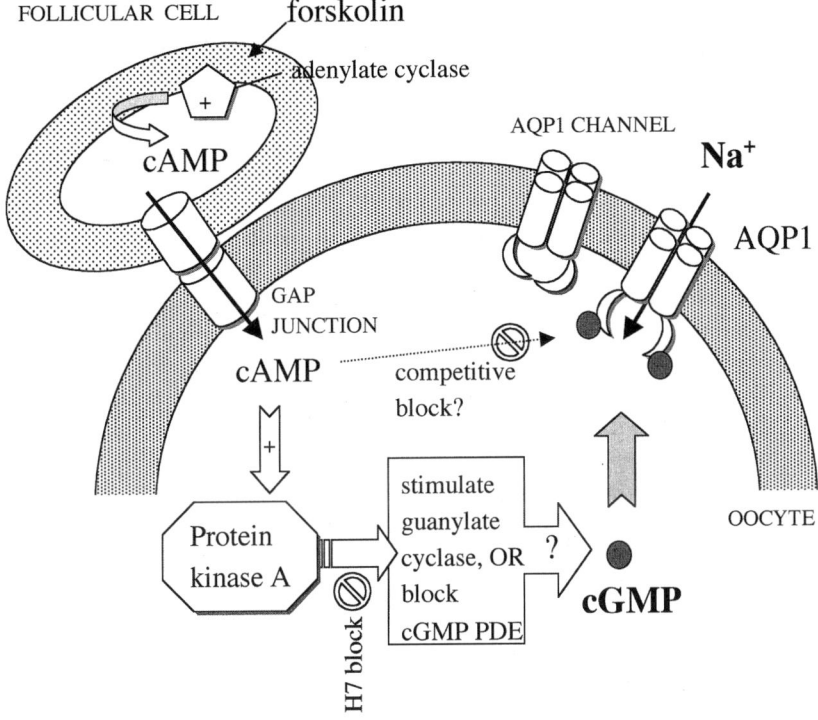

Fig. 6. Schematic diagram of a hypothetical signaling hierarchy leading to the activation of ion channel function in AQP1 channels expressed in *Xenopus* oocytes. Forskolin-sensitive adenylate cyclase is present in associated follicular cells and provides cAMP to the oocyte via gap junctions. The ionic conductance also is activated with direct injection of catalytic subunit of protein kinase A (PKA). The effects of forskolin and PKA are blocked by pre-treatment with the kinase inhibitor H7. The effect of PKA on cGMP levels is not known, but may occur by stimulating guanylate cyclase, by inhibiting cGMP phosphodiesterase (PDE), or by modulating other signaling cascades. Channel activation by cGMP is direct, not blocked by H7, and generated in excised patches removed from the cytoplasmic environment, indicating direct binding at the channel. Elevated cAMP may be inhibitory; thus the level of stimulation (or block) of ion channel activity by forskolin may depend on the balance achieved between cGMP and cAMP in a given oocyte.

CNG channels [63]. The effect of cGMP is *not* blocked by pre-treatment with the kinase inhibitor, H7; whereas, the effects of cAMP and PKA are blocked by H7. Activation of AQP1 channels by forskolin appears to rely on a complex balance between the level of activation of PKA, a positive inducer, and the level of cAMP, which appears to act as an antagonist (or a competitive weak agonist). The antagonistic effect of cAMP is evident in the ability of forskolin to inhibit the ionic conductance response, once it is maximally activated by cGMP (Yool, unpublished observations). Thus, the stimulatory action of cAMP appears to occur upstream of PKA; conversely, cGMP acts directly on the channel.

The pathway linking protein kinase A to the cGMP-regulated ionic conductance is unknown; however, there are several possibilities. PKA may activate the AQP1 conductance by:

(1) phosphorylating an unidentified accessory protein or by modulating other protein kinases;
(2) phosphorylating AQP1 at an atypical consensus sequence; or
(3) increasing cGMP concentration, either by enhancing guanylate cyclase or by decreasing cGMP phosphodiesterase activity.

Inter-dependent regulatory mechanisms for cGMP and cAMP are involved in *Xenopus* oocyte maturation [59,64]. "Crosstalk" or mutual regulation occurs between these signaling pathways [64,65]. The hierarchy in Fig. 6 represents a plausible scheme based on currently available evidence, and provides a set of testable predictions. Further studies are needed to dissect the signaling pathways involved in the physiological regulation of AQP1 channel function.

2.3. Aquaporin-6 ion channel activity regulated by pH

AQP6 is found in renal epithelia where it is associated with the membranes of intracellular vesicles [66]; however, in AQP6-expressing *Xenopus* oocytes, the protein is routed to plasma membrane providing a convenient means for functional analyses. Hg^{2+}, known as a blocker of water channels, stimulates both water permeability and ionic conductance (anionic) in AQP6-expressing *Xenopus* oocytes. The anionic conductance also is induced by external pH less than 5.5, and may have physiological significance in allowing for electroneutrality of proton transport at low pH [22].

AQP 6 shows a low permeability to water under standard physiological conditions, but within seconds after exposure to $HgCl_2$ undergoes an approximately 8-fold increase in osmotic water permeability [22]. The effect is accompanied by an increase in ionic permeability, yielding a nearly linear current–voltage relationship with a reversal potential at $-22\,mV$ in 100 mM NaCl saline. The currents attributed to AQP6 are similar to those described previously for AQP1 [43] in kinetics and near linear current-voltage properties, but differ in that AQP6 appears to be anion- rather than cation-selective. Mercury activates AQP6, but blocks AQP1 by covalent modification of a cysteine residue at position 189 [32] in a putative pore region (loop E) of the channel. Mutations of Cys 155 (in M4) and Cys 190 (in loop E) of AQP6 each caused a reduction in the Hg^{2+}-induced permeability to water and to ions, indicating that the effect of mercury is mediated by covalent modification of multiple cysteine residues in the AQP6 protein.

Low pH is suggested to be the natural activator of AQP6 channels [22]. The membrane conductance is inactive at pH 7.5, and induced reversibly by decreasing the external pH to 4.0. Although effects of acidic pH on native oocyte channels may be a concern, the response can be attributed at least in part to AQP6 proteins, in that mutation of a residue in the loop B region, Lys 72 to Glu, alters the ionic selectivity of the conductance.

2.4. Big Brain (Drosophila BIB) ion channel activity modulated by tyrosine kinases

Loss of the neurogenic gene for Big Brain (*bib*), a member of the MIP family, compromises cell–cell interaction (lateral inhibition) in early *Drosophila* embryogenesis and leads to an overabundance of neuroblasts and resultant neural hypertrophy at the expense of ventral epidermis [67]. The fact that *bib* has a substantial influence on cell fate indicates it must participate in some aspect of transmembrane signaling, but its mechanism of action has remained a mystery. A proposed ion channel function for BIB is consistent with results showing that the *Drosophila* protein in vivo is found in plasma membranes of cells that require neurogenic gene activity for their development, and that when over-expressed its signaling function can augment the activity of the Notch-Delta pathway [68]. In concept, BIB would function in a subset of epidermal cells to maintain the epidermal fate and to suppress the neural fate, by mediating the response to a lateral inhibition signal [68].

Studies of BIB channels heterologously expressed in *Xenopus* oocytes have shown that the protein functions as a tyrosine-kinase-regulated cationic channel [27]. Two-electrode voltage clamp recordings showed that BIB-expressing oocytes carry a cationic membrane conductance that increases approximately 20-fold over basal levels within 25 min of recording, whereas control oocytes showed little change in identical conditions. Pre-treatment with tyrosine kinase inhibitors increased the net conductance triggered by prick activation; conversely, stimulation of an endogenous tyrosine-kinase-linked receptor inhibited the conductance response [27]. These data support the intriguing speculation that the BIB channel conductance is reduced by factors that act via tyrosine-kinase-linked pathways, and that BIB might generate a neural-fate suppressing signal through a shift in resting membrane potential. In theory, the activation of BIB channels, known to be expressed in cells surrounding the neuroblast, should depolarize the membrane potential. Consistent with this idea, Goodman and Spitzer found in grasshopper embryos that the membrane potential of differentiated neuroblasts was -60 to $-80\,\text{mV}$, while the surrounding non-neural cells were relatively depolarized with potentials of -40 to $-60\,\text{mV}$ [69]. The role of BIB ion channel activity in regulating developmental cell fate remains to be demonstrated.

2.5. Other ion channels

The protein Nodulin-26 from soybean symbiosome nodules forms an ion channel when reconstituted in lipid bilayers, and increases water permeability when expressed in oocytes [70,71]. Single channels show unitary conductances of 3100 pS in 1 M KCl, and subconductance states ranging from 500 to 2500 pS [70]. Tris ion at the trans side of the bilayer appeared to act as a channel blocker. Possible ion channel functions of other MIP proteins remain to be evaluated.

The possibility that ion channel activity in oocytes results from upregulation of an endogenous channel must be considered [72,73]. However, this explanation seems unlikely given the differences in channel properties reported for the various MIP channels expressed in *Xenopus* oocytes. AQP1, AQP6 and BIB differ in their mechanisms of activation, ionic selectivity, and sensitivities to blocking agents. Furthermore, the properties of the native channels that are upregulated by high levels of exogenous protein expression differ from those reported for the MIP-mediated conductances. Thus, the *Xenopus* oocyte appears to be justified as a reasonable system for the analysis of AQP ion channel function, just as it has been proven to be a valuable tool in many studies of voltage-gated channels and ligand-gated receptors.

3. Physiological and pathophysiological significance of aquaporin channel activity

3.1. Regulatory pathways

Many integral membrane proteins are regulated by intracellular second messengers and modifiers that affect their activity or cellular localization. The first three AQPs cloned, AQP0, AQP1 and AQP2, have been well characterized in this regard.

Water permeability of AQP0 is modulated by pH, and by intracellular Ca^{2+}, with the latter thought to act via calmodulin [74]. Voltage-dependence of AQP0 channel function is altered following phosphorylation of Ser243 and Ser245 by protein kinase A [50]. AQP0 is also phosphorylated by protein kinase C both in vitro and in vivo; however, the biological consequences are unknown [75]. Among the possible functional consequences is a decreased permeability to water, as observed following phosphorylation of AQP4 by PKC [76].

Arginine-vasopressin regulates water reabsorption in renal collecting duct epithelial cells. Vital to its antidiuretic action in mammals is the insertion of AQP2 from intracellular vesicles into the apical membrane of epithelial cells. Binding of arginine-vasopression to V2 receptors on renal collecting duct epithelial cells initiates an increase in cAMP, activation of protein kinase A and phosphorylation of AQP2 at Ser 256, a PKA consensus site [77,78]. SNAP-23 in rat kidney co-localizes with AQP2, providing insight into the mechanisms that may govern hormone-dependent shuttling of AQP2 to the plasma membrane [79].

AQP1 responds to second messengers in a cell-type dependent manner. As with AQP2, intracellular vesicles containing AQP1 translocate and insert into the plasma membrane in response to elevated cAMP. This event occurs in some but not all cell types [80–82] and may or may not explain the increased water permeability of AQP1 in response to elevated cAMP [43,83]. PKA-dependent phosphorylation at an atypical consensus site in the carboxyl terminus of AQP1 has been implicated in this process [43,81]. PKA dependent phosphorylation of native AQP1 protein from renal preparations [81] or recombinant AQP1 from expressing oocytes yielded only weak or undetectable signals (Stamer, Yool and Regan, unpublished observations), suggesting that the effects of PKA on AQP1 function may be indirect or occur with low fidelity. In addition to affecting AQP1 water permeability or cellular location, recent data has shown that the direct binding of cGMP to a cytoplasmic domain gates a cationic current through AQP1 [13]. Interestingly, this effect appears to be modulated by cytoplasmic levels of other second messengers.

Modulation of AQP function was once thought to be limited to AQP2 and its role in antidiuresis. This stereotype has been challenged by further characterization of other AQP family members, revealing a myriad of responses to second messenger systems that are cell-type specific. Simple analyses of sequence from novel AQPs reveals the presence of consensus sites for various post-translational modifications, including consensus PKA sites in AQP5 and AQP9. However, whether these sites have functional effects remains to be demonstrated.

3.2. Role of AQP1 in epithelial and transmembrane transport

AQP1 channels are expressed in a specialized subset of mammalian tissues that require enhanced transmembrane water movement, including kidney (proximal tubule and descending Henle's loop), eye (ciliary epithelia, trabecular meshwork and canal of Schlemm), brain (choroid plexus), lung (peribronchial vasculature), and other tissues including lymph vessels and muscle [84]. At present, the role of a regulated ionic conductance of AQP1 channels with respect to the physiology of these various tissues is unknown. It will be important to carry out experiments on tissues in which AQP1 and other MIP channels are expressed to confirm the physiological relevance of the ion channel function, and to begin to investigate how the dual permeability of these channels to both water and ions might influence the processes of net water and salt movement across epithelial membranes.

As a first step towards conceptualizing roles for AQP1 ion channels, it is possible to ask whether the properties of the systems in which these channels are expressed are consistent with a gated cation channel component. It is important to state that these ideas are necessarily speculative in the absence of experimental data, but may serve as a framework for formulating testable hypotheses to assess the overall significance of these channels to the intact functioning organism. For example, in the eye, there are indications of cGMP-gated channel function, but the component potentially contributed by AQP1 remains to be determined. AQP1 channels are expressed in trabecular meshwork cells [85,86]. Bovine trabecular meshwork cells in culture analyzed by whole-cell patch clamp showed a robust outward conductance [87]. Approximately 35% of the conductance was blocked by charybdotoxin, an antagonist of calcium-dependent K^+ channels. 8Br-cGMP (1 mM) increased the outward current almost 300%; only half of this response was blocked by charybdotoxin. The identity of the insensitive component of the response was not determined, but in theory might include cGMP-activated AQP1 channels. Ciliary epithelial cells of the eye express AQP1, AQP4 and guanylate-cyclase ANP receptors. ANP stimulates cGMP production, and reduces intraocular pressure by altering trans-membrane water flux [83,88]. The role of AQP1 channels in these cGMP-mediated changes is currently unknown. Other hypothetical examples of possible AQP1-mediated contributions to absorption and secretion processes are proposed for choroid plexus and for the renal proximal tubule below.

3.3. Potential role in choroid plexus

AQP1 channels abundantly expressed in the apical membrane of choroid plexus are thought to serve in the production of cerebral spinal fluid (CSF) [89]. Choroid plexus produces about 90% of the total extracellular fluid in the brain. Given the rigid confines of the skull, an exact control of volume and osmotic composition of the CSF is essential for normal brain function [90], and several regulatory pathways converge on ion channels and transporters. Pathological changes in brain volume arise from a number of conditions: (i) Vasogenic edema, a breakdown of the blood–brain barrier (due to tumors, infarcts, hemorrhage or meningitis); (ii) Hydrocephalic edema due to obstruction of the outflow pathways; (iii) Cytotoxic brain swelling, resulting from conditions such as head trauma, ischemia, stroke and hypoxia; and (iv) Anisotonic cell volume changes, resulting from perturbations in extracellular osmolality associated with a number of diseases including diabetes, congestive heart failure, renal failure, or clinical manipulation of plasma osmolality [90].

A diagram of the signaling pathways in choroid plexus and the possible effects of stimulation on the transepithelial movement of salt and water is shown in Fig. 7. Unlike most epithelial cells, the $Na^+-K^+-ATPase$ transporter is located in the apical membrane in choroid plexus [91]. The $Na^+-K^+-ATPase$ pump provides the primary driving force for fluid secretion into the ventricle, as osmotically obliged water follows the transport of Na^+. Atrial natriuretic peptide receptors in choroid plexus epithelia couple to guanylate cyclase, stimulate cGMP generation [92,93] and inhibit CSF production [94]. Single channel studies have characterized K^+ and Cl^- channels in choroid plexus cells that are regulated by serotonin [95]. Activation of the serotonin receptor increases IP_3 and Ca^{2+},

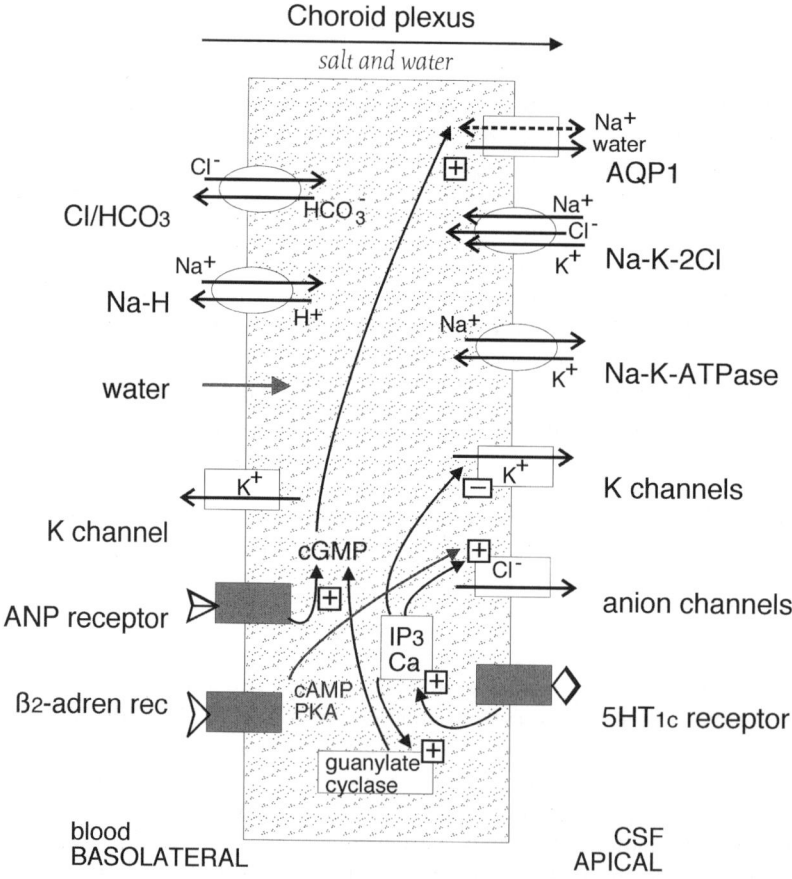

Fig. 7. Schematic diagram of some of the intracellular signaling pathways that may be involved in modulating the functional properties of choroid plexus during secretion of CSF. See text for details.

and stimulates guanylate cyclase [96]. An increase in cGMP induced by serotonin, ANP, or muscarinic receptors inhibits secretion; whereas agents that elevate cAMP stimulate secretion by enhancing apical Cl^- and HCO_3^- conductances (reviewed in Ref. [95]). The constitutively active $Na^+-K^+-2Cl^-$ co-transporter located in apical membrane has been proposed to function in reabsorption of K^+ from CSF [97]. Activation of a cGMP-gated cationic conductance in AQP1 channels by serotonin would be expected to increase the back leak of Na^+ across the apical membrane, and in combination with known inhibition of K^+ channels in the membrane would be consistent with the observed effect of serotonin in decreasing net secretion of CSF. The effects of cAMP also are consistent with the possible inhibition of AQP1 channels, thus stimulating secretion by decreasing Na^+ back leak in combination with the known stimulation of Cl^- channels. Further work is needed to detail the cellular and hormonal mechanisms that control solute and water movement across the choroid plexus under normal conditions and after the imposition of osmotic disturbances.

3.4. Potential role in renal proximal tubule

Approximately 60–70% of the filtered Na^+ and water is absorbed in the proximal nephron by an efficient process, estimated to move as many as 9 Na^+ ions per ATP expended. A small transepithelial voltage of 1–2 mV, positive in the lumen, is thought to drive the passive paracellular movement of Na^+ and Cl^- from lumen to blood, although it is difficult to distinguish precise quantities for paracellular versus transcellular flow [98]. A diagram of possible signaling pathways involving cGMP is shown in Fig. 8. Atrial natriuretic peptide (ANP) elevates cGMP and reduces Na^+-coupled transport in proximal tubules [99]. Proximal tubule expresses high levels of inducible nitric oxide synthase (iNOS) and soluble guanylate cyclase. Nitric oxide at high doses decreases sodium reabsorption 50–70%, thought to be due to inhibition of the Na^+-K^+-ATPase and the Na^+-H^+ exchanger. In contrast at low levels of nitric oxide, proximal Na^+ reabsorption is stimulated 30 50% via an unknown mechanism [100]. It is interesting to speculate that the AQP1 channels activated by cGMP might enable an enhanced Na^+ entry from the luminal side, promoting Na^+ reabsorption through activation of the basolateral Na^+-K^+-ATPase. Conversely, depolarization of the membrane potential by AQP1 ion channel activation might decrease Na^+ entry and contribute to the inhibitory response.

Model-based calculations of the conductance of the cGMP-activated AQP1 suggest that its impact on proximal tubule function could be physiologically relevant [41]. If an AQP1 non-selective cationic conductance was activated only in luminal membrane of proximal tubule, it could mediate a 45% increase in cell Na^+ entry (an 18% increase in overall epithelial Na^+ reabsorption), along with a -2 mV change in epithelial potential difference from a baseline of -0.4 mV. Concomitant peritubular activation of a similar conductance would slightly decrease these effects.

More experiments are needed to define the possible roles of AQP1, and to extrapolate the physiological significance from the oocyte to a native tissue. A cGMP-activated channel found in renal proximal tubule was found to show a unitary conductance of 150 pS

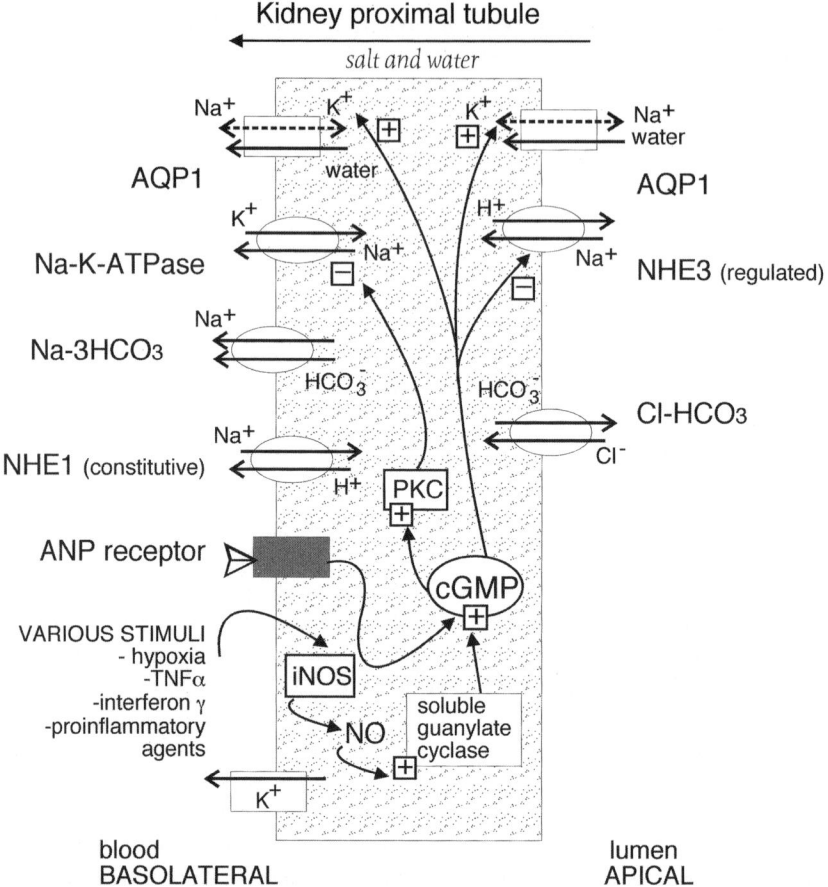

Fig. 8. Schematic diagram of some of the intracellular signaling pathways that may be involved in modulating the functional properties of renal proximal tubule during the reabsorption of salt and water from the filtrate. See text for details.

and was maximally activated by 1 mM cGMP, but not by cAMP [101]. Activation was not affected by kinase inhibitors, suggesting direct ligand binding. A decreased driving force for Na^+ entry with ANP was attributed to depolarization caused by activation of cGMP-gated channels. These findings parallel the properties of AQP1 channels that were characterized in oocytes [13], with one exception: Darvish and colleagues [101] suggested that their novel ion channel carried chloride, based on the apparent reversal potential. Ionic substitution experiments are needed to confirm the ionic selectivity of the cGMP-gated channels of proximal tubule.

In AQP1 knockout mice, the osmotic water permeability of the proximal tubule is decreased by approximately 80%, whereas water reabsorption is reduced by only 50% [102]. It is interesting to note that the back leak of Na^+ across the proximal tubule is also reduced in the AQP1 knockout mouse; altered paracellular flux is a suggested mechanism but these data also would be consistent with a loss of cGMP-stimulated AQP1 cation channels.

3.5. Disease states and ion channelopathies

With the establishment of ion channel function for several members of the MIP family, these proteins may join the ranks of ion channel types that underlie diseases collectively known as "ion channelopathies." A large number of inherited diseases have now been linked to specific mutations in genes encoding Na^+, K^+, Ca^{2+} and Cl^- channels, and include skeletal muscle myotonias, neurodegenerative diseases, ataxias, malignant hyperthermia, long QT syndrome in heart, and others [103]. The precise roles of ion channel function and dysfunction in the physiological processes controlled by AQPs remain to be identified. A link between a disease state and AQP4 expression is suggested from studies of the dystrophin-deficient *mdx* mouse, a model of muscular dystrophy. The observed reduction in expression of AQP4 channels in *mdx* mouse muscle and brain suggests a link between AQP4 and other components of the dystrophin complex, perhaps mediated by PDZ binding domains [104].

When expressed at high concentrations as in native vertebrate lens fiber cells, dysfunction of AQP0 channels is associated with cataract formation [105]. AQP0 is vital in maintaining the optical clarity of the crystalline lens; missense mutations in human AQP0 underlie inherited autosomal dominant cataract disease [106]. The essential function may be water permeability, structural organization, or another property (such as ion channel function) with a role that has not yet been identified.

AQP1 is present in red blood cells, kidney proximal tubule and thin descending limb of Henle, vascular endothelia, aortic vascular smooth muscle, choroid plexus, ciliary body and in other regions. The distribution of AQP1 protein suggested potential roles in pathophysiology of the eye, kidney, choroid plexus, developing lung, and in the vasculature [84]. AQP1 expression appears to be subject to regulatory mechanisms involving growth factor or steroid hormone signaling [107,108]. Insight into the essential renal role of AQP1 in the formation of concentrated urine came from transgenic knockout mice, which are grossly normal in survival, appearance and organ structure, but are susceptible to severe dehydration when deprived of free access to water [109]. Although there are rare cases of humans with homozygous mutations in AQP1, as identified by the absence of associated Colton blood group antigens [110], the low frequency of occurrence in humans suggests that AQP1 mutations may not be well tolerated and that AQP1 plays an essential physiological role. Although the rare affected humans are overtly normal, they suffer renal insufficiency as do the transgenic mice, and must ingest large volumes of water to mitigate rapid dehydration.

AQP6 is present in intracellular vesicles in renal epithelia [66] where it is proposed to play a role in enabling electroneutral transport of protons by a V-type H^+-ATPase at low pH levels that would preclude activity of the intracellular chloride channel ClC-5 [22]. The basis for the restricted distribution in renal epithelia is not yet understood. AQP6 shows a low water permeability [111] that suggests its major contribution is in mediating Cl^- fluxes; however, water permeability could be involved in volume changes associated with membrane fusion or other processes. Functional consequences of mutations in AQP6 may better define a physiological role in its native setting.

The gene for *big brain* (*bib*), cloned in 1990, was observed to have a significant sequence similarity with lens MIP AQP0 that prompted the suggestion that the Big Brain protein (BIB) might enable passage of small molecules across the membrane [112]. Mutations of *bib* in *Drosophila* impact the early development of the nervous system. *bib* is one of a group of neurogenic genes that include *Notch*, *Delta*, and others, but *bib* is unique in that its phenotype is less severe and it does not appear to show genetic interactions with the other neurogenic genes [113,114]. A recent study indicates that BIB is a cation channel modulated by tyrosine kinase and dephosphorylation [27], extending the previous suggestion of a possible channel function for BIB based on membrane localization, sequence similarity with MIP channels, and its unique features as a neurogenic gene [68,112]. Demonstration of an ion channel function for BIB in early nervous system development would induct the *bib* gene into the large group of genes associated with ion-channelopathy diseases. Candidates for a human homolog of *bib* are likely to be identified in the human genome project, and their role in influencing neural determination remains to be discovered.

4. Significance

Among the mammalian aquaporins, AQP0, AQP1 and AQP6 channels are not strictly dedicated to water transport, and can serve as ion channels. Related proteins from the MIP family including *Drosophila* Big Brain and soybean Nodulin 26 show ion channel function. Other AQPs may be found to serve as ion channels as well, when activated by an appropriate stimulus. AQP1 is co-expressed in tissues with guanylyl-cyclase-linked receptors for atrial natriuretic peptide (such as renal proximal tubule, eye ciliary epithelia, and choroid plexus), suggesting a physiological role for cGMP binding in altering the net movement of Na^+ and K^+ across membranes, and perhaps in modulating water flux through AQP1 channels.

Why would a five-pore channel exist to carry water and ions simultaneously? It is possible that the presence of a regulated ion channel may simply be a compact mechanism for combining cation or anion fluxes into the same molecule that mediates water flux, and that the two pathways coexist without any appreciable interaction at the molecular level. On the other hand, one might speculate that the coexistence of these functions at the single protein level suggests a capacity for useful interaction. Although solvent drag or pseudo-solvent drag resulting from local accumulation of ions on one face of the channel [115] has not been described for AQPs, it is possible that the phenomenon has not been readily apparent in the absence of ion channel activation. Solvent drag has been shown for other channels [115], and implicated in tissue transport physiology [116,117]. It will be interesting to determine whether any aspect of transepithelial water movement involves a solute-drag or pseudo-solute drag phenomenon through MIP channels. At the very least, activation of aquaporin ion channels should alter membrane potential, and thus influence other electrogenic transport processes in the same cell. The putative 5-pore design of ion-channel AQPs may reflect a clever economy in the biological strategies for controlling salt and water movement across cell membranes.

The regulation of cation permeability of AQP1 through a cGMP-dependent signaling cascade has potential relevance to the control of secretion and absorption of fluid in many tissues that express this channel at levels that are high enough to provide a measurable ion channel component. Insights into new clinical methods may take advantage of the presence of cGMP-coupled receptors for manipulating salt and water transport in tissues that express AQP1 channels. AQP1 may subserve mechanisms of regulation that would not be possible with a constitutively active and strictly water-selective channel. Derivatives of the quaternary amine blocker, TEA, may define a lead compound for the development of potentially significant agents for research, or perhaps for treating diseases involving fluid imbalance. Early developmental expression of AQP1 might exploit an ionic signaling capacity of these channels, perhaps in response to nitric oxide signaling and elevated cGMP. These areas of research are of interest in future studies.

The powerful combination of molecular biology and patch clamp electrophysiology revolutionized our understanding of voltage-gated ion channels, allowing the analysis of function at the single protein level. Electrophysiological characterization of AQPs initiates a new field of research on the relationships between structure and function that were not conceivable previously. Re-evaluation of a possible role in signaling may not be limited to a few MIP channels; it is possible that other members of the family may also be found to carry ion currents when activated by an appropriate stimulus.

In summary, the discovery of ion channel function in AQPs has potential significance to basic and clinical research involving the regulated control of water and ion fluxes across membranes. This review confirms the need for further investigation of possible ion channel properties in other mammalian AQPS. It is time to acknowledge new breadth in the fundamental properties for this important class of channels, and to revise a major paradigm for research in this field.

Acknowledgements

Thanks to Drs. Stephen Wright and Torsten Falk (University of Arizona) for comments, and to all the members of our research team. Supported by NIH R01 GM59986.

References

[1] Agre, P., Preston, G.M., Smith, B.L., Jung, J.S., Raina, S., Moon, C., Guggino, W.B., Nielsen, S., 1993. Aquaporin CHIP: the archetypal molecular water channel. Am. J. Physiol. 265, F463–F476.

[2] Macey, R.I., 1984. Transport of water and urea in red blood cells. Am. J. Physiol. Cell Physiol. 246, C195–C203.

[3] Gorin, M.B., Yancey, S.B., Cline, J., Revel, J.P., Horwitz, J., 1984. The major intrinsic protein (MIP) of the bovine lens fiber membrane: characterization and structure based on cDNA cloning. Cell 39, 49–59.

[4] Preston, G.M., Agre, P., 1991. Isolation of the cDNA for erythrocyte integral membrane protein of 28 kilodaltons: member of an ancient channel family. Proc. Natl Acad. Sci. USA 88, 11110–11114.

[5] Preston, G.M., Carroll, T.P., Guggino, W.B., Agre, P., 1992. Appearance of water channels in *Xenopus* oocytes expressing red cell CHIP28 protein. Science 256, 385–387.

[6] Jan, L.Y., Jan, Y.N., 1992. Structural elements involved in specific K^+ channel functions. Annu. Rev. Physiol. 54, 537–555.

[7] Reizer, J., Reizer, A., Saier, M.H. Jr, 1993. The MIP family of integral membrane channel proteins: Sequence comparisons, evolutionary relationships, reconstructed pathway of evolution, and proposed functional differentiation of the two repeated halves of the proteins. Crit. Rev. Biochem. Mol. Biol. 28, 235–257.

[8] Cheng, A., van Hoek, A.N., Yeager, M., Verkman, A.S., Mitra, A.K., 1997. Three-dimensional organization of a human water channel. Nature 387, 627–630.

[9] Walz, T., Hirai, T., Murata, K., Heymann, J.B., Mitsuoka, K., Fujiyoshi, Y., Smith, B.L., Agre, P., Engel, A., 1997. The three-dimensional structure of aquaporin-1. Nature 387, 624–627.

[10] Fu, D., Libson, A., Miercke, L.J.W., Weitzman, C., Nollert, P., Krucinski, J., Stroud, R.M., 2000. Structure of a glycerol-conducting channel and the basis for its selectivity. Science 290, 481–486.

[11] Goulding, E.H., Tibbs, G.R., Siegelbaum, S.A., 1994. Molecular mechanism of cyclic-nucleotide-gated channel activation. Nature 372, 369–374.

[12] Varnum, M.D., Black, K.D., Zagotta, W.N., 1995. Molecular mechanism for ligand discrimination of cyclic nucleotide-gated channels. Neuron 15, 619–625.

[13] Anthony, T.L., Brooks, H.L., Boassa, D., Leonov, S., Yanochko, G., Regan, J.W., Yool, A.J., 2000. Cloned human aquaporin-1 is a cyclic GMP-gated ion channel. Mol. Pharm. 57, 576–588.

[14] Saparov, S.M., Kozono, D., Rothe, U., Agre, P., Pohl, P., 2001. Water and ion permeation of aquaporin-1 in planar lipid bilayers: Major differences in structural determinants and stoichiometry. J. Biol. Chem. 276, 31515–31520.

[15] Kumar, V.D., Weber, I.T., 1992. Molecular model of the cyclic GMP-binding domain of the cyclic GMP-gated ion channel. Biochemistry 31, 4643–4649.

[16] Weber, I.T., Steitz, T.A., 1987. Structure of a complex of a catabolite gene activator protein and cyclic AMP refined at 2.5 Å resolution. J. Mol. Biol. 198, 311–326.

[17] Boadu, E., Sager, G., 1997. Binding characterization of a putative cGMP transporter in the cell membrane of human erythrocytes. Biochemistry 9, 10954–10958.

[18] Altenhofen, W., Ludwig, J., Eismann, E.W., Bonigk, W., Kaupp, U.B., 1991. Control of ligand specificity in cyclic nucleotide-gated channels from rod photoreceptors and olfactory epithelium. Proc. Natl Acad. Sci. USA 88, 9868–9872.

[19] Tibbs, G.R., Liu, D.T., Leypold, B.G., Siegelbaum, S.A., 1998. A state-independent interaction between ligand and a conserved arginine residue in cyclic nucleotide-gated channels reveals a functional polarity of the cyclic nucleotide binding site. J. Biol. Chem. 273, 4497–4505.

[20] Scott, S.-P., Tanaka, J.C., 1998. Three residues predicted by molecular modeling to interact with the purine moiety alter ligand binding and channel gating in cyclic nucleotide-gated channels. Biochemistry 37, 17239–17252.

[21] Jung, J.S., Preston, G.M., Smith, B.L., Guggino, W.B., Agre, P., 1994. Molecular structure of the water channel through aquaporin CHIP: The hourglass model. J. Biol. Chem. 269, 14648–14654.

[22] Yasui, M., Hazama, A., Kwon, T.-H., Nielsen, S., Guggino, W.B., Agre, P., 1999. Rapid gating and anion permeability of an intracellular aquaporin. Nature 402, 184–187.

[23] Ren, G., Reddy, V.S., Cheng, A., Melnyk, P., Mitra, A.K., 2001. Visualization of a water-selective pore by electron crystallography in vitreous ice. Proc. Natl Acad. Sci. 98, 1398–1403.

[24] Lagrée, V., Froger, A., Deschamps, S., Pellerin, I., Delamarche, C., Bonnec, G., Gouranton, J., Thomas, D., Hubert, J.-F., 1998. Oligomerization state of water channels and glycerol facilitators. J. Biol. Chem. 273, 33949–33953.

[25] Kuwahara, M., Shinbo, I., Sato, K., Terada, Y., Marumo, F., Sasaki, S., 1999. Transmembrane helix 5 is critical for the high water permeability of aquaporin. Biochemistry 38, 16340–16346.

[26] Mulders, S.M., Preston, G.M., Deen, P.M.T., Guggino, W.B., van Os, C.H., Agre, P., 1995. Water channel properties of major intrinsic protein of the lens. J. Biol. Chem. 270, 9010–9016.

[27] Yanochko, G.M., Yool, A.J., 2002. Regulated cationic channel function in *Xenopus* oocytes expressing *Drosophila* Big Brain. J. Neurosci. 22, 2530–2540.

[28] Borgnia, M.J., Agre, P., 2001. Reconstitution and functional comparison of purified GlpF and AqpZ, the glycerol and water channels from *Escherichia coli*. Proc. Natl Acad. Sci. 98, 2888–2893.

[29] Brooks, H.L., Regan, J.W., Yool, A.J., 2000. Inhibition of Aquaporin-1 water permeability by tetraethylammonium: Involvement of the loop E pore region. Mol. Pharm. 57, 1021–1026.

[30] MacKinnon, R., Yellen, G., 1990. Mutations affecting TEA blockade and ion permeation in voltage-activated K^+ channels. Science 250, 276–279.

[31] Heginbotham, L., Lu, Z., Abramson, T., MacKinnon, R., 1994. Mutations in the K^+ channel signature sequence. Biophys. J. 66, 1061–1067.

[32] Preston, G.M., Jung, J.S., Guggino, W.B., Agre, P., 1993. The mercury-sensitive residue at cysteine-189 in the CHIP28 water channel. J. Biol. Chem. 268, 17–20.

[33] Lagrée, V., Froger, A., Deschamps, S., Hubert, J.-F., Delamarche, C., Bonnec, G., Thomas, D., Gouranton, J., Pellerin, I., 1999. Switch from an aquaporin to a glycerol channel by two amino acids substitution. J. Biol. Chem. 274, 6817–6819.

[34] Mathai, J.C., Agre, P., 1999. Hourglass pore-forming domains restrict aquaporin-1 tetramer assembly. Biochemistry 38, 923–928.

[35] Shi, L.-B., Skach, W.R., Verkman, A.S., 1994. Functional independence of monomeric CHIP28 water channels revealed by expression of wild-type mutant heterodimers. J. Biol. Chem. 269, 10417–10422.

[36] van Hoek, A.N., Hom, M.L., Luthjens, L.H., de Jong, M.D., Dempster, J.A., van Os, C.H., 1991. Functional unit of 30 kDa for proximal tubule water channels as revealed by radiation inactivation. J. Biol. Chem. 266, 16633–16635.

[37] MacKinnon, R., 1991. Determination of the subunit stoichiometry of a voltage-activated potassium channel. Nature 350, 232–235.

[38] Lesage, F., Lazdunski, M., 2000. Molecular and functional properties of two-pore-domain potassium channels. Am. J. Physiol. Renal Physiol. 279, F793–F801.

[39] Lopez, G.A., Jan, Y.N., Jan, L.Y., 1994. Evidence that the S6 segment of the *Shaker* voltage-gated K^+ channel comprises part of the pore. Nature 367, 179–182.

[40] Liu, Y., Joho, R.H., 1998. A side chain in S6 influences both open-state stability and ion permeation in a voltage-gated K^+ channel. Pflugers Arch. Eur. J. Physiol. 435, 654–661.

[41] Yool, A.J., Weinstein, A.M., 2002. New roles for old holes: Ion channel function in Aquaporin 1. News in Physiol. Sci. 17, 68–72.

[42] Murata, K., Mitsuoka, K., Hirai, T., Walz, T., Agre, P., Heymann, J.B., Engel, A., Fujiyoshi, Y., 2000. Structural determinants of water permeation through aquaporin-1. Nature 407, 599–605.

[43] Yool, A.J., Stamer, W.D., Regan, J.W., 1996. Forskolin stimulation of water and cation permeability in aquaporin 1 water channels. Science 273, 1216–1218.

[44] Borgnia, M., Nielsen, S., Engel, A., Agre, P., 1999. Cellular and molecular biology of the aquaporin water channels. Annu. Rev. Biochem. 68, 425–458.

[45] van Os, C.H., Kamsteeg, E.-J., Marr, N., Deen, P.M.T., 2000. Physiological relevance of aquaporins: luxury or necessity? Pflügers Arch. Eur. J. Physiol. 440, 513–520.

[46] Chandy, G., Zampighi, G.A., Kreman, M., Hall, J.E., 1997. Comparison of the water transporting properties of MIP and AQP1. J. Membr. Biol. 159, 29–39.

[47] Zampighi, G., Hall, J.E., Kreman, M., 1985. Purified lens junction protein forms channels in planar lipid films. Proc. Natl Acad. Sci. USA 82, 8468–8472.

[48] Ehring, G.R., Zampighi, G., Horwitz, J., Bok, D., Hall, J.E., 1990. Properties of channels reconstituted from the major intrinsic protein of lens fiber membranes. J. Gen. Physiol. 96, 631–664.

[49] Modesto, E., Lampe, P.D., Ribeiro, M.C., Spray, D.C., Campos de Carvalho, A.C., 1996. Properties of chicken lens MIP channels reconstituted into planar lipid bilayers. J. Membr. Biol. 154, 239–249.

[50] Ehring, G.R., Lagos, N., Zampighi, G.A., Hall, J.E., 1991. Phosphorylation modulates the voltage dependence of channels reconstituted from the major intrinsic protein of lens fiber membranes. J. Membr. Biol. 126, 75–88.

[51] Kushmerick, C., Rice, S.J., Baldo, G.J., Haspel, H.C., Mathias, R.T., 1995. Ion, water and neutral solute transport in *Xenopus* oocytes expressing frog lens MIP. Exp. Eye Res. 61, 351–362.

[52] Nakhoul, N.L., Davis, B.A., Romero, M.F., Boron, W.F., 1998. Effect of expressing the water channel aquaporin-1 on the CO_2 permeability of *Xenopus* oocytes. Am. J. Physiol. Cell Physiol. 274, C543–C548.

[53] Prasad, G.V., Coury, L.A., Finn, F., Zeidel, M.L., 1998. Reconstituted aquaporin 1 water channels transport CO_2 across membranes. J. Biol. Chem. 273, 33123–33126.

[54] Abrami, L., Tacnet, F., Ripoche, P., 1996. Evidence for a glycerol pathway through aquaporin 1 (CHIP 28) channels. Pflügers Arch. Eur. J. Physiol. 430, 447–458.

[55] Yang, B., Verkman, A.S., 1997. Water and glycerol permeability of aquaporins 1-5 and MIP determined quantitatively by expression of epitope-tagged constructs in Xenopus oocytes. J. Biol. Chem. 272, 16140–16146.

[56] Yang, B., Fukuda, N., van Hoek, A.N., Matthay, M.A., Ma, T., Verkman, A.S., 2000. Carbon dioxide permeability of aquaporin-1 null mice and in reconstituted proteoliposomes. J. Biol. Chem. 275, 2686–2692.

[57] Raina, S., Preston, G.M., Guggino, W.B., Agre, P., 1995. Molecular cloning and characterization of an aquaporin cDNA from salivary, lacrimal, and respiratory tissue. J. Biol. Chem. 270, 1908–1912.

[58] Weber, W.-M., 1999. Ion currents of Xenopus laevis oocytes: state of the art. Biochim. Biphys. Acta 1421, 213–233.

[59] Cicirelli, M.F., Smith, L.D., 1985. Cyclic AMP levels during the maturation of Xenopus oocytes. Dev. Biol. 108, 254–258.

[60] Miledi, R., Woodward, R.M., 1989. Effects of defolliculation on membrane current responses of Xenopus oocytes. J. Physiol. 416, 601–621.

[61] Smith, A.A., Brooker, T., Brooker, G., 1987. Expression of rat mRNA coding for hormone-stimulated adenylate cyclase in Xenopus oocytes. FASEB J. 1, 380–387.

[62] Patil, R.V., Han, Z., Wax, M.B., 1997. Regulation of water channel activity of aquaporin 1 by arginine vasopressin and atrial natriuretic peptide. Biochem. Biophys. Res. Commun. 238, 392–396.

[63] Karpen, J.W., Zimmerman, A.L., Stryer, L., Baylor, D.A., 1988. Gating kinetics of the cyclic-GMP-activated channel of retinal rods: flash photolysis and voltage-jump kinetics. Proc. Natl Acad. Sci. USA 85, 1287–1291.

[64] Sandberg, K., Bor, M., Ji, H., Carvallo, P.M., Catt, K.J., 1993. Atrial natriuretic factor activates cyclic adenosine 3'-5'-monophosphate phosphodiesterase in Xenopus laevis oocytes and potentiates progesterone-induced maturation via cyclic guanosine 5'-monophosphate accumulation. Biol. Reprod. 49, 1074–1082.

[65] Houslay, M.D., Milligan, G., 1997. Tailoring cAMP-signalling responses through isoform multiplicity. Trends Biochem. Sci. 22, 217–224.

[66] Yasui, M., Kwon, T.-H., Knepper, M.A., Nielsen, S., Agre, P., 1999. Aquaporin-6 an intracellular vesicle water channel protein in renal epithelia. Proc. Natl Acad. Sci. USA 96, 5808–5813.

[67] Lehmann, R., Jiménez, F., Dietrich, U., Campos-Ortega, J.A., 1983. On the phenotype and development of mutants of early neurogenesis in Drosophila melanogaster. Roux's Arch. Dev. Biol. 192, 62–74.

[68] Doherty, D., Jan, L.Y., Jan, Y.N., 1997. The Drosophila neurogenic gene big brain, which encodes a membrane-associated protein, acts cell autonomously and can act synergistically with Notch and Delta. Development 124, 3881–3893.

[69] Goodman, C.S., Spitzer, N.C., 1979. Embryonic development of identified neurones: differentiation from neuroblast to neurone. Nature 280, 208–214.

[70] Weaver, C.D., Shomers, N.H., Louis, C.F., Roberts, D.M., 1994. Nodulin 26, a nodule specific symbiosome membrane protein from soybean, is an ion channel. J. Biol. Chem. 269, 17858–17862.

[71] Rivers, R.L., Dean, R.M., Chandy, G., Hall, J.E., Roberts, D.M., Zeidel, M.L., 1997. Functional analysis of nodulin 26, an aquaporin in soybean root nodule symbiosomes. J. Biol. Chem. 272, 16256–16261.

[72] Han, Z., Wax, M.B., Patil, R.V., 1998. Regulation of aquaporin-4 water channels by phorbol ester-dependent protein phosphorylation. J. Biol. Chem. 273, 6001–6004.

[73] Tzounopoulos, T., Maylie, J., Adelman, J.P., 1998. Induction of endogenous channels by high levels of heterologous membrane proteins in Xenopus oocytes. Biophys. J. 69, 904–908.

[74] Ebihara, L., 1996. Xenopus connexin38 forms hemi-gap-junctional channels in the nonjunctional plasma membrane of Xenopus oocytes. Biophys. J. 71, 742–748.

[75] Németh-Cahalan, K.L., Hall, J.E., 2000. pH and calcium regulate the water permeability of Aquaporin 0. J. Biol. Chem. 275, 6777–6782.

[76] Lampe, P.D., Johnson, R.G., 1989. Phosphorylation of MP26, a lens junction protein, is enhanced by activators of protein kinase, C. J. Membr. Biol. 107, 145–155.

[77] Kuwahara, M., Fushimi, K., Terada, Y., Bai, L., Marumo, F., Sasaki, S., 1995. cAMP-dependent phosphorylation stimulates water permeability of aquaporin-collecting duct water channel protein expressed in *Xenopus* oocytes. J. Biol. Chem. 270, 10384–10387.

[78] Christensen, B.M., Zelenina, M., Aperia, A., Nielsen, S., 2000. Localization and regulation of PKA-phosphorylated AQP2 response to V(2)-receptor agonist/antagonist treatment. Am. J. Physiol. 278, F29–F42.

[79] Inoue, T., Nielsen, S., Mandon, B., Terris, J., Kishore, B.K., Knepper, M.A., 1998. SNAP-23 in rat kidney: colocalization with aquaporin-2 in collecting duct vesicles. Am. J. Physiol. 275, F752–F760.

[80] Marinelli, R.A., Pham, L., Agre, P., Larusso, N.F., 1997. Secretin promotes osmotic water transport in rat cholangiocytes by increasing aquaporin-1 water channels in plasma membrane: Evidence for a secretin-induced vesicular translocation of aquaporin-1. J. Biol. Chem. 272, 12984–12988.

[81] Han, Z., Patil, R.V., 2000. Protein kinase A-dependent phosphorylation of aquaporin-1. Biochem. Biophys. Res. Comm. 273, 328–332.

[82] Katsura, T., Verbavatz, J.M., Farina, J., Ma, T., Ausiello, D.A., Verkman, A.S., Brown, D., 1995. Constitutive and regulated membrane expression of aquaporin 1 and aquaporin 2 water channels in stably transfected LLC-PK1 epithelial cells. Proc. Natl Acad. Sci. USA 92, 7212–7216.

[83] Han, Z., Wax, M.B., Patil, R.V., 1998. Potential role for AQPs and atrial natriuretic peptides in the aqueous humor dynamics. Exp. Eye Res. 67, 251–253.

[84] King, L.S., Agre, P., 1996. Pathophysiology of the aquaporin water channels. Annu. Rev. Physiol. 58, 619–648.

[85] Stamer, W.D., Seftor, R.E.B., Snyder, R.W., Regan, J.W., 1995. Cultured human trabecular meshwork cells express aquaporin-1 water channels. Curr. Eye Res. 14, 1095–1100.

[86] Stamer, W.D., Snyder, R.W., Smith, B.L., Agre, P., Regan, J.W., 1994. Localization of aquaporin CHIP in the human eye: implications in the pathogenesis of glaucoma and other disorders of ocular fluid balance. Invest. Ophthalmol. Vis. Sci. 35, 3867–3872.

[87] Stumpff, F., Strauss, O., Boxberger, M., Wiederholt, M., 1997. Characterization of maxi- K-channels in bovine trabecular meshwork and their activation by cyclic guanosine monophosphate. Invest. Ophthalmol. Vis. Sci. 38, 1883–1892.

[88] Millar, J.C., Shahidullan, M., Wilsor, W.S., 1997. Atriopeptin lowers aqueous humor formation and intraocular pressure and elevates ciliary cyclic GMP but lacks uveal vascular effects in the bovine perfused eye. J. Ocul. Pharmacol. Ther. 13, 1–11.

[89] Nielsen, S., Smith, B.L., Christensen, I., Agre, P., 1993. Distribution of the aquaporin CHIP in secretory and resorptive epithelia and capillary endothelia. Proc. Natl Acad. Sci. USA 90, 7275–7279.

[90] Strange, K., 1992. Regulation of solute and water balance and cell volume in the central nervous system. J. Am. Soc. Nephrol. 3, 12–27.

[91] Masuzawa, T., Ohta, T., Kawamura, M., Nakahara, N., Sato, F., 1984. Immunohistochemical localization of Na$^+$, K$^+$,-ATPase in the choroid plexus. Brain Res. 302, 357–362.

[92] Israel, A., Garrido, M.R., Barbella, Y., Becemberg, I., 1988. Rat atrial natriuretic peptide (99-126) stimulates guanylate cyclase activity in rat subfornical organ and choroid plexus. Brain Res. Bull. 20, 253–256.

[93] Tsutsumi, K., Niwa, M., Kawano, T., Ibaragi, M., Ozaki, M., Mori, K., 1987. Atrial natriuretic polypeptides elevate the level of cyclic GMP in the rat choroid plexus. Neurosci. Lett. 79, 174–178.

[94] Steardo, L., Nathanson, J.A., 1987. Brain barrier tissues: End organs for atriopeptins. Science 235, 470–473.

[95] Hung, B.C.P., Loo, D.D.F., Wright, E.M., 1993. Regulation of mouse choroid plexus apical Cl$^-$ and K$^+$ channels by serotonin. Brain Res. 617, 285–295.

[96] Dascal, N., Ifune, C., Hopkins, R., Snutch, T.P., Lübbert, J., Davidson, N., Simon, M.I., Lester, H.A., 1986. Involvement of a GTP-binding protein in mediation of serotonin and acetylcholine responses in *Xenopus* oocytes injected with rat brain messenger RNA. Mol. Brain Res. 1, 201–209.

[97] Wu, Q., Delpire, E., Hebert, S.C., Strange, K., 1998. Functional demonstration of Na$^+$-K$^+$-2Cl$^-$ cotransporter activity in isolated, polarized choroid plexus cells. Am. J. Physiol. 275, C1565–C1572.

[98] Greger, R., 2000. Physiology of renal sodium transport. Am. J. Med. Sci. 319, 51–62.

[99] Hammond, T.G., Yusufi, A., Knox, F.G., Dousa, T.P., 1985. Administration of atrial natriuretic factor inhibits sodium-coupled transport in proximal tubules. J. Clin. Invest. 75, 1983–1989.

[100] Liang, M., Knox, F.G., 2000. Production and functional roles of nitric oxide in the proximal tubule. Am. J. Physiol. Integ. Comp. Physiol. 278, R1117–R1124.

[101] Darvish, N., Winaver, J., Dagan, D., 1995. A novel cGMP-activated Cl⁻ channel in renal proximal tubules. Am. J. Physiol. 268, F323–F329.

[102] Schnermann, J., Chou, C.-L., Ma, T., Traynor, T., Knepper, M.A., Verkman, A.S., 1998. Defective proximal tubular fluid reabsorption in transgenic aquaporin-1 null mice. Proc. Natl Acad. Sci. USA 95, 9660–9664.

[103] Lehmann-Horn, F., Jurkat-Rott, K., 1999. Voltage-gated ion channels and hereditary disease. Physiol. Rev. 79, 1317–1372.

[104] Frigeri, A., Nicchia, G.P., Nico, B., Quondamatteo, F., Herken, R., Roncali, L., Svelto, M., 2001. Aquaporin-4 deficiency in skeletal muscle and brain of dystrophic mdx mice. FASEB J. 15, 90–98.

[105] Shiels, A., Bassnett, S., 1996. Mutations in the founder of the MIP gene family underlie cataract development in the mouse. Nat. Genet. 12, 212–215.

[106] Berry, V., Francis, P., Kaushal, S., Moore, A., Bhattacharya, S., 2000. Missense mutations in MIP underlie autosomal dominat "polymorphic" and lamellar cataracts linked to 12q. Nat. Genet. 25, 15–17.

[107] Lanahan, A., Williams, J.B., Sanders, L.K., Nathans, D., 1992. Growth factor-induced delayed early response genes. Mol. Cell Biol. 12, 3919–3929.

[108] King, L.S., Nielsen, S., Agre, P., 1996. Aquaporin-1 water channel protein in lung: ontogeny, steroid-induced expression, and distribution in rat. J. Clin. Invest. 97, 2183–2191.

[109] Ma, T., Yang, B., Gillespie, A., Carlson, E.J., Epstein, C.J., Verkman, A.S., 1998. Severely impaired urinary concentrating ability in transgenic mice lacking aquaporin-1 water channels. J. Biol. Chem. 273, 4296–4299.

[110] Preston, G.M., Smith, B.L., Zeidel, M.L., Moulds, J.J., Agre, P., 1994. Mutations in aquaporin-1 in phenotypically normal humans without functional CHIP water channels. Science 265, 1585–1587.

[111] Ma, T., Yang, B., Kuo, W.L., Verkman, A.S., 1993. cDNA cloning and gene structure of a novel water channel expressed exclusively in human kidney: evidence for a gene cluster of AQPs at chromosome locus 12q13. Genomics 35, 543–550.

[112] Rao, Y., Jan, L.Y., Jan, Y.N., 1990. Similarity of the product of the *Drosophila* neurogenic gene *big brain* to transmembrane channel proteins. Nature 345, 163–167.

[113] Rao, Y., Bodmer, R., Jan, L.Y., Jan, Y.N., 1992. The *big brain* gene of *Drosophila* functions to control the number of neuronal precursors in the peripheral nervous system. Development 116, 31–40.

[114] Brand, M., Campos-Ortega, J.A., 1988. Two groups of interrelated genes regulate early neurogenesis in Drosophila melanogaster. Roux's Arch. Dev. Biol. 197, 457–470.

[115] Pohl, P., Saparov, S.M., 2000. Solvent drag across gramicidin channels demonstrated by microelectrodes. Biophys. J. 78, 2426–2434.

[116] Wilson, R.W., Wareing, M., Green, R., 1997. The role of active transport in potassium reabsorption in the proximal convoluted tubule of the anaesthetized rat. J. Physiol. (Lond.) 500, 155–164.

[117] Zeuthen, T., 1994. Cotransport of K^+, Cl⁻ and H_2O by membrane proteins from choroid plexus epithelium of Necturus maculosus. J. Physiol. (Lond.) 478, 203–219.

[118] Ohtaka-Maruyama, C., Wang, X., Ge, H., Chepelinsky, A.B., 1998. Overlapping Sp1 and AP2 binding sites in a promoter element of the lens-specific MIP gene. Nucleic Acids Res. 26, 407–414.

[119] Li, J., Nielsen, S., Dai, Y., Lazowski, K.W., Christensen, E.I., Tabak, L.A., Baum, B.J., 1994. Examination of rat salivary glands for the presence of the aquaporin CHIP. Pflügers Arch. Eur. J. Physiol. 428(5-6), 455–460.

[120] Deen, P.M., Verdijk, M.A., Knoers, N.V., Wieringa, B., Monnens, L.A., van Os, C.H., van Oost, B.A., 1994. Requirement of human renal water channel aquaporin-2 for vasopressin-dependent concentration of urine. Science 264, 92–95.

[121] Yang, B., Ma, T., Xu, Z., Verkman, A.S., 1999. cDNA and genomic cloning of mouse aquaporin-2 functional analysis of an orthologous mutant causing nephrogenic diabetes insipidus. Genomics 57, 79–83.

[122] Ishibashi, K., Sasaki, S., Fushimi, K., Uchida, S., Kuwahara, M., Saito, H., Furukawa, T., Nakajima, K., Yamaguchi, Y., Gojobori, T., Marumo, F., 1994. Molecular cloning and expression of a member of the aquaporin family with permeability to glycerol and urea in addition to water expressed at the basolateral membrane of kidney collecting duct cells. Proc. Natl Acad. Sci. USA. 91, 6269–6273.

[123] Ma, T., Song, Y., Yang, B., Gillespie, A., Carlson, E.J., Epstein, C.J., Verkman, A.S., 2000. Nephrogenic diabetes insipidus in mice lacking aquaporin-3 water channels. Proc. Natl Acad. Sci. USA 97, 4386–4391.

[124] Jung, J.S., Bhat, R.V., Preston, G.M., Guggino, W.B., Baraban, J.M., Agre, P., 1994. Molecular characterization of an aquaporin cDNA from brain: candidate osmoreceptor and regulator of water balance. Proc. Natl Acad. Sci. USA 91, 13052–13056.

[125] Sobue, K., Yamamoto, N., Yoneda, K., Fujita, K., Miura, Y., Asai, K., Tsuda, T., Katsuya, H., Kato, T., 1999. Molecular cloning of two bovine aquaporin-4 cDNA isoforms and their expression in brain endothelial cells. Biochim. Biophys. Acta 1489, 393–398.

[126] Turtzo, L.C., Lee, M.D., Lu, M., Smith, B.L., Copeland, N.G., Gilbert, D.J., Jenkins, N.A., Agre, P., 1997. Cloning and chromosomal localization of mouse aquaporin 4: exclusion of a candidate mutant phenotype, ataxia. Genomics 41, 267–270.

[127] Krane, C.M., Towne, J.E., Menon, A.G., 1999. Cloning and characterization of murine Aqp5: evidence for a conserved aquaporin gene cluster. Mamm. Genome 10, 498–505.

[128] Ishibashi, K., Kuwahara, M., Gu, Y., Kageyama, Y., Tohsaka, A., Suzuki, F., Marumo, F., Sasaki, S., 1997. Cloning and functional expression of a new water channel abundantly expressed in the testis permeable to water, glycerol, and urea. J. Biol. Chem. 272, 20782–20786.

[129] Ishibashi, K., Kuwahara, M., Kageyama, Y., Tohsaka, A., Marumo, F., Sasaki, S., 1997. Cloning and functional expression of a second new aquaporin abundantly expressed in testis. Biochem. Biophys. Res. Commun. 237, 714–718.

[130] Koyama, Y., Yamamoto, T., Kondo, D., Funaki, H., Yaoita, E., Kawasaki, K., Sato, N., Hatakeyama, K., Kihara, I., 1997. Molecular cloning of a new aquaporin from rat pancreas and liver. J. Biol. Chem. 272, 30329–30333.

[131] Koyama, N., Ishibashi, K., Kuwahara, M., Inase, N., Ichioka, M., Sasaki, S., Marumo, F., 1998. Cloning and functional expression of human aquaporin-8 cDNA and analysis of its gene. Genomics 54, 169–172.

[132] Calamita, G., Spalluto, C., Mazzone, A., Rocchi, M., Svelto, M., 1999. Cloning, structural organization and chromosomal localization of the mouse aquaporin-8 water channel gene (Aqp8). Cytogenet. Cell Genet. 85, 237–241.

[133] Ishibashi, K., Kuwahara, M., Gu, Y., Tanaka, Y., Marumo, F., Sasaki, S., 1998. Cloning and functional expression of a new aquaporin (AQP9) abundantly expressed in the peripheral leukocytes permeable to water and urea, but not to glycerol. Biochem. Biophys. Res. Commun. 244, 268–274.

[134] Dhallan, R.S., Yau, K.W., Schrader, K.A., Reed, R.R., 1990. Primary structure and functional expression of a cyclic nucleotide-activated channel from olfactory neurons. Nature 347, 184–187.

[135] Bradley, J., Li, J., Davidson, N., Lester, H.A., Zinn, K., 1994. Heteromeric olfactory cyclic nucleotide-gated channels: A new subunit that confers increased sensitivity to cAMP. Proc. Natl Acad. Sci. USA 91, 8890–8894.

[136] Goulding, E.H., Ngai, J., Kramer, R.H., Colicos, S., Axel, R., Siegelbaum, S.A., Chess, A., 1992. Molecular cloning and single channel properties of the cyclic nucleotide-gated channel from catfish olfactory neurons. Neuron 8, 45–58.

[137] Kaupp, U.B., Niidome, T., Tanabe, T., Terada, S., Boenigk, W., Stuehmer, W., Cook, N.J., Kangawa, K., Matsuo, H., Hirose, T., Miyata, T., Numa, S., 1989. Primary structure and functional expression from complementary DNA of the rod photoreceptor cyclic GMP-gated channel. Nature 342, 762–766.

[138] Fortin, M.G., Morrison, N.A., Verma, D.P., 1987. Nodulin-26, a peribacteroid membrane nodulin is expressed independently of the development of the peribacteroid compartment. Nucleic Acids Res. 15, 813–824.

[139] Warmke, J., Drysdale, R., Ganetzky, B.A., 1991. A distinct potassium channel polypeptide encoded by the *Drosophila eag* locus. Science 252, 1560–1562.

[140] Sun, X.C., Allen, K.T., Xie, Q., Stamer, W.D., Bonanno, J.A., 2001. Effect of AQP1 expression level on CO2 permeability in bovine corneal endothelium. Invest. Ophthalmol. Vis. Sci. 42, 417–423.

Molecular and functional insights into voltage-gated calcium channels

Anthony Stea[a] and Terrance P. Snutch[b,*]

[a]University-College of the Fraser Valley, 33844 King Rd., Abbotsford, B.C., Canada V2S 7M8
[b]Biotechnology Laboratory, Rm 237-6174 University Blvd., University of British Columbia, Vancouver, B.C.,
Canada V6T 1Z3
*Correspondence address: Tel.: +1-604-822-6968; fax: +1-604-822-6470
E-mail: snutch@zoology.ubc.ca

1. Introduction

The first insights into voltage-gated calcium channels (Ca channels) came from work by Fatt, Ginsborg, Hagiwara, and coworkers in the 1950s [see Ref. 49] who determined that action potentials in crustacean muscles were dependent on the influx of calcium and not sodium ions. These experiments indicated that proteins other than voltage-gated sodium channels, well described from the squid axon, could play a role in membrane excitability. Since their pioneering work, a number of other important physiological roles have been determined for Ca channels. Ca channels are not only found ubiquitously in nervous, neuroendocrine, and muscle tissues but are also found in endocrine, epithelial, and endothelial tissues [15,76,84,98,168]. In smooth and cardiac muscle, Ca influx mediates calcium-induced calcium release from intracellular stores stimulating muscle contraction [6,29,59,84]. In skeletal muscle, Ca channels act to transfer the action potential electrical signal into a mechanical stimulus to evoke intracellular Ca release from ryanodine receptors which in turn initiates muscle contraction [84]. In endocrine and neuroendocrine cells, Ca influx through Ca channels is coupled to exocytosis of secretory granules. For example, insulin secretion in pancreatic β cells is dependent on several types of Ca channels [76]. In neurons, Ca channels play roles in neurotransmission, electrical excitability, gene expression, and modulation of signal tranduction events [2,15].

Based on physiological criteria, Ca channels have been broadly classified into high-voltage activated (HVA) or low-voltage activated (LVA) depending on the amount of depolarization required to first activate the channels [15,49,52]. In general HVA channels are activated at potentials > -30 mV while LVA channels are activated at lower potentials (e.g.   50 to   70 mV). HVA channels are essential for stimulating secretory events (e.g. synaptic neurotransmission), muscle contraction, and downstream signal transduction events, including gene expression [15,84]. LVA channels have been

Advances in Molecular and Cell Biology, Vol. 32, pages 381–406
ISSN: 1569-2558 / DOI: 10.1016/S1569-2558(03)32016-8

Table 1
Classification of voltage-gated calcium channels

Channel subtype	Major pore-forming α_1 subunits[a]	Pharmacological profile[b]			
		ω-CTX-GVIA	DHP's	ω-AGA-IVA	ω-CTX-MVIIC
L-type	α_{1S} (Ca$_V$1.1), α_{1C} (Ca$_V$1.2), α_{1D} (Ca$_V$1.3), α_{1F} [c] (Ca$_V$1.4)	–	✔	–	–
P/Q-type	α_{1A} (Ca$_V$2.1)	–	–	✔	✔
N-type	α_{1B} (Ca$_V$2.2)	✔	–	–	✔
Novel, R-type	α_{1E} (Ca$_V$2.3)	–	–	–	–
T-type	α_{1G} (Ca$_V$3.1), α_{1H} (Ca$_V$3.2), α_{1I} (Ca$_V$3.3)	–	–	–	–

[a]The molecular classification for Ca channels is based on the presence of distinct pore-forming α_1 subunits [28].
[b]Indicates a distinctive sensitivity to the drugs listed.
[c]α_{1F} is predicted to be a dihydropyridine-sensitive channel based on sequence similarity to the other L-type channels.

implicated in controlling the generation of patterns of excitability in the heart and brain [6,52]. HVA and LVA channels can be further subdivided according to their susceptability to block by various pharmacological agents (see Table 1). In particular, dihydropyridines (DHPs) are selective for one diverse group of HVA channels called L-type (Table 1 and Fig. 2; [15,51]). A peptide toxin isolated from the cone snail *Conus geographus* (ω-CgTx-GVIA) can irreversibly block N-type channels [99], and this blockade is a distinguishing characteristic of this class of HVA Ca channel (Table 1). P-type HVA channels are effectively blocked by nanomolar concentrations of the peptide toxin ω-agatoxin IVA isolated from the funnel web spider *Agelenopsis aperta* [89]. N and P/Q type channels can also be blocked by another peptide toxin, ω-CgTx-MVIIC, although at different affinities [1,50,78]. Although these agents have been commonly used to classify Ca channels in native tissues, many other types of molecules have been found to bind to Ca channels with varying affinities and are not always a reliable way of distinguishing between subtypes.

Over the last decade, the molecular identity of a heterogeneous family of Ca channels has been elucidated and has led to a better understanding of the diversity of form and function of these ion channel proteins. In this review, we will attempt to summarize the recent developments in understanding the molecular structure as well as the physiological and pathological roles of Ca channels.

2. Molecular classification and structure of voltage-gated Ca channels

All Ca channels contain a large ($\sim 180–270$ kDa) integral membrane protein called the α_1 subunit that contains the divalent cation selective pore, which allows calcium to enter cells during membrane depolarization [15]. To date, 10 distinctly different α_1 subunit genes have been isolated from mammals (Table 1, Fig. 2), with many homologues also described in non-mammals. HVA Ca channels are heteromeric proteins consisting of several subunits in addition to the pore-forming α_1 subunit (Fig. 1). Early purification and

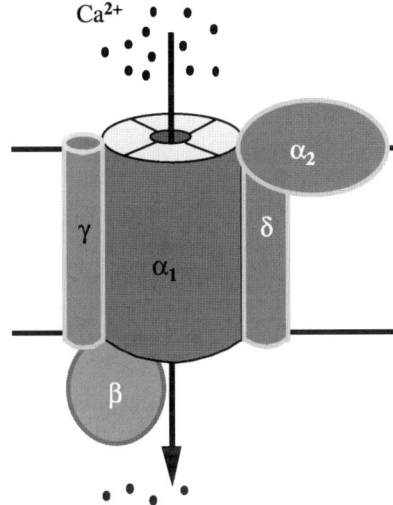

Fig. 1. Voltage-gated HVA Ca calcium channel complex. A generalized HVA Ca channel complex consists of the pore-forming α_1 subunit along with membrane attached γ and $\alpha_2-\delta$ subunits and the intracellular β subunit [14,15].

immunoprecipitation experiments revealed that the dihydropyridine receptor (L-type channel) from skeletal muscle contains five distinct subunits; α_1 ($\sim$180–270 kDa), α_2 ($\sim$140 kDa), β ($\sim$50–70 kDa), δ ($\sim$30 kDa), and γ ($\sim$25 kDa) subunits [14,15,126].

More recent studies have indicated that other types of HVA Ca channels also contain a similar subunit composition. Both the N-type and P/Q type HVA channels contain at least the α_1, $\alpha_2\delta$ and β subunits in the functional protein complex [78,118]. While the exact biochemical composition of T-type channels has not been determined, these channels may not contain associated $\alpha_2\delta$ or β subunits [68,69].

3. Molecular diversity of Ca channel α_1 subunits

The pore-forming α_1 subunit determines the unique properties amongst the various subtypes of HVA and LVA Ca channels. The first α_1 subunit (α_{1S}; Table 1; Fig. 2) was characterized from skeletal muscle by Tanabe and co-workers [137] who showed that the L-type dihydropyridine receptor that functions as the voltage sensor in excitation-contraction [84]. The α_{1S} cDNA coded for a 1873 amino acid protein with a predicted structure most similar to that of voltage-gated sodium channels [2,15]. The predicted structure consists of four repeating domains (I–IV), each with six membrane-spanning segments (S1–S6; see Fig. 3). Between S5 and S6 in each of the four domains is a P region that contains a highly conserved glutamate residue that forms the divalent cation selective pore [26]. The S4 segment contains several repeated positively charged residues and contributes to the voltage sensor of the channel (Fig. 3; [2]).

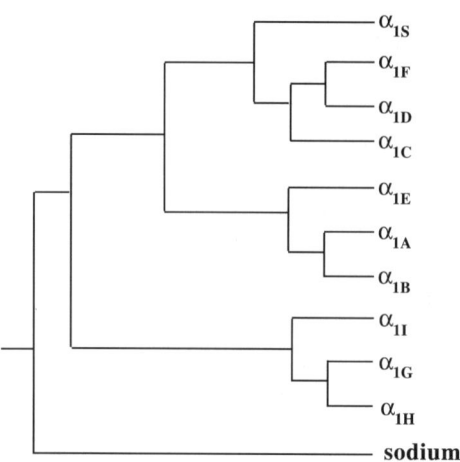

Fig. 2. Family tree of mammalian Ca channel α_1 subunits. Ten distinct mammalian α_1 subunits and the voltage-gated sodium channel were aligned (using domains III through IV) and compared using protein parsimony analysis [31]. Three major groupings were observed; the DHP-sensitive L-type Ca channels (α_{1S}, α_{1F}, α_{1D}, α_{1C}), the DHP-insensitive HVA Ca channels (α_{1E}, α_{1A}, α_{1B}) and the LVA Ca channels (α_{1I}, α_{1G}, α_{1H}). Genebank Accession numbers: α_{1S} – NP_000060; α_{1F} – NP_005174; α_{1D} – NP_000711; α_{1C} – NP_000710; α_{1E} – A37490; α_{1A} – NP_037050; α_{1B} – A45386; α_{1I} – Q9Z0Y8; α_{1G} – O54898; α_{1H} – O95180; sodium – AAA79965.

Sequence comparison of the 10 different types of mammalian α_1 subunits indicates three distinct groupings that can mostly be correlated to physiological and pharmacological subtypes of native Ca channels. The DHP-sensitive L-type Ca channels (α_{1S}, α_{1F}, α_{1D}, α_{1C}) form one group, the DHP-insensitive HVA Ca channels (α_{1E}, α_{1A}, α_{1B}) represent another group and a third group consists of the LVA Ca channels (α_{1I}, α_{1G}, α_{1H}).

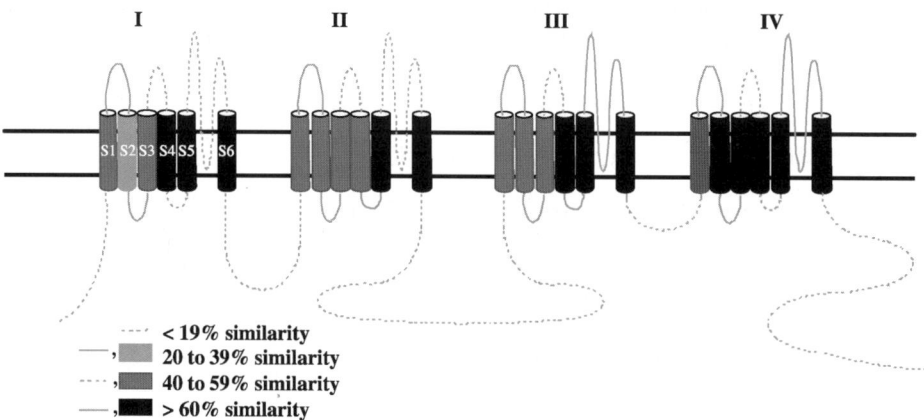

Fig. 3. Similarity of structural domains of mammalian Ca channels. A comparison of the amino acid similarity (based on either identical or conserved residues) in the various structural domains in the 10 different mammalian α_1 subunits (Fig. 2) showed the least similarity in the larger cytoplasmic domains while the greatest similarity was seen in the transmembrane domains (in particular domain IV).

Based upon overall sequence homology, the 10 mammalian α_1 subunits are most similar (20 to $> 60\%$) in the membrane spanning segments and least similar ($<20\%$) in the larger cytoplasmic domains. Of the membrane spanning segments, domain IV exhibits the greatest similarity ($>60\%$) and is also a major region for DHP-binding to the L-type Ca channels [51].

Four α_1 subunits (α_{1S}, α_{1D}, α_{1F}, α_{1C}) code for Ca channels classified as HVA L-type channels (Table 1). The major distinguishing characteristics of this group are activation at strongly depolarized potentials (>-20 mV), relatively slowly inactivating current waveforms and sensitivity to dihydropyridines [15,51,126]. The α_{1S} Ca channel is involved in excitation–contraction coupling in skeletal muscle and is not found in other tissues. The native α_{1S} subunit is associated with the β_{1a} subunit (along with $\alpha_2\delta$ and γ) and forms a complex at triad junctions where it couples membrane excitation to rapid release of Ca from intracellular stores [4]. A small segment (between amino acids 720–765) of the II–III loop of the α_{1S} subunit couples the Ca channel to the ryanodine receptor in skeletal muscle [40]. This type of Ca channel is also found in the plasma membrane of the myotubes where they mediate Ca influx into the cells [4]. The auxillary β_{1a} subunit appears to be essential for both of these physiological roles as skeletal myotubes lacking the β_{1a} subunit are both deficient in excitation–contraction (E–C) coupling and lack L-type Ca currents [7]. Transfection of a cardiac muscle and brain specific β_{2a} subunit restores the L-type current in myotubes but could not restore the E–C coupling (unlike the native skeletal muscle β_{1a} subunit; [7]). Dysgenic mice with skeletal myotubes lacking E–C coupling can be rescued by transfection of the α_{1S} subunit alone [97] or the α_{1C} L-type subunit [39], although the neuronal α_{1A} and α_{1B} subunits could not restore E–C coupling [39].

The α_{1C} subunit was first isolated from cardiac muscle [87], but has also been described in smooth muscle and brain (see review [126]). Ca channels containing α_{1C} are important in cardiac E–C coupling [87], secretion in endocrine cells [157], stimulation of gene expression [20], and modulating Ca homeostasis in the cell body of neurons [15]. The α_{1C} Ca channels are localized mainly to the cell body and proximal dendrites of neurons as determined by antibody staining (Fig. 4; [47]). Alternatively spliced forms of the α_{1C} subunit have been isolated which have distinct localization patterns in the brain [121]. The α_{1C} subunit is associated with the β_{2a} and α_2 subunits in cardiac muscle [36] and in brain may also associate with the prominant β_3 or β_4 subunits [118]. The association of the α_{1C} and β subunits appears to be critical for targeting of the Ca channels complex to the plasma membrane, as disruptions in this association prevent this transport [36,38]. Evidence suggests that a Src-3 motif in the β_{2a} subunit is critical for the membrane trafficking of the α_{1C} Ca channel [36].

Similar to α_{1C}, α_{1D} Ca channels are sensitive to DHPs and show $\sim 60\%$ amino acid identity between the α_1 subunits [154]. The α_{1D} L-type channels are broadly distributed through the central nervous system with expression patterns very similar to α_{1C} except for higher expression in the superior colliculus [47,81]. Both the α_{1D} and α_{1C} Ca channels are found predominantly on the cell bodies and proximal dendrites of neurons, where they likely play a role in controlling Ca influx into the soma of these cells (Fig. 4; [47]). It has been shown that the α_{1D} channels can be modulated by neurotransmitters differently than α_{1C} [30], leading to the conclusion that these L-type Ca channels have overlapping, but not

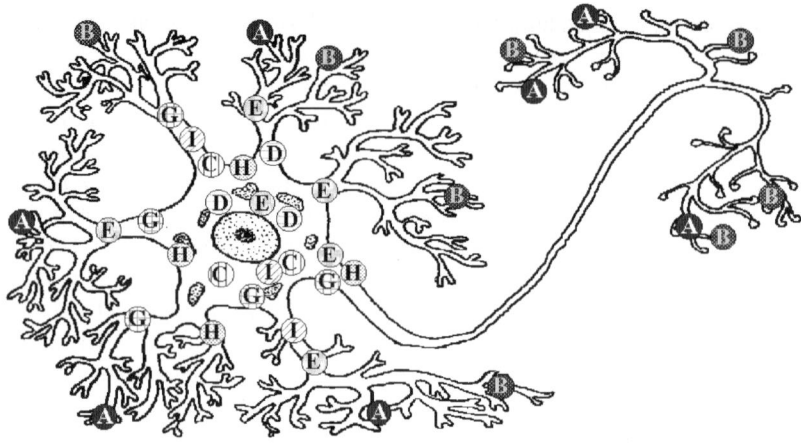

Fig. 4. Cellular localization of neuronal Ca channels. In this generalized neuron, antibody and toxin binding studies as well as physiological studies have determined the cellular localization of the major Ca channel isoforms. See text for details.

identical, roles in various neurons. The α_{1D} Ca channels are also found in subsets of neurons of the PNS (e.g. vasoactive intestinal peptide containing enteric neurons; [61]). Another important physiological role for the α_{1D} channels is in facilitating tonic release of neurotransmitter from hair cells in the cochlea [64]. These channels are the predominant L-type Ca channel in hair cells and have unusual physiological properties, including a lower threshold of activation and little Ca-dependent inactivation [64]. These properties may be due to the presence of an alternatively spliced form of the α_{1D} gene [65].

The sequence of the L-type α_{1F} subunit was obtained from analyses of human genomic DNA sequenced for the human genome project [132]. The α_{1F} is predicted to code for an L-type Ca channel based on similarity (50–60% identity) with the other L-type Ca channel subunits (Fig. 2; [132]). No physiological studies have been performed on this L-type variant but mutations in this gene have been linked to a type of stationary night blindness and it appears to be expressed solely in the neural retina [96,132].

The two major types of Ca channels that contribute to neurotransmitter release at many central and peripheral nervous system synapses are the P/Q- and N-types [135,153]. A clear illustration of the difficulties in distinguishing the subtypes of Ca channels based on physiological and pharmacological characteristics is found in the P and Q-type Ca channels. P-type channels were first studied in cerebellar Purkinje cells and were found to be selectively and potently blocked by ω-Aga-IVA [89]. Q-type currents were first described in cerebellar granule cells and had distinctly different inactivating kinetics compared to P-type, and were not potently blocked by ω-Aga-IVA but blocked by ω-CgTx-MVIIC [166]. In situ hybridization studies indicated distinct α_{1A} expression in cerebellar Purkinje cells [123,125] and this led to speculation that the α_{1A} gene coded for P-type channels. However, electrophysiological experiments carried out in *Xenopus* oocytes showed the α_{1A} currents inactivated more and were less sensitive to ω-Aga-IVA than native P-type channels and were more similar to native Q-type channels [115,125].

A study by Bourinet and colleagues [12] appears to have resolved this controversy. This group showed that the gene coding for the α_{1A} subunit is alternatively spliced and that one of the splice variants predominates in Purkinje cells and codes for slowly inactivating currents while another splice variant is expressed predominantly in cerebellar granule neurons and inactivated more rapidly like native Q-type channels [12]. Thus it appears the α_{1A} gene codes for both the P and Q-type Ca channels. Antibody localization has indicated that α_{1A} Ca channels are localized to the synaptic sites on dendrites and axons and show punctate patterns of localization on cell bodies where synapses likely occur (Fig. 4; [150]). This cellular localization pattern supports a major role of the α_{1A} Ca channel in stimulating neurotransmitter release at many central presynaptic terminals.

The α_{1B} gene codes for GVIA sensitive N-type channels [124,155], which are essential for neurotransmitter release at some presynaptic terminals in neurons (Fig. 4; [149]). Biochemical analysis of purified N-type channels revealed the presence of the α_{1B}, α_2-δ, β, and other associated proteins [158]. In adult brain α_{1B} Ca channels are mostly associated with the β_3 or β_4 subunits [118,142], while in early development the β_{1b} predominates [142]. A null mutation in the β_4 gene in the *lethargic* mouse causes an abnormal assembly of N-type channels resembling the immature state and may play a role in the epileptic phenotype [85]. Along with the typical Ca channel auxiliary subunits, α_{1B} channels also associate with synaptic vesicle proteins including synaptotagmin, syntaxin, and SNAP-25 [119]. These associations are critical for normal exocytotic processes to occur at the synapse [91] and can be modulated by phosphorylation [161]. Several studies have indicated that N-type channel diversity can be increased by alternative splicing of the α_{1B} subunit [77,80,128]. Subunit differences and alternative splicing might explain the diversity of native N-type currents seen in central and peripheral neurons.

The α_{1E} subunit is structurally related to the α_{1A} and α_{1B} subunits (Fig. 2) but has a cellular distribution pattern more similar to the α_{1C} and α_{1D} channel on the soma and proximal dendrites of neurons [151]. Of particular note, several studies also indicate a role for the α_{1E} R-type Ca channel in neurotransmission at a specific subset of presynaptic terminals [173,174,175]. Characterization of α_{1E} channels showed some physiological characteristics similar to T-type currents, including rapid inactivation and a relatively lower threshold for activation than typical HVA Ca channels [122]. The expression patterns of α_{1E} Ca channels in the brain also shows some similarities to T-type channels [122]. However, other studies indicated that this channel may be more like HVA channels (R-type) found in cerebellar granule neurons [107,140]. The α_{1E} channel shows permeation properties (Ca > Ba) typical of many T-type channels [11], although this is not diagnostic for all T-type currents [52]. Antisense oligonucleotides against α_{1E} can decrease LVA T-type current in atrial myocytes [106], but also reduces HVA R-type currents in cerebellar granule cells [107]. Alternative splicing of the α_{1E} gene [141] may generate Ca channels with physiological roles overlapping with "typical" HVA and LVA (T-type) channels in different tissues.

The last group of Ca channels (α_{1G}, α_{1H}, α_{1I}; Fig. 2) are collectively known as LVA (or T-type) channels based mainly on their voltage dependent properties including a low threshold for activation and rapid inactivation kinetics (see Ref. [52]). Pharmacological analyses have indicated that there are as yet no specific blockers for T-type Ca channels (Table 1; [52]), which has complicated the determination of the functional diversity of these channels in neurons. T-type Ca channels are thought to play essential roles in modifying the firing patterns in CNS neurons and may be important in local increases in intracellular calcium [52]. Physiological studies have shown that T-type channels are localized to the soma and dendrites of neurons consistent with their physiological roles (Fig. 4; [18,52]). The α_{1G}, α_{1H}, and α_{1I} subunits are more distantly related based on sequence homology to the other α_1 subunits (Fig. 2) but show the same four domain primary structure pattern (Fig. 3). Expression studies have shown that the α_{1G}, α_{1H}, and α_{1I} genes all code for channels with hyperpolarized activation thresholds very similar to typical T-type Ca channels in neurons (-60 to -50 mV; [19,70,86,103,156]). Both α_{1G} and α_{1H} code for Ca channels with fast activation and inactivation kinetics while α_{1I} shows slower kinetics similar to a subset of native T-type currents [70,86,103]. Unlike the other seven Ca channels described, the three T-type channels do not appear to require auxillary subunits (e.g. $\alpha_2\delta$ or β) for expression or modulation [68,70,103] and depletion of β subunits by antisense RNA in nodose sensory neurons did not affect native T-type current expression [69].

4. Molecular diversity of auxillary subunits

Along with the diverse types of α_1 subunits, there are also at least 4 genes (β_1, β_2, β_3, β_4) responsible for the diversity of β subunits (see review [126]). The coexpression of a β subunit is required for expression of certain α_1 subunits and also causes changes in voltage dependence and kinetics of the expressed currents [126]. A β interaction domain (BID) in the β subunit protein interacts with a conserved motif in the I–II loop of all α_1 subunits (α_1 interaction domain, AID; [110]) except for α_{1G}, α_{1H}, and α_{1I}. Another lower affinity interaction site in the carboxyl terminus of α_{1A} binds with the carboxyl terminus of β_4 [146]. It is thought that secondary interaction sites modulate the voltage-dependent and kinetic properties of the Ca channels [146] and play a role in plasma membrane targeting [37,38]. Early expression studies indicated that β subunit coexpression significantly enhanced whole cell currents and altered electrophysiological properties of the channels [115,124,125]. Furthermore, an increase in distribution of Ca channels to the plasma membrane was observed when α_1 subunits were coexpressed with β subunits [38]. Interestingly, it appears that the C-terminus of the α_{1C} subunit is responsible for the membrane-targeting interaction with the β subunit [37], supporting the argument of a second important interaction site between the α_1 and β subunits.

Localization studies have shown that all four types of β subunits are found in the brain [81] and immunostaining has shown that β_1, β_2, and β_3 are found mainly in the cell bodies while β_4 is localized to dendrites of neurons [75]. Mice homozygous for a null mutation in the β_1 gene die at birth from asphyxiation and have dramatically decreased levels of α_{1S} L-type Ca channels in the membrane of their myotubes [43]. Biochemical analysis of N-type and L-type Ca channels has shown that the β_3 and β_4 subunits associate preferentially with these subtypes in neurons [105,118]. β_3 knockout mice ($\beta_3^{-/-}$)showed no obvious physical or behavioural abnormalities even though they had slightly lower densities of L- and N-type Ca channels in sympathetic neurons [95], suggesting that β subunits have overlapping roles in the adult brain. The decrease in L- and N-type channels appeared to be compensated by modified P/Q-type channels, indicating that there may be some preferential binding of β_3 subunits with L- and N-type and β_4 subunits with P/Q-type Ca channels [95].

The α_2 and δ subunits were purified along with the dihydropyridine receptor [14,15] and the GVIA sensitive N-type Ca channels [158], and appear to be integral parts of HVA Ca channel complexes [126]. These subunits were found to be encoded by the same gene [57], but are post-translationally cleaved. The α_2 and δ subunits are joined by disulphide linkages and each part of the $\alpha_2\delta$ complex (Fig. 1) has distinct effects on the expressed Ca currents. The δ subunit anchors the $\alpha_2\delta$ complex to the membrane and to the α_1 subunit (Fig. 1; [44]). The α_2 is largely extracellular and appears essential for Ca channel current stimulation (Fig. 1; [44]). Functional interactions between the α_1 and $\alpha_2\delta$ require glycosylation of the $\alpha_2\delta$ complex [44]. There are four $\alpha_2\delta$ genes that share 30–60% amino acid identity [62,172]. The $\alpha_2\delta_1$ is present ubiquitously in brain, heart, and skeletal muscle tissues [57], while the $\alpha_2\delta_2$ appears to be most highly expressed in heart, pancreas and skeletal muscle [62,172]. The $\alpha_2\delta_3$ subunit complex is localized to the brain and shows the same current stimulation effects of the $\alpha_2\delta_1$ subunit [62]. One target of the novel anticonvulsant drug gabapentin is the $\alpha_2\delta$ subunit of Ca channels, and this interaction may underlie the therapeutic action of this agent [147].

Like the $\alpha_2\delta$, the γ subunit protein (~ 25 kDa) was first isolated from the L-type skeletal muscle dihydropyridine receptor [14,15] but in contrast, was not isolated from the purified N-type GVIA receptor from brain [158]. The mRNA for the skeletal muscle γ subunit isoform (γ_1) is only expressed in skeletal muscle tissue [56]. Mice with a targeted disruption of the γ_1 gene show no physical or behavioural differences from control littermates, though there are some small changes in the electrophysiological properties of the α_{1S} L-type Ca channel [34]. A second γ subunit isoform, stargazin (γ_2), has been identified from brain [73,172]. Stargazin (~ 36 kDa) has a longer COOH tail, is enriched in synaptic membranes and interacts with α_{1A}, P/Q-type channel in vitro [73]. Premature trancriptional termination mutations in this gene cause a seizure phenotype in mice (*stargazer* mice) with characteristic symptoms similar to absence epilepsy [73]. More recently six novel γ subunit genes have been identified. The γ_3 and γ_5 subunits are expressed solely in brain and γ_4, γ_6, γ_7 and γ_8 are highly expressed in a variety of tissues including brain and heart [63,172].

5. Structural determinants of Ca channel α_1 subunits

Many studies have given us insight into the specific regions of the α_1 subunits that are responsible for Ca channel functional properties. An early study by Tanabe and coworkers [138] showed that the α_{1S} II–III loop was essential for excitation–contraction coupling using chimeric α_1 subunits. Other studies examining α_1 subunit chimeras indicated that domain I appeared to play an important role in L-type Ca channel activation [139]. Further studies have indicated that the S3 transmembrane segment and S3–S4 linker of domain I appear to be critical in the activation properties of the α_{1S} and α_{1C} channels [94]. Activation is also dependent on the S4 segments of each domain, a region that has been shown to be the major voltage sensor region of voltage-gated ion channels [2]. The S4 segments move outwards when exposed to a depolarizing potential due to electrostatic interactions of the abundant positively charged residues (Arg or Lys every 3 amino acids [2]). Along with the role of the basic amino acids in the S4 segments conserved proline residues in I S4 and III S4 appear to be important in activation gating [159]. Replacement of these prolines with leucines caused a decrease in α_{1C} channel activation ($\downarrow$ single channel open probability) and the introduction of proline residues in II S4 and IV S4 caused an increase in channel activation [159]. The voltage dependence of activation in T-type channels is affected by a seven amino acid segment in the III–IV linker as determined by studying splice variants of α_{1G} [16].

The membrane-associated loop between the S5 and S6 segments of each domain (termed the P or SS1-SS2 region) has been determined to form the pore region of voltage-gated ion channels [2,83]. In Ca channels, the pore is selective for divalent ions such as Ca and barium over monovalent ions such as sodium. In HVA Ca channels the divalent selectivity is due to glutamate residues in the P region of each domain [160]. Mutation of individual glutamates in the EEEE locus resulted in an increase in monovalent ion permeability [83,160]. In the loop between S5 and S6 is another conserved region which shares homology to known Ca binding EF hand motifs [32]. Mutations in conserved residues in the EF hand motif affect the permeability of different divalent ions, suggesting that this region is important in the preferential permeation of barium ions over Ca ions, a characteristic of HVA channels that distinguishes them from LVA Ca channels [32].

Unlike most voltage-gated sodium channels, Ca channels have a large range of inactivation kinetics with some Ca channels showing little inactivation (P-type, L-types), while others inactivate very rapidly (T-types) (see review [131]). To identify specific regions in the α_1 subunit responsible for inactivation gating, chimeras were generated from slow and fast inactivating Ca channels [130]. From these studies, three distinct regions were shown to play a key role in the fast inactivation exhibited by α_{1E} channels: II S6, III S6, and the I–II linker [130]. When these regions were inserted into chimeric α_{1C} channels, the currents inactivated rapidly, typical of wild type α_{1E} channels [130]. The authors suggest that the I–II linker acts like an inactivation particle and interacts with the II S6 and III S6 regions [130]. This model for Ca channel fast inactivation has similarities to the mechanisms proposed for both sodium and potassium channel fast inactivation [131]. This model would also explain the dramatic effects of different β subunits on the inactivation of Ca channels [124,125,126], since the I–II linker contains a high affinity β-binding site [110].

6. Physiological roles and modulation of Ca channels

One important role for Ca channels is the release of neurotransmitters at neuronal and neuromuscular synapses (see review [88]). The major presynaptic Ca channels responsible for synaptic transmission in vertebrates are the N- and P/Q-type channels [88,135]. These Ca channels are localized to presynaptic terminals (Fig. 4; [149,150,151]), and when specific blockers such as ω-CgTx-GVIA or ω-Aga-IVA are added, neurotransmission is largely inhibited [135]. The N-type (α_{1B}) and P/Q-type (α_{1A}) channels bind to proteins associated with the synaptic vesicles and part of the exocytotic machinery at the synapse (Fig. 5).

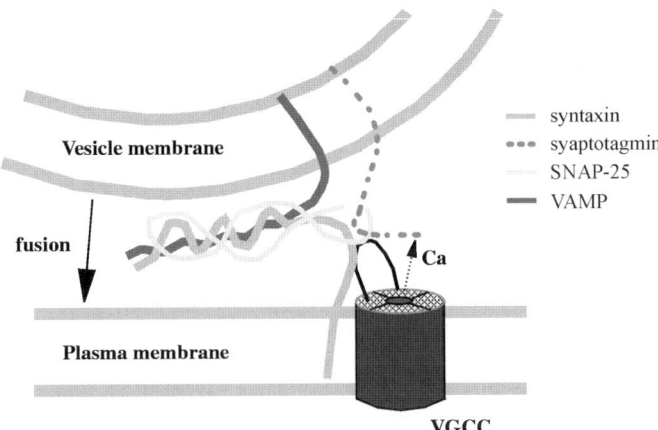

Fig. 5. Interaction of presynaptic Ca channels with SNARE proteins. The synprint site on the II–III loop of α_{1A} and α_{1B} Ca channels interacts with syntaxin, synaptotagmin, and SNAP-25 to facilitate exocytosis at the presynaptic membrane. Modified from Ref. [133], see text for details.

There has been direct evidence of binding of SNARE proteins (synaptic core complex) to the synaptic protein interaction (synprint) site on the II–III linker of the α_1 subunit of certain Ca channels [60,91,114,119,161]. Specifically, syntaxin, SNAP-25 and synapto-tagmin can bind to the synprint site in the α_{1A} and α_{1B} (Fig. 5; [60,91]). Syntaxin can modulate the activity of both N-type and P/Q-type Ca channels in vitro indicating a functional association between these proteins [8,134]. Also, when synprint peptides are injected into neurons they block the interaction between the Ca channels and the SNARE proteins and inhibit neurotransmission [91]. It is likely that the major advantage for the close association between Ca channels and the SNARE proteins is the ability to increase Ca-dependent fast exocytosis by allowing for rapid local increases in intracellular Ca. However, it appears that Ca independent signals, specifically voltage-dependent signals, can also be transferred from the Ca channels to the SNARE proteins and this Ca independent signal is essential for fast neurotransmitter release at synapses [91]. Other proteins that bind to the synprint site of P/Q-type Ca channels and to VAMP (synaptobrevin) are the cysteine string proteins (Csps), which may act as molecular chaparones to direct assembly or disassembly of the exocytotic complex [74]. One interesting study has shown that not only do α_{1A} P/Q-type Ca channels physically interact with the SNARE complex but may also direct the Ca-dependent transcription of part of that same complex [134]. The researchers found that discrete levels of Ca influx through α_{1A} P/Q-type but not other types of Ca channels was essential for expression of syntaxin-1A in both HEK cells and native cerebellar granule neurons [91].

The modification of synaptic strength is a key component of nervous communication in the brain and the basis for processes like long-term potentiation (LTP) or long-term depression (LTD) thought to be involved in learning and memory. Ca channels are targets for many types of modulators, including G-proteins and protein kinases, and these interactions likely underly synaptic modification. Both N-type and P/Q-type, but not L-type or T-type, Ca channels are inhibited by activation of G-protein coupled receptors such as muscarinic acetylcholine, GABA, opioid, somatostatin, prostaglandin, adrenergic, and adenosine receptors (see reviews [21,88,164]). The rapid G-protein dependent inhibition of the N-type α_{1B} and P/Q-type α_{1A} Ca channels is due to binding of $G_{\beta\gamma}$ subunits after dissociation from the G_{α} subunit during neurotransmitter receptor activation [48,53]. The main site of the 1:1 interaction of the $G_{\beta\gamma}$ with the α_1 subunit is the I–II intracellular loop (Fig. 6; [101]) specifically at two sites; one overlapping where the β subunit binds (AID site; [110]), and the other downstream from the AID site (Fig. 6; [163]). Other studies have identified a putative site of interaction on the carboxyl terminus of the α_1 subunit [112] and possibly other parts of domain I [167]. Previous studies had shown that native and exogenously expressed Ca channels can be modulated by protein kinase C and that the intracellular loop between domains I and II of the α_1 subunit may be involved in this process [127]. It has been shown that there is crosstalk between these modulators of the α_{1B} N-type channel in that PKC activation depresses the subsequent G-protein dependent inhibition of these channels [45,163]. The crosstalk region has been localized to a I–II loop site downstream of the AID sequence by mutation analysis, where inactivation of the PKC phosphorylation sites was associated with a reduction in G-protein inhibition [45]. Deletion of some of the carboxyl tail of α_{1B} also reduced G-protein inhibition lending credence to this being an area of interaction of G-proteins [45].

Another modulator of Ca channels is cAMP dependent protein kinase A (PKA), which was first implicated in affecting the cardiac L-type channels during stimulation of β-adrenergic receptors (see reviews [15,41]). PKA has been shown to affect both skeletal muscle (α_{1S}) and cardiac (α_{1C}) L-type Ca channels and is essential for the

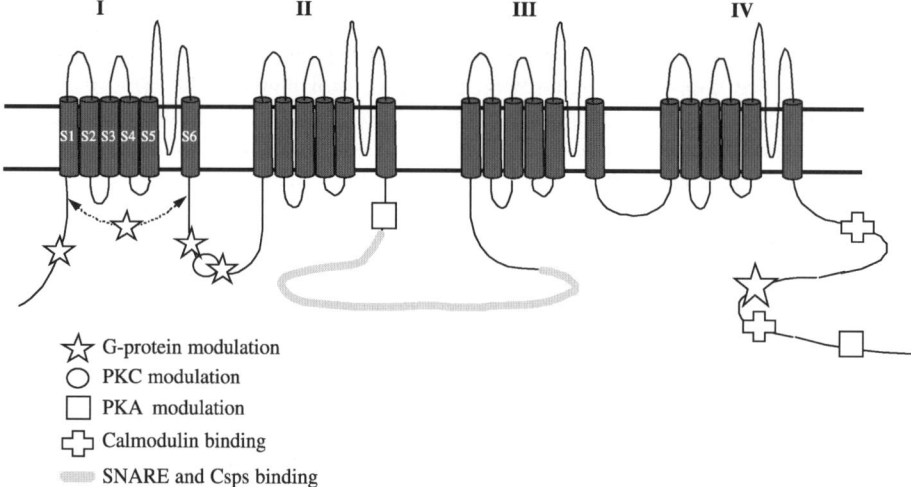

☆ G-protein modulation
○ PKC modulation
▢ PKA modulation
⊕ Calmodulin binding
▬ SNARE and Csps binding

Fig. 6. Sites of modulation of some intracellular effectors on the α_1 subunit of neuronal Ca channels. The G-protein $\beta\gamma$ complex interacts with the I–II loop and the COOH tail of α_{1A} and α_{1B}. PKC has been shown to phosphorylate the I–II linker of these Ca channels. PKA has been shown to interact with the COOH tail of α_{1C}. Two separate sites for calmodulin binding on the α_{1C} COOH tail have been found. SNARE proteins and cysteine string proteins (Csps) are thought to interact with the synprint site in the II–III loop of the α_1 subunit. See text for details.

voltage-dependent facilitation displayed by these channels [10,15]. A cAMP-dependent protein kinase anchoring protein (AKAP) is required to facilitate modulation of both the α_{1S} and α_{1C} subunits by PKA [35,42]. These AKAPs are thought to tether PKA enzyme to both the Ca channel and the membrane, and allow for rapid and efficient phosphorylation of channel. A key site of AKAP dependent PKC phosphorylation of the α_{1C} channel is a carboxyl terminal serine (Ser 1928) [35]. Modulation of the skeletal and cardiac muscle L-type Ca channels by neurotransmitter receptors which stimulate PKA could therefore enhance Ca influx into these cell types in order to replenish calcium stores after contraction events. In the brain, this type of modulation of α_{1C} L-type channels may allow for increased intracellular Ca concentrations, which are known to stimulate gene expression [20].

Besides PKA-dependent facilitation, L-type channels in particular also show Ca-dependent inactivation that decreases the flow of Ca ions into cells. It has been well documented that intracellular calcium levels are regulated by binding to calmodulin (CaM), and several groups have shown that calmodulin associates with Ca channels and is essential in the Ca-dependent inactivation of these channels [104,113,171]. Disruption of a carboxyl-terminal CaM-binding "IQ" motif (IQEYFRKFKKRK; [171]) prevents Ca-dependent inactivation of the α_{1C} channel [113,171]. Further evidence of this phenomenon was demonstrated by Peterson and collegues [104], who utilized CaM mutants that do not bind Ca. They found that cells coexpressioning the mutant CaM and the α_{1C} channel exhibited no Ca-dependent inactivation [104]. Mutations in the "IQ" motif which affect Ca-dependent inactivation also affect voltage-dependent faciliation of the α_{1C} L-type Ca channel [171], which may indicate an interaction between PKA phosphorylation and CaM binding. CaM binding does not appear to be specific to the α_{1C} Ca channel, as fusion proteins containing the "IQ" regions of α_{1A}, α_{1B}, and α_{1E} subunits also bind CaM [104]. Additionally, using a yeast two hybrid approach, another carboxyl terminal site (Fig. 6; calmodulin binding domain, CBD) further downstream from the IQ motif on the α_{1A} has been shown to bind to CaM intracellularly [71]. Unlike the α_{1C} L-type channels, the α_{1A} P/Q-type channel does not show pronounced Ca-dependent inactivation, although when CaM is inhibited the inactivation kinetics of the α_{1A} currents are modified slightly [71]. Of note, the CBD domain of the α_{1A} subunit is deleted in *leaner* mice and this may contribute to the ataxia, epilepsy and cerebellar degeneration of these mice [71].

Ca channels have been implicated in both contributing to LTP and LTD in the hippocampus [129]. These multi-faceted processes are activity-dependent increases (LTP) or decreases (LTD) in the synaptic strength that can last for at least an hour. There are two major types of LTP in the hippocampus – one that relies on Ca influx through NMDA receptors in the postsynaptic membrane and another that is independent of NMDA receptors and appears to be initiated in the presynaptic membrane (see review [17]). NMDA dependent LTP occurs in the CA1 region of the hippocampus and the rise in intracellular Ca is thought to stimulate Ca-dependent enzymes like PKC and CamKII [17]. In both cultured hippocampal neurons [9] and in the thalamo-amygdala pathway [148] LTP can be abolished by using dihydropyridines to specifically block L-type Ca channels. This suggests that L-type Ca channels in the postsynaptic membrane play a role in the Ca influx that is essential to LTP.

7. Ca channels and disease

Voltage-gated Ca channels play important roles in a number of human diseases. L-type Ca channels are essential in initiating and maintaining cardiac and smooth muscle contraction, which has made them the target for drugs used in the treatment of cardio-vascular disorders (see review [51]). Benzothiazepines (e.g. Diltiazem), phenyl-alkylamines (e.g. Verapamil), and dihydropyridines (e.g. Nifedipine) have been used to treat cardiovascular diseases such as hypertension and heart arrhythmias, with varying degrees of effectiveness. The α_1 subunit is the main target for these L-type blockers and studies have shown that amino acid residues (nine in particular) in domain III S5,S6 and IV S6 are crucial in dihydropyridine binding [51,120]. Further proof domain comes from studies that show that dihydropyridine-insensitive channels (e.g. α_{1A}, α_{1E}) can be made sensitive to this class of drug by mutations in the amino acid residues responsible for this binding [55,120].

Along with being the target for cardiovascular therapeutics, Ca channels have also been the target of drugs that may convey cellular protection after ischemia. Early Ca influx into cells after ischemic events is a major factor in stimulating apoptotic cascades which destroys heart cells after angina or neurons after a stroke [72]. Studies using animal models of brain ischemia have shown some neuroprotective effects of blockade of L- and T-type Ca channels with dihydropyridine antagonists or flunarizine [165]. Peptide toxin antagonists of other types of HVA channels have also shown some protective effects in animal models. For example, the specific N-type channel blocker ω-CgTx-GVIA (also called SNX-111; Table 1) protects rats after transient ischemia [102]. Additionally, the P/Q-type channel antagonist, ω-Aga-IVA [89], protects against brain injury after focal ischemia in rats [3]. Animal studies with Ca channels blockers have shown that these drugs show varying degrees of effectiveness in neuroprotection [72]. In human phase III clinical trials of the dihydropyridine L-type antagonist nimodipine and the less specific Ca channel antagonist, flunarizine, little efficacy in treatment of ischemic patients was observed [72]. There are several possible reasons for the variability in the animal studies and lack of effectiveness of Ca channel antagonists in these clinical trials. Firstly, it is thought that astrocytes in the brain play a beneficial role after ischemia by modulating extracellular ionic concentrations and releasing growth factors responsible for the health of the local neurons. It has been shown that α_{1C} L-type channels are upregulated in astrocytes after focal ischemia in animals [152] and this excess Ca influx may be essential in the healing process after ischemia by stimulating the release of neuroprotective growth factors. Ca channels blockers would prevent this Ca influx and actually inhibit the neuroprotective effects of the astrocytes [152]. A second possible reason for the varied effects of Ca channels blockers on ischemic injury may have to do with the effect of zinc toxicity in neurons. Zinc has been implicated in ischemic brain damage, and Ca channels blockers may exert some of the neuroprotective effects seen in animal models by preventing zinc influx through Ca channels and not by preventing Ca influx [72].

Epileptiform activity in the brain has also been linked with neuronal death and can also cause aberrant growth during development. For example, hippocampal mossy fibres develop postnatally and are altered in seizure-induced brain slices from premature rats [54]. The L-type Ca channel blocker nicardipine prevents this abnormal development and protects against neuronal injuries in this preparation, suggesting that increased levels of intracellular calcium through Ca channels during epileptiform activity is one of the contributing factors to abnormal development [54]. Some forms of epilepsy have been the direct target of therapeutic Ca channel antagonists, as these channels play important roles in modulating the electrical activity in the brain [52]. Many of the drugs used in the treatment of various forms of epilepsy have been implicated in blocking Ca channel function. Phenytoin, carbamazepine, and ethosuximide have all been shown to affect Ca channel function, with the L- and T-types being the major targets [52,117]. A relatively new anticonvulsant drug, gabapentin, has shown promise for the treatment of epilepsy [27], and has been shown to bind directly to the $\alpha_2\delta$ subunit of HVA Ca channels [147]. Also of note, in a rat model of absence epilepsy, small but significantly increased levels of α_{1G} and α_{1H} mRNA were observed compared to wild-type controls, which may account in part for the increased electrical activity typical of epileptic seizures [136].

Along with targeting Ca channels for drug therapy, mutations in the genes coding for Ca channels have also been implicated in a number of human nervous and muscular disorders (see review [79]). Human *incomplete X-linked congenital stationary night blindness* is thought to be caused by impaired synaptic neurotransmission from photoreceptor cells to neurons in the retina and has been linked to a variety of mutations in the L-type α_{1F} subunit (see Fig. 7; [5,132]). *Familial hemiplegic migraine* has been linked to seven mutations in the α_{1A} subunit gene (Fig. 7; [25,67,100]) all of which when introduced into α_{1A} cDNA cause some changes in the voltage-dependent or kinetic properties of the expressed α_{1A} current [46,66,67]. *Episodic ataxia-2* (EA-2) is a human disease that causes loss of

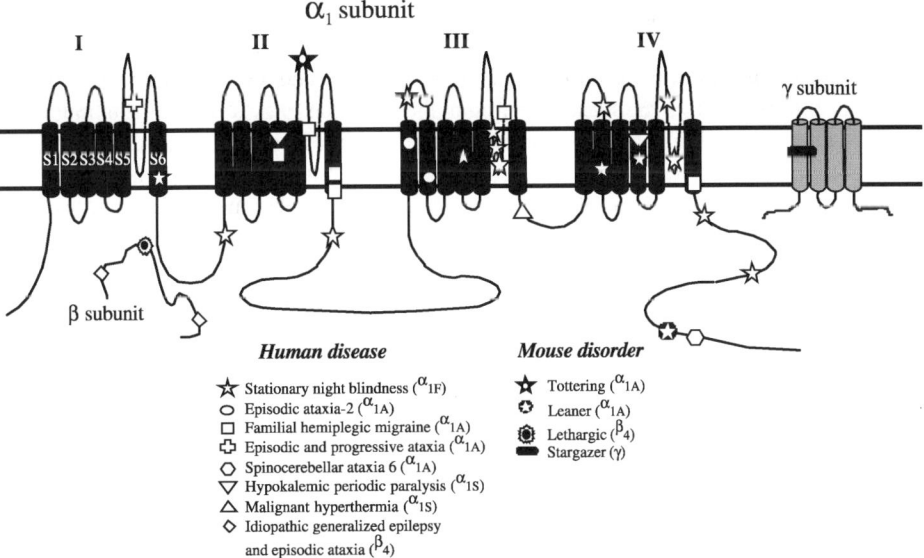

Fig. 7. Mutations in Ca channel subunits associated with inherited diseases. The mutations represented on the figure are associated with the labeled diseases. Human diseases are indicated with the open symbols. Adapted from Ref. [79], see text for details.

muscular coordination, dizziness, difficulty in speech, and cerebellar atrophy. This disease has also been linked to mutations in the α_{1A} gene (Fig. 7; [100]). *Autosomal dominant spinocerebellar ataxia 6* (SCA6) has similar symptoms to EA-2 and is also linked to mutations in α_{1A} (Fig. 7; [169]). However, unlike the other two α_{1A}-linked diseases this is a progressive disorder with onset at 40–50 years of age. Like other neurodegenerative disorders (e.g. Huntington's disease) there is the presence of expanded triplet repeats (normal individuals = 4–16 CAG's; SCA6 affected individuals = 21–27 CAG's) although the expansion in SCA6 is limited compared to Huntington's [169]. A fourth α_{1A}-linked human disease is *Episodic and progessive ataxia* which shows overlapping symptoms with SCA6 and EA-2 but has an earlier onset (5 to 15 years) of the disease [162]. The four α_{1A}-linked human diseases all show some cerebellar degeneration as the disease progresses and since the α_{1A} gene is highly expressed in the cerebellum [123,125], alterations in the functioning of α_{1A} P/Q-type Ca channels may contribute to this degeneration [79].

Two mutant strains of mice also show neurological symptoms linked to α_{1A} gene mutations. *Tottering* (*tg*) mice show mild ataxia and display seizures that are due to a single point mutation in the IIS5–S6 linker (Fig. 7; [23,33]). Ca channel current densities in the mutant mice are significantly reduced compared to wild type mice [145]. *Leaner* (*tg*^{la}) mice also have defects in the α_{1A} gene (Fig. 7; [23,33]) which results in decreased whole-cell Ca current densities in Purkinje cells [22]. Unlike *tg* mice, *leaner* mice show severe ataxia and pronounced cerebellar degeneration similar to some of the human ataxic diseases above.

Mutations in Ca channels involved in excitation–contraction coupling have been linked to human muscular diseases. *Hypokalemic periodic paralysis* causes episodic muscle weakness and has been linked to mutations in the voltage sensor region (S4) of the α_{1S} subunit gene (Fig. 7; [58,111]). *Malignant hyperthermia* (MH) episodes are characterized by rigidity, metabolic rise, fever and tachycardia [79]. Certain forms of MH appear to be due to a missense mutation in the α_{1S} gene [92]. It is likely that mutations in α_{1S} alters coupling of the dihydropyridine receptor to the ryanodine receptors preventing proper intracellular Ca homeostasis and affecting muscle contraction [79].

Mutations in the β_4 subunit of the HVA Ca channel complex cause a murine disorder, *lethargic* (*lh*), which has symptoms which mimic the human α_{1A} diseases (e.g. ataxia and seizures) but does not cause neuronal degeneration (Fig. 7; [13]). In the *lethargic* brain, the β_4 subunit protein is not detectable, and α_{1B} N-type Ca channel expression is decreased [85]. In contrast, in normal brain tissue β_4 is present at high levels [81,105,118]. Interestingly, there may be some redundancy in the roles of the β subunits and some compensatory cellular changes that occur, as the β_{1b} subunit is upregulated in *lethargic* mice [85].

In *Stargazer* (*stg*) mice, the γ_2 subunit is not expressed properly due to an early transposon insertion in the first intron of the gene, and this misexpression is responsible for the ataxic gate and epileptic seizures displayed by these mice [73].

Another disorder that is associated with Ca channels is a paraneoplastic nervous system disease called *Lambert–Eaton myasthenic syndrome* (LEMS; see review [109]). LEMS is an autoimmune disorder in which antibodies to Ca channels are elicited presumably against these proteins produced by small cell lung carcinoma (SCLC) cells associated with this disease [109]. The site of pathology of LEMS is the presynaptic terminal of the cholinergic synapse which causes muscle weakness, impotence, and a lack of reflexes [109]. The major Ca channel target of these autoantibodies appears to be the α_{1A} P/Q-type channel [144], and HEK cells specifically expressing α_{1A} Ca channels show reduced Ca influx when exposed to LEMS serum [108]. Other possible targets of the LEMS autoantibodies include the Ca channel β subunit [143] or the N-type Ca channel [93]. Both P/Q-type and N-type channels are present at the presynaptic terminal of the human neuromuscular junction and the LEMS autoantibodies presumably cross-link these Ca channels and limit their function preventing transmitter release and inhibiting muscle contraction [109].

8. Future directions of research

One of the controversial aspects of research into Ca channel function has been the elucidation of the molecular structure–function relationships of the various domains. The putative four domain structure of Ca channels (Fig. 3) is largely based on homology with other ion channels, hydrophobicity patterns, biochemical analysis, and mutation studies [2,15]. However, when these same methods were used for predicting 3-D structure of other membrane proteins (e.g. acetylcholine receptor) and compared to the actual 3-D structure as resolved by X-ray crystallography, a number of errors were noted [90]. The first 3-D structure of a voltage-gated ion channel was determined for a small bacterial potassium channel (KcsA). The structure showed that the transmembrane domains adopted an inverted teepee structure [82], which is very different than the rigid vertical domains shown in Ca channel predictions [24,82]. However, it must be noted that the KcsA channel is a two-transmembrane, one domain channel quite different than the six-transmembrane, four domain α_1 subunit of the Ca channels. Recently, a low-resolution 3-D structure for a voltage-gated sodium channel from the electric eel was determined by cryo-electron microscopy and image reconstruction techniques [116]. An analysis of the structure shows the predicted extracellular and intracellular domains but also shows an unexpected finding of four peripheral transmembrane pores instead of the predicted one central pore [116]. It will be of interest to see if this technique can be applied to the very structurally similar Ca channels and to evaluate predictions of voltage activation, inactivation, subunit interactions, and accessory protein binding.

The majority of Ca channel α_1 subunits have been isolated from mammals (for reviews see [15,126]) but over the last 5 years Ca channels from other classes and phyla have been determined and allow a primary sequence comparison (from Genbank). When the L-type Ca channels from different phyla are compared (using protein parsimony analysis, [31]) one sees a distinct pattern of relatedness similar to the phylogenetic patterns of the identified groups (Fig. 8; see also Ref. [170]). The greatest divergence is seen between the putative α_1 subunit isolated from the yeast (Kingdom: Fungi) compared to the other α_1 subunits (Kingdom: Animalia). The next three most distinct groupings consist of α_1 subunits isolated from the Phyla Arthropoda, Cnidaria, and Nematoda thought to diverge during the Cambrian explosion. All the members of our phylum (Chordata) show the most similarity (Fig. 8) with the most distant being that of the subphylum Urochordata (ascidian) compared to the others, which are members of the subphylum Vertebrata. The presence of at least four

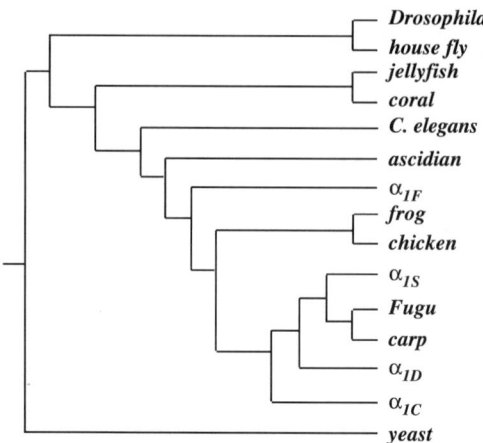

Fig. 8. Phylogenetic comparison of L-type Ca channel α_1 subunits. A protein parsimony analysis [31] of the aligned fourth domain of the L-type α_1 subunits showed divergences that mimic the phylogenetic relationships of the organisms shown. Genebank Accession numbers: *Drosophila* – AAA81883; housefly – S41742; jellyfish – AAC63050; coral – AAD11470; *C. elegans* – AAC47755; ascidian – BAA34927; α_{1F} – NP_005174; frog – AAC36126; chicken – AAC08304; α_{1S} – NP_000060; *Fugu* – AAC15583; carp – A37860; α_{1D} – NP_000711; α_{1C} – NP_000710; yeast – S64540.

distinct types (α_{1S}, α_{1C}, α_{1D}, α_{1F}) of mammalian L-type Ca channels and the divergence exhibited by these subunits suggests that these channels separated fairly early in the evolutionary process in vertebrates (Fig. 8). This suggests that comparison of sequences from different species may help identify conserved regions with important functional roles.

References

[1] Adams, M.E., Myers, R.A., Imperial, J.S., Olivera, B.M., 1993. Toxityping rat brain calcium channels with omega-toxins from spider and cone snail venoms. Biochemistry 32(47), 12566–12570.

[2] Armstrong, C.M., Hille, B., 1998. Voltage-gated ion channels and electrical excitability. Neuron 20(3), 371–380.

[3] Asakura, K., Matsuo, Y., Kanemasa, T., Ninomiya, M., 1997. P/Q-type Ca2+ channel blocker omega-agatoxin IVA protects against brain injury after focal ischemia in rats. Brain Res. 776(1–2), 140–145.

[4] Beam, K.G., Adams, B.A., Niidome, T., Numa, S., Tanabe, T., 1992. Function of a truncated dihydropyridine receptor as both voltage sensor and calcium channel. Nature 360, 169–171.

[5] Bech-Hansen, N.T., Naylor, M.J., Maybaum, T.A., Pearce, W.G., Koop, B., Fishman, G.A., Mets, M., Musarella, M.A., Boycott, K.M., 1998. Loss-of-function mutations in a calcium-channel alpha1-subunit gene in Xp11.23 cause incomplete X-linked congenital stationary night blindness. Nat. Genet. 19(3), 264–267.

[6] Bers, D.M., Perez-Reyes, E., 1999. Ca channels in cardiac myocytes: structure and function in Ca influx and intracellular Ca release. Cardiovasc. Res. 42(2), 339–360.

[7] Beurg, M., Sukhareva, M., Ahern, C.A., Conklin, M.W., Perez-Reyes, E., Powers, P.A., Gregg, R.G., Coronado, R., 1999. Differential regulation of skeletal muscle L-type Ca2+ current and excitation–contraction coupling by the dihydropyridine receptor beta subunit. Biophys. J. 76(4), 1744–1756.

[8] Bezprozvanny, I., Scheller, R.H., Tsien, R.W., 1995. Functional impact of syntaxin on gating of N-type and Q-type calcium channels. Nature 378(6557), 623–626.

[9] Bi, G.Q., Poo, M.M., 1998. Synaptic modifications in cultured hippocampal neurons: dependence on spike timing, synaptic strength, and postsynaptic cell type. J. Neurosci. 18(24), 10464–10472.

[10] Bourinet, E., Charnet, P., Tomlinson, W.J., Stea, A., Snutch, T.P., Nargeot, J., 1994. Voltage-dependent facilitation of a neuronal alpha 1C L-type calcium channel. EMBO J. 13(21), 5032–5039.

[11] Bourinet, E., Zamponi, G.W., Stea, A., Soong, T.W., Lewis, D.A., Jones, L.P., Yue, D.T., Snutch, T.P., 1996. The α_{1E} calcium channel exhibits permeation properties similar to low-voltage-activated calcium channels. J. Neurosci. 16, 4983–4993.

[12] Bourinet, E., Soong, T.W., Sutton, K., Slaymaker, S., Mathews, E., Monteil, A., Zamponi, G.W., Nargeot, J., Snutch, T.P., 1999. Splicing of alpha 1A subunit gene generates phenotypic variants of P- and Q-type calcium channels. Nat. Neurosci. 2(5), 407–415.

[13] Burgess, D.L., Jones, J.M., Meisler, M.H., Noebels, J.L., 1997. Mutation of the Ca2+ channel beta subunit gene Cchb4 is associated with ataxia and seizures in the lethargic (lh) mouse. Cell 88(3), 385–392.

[14] Campbell, K.P., Leung, A.T., Sharp, A.H., 1988. The biochemistry and molecular biology of the dihydropyridine-sensitive calcium channel. Trends Neurosci. 11, 425–430.

[15] Catterall, W.A., 2000. Structure and regulation of voltage-gated Ca2+ channels. Annu. Rev. Cell Dev. Biol. 16, 521–555.

[16] Chemin, J., Monteil, A., Bourinet, E., Nargeot, J., Lory, P., 2001. Alternatively spliced alpha(1G) (Ca(V)3.1) intracellular loops promote specific T-type Ca(2+) channel gating properties. Biophys. J. 80(3), 1238–1250.

[17] Chen, C., Tonegawa, S., 1997. Molecular genetic analysis of synaptic plasticity, activity-dependent neural development, learning, and memory in the mammalian brain. Annu. Rev. Neurosci. 20, 157–184.

[18] Craig, P.J., Beattie, R.E., Folly, E.A., Banerjee, M.D., Reeves, M.B., Priestley, J.V., Carney, S.L., Sher, E., Perez-Reyes, E., Volsen, S.G., 1999. Distribution of the voltage-dependent calcium channel alpha1G subunit mRNA and protein throughout the mature rat brain. Eur. J. Neurosci. 11(8), 2949–2964.

[19] Cribbs, L.L., Lee, J.H., Yang, J., Satin, J., Zhang, Y., Daud, A., Barclay, J., Williamson, M.P., Fox, M., Rees, M., Perez-Reyes, E., 1998. Cloning and characterization of alpha1H from human heart, a member of the T-type Ca2+ channel gene family. Circ. Res. 83(1), 103–109.

[20] Deisseroth, K., Heist, E.K., Tsien, R.W., 1998. Translocation of calmodulin to the nucleus supports CREB phosphorylation in hippocampal neurons. Nature 392(6672), 198–202.

[21] Dolphin, A.C., 1998. Mechanisms of modulation of voltage-dependent calcium channels by G proteins. J. Physiol. (Lond.) 506(Pt 1), 3–11.

[22] Dove, L.S., Abbott, L.C., Griffith, W.H., 1998. Whole-cell and single-channel analysis of P-type calcium currents in cerebellar Purkinje cells of leaner mutant mice. J. Neurosci. 18(19), 7687–7699.

[23] Doyle, J., Ren, X., Lennon, G., Stubbs, L., 1997. Mutations in the Cacnl1a4 calcium channel gene are associated with seizures, cerebellar degeneration, and ataxia in tottering and leaner mutant mice. Mamm. Genome. 8(2), 113 120.

[24] Doyle, D.A., Morais Cabral, J., Pfuetzner, R.A., Kuo, A., Gulbis, J.M., Cohen, S.L., Chait, B.T., MacKinnon, R., 1998. The structure of the potassium channel: molecular basis of K + conduction and selectivity. Science 280(5360), 69–77.

[25] Ducros, A., Denier, C., Joutel, A., Vahedi, K., Michel, A., Darcel, F., Madigand, M., Guerouaou, D., Tison, F., Julien, J., Hirsch, E., Chedru, F., Bisgard, C., Lucotte, G., Despres, P., Billard, C., Barthez, M.A., Ponsot, G., Bousser, M.G., Tournier-Lasserve, E., 1999. Recurrence of the T666M calcium channel CACNA1A gene mutation in familial hemiplegic migraine with progressive cerebellar ataxia. Am. J. Hum. Genet. 64(1), 89–98.

[26] Ellinor, P.T., Yang, J., Sather, W.A., Zhang, J.F., Tsien, R.W., 1995. Ca2+ channel selectivity at a single locus for high-affinity Ca2+ interactions. Neuron 15(5), 1121–1132.

[27] Emilien, G., Maloteaux, J.M., 1998. Pharmacological management of epilepsy. Mechanism of action, pharmacokinetic drug interactions, and new drug discovery possibilities. Int. J. Clin. Pharmacol. Ther. 36(4), 181–194.

[28] Ertel, E.A., Campbell, K.P., Harpold, M.M., Hofmann, F., Mori, Y., Perez-Reyes, E., Schwartz, A., Snutch, T.P., Tanabe, T., Birnbaumer, L., Tsien, R.W., Catterall, W.A., 2000. Nomenclature of voltage-gated calcium channels. Neuron 25(3), 533–535.

[29] Farrugia, G., 1999. Ionic conductances in gastrointestinal smooth muscles and interstitial cells of Cajal. Annu. Rev. Physiol. 61, 45–84.

[30] Fass, D.M., Takimoto, K., Mains, R.E., Levitan, E.S., 1999. Tonic dopamine inhibition of L-type Ca^{2+} channel activity reduces alpha1D Ca^{2+} channel gene expression. J. Neurosci. 19(9), 3345–3352.

[31] Felsenstein, J., 1993. PHYLIP (Phylogeny Inference Package) version 3.5c. Distributed by the author, Department of Genetics, University of Washington, Seattle, http://bioweb.pasteur.fr/seqanal/interfaces/protpars-simple.html.

[32] Feng, Z.P., Hamid, J., Doering, C., Jarvis, S.E., Bosey, G.M., Bourinet, E., Snutch, T.P., Zamponi, G.W., 2001. Amino acid residues outside of the pore region contribute to N-type calcium channel permeation. J. Biol. Chem. 276(8), 5726–5730.

[33] Fletcher, C.F., Lutz, C.M., O'Sullivan, T.N., Shaughnessy, J.D. Jr., Hawkes, R., Frankel, W.N., Copeland, N.G., Jenkins, N.A., 1996. Absence epilepsy in tottering mutant mice is associated with calcium channel defects. Cell 87(4), 607–617.

[34] Freise, D., Held, B., Wissenbach, U., Pfeifer, A., Trost, C., Himmerkus, N., Schweig, U., Freichel, M., Biel, M., Hofmann, F., Hoth, M., Flockerzi, V., 2000. Absence of the gamma subunit of the skeletal muscle dihydropyridine receptor increases L-type Ca^{2+} currents and alters channel inactivation properties. J. Biol. Chem. 275(19), 14476–14481.

[35] Gao, T., Yatani, A., Dell'Acqua, M.L., Sako, H., Green, S.A., Dascal, N., Scott, J.D., Hosey, M.M., 1997. cAMP-dependent regulation of cardiac L-type Ca^{2+} channels requires membrane targeting of PKA and phosphorylation of channel subunits. Neuron 19(1), 185–196.

[36] Gao, T., Chien, A.J., Hosey, M.M., 1999. Complexes of the alpha1C and beta subunits generate the necessary signal for membrane targeting of class C L-type calcium channels. J. Biol. Chem. 274(4), 2137–2144.

[37] Gao, T., Buenemann, M., Gerhardstein, B.L., Ma, H., Hosey, M.M., 2000. Role of the C-terminus of the alpha 1C (CaV1.2) subunit in membrane targeting of cardiac L-type calcium channels. J. Biol. Chem. 275(33), 25436–25444.

[38] Gerster, U., Neuhuber, B., Groschner, K., Striessnig, J., Flucher, B.E., 1999. Current modulation and membrane targeting of the calcium channel alpha1C subunit are independent functions of the beta subunit. J. Physiol. (Lond.) 517(Pt 2), 353–368.

[39] Grabner, M., Dirksen, R.T., Beam, K.G., 1998. Tagging with green fluorescent protein reveals a distinct subcellular distribution of L-type and non-L-type Ca^{2+} channels expressed in dysgenic myotubes. Proc. Natl Acad. Sci. USA 95(4), 1903–1908.

[40] Grabner, M., Dirksen, R.T., Suda, N., Beam, K.G., 1999. The II–III loop of the skeletal muscle dihydropyridine receptor is responsible for the bi-directional coupling with the ryanodine receptor. J. Biol. Chem. 274(31), 21913–21919.

[41] Gray, P.C., Scott, J.D., Catterall, W.A., 1998. Regulation of ion channels by cAMP-dependent protein kinase and A-kinase anchoring proteins. Curr. Opin. Neurobiol. 8(3), 330–334.

[42] Gray, P.C., Johnson, B.D., Westenbroek, R.E., Hays, L.G., Yates, J.R. III, Scheuer, T., Catterall, W.A., Murphy, B.J., 1998. Primary structure and function of an A kinase anchoring protein associated with calcium channels. Neuron 20(5), 1017–1026.

[43] Gregg, R.G., Messing, A., Strube, C., Beurg, M., Moss, R., Behan, M., Sukhareva, M., Haynes, S., Powell, J.A., Coronado, R., Powers, P.A., 1996. Absence of the beta subunit (cchb1) of the skeletal muscle dihydropyridine receptor alters expression of the alpha 1 subunit and eliminates excitation–contraction coupling. Proc. Natl Acad. Sci. USA 93(24), 13961–13966.

[44] Gurnett, C.A., Felix, R., Campbell, K.P., 1997. Extracellular interaction of the voltage-dependent Ca^{2+} channel alpha2delta and alpha1 subunits. J. Biol. Chem. 272(29), 18508–18512.

[45] Hamid, J., Nelson, D., Spaetgens, R., Dubel, S.J., Snutch, T.P., Zamponi, G.W., 1999. Identification of an integration center for cross-talk between protein kinase C and G protein modulation of N-type calcium channels. J. Biol. Chem. 274(10), 6195–6202.

[46] Hans, M., Luvisetto, S., Williams, M.E., Spagnolo, M., Urrutia, A., Tottene, A., Brust, P.F., Johnson, E.C., Harpold, M.M., Stauderman, K.A., Pietrobon, D., 1999. Functional consequences of mutations in the human alpha1A calcium channel subunit linked to familial hemiplegic migraine. J. Neurosci. 19(5), 1610–1619.

[47] Hell, J.W., Westenbroek, R.E., Warner, C., Ahlijanian, M.A., Prystay, W., Gilbert, M.M., Snutch, T.P., Catterall, W.A., 1993. Identification and differential subcellular localization of the neuronal class C and class D L-type calcium channel α_1 subunits. J. Cell Biol. 123, 949–962.

[48] Herlitze, S., Garcia, D.E., Mackie, K., Hille, B., Scheuer, T., Catterall, W.A., 1996. Modulation of Ca^{2+} channels by G-protein beta gamma subunits. Nature 380(6571), 258–262.

[49] Hille, B., 1992. Ionic Channels of Excitable Membranes. Sinauer Associates Inc., Sunderland, Mass.

[50] Hillyard, D.R., Monje, V.D., Mintz, I.M., Bean, B.P., Nadasdi, L., Ramachandran, J., Miljanich, G., Azimi-Zoonooz, A., McIntosh, J.M., Cruz, L.J., 1992. A new Conus peptide ligand for mammalian presynaptic Ca^{2+} channels. Neuron 9(1), 69–77.

[51] Hockerman, G.H., Peterson, B.Z., Johnson, B.D., Catterall, W.A., 1997. Molecular determinants of drug binding and action on L-type calcium channels. Annu. Rev. Pharmacol. Toxicol. 37, 361–396.

[52] Huguenard, J.R., 1996. Low-threshold calcium currents in central nervous system neurons. Annu. Rev. Physiol. 58, 329–348.

[53] Ikeda, S.R., 1996. Voltage-dependent modulation of N-type calcium channels by G-protein beta gamma subunits. Nature 380(6571), 255–258.

[54] Ikegaya, Y., 1999. Abnormal targeting of developing hippocampal mossy fibers after epileptiform activities via L-type Ca^{2+} channel activation in vitro. J. Neurosci. 19(2), 802–812.

[55] Ito, H., Klugbauer, N., Hofmann, F., 1997. Transfer of the high affinity dihydropyridine sensitivity from L-type to non-L-type calcium channel. Mol. Pharmacol. 52(4), 735–740.

[56] Jay, S.D., Ellis, S.B., McCue, A.F., Williams, M.E., Vedvick, T.S., Harpold, M.M., Campbell, K.P., 1990. Primary structure of the γ subunit of the DHP-sensitive calcium channel from skeletal muscle. Science 248, 490–492.

[57] Jay, S.D., Sharp, A.H., Kahl, S.D., Vedvick, T.S., Harpold, M.M., Campbell, K.P., 1991. Structural characterization of the dihydropyridine-sensitive calcium channel α_2-subunit and the associated δ peptides. J. Biol. Chem. 266, 3287–3293.

[58] Jurkat-Rott, K., Lehmann-Horn, F., Elbaz, A., Heine, R., Gregg, R.G., Hogan, K., Powers, P.A., Lapie, P., Vale-Santos, J.E., Weissenbach, J., et al., 1994. A calcium channel mutation causing hypokalemic periodic paralysis. Hum. Mol. Genet. 3(8), 1415–1419.

[59] Karaki, H., Ozaki, H., Hori, M., Mitsui-Saito, M., Amano, K., Harada, K., Miyamoto, S., Nakazawa, H., Won, K.J., Sato, K., 1997. Calcium movements, distribution, and functions in smooth muscle. Pharmacol. Rev. 49(2), 157–230.

[60] Kim, D.K., Catterall, W., 1997. Ca^{2+}-dependent and -independent interactions of the isoforms of the alpha1A subunit of brain Ca^{2+} channels with presynaptic SNARE proteins. Proc. Natl Acad. Sci. USA 94(26), 14782–14786.

[61] Kirchgessner, A.L., Liu, M.T., 1999. Differential localization of Ca^{2+} channel alpha1 subunits in the enteric nervous system: presence of alpha1B channel-like immunoreactivity in intrinsic primary afferent neurons. J. Comp. Neurol. 409(1), 85–104.

[62] Klugbauer, N., Lacinova, L., Marais, E., Hobom, M., Hofmann, F., 1999. Molecular diversity of the calcium channel alpha2delta subunit. J. Neurosci. 19(2), 684–691.

[63] Klugbauer, N., Dai, S., Specht, V., Lacinova, L., Marais, E., Bohn, G., Hofmann, F., 2000. A family of gamma-like calcium channel subunits. FEBS Lett. 470(2), 189–197.

[64] Kollmar, R., Montgomery, L.G., Fak, J., Henry, L.J., Hudspeth, A.J., 1997. Predominance of the alpha1D subunit in L-type voltage-gated Ca^{2+} channels of hair cells in the chicken's cochlea. Proc. Natl Acad. Sci. USA 94(26), 14883–14888.

[65] Kollmar, R., Fak, J., Montgomery, L.G., Hudspeth, A.J., 1997. Hair cell-specific splicing of mRNA for the alpha1D subunit of voltage-gated Ca^{2+} channels in the chicken's cochlea. Proc. Natl Acad. Sci. USA 94(26), 14889–14893.

[66] Kraus, R.L., Sinnegger, M.J., Glossmann, H., Hering, S., Striessnig, J., 1998. Familial hemiplegic migraine mutations change alpha1A Ca^{2+} channel kinetics. J. Biol. Chem. 273(10), 5586–5590.

[67] Kraus, R.L., Sinnegger, M.J., Koschak, A., Glossmann, H., Stenirri, S., Carrera, P., Striessnig, J., 2000. Three new familial hemiplegic migraine mutants affect P/Q-type Ca(2+) channel kinetics. J. Biol. Chem. 275(13), 9239–9243.

[68] Lacinova, L., Klugbauer, N., Hofmann, F., 1999. Absence of modulation of the expressed calcium channel alpha1G subunit by alpha2delta subunits. J. Physiol. (Lond.) 516(Pt 3), 639–645.

[69] Lambert, R.C., Maulet, Y., Mouton, J., Beattie, R., Volsen, S., De Waard, M., Feltz, A., 1997. T-type Ca^{2+} current properties are not modified by Ca^{2+} channel beta subunit depletion in nodosus ganglion neurons. J. Neurosci. 17(17), 6621–6628.

[70] Lee, J.H., Daud, A.N., Cribbs, L.L., Lacerda, A.E., Pereverzev, A., Klockner, U., Schneider, T., Perez-Reyes, E., 1999. Cloning and expression of a novel member of the low voltage-activated T-type calcium channel family. J. Neurosci. 19(6), 1912–1921.

[71] Lee, A., Wong, S.T., Gallagher, D., Li, B., Storm, D.R., Scheuer, T., Catterall, W.A., 1999. Ca^{2+}/calmodulin binds to and modulates P/Q-type calcium channels. Nature 399(6732), 155–159.

[72] Lee, J.M., Zipfel, G.J., Choi, D.W., 1999. The changing landscape of ischaemic brain injury mechanisms. Nature 399(6738 Suppl), A7–A14.

[73] Letts, V.A., Felix, R., Biddlecome, G.H., Arikkath, J., Mahaffey, C.L., Valenzuela, A., Bartlett, F.S. II, Mori, Y., Campbell, K.P., Frankel, W.N., 1998. The mouse stargazer gene encodes a neuronal Ca^{2+}-channel gamma subunit. Nat. Genet. 19(4), 340–347.

[74] Leveque, C., Pupier, S., Marqueze, B., Geslin, L., Kataoka, M., Takahashi, M., De Waard, M., Seagar, M., 1998. Interaction of cysteine string proteins with the alpha1A subunit of the P/Q-type calcium channel. J. Biol. Chem. 273(22), 13488–13492.

[75] Lie, A.A., Blumcke, I., Volsen, S.G., Wiestler, O.D., Elger, C.E., Beck, H., 1999. Distribution of voltage-dependent calcium channel beta subunits in the hippocampus of patients with temporal lobe epilepsy. Neuroscience 93(2), 449–456.

[76] Ligon, B., Boyd, A.E. III, Dunlap, K., 1998. Class A calcium channel variants in pancreatic islets and their role in insulin secretion. J. Biol. Chem. 273(22), 13905–13911.

[77] Lin, Z., Haus, S., Edgerton, J., Lipscombe, D., 1997. Identification of functionally distinct isoforms of the N-type Ca^{2+} channel in rat sympathetic ganglia and brain. Neuron 18, 153–166.

[78] Liu, H., De Waard, M., Scott, V.E.S., Gurnett, C.A., Lennon, V.A., Campbell, K.P., 1996. Identification of three subunits of the high affinity omega-conotoxin MVIIC-sensitive Ca^{2+} channel. J. Biol. Chem. 271(23), 13804–13810.

[79] Lorenzon, N.M., Beam, K.G., 2000. Calcium channelopathies. Kidney Int. 57(3), 794–802.

[80] Lu, Q., Dunlap, K., 1999. Cloning and functional expression of novel N-type Ca(2+) channel variants. J. Biol. Chem. 274(49), 34566–34575.

[81] Ludwig, A., Flockerzi, V., Hofmann, F., 1997. Regional expression and cellular localization of the alpha1 and beta subunit of high voltage-activated calcium channels in rat brain. J. Neurosci. 17(4), 1339–1349.

[82] MacKinnon, R., Cohen, S.L., Kuo, A., Lee, A., Chait, B.T., 1998. Structural conservation in prokaryotic and eukaryotic potassium channels. Science 280(5360), 106–109.

[83] McCleskey, E.W., 1999. Calcium channel permeation: a field in flux. J. Gen. Physiol. 113(6), 765–772.

[84] McDonald, T.F., Pelzer, S., Trautwein, W., Pelzer, D.J., 1994. Regulation and modulation of calcium channels in cardiac, skeletal, and smooth muscle cells. Physiol. Rev. 74(2), 365–507.

[85] McEnery, M.W., Copeland, T.D., Vance, C.L., 1998. Altered expression and assembly of N-type calcium channel alpha1B and beta subunits in epileptic lethargic (lh/lh) mouse. J. Biol. Chem. 273(34), 21435–21438.

[86] McRory, J.E., Santi, C.E., Hamming, K.S.C., Mezeyova, J., Sutton, K.G., Baillie, D.L., Stea, A., Snutch, T.P., 2001. Molecular and functional characterization of a family of rat brain T-type calcium channels. J. Biol. Chem. 276, 3999–4011.

[87] Mikami, A., Imoto, K., Tanabe, T., Niidome, T., Mori, Y., Takeshima, H., Narumiya, S., Numa, S., 1989. Primary structure and functional expression of the cardiac dihydropyridine-sensitive calcium channel. Nature 340, 230–233.

[88] Miller, R.J., 1998. Presynaptic receptors. Annu. Rev. Pharmacol. Toxicol. 38, 201–227.

[89] Mintz, I.M., Venema, V.J., Swiderek, K.M., Lee, T.D., Bean, B.P., Adams, M.E., 1992. P-type calcium channels blocked by the spider toxin ω-Aga-IVA. Nature 355, 827–829.

[90] Miyazawa, A., Fujiyoshi, Y., Stowell, M., Unwin, N., 1999. Nicotinic acetylcholine receptor at 4.6 A resolution: transverse tunnels in the channel wall. J. Mol. Biol 288(4), 765–786.

[91] Mochida, S., Yokoyama, C.T., Kim, D.K., Itoh, K., Catterall, W.A., 1998. Evidence for a voltage-dependent enhancement of neurotransmitter release mediated via the synaptic protein interaction site of N-type Ca^{2+} channels. Proc. Natl Acad. Sci. USA 95(24), 14523–14528.

[92] Monnier, N., Procaccio, V., Stieglitz, P., Lunardi, J., 1997. Malignant-hyperthermia susceptibility is associated with a mutation of the alpha 1-subunit of the human dihydropyridine-sensitive L-type voltage-dependent calcium-channel receptor in skeletal muscle. Am. J. Hum. Genet. 60(6), 1316–1325.

[93] Motomura, M., Lang, B., Johnston, I., Palace, J., Vincent, A., Newsom-Davis, J., 1997. Incidence of serum anti-P/O-type and anti-N-type calcium channel autoantibodies in the Lambert–Eaton myasthenic syndrome. J. Neurol. Sci. 147(1), 35–42.

[94] Nakai, J., Adams, B.A., Imoto, K., Beam, K.G., 1994. Critical roles of the S3 segment and S3–S4 linker of repeat I in activation of L-type calcium channels. Proc. Natl Acad. Sci. USA 91(3), 1014–1018.

[95] Namkung, Y., Smith, S.M., Lee, S.B., Skrypnyk, N.V., Kim, H.L., Chin, H., Scheller, R.H., Tsien, R.W., Shin, H.S., 1998. Targeted disruption of the Ca^{2+} channel beta3 subunit reduces N- and L-type Ca^{2+} channel activity and alters the voltage-dependent activation of P/Q-type Ca^{2+} channels in neurons. Proc. Natl Acad. Sci. USA 95(20), 12010–12015.

[96] Naylor, M.J., Rancourt, D.E., Bech-Hansen, N.T., 2000. Isolation and characterization of a calcium channel gene, Cacna1f, the murine orthologue of the gene for incomplete X-linked congenital stationary night blindness. Genomics 66(3), 324–327.

[97] Neuhuber, B., Gerster, U., Doring, F., Glossmann, H., Tanabe, T., Flucher, B.E., 1998. Association of calcium channel alpha1S and beta1a subunits is required for the targeting of beta1a but not of alpha1S into skeletal muscle triads. Proc. Natl Acad. Sci. USA 95(9), 5015–5020.

[98] Nilius, B., Viana, F., Droogmans, G., 1997. Ion channels in vascular endothelium. Annu. Rev. Physiol. 59, 145–170.

[99] Olivera, B.M., Gray, W.R., Zeikus, R., McIntosh, J.M., Varga, J., Rivier, J., de Santos, V., Cruz, L.J., 1985. Peptide neurotoxins from fish-hunting cone snails. Science 230(4732), 1338–1343.

[100] Ophoff, R.A., Terwindt, G.M., Vergouwe, M.N., van Eijk, R., Oefner, P.J., Hoffman, S.M., Lamerdin, J.E., Mohrenweiser, H.W., Bulman, D.E., Ferrari, M., Haan, J., Lindhout, D., van Ommen, G.J., Hofker, M.H., Ferrari, M.D., Frants, R.R., 1996. Familial hemiplegic migraine and episodic ataxia type-2 are caused by mutations in the Ca^{2+} channel gene CACNL1A4. Cell 87(3), 543–552.

[101] Page, K.M., Stephens, G.J., Berrow, N.S., Dolphin, A.C., 1997. The intracellular loop between domains I and II of the B-type calcium channel confers aspects of G-protein sensitivity to the E-type calcium channel. J. Neurosci. 17(4), 1330–1338.

[102] Perez-Pinzon, M.A., Yenari, M.A., Sun, G.H., Kunis, D.M., Steinberg, G.K., 1997. SNX-111, a novel, presynaptic N-type calcium channel antagonist, is neuroprotective against focal cerebral ischemia in rabbits. J. Neurol. Sci. 153(1), 25–31.

[103] Perez-Reyes, E., Cribbs, L.L., Daud, A., Lacerda, A.E., Barclay, J., Williamson, M.P., Fox, M., Rees, M., Lee, J.H., 1998. Molecular characterization of a neuronal low-voltage-activated T-type calcium channel. Nature 391(6670), 896–900.

[104] Peterson, B.Z., DeMaria, C.D., Adelman, J.P., Yue, D.T., 1999. Calmodulin is the Ca^{2+} sensor for Ca^{2+}-dependent inactivation of L-type calcium channels. Neuron 22(3), 549–558.

[105] Pichler, M., Cassidy, T.N., Reimer, D., Haase, H., Kraus, R., Ostler, D., Striessnig, J., 1997. Beta subunit heterogeneity in neuronal L-type Ca^{2+} channels. J. Biol. Chem. 272(21), 13877–13882.

[106] Piedras-Renteria, E.S., Chen, C.C., Best, P.M., 1997. Antisense oligonucleotides against rat brain alpha1E DNA and its atrial homologue decrease T-type calcium current in atrial myocytes. Proc. Natl Acad. Sci. USA 94(26), 14936–14941.

[107] Piedras-Renteria, E.S., Tsien, R.W., 1998. Antisense oligonucleotides against alpha1E reduce R-type calcium currents in cerebellar granule cells. Proc. Natl Acad. Sci. USA 95(13), 7760–7765.

[108] Pinto, A., Gillard, S., Moss, F., Whyte, K., Brust, P., Williams, M., Stauderman, K., Harpold, M., Lang, B., Newsom-Davis, J., Bleakman, D., Lodge, D., Boot, J., 1998. Human autoantibodies specific for the alpha1A calcium channel subunit reduce both P-type and Q-type calcium currents in cerebellar neurons. Proc. Natl Acad. Sci. USA 95(14), 8328–8333.

[109] Posner, J.B., Dalmau, J., 1997. Paraneoplastic syndromes. Curr. Opin. Immunol. 9(5), 723–729.

[110] Pragnell, M., De Waard, M., Mori, Y., Tanabe, T., Snutch, T.P., Campbell, K.P., 1994. Calcium channel β subunit binds to a conserved motif in the I–II cytoplasmic linker of the α_1 subunit. Nature 368, 68–70.

[111] Ptacek, L.J., Tawil, R., Griggs, R.C., Engel, A.G., Layzer, R.B., Kwiecinski, H., McManis, P.G., Santiago, L., Moore, M., Fouad, G., et al., 1994. Dihydropyridine receptor mutations cause hypokalemic periodic paralysis. Cell 77(6), 863–868.

[112] Qin, N., Platano, D., Olcese, R., Stefani, E., Birnbaumer, L., 1997. Direct interaction of gbetagamma with a C-terminal gbetagamma-binding domain of the Ca^{2+} channel alpha1 subunit is responsible for channel inhibition by G protein-coupled receptors. Proc. Natl Acad. Sci. USA 94(16), 8866–8871.

[113] Qin, N., Olcese, R., Bransby, M., Lin, T., Birnbaumer, L., 1999. Ca^{2+}-induced inhibition of the cardiac Ca^{2+} channel depends on calmodulin. Proc. Natl Acad. Sci. USA 96(5), 2435–2438.

[114] Rettig, J., Sheng, Z.H., Kim, D.K., Hodson, C.D., Snutch, T.P., Catterall, W.A., 1996. Isoform-specific interaction of the alpha1A subunits of brain Ca^{2+} channels with the presynaptic proteins syntaxin and SNAP-25. Proc. Natl Acad. Sci. USA 93(14), 7363–7368.

[115] Sather, W.A., Tanabe, T., Zhang, J.-F., Mori, Y., Adams, M.E., Tsien, R.W., 1993. Distinctive biophysical and pharmacological properties of class A (BI) calcium channel α_1 subunits. Neuron 11, 291–303.

[116] Sato, C., Ueno, Y., Asai, K., Takahashi, K., Sato, M., Engel, A., Fujiyoshi, Y., 2001. The voltage-sensitive sodium channel is a bell-shaped molecule with several cavities. Nature 409(6823), 1047–1051.

[117] Schumacher, T.B., Beck, H., Steinhauser, C., Schramm, J., Elger, C.E., 1998. Effects of phenytoin, carbamazepine, and gabapentin on calcium channels in hippocampal granule cells from patients with temporal lobe epilepsy. Epilepsia 39(4), 355–363.

[118] Scott, V.E., De Waard, M., Liu, H., Gurnett, C.A., Venzke, D.P., Lennon, V.A., Campbell, K.P., 1996. Beta subunit heterogeneity in N-type Ca^{2+} channels. J. Biol. Chem. 271(6), 3207–3212.

[119] Sheng, Z.H., Yokoyama, C.T., Catterall, W.A., 1997. Interaction of the synprint site of N-type Ca^{2+} channels with the C2B domain of synaptotagmin I. Proc. Natl Acad. Sci. USA 94(10), 5405–5410.

[120] Sinnegger, M.J., Wang, Z., Grabner, M., Hering, S., Striessnig, J., Glossmann, H., Mitterdorfer, J., 1997. Nine L-type amino acid residues confer full 1,4-dihydropyridine sensitivity to the neuronal calcium channel alpha1A subunit. Role of L-type Met1188. J. Biol. Chem. 272(44), 27686–27693.

[121] Snutch, T.P., Tomlinson, W.J., Leonard, J.P., Gilbert, M.M., 1991. Distinct calcium channels are generated by alternative splicing and are differentially expressed in the mammalian CNS. Neuron 7, 45–57.

[122] Soong, T.W., Stea, A., Hodson, C.D., Dubel, S.J., Vincent, S.R., Snutch, T.P., 1993. Structure and functional expression of a member of the low voltage-activated calcium channel family. Science 260, 1133–1136.

[123] Starr, T.V.B., Prystay, W., Snutch, T.P., 1991. Primary structure of a calcium channel that is highly expressed in the rat cerebellum. Proc. Natl. Acad. Sci. USA 88, 5621–5625.

[124] Stea, A., Dubel, S.J., Pragnell, M., Leonard, J.P., Campbell, K.P., Snutch, T.P., 1993. A β subunit normalizes the electrophysiological properties of a cloned N-type Ca channel α1 subunit. Neuropharmacology 32, 1103–1116.

[125] Stea, A., Tomlinson, W.J., Soong, T.W., Bourinet, E., Dubel, S.J., Vincent, S.R., Snutch, T.P., 1994. Localization and functional properties of a rat brain α_{1A} calcium channel reflect similarities to neuronal Q- and P-type channels. Proc. Natl. Acad. Sci. USA 91, 10567–10580.

[126] Stea, A., Soong, T.W., Snutch, T.P., 1995. Voltage-Gated Calcium Channels. In: Alan North, R. (Ed.), Handbook of Receptors and Channels; Ligand- and Voltage-Gated Ion Channels. CRC Press Inc., Boca Raton, Florida, pp. 113–152.

[127] Stea, A., Soong, T.W., Snutch, T.P., 1996. Determinants of PKC-dependent modulation of a family of neuronal calcium channels. Neuron 15, 929–940.

[128] Stea, A., Dubel, S.J., Snutch, T.P., 1999. Alpha 1B N-type calcium channel isoforms with distinct biophysical properties. Ann. N Y Acad. Sci. 868, 118–130.

[129] Stevens, C.F., 1998. A million dollar question: does LTP = memory ? Neuron 20(1), 1–2.

[130] Stotz, S.C., Hamid, J., Spaetgens, R.L., Jarvis, S.E., Zamponi, G.W., 2000. Fast inactivation of voltage-dependent calcium channels. A hinged-lid mechanism? J. Biol. Chem. 275(32), 24575–24582.

[131] Stotz, S.C., Zamponi, G.W., 2001. Structural determinants of fast inactivation of high voltage-activated Ca^{2+} channels. Trends Neurosci. 24(3), 176–182.

[132] Strom, T.M., Nyakatura, G., Apfelstedt-Sylla, E., Hellebrand, H., Lorenz, B., Weber, B.H., Wutz, K., Gutwillinger, N., Ruther, K., Drescher, B., Sauer, C., Zrenner, E., Meitinger, T., Rosenthal, A., Meindl, A., 1998. An L-type calcium-channel gene mutated in incomplete X-linked congenital stationary night blindness. Nat Genet. 19(3), 260–263.

[133] Sutton, R.B., Fasshauer, D., Jahn, R., Brunger, A.T., 1998. Crystal structure of a SNARE complex involved in synaptic exocytosis at 2.4 A resolution. Nature 395(6700), 347–353.

[134] Sutton, K.G., McRory, J.E., Guthrie, H., Murphy, T.H., Snutch, T.P., 1999. P/Q-type calcium channels mediate the activity-dependent feedback of syntaxin-1A. Nature 401(6755), 800–804.

[135] Takahashi, T., Momiyama, A., 1993. Different types of calcium channels mediate central synaptic transmission. Nature 366, 156–158.

[136] Talley, E.M., Solorzano, G., Depaulis, A., Perez-Reyes, E., Bayliss, D.A., 2000. Low-voltage-activated calcium channel subunit expression in a genetic model of absence epilepsy in the rat. Brain Res. Mol. Brain Res. 75(1), 159–165.

[137] Tanabe, T., Takeshima, H., Mikami, A., Flockerzi, V., Takahashi, H., Kangawa, K., Kojima, M., Matsuo, H., Hirose, T., Numa, S., 1987. Primary structure of the receptor for calcium channel blockers from skeletal muscle. Nature 328, 313–318.

[138] Tanabe, T., Beam, K.G., Adams, B.A., Niidome, T., Numa, S., 1988. Regions of the skeletal muscle dihydropyridine receptor critical for excitation–contraction coupling. Nature 346, 567–569.

[139] Tanabe, T., Adams, B.A., Numa, S., Beam, K.G., 1991. Repeat I of the dihydropyridine receptor is critical in determining calcium channel activation kinetics. Nature 352, 800–803.

[140] Tottene, A., Volsen, S., Pietrobon, D., 2000. Alpha(1E) subunits form the pore of three cerebellar R-type calcium channels with different pharmacological and permeation properties. J. Neurosci. 20(1), 171–178.

[141] Vajna, R., Schramm, M., Pereverzev, A., Arnhold, S., Grabsch, H., Klockner, U., Perez-Reyes, E., Hescheler, J., Schneider, T., 1998. New isoform of the neuronal Ca^{2+} channel alpha1E subunit in islets of Langerhans and kidney – distribution of voltage-gated Ca^{2+} channel alpha1 subunits in cell lines and tissues. Eur. J. Biochem. 257(1), 274–285.

[142] Vance, C.L., Begg, C.M., Lee, W.L., Haase, H., Copeland, T.D., McEnery, M.W., 1998. Differential expression and association of calcium channel alpha1B and beta subunits during rat brain ontogeny. J. Biol. Chem. 273(23), 14495–14502.

[143] Verschuuren, J.J., Dalmau, J., Tunkel, R., Lang, B., Graus, F., Schramm, L., Posner, J.B., Newsom-Davis, J., Rosenfeld, M.R., 1998. Antibodies against the calcium channel beta-subunit in Lambert–Eaton myasthenic syndrome. Neurology 50(2), 475–479.

[144] Voltz, R., Carpentier, A.F., Rosenfeld, M.R., Posner, J.B., Dalmau, J., 1999. P/Q-type voltage-gated calcium channel antibodies in paraneoplastic disorders of the central nervous system. Muscle Nerve. 22(1), 119–122.

[145] Wakamori, M., Yamazaki, K., Matsunodaira, H., Teramoto, T., Tanaka, I., Niidome, T., Sawada, K., Nishizawa, Y., Sekiguchi, N., Mori, E., Mori, Y., Imoto, K., 1998. Single tottering mutations responsible for the neuropathic phenotype of the P-type calcium channel. J. Biol. Chem. 273(52), 34857–34867.

[146] Walker, D., Bichet, D., Campbell, K.P., De Waard, M., 1998. A beta 4 isoform-specific interaction site in the carboxyl-terminal region of the voltage-dependent Ca^{2+} channel alpha 1A subunit. J. Biol. Chem. 273(4), 2361–2367.

[147] Wang, M., Offord, J., Oxender, D.L., Su, T.Z., 1999. Structural requirement of the calcium-channel subunit alpha2delta for gabapentin binding. Biochem. J. 342(Pt 2), 313–320.

[148] Weisskopf, M.G., Bauer, E.P., LeDoux, J.E., 1999. L-type voltage-gated calcium channels mediate NMDA-independent associative long-term potentiation at thalamic input synapses to the amygdala. J Neurosci. 19(23), 10512–10519.

[149] Westenbroek, R.E., Hell, J.W., Warner, C., Dubel, S.J., Snutch, T.P., Catterall, W.A., 1992. Biochemical properties and subcellular distribution of an N-type calcium channel alpha 1 subunit. Neuron 9(6), 1099–1115

[150] Westenbroek, R.E., Sakurai, T., Elliott, E.M., Hell, J.W., Starr, T.V., Snutch, T.P., Catterall, W.A., 1995. Immunochemical identification and subcellular distribution of the alpha 1A subunits of brain calcium channels. J. Neurosci. (10), 6403–6418.

[151] Westenbroek, R.E., Hoskins, L., Catterall, W.A., 1998. Localization of Ca^{2+} channel subtypes on rat spinal motor neurons, interneurons, and nerve terminals. J. Neurosci. 18(16), 6319–6330.

[152] Westenbroek, R.E., Bausch, S.B., Lin, R.C., Franck, J.E., Noebels, J.L., Catterall, W.A., 1998. Upregulation of L-type Ca^{2+} channels in reactive astrocytes after brain injury, hypomyelination, and ischemia. J. Neurosci. 18(7), 2321–2334.

[153] Wheeler, D.B., Randall, A., Tsien, R.W., 1994. Role of N-type and Q-type Ca channels in supporting hippocampal synaptic transmission. Science 264, 107–111.

[154] Williams, M.E., Feldman, D.H., McCue, A.F., Brenner, R., Velicelebi, G., Ellis, S.D., Harpold, M.M., 1992. Structure and functional expression of α1, α2, and β subunits of a novel human neuronal calcium channel subtype. Neuron 8, 71–84.

[155] Williams, M.E., Brust, P.F., Feldman, D.H., Patthi, S., Simerson, S., Maroufi, A., McCue, A.F., Velicelebi, G., Ellis, S.B., Harpold, M., 1992. The structure and functional expression of an ω-conotoxin-sensitive human N-type calcium channel. Science 257, 389–395.

[156] Williams, M.E., Washburn, M.S., Hans, M., Urrutia, A., Brust, P.F., Prodanovich, P., Harpold, M.M., Stauderman, K.A., 1999. Structure and functional characterization of a novel human low-voltage activated calcium channel. J. Neurochem. 72(2), 791–799.

[157] Wiser, O., Trus, M., Hernandez, A., Renstroïn, E., Barg, S., Rorsman, P., Atlas, D., 1999. The voltage sensitive Lc-type Ca^{2+} channel is functionally coupled to the exocytotic machinery. Proc. Natl Acad. Sci. USA 96(1), 248–253.

[158] Witcher, D.R., De Waard, M., Sakamoto, J., Franzini-Armstrong, C., Pragnell, M., Kahl, S.D., Campbell, K.P., 1993. Subunit identification and reconstitution of the N-type Ca channel complex purified from brain. Science 261, 486–489.

[159] Yamaguchi, H., Muth, J.N., Varadi, M., Schwartz, A., Varadi, G., 1999. Critical role of conserved proline residues in the transmembrane segment 4 voltage sensor function and in the gating of L-type calcium channels. Proc. Natl Acad. Sci. USA 96(4), 1357–1362.

[160] Yang, J., Ellinor, P.T., Sather, W.A., Zhang, J.-F., Tsien, R.W., 1993. Molecular determinants of Ca selectivity and ion permeation in L-type Ca channels. Nature 366, 158–161.

[161] Yokoyama, C.T., Sheng, Z.-H., Catterall, W.A., 1997. Phosphorylation of the synaptic protein interaction site on N-type calcium channels inhibits interactions with SNARE proteins. J. Neurosci. 17(18), 6929–6938.

[162] Yue, Q., Jen, J.C., Nelson, S.F., Baloh, R.W., 1997. Progressive ataxia due to a missense mutation in a calcium-channel gene. Am. J. Hum. Genet. 61(5), 1078–1087.

[163] Zamponi, G.W., Bourinet, E., Nelson, D., Nargeot, J., Snutch, T.P., 1997. Crosstalk between G proteins and protein kinase C mediated by the calcium channel alpha1 subunit. Nature 385(6615), 442–446.

[164] Zamponi, G.W., Snutch, T.P., 1998. Modulation of voltage-dependent calcium channels by G proteins. Curr. Opin. Neurobiol. 8(3), 351–356.

[165] Zapater, P., Moreno, J., Horga, J.F., 1997. Neuroprotection by the novel calcium antagonist PCA50938, nimodipine and flunarizine, in gerbil global brain ischemia. Brain 1; 772(1–2), 57–62.

[166] Zhang, J.F., Randall, A.D., Ellinor, P.T., Horne, W.A., Sather, W.A., Tanabe, T., Schwarz, T.L., Tsien, R.W., 1993. Distinctive pharmacology and kinetics of cloned neuronal Ca^{2+} channels and their possible counterparts in mammalian CNS neurons. Neuropharmacology 32(11), 1075–1088.

[167] Zhang, J.F., Ellinor, P.T., Aldrich, R.W., Tsien, R.W., 1996. Multiple structural elements in voltage-dependent Ca^{2+} channels support their inhibition by G proteins. Neuron 17(5), 991–1003.

[168] Zhang, M.I., O'Neil, R.G., 1999. The diversity of calcium channels and their regulation in epithelial cells. Adv. Pharmacol. 46, 43–83.

[169] Zhuchenko, O., Bailey, J., Bonnen, P., Ashizawa, T., Stockton, D.W., Amos, C., Dobyns, W.B., Subramony, S.H., Zoghbi, H.Y., Lee, C.C., 1997. Autosomal dominant cerebellar ataxia (SCA6) associated with small polyglutamine expansions in the alpha 1A-voltage-dependent calcium channel. Nat. Genet. 15(1), 62–69.

[170] Zoccola, D., Tambutte, E., Senegas-Balas, F., Michiels, J.F., Failla, J.P., Jaubert, J., Allemand, D., 1999. Cloning of a calcium channel alpha1 subunit from the reef-building coral, *Stylophora pistillata*. Gene 227(2), 157–167.

[171] Zuhlke, R.D., Pitt, G.S., Deisseroth, K., Tsien, R.W., Reuter, H., 1999. Calmodulin supports both inactivation and facilitation of L-type calcium channels. Nature 399(6732), 159–162.

[172] Arikkath, J. and Campbell, K.P., 2003. Auxiliary subunits: essential components of the voltage-gated calcium channel complex. Current Opinion in Neurobiology 13, 298–307.

[173] Dietrich, D., Kirschstein, T., Kukley, M., Pereverzev, A., von der Brelie, C., Schneider, T. and Beck, H., 2003. Functional specialization of presynaptic Cav2.3 Ca^{2+} channels. Neuron 39(3), 483–496.

[174] Wu, L.G., Borst, J.G. and Sakmann, B., 1998. R-type Ca^{2+} currents evoke transmitter release at a rat central synapse. Proc. Natl. Acad. Sci. USA 95(8), 4720–4725.

[175] Wu, L.G., Westenbroek, R.E., Borst, J.G., Catterall, W.A. and Sakmann, B., 1999. Calcium channel types with distinct presynaptic localization couple differentially to transmitter release in single calyx-type synapses. J. Neurosci. 19(2), 726–736.

Ion channels and sperm function

Ricardo Felix,[a] Ignacio López-González,[b] Carlos Muñoz-Garay[b] and Alberto Darszon[b,*]

[a]*Department of Physiology, Biophysics and Neuroscience, Cinvestav-IPN, Mexico City, Mexico*
[b]*Department of Genetics and Molecular Physiology, Institute of Biotechnology, UNAM, Cuernavaca, Mexico*
[*]Correspondence address: Departamento de Genética del Desarrollo y Fisiología Molecular, Instituto de Biotecnología, UNAM, Avenida Universidad 2001, Col. Chamilpa, CP 62100, Cuernavaca, Mor., Mexico, Tel.: +1-525-622-7611; fax: +1-5273-17-23-88. E-mail: darszon@ibt.unam.mx

1. Importance of ionic fluxes in sperm physiology

Development of multicellular organisms usually starts when a mature sperm fertilizes an egg. The sperm must reach the egg, fuse with it and deliver its precious cargo consisting of a haploid nucleus to form the zygote. This initiates a complex process that results in an embryo, which will develop into an organism made from many different cell types organized into tissues and organs [1].

Prior to fertilization, ejaculated sperm from mammals must undergo a maturational process in order to fertilize an egg. Sperm become "fertilization competent" as they reside in the female reproductive tract where they complete a series of poorly understood changes, globally known as capacitation. This includes metabolic changes and alterations in the permeability of the plasma membrane to ions [2,3]. Although the cell machinery that elicits these cellular changes requires several elements, as we shall see later, recent data indicate the importance of voltage-gated K^+ channels during capacitation [4,5].

After gametes go through these maturational processes, their encounter requires that the sperm penetrate the surface coats that surround the egg. This is facilitated by the acrosome reaction (AR), a Ca^{2+}-dependent exocytotic event involving the fusion of the plasma and outer acrosomal membranes of the sperm head. Fusion exposes the sperm's inner acrosomal membrane and releases acrosomal enzymes needed for penetration and fusion to the egg plasma membrane [1,6]. In spite of its importance in reproduction, many fundamental questions regarding the sperm AR remain unanswered. However, diverse studies using sea urchin and mammalian sperm as model systems indicate that a drastic increase in the concentration of intracellular Ca^{2+} ($[Ca^{2+}]_i$) is one of the critical events during the AR [1,7,8].

In this manner, ion fluxes appear to be key elements in the dialogue between sperm, its environment, and the egg. Though a variety of molecules are implicated in these ion

Advances in Molecular and Cell Biology, Vol. 32, pages 407–431
ISSN: 1569-2558/DOI: 10.1016/S1569-2558(03)32017-X

transport events [3,9], in this chapter we shall focus on ionic fluxes through plasma membrane pores (known as ion channels) [10] in mammalian spermatogenic cells and sperm, and discuss some of the functional consequences of their activation.

2. Electrophysiological approaches to sperm ion channel function

Sperm are among the most highly specialized cell types described. Much of our present knowledge of sperm function derives from early studies in invertebrates [11,12]. Understanding how sperm ion channels participate in fertilization required combining several experimental strategies, including planar bilayer techniques, in vivo measurements of membrane potential, $[Ca^{2+}]_i$ and intracellular pH (pH_i) using fluorescent probes.

2.1. Planar bilayers

The size, complex geometry and highly differentiated and motile nature of sperm have made it very difficult to systematically characterize their channels with conventional electrophysiological methods. However, the availability of large quantities of sperm allows, in principle, the isolation and characterization of plasma membrane fractions from the different regions of the cell. A mature sea urchin male can provide up to $\sim 5 \times 10^{10}$ sperm, while a mature mouse has around 10^8 sperm. The isolated sperm plasma membrane vesicles can be reassembled in various model systems to study sperm ion channels [12]. K^+ channels were first recorded in bilayers made at the tip of patch pipettes containing isolated sperm flagellar membranes [13]. Later a high conductance, voltage-dependent, multi-state Ca^{2+} channel was characterized in black lipid planar bilayers (BLMs) containing isolated *Strongylocentrotus purpuratus* sperm plasma membranes. This channel is insensitive to verapamil and nisoldipine, but blocked by Cd^{2+} and Co^{2+} at concentrations, which in sperm inhibit the AR induced by the physiological agonist [14] (a fucose sulfate glycoconjugate present in the egg jelly layer that surrounds the egg). More recently, a mildly selective K^+ channel, directly modulated by cAMP, was studied in planar lipid bilayers with incorporated flagellar sperm membranes [15]. This experimental strategy also revealed the presence of Cl^- channels in sea urchin sperm [16].

Planar bilayer work incorporating mammalian sperm plasma membranes has also documented the presence of Ca^{2+} channels [17,18]. A 10 pS Ca^{2+} channel, blocked by nitrendipine, a blocker of the sperm AR, was detected in BLMs containing boar sperm plasma membranes. This channel is weakly activated by voltage and selects poorly between monovalent and divalent cations but is activated by the agonist Bay K 8644 [19]. In addition, incorporation of mouse sperm plasma membranes into BLMs revealed the presence of several ion channels including an anion channel, a cation selective channel ($PNa^+/PK^+ = 2.5$) and a high-conductance-Ca^{2+} selective channel ($PCa^{2+}/PNa^+ = 4$) similar to the one described from sea urchin sperm plasma membranes [15]. TEA^+-sensitive K^+ channels and poorly selective cation channels have also been recorded in bilayers containing rat sperm plasma membranes [20].

The transfer of ion channels from live sperm to BLMs represents an additional possibility to circumvent the sperm size limitation [21]. The probability of ion channel transfer is at least doubled by the AR, both in sea urchin and in mouse sperm. A Ca^{2+} permeable channel with several subconductance states, similar to the one described from mouse and sea urchin sperm plasma membranes, has been observed upon the transfer of ion channels from live sperm to BLMs [21].

2.2. Patch-clamp recordings of single ion channels in sperm

More recently, because of its medical, sociological and economic implications, mammalian sperm physiology has captivated the attention of a broader group of investigators. New molecular tools that allow the study of ion channels at the single cell level have led to rapid progress in our understanding of how an unexpected diversity of ion channels deeply influences mammalian sperm physiology [7,8,22].

The patch-clamp technique has revolutionized the study of cellular physiology by allowing the study of individual ion channels in many cell types [10,23,24]. However, applying patch-clamp techniques directly to mature sperm is not an easy task. Indeed, ionic currents have been documented only rarely from these cells using this method. Initially, single K^+ channels were recorded directly from sea urchin sperm heads using the patch-clamp technique [25]. Sea urchin sperm were then swollen in diluted sea water to improve the success rate of patch formation (from $<1\%$ in non-swollen cells to $<20\%$) [26]. The analysis of swollen sea urchin sperm opened up new possibilities to directly study ion channel regulation, and revealed the presence of a K^+-selective channel activated by speract, a decapeptide from the outer layer of the sea urchin egg [26]. Weyand et al. were able to record "on cell" cGMP modulated currents in mouse and human sperm (see Section 4.6) [67]. In addition, Espinosa et al. (1998) recorded cation and anion-selective single channel currents in cell-attached and excised patches from mammalian sperm heads for the first time [27]. The anion channels were blocked by micromolar niflumic acid, an anion channel antagonist, which at these concentrations also inhibited the AR [27]. These findings suggested that anion channels participate in the sperm–egg dialogue.

2.3. Whole-cell patch-clamp in spermatogenic cells

Fortunately spermatogenic cells, the precursors of sperm, are larger and synthesize the ion channels that will end up in mature sperm. Therefore, spermatogenic cells constitute an excellent model system to study sperm ion channels with patch-clamp techniques [28–30]. Furthermore, combining the information from patch-clamp recordings with new molecular biological techniques in spermatogenic cells is revealing new insights about the role of sperm ion channels in the AR. Since sperm are transcriptionally and translationally inert [31], ion channels that are utilized in sperm must therefore be synthesized during spermatogenesis. This complex and highly co-ordinated process by which spermatogonia proliferate and differentiate, leads to the production of mature sperm. It is possible to isolate significant amounts of cells from each stage of spermatogenesis. In adult rats (>80 days old) it has been documented that late primary spermatocytes and early spermatids are most abundant ($\sim80\%$ of dissociated cells) [32]. Pachytene spermatocytes, and round and condensing spermatids are at the later stages of differentiation and much larger than sperm, and therefore easier to analyse using the patch-clamp technique.

Indeed, a wide variety of voltage-gated ion channels can be detected in mouse spermatogenic cells by using whole-cell patch-clamp methods. As illustrated in Fig. 1, when a depolarizing voltage step is delivered from a holding potential of -90 mV, a fast transient inward current is seen followed by a sustained outward current (panel A). Both inward and outward currents result from the combination of several ionic currents. Using ion substitution, pharmacological reagents and appropriate voltage protocols, it is possible to isolate the different ionic components of these currents. To measure the inward current, outward currents are eliminated by substituting Cs^+ for K^+ in the patch pipette solution and by adding millimolar concentrations of tetraethylamonium (TEA^+), a K^+ channel blocker, to the extracellular solution. As illustrated in panel B of Fig. 1, the remaining inward current rapidly activates and then inactivates. As we shall discuss later, this transient inward current represents a Ca^{2+} current of the T-type. To measure outward currents, the transient inward current (I_{Ca}) is blocked with micromolar concentrations of Ni^{2+} or Cd^{2+}, two well-known effective blockers of voltage-gated Ca^{2+} channels. Experiments in which the intra- and extracellular K^+ concentrations are varied suggest that K^+ is the major charge carrier of the outward current. Fig. 1 (panel C) shows a typical non-inactivating K^+ outward current (called delayed rectifying current) recorded in mouse synplasts of round and condensing spermatids in response to a voltage step of $+70$ mV from a holding potential of -90 mV. Synplasts are multinucleated cells from the same stage of spermatogenesis. In addition to this outward current, recent studies have shown that at least another K^+ current is present in mouse spermatogenic cells. At hyperpolarizing voltage steps, a substantial population of spermatogenic cells display a rapidly activating and sustained inwardly rectifying K^+ current (Fig. 1, panel D) that shows strong rectification around the K^+ equilibrium potential. The magnitude of this current is also dependent on external K^+ concentration and the channels are apparently highly selective for K^+ [33].

Chloride currents can be also isolated from spermatogenic cells using both ion substitution and pharmacological agents [27]. Cl^- channels activate at voltages above -50 to -40 mV and the isolated currents through these channels do not decay during maintained depolarizing voltage steps (Fig. 1, panel E). Lastly, as depicted in panel F no voltage-gated Na^+ channels have been detected thus far in spermatogenic cells using the patch-clamp technique.

3. K^+ channels involved in sperm physiology

K^+ channels appear to be important in sperm physiology, particularly during capacitation. This process is accompanied by a plasma membrane hyperpolarization, which is thought to be mediated by an increased contribution of K^+ permeability to the transmembrane potential [4,5]. Voltage-gated K^+ channels constitute a superfamily of transmembrane proteins that include several members with distinct permeability and gating properties, determined mainly by the type of ion-conducting pore subunit (α) that the channel contains [10]. The structure of the α subunits is modelled as six hydrophobic

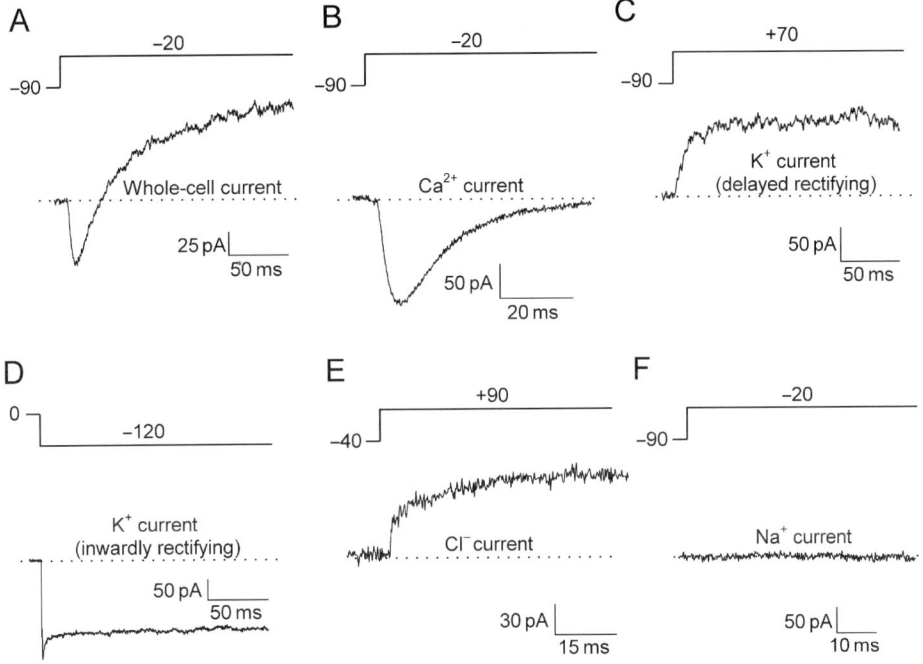

Fig. 1. Whole-cell currents in spermatogenic cells. (A) Typical patch-clamp whole-cell currents recorded in control KCl-based (pipette), and NaCl-based (bath), respectively. Pulse protocol: 70 mV for 40 ms, from a holding potential of -90 mV. A transient inward current followed by a slowly developing outward current is produced when the membrane potential is made more positive than -60 mV. (B) In bath solutions containing elevated (10 mM) Ca^{2+} or Ba^{2+} and adding outward K^+ current blockers it is possible to isolate currents through voltage-gated Ca^{2+} channels of the T-type. This current has pharmacological attributes consistent with those of the putative rapid channel responsible for Ca^{2+} influx mediating the AR. Likewise, in KCl-based solutions after blocking the current through voltage-gated Ca^{2+} channels, mammalian spermatogenic cells show a slowly developing outward current with negligible initial inward current. This current has been identified as membrane-potential-dependent K^+ current. More recent studies have shown that at least two major types of K^+ currents are present in isolated mammalian spermatogenic cells. One is due to a voltage-gated channel that is blocked by Cs^+ and other K^+ channel blockers (Ba^{2+} and TEA$^+$), which is similar to the family of delayed rectifying currents found in excitable cells (C), and the other is due to a pH-sensitive inwardly rectifying K^+ channel (D). These two currents are probably responsible for setting the ~ -50 mV resting potential reported for mammalian sperm. The inwardly rectifying K^+ channels may also produce hyperpolarization under physiological conditions and contribute to the cellular changes that give rise to the capacitated state in mature sperm. (E) In spermatogenic cells, voltage steps induce a sustained membrane current in the absence of K^+ and Ca^{2+} ions as charge carriers. Internal EGTA and the Cl^- channel blocker niflumic acid can suppress this current. The blocker is effective also in inhibiting the sperm AR suggesting that Cl^- channels participate in this event. It has been suggested that Cl^- channel activation may result in plasma membrane depolarization and thereby exert at least a partial control over voltage-gated sperm Ca^{2+} channels and Ca^{2+} influx important to the AR. (F) No electrophysiological evidence has been provided thus far regarding the expression of voltage-gated Na^+ channels in mammalian spermatogenic cells. All traces displayed correspond to recordings obtained from mouse cells in the round and condensing spermatid stages of spermatogenesis.

segments (S1–S6) embedded in the plasma membrane with the N- and C-terminal domains of the protein positioned intracellularly. One transmembrane segment (S4) contains a unique array of positive charges that function as the voltage sensor of the channel. The region separating segments S5 and S6 may contain two additional transmembrane segments that together form the pore of the channel [10].

Unlike functional Na^+ and Ca^{2+} channel α subunits that contain four sets of S1–S6 membrane spanning regions coded as parts of a single gene product linked by hydrophilic loops, most functional K^+ channels are formed from four α subunits that constitute a hetero- or homotetrameric structure [10]. However, there is evidence that some K^+ channels more distantly related to the superfamily of voltage-gated channels (e.g. the inwardly rectifying and the G-protein regulated inwardly rectifying K^+ channels, IRKs and GIRKs, respectively) possess only two membrane-spanning segments that resemble the S5 and S6 regions of the α subunit. Even though K^+ channel α subunits are capable of conducting ions by themselves, some channels include an ancillary cytoplasmic protein called β that apparently plays modulatory, structural, and/or stabilizing roles. Most traditional voltage-gated K^+ channels have not been shown to possess additional transmembrane-containing subunits, although a family of membrane proteins including I_sK (also called minK) may fit into this category [70].

3.1. Molecular identity of K^+ channels in spermatiogenic cells and sperm

K^+ channels constitute by far the largest and most diverse family of ion channels. They contribute to the resting potential and help regulate the degree of excitability of numerous cell types. In spermatogenic cells, initial studies showed that the macroscopic phenotype of K^+ current (I_K) was a delayed rectifying K^+-selective current that was blocked by TEA^+ [28]. However, more recent studies have shown that sperm I_K actually consists of different types of K^+ channels. Indeed a number of members of the mammalian K^+ channel α subunit family have been identified and cloned from mammalian testis. Initially, a novel gene abundantly expressed in spermatocytes encoding a unique type of K^+ channel regulated by both pH_i and membrane voltage (called mSlo3) was cloned from a mouse testis cDNA library and expressed in *Xenopus* oocytes [71]. Molecular studies showed that Slo3 is primarily expressed in testis in both mice and humans and its sequence is similar to Slo1, the large conductance, Ca^{2+}- and voltage-gated K^+ channel [71]. Its functional expression in spermatogenic cells and sperm remains to be determined.

Likewise $K_{ir}5.1$, a member of the inward rectifier K^+ channel superfamily, was detected in RT-PCR experiments using rat testis. The function of this channel remains unknown since it does not form functional channels when expressed in *Xenopus* oocytes. However, its temporal pattern of expression suggests that it may be involved in spermatogenesis [75]. Using specific antibodies, $K_{ir}5.1$ was found in seminiferous tubules of rat testis and, particularly, in spermatogonia, primary and secondary spermatocytes, spermatids and in the head and body of sperm. The expression of $K_{ir}5.1$ increases with age during all stages of sperm development. It reaches a peak in 2-month-old rats and its expression decreases in 3-month-old animals where it is detected mainly in mature sperm [75].

The expression of a third class of K^+ channel has been suggested from functional studies in *Xenopus* oocytes injected with RNAs from rat spermatogenic cells. ATP, a factor known to induce sperm activation, induced a K^+-dependent outwardly rectifying whole-cell current in RNA-injected oocytes that was inhibited by K^+ channel blockers charybdotoxin (CTX) and TEA^+ [76]. This current can be also elicited by the Ca^{2+} ionophore ionomycin [77], suggesting that a Ca^{2+}-activated K^+ channel (known also as Maxi-K type channels due to its large conductance) mediates it. Thereafter, using immunofluorescent detection of sperm-bound biotinylated CTX, the presence of this channel was identified over the surface of both heads and tails of unfixed rat epididymal sperm [78]. In addition, based on RT-PCR experiments, Jacob and colleagues amplified sequences homologous to K^+ channels of the delayed rectifier type in primary spermatocytes and post-meiotic elongating spermatids. Though an assembled sequence of 2693 base pairs with $>90\%$ homology to a delayed rectifier K^+ channel Kv1.3 has been obtained [78], electrophysiological evidence for the expression of these channels in spermatozoa has not been provided thus far.

More recently, by using immunoconfocal and electron microscopy we have shown that a minimum of four different classes of K^+ channels (Kv1.1, Kv1.2, Kv3.1 and GIRK1) are present and regionally distributed over the surface of mouse epididymal sperm [145]. In addition, the use of reverse transcription-polymerase chain reaction on RNA from mouse spermatogenic cells allowed the amplification of multiple transcripts corresponding to the channels identified by immunocytochemistry. Coincident with these findings, whole-cell patch-clamp recordings have indicted the functional expression of at least two different outwardly rectifying K^+ currents in spermatogenic cells [145].

Lastly, two novel members of the hyperpolarization-activated and cylic-nucleotide-gated channel (HCN) family have been identified and characterized from sea urchin and human testis [72,73]. Heterologous expression of the channel cloned from the sea urchin, called SPIH, gives rise to weakly K^+-selective hyperpolarization-activated currents that are gated by cAMP [72]. These properties are similar to those described previously for a native K^+ channel detected in planar bilayers with incorporated *S. purpuratus* sperm plasma membranes [15]. The functional relevance of these channels is unclear. It is thought that they might become activated in vivo by a membrane hyperpolarization and a concomitant rise in cAMP levels induced by speract, a sperm-activating peptide derived from the egg jelly coat [26,74], and could thereby control the waveform of flagellar beating. HCN channels usually participate in determining the rhythmic activity of neurons; however, as mentioned above, a second member of this superfamily of proteins (hHCN4) has been found abundantly expressed in human testis [73]. It has been hypothesized that hHCN4 may represent the mammalian equivalent to the HCN channel in the flagellum of sea urchin sperm and may serve the same role in the generation of rhythmic activity that controls the waveform of flagellar beating [73].

3.2. Functional analysis of native K^+ channels in sperm and spermatogenic cells

Though biochemical and molecular biology data suggest the presence of different K^+ channels in both mammalian testis and spermatozoa [71,73,75–78], initial electrophysiological studies detected only one functional type of K^+ channels in rodent spermatogenic cells [28]. Recent patch-clamp experiments in mouse spermatogenic cells confirmed the presence of a delayed rectifier K^+ outward current that matched the properties of the native whole-cell outward K^+ current originally described by Hagiwara and Kawa [145]. However, in a substantial population of spermatogenic cells ($>60\%$) a rapidly activating inward current could be elicited at hyperpolarizing voltages, whose amplitude was sustained throughout the duration of the voltage pulse (Fig. 1D) and showed strong rectification around the K^+ equilibrium potential. The magnitude of this current was strongly dependent on the external K^+ concentration and the channels were highly selective for K^+ over other monovalent cations [33]. Addition of micromolar concentrations of Cs^+ or Ba^{2+} effectively blocked the current, and more importantly, cytosolic acidification reversibly inhibited it.

Interestingly enough, the native inwardly rectifying K^+ channels of spermatogenic cells share biophysical and pharmacological properties with cloned members of the $K_{ir}2$ and $K_{ir}4$ superfamilies previously expressed in heterologous systems [70]. There is a strong correlation between structure and function within these subfamilies. $K_{ir}2$ channels are inward rectifiers and are subject to modulation by various effectors including pH_o, intracellular ATP, PKC activity, G-protein, and Mg^{2+} [79], while $K_{ir}4$ channel activity is subject to modulation by pH_i [80]. $K_{ir}4.1$ has additionally been shown to form novel functional heteromultimeric channels with $K_{ir}5.1$ [75,81], a channel protein highly expressed in spermatozoa and spermatogenic cells which does not form functional homomeric channels when expressed in heterologous systems. Although the functional relevance of $K_{ir}5.1$ is not yet clear, it has been proposed recently that it may play an important role in conferring pH sensitivity to $K_{ir}4.1$ recombinant channels [82]. Further studies will be needed to determine whether K_{ir} channels form heteromultimers in native tissue and whether the inwardly rectifying currents expressed in spermatogenic cells are homo or heterotetramers and what is their molecular identity.

4. Ca^{2+} channels involved in mammalian sperm physiology

Ca^{2+} channels are involved in orchestrating diverse physiological processes including, among many others, spermatogenesis and the sperm AR. During spermatogenesis, an increment of $[Ca^{2+}]_i$ occurs and appears to be important for the proliferation, differentiation and maturation of spermatozoa [34]. In addition, at least two different Ca^{2+} permeable channels participate in the increase in $[Ca^{2+}]_i$ that is essential for the AR [7,8,22].

4.1. Electrophysiological and optical characterization of Ca^{2+} channels in spermatogenic cells and sperm

Measurements in single sperm loaded with fluorescent ion probes indicate that the zona pellucida (ZP) triggers $[Ca^{2+}]_i$ increases that precede exocytosis [110,112,142]. These probes have revealed two phases of $[Ca^{2+}]_i$ increase induced by ZP3, a ZP glycoprotein, consistent with the participation of at least two different types of Ca^{2+}

channels in the mammalian sperm AR [112]. Addition of ZP3 transiently elevates $[Ca^{2+}]_i$ to micromolar levels within 40–50 ms, it then relaxes to resting values within the next 200 ms [5]. This transient has pharmacological and kinetic properties that are consistent with those of T-type Ca^{2+} channels. T-type Ca^{2+} currents are the main voltage dependent Ca^{2+} currents present during the later stages of rodent spermatogenesis [29,30]. These currents, the AR, and the transient increase in $[Ca^{2+}]_i$ are inhibited by micromolar concentrations of dihydropyridines (DHPs), pimozide and Ni^{2+}. Though the fast transient increase in mouse sperm $[Ca^{2+}]_i$ is probably due mainly to a ZP3-dependent activated T-type Ca^{2+} channel [7,136], the molecular identity of the AR relevant voltage-gated Ca^{2+} channels remains to be established [35,53,54]. Moreover, a slower and sustained $[Ca^{2+}]_i$ increase, necessary for the AR, follows the fast transient. The sustained $[Ca^{2+}]_i$ elevation induced by ZP3 involves releasing Ca^{2+} from an IP3-sensitive intracellular store [57] and subsequent Ca^{2+} entry through store-operated Ca^{2+} channels (SOCs) in the plasma membrane [65,113] (see Section 4.5).

4.2. Molecular composition of voltage-gated Ca^{2+} channels

Mounting functional and pharmacological evidence indicates that voltage-gated Ca^{2+} channels constitute a fundamental pathway in the plasma membrane controlling Ca^{2+} entry and modulating intracellular ionized Ca^{2+} levels in spermatozoa [7,8,22,35]. Therefore, understanding the molecular composition and distribution of these channels is important for the comprehension of several reproductive phenomena including sperm AR. In general, two different types of voltage-gated Ca^{2+} channels have thus far been identified in mammalian cells on the basis of voltage-activation threshold: low voltage- and high voltage-activated channels (LVA and HVA, respectively). Four subtypes of HVA Ca^{2+} channels (named L, N, P/Q, and R) and one subtype of low voltage-activated Ca^{2+} channels (known as T) have been defined [36]. HVA-channels consist of four subunits: α_1, β, $\alpha_2\delta$ and γ. Molecular cloning has revealed seven HVA-channel α_1 genes ($Ca_v1.1$–1.4 and $Ca_v2.1$ 2.3) coding for proteins responsible for ion conduction, voltage sensing, and binding of specific drugs and toxins [37]. Recently, the cloning of three novel α_1 subunits ($Ca_v3.1$–3.3) by Perez-Reyes and colleagues has given the initial insight into the molecular structure of the LVA Ca^{2+} channels [38–40]. In addition, a number of β subunits ($Ca_v\beta_1$-β_4) have also been cloned. These β subunits do not transverse the plasma membrane, but interact directly with the α_1 pore-forming subunit and appear to be important for expression of the kinetic characteristics of the channel [41]. Some of them have been involved also in the intracellular trafficking of the α_1 subunits [42]. Conversely, much less is known about the other auxiliary subunits, $Ca_v\alpha_2\delta$ and $Ca_v\gamma$ [37,43,44].

4.3. Molecular identification of voltage-gated Ca^{2+} channels subunits in spermatogenic cells and sperm

Despite the crucial role of voltage-gated Ca^{2+} channels in the physiology of spermatozoa, their definitive identification remains elusive [45,46]. The size and complex nature of sperm has precluded their systematic electrophysiological characterization.

In addition, the inability of sperm to synthesize proteins impedes the use of standard molecular approaches to learn about their ion channels. For these reasons, as mentioned earlier, more recent efforts have focused on germ-line cells from which sperm arise. Though mainly only T-type Ca^{2+} currents have been detected in immature spermatogenic cells from rat [28] and mouse [29,30,47,48], transcripts for a number of voltage-gated Ca^{2+} channel α_1 subunits have been identified in these cells. These include: $Ca_v2.1$ and $Ca_v2.3$ [29], $Ca_v1.2$ [49,50] as well as $Ca_v3.1$ and $Ca_v3.2$ [50]. The first three α_1 code for HVA- and the later two for LVA or T-type channels. In addition, PCR analysis has demonstrated the expression of $Ca_v3.2$ subunits in human testis [141].

4.4. Immunolocalization of voltage-gated Ca^{2+} channels in spermatogenic cells and sperm

Immunocytochemical evidence for α_1 protein expression was only recently provided. The findings indicate that four Ca^{2+} channel α_1 subunits ($Ca_v1.2$, and $Ca_v2.1-2.3$) are present and regionally localized in mammalian mature sperm [51–53]. In addition, using specific antibodies we showed that $Ca_v1.2$ and $Ca_v2.1$ subunits are expressed not only in sperm but also in spermatogenic cells, and provided evidence for the presence of the four $Ca_v\beta$ auxiliary subunits in both cell types [54]. A summary on the information regarding the expression of distinct Ca^{2+} channel proteins is given in Table 1.

4.5. Intracellular and capacitative Ca^{2+} channels

Many cells are able to release sequestered Ca^{2+} from pools of the endoplasmic reticulum (ER). The membranes of these pools contain both a sarco/endoplasmic reticulum Ca^{2+} ATPase (SERCA) pump and an inositol-1,4,5-trisphosphate (IP3)-gated Ca^{2+} channel (IP3 receptor or IP3R) which mediate Ca^{2+} uptake from and

Table 1
Ca^{2+} channel expression in mammalian spermatogenic cells and sperm

	Name	Current type	Former name	mRNA	Protein	References
High voltage-activated Ca^{2+} channels (HVA)	$Ca_v1.1$ ($\alpha1.1$)	L	α_{1S}	ND	ND	
	$Ca_v1.2$ ($\alpha1.2$)	L	α_{1C}	+	+	[49–54]
	$Ca_v1.3$ ($\alpha1.3$)	L	α_{1D}	−	−	
	$Ca_v1.4$ ($\alpha1.4$)	L	α_{1F}	ND	ND	
	$Ca_v2.1$ ($\alpha2.1$)	P/Q	α_{1A}	+	+	[29]
	$Ca_v2.2$ ($\alpha2.2$)	N	α_{1B}	+	+	[53]
	$Ca_v2.3$ ($\alpha2.3$)	R	α_{1E}	+	+	[29,53]
Low voltage-activated Ca^{2+} channels (LVA)	$Ca_v3.1$ ($\alpha3.1$)	T	α_{1G}	+	+	[50]
	$Ca_v3.2$ ($\alpha3.2$)	T	α_{1H}	+	ND	[50,141]
	$Ca_v3.3$ ($\alpha3.3$)	T	α_{1I}	ND	ND	

ND = not determined.

release into the cytosol, respectively. Extracellular ligands evoke Ca^{2+} release from these pools through receptor-mediated activation of phospholipase C and the resultant production of IP3. Although mature sperm do not possess ER, there is evidence that the acrosome may function as a Ca^{2+} store. Using specific antibodies it has been shown that all three known types of IP3Rs are expressed in sperm [55–57], and suggest that immature IP3Rs may accumulate in the Golgi complex of spermatogenic cells in preparation for their further localization in the acrosome of mature sperm [57]. It has been suggested that IP3Rs in the acrosome become functional and participate in the AR [55,57].

In addition to the IP3 receptors, there is a second class of sperm Ca^{2+} release channel called ryanodine receptors (RyRs). Although these channels were originally discovered in skeletal and cardiac muscles where they play crucial roles in excitation–contraction coupling [58], the expression of RyRs in spermatogenic cells and in mouse mature sperm have been documented using specific antibodies [55,59]. Notably, RyR type 3 expression is restricted to the apical tip of the sperm head, and segregated to the principal piece of the flagellum. This pattern of expression resembles that of the α_1 subunit of the $Ca_v1.2\ Ca^{2+}$ channels [54], suggesting the formation of signaling complexes comparable to that of skeletal muscle and myocardial cells, though experimental evidence for this possibility is lacking.

The emptying of the intracellular Ca^{2+} pool is generally followed by a Ca^{2+} influx from the extracellular medium [60]. The channels involved in this process are known as capacitative entry channels or SOCs. Although the molecular identity of SOCs has not been unequivocally determined, it has been suggested that some of the *Drosophila* photoreceptor *trp* genes might encode SOCs. Seven mammalian *trp* homologs have thus far been identified. Expression experiments do indicate that only some of these clones encode SOCs [60,61]. Evidence is emerging that SOCs are not simply homeostatic regulators that mediate the refilling of intracellular Ca^{2+} pools, but rather participate in the control of such diverse processes as secretion, apoptosis, cell cycle control, proliferation and gene transcription [60].

Interestingly, thapsigargin promotes Ca^{2+} release from sperm internal stores, probably the acrosome, increases $[Ca^{2+}]_i$ and induces sperm AR [62,63]. Mouse spermatogenic cells, as well as immobile testicular sperm possess SOCs which have been proposed to be responsible for a sustained elevation in $[Ca^{2+}]_i$ necessary for the AR [64]. Indeed, this ER Ca^{2+}-ATPase inhibitor stimulates external Ca^{2+} uptake with similar kinetics and sensitivity to Ni^{2+} and DHPs as the second phase of Ca^{2+} influx induced by ZP3 [113]. Furthermore, very recently Jungnickel and co-workers showed that *trp2* localizes to the sperm head and is important for the prolonged Ca^{2+} influx that drives the mammalian sperm AR [65]. In addition, preliminary evidence from our laboratory suggests the expression of seven of *Drosophila*'s transient receptor potential homolog genes (*trp1*–7) and three of their protein products (Trp1, Trp3 and Trp6) in mouse sperm. Immunoconfocal analysis showed that Trp6 is present in the post-acrosomal region and could be involved in sperm AR. On the other hand, Trp1 and Trp3 are confined to the flagellum, suggesting that they may participate in other Ca^{2+}-dependent events such as motility and/or hyperactivation [143].

4.6. Cyclic-nucleotide-gated channels

Cyclic-nucleotides are key elements of cGMP- and cAMP-signaling in sperm. They have been implicated in several cellular processes including the AR [12]. Interestingly, a cyclic nucleotide-gated (CNG) channel was identified and cloned in mammalian sperm by Weyand and co-workers in 1994 [67]. Heterologous expression showed that sperm CNG channels are directly opened by either cAMP or cGMP, being more sensitive to cGMP, and permeable to Ca^{2+} ions [67,68]. Similar to other Ca^{2+} channels, the areas of CNG channel expression coincide with the segmentation of the flagellum based on morphological criteria. Hence, molecular studies have shown that in mature sperm the ion-conducting α subunit of CNG channels is observed along the entire flagellum. In contrast, the different ancillary β subunit isoforms are heterogeneously distributed in the flagellum, suggesting that different forms of CNG channels may coexist in this region of the cell. Differential subunit association caused by regional segregation may give rise to distinct CNG channel subtypes that differ in Ca^{2+} permeability. These CNG channel subtypes could generate particular Ca^{2+} microdomains along the flagellum, thereby providing the molecular basis for control of flagellar bending waves [68].

More recently, the cloning and expression of a unique sperm cation channel (CatSper) has been reported [69,144]. Although CatSper structure resembles a six-transmembrane-spanning repeat of the voltage-gated ion channels discussed earlier, functional studies shows that it is sensitive to cell-membrane-permeant cAMP and cGMP analogues. CatSper RNA expression is restricted to the testis and the subcellular localization of the protein seems to be confined to the sperm flagellum (see also Section 6) [69].

5. Physiological relevance of ion channels in sperm and spermatogenic cells

5.1. Capacitation

Mammalian sperm are morphologically differentiated after leaving the testis, but still have to acquire progressive motility and the ability to fertilize an egg. During epididymal transit, sperm attain the ability to move progressively; however, they are still fertilization-incompetent. Sperm gain the capacity to fertilize in the female reproductive tract by a poorly understood process called capacitation [2,3,6]. The ability of sperm to undergo the AR in response to physiological stimuli (ZP, progesterone) can be taken as the goal of capacitation, though fertilization is its endpoint.

Studies of sperm loaded with ion-selective fluorescent probes indicate that during capacitation sperm pH_i increases, membrane potential hyperpolarizes and $[Ca^{2+}]_i$ is elevated from $50-100$ nM to $125-175$ nM [2-6,22]. In addition, capacitation is accompanied by several metabolic alterations, including a time-dependent and cAMP-regulated increase in the protein tyrosine phosphorylation of a subset of proteins [2,3,83,84] and changes in the distribution and composition of plasma membrane lipids and phospholipids [2,3]. Three elements appear to be necessary to accomplish capacitation: Ca^{2+}, $NaHCO_3$ and serum albumin [83-85].

The requirement for serum albumin in capacitation has been related to its ability to remove cholesterol from the membrane [86-88]. It has been proposed that cholesterol efflux then leads to changes in the membrane architecture and fluidity that give rise to the capacitated state [89,90]. More recently, whole-cell patch-clamp experiments indicate that BSA can regulate Ca^{2+} T-currents in spermatogenic cells in a cholesterol-independent manner, showing alternative ways in which this protein may regulate capacitation and/or the AR [91]. These findings suggest that voltage-gated Ca^{2+} channels may participate in promoting sperm capacitation. Much work is needed still to understand how sperm $[Ca^{2+}]_i$ is controlled during capacitation.

The membrane potential of uncapacitated sperm populations is relatively depolarized (~ -50 mV) [4,92] keeping Ca^{2+} influx through T-channels minimal due to inactivation [5,47]. Under these conditions, a further depolarization would not enhance T-currents and would not initiate AR. However, during capacitation sperm membrane potential hyperpolarizes to ~ -70 mV due to an enhanced contribution of K^+ permeability [5,33]. This hyperpolarization may relieve T-channel inactivation, increasing their availability to open after a depolarization. Thus, a hyperpolarization may act to prime T-channels for subsequent activation by ZP3 [5].

In spite of its importance, it is not known how K^+ permeability is regulated during capacitation. It is reasonable to invoke protein phosphorylation as a means to modulate K^+ channels [70]. On the other hand, it is worth considering that an inwardly rectifying K^+ channel in spermatogenic cells is modulated by pH_i. This opens the possibility that this K_{ir} may be involved in setting the cell's resting potential [33]. Furthermore, the pH_i changes that occur during capacitation could activate K_{ir} channels and thus influence this process under physiological conditions. Before capacitation mammalian sperm pH_i is relatively acidic compared to other somatic cells [93] and may impose a functionally quiescent state. This may contribute to maintain the uncapacitated state [94], prolong sperm viability during storage in the epididymis and suppress spontaneous AR [85]. In addition, an acidic pH_i may act as a negative regulator of sperm K_{ir} channels thereby maintaining depolarized membrane potential and indirectly preventing unregulated Ca^{2+} entry, and thus AR. In mature sperm pH_i increases during capacitation by more than 0.2 pH units [93], a change sufficient to induce a 0.5 to 3-fold increase in the open probability of some K_{ir} channels [95,96]. Two acid efflux pathways are mainly responsible for pH_i regulation – one mechanism dependent upon extracellular Na^+, Cl^-, and HCO_3^-, and the other characterized by its sensitivity to arylaminobenzoates [93]. Hence, under physiological conditions, an increase in pH_i would activate inwardly rectifying K^+ channels, permitting K^+ ions to flow out of the cell, driving the potential towards the K^+ reversal potential [97] and hyperpolarizing the sperm (Fig. 2).

Lastly, glucose has been reported to be beneficial to human sperm for optimal capacitation and fertilization, though it is unclear whether glucose is required for providing extra metabolic energy through glycolysis, or for generating some other metabolic product. Recent experimental evidence support the idea that ATP is required for protein tyrosine phosphorylation events that take place during sperm capacitation [98–100], and more specifically, the ATP produced by a compartmentalized glycolytic pathway in the principal piece of the flagellum [100]. Glucose, however, might be used in alternative metabolic pathways. In this regard, it should be noted that besides ion channels, electrogenic transport systems may also contribute to the plasma membrane potential. Hence, it has been recently shown that an increase in ATP synthesis, generated by

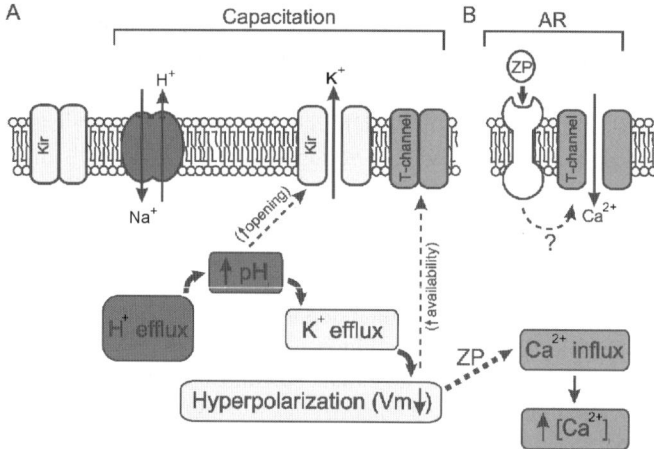

Fig. 2. Model of functional events during capacitation. (A) In vitro capacitation is accompanied by a pH$_i$ increase due to the activation of two acid efflux mechanisms. This cytosolic alkalization results in the activation of inwardly rectifying K$^+$ channels in sperm plasma membrane leading to membrane hyperpolarization. Sperm transmembrane potential (V_m) can become as negative as -80 mV, a value sufficient to relieve steady-state, voltage-dependent inactivation of the voltage-gated Ca^{2+} currents. (B) Ca^{2+} influx through these T-channels activated by ZP (arrow) via an unknown mechanism (question mark) during fertilization may drive AR. (*For a colored version of this figure, see plate section, page 468.*)

glycolysis, results in a stimulation of the Na$^+$/K$^+$-ATPase that hyperpolarizes the plasma membrane in mature human sperm and may contribute to the capacitated state of the cells [101].

5.2. The acrosome reaction

The ZP, a thick extracellular glycoprotein coat surrounding the egg, is the main mediator of the sperm AR in mammals. ZP3 is the murine ZP sulfated glycoprotein that induces the AR. As mentioned earlier, the AR is a Ca^{2+}-dependent exocytotic reaction involving the fusion of the plasma and outer acrosomal membranes of the sperm head. Fusion exposes the sperm's inner acrosomal membrane, whose surface contains proteases and/or glycosidases that allow its penetration through ZP to reach the egg plasma membrane [1,85].

SNARE proteins, soluble N-ethylmalimide-sensitive attachment protein receptors [102] appear to play a role in the fusion events leading to the AR [103–105]. Interestingly, HVA Ca^{2+} channels interact with proteins of the synaptic machinery [106] and both interact with calmodulin [107,108], though this interaction has not been documented in sperm yet.

Being a crucial step in fertilization, acrosomal exocytosis is a very dynamic and complex process. Multiple, concerted and co-operative interactions between ZP3 and the sperm surface may be needed to achieve AR. How this interaction conveys information to initiate signal transduction in mammalian sperm is not known. However, it is generally accepted that a rise in [Ca^{2+}]$_i$ is an essential step in the ZP3 signaling path leading to the

AR. It is also well known that external Ca^{2+} is required for the physiological AR to occur [1,85], and that ZP induces $[Ca^{2+}]_i$ and pH_i increases that precede acrosomal exocytosis [7,22]. Interestingly, pertussis toxin (PTX), an inactivator of G_i proteins, inhibits the AR [109], and the $[Ca^{2+}]_i$ and pH_i changes that accompany it [110]. Although the pH_i increase appears to be the G_i-regulated step in the ZP-induced AR [22], it is not known if sperm ion channels are regulated by ZP3-activated G_i proteins or how pH_i and $[Ca^{2+}]_i$ are intimately related and finely tuned in mammalian sperm.

As in the sea urchin sperm [111], at least two different types of Ca^{2+} permeable channels are involved in the mammalian sperm AR [112] (Fig. 3); one necessary for a fast (ms) transient change in $[Ca^{2+}]_i$ and another to achieve a sustained $[Ca^{2+}]_i$ elevation [5,113]. Growing evidence indicates that the fast response is mediated by a voltage-gated Ca^{2+} channel [5,7,22]. On the other hand, micromolar concentrations of DHPs are needed to inhibit the AR and the increase in $[Ca^{2+}]_i$ associated to it. Depolarizing conditions, at elevated pH_i, open Ca^{2+} channels sensitive to micromolar concentrations of DHPs that bypass the inhibition of the ZP3-induced exocytosis produced by PTX. While L-type

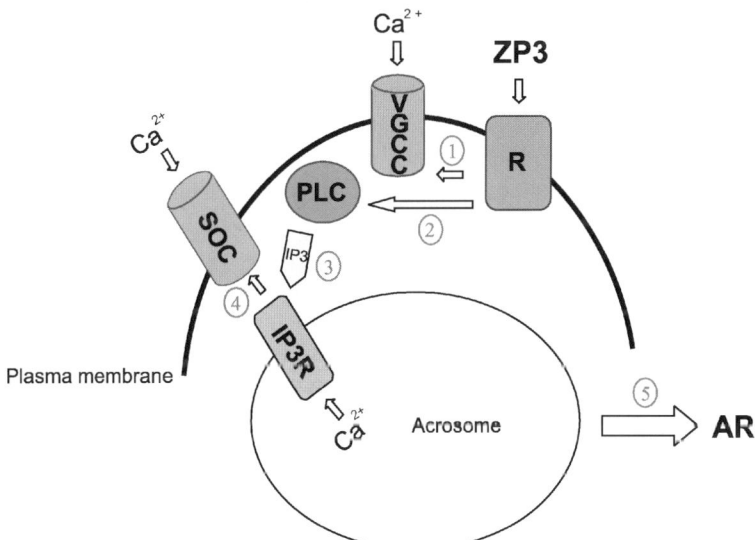

Fig. 3. Working model of the AR. Based on the available experimental evidence, the following model for sperm $[Ca^{2+}]_i$ control during AR may be proposed. Acting through membrane receptors (R), one of which could be a ligand-gated channel that depolarizes sperm, ZP3 activates two main signal transducers: (1) an LVA-type, voltage-gated Ca^{2+} channel (VGCC) opens and produces a fast transient Ca^{2+} influx within 40–50 ms that inactivates within an additional 150–200 ms. (2) ZP3 signal transduction activates a phospholipase C (PLC), which results in IP3 production. IP3 opens sperm IP3 receptors (IP3R), leading to a depletion of intracellular Ca^{2+} stores (3). The emptying of these stores generates a gating signal that couples store depletion to the opening of the store-operated permeation pathway (SOC) in the plasma membrane (4). The nature of this gating signal is unknown. Other modulators of sperm physiology may regulate Ca^{2+} conductance through this pathway. For instance, ZP3 transiently increases sperm pH_i with the same time course as the sustained $[Ca^{2+}]_i$ elevation and mouse SOCs from spermatogenic cells may be controlled by pH_i. (5) Finally, Ca^{2+} entering through SOCs induces secretion and triggers the final stages of acrosome reaction. (*For a colored version of this figure, see plate section, page 469.*)

Ca^{2+} channels are blocked by submicromolar concentrations of these blockers, tens of micromolar are required to inhibit T-type Ca^{2+} channels [29,30]. On the basis of the pharmacological profiles of these voltage-gated Ca^{2+} channels it has been suggested that a T-type Ca^{2+} channel ($Ca_v3.1$ and/or $Ca_v3.2$) participates in the mouse sperm AR [5,29,30, 50,141]. However, it is unclear how ZP opens voltage-gated Ca^{2+} channels at the present time. Although ZP or ZP3 have been reported to induce a 30 mV depolarization in bovine or mouse sperm [66], this membrane potential change seems too slow (many seconds) to activate T-type Ca^{2+} channels. In addition, recent studies suggest the activation of N- and R-type Ca^{2+} channels in mouse sperm in response to membrane depolarizations evoked by high K^+ [53], though no experimental evidence for the contribution of these channels to the AR has been provided thus far.

The fast transitory elevation in $[Ca^{2+}]_i$ is followed by a much slower one, which remains high while ZP is present. In mouse sperm this slow response requires at least ~ 30 s to develop and AR occurs only after the high sustained $[Ca^{2+}]_i$ is reached [22,66]. The kinetic characteristics of the slow sustained elevation in $[Ca^{2+}]_i$ are incompatible with the properties of T-type Ca^{2+} channels [114], therefore, at least another pathway for Ca^{2+} is necessarily involved in triggering the AR. Antagonists of LVA Ca^{2+} channels added before ZP3 also inhibit the sustained elevation in $[Ca^{2+}]_i$ [66]. These results indicate that the transient increase in $[Ca^{2+}]_i$ due to the ZP-induced activation of T-type channels is necessary to open a second Ca^{2+} pathway that keeps $[Ca^{2+}]_i$ elevated to allow AR [3,22].

ZP3 signaling may lead to the sustained elevation of $[Ca^{2+}]_i$ through the release of Ca^{2+} from an IP3-sensitive intracellular store [57] by a mechanism requiring prior Ca^{2+} influx through voltage-gated Ca^{2+} channels. There are several isoforms of phospholipase C in sperm [115] and ZP3 stimulates IP3 production in these cells [116]. IP3 receptors are present in acrosomal membranes [55,57] and, in digitonin-permeabilized mouse sperm exogenous IP3 causes $^{45}Ca^{2+}$ efflux from a non-mitochondrial pool [57]. Agents that are expected to promote Ca^{2+} release from ER pools like thapsigargin, promote elevations of $[Ca^{2+}]_i$ and AR in sperm [62,63]. These agents are thought to act on the sperm acrosome vesicle. Mouse spermatogenic cells, as well as immobile testicular sperm possess a SOC which was proposed to be responsible for the sustained $[Ca^{2+}]_i$ elevation necessary for the AR [64].

Mature mouse sperm do possess SOCs. Thapsigargin activates a Ca^{2+} uptake mechanism that requires extracellular Ca^{2+} and is inhibited by Ca^{2+}-entry antagonist, such as La^{3+} or Ni^{2+}. ZP3 also produces a sustained elevation in sperm $[Ca^{2+}]_i$. This late phase of Ca^{2+} influx shares several features with the sperm SOC, including a similar time course and magnitude of $[Ca^{2+}]_i$ increase, and a similar sensitivity to inhibition by Ni^{2+} and DHPs [113]. How voltage-gated Ca^{2+} channels, Ca^{2+} release from intracellular stores and AR are co-ordinated is not well understood. Furthermore, the molecular identity of this store-operated Ca^{2+} channel has to be established.

Lastly, it has been shown that progesterone promotes AR in sperm of several species [117]. Since this steroid activates phospholipase C [117], it may also induce Ca^{2+} depletion from sperm stores and activate the same SOC. This would explain why progesterone and ZP3 activate distinct upstream signaling elements [118,119] but can act co-operatively in driving acrosomal secretion [120].

6. Ion channel malfunction and infertility

Male infertility can be classified into several categories including spermatogenic disorders, obstruction of the seminal tract, inflammatory processes and sexual disorders. Inasmuch as the cause of spermatogenic disorders has not been identified clearly, idiopathic spermatogenic disorders accounts for more than 50% of all cases [121]. Abnormal human spermatogenesis and/or sperm fertilization incapability is caused by a variety of genetic and acquired conditions that affect mammalian fertility. Some reports have suggested that such spermatogenic disorders could involve malfunction of ion channels [122,123]. However, the precise contribution of ion channels to the etiology of this condition is difficult to dissect because alterations in many other proteins that are important in spermatogenesis, capacitation and/or AR can affect ion channel activity. For example, mutations in androgen receptors or tyrosine kinases could inhibit ionic fluxes through voltage-gated ion channels due to improper or lack of modulation.

6.1. Molecular genetic changes in ion channels and male infertility

In spite of the complexity of the system, specific ion channel mutants or transgenic animals can reveal important information regarding ion channel participation in sperm physiology and fertilization. Recently it was shown that sperm from mice mutants in the glycine receptor/Cl$^-$ channel (GlyR) α or β subunits (*spasmodic* and *spastic*) are deficient in their ability to undergo the AR triggered in vitro by glycine or by mouse egg ZP. Furthermore, a monoclonal antibody against GlyR blocked the ZP-induced AR in normal mouse sperm. These results suggest that the sperm GlyR plays an important role in the ZP-initiated AR [124]. They are also consistent with previous findings showing that certain compounds that inhibit single mouse sperm Cl$^-$ channel activity and macroscopic Cl$^-$ currents in mouse spermatogenic cells, block the ZP-induced AR [27,50]. Moreover, massive degeneration of male germ cells has been observed in mice deficient in plasma membrane Cl$^-$ channels, presumably due to alterations in the ionic homeostasis of these cells and defects in transepithelial transport by Sertoli cells an essential partner of spermatogenic cells in differentiation [125].

As mentioned earlier, two sperm-specific membrane proteins (CatSper 1 and 2) have been recently discovered [69,144]. These novel proteins are ion channel α$_1$ subunits resembling the six-transmembrane-segment protein voltage-gated K$^+$ channels. However, the predicted ion selectivity characteristics are the same as those expected for a Ca^{2+} channel. Notably, these channels are expressed by meiotic and post-meiotic spermatogenic cells, but not by other cells, and are present on the sperm flagellum, suggesting a role in the regulation of sperm motility. Notably, targeted disruption of mouse CatSper gene results in male sterility, due mainly to the incapability of sperm to maintain normal patterns of motility and their inability to penetrate the egg's ZP [69].

6.2. Environmental pathogenesis of reduced fertility

As discussed above, multiple Ca^{2+} and K^+ channel isoforms have been identified in mammalian spermatogenic cells and sperm [3]. It is generally accepted that heavy metal ions exert little effect on voltage-gated K^+ channels, but significantly inhibit ion flow through voltage-activated Ca^{2+} channels [126,127]. Consequently, the presence of these toxic ions in the male reproductive tract could affect male fertility by blocking the Ca^{2+} channels, which are sensitive to Cd^{2+} and Pb^{2+}, but can also affect Pb^{2+}-sensitive K^+ channels before causing heavy metal intoxication [128]. It has been proposed that at high concentrations heavy metals can permeate through sperm Ca^{2+} or K^+ channels [127, 129–132]. Permeation of Zn^{2+}, Cd^{2+}, Ni^{2+}, Co^{2+} and Mn^{2+} into the male reproductive tract through voltage-gated Ca^{2+} channels [35,46,112], can alter not only membrane polarization per se, but affect also intracellular Ca^{2+} homeostasis and therefore modify several metabolic pathways in spermatozoa.

Although exposure to high levels of heavy metals leads to a reduction in human male fertility, there are few studies about the effects of chronic low-level environmental exposures to such toxic ions on male reproductive health [133,134]. A recent prospective study revealed that a significant fraction of men with unexplained infertility had high blood and semen levels of Pb^{2+}. More importantly, sperm dysfunction was observed in these patients in the absence of other outward signs such as abnormal semen parameters or altered serum hormone levels [128]. Earlier it was mentioned that Pb^{2+} might enter cells through voltage-gated Ca^{2+} channels. Nifedipine, a potent Ca^{2+} channel antagonist, inhibits Pb^{2+} influx in different somatic cells [129,135]. Interestingly, patch-clamp and fluorescent dye studies have revealed the presence of nifedipine-sensitive Ca^{2+} currents in spermatogenic cells [29,30,47,136] and in mature sperm (Fig. 1) [8,22]. These observations suggest that Ca^{2+} channels contribute, at least in part, to Pb^{2+} influx in spermatozoa. Therefore, the action of Pb^{2+} in infertility might not only be related to Ca^{2+} channel blockade, but to its effects on other molecules of the fertilization machinery inside the cell.

Lastly, varicocele, a clinical condition found to be the cause of infertility in about 17% of the males who are treated for infertility, has also been associated with deficits in ion transport. Incompetent or inadequate valves within the veins along the spermatic cord cause varicocele. The abnormal valves obstruct healthy blood flow causing dilation of the veins. Interestingly, among other clinical signs, infertile men with varicocele present abnormally high levels of Cd^{2+} in seminal plasma. This increased Cd^{2+} accumulation was observed even in patients who were not occupationally exposed to Cd^{2+} and who did not smoke cigarettes, another source of Cd^{2+} [35,137]. Moreover, the ZP-induced sperm AR of fertile donors is deficient in the presence of Cd^{2+} concentrations equivalent to those observed in infertile men with varicocele. These observations suggest a causal relationship between Cd^{2+} exposure and varicocele-related AR insufficiency [35,138]. Likewise in the case of Pb^{2+}, Ca^{2+} channel blockers also inhibit Cd^{2+} uptake within somatic cells [139,140]. In mammalian sperm, the sites of Cd^{2+} entrance and Ca^{2+} channels were co-localized in the sperm head. These observations lead to the suggestion that AR inefficiency in varicocele patients may be due to defective Ca^{2+} influx through voltage-gated Ca^{2+} channels.

Acknowledgements

R.F. and A.D. were supported by CONACyT-México (grants 31735-N and 27707-N, respectively), and A.D. by DGAPA (grant IN201599). This chapter was written in 2001, while A.D. was in sabbatical at the University of Newcastle upon Tyne in the laboratory of Dr. Michael Whitaker. The support of The Wellcome Trust and BBSRC is acknowledged.

References

[1] Wassarman, P.M., 1999. Fertilization in animals. Dev. Genet. 25, 83–86.

[2] Visconti, P.E., Galantino-Homer, H., Moore, G.D., Bailey, J.L., Ninag, X., Fornes, M., Kopf, G.S., 1998. The molecular basis of sperm capacitation. J. Androl. 19, 242–248.

[3] Baldi, E., Luconi, M., Bonaccorsi, L., Muratori, M., Forti, G., 2000. Intracellular events and signaling pathways involved in sperm acquisition of fertilizing capacity and acrosome reaction. Front. Biosci. 5, E110–E123.

[4] Zeng, Y., Clark, E.N., Florman, H.M., 1995. Sperm membrane potential: hyperpolarization during capacitation regulates zona pellucida-dependent acrosomal secretion. Dev. Biol. 171, 554–563.

[5] Arnoult, C., Kazam, I.G., Visconti, P.E., Kopf, G.S., Villaz, M., Florman, H.M., 1999. Control of the low voltage-activated calcium channel of mouse sperm by egg ZP3 and by membrane hyperpolarization during capacitation. Proc. Natl Acad. Sci. USA 96, 6757–6762.

[6] Flesch, F.M., Gadella, B.M., 2000. Dynamics of the mammalian sperm plasma membrane in the process of fertilization. Biochim. Biophys. Acta 1469, 197–235.

[7] Darszon, A., Labarca, P., Nishigaki, T., Espinosa, F., 1999. Ion channels in sperm physiology. Physiol. Rev. 79, 481–510.

[8] Publicover, S.J., Barratt, C.L., 1999. Voltage-operated Ca^{2+} channels and the acrosome reaction: which channels are present and what do they do? Hum. Reprod. 14, 873–879.

[9] Baldi, E., Luconi, M., Bonaccorsi, L., Krausz, C., Forti, G., 1996. Human sperm activation during capacitation and acrosome reaction: role of calcium, protein phosphorylation and lipid remodelling pathways. Front. Biosci. 1, d189–d205.

[10] Hille, B., 1992. Ionic channels of excitable membranes. Sinauer Associates Inc., Sunderland.

[11] Ohlendieck, K., Lennarz, W.J., 1995. Role of the sea urchin egg receptor for sperm in gamete interactions. Trends Biochem. Sci. 20, 29–33.

[12] Darszon, A., Liévano, A., Beltrán, C., 1996. Ion channels: key elements in gamete signaling. Curr. Top. Dev. Biol. 34, 117–167.

[13] Liévano, A., Sánchez, J.A., Darszon, A., 1985. Single-channel activity of bilayers derived from sea urchin sperm plasma membranes at the tip of a patch-clamp electrode. Dev. Biol. 112, 253–257.

[14] Liévano, A., Vega-Saenz de Miera, E.C., Darszon, A., 1990. Ca^{2+} channels from the sea urchin sperm plasma membrane. J. Gen. Physiol. 95, 273–296.

[15] Labarca, P., Santi, C., Zapata, O., Morales, E., Beltrán, C., Liévano, A., Darszon, A., 1996. A cAMP regulated K^+-selective channel from the sea urchin sperm plasma membrane. Dev. Biol. 174, 271–280.

[16] Morales, E., de la Torre, L., Moy, G.W., Vacquier, V.D., Darszon, A., 1993. Anion channels in the sea urchin sperm plasma membrane. Mol. Reprod. Dev. 36, 174–182.

[17] Cox, T., Peterson, R.N., 1989. Identification of calcium conducting channels in isolated boar sperm plasma membranes. Biochem. Biophys. Res. Commun. 161, 162–168.

[18] Cox, T., Campbell, P., Peterson, R.N., 1991. Ion channels in boar sperm plasma membranes: characterization of a cation selective channel. Mol. Reprod. Dev. 30, 135–147.

[19] Tiwari-Woodruff, S.K., Cox, T.C., 1995. Boar sperm plasma membrane Ca^{2+}-selective channels in planar lipid bilayers. Am. J. Physiol. 268, C1284–C1294.

[20] Chan, H.C., Zhou, T.S., Fu, W.O., Wang, W.P., Shi, Y.L., Wong, P.Y., 1997. Cation and anion channels in rat and human spermatozoa. Biochim. Biophys. Acta 1323, 117–129.

[21] Beltrán, C., Darszon, A., Labarca, P., Liévano, A., 1994. A high-conductance voltage-dependent multistate Ca^{2+} channel found in sea urchin and mouse spermatozoa. FEBS Lett. 338, 23–26.

[22] Florman, H.M., Arnoult, C., Kazam, I.G., Li, C., O'Toole, C.M., 1998. A perspective on the control of mammalian fertilization by egg-activated ion channels in sperm: a tale of two channels. Biol. Reprod. 59, 12–16.

[23] Hamill, O.P., Marty, A., Neher, E., Sakmann, B., Sigworth, F.J., 1981. Improved patch-clamp techniques for high-resolution current recording from cells and cell-free membrane patches. Pflugers Arch. 391, 85–100.

[24] Sakmann, B., Neher, E., 1995. Single-channel recording. Plenum Press, New York.

[25] Guerrero, A., Sánchez, J.A., Darszon, A., 1987. Single-channel activity in sea urchin sperm revealed by the patch-clamp technique. FEBS Lett. 220, 295–298.

[26] Babcock, D.F., Bosma, M.M., Battaglia, D.E., Darszon, A., 1992. Early persistent activation of sperm K^{+} channels by the egg peptide speract. Proc. Natl Acad. Sci. USA 89, 6001–6005.

[27] Espinosa, F., de la Vega-Beltrán, J.L., López-González, I., Delgado, R., Labarca, P., Darszon, A., 1998. Mouse sperm patch-clamp recordings reveal single Cl^{-} channels sensitive to niflumic acid, a blocker of the sperm acrosome reaction. FEBS Lett. 426, 47–51.

[28] Hagiwara, S., Kawa, K., 1984. Calcium and potassium currents in spermatogenic cells dissociated from rat seminiferous tubules. J. Physiol. 356, 135–149.

[29] Liévano, A., Santi, C.M., Serrano, C.J., Treviño, C.L., Bellve, A.R., Hernández-Cruz, A., Darszon, A., 1996. T-type Ca^{2+} channels and $\alpha 1E$ expression in spermatogenic cells, and their possible relevance to the sperm acrosome reaction. FEBS Lett. 388, 150–154.

[30] Arnoult, C., Cardullo, R.A., Lemos, J.R., Florman, H.M., 1996. Activation of mouse sperm T-type Ca^{2+} channels by adhesion to the egg zona pellucida. Proc. Natl Acad. Sci. USA 93, 13004–13009.

[31] Hecht, N.B., 1998. Molecular mechanisms of male germ cell differentiation. Bioessays 20, 555–561.

[32] Bellvé, A.R., 1998. Introduction: the male germ cell; origin, migration, proliferation and differentiation. Semin. Cell Dev. Biol. 9, 379–391.

[33] Muñoz-Garay, C., de la Vega-Beltrán, J.L., Delgado, R., Labarca, P., Felix, R., Darszon, A., 2001. Inwardly-rectifying K^{+} channels in spermatogenic cells: functional expression and implication in sperm capacitation. Dev. Biol. 234, 261–274.

[34] Abou-Haila, A., Tulsiani, D.R., 2000. Mammalian sperm acrosome: formation, contents, and function. Arch. Biochem. Biophys. 379, 173–182.

[35] Benoff, S., 1998. Voltage dependent calcium channels in mammalian spermatozoa. Front. Biosci. 3, d1220–d1240.

[36] De Waard, M., Gurnett, C.A., Campbell, K.P., 1996. Structural and functional diversity of voltage-activated calcium channels. In: Narahashi, T. (Ed.), Ion Channels, Vol. IV. Plennum Press, New York, pp. 41–87.

[37] Walker, D., De Waard, M., 1998. Subunit interaction sites in voltage-dependent Ca^{2+} channels: role in channel function. Trends Neurosci. 21, 148–154.

[38] Perez-Reyes, E., Cribbs, L.L., Daud, A., Lacerda, A.E., Barclay, J., Williamson, M.P., Fox, M., Rees, M., Lee, J.H., 1998. Molecular characterization of a neuronal low-voltage-activated T-type calcium channel. Nature 391, 896–900.

[39] Cribbs, L.L., Lee, J.H., Yang, J., Satin, J., Zhang, Y., Daud, A., Barclay, J., Williamson, M.P., Fox, M., Rees, M., Perez-Reyes, E., 1998. Cloning and characterization of α_{1H} from human heart, a member of the T-type Ca^{2+} channel gene family. Circ. Res. 83, 103–109.

[40] Lambert, R.C., McKenna, F., Maulet, Y., Talley, E.M., Bayliss, D.A., Cribbs, L.L., Lee, J.H., Perez-Reyes, E., Feltz, A., 1998. Low-voltage-activated Ca^{2+} currents are generated by members of the CavT subunit family ($\alpha_{1G/H}$) in rat primary sensory neurons. J. Neurosci. 18, 8605–8613.

[41] Birnbaumer, L., Qin, N., Olcese, R., Tareilus, E., Platano, D., Costantin, J., Stefani, E., 1998. Structures and functions of calcium channel beta subunits. J. Bioenerg. Biomembr. 30, 357–375.

[42] Chien, A.J., Hosey, M.M., 1998. Post-translational modifications of beta subunits of voltage-dependent calcium channels. J. Bioenerg. Biomembr. 30, 377–386.

[43] Felix, R., 1999. Voltage-dependent Ca^{2+} channel $\alpha 2\delta$ auxiliary subunit: structure, function and regulation. Receptors Channels 6, 351–362.

[44] Catterall, W.A., 2000. Structure and regulation of voltage-gated Ca^{2+} channels. Annu. Rev. Cell Dev. Biol. 16, 521–555.

[45] Babcock, D.F., Pfeiffer, D.R., 1987. Independent elevation of cytosolic $[Ca^{2+}]$ and pH of mammalian sperm by voltage-dependent and pH-sensitive mechanisms. J. Biol. Chem. 262, 15041–15047.

[46] Florman, H.M., Corron, M.E., Kim, T.D., Babcock, D.F., 1992. Activation of voltage-dependent calcium channels of mammalian sperm is required for zona pellucida-induced acrosomal exocytosis. Dev. Biol. 152, 304–314.

[47] Santi, C.M., Darszon, A., Hernández-Cruz, A., 1996. A dihydropyridine-sensitive T-type Ca^{2+} current is the main Ca^{2+} current carrier in mouse primary spermatocytes. Am. J. Physiol. 271, C1583–C1593.

[48] Arnoult, C., Lemos, J.R., Florman, H.M., 1997. Voltage-dependent modulation of T-type calcium channels by protein tyrosine phosphorylation. EMBO J. 16, 1593–1599.

[49] Goodwin, L.O., Leeds, N.B., Hurley, I., Mandel, F.S., Pergolizzi, R.G., Benoff, S., 1997. Isolation and characterization of the primary structure of testis-specific L-type calcium channel: implications for contraception. Mol. Hum. Reprod. 3, 255–268.

[50] Espinosa, F., López-González, I., Serrano, C.J., Gasque, G., de la Vega-Beltrán, J.L., Trevino, C.L., Darszon, A., 1999. Anion channel blockers differentially affect T-type Ca^{2+} currents of mouse spermatogenic cells, α1E currents expressed in Xenopus oocytes and the sperm acrosome reaction. Dev. Genet. 25, 103–114.

[51] Goodwin, L.O., Leeds, N.B., Hurley, I., Cooper, G.W., Pergolizzi, R.G., Benoff, S., 1998. Alternative splicing of exons in the $α_1$ subunit of the rat testis L-type voltage-dependent calcium channel generates germ line-specific dihydropyridine binding sites. Mol. Hum. Reprod. 4, 215–226.

[52] Westenbroek, R.E., Babcock, D.F., 1999. Discrete regional distributions suggest diverse functional roles of calcium channel $α_1$ subunits in sperm. Dev. Biol. 207, 457–469.

[53] Wennemuth, G., Westenbroek, R.E., Xu, T., Hille, B., Babcock, D.F., 2000. $Ca_v2.2$ and $Ca_v2.3$ (N- and R-type) Ca^{2+} channels in depolarization-evoked entry of Ca^{2+} into mouse sperm. J. Biol. Chem. 275, 21210–21217.

[54] Serrano, C.J., Trevino, C.L., Felix, R., Darszon, A., 1999. Voltage-dependent Ca^{2+} channel subunit expression and immunolocalization in mouse spermatogenic cells and sperm. FEBS Lett. 462, 171–176.

[55] Treviño, C.L., Santi, C.M., Beltrán, C., Hernández-Cruz, A., Darszon, A., Lomelí, H., 1998. Localisation of inositol trisphosphate and ryanodine receptors during mouse spermatogenesis: possible functional implications. Zygote 6, 159–172.

[56] Kuroda, Y., Kaneko, S., Yoshimura, Y., Nozawa, S., Mikoshiba, K., 1999. Are there inositol 1,4,5-triphosphate (IP3) receptors in human sperm? Life Sci. 65, 135–143.

[57] Walensky, L.D., Snyder, S.H., 1995. Inositol 1,4,5-trisphosphate receptors selectively localized to the acrosomes of mammalian sperm. J. Cell Biol. 130, 857–869.

[58] Tanabe, T., Beam, K.G., Adams, B.A., Niidome, T., Numa, S., 1990. Regions of the skeletal muscle dihydropyridine receptor critical for excitation–contraction coupling. Nature 346, 567–569.

[59] Giannini, G., Conti, A., Mammarella, S., Scrobogna, M., Sorrentino, V., 1995. The ryanodine receptor/calcium channel genes are widely and differentially expressed in murine brain and peripheral tissues. J. Cell Biol. 128, 893–904.

[60] Putney, J.W. Jr., McKay, R.R., 1999. Capacitative calcium entry channels. Bioessays 21, 38–46.

[61] Boulay, G., Brown, D.M., Qin, N., Jiang, M., Dietrich, A., Zhu, M.X., Chen, Z., Birnbaumer, M., Mikoshiba, K., Birnbaumer, L., 1999. Modulation of Ca^{2+} entry by polypeptides of the inositol 1,4, 5-trisphosphate receptor (IP3R) that bind transient receptor potential (TRP): evidence for roles of TRP and IP3R in store-depletion-activated Ca^{2+} entry. Proc. Natl Acad. Sci. USA 96, 14955–14960.

[62] Meizel, S., Turner, K.O., 1993. Initiation of the human sperm acrosome reaction by thapsigargin. J. Exp. Zool. 267, 350–355.

[63] Blackmore, P.F., 1993. Thapsigargin elevates and potentiates the ability of progesterone to increase intracellular free calcium in human sperm: possible role of perinuclear calcium. Cell Calcium 14, 53–60.

[64] Santi, C.M., Santos, T., Hernández-Cruz, A., Darszon, A., 1998. Properties of a novel pH-dependent Ca^{2+} permeation pathway present in male germ cells with possible roles in spermatogenesis and mature sperm function. J. Gen. Physiol. 112, 33–53.

[65] Jungnickel, M.K., Marrero, H., Birnbaumer, L., Lemos, J.R., Florman, H.M., 2001. Trp2 regulates entry of Ca^{2+} into mouse sperm triggered by egg ZP3. Nat. Cell Biol. 3, 499–502.

[66] Arnoult, C., Zeng, Y., Florman, H.M., 1996. ZP3-dependent activation of sperm cation channels regulates acrosomal secretion during mammalian fertilization. J. Cell Biol. 134, 637–645.

[67] Weyand, I., Godde, M., Frings, S., Weiner, J., Muller, F., Altenhofen, W., Hatt, H., Kaupp, U.B., 1994. Cloning and functional expression of a cyclic-nucleotide-gated channel from mammalian sperm. Nature 368, 859–863.

[68] Wiesner, B., Weiner, J., Middendorff, R., Hagen, V., Kaupp, U.B., Weyand, I., 1998. Cyclic nucleotide-gated channels on the flagellum control Ca^{2+} entry into sperm. J. Cell Biol. 142, 473–484.

[69] Ren, D., Navarro, B., Perez, G., Jackson, A.C., Hsu, S., Shi, Q., Tilly, J.L., Clapham, D.E., 2001. A sperm ion channel required for sperm motility and male fertility. Nature 413, 603–609.

[70] Coetzee, W.A., Amarillo, Y., Chiu, J., Chow, A., Lau, D., McCormack, T., Moreno, H., Nadal, M.S., Ozaita, A., Pountney, D., Saganich, M., Vega-Saenz de Miera, E., Rudy, B., 1999. Molecular diversity of K^+ channels. Ann. NY Acad. Sci. 868, 233–285.

[71] Schreiber, M., Wei, A., Yuan, A., Gaut, J., Saito, M., Salkoff, L., 1998. Slo3, a novel pH-sensitive K^+ channel from mammalian spermatocytes. J. Biol. Chem. 273, 3509–3516.

[72] Gauss, R., Seifert, R., Kaupp, U.B., 1998. Molecular identification of a hyperpolarization-activated channel in sea urchin sperm. Nature 393, 583–587.

[73] Seifert, R., Scholten, A., Gauss, R., Mincheva, A., Lichter, P., Kaupp, U.B., 1999. Molecular characterization of a slowly gating human hyperpolarization-activated channel predominantly expressed in thalamus, heart, and testis. Proc. Natl Acad. Sci. USA 96, 9391–9396.

[74] Hansbrough, J.R., Garbers, D.L., 1981. Speract –purification and characterization of a peptide associated with eggs that activates spermatozoa. J. Biol. Chem. 256, 1447–1452.

[75] Salvatore, L., D'Adamo, M.C., Polishchuk, R., Salmona, M., Pessia, M., 1999. Localization and age-dependent expression of the inward rectifier K^+ channel subunit Kir 5.1 in a mammalian reproductive system. FEBS Lett. 449, 146–152.

[76] Chan, H.C., Wu, W.L., Sun, Y.P., Leung, P.S., Wong, T.P., Chung, Y.W., So, S.C., Zhou, T.S., Yan, Y.C., 1998. Expression of sperm Ca^{2+}-activated K^+ channels in *Xenopus* oocytes and their modulation by extracellular ATP. FEBS Lett. 438, 177–182.

[77] Wu, W.L., So, S.C., Sun, Y.P., Zhou, T.S., Yu, Y., Chung, Y.W., Wang, X.F., Bao, Y.D., Yan, Y.C., Chan, H.C., 1998. Functional expression of a Ca^{2+}-activated K^+ channel in *Xenopus* oocytes injected with RNAs from the rat testis. Biochim. Biophys. Acta 1373, 360–365.

[78] Jacob, A., Hurley, I.R., Goodwin, L.O., Cooper, G.W., Benoff, S., 2000. Molecular characterization of a voltage-gated potassium channel expressed in rat testis. Mol. Hum. Reprod. 6, 303–313.

[79] Reimann, F., Ashcroft, F.M., 1999. Inwardly rectifying potassium channels. Curr. Opin. Cell Biol. 11, 503–508.

[80] Tsai, T.D., Shuck, M.E., Thompson, D.P., Bienkowski, M.J., Lee, K.S., 1995. Intracellular H^+ inhibits a cloned rat kidney outer medulla K^+ channel expressed in *Xenopus* oocytes. Am. J. Physiol. 268, C1173–C1178.

[81] Pessia, M., Tucker, S.J., Lee, K., Bond, C.T., Adelman, J.P., 1996. Subunit positional effects revealed by novel heteromeric inwardly rectifying K^+ channels. EMBO J. 15, 2980–2987.

[82] Tucker, S.J., Imbrici, P., Salvatore, L., D'Adamo, M.C., Pessia, M., 2000. pH Dependence of the inwardly rectifying potassium channel, Kir5.1, and localization in renal tubular epithelia. J. Biol. Chem. 275, 16404–16407.

[83] Visconti, P.E., Bailey, J.L., Moore, G.D., Pan, D., Olds-Clarke, P., Kopf, G.S., 1995. Capacitation of mouse spermatozoa. I. Correlation between the capacitation state and protein tyrosine phosphorylation. Development 121, 1129–1137.

[84] Visconti, P.E., Moore, G.D., Bailey, J.L., Leclerc, P., Connors, S.A., Pan, D., Olds-Clarke, P., Kopf, G.S., 1995. Capacitation of mouse spermatozoa. II. Protein tyrosine phosphorylation and capacitation are regulated by a cAMP-dependent pathway. Development 121, 1139–1150.

[85] Yanagimachi, R., 1994. Mammalian fertilization. In: Knobil, E., Neil, J.D. (Eds.), The Physiology of Reproduction. Raven Press, New York, pp. 189–317.

[86] Davis, B.K., Byrne, R., Hungund, B., 1979. Studies on the mechanism of capacitation. II. Evidence for lipid transfer between plasma membrane of rat sperm and serum albumin during capacitation in vitro. Biochim. Biophys. Acta 558, 257–266.

[87] Go, K.J., Wolf, D.P., 1985. Albumin-mediated changes in sperm sterol content during capacitation. Biol. Reprod. 32, 145–153.

[88] Langlais, J., Kan, F.W., Granger, L., Raymond, L., Bleau, G., Roberts, K.D., 1988. Identification of sterol acceptors that stimulate cholesterol efflux from human spermatozoa during in vitro capacitation. Gamete Res. 20, 185–201.

[89] Martínez, P., Morros, A., 1996. Membrane lipid dynamics during human sperm capacitation. Front. Biosci. 1, d103–d117.

[90] Visconti, P.E., Galantino-Homer, H., Ning, X., Moore, G.D., Valenzuela, J.P., Jorgez, C.J., Alvarez, J.G., Kopf, G.S., 1999. Cholesterol efflux-mediated signal transduction in mammalian sperm. β-Cyclodextrins initiate transmembrane signaling leading to an increase in protein tyrosine phosphorylation and capacitation. J. Biol. Chem. 274, 3235–3242.

[91] Espinosa, F., López-González, I., Munoz-Garay, C., Felix, R., de la Vega-Beltrán, J.L., Kopf, G.S., Visconti, P.E., Darszon, A., 2000. Dual regulation of the T-type Ca^{2+} current by serum albumin and beta-estradiol in mammalian spermatogenic cells. FEBS Lett. 475, 251–256.

[92] Espinosa, F., Darszon, A., 1995. Mouse sperm membrane potential: changes induced by Ca^{2+}. FEBS Lett. 372, 119–125.

[93] Zeng, Y., Oberdorf, J.A., Florman, H.M., 1996. pH regulation in mouse sperm: identification of Na^+-, Cl^--, and $HCO3^-$-dependent and arylaminobenzoate-dependent regulatory mechanisms and characterization of their roles in sperm capacitation. Dev. Biol. 173, 510–520.

[94] Parrish, J.J., Susko-Parrish, J.L., First, N.L., 1989. Capacitation of bovine sperm by heparin: inhibitory effect of glucose and role of intracellular pH. Biol. Reprod. 41, 683–699.

[95] Choe, H., Zhou, H., Palmer, L.G., Sackin, H., 1997. A conserved cytoplasmic region of ROMK modulates pH sensitivity, conductance, and gating. Am. J. Physiol. 273, F516–F529.

[96] Qu, Z., Zhu, G., Yang, Z., Cui, N., Li, Y., Chanchevalap, S., Sulaiman, S., Haynie, H., Jiang, C., 1999. Identification of a critical motif responsible for gating of Kir2.3 channel by intracellular protons. J. Biol. Chem. 274, 13783–13789.

[97] Johns, D.C., Marx, R., Mains, R.E., O'Rourke, B., Marban, E., 1999. Inducible genetic suppression of neuronal excitability. J. Neurosci. 19, 1691–1697.

[98] Galantino-Homer, H.L., Visconti, P.E., Kopf, G.S., 1997. Regulation of protein tyrosine phosphorylation during bovine sperm capacitation by a cyclic adenosine $3'5'$-monophosphate-dependent pathway. Biol. Reprod. 56, 707–719.

[99] Urner, F., Leppens-Luisier, G., Sakkas, D., 2001. Protein tyrosine phosphorylation in sperm during gamete interaction in the mouse: the influence of glucose. Biol. Reprod. 64, 1350–1357.

[100] Travis, A.J., Jorgez, C.J., Merdiushev, T., Jones, B.H., Dess, D.M., Diaz-Cueto, L., Storey, B.T., Kopf, G.S., Moss, S.B., 2001. Functional relationships between capacitation-dependent cell signaling and compartmentalized metabolic pathways in murine spermatozoa. J. Biol. Chem. 276, 7630–7636.

[101] Guzman-Grenfell, A.M., Bonilla-Hernandez, M.A., Gonzalez-Martinez, M.T., 2000. Glucose induces a Na^+,K^+-ATPase dependent transient hyperpolarization in human sperm. 1. Induction of changes in plasma membrane potential by the proton ionophore CCCP. Biochim. Biophys. Acta 1464, 188–198.

[102] Jahn, R., Sudhof, T.C., 1999. Membrane fusion and exocytosis. Annu. Rev. Biochem. 68, 863–911.

[103] Schulz, J.R., Sasaki, J.D., Vacquier, V.D., 1998. Increased association of synaptosome-associated protein of 25 kDa with syntaxin and vesicle-associated membrane protein following acrosomal exocytosis of sea urchin sperm. J. Biol. Chem. 273, 24355–24359.

[104] Ramalho-Santos, J., Moreno, R.D., Sutovsky, P., Chan, A.W., Hewitson, L., Wessel, G.M., Simerly, C.R., Schatten, G., 2000. SNAREs in mammalian sperm: possible implications for fertilization. Dev. Biol. 223, 54–69.

[105] Michaut, M., Tomes, C.N., De Blas, G., Yunes, R., Mayorga, L.S., 2000. Calcium-triggered acrosomal exocytosis in human spermatozoa requires the coordinated activation of Rab3A and N-ethylmaleimide-sensitive factor. Proc. Natl Acad. Sci. USA 97, 9996–10001.

[106] Stanley, E.F., 1997. The calcium channel and the organization of the presynaptic transmitter release face. Trends Neurosci. 20, 404–409.

[107] Lee, A., Scheuer, T., Catterall, W.A., 2000. Ca^{2+}/calmodulin-dependent facilitation and inactivation of P/Q-type Ca^{2+} channels. J. Neurosci. 20, 6830–6838.

[108] Peters, C., Mayer, A., 1998. Ca^{2+}/calmodulin signals the completion of docking and triggers a late step of vacuole fusion. Nature 396, 575–580.

[109] Endo, Y., Lee, M.A., Kopf, G.S., 1988. Characterization of an islet-activating protein-sensitive site in mouse sperm that is involved in the zona pellucida-induced acrosome reaction. Dev. Biol. 129, 12–24.

[110] Florman, H.M., Tombes, R.M., First, N.L., Babcock, D.F., 1989. An adhesion-associated agonist from the zona pellucida activates G protein-promoted elevations of internal Ca^{2+} and pH that mediate mammalian sperm acrosomal exocytosis. Dev. Biol. 135, 133–146.

[111] Guerrero, A., Darszon, A., 1989. Evidence for the activation of two different Ca^{2+} channels during the egg jelly-induced acrosome reaction of sea urchin sperm. J. Biol. Chem. 264, 19593–19599.

[112] Florman, H.M., 1994. Sequential focal and global elevations of sperm intracellular Ca^{2+} are initiated by the zona pellucida during acrosomal exocytosis. Dev. Biol. 165, 152–164.

[113] O'Toole, C.M., Arnoult, C., Darszon, A., Steinhardt, R.A., Florman, H.M., 2000. Ca^{2+} entry through store-operated channels in mouse sperm is initiated by egg ZP3 and drives the acrosome reaction. Mol. Biol. Cell 11, 1571–1584.

[114] Bean, B.P., McDonough, S.I., 1998. Two for T. Neuron 20, 825–828.

[115] Vanha-Perttula, T., Kasurinen, J., 1989. Purification and characterization of phosphatidylinositol-specific phospholipase C from bovine spermatozoa. Int. J. Biochem. 21, 997–1007.

[116] Tomes, C.N., McMaster, C.R., Saling, P.M., 1996. Activation of mouse sperm phosphatidylinositol-4,5 bisphosphate-phospholipase C by zona pellucida is modulated by tyrosine phosphorylation. Mol. Reprod. Dev. 43, 196–204.

[117] Thomas, P., Meizel, S., 1989. Phosphatidylinositol 4,5-bisphosphate hydrolysis in human sperm stimulated with follicular fluid or progesterone is dependent upon Ca^{2+} influx. Biochem. J. 264, 539–546.

[118] Tesarik, J., Moos, J., Mendoza, C., 1993. Stimulation of protein tyrosine phosphorylation by a progesterone receptor on the cell surface of human sperm. Endocrinology 133, 328–335.

[119] Murase, T., Roldan, E.R., 1996. Progesterone and the zona pellucida activate different transducing pathways in the sequence of events leading to diacylglycerol generation during mouse sperm acrosomal exocytosis. Biochem. J. 320, 1017–1023.

[120] Roldan, E.R., Murase, T., Shi, Q.X., 1994. Exocytosis in spermatozoa in response to progesterone and zona pellucida. Science 266, 1578–1581.

[121] Namiki, M., 2000. Genetic aspects of male infertility. World J. Surg. 24, 1176–1179.

[122] Yeung, C.H., Sonnenberg-Riethmacher, E., Cooper, T.G., 1999. Infertile spermatozoa of c-ros tyrosine kinase receptor knockout mice show flagellar angulation and maturational defects in cell volume regulatory mechanisms. Biol. Reprod. 61, 1062–1069.

[123] Stuhrmann, M., Dork, T., 2000. CFTR gene mutations and male infertility. Andrologia 32, 71–83.

[124] Sato, A., Son, J.-H., Tucker, R.P., Meizel, S, 2000. The zona pellucida-initiated acrosome reaction: defect due to mutations in the sperm glycine receptor/Cl^- channel. Dev. Biol. 227, 211–218.

[125] Bosl, M.R., Stein, V., Hubner, C., Zdebik, A.A., Jordt, S.E., Mukhopadhyay, A.K., Davidoff, M.S., Holstein, A.F., Jentsch, T.J., 2001. Male germ cells and photoreceptors, both dependent on close cell–cell interactions, degenerate upon ClC-2 Cl(-) channel disruption. EMBO J. 20, 1289–1299.

[126] Audesirk, G., 1993. Electrophysiology of lead intoxication: effects on voltage-sensitive ion channels. Neurotoxicology 14, 137–147.

[127] Busselberg, D., Platt, B., Michael, D., Carpenter, D.O., Haas, H.L., 1994. Mammalian voltage-activated calcium channel currents are blocked by Pb^{2+}, Zn^{2+}, and Al^{3+}. J. Neurophysiol. 71, 1491–1497.

[128] Benoff, S., Jacob, A., Hurley, I.R., 2000. Male infertility and environmental exposure to lead and cadmium. Hum. Reprod. Update 6, 107–121.

[129] Tomsig, J.L., Suszkiw, J.B., 1991. Permeation of Pb^{2+} through calcium channels: fura-2 measurements of voltage- and dihydropyridine-sensitive Pb^{2+} entry in isolated bovine chromaffin cells. Biochim. Biophys. Acta 1069, 197–200.

[130] Audesirk, G., Audesirk, T., 1993. The effects of inorganic lead on voltage-sensitive calcium channels differ among cell types and among channel subtypes. Neurotoxicology 14, 259–265.

[131] Busselberg, D., Pekel, M., Michael, D., Platt, B., 1994. Mercury (Hg^{2+}) and zinc (Zn^{2+}): two divalent cations with different actions on voltage-activated calcium channel currents. Cell Mol. Neurobiol. 14, 675–687.

[132] Platt, B., Busselberg, D., 1994. Combined actions of Pb^{2+}, Zn^{2+}, and Al^{3+} on voltage-activated calcium channel currents. Cell Mol. Neurobiol. 14, 831–840.

[133] Sallmen, M., Lindbohm, M.L., Anttila, A., Taskinen, H., Hemminki, K., 2000. Time to pregnancy among the wives of men occupationally exposed to lead. Epidemiology 11, 141–147.

[134] Sallmen, M., Lindbohm, M.L., Nurminen, M., 2000. Paternal exposure to lead and infertility. Epidemiology 11, 148–152.

[135] Schulte, S., Muller, W.E., Friedberg, K.D., 1995. In vitro and in vivo effects of lead on specific 3H-MK-801 binding to NMDA-receptors in the brain of mice. Neurotoxicology 16, 309–317.

[136] Arnoult, C., Villaz, M., Florman, H.M., 1998. Pharmacological properties of the T-type Ca^{2+} current of mouse spermatogenic cells. Mol. Pharmacol. 53, 1104–1111.

[137] Chia, S.E., Ong, C.N., Tsakok, F.M., 1994. Effects of cigarette smoking on human semen quality. Arch. Androl. 33, 163–168.

[138] Benoff, S., Hurley, I.R., Barcia, M., Mandel, F.S., Cooper, G.W., Hershlag, A., 1997. A potential role for cadmium in the etiology of varicocele-associated infertility. Fertil. Steril. 67, 336–347.

[139] Souza, V., Bucio, L., Gutierrez-Ruiz, M.C., 1997. Cadmium uptake by a human hepatic cell line (WRL-68 cells). Toxicology 120, 215–220.

[140] Weidner, W.J., Sillman, A.J., 1997. Low levels of cadmium chloride damage the corneal endothelium. Arch. Toxicol. 71, 455–460.

[141] Son, W.Y., Lee, J.H., Lee, J.H., Han, C.T., 2000. Acrosome reaction of human spermatozoa is mainly mediated by alpha1H T-type calcium channels. Mol. Hum. Reprod. 6, 893–897.

[142] Storey, B.T., Hourani, C.L., Kim, J.B., 1992. A transient rise in intracellular Ca^{2+} is a precursor reaction to the zona pellucida-induced acrosome reaction in mouse sperm and is blocked by the induced acrosome reaction inhibitor 3-quinuclidinyl benzilate. Mol. Reprod. Dev. 32, 41–50.

[143] Treviño, C.L., Serrano, C.J., Beltrán, C., Felix, R., Darszon, A., 2001. Identification of mouse *trp* homologues and lipid rafts from spermatogenic cells and sperm. FEBS Lett. 509, 119–125.

[144] Quill, T.A., Ren, D., Clapham, D.E., Garbers, D.L., 2001. A voltage-gated ion channel expressed specifically in spermatozoa. Proc. Natl Acad. Sci. USA 98, 12527–12531.

[145] Felix, R., Serrano, C.J., Treviño, C.L., Muñoz-Garay, C., Bravo, A., Navarro, A., Pacheco, J., Tsutsumi, V., Darszon, A., 2002. Identification of distinct K^+ channels in mouse spermatogenic cells and sperm. Zygote 10, 183–188.

Ion channels on intracellular organelles

Leonard K. Kaczmarek[a],* and Elizabeth A. Jonas[b]

[a]Departments of Pharmacology, Cellular and Molecular Physiology, Yale University School of Medicine,
333 Cedar Street, New Haven, CT 06520, USA
[b]Internal Medicine, Yale University School of Medicine, 333 Cedar Street, New Haven, CT 06520, USA
*Correspondence address: Tel.: +1-203-785-4500; fax: +1-203-785-7670
E-mail: Leonard.Kaczmarek@yale.edu

1. Introduction

Ion channels in the plasma membrane are the prime determinants of the electrical properties of excitable cells. In non-excitable cells, plasma membrane channels regulate the flux of ions into cells and across epithelia. Because of their accessibility to patch clamp techniques, the function and regulation of plasma membrane ion channels have been widely investigated, and the molecular identities of most such channels are now known. Nevertheless, at any given time, the activity of channels in the plasma membrane may represent only a small component of the channel activity within a cell. The surface area of the plasma membrane is estimated to be only 2–5% of the total membrane area of a cell [1]. Intracellular membranes include those of rough and smooth endoplasmic reticulum (ER), nucleus, secretory granules and vesicles, mitochondria, subsurface cisternae, lysosomes and Golgi. It is known that membranes of such intracellular organelles also contain ion channels, and that these channels play important roles in cell function.

One role for ion channels on intracellular membranes that is relatively well understood is their management of intracellular calcium. Ryanodine receptors and inositol trisphosphate receptors regulate the release of calcium from internal stores for cellular functions such as muscle contraction and secretion, and, in response to external signals, their activity can initiate oscillations and traveling waves of cytoplasmic calcium. Mitochondrial channels, some of whose molecular identities have not been fully established, also play key roles in the regulation of cytoplasmic calcium levels. There is an important distinction between the regulation of calcium influx across the plasma membrane and that from intracellular stores. As we shall see in this review, channels in the endoplasmic reticulum and mitochondrial membranes can be profoundly influenced by the levels of calcium stored within these organelles. Because the stimulation of cells such as neurons produces changes in the loading of calcium stores, these channels provide a mechanism for the integration of the recent history of stimulation.

Advances in Molecular and Cell Biology, Vol. 32, pages 433–458
ISSN: 1569-2558 / DOI: 10.1016/S1569-2558(03)32018-1

The rapid release of calcium across membranes is not the only function of internal channels. The movement of metabolites such as ATP, and even of large proteins, across internal membranes is associated with the activation of ion channels in these membranes. A description of all the known types of organellar ion channels is beyond the scope of this chapter. We shall, however, review some of the most widely studied channels. A major part of the review will discuss the properties of channels located in the outer and inner membranes of mitochondria, which control not only the handling of calcium, but also determine whether a cell survives or undergoes programmed cell death in response to external signals. For each of the channels, we shall focus on their potential role in neurons, where they may contribute to the modulation of synaptic transmission.

2. Techniques for the investigation of intracellular ion channels

Because of the inaccessibility of internal membranes for electrophysiological techniques such as microelectrode recording or conventional patch clamp techniques, a variety of experimental approaches have been adopted to analyze the biophysical properties of channels on such membranes. These include the reconstitution of channels into lipid bilayers, the patch clamping of isolated organelles, the use of cells that have been skinned of their plasma membrane, and a modified patch clamp approach that allows the clamping of organelles in intact cells.

Membrane fractions enriched in specific organelles can be reconstituted into planar lipid bilayers (Fig. 1a, [2]). This has been the major approach that has been used to characterize the electrophysiological characteristics of most intracellular channels. Because internal membranes represent the majority of the membrane of most cells, intracellular channels are well represented in even crude fractions. Reconstitution into spherical liposomes has also been used, and is particularly appropriate for the measurement of the flux of tracers such as radioisotopes or fluorescent indicators that permeate the reconstituted channel. These approaches allow for a rigorous investigation

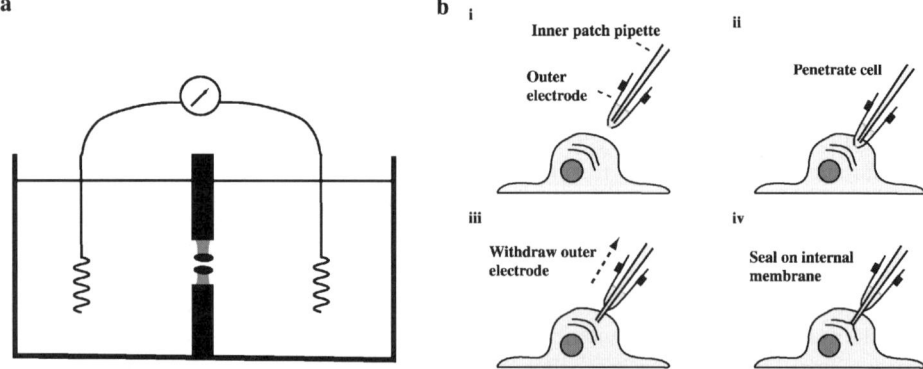

Fig. 1. Two techniques for the investigation of intracellular ion channels. a. Reconstitution into lipid bilayers. b. A series of steps in the formation of giga-ohm seals on internal membranes in intact cells [7].

of the permeation and gating properties of the reconstituted channels. Because the artificial lipids used for reconstitution differ from those of normal cellular membranes, however, these techniques isolate the channel from its normal environment, which includes both its surrounding lipids and many of the regulatory proteins with which the channel may normally interact.

Conventional patch clamp techniques can be applied to isolated organelles. This approach has been particularly useful for nuclei and mitochondria [3–6]. In the latter case, patching of the inner mitochondrial membrane has been made possible by first rupturing the outer mitochondrial membrane using hypo-osmotic solutions. In some cells, it has also been possible to remove the plasma membrane of intact cells, allowing direct access of patch pipettes to the sacroplasmic reticulum.

It is also possible to record directly from intracellular channels in intact living cells. A variant of the patch clamp technique has been developed in which the patch pipette is protected from contact with the plasma membrane during penetration of the cells (Fig. 1b, [7,8]). The patch electrode is contained within an outer, large bore microelectrode. The concentric electrodes can be manipulated past the plasma membrane, after which the outer electrode is withdrawn, exposing the inner tip. Negative pressure causes the inner tip to form a high resistance seal on an intracellular organelle. Single channel data can be gathered either on the organelle, or after excision of the patch into the cytoplasm or bath. This technique has been used successfully in a number of cell types including neurons of *Aplysia*, the squid giant presynaptic terminal, CHO cells, rat microglia, and *Xenopus* oocytes [7,8].

2.1. Ryanodine receptors

Many cells possess mechanisms that allow a small local elevation in cytoplasmic calcium levels, for example through transient opening of plasma membrane calcium channels, to become amplified, resulting in a global elevation of calcium throughout the entire cell or in waves of cytoplasmic calcium that propagate throughout the cell [9–11]. Such amplification is produced by the process of calcium-induced calcium release (CICR), which results from the opening of calcium-permeable channels on endoplasmic reticulum (ER). The initial stimulus for the opening of such channels may be either the local influx of calcium across the plasma membrane or the local release of calcium from an intracellular store. The opening of the channels on the ER membrane causes the release of calcium from within the ER, resulting in further elevation of cytoplasmic calcium. The predominant calcium release channel in many cells is the ryanodine receptor. Ryanodine receptors were first described in skeletal muscle, where their activation in the sarcoplasmic reticulum provides the calcium elevation necessary for contraction of muscle fibers [12–15].

The ryanodine receptor is the largest ion channel protein complex known [16–18]. It comprises a tetramer of subunits each of which consists of about 5000 amino acids. The channels formed by this complex are non-selective cation channels with a high unitary conductance. There are three isoforms of ryanodine receptors (Ryr1, Ryr2, and Ryr3), and while all three have been found in a wide variety of tissues [19,20], they have different tissue and cell-type distributions. For example, Ryr1 is predominant in skeletal muscle, while Ryr2 and Ryr3 are primarily localized to heart and brain. Furthermore, within the nervous system, Ryr2 is found throughout the brain, while Ryr3 is found at high levels in the CA1 region of the hippocampus, and Ryr1 is selectively expressed in cerebellar Purkinje cells [19].

There are two distinct mechanisms for the activation of ryanodine receptors [21,22]. The first of these does not depend directly on the sensitivity of the channel to calcium, but rather to its physical coupling to calcium channels in the plasma membrane. In skeletal muscle, ryanodine receptors are localized to specialized regions of the sacroplasmic reticulum called terminal cisternae. These regions of the sarcoplasmic reticulum are physically linked to infoldings of the plasma membrane termed t-tubules. At these sites, the ryanodine receptor binds two other proteins, triadin and a related protein junctin, which may be important for localization to the terminal cisternae. At these specialized contacts between the intracellular and plasma membranes, there is a direct coupling of the ryanodine receptor to the α-1 subunit of the skeletal muscle voltage-gated calcium channel. Depolarization of the plasma membrane directly triggers the opening of the ryanodine receptor, causing the release of calcium ions from the sarcoplasmic reticulum into the cytosol. The coupling between the two proteins appears to be mediated through the intracellular loop between domains II and III of the calcium channel α-1 subunit. In lipid bilayer experiments, application of peptides from this region of the calcium channel has been found to activate the ryanodine receptors [23,24].

In contrast to the situation in skeletal muscle, in many other tissues the ryanodine receptor acts as a true CICR channel [9,25]. Activation of this receptor in these tissues occurs as a result of an elevation of cytosolic calcium, producing a positive feedback mechanism for calcium elevation. Normal physiological activation probably occurs upon elevation of cytosolic calcium to concentrations of $1-100$ μM.

Ryanodine receptors appear to be organized into discrete clusters on intracellular membranes. Openings of ryanodine receptors in such clusters can occur spontaneously, in the absence of external stimuli [9,25,26]. Such openings give rise to localized elevations of cytoplasmic calcium that are termed calcium "sparks". These events, which were first described in ventricular myocytes, are likely to represent the simultaneous opening of all the closely linked ryanodine receptors in the cluster. A spark may also be triggered by local elevation of cytoplasmic calcium, such as occurs on the opening of a single plasma membrane calcium channel. In response to larger or more global elevations of cytoplasmic calcium, however, the calcium that is released from one cluster may be sufficient to activate neighboring clusters of ryanodine receptors resulting, in a more regenerative release and providing the substrate for propagating waves of calcium release.

Two pharmacological agents, ryanodine and caffeine, have been key tools in the analysis of ryanodine receptors and are commonly used to manipulate their activity. Caffeine is a potent agonist of the receptor in artificial membranes, and treatment of intact cells with this agent can lead to rapid and complete depletion of calcium stores in sarcoplasmic reticulum. Ryanodine binds to the receptors slowly and irreversibly. Ryanodine has a biphasic effect. Low concentrations ($1\ nM-10$ μM) activate the channel by causing it to enter a prolonged subconductance state, while application of higher concentrations block the channel [15,22].

The probability of opening of the ryanodine receptor depends not only on the level of cytoplasmic calcium but also on calcium levels at the lumenal side of the channel [22,27]. In lipid bilayers, increasing calcium on the lumenal side increases opening probability. Similarly, in intact cells, the frequency of calcium sparks depends on the calcium load within the lumen of the sarcoplasmic reticulum. As calcium levels in the lumen increase, the amplitude as well as the frequency of sparks increases and can trigger macroscopic waves of calcium that propagate throughout the cell. Calcium in the lumen of the ER is bound to calsequestrin, a low affinity, high capacity calcium-binding protein that serves to buffer calcium levels in the ER, and there is evidence that calsequestrin may directly regulate the gating of the ryanodine receptor [21]. As we shall see later, calcium release from its other major storage organelle, the mitochondrion, is also sensitive to levels of calcium loading.

The activity of ryanodine receptors can be regulated by a wide variety of signaling pathways, either by phosphorylation or by direct binding to the receptors [21]. For example, the activity of the ryanodine receptor can be modulated by phosphorylation through the Ca-calmodulin-dependent protein kinase II, the cyclic AMP-dependent protein kinase, and by binding to calmodulin and to FK-506 binding proteins (FKBPs). As described above, the activation of physically adjacent ryanodine receptors may occur co-operatively (e.g. during a spark) and there is evidence that the FKBPs represent part of the mechanical coupling between adjacent ryanodine receptor tetramers. Such coupling through FKBPs may act in concert with CICR release to ensure complete activation of ryanodine receptors following cellular stimulation.

2.2. Ryanodine receptors in neurons

The complex three-dimensional structure of neurons is paralleled by an equally rich elaboration of internal membranes [10,28,29]. The ER in these cells extends from the soma into dendrites, where it enters many dendritic spines to form a structure known as the spine apparatus, which is comprised of a series of closely spaced stacks of endoplasmic reticular membrane. It also enters the axon, where it forms an internal pathway of connecting tubular membranes extending toward the synapse. Individual synaptic endings vary considerably in the extent to which ER is represented close to neurotransmitter release sites [8,28,29]. Mitochondria are the predominant calcium storage organelles in many synaptic terminals, with relatively less representation from ER. ER from the axon can, however, make very close functional contacts with these synaptic mitochondria [28].

In the brain, the ryanodine receptors are localized predominantly to the somata, where they are found on subsurface cisternae derived from the smooth ER (Fig. 2, [10]). The subsurface cisternae are flattened stacks of internal membrane abutting the inner face of the plasma membrane, from which they are separated by a gap of no more than 5–8 nm. This unique morphological specialization suggests a special relationship of these internal stores with the plasma membrane. The subsurface cisternae are not only the major location of ryanodine receptors, but may also contain high levels of voltage-dependent channels such as the Kv2.1 potassium channel, whose function is more normally associated with the plasma membrane [30]. The explanation for the presence of plasma membrane channels

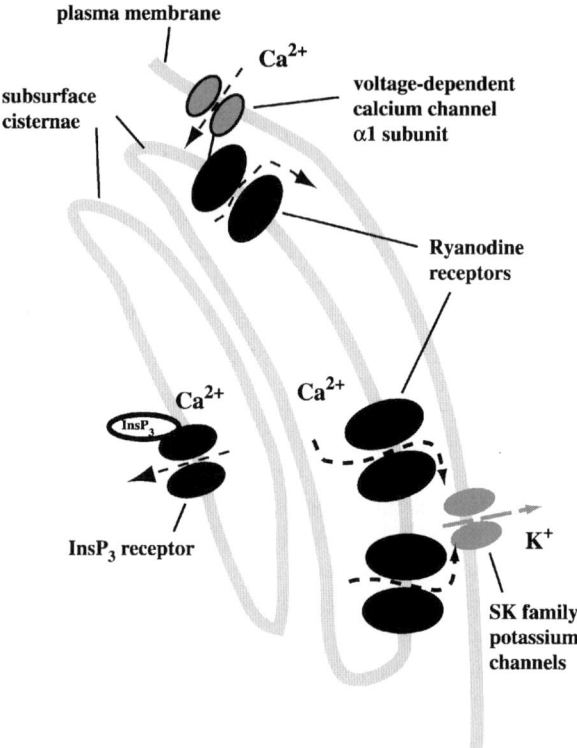

Fig. 2. Putative relationships between intracellular calcium release channels on subsurface cisternae and plasma membrane calcium and potassium channels in the plasma membrane of neurons.

on these organelles is not known. It is possible, however, that it represents a late stage in their constitutive secretion into the plasma membrane, or that such channels "shuttle" between the plasma membrane and internal membranes.

There is evidence in cultured neurons that as in skeletal muscle, there is close functional connection between L-type calcium channels and ryanodine receptors. Accordingly, the activation of calcium channels during action potentials triggers the release of calcium from ER stores [31,32]. The occurrence of action potentials is not the only trigger for the opening of ryanodine receptors. Activation of ionotropic receptors such as NMDA receptors that are permeable to calcium ions can also result in the opening of ryanodine receptors [33].

Ryanodine receptors may be required for the modulation of certain plasma membrane ion channels by neurotransmitters [34]. For example, in cultured cerebellar granule cells, stimulation of the metabotropic glutamate receptor enhances the opening of L-type calcium channels. This effect appears to require activation of the ryanodine receptor and can be abolished by ryanodine. Interestingly, the effects of receptor stimulation appear to persist even after isolation of the glutamate receptor from the calcium channel by excision of the patches, and ryanodine can reverse the activation of the calcium channels even in these excised patches [35]. Such experiments lend additional support to the concept of a close physical and regulatory link between the intracellular ryanodine receptor channels and the plasma membrane calcium channels.

One important action of ryanodine receptors in certain neurons is to enhance the amplitude of neurotransmitter release at presynaptic endings. In axons and nerve terminals of the basket cells that synapse onto Purkinje cells of the cerebellum, spontaneous openings of clusters of ryanodine receptors produce localized transients analogous to the sparks of skeletal muscle [36]. These calcium transients, which are sensitive to ryanodine and to depletion of intracellular calcium stores, can trigger the release of multiple neurotransmitter-containing vesicles, giving rise to spontaneous large amplitude inhibitory postsynaptic potentials that can be recorded in the Purkinje cells. These multi-vesicular miniature events have been termed "maximinis", by analogy with the miniature potentials that represent the release of the contents of only a single vesicle [36,37].

Another major consequence of the release of calcium through ryanodine receptors for neuronal excitability is the activation of calcium-dependent potassium channels in the plasma membrane. In dorsal root ganglion cells and sympathetic ganglion cells, spontaneous miniature outward currents can be recorded and these have been termed SMOCs [38,39]. The frequency of these events can be increased by caffeine, and it is likely that they represent the neuronal equivalent of calcium sparks, i.e. that they that are triggered by the co-operative release of calcium through a single cluster of ryanodine receptors on the subsurface cisternae. More extensive activation of calcium-dependent potassium currents occurs during repetitive stimulation of action potentials, which produces a slow afterhyperpolarization (sAHP) that lasts for many seconds in sympathetic ganglion cells and many other neurons. This sAHP serves to inhibit further neuronal firing. Although the sAHP results from the activation of small-conductance calcium-activated potassium channels, it appears to be calcium released from the ER, rather than calcium entering across the plasma membrane, that directly activates these channels. The sAHP can be inhibited by ryanodine and prolonged by caffeine [40–42]. The delayed onset and slow time course of the sAHP also strongly suggests that the activation of potassium channels is indirect [10].

The activation of plasma membrane calcium-activated potassium channels that may be closely associated with the subsurface cisternae is not limited to stimuli that trigger repetitive firing. Activation of CICR can also be triggered by direct activation of metabotropic glutamate receptors or by the stimulation of synaptic pathways that activate these receptors. For example, in dopaminergic neurons of the ventral tegmental area, activation of metabotropic glutamate receptors stimulates a slow inhibitory potential that is blocked by ryanodine [43].

2.3. Inositol trisphosphate receptor channels

The second major channel that is responsible for calcium release from ER is the inositol trisphosphate (InsP$_3$) receptor channel [44–47]. Like the ryanodine receptor, it is a tetramer, but its subunits are composed of about 2700 amino acids, approximately half the size of those for the ryanodine receptors. It is possible to speculate that its smaller size reflects the lack of sites for interaction with other large protein complexes such as plasma membrane calcium channels. Its function in calcium release appears to be very similar to that of ryanodine receptor except that it functions as a channel only when bound to the cytoplasmic second messenger InsP$_3$, which is produced by neurotransmitters and neuropeptides whose receptors stimulate the hydrolysis of phosphatidylinositol 4,5-bisphosphate (PtdIns(4,5)P$_2$) [48–50].

There are three known isoforms of the InsP$_3$ receptor, which differ in subtle ways in their sensitivity to InsP$_3$ and to cytoplasmic calcium [51]. Both the type I and type II InsP$_3$ receptors are found at high levels in the brain, while the type III receptor has a widespread distribution. Like the ryanodine receptors, the InsP$_3$ receptors have binding sites for multiple regulatory proteins including calmodulin and FKBPs. In addition, the InsP$_3$ receptors have binding sites for ATP, and, at low concentrations in the presence of InsP$_3$, this nucleotide stimulates the activity of the InsP$_3$ receptor channel.

When it is bound to InsP$_3$, the InsP$_3$ receptor acts as a CICR channel. The activation of type I InsP$_3$ receptors by cytoplasmic calcium displays a bell-shaped dose-response curve, while in lipid bilayers, the type II and type III receptors are activated monotonically by increasing calcium [51]. The normal resting calcium level within cells is on the rising phase of this curve, such that, in the presence of InsP$_3$, small local elevations of cytoplasmic calcium promote channel opening, resulting in further release of calcium from internal stores. Many of the spatial factors that determine the response of ryanodine receptors also pertain to the InsP$_3$ receptors. For example, groups of InsP$_3$ receptors may form clusters that can be activated synchronously to generate "puffs" of calcium release that are analogous to the calcium sparks produced by ryanodine receptors. More intense stimulation may produce co-operative activation of multiple local clusters, resulting in a regenerative release or propagating waves of calcium release [52,53].

InsP$_3$ receptors are very widely distributed in the brain. In many locations, they appear to be co-localized with ryanodine receptors [54,55], although as described above, ryanodine receptors may be preferentially targeted to subsurface cisternae at the soma, whereas in many cells, there is a more widespread localization of InsP$_3$ receptors to ER membranes within dendrites and axons. Moreover InsP$_3$ receptors are not restricted to the ER. It has been possible to record the activity of InsP$_3$ receptors on isolated nuclear membranes [3,6], where their properties generally match those of InsP$_3$ receptors reconstituted into lipid bilayers. Their presence has also been reported on secretory granules of endocrine cells, where they may play a role in localization of intracellular calcium signals responsible for secretion [56].

As described above, the ER of neurons is thought to represent a continuous network of intracellular membrane that can extend from the soma into dendrites and into axons. In dendrites, it can extend into the spine apparatus, where it may be closely associated with synaptic inputs. The stimulation of synaptic inputs that trigger InsP$_3$ formation can produce very localized calcium signals that remain restricted to a single spine or to multiple spines on adjacent dendritic shafts [57]. The continuous nature of ER suggests, however, that regenerative release of calcium, triggered at one location by activation of InsP$_3$ receptors and ryanodine receptors, may, under some circumstances, propagate regeneratively throughout the cell. Indeed the extensive nature of this organelle has led to the concept that it comprises a "neuron within a neuron", i.e. a signaling system distinct from the plasma membrane that can send regenerative messages between soma, dendrites and synapses [10].

There are two aspects of signals generated by the InsP$_3$ and ryanodine receptors that have particular significance for neuronal activity. First, the InsP$_3$ receptor is maximally activated by a cytoplasmic elevation of both calcium and InsP$_3$ levels. It has, therefore, the potential to act as a detector for the simultaneous activation of two independent synaptic inputs, one of which stimulates receptors coupled to PtdIns(4,5)P$_2$ hydrolysis and the other that triggers neuronal firing resulting in calcium influx [10,58]. For example, while the activation of either pathway may be insufficient to trigger a global release of calcium, the simultaneous activation of both inputs could trigger regenerative calcium release that propagates throughout the cell. A role for intracellular calcium in coincidence detection has been proposed for the induction of long-term depression by simultaneous activation of parallel fibers and climbing fibers onto Purkinje cells of the cerebellum [59]. Secondly, as described earlier, the activity of the ryanodine and InsP$_3$ receptors is highly sensitive to the levels of calcium within the lumen of the ER [60–63]. The loading of these calcium stores is very sensitive to the recent firing pattern of a neuron. For example, calcium entry during repetitive trains of action potentials may result in the loading of these stores [64]. This, in turn, may enhance the sensitivity of the CICR system, so that weak stimuli that were previously ineffective may now generate waves of regenerative calcium release.

2.4. Mitochondrial channels

Ion channels in mitochondria and in the ER work together to regulate intracellular calcium levels. Indeed, in many cells, mitochondria make numerous close contacts with the ER. At these contact sites, the opening of InsP$_3$ receptors provides a direct pathway for the transfer of calcium from the ER to the mitochondria, without significant diffusion into bulk cytoplasm [65]. Moreover, the uptake of calcium into mitochondria sets the threshold for regenerative CICR by ER [66]. In secretory cells such as neurons and endocrine cells, mitochondria are particularly important for uptake of calcium following the stimulation of neurotransmitter or neuropeptide release [67–70]. The interaction of mitochondria with ER stores is also essential for generating slow waves of calcium release that produce the rhythmic activity of pacemaker cells in the gastrointestinal tract [71].

In addition to their role in calcium dynamics, the opening and closing of channels in mitochondrial membranes is closely linked both to everyday housekeeping functions such as the synthesis and export of ATP, and to biochemical pathways that determine the survival or death of a cell in response to external factors.

2.5. Channels in the inner mitochondrial membrane

Mitochondria possess a double layer of membranes (Fig. 3). The outer membrane was, for many years, considered to represent a rather leaky sieve. As we shall see below, this view is currently changing, and evidence exists that ion channels in the outer membrane are tightly regulated [4,8,72]. In contrast, it has long been known that the permeability

Inner membrane channels

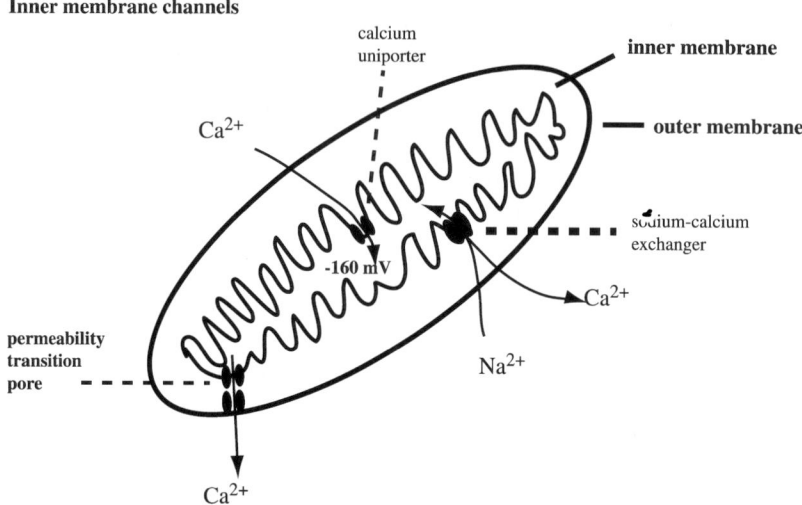

Fig. 3. Channels and transporters known to regulate uptake and release of calcium across the inner mitochondrial membrane.

of the inner membrane is central to the basic functions of a mitochondrion. The inner mitochondrial membrane separates the mitochondrial matrix from the intermembrane space, and it is the proton gradient across this membrane that drives respiration [73]. The membrane potential across the inner membrane is approximately -160 mV with respect to the cytoplasm, and this provides the driving force for entry of calcium into the mitochondrial matrix from the intermembrane space during sustained elevations of cytoplasmic calcium [74–76]. At rest, the matrix level of free calcium is similar to that of the cytosol [77,78], but during a cytosolic calcium load such as occurs during depolarization in neurons, total mitochondrial calcium can reach levels up to 80 times higher than prestimulation levels. The time course of changes in mitochondrial calcium following depolarization of neurons has been investigated at the level of electron microscopy [79]. Using X-ray microanalysis, total calcium in the mitochondria is seen to rise during neuronal activity, and revert to a normal low value within 5 min. Calcium entry across the inner mitochondrial membrane occurs through a channel that has been termed the calcium uniporter, as well as through other uptake mechanisms [80]. Relatively selective blockade of this calcium channel can be produced by the dye Ruthenium Red. The molecular identity of the uniporter has, however, not yet been established.

It is believed that in order to maintain the proton gradient necessary for respiration, the ionic permeability of inner membrane must normally be low. Perhaps for this reason the calcium uniporter has not been identified in patch clamp recordings, and evidence for its existence and physiological role has largely accrued through flux studies and the use of indicator dyes [68,81–84]. Nevertheless, both flux experiments and direct patch clamp recordings (which typically use isolated mitoplasts stripped of their outer membranes) have revealed a number of other channel types in the inner membrane [4,5,86]. Although the function of most of these channels has yet to be discovered, they include a potassium channel that is inhibited by both ATP and ADP, and whose pharmacology closely matches that of plasma membrane K_{ATP} channels [87,88]. It has been suggested that this channel plays a role in mitochondrial division [86].

Once calcium has entered the matrix of the mitochondrion, it triggers an increase in the activity of several enzymes involved in respiration [89]. In addition, as we shall see below, the subsequent release of calcium back into the cytoplasm also plays an important role in cellular events, including the potentiation of synaptic transmission in nerve terminals [8,69,90]. One prominent physiological route for the transport of calcium out of the matrix is through the activity of a sodium/calcium exchanger. This appears to be the major pathway for mitochondrial calcium release in certain neurons, where release is blocked by a specific inhibitor of the exchanger or by intracellular solutions lacking sodium ions [91,92]. An alternative pathway for the release of calcium is, however, the activation of the mitochondrial permeability transition pore (mPTP), a very large conductance channel on the inner mitochondrial membrane.

2.6. Mitochondrial permeability transition pore

Under conditions of high calcium influx into mitochondria, the accumulation of calcium in the matrix can trigger a rapid and large increase in permeability of the inner membrane [93–95]. The inner membrane then becomes permeable to solutes of molecular weight up to about 1.5 kDa. Such an increase in permeability completely dissipates the negative potential gradient across the membrane, so that calcium ions leave the matrix. Persistent activation of this permeability transition may lead to apoptosis and to oxidative cell death. Evidence suggests, however, that the permeability transition can also occur in a transient low-conductance mode [94]. Such transient activation produces a reversible depolarization of the inner membrane and the release of mitochondrial calcium. In intact cells, the mitochondrial calcium release that is produced by this form of transient permeability transition has been shown to amplify signals generated by InsP_3-induced calcium release from the ER [84].

Patch clamp studies of the inner mitochondrial membrane have revealed a large conductance channel (1.3 nS) with multiple lower conductances from 30 pS on up [85,86, 93]. Pharmacological and biophysical data clearly indicate that this channel, sometimes termed the mitochondrial megachannel, is the channel that underlies the permeability transition. The activity of the channel depends on an elevation of calcium levels in the matrix and its behavior is consistent with the rapid calcium release from mitochondria detected in physiologic studies [81,95]. Both the activity of the channel and the permeability transition have a rich pharmacology. Channel activity is sensitive to voltage [96] and can be inhibited by agents such as propranolol, amiodarone and quinine, as well as certain anesthetics and benzodiazepines [4]. One agent that appears to be a particularly effective inhibitor of the permeability transition is cyclosporin A [97].

Although the molecular identity of the permeability transition channel is not certain, evidence suggests that the adenine nucleotide transporter (ANT) of the inner membrane may be a component of the channel [98]. The permeability transition is inhibited by bongkrekic acid and activated by actractylate, agents that bind to this transporter.

Moreover, when the ANT is reconstituted into liposomes, it generates large-conductance channel activity that resembles activity recorded in the inner membrane [99]. Nevertheless, the pharmacology of these channels does not fully match that of the mitochondrial permeability transition pores, suggesting that, at the very least, additional mitochondrial proteins contribute to the regulated channel. A current model, based on a variety of experimental approaches, is that the permeability transition pore is a complex consisting of ANT and the outer mitochondrial membrane channel VDAC (see below), which together with several other proteins including creatine kinase and hexokinase, provide a regulated pore that links the matrix to the cytoplasm.

2.7. Channels in the outer mitochondrial membrane

As mentioned above, the outer mitochondrial membrane was for many years considered to be a leaky "sieve", which allows the free passage of ions and metabolites, but which excludes proteins, providing for the retention of enzymes such as cytochrome *c* in the intermembrane space (Fig. 4). This view was based largely on the properties of mitochondria after biochemical isolation. Studies with intact cells, however, now indicate that the permeability of the outer membrane is also tightly regulated [72].

2.8. Voltage-dependent anion channel (VDAC)

The most prominent channel seen in the outer mitochondrial membrane is the VDAC [100]. A single VDAC channel is probably formed from a single VDAC protein of 30 kDa and is most likely present in mitochondrial membrane as a monomer. The biophysical characteristics of VDAC have been determined by reconstituting the channel into planar phospolipid membranes. It has a very large unitary conductance, and a distinctive voltage

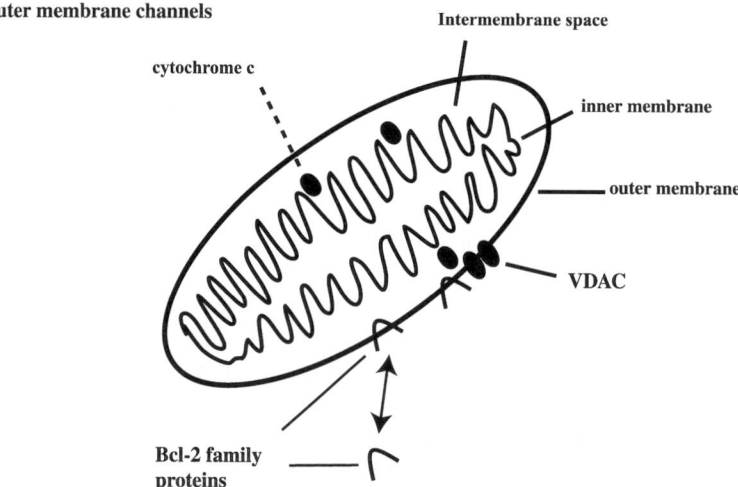

Fig. 4. Some of the channels of the outer mitochondrial membrane.

dependence. The channel properties of the protein are well conserved in eukaryotes, and include an open channel conductance of ~4 nS in 1 M KCl, and an open channel selectivity of 2:1 for chloride over potassium ions. The size of the open pore has been estimated to be between 2.4 and 3.0 nm. The VDAC channel characteristically opens at potentials between −40 and +40 mV but tends to "close" at all other voltages. In contrast to most other ion channels, however, the "closed" state does in fact conduct ions, and comprises multiple lower-conductance states with conductances of several hundred picosiemens. The 'closed' state of VDAC is relatively cation selective.

The physiological role of the VDAC channel is likely to be the control of metabolite flux across the outer membrane. For example, ATP, succinate, and creatine phosphate readily pass through the VDAC channel in its open configuration, but their passage is greatly reduced when VDAC enters its closed state [72,101,102]. In contrast, the movement of sodium ions through VDAC incorporated into lipid bilayers is little affected by the state of the channel.

It is likely that the gating of the native VDAC channel in the outer membrane is tightly regulated by metabolites and other mitochondrial proteins, and that this regulation is lost on reconstitution into lipid bilayers. For example, an endogenous modulator of VDAC, which most likely resides in the intermembrane space, has been described [100]. The nature of this modulator has not yet been characterized. When this modulator is added at very low concentrations to VDAC reconstituted in planar lipid membranes, however, it binds tightly and maintains the channel in a closed state, allowing it to open to a small conductance only at high voltages. Another characteristic behavior of VDAC is its sensitivity on the lumenal side to mitochondrial creatine kinase, an enzyme located in the intermembrane space and concentrated at contact sites between the inner and outer membranes. In lipid bilayers, the conductance of VDAC was reduced by half when exposed to creatine kinase [103].

The nucleotide NADH also has significant effects on VDAC channel function. It doubles the voltage dependence of the channel with an estimated Kd in the low micromolar range [104]. Consistent with its effect on VDAC, exposure of isolated mitochondria to NADH can result in a 6-fold reduction in the permeability of the outer membrane to ADP. Although there as yet exist no specific inhibitors of VDAC activity, exposure of the channel to polyanions such as dextran sulfate and Konig's polyanion, like NADH, favor closure of the channel by increasing the steepness of its voltage dependence [105,106].

2.9. Bcl-2 family channels

In addition to regulating cellular metabolism and calcium signaling, channels in mitochondrial membranes appear to play a key role in determining whether a cell will live or die in response to changes in its environment. Apoptosis is a programmed form of cell death that is regulated genetically and produces the normal elimination of cells that occurs in many tissues, both during development and in mature organisms [107,108]. In the major pathway for apoptosis, certain proteins that are normally resident in the intermembrane space are released from this space into the cytoplasm. These proteins include cytochrome c (Fig. 4) and apoptosis-inducing factor. Once in the cytoplasm these proteins bind to a protease termed caspase-9, activating its proteolytic activity. This, in turn, activates further proteolytic enzymes, eventually leading to the cleavage of cellular DNA [109].

The pathway by which cytochrome c passes through the outer mitochondrial membrane is not understood. VDAC has been proposed as a candidate for a conduit across the membrane, but the largest estimated pore size of VDAC in its open state may be too small to allow cytochrome c to pass [100,110]. Passage of cytochrome c is known to be regulated by the Bcl-2 family of proteins, many of which are believed to function as ion channels in the outer mitochondrial membrane [111].

Bcl-2 family proteins come in two flavors. The prototypical member, Bcl-2 itself, is traditionally thought to be anti-apoptotic. Thus, its presence in cells protects them against apoptotic signals such as growth factor deprivation, ultraviolet and gamma-radiation, heat shock, tumor necrosis factor, calcium ionophores, viral infection, and agents that promote free radicals [109]. Another member of the anti-apoptotic group is Bcl-x_L. In contrast, other members of this family, including Bcl-x_S, Bax, Bid and Bak, promote or trigger apoptosis when they are expressed in cells [108]. Many of these proteins share four homologous domains denoted BH1, 2, 3, and 4 (Fig. 5). The BH3 domain is present in all family members, and is thought to be a key domain responsible for cell death [112–115]. BH4 is found at the N-terminus of anti-apoptotic proteins [112,116]. Indeed the anti-apoptotic Bcl-x_L protein can be converted into a pro-apoptotic form through a caspase-induced cleavage of the BH4 domain, which exposes the BH3 domain [117] (Fig. 5).

It is known that Bcl-2, Bcl-x_L, and Bax are localized to the outer mitochondrial membrane via a C terminal ~ 20 amino acid residue hydrophobic tail [118,119]. Bcl-2 and Bcl-x_L can also be found at other sites, including the nuclear envelope and the ER. Immuno-electron microscopy has suggested that Bcl-2 may be preferentially localized to zones of adhesion joining the outer and inner mitochondrial membranes. Association of Bcl-2 with these membranes implies that Bcl-2 may regulate protein and/or ion import.

The structure of Bcl-x_L has been determined by X-ray crystallography and NMR spectroscopy [112,120]. The arrangement of alpha-helices in Bcl-x_L closely resembles the membrane translocation domain of bacterial toxins, in particular diphtheria toxin and the colicins, which form channels in cellular membranes. Based on this homology to the

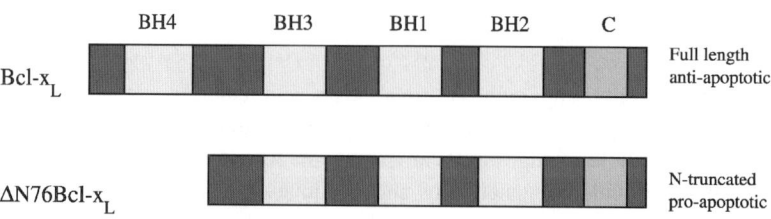

Fig. 5. BH domains 1–4 and the C-terminal anchoring domain (C) in the anti-apoptotic protein Bcl-x_L and its pro-apoptotic product ΔN76 Bcl-x_L [117].

colicins, the pore-forming region of Bcl-2 proteins is believed to lie between domains BH1 and BH2, in a region with two alpha helices, which are predicted to form a transmembrane channel [120].

Bcl-x$_L$, Bcl-2, Bax, and cleaved BID all form ion channels when reconstituted into planar lipid bilayers [115,121,122]. The ion channels formed by these proteins tend to exhibit multiple conductance states and the probability of channel opening or formation is modulated by changes in membrane potential or pH. The channels are relatively non-selective, although Bcl-2 and Bcl-x$_L$ channels have a preference for monovalent cations. Mutations that exchange the channel-forming region of Bcl-2 and Bcl-x$_L$ with that of Bax exhibit a reduced ability to protect cells from death; this seems to be correlated with a reduced ability to form channels at negative transmembrane potentials [122]. Bcl-x$_L$ and its pro-apoptotic truncated form have also been shown to induce channel activity when applied to intact mitochondria in squid presynaptic terminals using the intracellular patch technique (Fig. 1). Interestingly, although both proteins share the same pore-forming domain (Fig. 5), the pro-apoptotic form induces unitary currents with conductances of up to several nanosiemens, while the anti-apoptotic form produces much smaller channel activity of only a few hundred picosiemens conductance [123,124].

The role of the channel activity of the pro-apoptotic Bcl-2 proteins in the release of cytochrome *c* from the intermembrane space in not clear. It is possible that the pores formed by these proteins could themselves allow the passage of cytochrome *c* across the outer membrane. It has also been suggested that pro-apoptotic molecules such as Bax, cleaved Bcl-x$_L$ or BID, by activating or linking to VDAC, form a multi-protein complex that can conduct cytochrome *c*. The general idea that Bcl-2 proteins interact with some other endogenous channel has been supported by several findings. The application of peptides containing the BH3 domain alone has been reported to be sufficient to cause release of cytochrome *c* from isolated mitochondria [125]. Because of its sequence and structure, the BH3 domain alone is unlikely to form an ion channel. It, therefore, either destabilizes the membrane, causing it to form a "lipidic pore" that releases cytochrome *c*, or it increases the conductance of an endogenous outer membrane channel such as VDAC or one of the Bcl-2 family proteins already resident in the outer membrane. Furthermore, Bcl-2 family proteins have been found to co-immunoprecipitate with VDAC [126]. When VDAC was incorporated into large liposomes, Bax and Bak were reported to increase the permeability of the liposomes by opening VDAC, while Bcl-x$_L$ was found to decrease VDAC activity [126,127].

The role of the channel activity of the anti-apoptotic Bcl-2 proteins is also not understood. The widespread expression of Bcl-x$_L$ in adult brain suggests that Bcl-x$_L$ has roles beyond protecting neurons from developmental cell death [128], and suggests that control of the permeability of the outer membrane by Bcl-2 family proteins may occur during normal activity. For example, preliminary experiments in which Bcl-x$_L$ has been injected into presynaptic terminals, which are devoid of nuclei and where traditional apoptosis cannot occur, have indicated that this protein can alter the strength of synaptic transmission [129]. Moreover, recent findings suggest that the permeability of the outer membrane to metabolites such as ATP and creatine phosphate is enhanced by the anti-apoptotic proteins Bcl-x$_L$ and Bcl-2 [72].

2.10. Role of mitochondrial channels in the regulation of neurotransmitter release

In neurons, as in other cells types, mitochondria are found throughout the cell. A particularly high density of mitochondria, however, is found in presynaptic terminals, particularly those of neurons that are capable of firing at high rates. Many such synaptic endings are also relatively devoid of other organelles such as ER, which play significant roles in calcium signaling in other parts of the cell such as its axon, dendrites and soma. At some synapses, mitochondria are tethered by filamentous structures to the presynaptic membrane adjacent to active zones, where vesicle fusion occurs [130,131]. Vesicular structures can also be found attached to these filaments. These morphological specializations, termed mitochondrial adherens complexes (MACs) suggest that the activity of mitochondria is closely coupled to the release of neurotransmitters.

A period of repetitive stimulation of a presynaptic terminal generally produces a change in the amount of neurotransmitter release that can be evoked by a subsequent single action potential. Several phases of such short-term synaptic plasticity have been described. One of these, which produces a marked enhancement of release for many tens to hundreds of seconds following a brief tetanus, has been termed post-tetanic potentiation [132–134]. The induction of post-tetanic potentiation requires that cytoplasmic calcium levels within synaptic terminals remain elevated following the tetanus. Such prolonged elevation of calcium following the tetanus has been termed "residual calcium", and appears to result from the slow re-release of calcium that accumulates in synaptic mitochondria during the tetanus. The residual tail of calcium can be eliminated by mitochondrial uncoupling agents, and, in bullfrog sympathetic neurons and crayfish neuromuscular junctions, these agents also eliminate post-tetanic potentiation (Fig. 6, [68,69]).

The intracellular patch clamp technique has been used to measure the activity of ion channels in mitochondrial membranes within the squid giant presynaptic terminal following trains of action potentials similar to those that produce post-tetanic potentiation [8]. Electron microscopic studies of this synapse have indicated that mitochondria are the predominant organelles likely to be contacted by the internal patch pipette. Stimulation of a presynaptic tetanus produces two prominent forms of synaptic plasticity. One is a profound depression seen at high frequencies of firing, which results in failure of the postsynaptic action potential. This synaptic depression can be relieved by partially reducing calcium concentration in the external medium, revealing an underlying post-tetanic potentiation [135,136].

Prior to stimulation of the squid synapse, spontaneous small conductance channel activity is usually detected in mitochondrial patch recordings, although, in a very small number of experiments, openings comparable to those expected for VDAC or permeability transition pores can also be detected [8]. A brief tetanus, however, produces an approximately 60-fold increase in conductance of the mitochondrial membrane. At the end of the tetanus, the enhanced activity continues for tens of seconds, and, in many cases, increases in amplitude before returning to control (Fig. 7). This evoked increase in mitochondrial conductance requires calcium influx into the terminal, and is not detected in calcium-free media. The time course of the change in conductance generally matches that of both residual calcium [137] and of post-tetanic potentiation at this synapse [136].

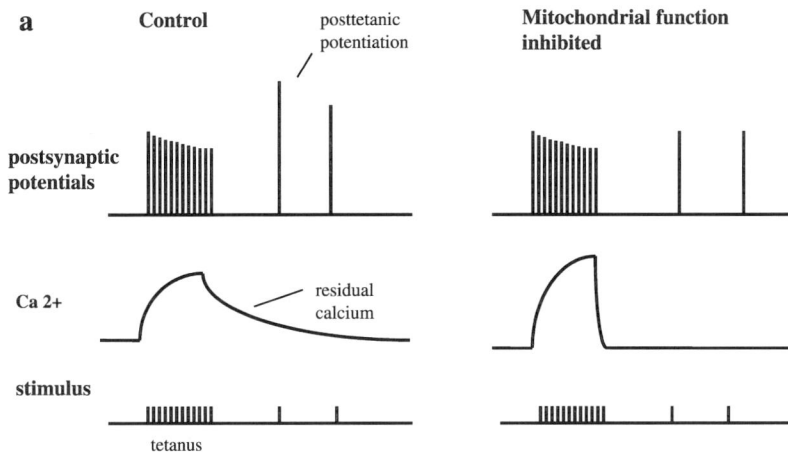

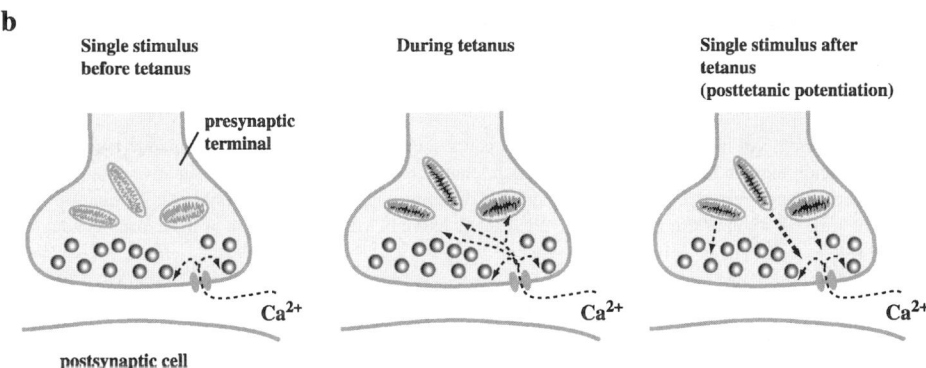

Fig. 6. Role proposed for mitochondrial calcium uptake in post-tetanic potentiation of synaptic responses. a. Time course of presynaptic calcium transients and size of postsynaptic potentials during and after a tetanus. The scheme depicts responses before and after inhibition of mitochondrial function using uncoupling agents or inhibitors of the calcium uniporter. b. Diagram depicting loading and subsequent unloading of calcium in mitochondria during the potentiation of synaptic responses.

Moreover, a mitochondrial uncoupling agent, the proton ionophore *p*-trifluoromethoxyphenylhydrazone (FCCP), abolished both post-tetanic potentiation and the evoked conductance change on mitochondrial membranes.

Because, in the above experiments, no special steps were taken to break the outer mitochondrial membrane, it is probable that the synaptically evoked activity represents a conductance change in the outer membrane, perhaps related to the activity of VDAC or Bcl-2 family channels. If this is the case, it may represent the activation of a pathway for the movement of metabolites such as ATP and calcium ions, to reach the cytoplasm from the intermembrane space. Because of the sensitivity of the conductance change to calcium entry and to uncoupling agents, however, it is not possible to exclude a contribution of inner membrane channels such as the permeability transition pore. Moreover, points of

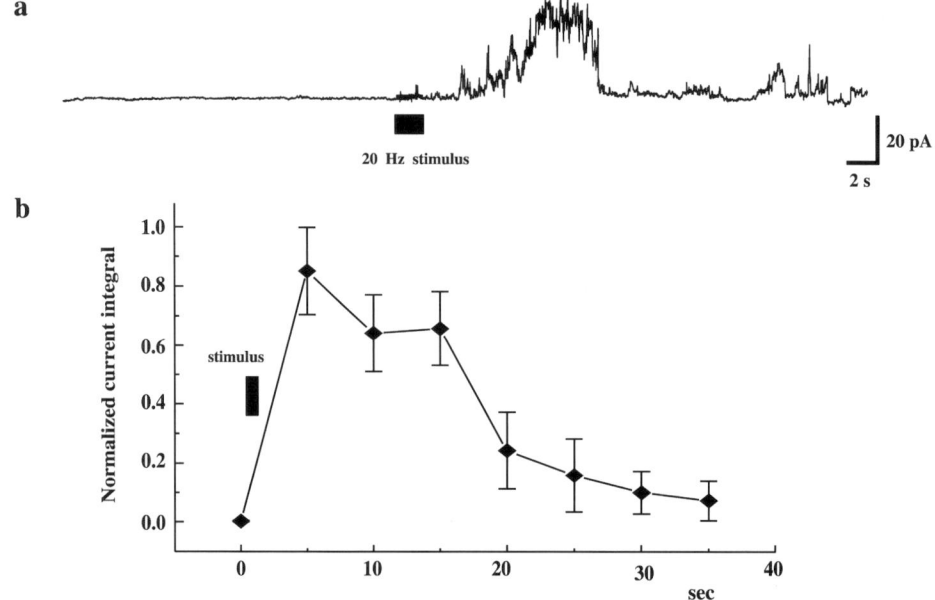

Fig. 7. An increase in the conductance of mitochondrial membranes evoked by a presynaptic tetanus can be recorded using the intracellular membrane patch clamp technique [8]. a. Representative example of a conductance change recorded on a mitochondrial membrane within the giant presynaptic terminal of squid following a 20 Hz stimulus train given to the presynaptic axon. b. Mean time course of the evoked conductance change [8].

contact exist between the outer and inner mitochondrial membranes, where coupling of inner and outer channels could occur [4].

One attractive hypothesis for the activation of outer mitochondrial membrane channels during neuronal activity is that it represents the activity of Bcl-2 family proteins that eventually determine the survival or death of synaptic connections. For example, although the outer mitochondrial membrane channel Bcl-x_L inhibits neuronal death induced by withdrawal of growth factors [138,139], the increased expression of Bcl-x_L subtypes in the mature nervous system suggests that it plays a role beyond the early developmental stages characterized by rapid neuronal growth and loss [128]. Thus it is possible that activation of Bcl-x_L channels during normal synaptic activity may influence synaptic events, and the long-term stability or elimination of a synapse, without influencing the survival of the cell itself.

2.11. Channels on secretory granules

In addition to the channels we have considered in this review, there exist other intracellular ion channels. For example, secretory granules contain ion channels. One of these is a large-conductance, calcium-dependent cation channel whose function is unknown, but may be important for releasing calcium for calcium-dependent fusion, or may contribute to the pore through which neurotransmitter gets released into the extracellular space [140–142]. As pointed out earlier, InsP₃ receptors have also been reported on secretory granules [56].

Evidence that granules in secretory terminals release calcium during the process of neurosecretion is further supported by work in peptidergic neurons. In certain *Aplysia* neurons, insulin triggers neuropeptide secretion by causing the release of calcium from an intracellular store, without causing a depolarization of the plasma membrane or producing calcium influx. Pharmacological and imaging experiments indicate that this store is unaffected by agents that act on ER or mitochondrial stores, and that it probably represents calcium loaded into granules [143,144]. The identity of the ion channel or transporter that releases calcium from the granules during secretion is not yet known.

The finding that different types of stores (ER, mitochondria and granules) can all be used in different cell types for similar processes, for example the potentiation of neurotransmitter or neuropeptide release, suggests that there exists evolutionary convergence of the function of these intracellular organelles and their ion channels. The mechanism by which a particular cell selects different calcium stores and release channels for such a purpose is, however, not yet understood.

2.12. Other channels on intracellular membranes

It should be pointed out that there exist many proteins that can masquerade as ion channels although their true function is not related to the passage of ions across membranes. Examples of such proteins are those that ferry other proteins across lipid membranes. For example, mitochondria import proteins from the surrounding cytosol [145,146]. Proteins are selected from a vast array of cytosolic proteins by recognition of a signal sequence. They are then transported across the outer mitochondrial membrane in an ATP-dependent fashion through a protein import pore known in yeast as Tom 40 [147]. Tom 40 is an integral membrane protein that exists as part of a large complex containing multiple subunits. In lipid bilayers Tom 40 forms a voltage dependent, cation-selective channel that has a conductance of about 360 pS. Thus, Tom 40 and many other proteins that access both sides of a biological membrane may be able to provide an ion conduction pathway in the absence of their transport substrates and their physiological control mechanisms. It is not yet known, however, whether ion flux through these proteins normally occurs in vivo.

2.13. Summary

One of the major roles for intracellular ion channels is the regulation of intracellular calcium levels. The ryanodine receptors and InsP₃ receptors on ER membranes, together with calcium channels in mitochondrial membranes, allow cells to store the calcium that enters through the plasma membrane during stimulation, and to release it at later times. The activity of these channels provides mechanisms for propagating waves of calcium throughout cells and for ensuring regenerative all-or-none responses to external stimuli. Because uptake of calcium into the ER alters the sensitivity of the calcium release channels, the degree of loading of calcium stores may provide a short-term memory trace for the recent pattern of stimulation of a cell. A similar integration of the pattern of recent stimulation is reflected in the degree of calcium uptake into mitochondria. The opening of ion channels in mitochondrial membranes also has the potential to change the fate of a cell and to determine its subsequent long-term stability in the face of further stimulation.

References

[1] Alberts, B., Bray, D., Lewis, J., Raff, M., Roberts, K., Watson, J.D., 1994. Molecular Biology of the Cell. Garland Publishing, New York.
[2] Miller, C., 1982. Reconstitution of ion channels in planar bilayer membranes: a five year progress report. Commun. Mol. Cell Biophys. 1, 413–428.
[3] Mak, D.O.D., Foskett, J.K., 1998. Effects of divalent cations on single-channel conduction properties of Xenopus IP3 receptor. Am. J. Physiol. 275, C179–C188.
[4] Kinnally, K.W., Tedeschi, H., 1994. Mitochondrial channels: an integrated view. In: Forte, M., Colombini, M. (Eds.), Molecular Biology of Mitochondrial Transport System. Spreinger-Verlag, Berlin.
[5] Sorgato, M.C., Keller, B.U., Stuhmer, W., 1987. Patch-clamping of the inner mitochondrial membrane reveals a voltage-dependent ion channel. Nature 330, 498–500.
[6] Stehno-Bittel, L., Luckhoff, A., Clapham, D.E., 1995. Calcium release from the nucleus by InsP3 receptor channels. Neuron 14, 163–167.
[7] Jonas, E.A., Knox, R.J., Kaczmarek, L.K., 1997. Giga-ohm seals on intracellular membranes: a technique for studying intracellular ion channels in intact cells. Neuron 19, 7–13.
[8] Jonas, E.A., Buchanan, J., Kaczmarek, L.K., 1999. Prolonged activation of mitochondrial conductances during synaptic transmission. Science 286, 1347–1350.
[9] Berridge, M.J., 1997. Elementary and global aspects of calcium signaling. J. Physiol. 499, 291–306.
[10] Berridge, M.J., 1998. Neuronal calcium signaling. Neuron 21, 13–26.
[11] Verkhratsky, A., Shmigol, A., 1996. Calcium-induced calcium release in neurones. Cell Calcium 19, 1–14.
[12] McPherson, P.S., Campbell, K.P., 1993. Characterization of the major brain form of the ryanodine receptor/Ca^{2+} release channel. J. Biol. Chem. 268, 19785–19790.
[13] Fill, M., Coronado, R., 1988. Ryanodine receptor channel of sarcoplasmic reticulum. Trends Neurosci. 10, 453–457.
[14] Sitsapesan, R., McGarry, S.J., Williams, A.J., 1995. Cyclic ADP-ribose, the ryanodine receptor and Ca^{2+} release. Trends Pharmacol. Sci. 16, 386–391.
[15] Striggow, F., Ehrlich, B.E., 1996. Ligand-gated calcium channels inside and out. Curr. Opin. Cell Biol. 8, 490–495.
[16] Otsu, K., Willard, H.F., Khanna, V.K., Zorzato, F., Green, N.M., MacLennan, D.H., 1990. Molecular cloning of cDNA encoding the Ca^{2+} release channel (ryanodine receptor) of rabbit cardiac muscle sarcoplasmic reticulum. J. Biol. Chem. 265, 13472–13483.
[17] Nakai, J., Imagawa, T., Hakamat, Y., Shigekawa, M., Takeshima, H., Numa, S., 1990. Primary structure and functional expression from cDNA of the cardiac ryanodine receptor/calcium release channel. FEBS Lett. 271, 169–177.
[18] Wagenknecht, T., Grassucci, R., Frank, J., Saito, A., Inui, M., Fleischer, S., 1989. Three-dimensional architecture of the calcium channel/foot structure of sarcoplasmic reticulum. Nature 338, 167–170.
[19] Giannini, G., Conti, A., Mammarella, S., Scrobogna, M., Sorrentino, V., 1995. The ryanodine receptor/calcium channel genes are widely and differentially expressed in murine brain and peripheral tissues. J. Cell Biol. 128, 893–904.
[20] Bennett, D.L., Cheek, T.R., Berridge, M.J., De Smedt, H., Parys, J.B., Missiaen, L., Bootman, M.D., 1996. Expression and function of ryanodine receptors in nonexcitable cells. J. Biol.Chem. 271, 6356–6362.

[21] Mackrill, J.J., 1999. Protein–protein interactions in intracellular Ca^{2+}-release channel function. Biochem. J. 337, 345–361.

[22] Bers, M.B., Perez-Reyes, E., 1999. Ca channels in cardiac myocytes: structure and function in Ca influx and intracellular Ca release. Cardiovasc. Res. 42, 339–360.

[23] Lu, X., Xu, L., Meissner, G., 1994. Activation of the skeletal muscle calcium release channel by a cytoplasmic loop of the dihydropyridine receptor. J. Biol. Chem. 269, 6511–6516.

[24] Lu, X., Xu, L., Meissner, G., 1995. Phosphorylation of dihydropyridine receptor II–III loop peptide regulates skeletal muscle calcium release channel function. Evidence for an essential role of the beta-OH group of Ser687. J. Biol. Chem. 270, 18459–18464.

[25] Bootman, M.D., Berridge, M.J., 1995. The elementary principles of calcium signaling. Cells 83, 675–678.

[26] Cheng, H., Lederer, W.J., Cannell, M.B., 1993. Calcium sparks: elementary events underlying excitation–contraction coupling in heart muscle. Science 262, 740–744.

[27] Satoh, H., Blatter, L.A., Bers, D.M., 1997. Effects of $[Ca^{2+}]i$, SR Ca^{2+} load, and rest on Ca^{2+} spark frequency in ventricular myocytes. Am. J. Physiol. 272, H657–H668.

[28] McGraw, C.F., Somlyo, A.V., Blaustein, M.P., 1980. Localization of calcium in presynaptic nerve terminals; an ultrastructural and electron microproble analysis. J. Cell Biol. 85, 228–241.

[29] Westrum, L.E., Gray, E.G., 1986. New observations on the substructure of the active zone of brain synapses and motor endplates. Proc. R. Soc. Lond. B 229, 29–38.

[30] Du, J., Tao-Cheng, J.-H., Zerfas, P., McBain, C.J., 1998. The K + channel Kv2.1, is apposed to astrocytic processes and is associated with inhibitory postsynaptic membranes in hippocampal and cortical principal neurons and inhibitory inerneurons. Neuroscience 84, 37–48.

[31] Llano, I., DiPolo, R., Marty, A., 1994. Calcium-induced calcium release in cerebellar Purkinje cells. Neuron 12, 663–673.

[32] Fisher, T.E., Levy, S., Kaczmarek, L.K., 1994. Transient changes in intracellular calcium associated with a prolonged increase in excitability in neurons of *Aplysia Californica*. J. Neurophysiol. 71, 1254–1257.

[33] Tsai, T.D., Barish, M.D., 1995. Imaging of caffeine-inducible release of intracellular calcium in cultured embryonic mouse telencephalic neurons. J. Neurobiol. 27, 252–265.

[34] Fagni, L., Chavis, P., Ango, F., Bochaert, J., 2000. Complex interactions between mGluRs, intracellular Ca^{2+} stores and ion channels in neurons. Trends Neurosci. 23, 80–88.

[35] Chavis, P., Fagni, L., Lansman, J.B., Bockaert, J., 1996. Functional coupling between ryanodine receptors and L-type calcium channels in neurons. Nature 382, 719–722.

[36] Llano, I., Gonzalez, J., Caputo, C., Lai, F.A., Blayney, L.M., Tan, Y.P., Marty, A., 2000. Presynaptic calcium stores underlie large-amplitude miniature IPSCs and spontaneous calcium transients. Nature Neurosci. 3, 1256–1265.

[37] Xu-Friedman, M.A., Regehr, W.G., 2000. Maximinis. Nature Neurosci. 3, 1229–1230.

[38] Mathers, D.A., Barker, J.L., 1984. Spontaneous voltage and current fluctuations in tissue cultured mouse dorsal root ganglion cells. Brain Res. 293, 25–47.

[39] Marrion, N.V., Adams, P.R., 1992. Release of intracellular calcium and modulation of membrane currents by caffeine in bullfrog sympathetic neurons. J. Physiol. 445, 515–535.

[40] Kawai, T., Watanabe, M., 1989. Effects of ryanodine on the spike after-hyperpolarization in sympathetic neurones of the rat superior cervical ganglion. Pflugers Arch. 413, 470–475.

[41] Fujimoto, S., Yamamoto, K., Kuba, K., Morita, K., Kato, E., 1980. Calcium localization in the sympathetic ganglion of the bullfrog and effects of caffeine. Brain Res. 202, 21–32.

[42] Davies, P.J., Ireland, D.R., McLachlan, E.M., 1996. Sources of Ca^{2+} for different Ca^{2+}-activated K^+ conductances in neurons of the rat superior cervical ganglion. J. Physiol. 495, 353–366.

[43] Fiorello, C.D., Williams, J.T., 1998. Glutamate activates an inhibitory postsynaptic potential in dopamine neurons. Nature 394, 78–82.

[44] Taylor, C.W., 1998. Inositol trisphosphate receptors: Ca^{2+}-modulated intracellular Ca^{2+} channels. Biochim. Biophys. Acta. 1436, 19–33.

[45] Bezprozvanny, I., Watras, J., Ehrlich, B.E., 1991. Bell-shaped calcium-response curves of Ins(145)P3-and calcium-gated channels from endoplasmic reticulum of cerebellum. Nature 351, 751–754.

[46] Watras, J., Bezprozvanny, I., Ehrlich, B.E., 1991. Inositol 1,4,5-trisphosphate-gated channels in cerebellum: presence of multiple conductance states. J. Neurosci. 11, 3239–3245.

[47] Hagar, R.E., Burgstahler, A.D., Nathanson, M.H., Ehrlich, B.E., 1998. Type III InsP3 receptor channel stays open in the presence of increased calcium. Nature 396, 81–84.

[48] Dawson, A.P., 1997. How do IP3 receptors work? Curr. Biol. 7, R544–R547.

[49] Parys, J.B., Bezprozvanny, I., 1995. The inositol trisphosphate receptor of xenopus oocytes. Cell Calcium 18, 353–363.

[50] Yoshida, Y., Imai, S., 1997. Structure and function of inositol 1,4,5-trisphosphate receptor. Jpn. J. Pharmacol. 74, 125–137.

[51] Thrower, E.C., Hagar, R.E., Ehrlich, B.E., 2001. Regulation of Ins(1,4,5)P3 receptor isoforms by endogenus modulators. Trends Pharm. Sci. 22, 580–586.

[52] Marchant, J., Callamaras, N., Parker, I., 1999. Initiation of IP$_3$-mediated Ca^{2+} waves in Xenopus oocytes. EMBO J. 18, 5285–5299.

[53] Sun, X.P., Callamaras, N., Marchant, J.S., Parker, I., 1998. A continuum of InsP3-mediated elementary Ca^{2+} signalling events in Xenopus oocytes. J. Physiol. 509, 67–80.

[54] Ross, C.A., Meldolesi, J., Milner, T.A., Satoh, T., Supattapone, S., Snyder, S.H., 1989. Inositol 1,4,5-trisphosphate receptor localized to endoplasmic reticulum in cerebellar Purkinje neurons. Nature 339, 468–470.

[55] Takei, K., Stukenbrok, H., Metcalf, A., Mignery, G.A., Sudhof, T.C., Volpe, P., De Camilli, P., 1992. Ca^{2+} stores in Purkinje neurons: endoplasmic reticulum subcompartments demonstrated by the heterogeneous distribution of the InsP3 receptor, Ca^{2+}-ATPase, and calsequestrin. J. Neurosci. 12, 489–505.

[56] Gerasimenko, O.V., Gerasimenko, J.V., Belan, P.V., Petersen, O.H., 1996. Inositol Trisphosphate and cyclic ADP-ribose-mediated release of Ca^{2+} from single isolated pancreatic zymogen granules. Cell 84, 473–480.

[57] Finch, E.A., Augustine, G.J., 1998. Local calcium signalling by inositol-1,4,5-trisphosphate in Purkinje cell dendrites. Nature 396, 753–756.

[58] Simpson, P.B., Challis, R.A.J., Nahorski, S.R., 1995. Neuronal Ca^{2+} stores: activation and function. Trends Neurosci. 18, 299–306.

[59] Wang, S.S.-H., Denk, W., Hausser, M., 2000. Coincidence detection in single dendritic spines mediated by calcium release. Nature Neurosci. 3, 1266–1273.

[60] Friel, D.D., Tsien, R.W., 1992. A caffeine and ryanodine-sensitive Ca^{2+} store in bullfrog sympathetic neurons modulates the effects of Ca^{2+} entry on [Ca^{2+}]$_i$. J. Physiol. 450, 217–246.

[61] Hernandez-Cruz, A., Escobar, A.L., Jimenez, N., 1997. Ca^{2+}-induced Ca^{2+} release phenomena in mammalian sympathetic, neurons are critically dependent on the rate of rise of trigger Ca^{2+}. J. Gen. Physiol. 109, 147–167.

[62] Usachev, Y.M., Thayer, S.A., 1999. Ca^{2+} influx in resting rat sensory neurones that regulates and is regulated by ryanodine-sensitive Ca^{2+} stores. J. Physiol. 519, 115–130.

[63] Usachev, Y.M., Thayer, S.A., 1997. All-or-none Ca^{2+} release from intracellular stores triggered by Ca^{2+} influx through voltage-gated Ca^{2+} channels in rat sensory neurons. J. Neurosci. 17, 7404–7414.

[64] Jaffe, D.B., Brown, T.H., 1994. Metabotropic glutamate receptor activation induces calcium waves within hippocampal dendrites. J. Neurophysiol. 72, 471–474.

[65] Rizzuto, R., Pinton, P., Carrington, W., Fay, F.S., Fogarty, K.E., Lifshitz, L.M., Tuft, R.A., Pozzan, T., 1998. Close contacts with the endoplasmic reticulum as determinants of mitochondrial Ca^{2+} responses. Science 280, 1763–1766.

[66] Hajnoczky, G., Hager, R., Thomas, A.P., 1999. Mitochondria suppress local feedback activation of inositol 1,4,5-trisphosphate receptors by Ca^{2+}. J. Biol. Chem. 274, 14157–14162.

[67] Herrington, J., Park, Y.B., Babcock, D.F., Hille, B., 1996. Dominant role of mitochondria in clearance of large Ca^{2+} loads from rat adrenal chromaffin cells. Neuron 16, 219–228.

[68] Friel, D.D., Tsien, R.W., 1994. An FCCP-sensitive Ca^{2+} store in bullfrog sympathetic neurons and its participation in stimulus-evoked changes in [Ca^{2+}]i. J. Neurosci. 14, 4007–4024.

[69] Tang, Y.-g., Zucker, R.S., 1997. Mitochondrial involvement in post-tetanic potentiation of synaptic transmission. Neuron 18, 483–491.

[70] Peng, Y.-Y., 1998. Effects of mitochondrion on calcium transients at intact presynaptic terminals depends on frequency of firing. J. Neurophysiol. 80, 186–198.

[71] Ward, S.M., Ordog, T., Koh, S.D., Abu Baker, S., Jun, J.Y., Amberg, G., Monaghan, K., Sanders, K.M., 2000. Pacemaking in interstitial cells of Cajal depends upon calcium handling by endoplasmic reticulum and mitochondria. J. Physiol. 525, 355–361.

[72] Vander Heiden, M.G., Chandel, N.S., Li, X.X., Schumacker, P.T., Colombini, M., Thompson, C.B., 2000. Outer mitochondrial membrane permeability can regulate coupled respiration and cell survival. Proc. Natl Acad. Sci. USA 97, 4666–4671.

[73] Nicholls, D.G., Ferguson, S.J., 1992., Bioenergetics, Vol. 2. Academic Press, San Diego.

[74] Lehninger, A.L., 1970. Mitochondria and calcium ion transport. Biochem. J. 119, 129–138.

[75] Mitchell, P., 1961. Coupling of phosphorylation to electron and hydrogen transfer by a chemi-osmotic type of mechanism. Nature 191, 144–148.

[76] Gunter, T.E., Pfeiffer, D.R., 1990. Mechanisms by which mitochondria transport calcium. Am. J. Physiol. 258, C755–C786.

[77] Miyata, H., Silverman, H.S., Sollott, S.J., Lakatta, E.G., Stern, M.D., Hansford, R.G., 1991. Measurement of mitochondrial free Ca^{2+} concentration in living single rat cardiac myocytes. Am. J. Physiol. 261, H1123–H1134.

[78] Rizzuto, R., Simpson, A.W., Brini, M., Pozzan, T., 1992. Rapid changes of mitochondrial Ca^{2+} revealed by specifically targeted recombinant aequorin. Nature 358, 325–327.

[79] Pivovarova, N.B., Hongpaisan, J., Andrews, S.B., Friel, D.D., 1999. Depolarization-induced mitochondrial Ca accumulation in sympathetic neurons: spatial and temporal characteristics. J. Neurosci. 19, 6372–6384.

[80] Sparagna, G.C., Gunter, K.K., Sheu, S.-S., Gunter, T.E., 1995. Mitochondrial calcium uptake from physiological-type pulses of calcium. J. Biol. Chem. 270, 27510–27515.

[81] Hunter, D.R., Haworth, R.A., Southard, J.H., 1976. Relationship between configuration, function, and permeability in calcium-treated mitochondria. J. Biol. Chem. 251, 5069–5077.

[82] Loew, L.M., Carrington, W., Tuft, R.A., Fay, F.S., 1994. Physiological cytosolic Ca^{2+} transients evoke concurrent mitochondrial depolarizations. Proc. Natl Acad. Sci. USA 91, 12579–12583.

[83] Babcock, D.F., Hille, B., 1998. Mitochondrial oversight of cellular Ca^{2+} signaling. Curr. Opin. Neurobiol. 8, 398–404.

[84] Ichas, F., Jouaville, L.S., Mazat, J.P., 1997. Mitochondria are excitable organelles capable of generating and conveying electrical and calcium signals. Cell 89, 1145–1153.

[85] Sorgato, M.C., Moran, O., 1993. Channels in mitochondrial membranes: knowns, unknowns, and prospects for the future. Crit. Rev. Biochem. Mol. Biol. 18, 127–171.

[86] Bernardi, P., 1999. Mitochondrial transport of cations: channels, exchangers and permeability transition. Physiol. Rev. 79, 1127–1155.

[87] Inoue, I., Hagase, H., Kishi, K., Higuti, T., 1991. ATP-sensitive K^+ channel in the mitochondrial inner membrane. Nature 352, 244–247.

[88] Jaburek, M., Yarov Yarovoy, V., Paucek, P., Garlid, K.D., 1998. State-dependent inhibition of the mitochondrial K_{ATP} channel by glybureide and 5-hydroxydecanoate. J. Biol. Chem. 273, 13578–13582.

[89] Hansford, R.G., 1994. Physiological role of mitochondrial Ca^{2+} transport. J. Bioeng. Biomembr. 26, 495–508.

[90] Kaczmarek, L.K., 2000. Mitochondrial memory banks; calcium stores keep a record of neuronal stimulation. J. Gen. Physiol. 115, 347–350.

[91] Colegrove, S.L., Albrecht, M.A., Friel, D.D., 2000. Quantitative analysis of mitochondrial Ca^{2+} uptake and release pathways in sympathetic neurons. Reconstruction of the recovery after depolarization-evoked $[Ca^{2+}](i)$ elevations. J. Gen. Physiol. 115, 371–388.

[92] Colegrove, S.L., Albrecht, M.A., Friel, D.D., 2000. Dissection of mitochondrial Ca^{2+} uptake and release fluxes in situ after depolarization-evoked $[Ca^{2+}](i)$ elevations in sympathetic neurons. J. Gen. Physiol. 115, 351–370.

[93] Petronilli, V., Szabo, I., Zoratti, M., 1989. The inner mitochondrial membrane contains ion-conducting channels similar to those found in bacteria. FEBS Lett. 259, 137–143.

[94] Zoratti, M., Szabo, I., 1995. The mitochondrial permeability transition. Biochim. Biophys. Acta 1241, 139–176.

[95] Szabo, I., Bernardi, P., Zoratti, M., 1992. Modulation of the mitochondrial megachannel by divalent cations and protons. J. Biol. Chem. 267, 2940–2946.

[96] Zorov, D.B., Kinnally, K.W., Perini, S., Tedeschi, H., 1992. Multiple conductance levels in rat heart inner mitochondrial membranes studied by patch clamping. Biochim. Biophys. Acta 1105, 263–270.

[97]   Altschuld, R.A., Hohl, C.M., Castillo, L.C., Garleb, A.A., Starling, R.C., Brierley, G.P., 1992. Cyclosporin
       inhibits mitochondrial calcium efflux in isolated adult rat ventricular cardiomyocytes. Am. J. Physiol. 262,
       H1699–H1704.
[98]   Halestrap, A.P., Davidson, A.M., 1990. Inhibition of Ca^{2+} induced large-amplitude swelling of liver and
       heart mitochondria by cyclosporin is probably caused by the inhibitor binding to mitochondrial-matrix
       peptidly-prolyl cis–trans isomerase and preventing it interacting with the adenine nucleotide translocase.
       Biochem. J. 268, 153–160.
[99]   Brustovetsky, N., Klingenberg, M., 1996. Mitochondrial ADP/ATP carrier can be reversibly converted
       into a large channel by Ca^{2+}. Biochemistry 35, 8483–8488.
[100] Colombini, M., Blachly-Dyson, E., Forte, M., 1996. VDAC, a channel in the outer mitochondrial
       membrane. Ion Channels 4, 169–202.
[101] Rostovtseva, T., Colombini, M., 1996. ATP flux is controlled by a voltage-gated channel from the
       mitochondrial outer membrane. J. Biol. Chem. 271, 28006–28008.
[102] Hodge, T., Colombini, M., 1997. Regulation of metabolite flux through voltage-gating of VDAC channels.
       J. Membr. Biol. 157, 271–279.
[103] Brdiczka, D., Kaldis, P., Wallimann, T., 1994. In vitro complex formation between the octamer of
       mitochondrial creatine kinase and porin. J. Biol. Chem. 269, 27640–27644.
[104] Zizi, M., Forte, M., Blachly-Dyson, E., Colombini, M., 1994. NADH regulates the gating of VDAC, the
       mitochondrial outer membrane channel. J. Biol. Chem. 269, 1614–1616.
[105] Mangan, P.S., Colombini, M., 1987. Ultrasteep voltage dependence in a membrane channel. Proc. Natl
       Acad. Sci. USA 84, 4896–4900.
[106] Wunder, U.R., Colombini, M., 1991. Patch clamping VDAC in liposomes containing whole mitochondrial
       membranes. J. Membr. Biol. 123, 83–91.
[107] Afford, S., Randhawa, S., 2000. Apoptosis. Mol. Pathol. 53, 55–63.
[108] Schendel, S.L., Montal, M., Reed, J.C., 1998. Bcl-2 family proteins as ion-channels. Cell Death Differ. 5,
       372–380.
[109] Green, D.R., Reed, J.C., 1998. Mitochondria and apoptosis. Science 281, 1309–1312.
[110] Vander Heiden, M.G., Thompson, C.B., 1999. Bcl-2 proteins: regulators of apoptosis or of mitochondrial
       homeostasis. Nature Cell Biol. 1, E209–E216.
[111] Reed, J.C., 1997. Double identity for proteins of the Bcl-2 family. Nature 387, 773–776.
[112] Kelekar, A., Thompson, C.B., 1998. Bcl-2-family proteins: the role of the BH3 domain in apoptosis.
       Trends Cell Biol. 8, 324–330.
[113] Chittenden, T., Flemington, C., Houghton, A.B., Ebb, R.G., Gallo, G.J., Elangovan, B., Chinnadurai, G.,
       Lutz, R.J., 1995. A conserved domain in Bak, distinct from BH1 and BH2, mediates cell death and protein
       binding functions. EMBO J. 14, 5589–5596.
[114] Wang, K., Gross, A., Waksman, G., Korsmeyer, S.J., 1998. Mutagenesis of the BH3 domain of BAX
       identifies residues critical for dimerization and killing. Mol. Cell Biol. 18, 6083–6089.
[115] Schendel, S.L., Azimov, R., Pawlowski, K., Godzik, A., Kagan, B.L., Reed, J.C., 1999. Ion channel
       activity of the BH3 only Bcl-2 family member, BID. J. Biol. Chem. 274, 21932–21936.
[116] Kroemer, G., 1997. The proto-oncogene Bcl-2 and its role in regulating apoptosis. Nat. Med. 3,
       614–620.
[117] Clem, R.J., Cheng, E.H., Karp, C.L., Kirsch, D.G., Ueno, K., Takahashi, A., Kastan, M.B., Griffin, D.E.,
       Earnshaw, W.C., Veliuona, M.A., Hardwick, J.M., 1998. Modulation of cell death by Bcl-XL through
       caspase interaction. Proc. Natl Acad. Sci. USA 95, 554–559.
[118] Krajewski, S., Tanaka, S., Takayama, S., Schibler, M.J., Fenton, W., Reed, J.C., 1993. Investigation of the
       subcellular distribution of the bcl-2 oncoprotein: residence in the nuclear envelope, endoplasmic
       reticulum, and outer mitochondrial membranes. Cancer Res. 53, 4701–4714.
[119] González-Garcia, M., Perez-Ballestro, R.P., Ling, L., Duan, L., Boise, L.H., Thompson, C.B., Núñez, G.,
       1994. bcl-xl is the major mRNA form expressed during murine development and its product localizes to
       mitochondria. Development 120, 3033–3042.
[120] Muchmore, S.W., Sattler, M., Liang, H., Meadows, R.P., Harlan, J.E., Yoon, H.S., Nettesheim, D., Chang,
       B.S., Thompson, C.B., Wong, S.L., Ng, S.L., Fesik, S.W., 1996. X-ray and NMR structure of human
       Bcl-x$_L$, an inhibitor of programmed cell death. Nature 381, 335–341.

[121] Schlesinger, P.H., Gross, A., Yin, X.M., Yamamoto, K., Saito, M., Waksman, G., Korsmeyer, S.J., 1997. Comparison of the ion channel characteristics of proapoptotic BAX and antiapoptotic BCL-2. Proc. Natl Acad. Sci. USA 94, 11357–11362.

[122] Minn, A.J., Velez, P., Schendel, S.L., Liang, H., Muchmore, S.W., Fesik, S.W., Fill, M., Thompson, C.B., 1997. Bcl-x_L forms an ion channel in synthetic lipid membranes. Nature 385, 353–357.

[123] Jonas, E., Hickman, J.A., Zimmerberg, J., Basanez, G., Zhang, J., Hardwick, J.M., Kaczmarek, L.K., 2000. Bcl-2 family proteins activate mitochondrial ion channels in presynaptic terminals. Soc. Neurosci. Abstr. 26, 359.

[124] Kaczmarek, L.K., Jonas, E.A., 2000. Regulation of synaptic stability by Bcl-2 family proteins. CNS Drug Rev. 6, 30–31.

[125] Luo, X., Budihardjo, I., Zou, H., Slaughter, C., Wang, X., 1998. Bid, a Bcl2 interacting protein, mediates cytochrome c release from mitochondria in response to activation of cell surface death receptors. Cell 94, 481–490.

[126] Shimizu, S., Narita, M., Tsujimoto, Y., 1999. Bcl-2 family proteins regulate the release of apoptogenic cytochrome c by the mitochondrial channel VDAC. Nature 399, 483–487.

[127] Shimizu, S., Tsujimoto, Y., 2000. Proapoptotic BH3-only Bcl-2 family members induce cytochrome c release, but not mitochondrial membrane potential loss, and do not directly modulate voltage-dependent anion channel activity. Proc. Natl Acad. Sci. USA 97, 577–582.

[128] Frankowski, H., Missotten, M., Fernandez, P.A., Martinou, I., Michel, P., Sadoul, R., Martinou, J.C., 1995. Function and expression of the Bcl-x gene in the developing and adult nervous system. Neuroreport 6, 1917–1921.

[129] Hoit, D., Hickman, J.A., Zhang, J., Ivanovska, I., Hardwick, J.M., Kaczmarek, L.K., Jonas, E.A., 2001. Modulation of mitochondrial conductance and synaptic transmission by Bcl-x_L. Biophys. J. 80, 239a.

[130] Tolbert, L.P., Morest, D.K., 1982. The neuronal architecture of the anteroventral cochlear nucleus of the cat in the region of the cochlear nerve root: electron microscopy. Neuroscience 7, 3053–3067.

[131] Rowland, K.C., Irby, N.K., Spirou, G.A., 2000. Specialized synapse-associated structures within the calyx of Held. J. Neurosci. 20, 9135–9144.

[132] Zucker, R.S., 1989. Short-term synaptic plasticity. Ann. Rev. Neurosci. 12, 13–31.

[133] Seward, E.P., Chernevskaya, N.I., Nowycky, M.C., 1996. Ba^{2+} ions evoke two kinetically distinct patterns of exocytosis in chromaffin cells, but not in neurohypophysial nerve terminals. J. Neurosci. 6, 1370–1379.

[134] Neher, E., Zucker, R.S., 1993. Multiple calcium-dependent processes related to secretion in bovine chromaffin cells. Neuron 10, 21–30.

[135] Charlton, M.P., Smith, S.J., Zucker, R.S., 1982. Role of presynaptic calcium ions and channels in synaptic facilitation and depression at the squid giant synapse. J. Physiol. 323, 173–193.

[136] Swandulla, D., Hans, M., Zipser, K., Augustine, G.J., 1991. Role of residual calcium in synaptic depression and posttetanic potentiation: fast and slow calcium signaling in nerve terminals. Neuron 7, 915–926.

[137] Smith, S.J., Buchanan, J., Osses, L.R., Charlton, M.P., Augustine, G.J., 1993. The spatial distribution of calcium signals in squid presynaptic terminals. J. Physiol. 472, 573–593.

[138] Gonzalez-Garcia, M., Garcia, I., Ding, L., O'Shea, S., Boise, L.H., Thompson, C.B., Nunez, G., 1995. bcl-x is expressed in embryonic and postnatal neural tissues and functions to prevent neuronal cell death. Proc. Natl Acad. Sci. USA 92, 4304–4308.

[139] Blomer, U., Kafri, T., Randolph-Moore, L., Verma, I.M., Gage, F.H., 1998. Bcl-xL protects adult septal cholinergic neurons from axotomized cell death. Proc. Natl Acad. Sci. USA 95, 2603–2608.

[140] Breckenridge, L.J., Almers, W., 1987. Currents through the fusion pore that forms during exocytosis of a secretory vesicle. Nature 328, 814–817.

[141] Lee, C.J., Dayanithi, G., Nordmann, J.J., Lemos, J.R., 1992. Possible role during exocytosis of a Ca^{2+}-activated channel in neurohypophysial granules. Neuron 8, 335–342.

[142] Yakir, N., Rahamimoff, R., 1995. The non-specific ion channel in Torpedo ocellata fused synaptic vesicles. J. Physiol. 485, 683–697.

[143] Jonas, E.A., Knox, R.J., Smith, T.C.M., Wayne, N.L., Connor, J.A., Kaczmarek, L.K., 1997. Regulation by insulin of a unique neuronal Ca^{2+} pool and neuropeptide secretion. Nature 385, 343–346.

[144] Magoski, N.M., Knox, R.I., Kaczmarek, L.K., 2000. Activation of a Ca^{2+}-permeable cation channel produces a prolonged attenuation of intracellular Ca^{2+} release in *Aplysia* bag cell neurons. J. Physiol. 522, 271–283.

[145] Kunkele, K.P., Heins, S., Dembowski, M., Nargang, F.E., Benz, R., Thieffry, M., Walz, J., Lill, R., Nussberger, S., Neupert, W., 1998. The preprotein translocation channel of the outer membrane of mitochondria. Cell 93, 1009–1019.
[146] Schatz, G., 1998. The doors to organelles. Nature 395, 439–440.
[147] Hill, K., Mode, K., Ryan, M.T., Dietmeier, K., Martin, F., Wagner, R., Pfanner, N., 1998. Tom40 forms the hydrophilic channel of the mitochondrial import pore for preproteins. Nature 395, 516–521.

Colour Plates

Advances in Molecular and Cell Biology, Vol. 32, pages 459–469
© 2004 Elsevier B.V. All rights of reproduction in any form reserved.
ISSN: 1569-2558 / DOI: 10.1016/S1569-2558(03)32023-5

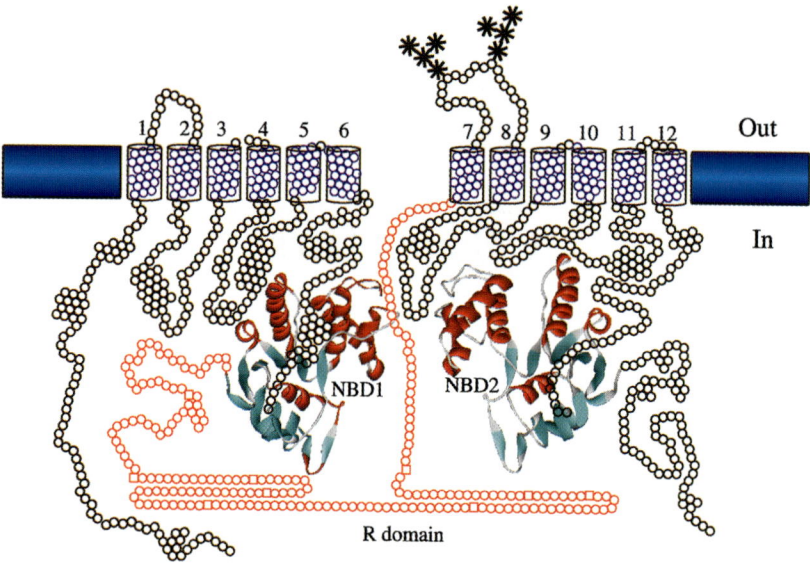

Fig. 1.

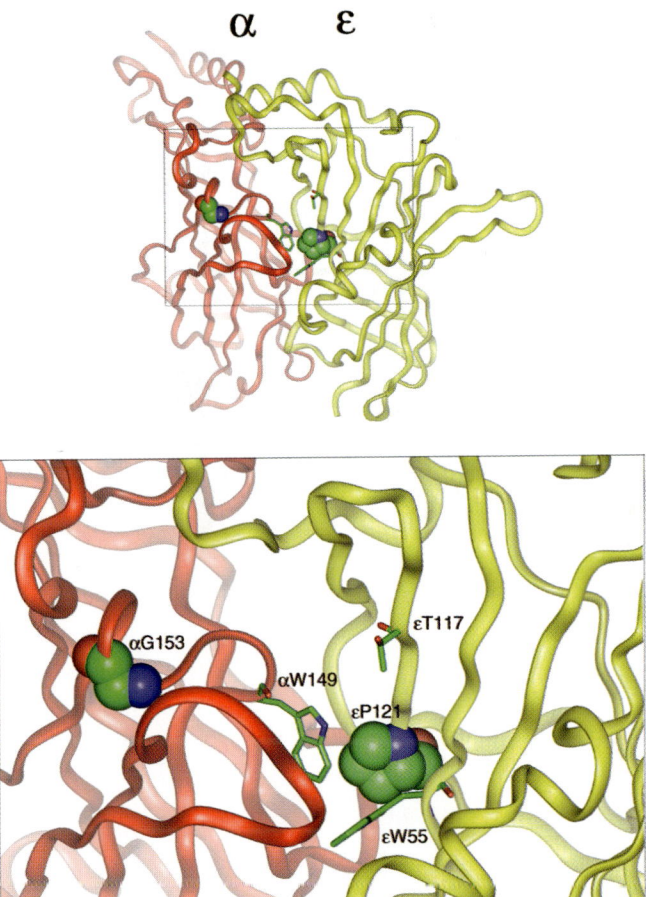

Fig. 4.

Colour Plates

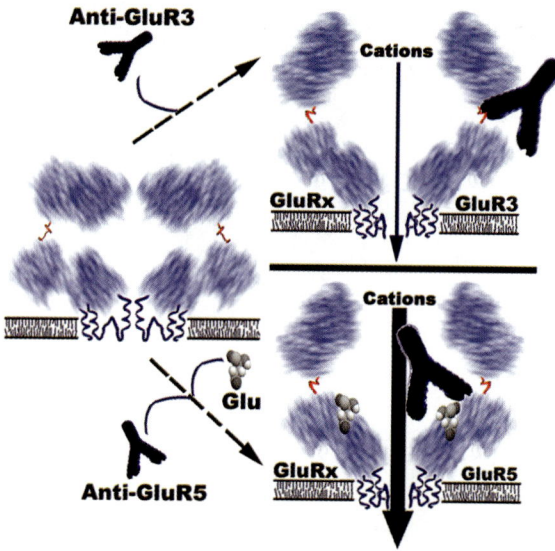

Fig. 2.

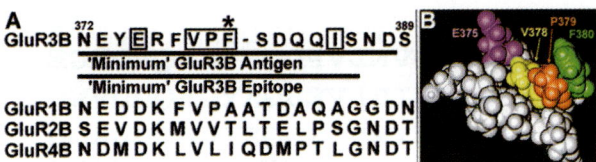

Fig. 3.

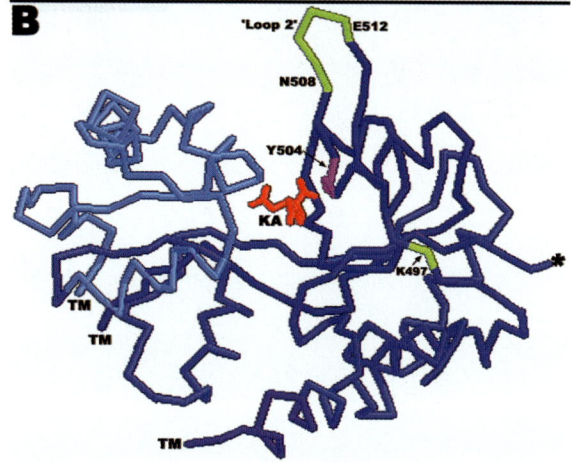

Fig. 4.

A

B

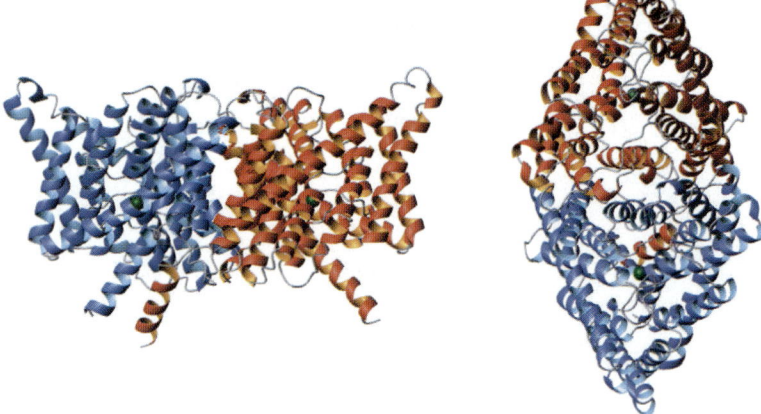

Fig. 6.

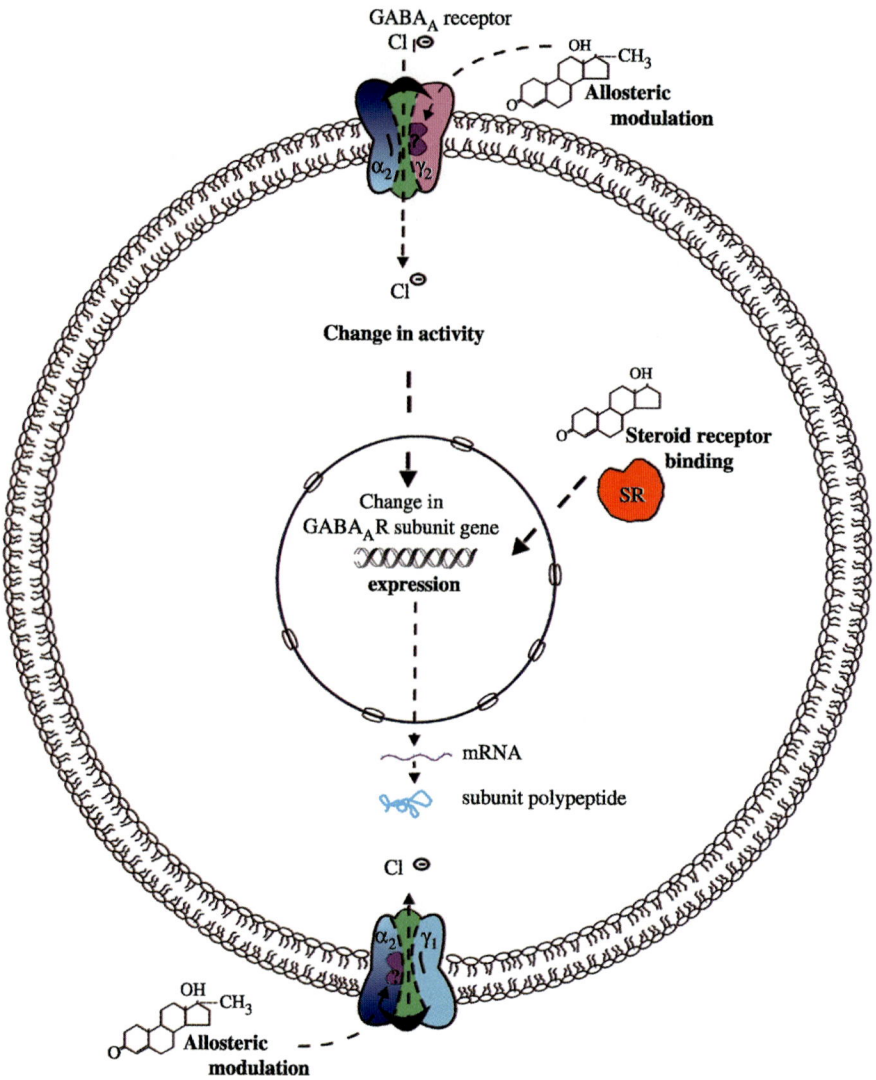

Fig. 3.

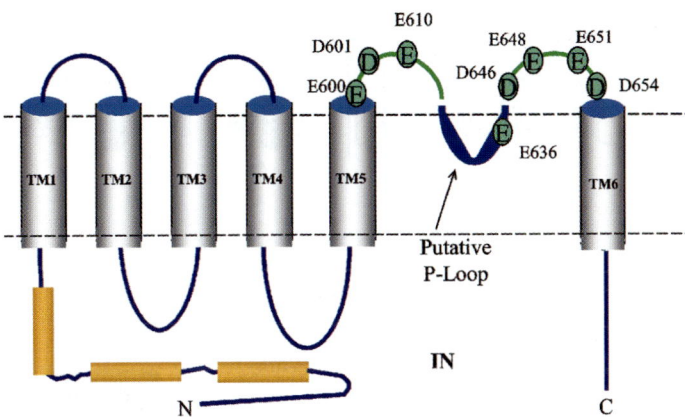

Fig. 1.

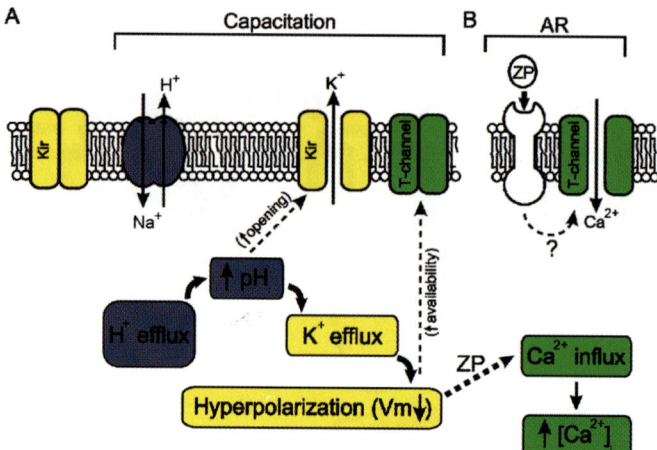

Fig. 2.

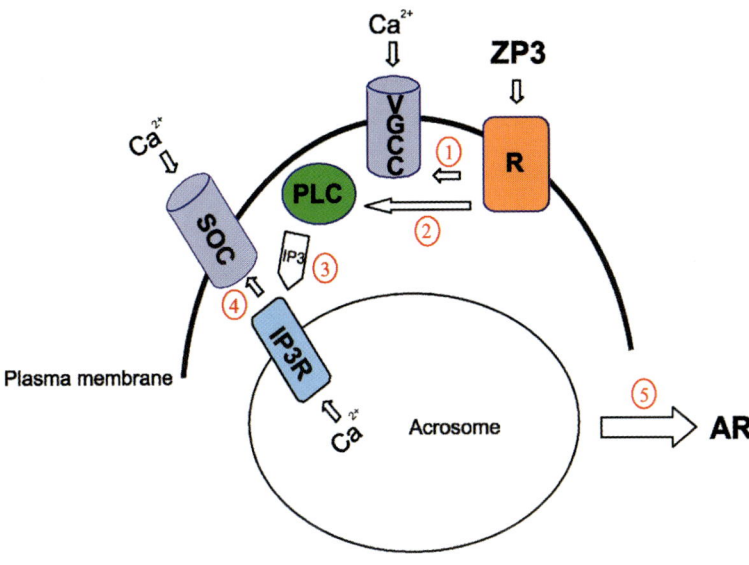

Fig. 3.

List of Contributors

Pascal Barbry

Institut de Pharmcologie Moleculaire et Cellulaire
CNRS UPR 411
660 route des Lucioles
06560 Sophia Antipolis, France

Stuart Bevan

Novartis Institute for Medical Sciences
5 Gower Place
London WC1E6BN, UK

Marie-Christine Broillet

Institute of Pharmacology and Toxicology
University of Lausanne
CH-1005 Lausanne, Switzerland

John H. Caldwell

Department of Cellular and Structural Biology
University of Colorado Health Sciences Center
Denver, Colorado 80262, USA

Noel G. Carlson

Geriatric Research Education and Clinical Center
Salt Lake City Veterans Administration Medical Center and
    University of Utah School of Medicine
Salt Lake City, Utah 84112-5330, USA

Sylvie Coscoy

Laboratoire de Physico-Chimie
Institut Curie
CNRS UMR 168
26 rue d'Ulm
75005 Paris, France

Alberto Darszon

Department of Genetics and Molecular Physiology
Institute of Biotechnology
UNAM
Cuernavaca, Mexico

Andrew G. Engel

Muscle Research Laboratory
Department of Neurology
Mayo Foundation
Rochester, Minnesota 55905, USA

Christoph Fahlke

RWTH Aachen
Institute of Physiology
Aachen, Germany

Centro de Estudios Cientificos
Avenida Prat 514
Valdiva, Chile

Ricardo Felix   Department of Physiology, Biophysics, and Neuroscience
                                 Cinvestav-IPN
                                 Mexico City, Mexico

Stuart Firestein   Department of Biological Sciences
                                 Columbia University
                                 New York, New York 10027, USA

Alfredo Fort   Department of Neuroscience
                                 Albert Einstein College of Medicine
                                 1300 Morris Park Avenue
                                 Bronx, New York 10461, USA

Lorise C. Gehring   Geriatric Research Education and Clinical Center
                                 Salt Lake City Veterans Administration Medical Center and
                                     University of Utah School of Medicine
                                 Salt Lake City, Utah 84112-5330, USA

John W. Hanrahan   Department of Physiology
                                 McGill University
                                 3655 Prom. Sir-William-Osler
                                 Montreal, Quebec
                                 Canada H3G1Y6

Leslie P. Henderson   Departments of Physiology and Biochemistry
                                 Dartmouth Medical School
                                 Hanover, New Hampshire 03755, USA

Matthew G. Hopperstad   Department of Neuroscience
                                 Albert Einstein College of Medicine
                                 1300 Morris Park Avenue
                                 Bronx, New York 10461, USA

Elizabeth A. Jones   Department of Internal Medicine
                                 Yale University School of Medicine
                                 333 Cedar Street
                                 New Haven, Connecticut 06520, USA

Juan Carlos Jorge   Department of Anatomy
                                 Medical Sciences Campus
                                 University of Puerto Rico
                                 San Juan, Puerto Rico

Leonard K. Kaczmarek   Department of Pharmacology
                                 Department of Cellular and Molecular Physiology
                                 Yale University School of Medicine
                                 333 Cedar Street
                                 New Haven, Connecticut 06520, USA

Joseph C. Koster   Department of Cell Biology and Physiology
                                 Washington University School of Medicine

660 South Euclid Avenue
St. Louis, Missouri 63110, USA

S. Rock Levinson

Department of Physiology and Biophysics
University of Colorado Health Sciences Center
Denver, Colorado 80262, USA

Richard S. Lewis

Department of Molecular and Cellular Physiology
Stanford University School of Medicine
Stanford, CA 94305, USA

Ignacio Lopez-Gonzalez

Department of Genetics and Molecular Physiology
Institute of Biotechnology
UNAM
Cuernavaca, Mexico

Joseph F. Margiotta

Department of Anatomy and Neurobiology
Medical College of Ohio
Toledo, Ohio 43614, USA

Bess A. Marshall

Department of Cell Biology and Physiology
Washington University School of Medicine
660 South Euclid Avenue
St. Louis, Missouri 63110, USA

David P. McCobb

Department of Neurobiology and Behavior
Cornell University
Ithaca, New York 14583, USA

Erin L. Meyer

Geriatric Research Education and Clinical Center,
Salt Lake City Veterans Administration Medical Center and
    University of Utah School of Medicine
Salt Lake City, Utah 84112-5330, USA

Carlos Munoz-Garay

Department of Genetics and Molecular Physiology
Institute of Biotechnology
UNAM
Cuernavaca, Mexico

Colin G. Nichols

Department of Cell Biology and Physiology
Washington University School of Medicine
660 South Euclid Avenue
St. Louis, Missouri 63110, USA

Kinji Ohno

Muscle Research Laboratory
Department of Neurology
Mayo Foundation
Rochester, Minnesota 55905, USA

Murali Prakriya   Department of Molecular and Cellular Physiology
Stanford University School of Medicine
Stanford, CA 94305, USA

Phyllis C. Pugh   Department of Anatomy and Neurobiology
Medical College of Ohio
Toledo, Ohio 43614, USA

Allison J. Reeve   Novartis Institute for Medical Sciences
5 Gower Place
London WC1E6BN, UK

Scott W. Rogers   Geriatric Research Education and Clinical Center
Salt Lake City Veterans Administration Medical Center and
    University of Utah School of Medicine
Salt Lake City, Utah 84112-5330, USA

Show-Ling Shyng   Center for Research on Occupational and
    Environmental Toxicology
Oregon Health Sciences University
Portland, Oregon 97201, USA

Steven M. Sine   Receptor Biology Laboratory
Department of Physiology and Biophysics
Mayo Foundation
Rochester, Minnesota 55905, USA

Terrance P. Snutch   Biotechnology Laboratory
Rm 237-6174 University Blvd.
University of British Columbia
Vancouver, B.C.
Canada V6T 1Z3

David C. Spray   Department of Neuroscience
Albert Einstein College of Medicine
1300 Morris Park Avenue
Bronx, New York 10461, USA

Miduturu Srinivas   Department of Neuroscience
Albert Einstein College of Medicine
1300 Morris Park Avenue
Bronx, New York 10461, USA

W. Daniel Stamer   Department of Pharmacology and Opthalmology
University of Arizona College of Medicine
Tucson, Arizona 85724, USA

Anthony Stea   University-College of the Fraser Valley
33844 King Road
Abbotsford, British Columbia V2S 7M8
Canada

Hai-Long Wang   Receptor Biology Laboratory
Department of Physiology and Biophysics
Mayo Foundation
Rochester, Minnesota 55905, USA

Andrea J. Yool   Departments of Physiology and Pharmacology
P.O. Box 24501
University of Arizona College of Medicine
Tucson, Arizona 85724-5051, USA